中国工程院院士文集

岑可法文集

岑可法 著

科学出版社
北京

内 容 简 介

岑可法院士长期从事能源高效清洁利用领域的教学和研究工作。本文集共分三篇，分别选录了岑可法院士自20世纪60年代以来在各学术期刊公开发表的论文、出版的著作及学术报告的代表性作品。内容涉及化石燃料的能源高效清洁利用、废弃物能源化处置新技术、生物质能利用技术、洁净煤燃烧高效裂解及气化技术、能源利用过程中多种污染物协同脱除技术、工程气固多相流和计算机辅助优化试验（CAT）及先进激光诊断技术等方面的研究成果，论文以全文形式呈现，著作以详细摘要的形式呈现。本书主要内容涵盖了岑可法院士所获得的18项国家级奖励的主要研究成果，体现了他为我国能源与环境的可持续发展作出的创造性的贡献。

本文集可供能源工程、环境保护及碳减排领域的科研人员与高校师生参阅。

图书在版编目（CIP）数据

岑可法文集 / 岑可法著. -- 北京：科学出版社，2024. 12. --（中国工程院院士文集）. -- ISBN 978-7-03-080702-1

Ⅰ．TK01-53

中国国家版本馆 CIP 数据核字第 20240R0D12 号

责任编辑：范运年 / 责任校对：王萌萌
责任印制：师艳茹 / 封面设计：陈 敬

科 学 出 版 社 出版
北京东黄城根北街 16 号
邮政编码：100717
http://www.sciencep.com

三河市春园印刷有限公司印刷
科学出版社发行　各地新华书店经销
*

2024 年 12 月第 一 版　开本：787×1092 1/16
2024 年 12 月第一次印刷　印张：35 1/2
字数：835 000

定价：268.00 元
（如有印装质量问题，我社负责调换）

《中国工程院院士文集》总序

二〇一二年暮秋，中国工程院开始组织并陆续出版《中国工程院院士文集》系列丛书。《中国工程院院士文集》收录了院士的传略、学术论著、中外论文及其目录、讲话文稿与科普作品等。其中，既有早年初涉工程科技领域的学术论文，亦有成为学科领军人物后，学术观点日趋成熟的思想硕果。卷卷《文集》在手，众多院士数十载辛勤耕耘的学术人生跃然纸上，透过严谨的工程科技论文，院士笑谈宏论的生动形象历历在目。

中国工程院是中国工程科学技术界的最高荣誉性、咨询性学术机构，由院士组成，致力于促进工程科学技术事业的发展。作为工程科学技术方面的领军人物，院士们在各自的研究领域具有极高的学术造诣，为我国工程科技事业发展作出了重大的、创造性的成就和贡献。《中国工程院院士文集》既是院士们一生事业成果的凝练，也是他们高尚人格情操的写照。工程院出版史上能够留下这样丰富深刻的一笔，与有荣焉。

我向来以为，为中国工程院院士们组织出版《院士文集》之意义，贵在"真善美"三字。他们脚踏实地，放眼未来，自朴实的工程技术升华至引领学术前沿的至高境界，此谓其"真"；他们热爱祖国，提携后进，具有坚定的理想信念和高尚的人格魅力，此谓其"善"；他们治学严谨，著作等身，求真务实，科学创新，此谓其"美"。《院士文集》集真善美于一体，辩而不华，质而不俚，既有"居高声自远"之澹泊意蕴，又有"大济于苍生"之战略胸怀，斯人斯事，斯情斯志，令人阅后难忘。

读一本文集，犹如阅读一段院士的"攀登"高峰的人生。让我们翻开《中国工程院院士文集》，进入院士们的学术世界。愿后之览者，亦有感于斯文，体味院士们的学术历程。

徐匡迪
二〇一二年

序　言

我与岑可法院士相知已有一个甲子有余，记得在我考上天津大学那一年，在我父亲任教的太原工学院宣传栏上看到岑可法讲学的通知，当时我读不准"岑"的发音，便问与岑先生从事同专业的、比他整整大一轮、请他来太原工学院讲学的父亲，父亲告诉我"岑"作为姓氏的正确读音并告诉我岑可法老师是留苏博士，刚回国不久，还不到30岁，在专业上就很有造诣。从那时起，岑可法就在我脑中留下深刻印象。后来，我们见面相识，并共同从事煤炭清洁高效利用的科研教学和战略研究。在我们长期的共事与合作中，他那种对事业锲而不舍的追求精神，"宁谢纷华甘淡泊"潜心育后人的伯乐美德，一直深深地感染与鞭策着我，使我从几十年前的肤浅相知变成崇高敬意。

岑可法院士是我国能源与环境领域的著名战略科学家。他长期从事煤炭的高效清洁利用、燃烧与污染控制、气固多相流动、热能工程测试技术、新能源等前沿领域的研究，在煤粉燃烧、水煤浆燃烧、流化床燃烧、煤的气化与液化、生物质能利用，气固多相流动与分离、电站锅炉与工业窑炉的节能与环保，以及能源与环境领域中的多种测试技术等方面，均卓有建树，为我国能源与环境的可持续发展做出了系统的、创造性的贡献。

在科学研究方面，岑可法教授特别重视基础研究与工程应用的紧密结合。他一贯主张从事科学研究既不能纸上谈兵、忽视工程实际，又不能仅仅满足于解决工程中的一般性问题而就事论事，必须把两者有机地结合起来。即应从工程实践中提炼出具有共性和基础性的科学问题开展深入研究，探索并掌握其中的基本规律，再用来指导并解决实际工程问题，才能取得既有学术价值又有实用意义的创新性成果。在长达数十年的科研生涯中，他始终不渝地坚持这一理念，并且身体力行、一以贯之。他所取得的丰硕成果，充分证明了这是一条成功的经验，值得广大科技工作者，特别是青年学者学习和借鉴。

岑可法院士不仅是一位杰出的科学家，而且是一位优秀的教育家。他以身作则，育人为先，始终把为国家培养高质量的人才视为己任，并且倾注了毕生心血。他特别重视对学生独立思考能力与创新精神的培养，总是要求学生不满足于掌握书本知识和一般性地完成课题任务，而要充分发挥自己的主观能动性，敢于探索、勇于创新，努力提出自己的新见解、新思路和新方法。他十分注重对学生综合素质的全面培养，总是教导学生学风正派，力戒浮躁，要老老实实做人、踏踏实实做事、扎扎实实做学问，在业务学习、科研工作、思想品德、生活作风等各个方面都要从严要求自己，不断取得新的进步。他还经常教导学生，要正确处理好个人、集体与国家三者之间的关系，始终把国家的利益放在首位。他的一言一行，都为学生树立了学习的榜样，一批又一批优秀的人才在他的精心培养下脱颖而出，迅速成长为我国能源与环境领域的骨干和栋梁。

在长期的科研与教育中，岑可法院士十分重视团队的作用，始终坚持并大力倡导"求是、团结、创新"的团队精神。他反复强调，任何一项重大科技成果的取得，都绝非个别人单枪匹马所能完成，必须依靠集体的智慧和艰苦努力与不懈奋斗，才能攻克科学堡垒、攀登科技高峰。他自己就是这样身体力行的。作为一位学术团队的领头人，他深学佩冠，厚德服群，具有宽广的胸怀和高尚的情操，总是乐于并善于为团队成员的脱颖而出和迅速成长提供各种有利条件与良好环境，因而能够吸引并凝聚一大批优秀学者团结在他的周围，形成一个特别能战斗的团队，共同为发展我国的能源与环境科学事业而努力奋斗。

岑可法院士先后承担了一大批国家与省部级重大科技项目，如 973 计划、863 计划和国家自然科学基金重点项目等，取得了一系列重大创新性成果。多年来，他带领的团队先后荣获国家"三大奖"18 项。他个人还获得"国家有突出贡献中青年科技专家""全国高等学校先进科技工作者""全国五一劳动奖章""浙江省劳动模范"等荣誉称号，是我国能源与环境科学领域公认的一位德高望重的学术大师。

为了系统地总结并展示岑可法院士及其学术团队所取得的丰硕成果与宝贵经验，为我国能源与环境科学事业的发展提供有益的借鉴与参考，在他从事科研和教学工作七十年之际，编辑出版《岑可法文集》。这是一件十分有意义的工作，该文集既是岑可法院士一生事业成果的凝练，也是他高尚人格情操的写照。我对此表示衷心的祝贺，"喜看后生多可亲，从来学术重传薪。老干扶持燃真火，天道酬勤还在精"，深信该文集的出版必将激励更多志同道合的后薪人才投身我国能源与环境科学事业，并为之做出新的贡献。

有幸为该文集作序，又一次获得向岑可法院士学习的机会。为人师表，人生如炬。智山慧海，真火已燃。愿随前薪做后薪，这是向岑可法院士学习的最好体现，也是该文集编辑出版的初衷。是为序。

最后，衷心祝愿岑可法院士健康长寿、永葆学术青春，为我国的能源与环境科学事业的发展和高质量人才培养继续作出贡献！

谢克昌

2024 年 11 月

前　言

公元二○二五年是岑可法院士从事教学科研工作70周年，也是他步入90后的第一年。谨以此书以致庆贺，同时亦希望能管窥岑可法院士七十年科研教学工作的成就。

岑可法院士是我国能源与环境领域的著名科学家和教育家。他长期从事煤炭的高效清洁利用、新能源、燃烧与污染控制、气固多相流动、热能工程测试技术等前沿领域的教学和科研工作，为我国能源与环境的可持续发展做出了系统的、创造性的贡献。

从抗战时期的一路流亡、一路学习，到留学苏联学成归国，再到美国访学，拒绝高薪聘请，七十年的科研教学工作充分体现了他胸怀祖国、服务人民的爱国精神。

岑可法院士特别重视创新，重视基础研究与工程应用的紧密结合。他先后承担了一大批国家与省部级重大科技项目，如国家863计划和国家自然科学基金重点项目等，取得了一系列重大创新性成果，先后荣获国家"三大奖"18项。这充分体现了他勇攀高峰、敢为人先的创新精神。

岑可法院士不仅是一位杰出的科学家，而且是一位优秀的教育家。他特别重视对学生独立思考能力与创新精神的培养。他十分注重对学生综合素质的全面培养，总是教导学生学风正派，力戒浮躁，要老老实实做人，踏踏实实做事。他还经常在论文和奖励署名等上面把自己往后排，体现了他甘为人梯、奖掖后学的育人精神。一批又一批优秀的人才在他的精心培养下脱颖而出，迅速成长为我国能源与环境领域的骨干和栋梁。

岑可法院士十分重视团队的作用，始终坚持并大力倡导"求是、团结、创新"的团队精神。他反复强调，任何一项重大科技成果的取得，都绝非个别人单枪匹马所能完成，必须依靠集体的智慧和艰苦努力与不懈奋斗，才能攻克科学堡垒、攀登科技高峰。他自己就是这样身体力行的。作为一位学术团队的领头人，他总是乐于并善于为团队成员的迅速成长提供各种有利条件与良好环境，体现了他集智攻关、团结协作的协同精神。

岑可法院士七十年的科研教学工作充分体现了他追求真理、严谨治学的求实精神。他个人还获得"国家有突出贡献中青年科技专家""全国高等学校先进科技工作者""全国五一劳动奖章""浙江省劳动模范"等荣誉称号，是我国能源与环境科学领域公认的一位德高望重的学术大师。

作为岑可法院士培养的第二位博士，我师从岑可法院士的时间可以追溯到1982年初的本科毕业设计。从那时开始，我的硕士、博士一直师从岑可法院士，参加工作后一直在岑可法院士的具体指导下进行科研教学工作，目前已逾四十载。听闻工程院正在组织院士文集的编撰工作，我自告奋勇承担了本文集的策划、选稿、组织工作，在岑院士的具体指导下，与其他团队成员一起几轮讨论，最终形成了本文集。文集从岑院士发表的三千余篇学术论文中精选了不同时期、不同科研方向的四十篇论文全文收入本文集，文集还收录了从岑院士公开出版的三十一本著作中筛选出的二十本著作的详细摘要，此外，文集也收录了根据岑院士公开发表的两篇关于人才培养方面的演讲整理的报告。

感谢倪明江、高翔、严建华、樊建人、施正伦、吴学成、周昊、刘建忠、方梦祥、李晓东、王勤辉、池涌、王智化、周劲松、肖刚、郑成航、程乐鸣、蒋旭光、俞自涛对文集作出的贡献，无论是参与论文和著作的几轮选择和著作详细摘要的撰写都体现出他们对本文集作出的贡献。感谢浙江大学能源工程学院陈浩和罗坤等领导对本书的支持。感谢王诗依、汪晓彤、邵焕霞等对本书的付出。

感谢谢克昌院士为本书作序言和对本文集总体策划的指导，特别感谢谢克昌院士两代人对岑可法院士的情谊和支持。

特别感谢岑可法院士夫人孙慧珍女士对岑院士无微不至的照顾，以及对本书成书过程的关心和支持。

七十年教学科研成果，不是一本文集所能包涵的，但其中体现了我国能源与环境科学事业的发展，希望本书能够为能源与环境领域的教学科研工作者提供有益的借鉴与参考。

骆仲泱甲辰秋于求是园
岑可法审定

目 录

《中国工程院院士文集》总序
序言
前言

第一篇 论 文

自由射流，圆柱及管簇后尾迹紊流结构的试验研究 ... 3
旋风燃烧室内气流紊流结构的研究 ... 25
气流紊流结构的研究方法 ... 62
劣质煤沸腾燃烧过程动力特性和双床并联运行沸腾炉提高燃烧效率的试验研究 ... 127
油煤混合燃料的试验研究 ... 134
水煤浆在沸腾床内燃烧过程的初步研究 ... 141
水煤浆滴燃烧过程的简化数学模型 ... 147
煤及水煤浆燃烧过程中 NO_x 和 SO_2 生成的研究 ... 151
脉冲沸腾床流体动力特性的试验研究 ... 159
应用脉冲鼓风来提高沸腾炉燃烧效率的研究 ... 163
煤粉颗粒在气流中的受力分析及其运动轨迹的研究 ... 173
洗选煤泥沸腾燃烧技术的研究 ... 184
水煤浆燃烧技术的研究 ... 187
A NUMERICAL MODEL FOR THE TURBULENT FLUCTUATION AND
  DIFFUSION OF GAS-PARTICLE FLOWS AND ITS APPLICATION
  IN THE FREEBOARD OF A FLUIDIZED BED ... 191
循环流化床内颗粒运动的预测与测量 ... 205
低倍率中温分离型循环流化床锅炉的设计 ... 213
数值试验(CAT)在大型电站锅炉设计及调试中应用的前景 ... 225
煤炭洁净综合利用技术的研究与前景 ... 234
新颖的热、电、燃气三联产装置 ... 241
流化床锅炉床下热烟气点火启动的理论及试验研究 ... 245
Experimental study of a finned tubes impact gas-solid separator for CFB boilers ... 254
Experimental studies on municipal solid waste pyrolysis in a laboratory-scale
  rotary kiln ... 265

循环流化床锅炉炉膛热力计算 … 275
Experimental research on solid circulation in a twin fluidized bed system … 281
中国能源与环境可持续发展问题的探讨（一） … 289
Modulations on turbulent characteristics by dispersed particles in gas-solid jets … 295
高效低污染燃烧及气化技术的最新研究进展 … 312
Simultaneous removal of NO_x, SO_2 and Hg in nitrogen flow in a narrow reactor by ozone injection: Experimental results … 319
On coherent structures in a three-dimensional transitional plane jet … 326
An experimental investigation of a natural circulation heat pipe system applied to a parabolic trough solar collector steam generation system … 337
Nitrogen oxide absorption and nitrite/nitrate formation in limestone slurry for WFGD system … 346
Pilot-scale investigation on slurrying, combustion, and slagging characteristics of coal slurry fuel prepared using industrial wasteliquid … 354
基于煤炭分级转化的发电技术前景 … 366
煤炭清洁发电技术进展与前景 … 371
Emerging energy and environmental applications of vertically-oriented graphenes … 378
Formation, Measurement, and Control of Dioxins from the Incineration of Municipal Solid Wastes: Recent Advances and Perspectives … 392
Influences of Coal Type and Particle Size on Soot Measurement by Laser-Induced Incandescence and Soot Formation Characteristics in Laminar Pulverized Coal Flames … 414
A solar micro gas turbine system combined with steam injection and ORC bottoming cycle … 425
Astigmatic dual-beam interferometric particle imaging for metal droplet 3D position and size measurement … 437
Nonlinear Dynamic Characteristics of Turbulent Non-Premixed Acoustically Perturbed Swirling Flames … 442

第二篇 著 作

锅炉燃烧试验研究方法及测量技术 … 457
工程气固多相流动的理论及计算 … 462
燃烧流体力学 … 466
锅炉和热交换器的积灰、结渣、磨损和腐蚀的防止原理与计算 … 470
煤浆燃烧、流动、传热和气化的理论与应用技术 … 474

循环流化床锅炉理论、设计与运行 …………………………………………………………478
气固分离理论及技术 ………………………………………………………………………484
大型电站锅炉安全及优化运行技术 ………………………………………………………487
高等燃烧学 …………………………………………………………………………………491
煤的热电气多联产技术及工程实例 ………………………………………………………495
燃烧理论与污染控制 ………………………………………………………………………499
洗煤泥及污泥焚烧技术与工程实例 ………………………………………………………503
臭氧氧化烟气中多种污染物同时脱除技术 ………………………………………………507
先进清洁煤燃烧与气化技术 ………………………………………………………………511
可燃固体废弃物能源化利用技术 …………………………………………………………515
推动能源生产和消费革命的支撑与保障 …………………………………………………520
基于智能计算的燃烧优化 …………………………………………………………………524
太阳能 ………………………………………………………………………………………529
烟气多种污染物高效脱除技术原理及应用 ………………………………………………533
循环流化床锅炉数值优化设计与运行 ……………………………………………………537

第三篇 报 告

研究型大学要培养多层次的现代工程师 …………………………………………………543
培养高质量博士的探索与实践 ……………………………………………………………546

附录 大事记 …………………………………………………………………………………551

第一篇 论　　文

自由射流，圆柱及管簇后尾迹紊流结构的试验研究

岑可法

摘 要

本文分两部份，第一部份阐述了自由射流的紊流结构，发现紊流参数的分布和平均参数的分布有着显著的不同，对此给予一定的分析。根据试验结果对 Prandtl 的新旧理论进行了验证，发现紊流微团"混合长度"和紊流粘度系数在射流横截面基本上是同一数量级，并近似保持不变，而沿射流纵向则线性地增长。文章对 Абрамович 的平均速度分布公式的实质进行了讨论。第二部份给出了射流绕圆柱及错列管簇后尾迹的紊流结构，对试验结果提出了详细的分析，证明非流线型物体是紊流脉动"发生器"。本文主要测量了二元脉动参数，部份地测量了三元脉动参数。

一、引 言

自由紊流射流及绕非流线型流体流动是实现及组织燃烧过程的两种基本空气动力方式，因此长期来各国学者进行了比较详尽的理论和试验研究（文献 1、3）。但是过去的研究主要集中在平均空气动力参数方面，而对这两种流动方式的紊流结构往往因理论和试验技术上的困难，进行的工作不多，特别是对于三元流动的情况尤感困难。可是对燃烧技术工作者来说，详尽地掌握它们的紊流结构有很大实际意义：一方面，紊流结构直接影响到燃料的着火及火焰底稳定性问题；另一方面，它也决定燃料的燃烬过程。本文目的在于对自由射流及绕非流线型流体后尾迹的紊流结构进行初步的试验研究，并对 Prandtl 的自由射流紊流机理假说作进一步的分析。

二、自由射流的紊流结构

§1. 问题的提出

由喷嘴喷向无限空间的射流很明显地分为两段：初始段及主体段（图 1）。在初始段内又分为两个区域：中间的核心区（具有不变的平均速度侧形）及边界层。如果喷嘴出口处具有均匀的平均速度，那么，射流离开喷嘴后由于与无限空间的介质进行强烈混合，能量不断向外传递及散逸，因此核心区会愈来愈小，而边界层则愈来愈厚。在主体段内，核心区完全消失，即边界层已扩展到射流轴心。根据自由射流这种特性，我们已经可以推测到：在射流边界层上一定有着强烈的紊流混合和脉动，在核心区则保持着较为平稳的管内流动紊流结构。至于其数量上的关系，早在 1943 年开始先后在英美及苏联进行过

不少研究(详细总结见文献1,2,3)。可是英美学者的研究主要集中在射流的主体段,并且多数只测量了两元脉动参数。但是从解决着火及火焰稳定的角度出发,研究射流起始段的紊流结构就显得特别重要。在1957年,АнтоноваГ.С.提出了等温及不等温自由射流紊流结构的论文(文献4),其中对初始段进行了研究,但是根据分析(文献10),该文作者所使用的测量方法准确度较低,其误差一般不少于20%～30%(其中理论计算误差15%～20%测量误差5%～15%),并且也只讨论了两元脉动问题。在1961年出现了Солнцев В.П.的文章,但他只研究了射流起始段核心区,并只进行了一元脉动的测量(文献6)。因此,我们把任务集中在用比较准确的测量方法研究紊流射流的初始段,并对三元脉动进行了部份的测量。

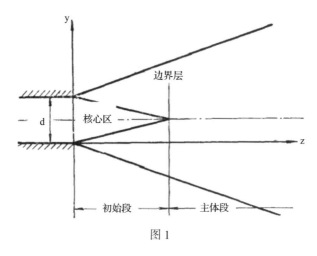

图 1

§2. 研究方法及试验设备

采用 V 一型感应元件的热电风计法使我们有可能对三元脉动进行比较准确的测量(方法详见文献10)。自由射流喷嘴采用80毫米直径的技术管道,这比上述作者采用磨制光滑、带有稳定栅栏的风洞更合乎燃烧技术实际情况。一般在技术管道所具有的起始紊流强度比风洞高2～3倍(根据测量,在技术管道 ε_0=4%～6%)。出口风速变化由0至40米/秒;测量了自由射流初始段下列五个截面:

$$\frac{x}{d}=0.325, 0.625, 1.25, 1.875, 3.12$$

§3. 试验结果综述

图2示出平均速度、脉动速度、纵向及横向紊流强度场沿射流初始段的分布。

值得注意的是下面几点事实:第一、和平均速度不同,在射流核心内不论沿横向及纵向分布的紊流参数均不保持为常量。因此严格来说,对紊流参数根本不存在核心区。反映在图3中则更明显了,在核心区内,紊流强度沿射流纵向不断增加,愈靠近主体段时增长速度愈快。第二、在初始段边界层内(混合区)和平均参数不断降低相反,紊流参数不断升高,至最大点后向外边界降低。最大点位置基本上和出口喷嘴直径的坐标相重合。混合最强烈地段既不可能在内边界(这里是弱扰动区),也不可能在外边界上(这里速度趋于零),而应位于两者之间。第三、作为表征脉动大小的紊流强度(或称为卡门数)采用下式表示:

$$\varepsilon = \frac{W'}{\overline{W}} 100\% \qquad (注)$$

我们知道,对自由射流平均速度而言:$W_x \gg W_y \approx W_z$。但对脉动速度则情况不同,虽然纵向脉动数值最大,但三个方向脉动速度值基本上处于同一数量级;即 $W'_x \approx W'_y \approx W'_z$(见图 4)。

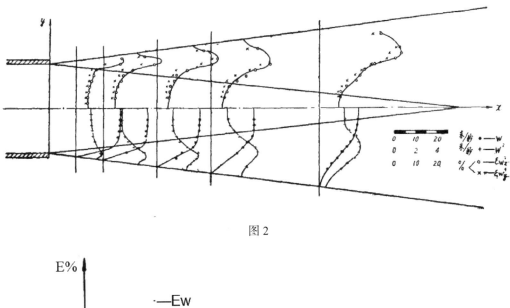

图 2

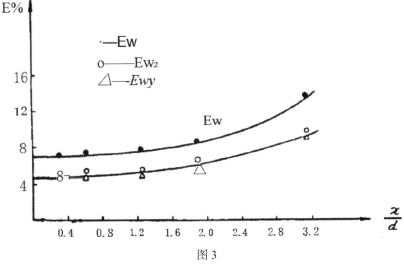

图 3

注:为简单起见,这里和以后符号 W' 代表脉动速度的平方平均开方值,即 $\sqrt{\overline{W'^2}}$

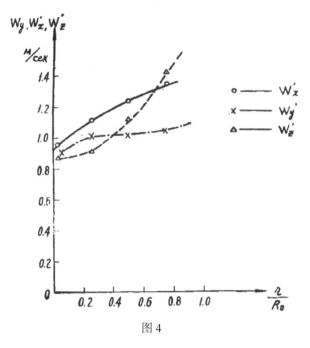

图 4

如果把各截面所测量数据用无因次坐标 y/b 来总结(这里 b 代表射流宽度),可以发现,在初始段核心区和边界层上紊流参数基本是自模的(图 5、6)。АнтоноваГ.С. 亦得出过同样的结论(图 5 中虚线所示)。应当指出,Антонова 的数据比我们高 30%～50%;例如在边界层中最大纵向紊流强度按我们数据为 $\varepsilon_x = 16\%$;按 Антонова 数据为 24%,这是因为她采用了误差较大的单丝感应元件来测量两元脉动的结果。我们的数据和英美学者数据是相近的,例如 Corrsin 氏在测量圆形射流时,当 $\frac{x}{d}=5$,最大的 $\varepsilon_x = 18\%$(文献 1)。数据比我们略高的原因是 $\frac{x}{d}=5$ 时已靠近主体段,因而应有较高的紊流脉动值。

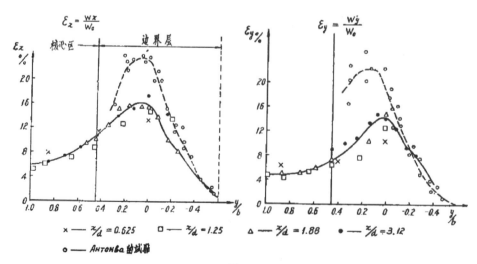

图 5

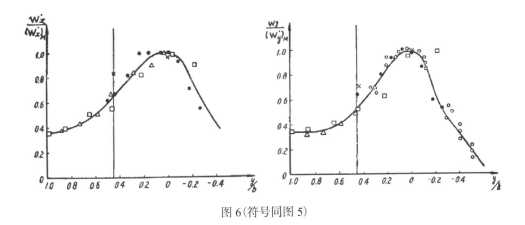

图 6（符号同图 5）

方向关联系数 $R_{W_x'W_y'} = \dfrac{\overline{W_x'W_y'}}{\sqrt{\overline{W_x'^2}}\sqrt{\overline{W_y'^2}}}$ 在射流起始段具有独特的变化规律（图 7）；在射流核心处（具有最低值）向边界层不断增加，但在边界层中却近似保持为常量，而且比核心区大 2～3 倍。这在目前还未能找出满意的解释，可能是因为在边界层中紊流摩擦应力 $\tau = -\rho \overline{W_x'W_y'}$ 比核心区高得多的缘故。同时测量结果表明，在射流的初始段不同方位的方向关联系数近似相等（图 8），并且在不同截面处有不同的数值，即没有出现自模现象。不过其平均值（不论是方向关联系数或是紊流强度）基本上都不随雷诺数变化（图 9、10），这点和平均参数自模有相同之处。至于在射流中紊流脉动频率问题，现有数据范围变化很大，并且发现和喷嘴出口直径、射流起始紊流强度等参数有关。测量结果表明，随着射流往前发展脉动频率不断降低。例如：对我们所研究出口直径为 80mm 的射流，在初始段（$\dfrac{x}{d}$ =0.625）内平均脉动频率为 199 赫，到达主体段时则降为 117 赫。知道平均脉动频率后，就有可能根据下式近似求得欧拉紊流标尺：

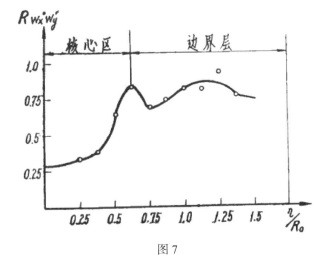

图 7

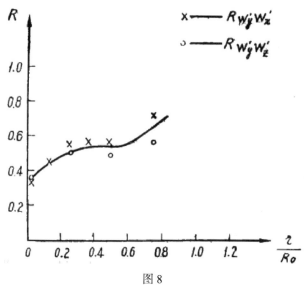

图 8

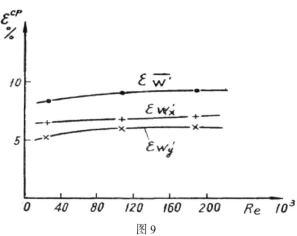

图 9

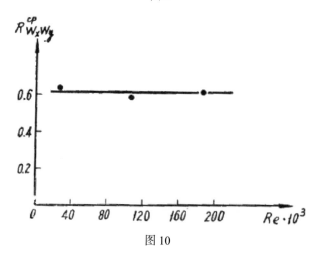

图 10

$$L = k\frac{\overline{w}}{f}$$

式中，k 为试验系数，对自由射流可近似取为 0.0268（文献 6）；$\overline{w}$、f 为气流平均速度及脉动频率。

这样可以得出初始段 L=2.7mm，在主体段为 4.62mm，这和试验所得的 Prandtl "混合长度" l 在数量上是同级的，并且沿射流的发展 L 不断增加（见图 12）。

§4. 有关自由射流紊流结构几个假定的讨论

由于紊流机理（即使最简单的各向同性紊流机理）至今还没有完全弄清楚，因此更不可能对复杂的紊流流动得出准确的理论解。为了满足工程上日益发展的需要，学者们从物理模型出发拟定了一系列半经验的紊流理论。其中最有名的是 Prandtl 底紊流微团动量转移理论和 Taylor 底旋涡转移理论。特别是 Prandtl 理论应用于解决射流工程问题更富成效（文献 1、2、3），直至今天仍保持着很大的实用价值。

1) 关于 Prandtl 旧理论有关假说讨论

早在 1925 年 Prandtl 为了解决工程问题把研究紊流运动的困难归结到所谓紊流微团"混合长度" l 上去，此时脉动速度可写成：

$$w'_x \approx w'_y = l\left(\frac{\mathrm{d}w}{\mathrm{d}y}\right)$$

则紊流摩擦应力等于：

$$\tau_{xy} = -\rho\overline{w'_x w'_y} = \pm\rho l^2 \frac{\mathrm{d}w}{\mathrm{d}y}\left|\frac{\mathrm{d}w}{\mathrm{d}y}\right| \tag{1}$$

如果我们得知 l 的变化规律，那么描述紊流运动的雷诺方程原则上能得出最终理论解答。显然，"混合长度" l 的变化规律只能根据流动的物理模型作出假定，当然对不同的流动工况应假定不同的变化规律。对于自由射流，Prandtl 根据平均参数沿射流不同截面自模（相似）的试验事实出发，假定混合长度沿射流横截面不变，而随着射流向前发展成正例地增加，即 $l \neq f(y)$ 而 $l=cx$。这样就得出了射流一系列的理论解答（见文献 3），并且和大量的速度场试验相符合，但这并不说明关于紊流模型的假定就是正确的。这方面各国学者持有不同的意见：例如 Pai Shi—I 根据 Corrin 及 Lilpmann 紊流参数测量的数据进行换算，认为沿射流横截面混合长度并没有保持为常量（文献 2），不过作者可能主要根据射流主体段的数据整理所得的结果。根据我们的测量证明，在射流初始段的边界层中不论是纵向或横向混合长度沿射流横截面分布在误差 20%范围内可以认为 Prandtl 的假定是正确的，即基本上保持为常量（图 11，А、Б），沿射流纵向混合长度线性增长的假定则比较准确（图 12）。根据试验我们可以总结成：

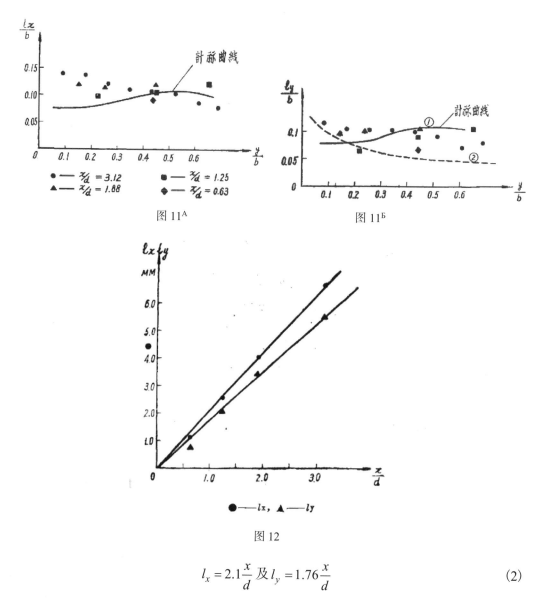

图 11ᴀ   图 11Б

图 12

$$l_x = 2.1\frac{x}{d} \text{ 及 } l_y = 1.76\frac{x}{d} \tag{2}$$

这些数据只对由技术管道喷出的射流才合用，过去很多学者的试验研究都是在风洞中进行的，在射流喷出以前往往用各种栅网对气流加以"过滤"、"稳定"，因而使得紊流标尺（在数值上与混合长度同数量级）随着栅网孔径减少而减少，或许正是由于这样使得他们的测量结果和 Prandtl 的假定偏差较大。

2）关于 Prandtl 新理论有关假说的讨论

Prandtl 关于混合长度假说对某些流动情况（例如旋转射流等）的讨论在数学上带来了较大的困难，因为按照式（1）在解雷诺方程时包含有速度一阶导数的平方项，为了克服上述困难，于 1942 年 Prandtl 提出了新的假说。此时他不假定混合长度沿射流横截面是常量，而是假定紊流粘度系数 A_T，在横截面中保持不变，这样紊流切应力就只与速度一阶导数一次方成比例，理由如下：

用紊流黏度系数表示的紊流切应力式可写成

$$\tau_{xy} = -\rho \overline{w'_x w'_y} = A_T \frac{dw}{dy} \tag{3}$$

和(1)式比较可以发现

$$A_T = \rho l^2 \left(\frac{dw}{dy} \right) \tag{4}$$

由于在横截面速度侧形相似可以认为

$$d\frac{W}{W_M} \bigg/ d\frac{y}{b} = \text{idem} \ \ \text{及} \ \ \frac{l}{b} = \text{idem}$$

则式(4)可改写成

$$A_T = \rho l^2 \left(\frac{dw}{dy} \right) = \rho b W_M \left(\frac{l}{b} \right)^2 \frac{d\left(\frac{W}{W_M}\right)}{d\left(\frac{y}{b}\right)} = k\rho b W_M \tag{5}$$

这里 W_M 为射流横截面上最大速度；k 为比例常数。

对于自由射流来说，式(5)表明 A_T 在横截面上保持为常量，因为(5)式中各参数均与 y 轴无关，但 A_T 沿纵向变化则取决于 bW_M 乘积，Л.Г.Лойцянский 教授及其学派更进一步发展了 Prandtl 的理论，认为对射流来说 $bW_M=\text{const}$，因为一般对射流的解可以得出 $b \sim x^n$，而 $W_M \sim \frac{1}{x^n}$，故 A_T 不但在横向而且在纵向均保持为常量，即可作为一物理常数看待（文献 7）。这样有关流体力学对层流的理论解均能适用于紊流工况，此时只不过黏度系数有不同的数值而已。于是 Лойцянский 就把这样的假说推广到旋转射流中去，当然这样的推论在解决紊流问题时带来很大的方便，但是不能不对他们的假设进行分析，特别是进行试验的验证。

a) 如果认为 Prandtl 的旧理论是对的话，即位移径在射流横截面上近似保持为常量，则按公式(4)可知在横截面上

$$A_T = \rho l^2 \left(\frac{dw}{dy} \right) = \text{const}$$

的实质是要求 $\frac{dw}{dy} \approx \text{const}$，亦即

$$\frac{dw}{dy} \approx \frac{\omega_M - \omega_1}{b}$$

这里 ω_M、ω_1 分别为射流核心及边界层外的气流速度。

亦即近似把速度在边界层变化规律取成线性(文献 13),和试验比较可知,这在边界层起始及结束段内误差是较大。

b) Лойцянский 认为 A_T 不但沿横向,而且在纵向均保持为常量,在自由射流初始段,这样的假想是值得商榷的,因为在初始段内 $\omega_M = \omega_0$。即为射流出口速度是个不变值,而 $b \sim k'x$.随 x 增加而增加,因 $A_T = K\rho b\omega_M = KK'\rho x\omega_0 \neq \text{const}$,而近似和 x 成正比例,我们的试验也证明了这一点,(见图 14)在自由射流主体段,$b\omega_M = \text{const}$(即 $A_T = \text{const}$)是比较符合实际情况,但是对于旋转气流则必须作进一步的研究,此时,ω_M 和坐标 x 是个比较复杂的函数。根据文献 7,$\omega_M = f\left(\dfrac{1}{x}\cdots\dfrac{1}{x^2},\cdots\right)$,此时 A_T 是否沿旋转射流横截面及长度保持不变是值得深入研究的,根据我们的初步试验,(文献 14),对于半自由旋转射流,A_T 至少沿流横截面是个变量,而且变化颇大(在边界层中尤为显著),Д.Н.Ляховский 采用 $A_T = \text{const}$ 的假定得出平均速度场分布理论公式,应用到喷燃器的旋转射流中(见文献 8),并且由所得速度场反过来求解 A_T 值,我们认为,这样计算 A_T 值的方法是缺乏足够根据的,所得的结果不能代表真正的 A_T 值,而只能是某个试验的转换系数。

c) 从紊流微观结构出发,所谓紊流粘度系数用拉格朗日法表示时:

$$A_T = w'L_1 = \overline{w'^2}\int_0^\infty R_\tau d\tau \tag{6}$$

或用欧拉法表示时:
$$A_T = w'L_2 = w'\int_0^\infty R_y dy \tag{7}$$

这里,L_1 和 L_2 为拉格朗日和欧拉法表示的紊流标尺,这里它们认为是同一数量级;R_τ、R_y 为时间关联系数及横向座标关联系数。

根据 Prandtl 假说 A_T 沿射流横截面是常量,则从微观上要求脉动速度与紊流标尺的乘积为常量;这只有在两参数均为常量或两参数变化规律刚相反时才有可能。上述试验证明,紊流标尺(如果在数量上和混合长度相近似的话)变化不会很大,而脉动速度在射流边界层中则变化较大(见图 6),因此在射流横截面上 A_T 保持常量的假说比混合长度保持常量的假说"微观"误差较大。试验证明了上述分析。在图 13 中列出在起始段不同横截面上无因次紊流粘度系数的分布,试验点很分散,公式(5)中系数 k 沿横截面分布变化很大(由 0.5 变至 1.6),作为极粗略的近似,式(5)可写成:

$$A_T = 1.05 \times 10^{-3} \rho b W_M \tag{8}$$

关于紊流粘度系数 A_T 在初始段沿射流向前推进的变化规律示于图 14,试验发现,A_T 沿纵向发展并不保持为常量,而是按规律:

$$A_T = ckx = f(x) \tag{9}$$

根据试验可粗略表示为

$$A_T = 5 \times 10^{-3} \left(\dfrac{x}{d}\right)^{0.8-1} \text{米}^2/\text{秒} \tag{10}$$

由此可见 Лойцянский 把 A_T=const 推广到旋转射流是缺乏足够根据的（见文献14）。

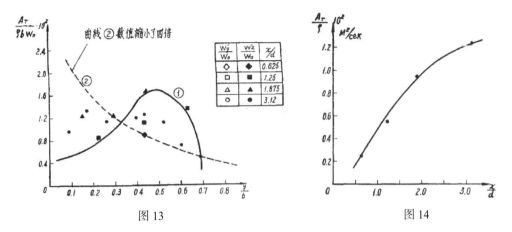

图 13　　　　　　　　　　图 14

3）对 Г.Н.Абрамович 关于自由射流速度场分布假定的讨论

Г.Н.Абрамович 教授根据 Prandtl 的假说，总结了大量试验拟定了自由射流全面的工程计算方法，普遍被应用于解决各种实际问题。关于速度分布 Абрамович 认为在射流边界层和射流主体段上是自模的，并采用了 Schlichting 关于射流绕圆柱形物体较远处的速度场分布理论公式作为射流速度场分布公式，即：

$$w_X = w_M \left[1 - \left(\frac{y}{b} \right)^{3/2} \right]^2 \tag{11}$$

公式(11)对求解平均速度的准确度是无可怀疑的，但它能否符合射流紊流结构的物理模型又是另一回事了！下面提出一些初步的分析。

我们知道对于紊流自由射流可以把复杂的雷诺方程及连续方程式简化成（见文献3）：

$$\omega_x \frac{\partial \omega_x}{\partial x} + \omega_y \frac{\partial \omega_x}{\partial y} = \frac{1}{\rho} \frac{\partial \tau_{xy}}{\partial y} = \frac{1}{\rho} A_T \frac{\partial^2 \omega_x}{\partial y^2} + \frac{1}{\rho} \frac{\partial A_T}{\partial y} \frac{\partial \omega_x}{\partial y} \tag{12}$$

$$\frac{\partial \omega_x}{\partial x} + \frac{\partial \omega_y}{\partial y} = 0 \tag{13}$$

解上方程式组求紊流参数应满足下列边界条件：

A）在射流轴心处　　　　$y=0$；$\tau_{xy}=0$

B）在射流边界层外　　　$y \geq b$；$\tau_{xy}=0$

第一种情况：最简单情况，可近似认为 $W_y=0$，射流宽度 b 与喷嘴距离成正比，即 $b=kx$，k 为试验常数，对圆形射流可取为 0.258。

设　　　　$$\frac{y}{b} = \frac{y}{kx} = \eta \tag{14}$$

并利用第一边界条件（$y=0$，$\eta=0$，$\tau_{xy}=0$）积分方程式(13)可得：

$$\tau_{xy} = -\rho \overline{\omega'_x \omega'_y} = \frac{\rho \omega_M^2 k}{1540}\left[3390\eta^4 + 660\eta^7 - 280\eta^{\frac{11}{2}} - 1848\eta^{\frac{5}{2}}\right]\left[\frac{公斤}{米^2}\right] \quad (15)$$

同理由式(3)、(4)及(6)得出

$$A_T = \frac{\rho b \omega_M k}{4620} \frac{\left(1848\eta^{\frac{5}{2}} - 3390\eta^4 + 280\eta^{\frac{11}{2}} - 660\eta^7\right)}{\eta^{\frac{1}{2}}\left(\eta^{\frac{3}{2}} - 1\right)}\left[\frac{公斤 \cdot 秒}{米^2}\right] \quad (16)$$

$$l = \sqrt{\frac{kb^2\left(1848\eta^{\frac{5}{2}} - 3390\eta^4 + 280\eta^{\frac{11}{2}} - 660\eta^7\right)}{13860\eta\left(\eta^{\frac{3}{2}} - 1\right)^2}}\,[米] \quad (17)$$

$$\frac{\omega'}{\omega_M} = \sqrt{\frac{k}{1540}\left(1848\eta^{\frac{5}{2}} - 3390\eta^4 + 280\eta^{\frac{11}{2}} - 660\eta^7\right)}\,\% \quad (18)$$

第二种情况：考虑 $\omega_y \neq 0$。由式(11)、(14)可知，速度分布只与新设函数 η 有关，因而上述偏微分方程可改写成常微分方程系。解方程式(13)可得

$$\frac{\omega_y}{\omega_M} = k\left(\frac{3}{4}\eta^4 - \frac{6}{5}\eta^{\frac{5}{2}}\right) + c$$

在轴心处 $y=0$；$\eta=0$，$\omega_y=0$，∴ $c=0$，可得

$$\omega_y = k\omega_M \eta^{\frac{5}{2}}\left(\frac{3}{4}\eta^{\frac{3}{2}} - \frac{6}{5}\right) \quad (19)$$

代入式(12)并考虑到式(3)、(4)、(6)，同时如果采用第二边界条件：$y=b$；$\eta=1$，$\tau_{xy}=0$。则可分别积分得出：

$$\tau_{xy} = k\omega_M\left(\frac{2}{5}\eta^{\frac{5}{2}} + \frac{21}{110}\eta^{\frac{11}{2}} - \frac{9}{20}\eta^4 - \frac{1}{28}\eta^7 - \frac{81}{770}\right) \quad (20)$$

$$A_T = \frac{\rho b k \omega_M\left(\frac{2}{5}\eta^{\frac{5}{2}} + \frac{21}{110}\eta^{\frac{11}{2}} - \frac{9}{20}\eta^4 - \frac{1}{28}\eta^7 - \frac{81}{770}\right)}{\eta^{\frac{1}{2}}\left(\eta^{\frac{3}{2}} - 1\right)} \quad (21)$$

$$l = \sqrt{\frac{b^2 k \left(\frac{2}{5}\eta^{\frac{5}{2}} + \frac{21}{110}\eta^{\frac{11}{2}} - \frac{9}{20}\eta^4 - \frac{1}{28}\eta^7 - \frac{81}{770}\right)}{3\eta\left(\eta^{\frac{3}{2}} - 1\right)^2}} \quad (22)$$

$$\frac{\omega'}{\omega_M} = \sqrt{3k\left(\frac{2}{5}\eta^{\frac{5}{2}} + \frac{21}{110}\eta^{\frac{11}{2}} - \frac{9}{20}\eta^4 - \frac{1}{28}\eta^7 - \frac{81}{770}\right)} \quad (23)$$

讨论：a) 所得的两种解无法同时满足两个边界条件。例如公式(15)能满足第一边界条件而不能满足第二边界条件，公式(20)则相反。因此，对于式(15)可以发现 $\eta=0.68$ 时，所有紊流参数等于零，当 $\eta>0.68$ 时则变成负值，这明显不符合射流外边界层的物理意义；对于式(20)，所得的紊流参数随 η 减少而单值地增加，在轴心处 ($\eta=0$) 达到最大值，这也明显不符合内边界层的物理意义。可见由式(11)得出的紊流参数公式(15)至(23)在实际中不能应用。

b) 由式(17)及(22)可知，$l = bf(\eta) = bf\left(\frac{y}{b}\right)$；当 $\frac{y}{b}=$const 时，$l = b \cdot $const$ = $const$ \cdot x$，即混合长度随射流发展成正比地增长，这和图12所示试验结果相符。同样由(16)及(17)式可知，$A_T = bf(\eta) = $const$ \cdot x$，这存有一定偏差(和式(10)比较)。

c) 由公式(15)至(18)计算所得的无因次紊流参数沿横截面分布和试验比较在数量级上基本是符合的(见图11、13、15、16中曲线(1))，但是变化规律有较大的偏差，特别是当 $\eta>0.7$ 时。在计算时考虑到实际情况对第一边界条件进行了修正，即当 $y=0$；$\eta=0$ 时 $\tau_{xy} = \tau_{xy}^0 \neq 0$，$\tau_{xy}^0$ 为射流起始段核心区切应力，因为测量发现，在射流核心区 τ_{xy} 并不等于零而有一定值，其大小应与起始脉动强度有关，在我们的试验中此值等于：$\tau_{xy}^0/\rho W_M^2 \approx 4\times10^{-3}$，在图15中示出修正前后的比较(曲线(1)与(3))。

d) 由第二边界条件所决定的公式(20)至(23)中，其计算结果与试验比较不但变化规律不同，而且对 τ_{xy} 及 A_T 值在数量级上比试验高四倍，而紊流强度及混合长度则得出偏低的结果(见图11、13、15、16中曲线(2))。

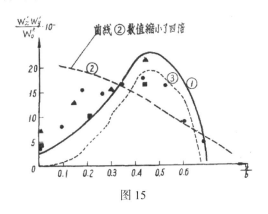

图15

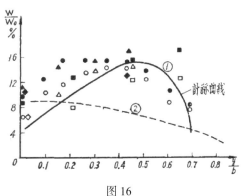

图16

e) 根据上述计算和试验比较可以得出下面几个结论：

第一、就普遍意义来说，由于紊流理论目前发展得还不够完善，因此往往只能根据一定物理模型提出紊流机理的假定，由此而得出速度场计算公式从理论上来说不能是很准确的，必须经过大量试验的校核。

第二、即使得出的速度场计算公式和试验比较非常吻合，并不能由此得出结论，证明推演速度场计算公式时所采用的紊流机理模型是正确的。例如 Prandtl 的动量转移理论和 Taylor 的旋涡转移理论两者在物理模型上显然是不同的（文献 1、2），但是所得出自由射流速度场分布的理论公式一样，而且和试验比较也得出令人满意的结果，但是这并不能证明 Prandtl 及 Taylor 的紊流模型都同样正确，很明显，如果把 Prandtl 理论应用于不等温射流中和试验比较则相差较大。由此可见，紊流机理模型的是否正确只能通过直接测量加以证明。同时必须指出，一定的"紊流模型"只能对特定的某些流动情况才有效，而不能任意加以推广。例如 Prandtl 的新旧理论对自由射流是近似符合的（在数量级意义上），但对旋转气流边界层则不能应用。

第三、根据一定的紊流模型，应用被大量试验证实的经验公式去求解紊流参数时，所得的结果如果没有准确的试验检查即使在定性上也不能推荐使用。Абрамович 式(11) 是公认被无数试验证实的公式，当应用来求解紊流参数时虽然数量级和试验相差不远，但变化规律却不大相同。这说明为什么在紊流理论中试验研究占有如此重要的地位。

三、射流在非流线型物体（圆柱和管簇）后尾迹的紊流结构

§1. 问题的提出

在自由射流中放置非流线形物体时，在物体后面出现有紊流尾迹，完全改变了自由射流的紊流结构，使紊流混合和脉动大为加强，因此非流线形物体广泛应用在燃烧技术及强化热质交换的过程中。这里我们选择了圆柱和管簇作为研究对象。根据文献记载：早在 1949 年美国学者 A.Townsend 对圆柱后紊流尾迹进行过详细研究，在同时期，苏联学者 Ю.Г.Захаров 也对圆柱后尾迹的一元脉动进行过测量，直至最近 Л Н.Уханова 把圆柱放在扩散喷咀内，研究有微小纵向压力梯度对紊流尾迹的影响（文献 9）。不过，上述作者都主要研究了远离圆柱后的情况（测量最近截面不小于 30d），至于管簇后尾迹，除了作为风洞平稳气流的栅网测量外，则研究更小，但是对解决稳定燃烧、强化热质交换来说主要兴趣集中在非流线形物体后的紊流结构，这就决定了我们的研究任务。总共分为三个试验方案：

第一方案：直径为 30mm 的圆柱，放在射流初始段，射流的绕流速度为

W=5,8,20,35,40 米/秒，测量的范围在 $\frac{x}{d}=1\sim 6.5$。

第二方案：直径为 80mm 圆柱组成两排错列管簇，横向管距 $\frac{s_1}{d}=3.5$，纵向管距

$\dfrac{s_2}{d}=1.5$，射流绕流速度为 $W=8,20,40$ 米/秒，测量范围在 $\dfrac{x}{d}=0.5$ 至 6.5。

第三方案：直径为 8mm 组成两排错列管簇，$\dfrac{s_1}{d}=2$，$\dfrac{s_2}{d}=1.5$，$W=20$ 米/秒。

主要测量了两元脉动，部分地测量了三元脉动。对单个圆柱的紊流尾迹作了比较详细的试验，因为它是组成管簇的基础，而对管簇后尾迹的测量主要是探讨各单个圆柱间的相互影响。

§2. 射流在圆柱后尾迹的紊流结构

当射流绕流圆柱时，在和气流方向相差 20°的圆柱表面附近，出现了边界层的脱离现象，之后在圆柱后形成了强烈的旋涡区，部份气流作反向流动，因此可以想象，绕流圆柱后，原来的射流紊流结构必然受到破坏，气流必然更加"紊流化"，在图 17 中得到很好的证明；这里列出了在圆柱后尾迹三元脉动紊流强度的变化，试验是按公式 $\varepsilon=\dfrac{W'}{\overline{W}}$ 总结的，$\overline{W}$ 代表该点的气流平均矢量速度。在圆柱前面，气流脉动完全和自由射流一样，一般 $\varepsilon=5\%\sim7\%$，但绕流圆柱之后，紊流脉动立即增高了 6～7 倍，在尾迹处，虽然平均速度很小或甚至是负的(逆流)，而且由尾迹边缘至轴心变化很大——不断地减少，但紊流强度在尾迹整个范围内却保持为常量，在尾迹外部，则保持为自由射流的紊流特性。距离"脉动源"(圆柱)愈远，尾迹紊流强度不断衰减，而自由射流的紊流强度却不断增加，因而在横截面上整个脉动场就愈来愈均匀。试验表明，在尾迹处纵向和横向间的脉动规律和射流也有显著的不同。在自由射流中，纵向脉动 W'_x 是大于横向 W'_y，因为射流是以前进运动为主的，但在圆柱后尾迹处，W'_y 则明显地大于 W'_x (图18)。我们认为，这是因为在尾迹后出现了强大的环流旋涡、因而大大地增强了横向脉动。气流脉动速度在尾迹后的变化和平均速度类似——在尾迹中心处达到最小值，但是在图 18 中可以发现，虽然在圆柱后平均速度比圆柱前射流速度少近十倍，但前者脉动速度的绝对值却比后者大一倍。可见，在这里气流运动速度很慢，而紊流脉动及混合却非常强烈，因而对稳定燃烧做成了极为有利的条件。在图 19 中列出脉动速度在尾迹处的发展过程，其特点是在尾迹边界处出现有两个脉动速度"高峰"，这里的热质交换进行最强烈，例如流过圆柱的

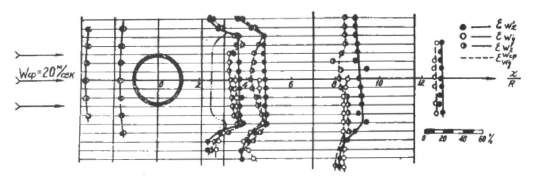

图 17

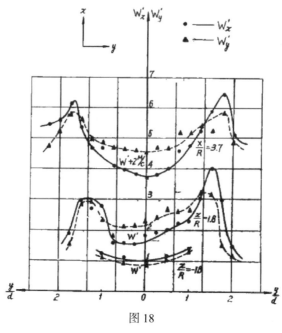

图 18

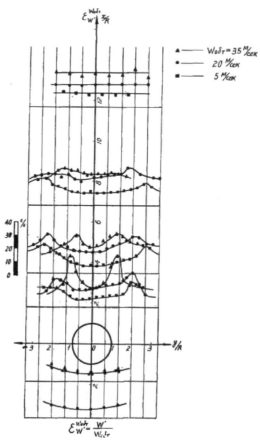

图 19

强迫对流换热试验证明，正是在这里出现有局部的最高放热系数值。同时脉动"高峰"随射流发展还不断向外"膨胀"，即尾迹影响范围不断扩大，尾迹内脉动速度不断被拉平，当 $\frac{x}{d}=6$ 时，脉动"高峰"实际上已消失，而脉动速度则趋于均匀，并且渐渐衰减；但是不是直接在圆柱后出现最高的紊流脉动呢？图20给出了否定的答复，原来脉动最强烈的截面并不直接在圆柱后，而是在其后一定距离处，这个距离取决于绕流圆柱的雷诺数，并随之增大而缩短，具体数据列于下表：

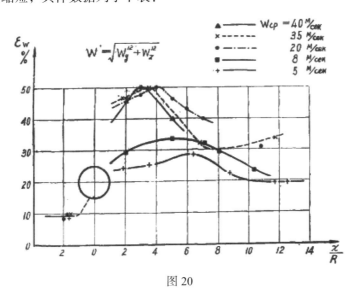

图 20

绕流速度 ω 米/秒	5	8	20	35	35
雷诺数 Re	10000	16000	40000	70000	80000
最大脉动截面位置 $\frac{x}{d}$	3.25	2.5	2.0	1.75	1.5
最大紊流强度 ε_{max}%	28.1	33.5	50.0	49.7	50.5

由试验数据可见：在较低的雷诺数下圆柱尾迹后紊流脉动增长不多，且随 Re 值提高而增加，当 $Re \geqslant 40000$ 时达到稳定值，紊流强度保持在50%左右。关于紊流参数自模的问题，Л. Н.Уханова 曾经作出这样的结论，认为在离圆柱较远处，紊流参数是自模的，具有通用的紊流参数分布曲线。但是令我们感兴趣离圆柱不远处的尾迹范围内既没有通用性的紊流参数侧形，又没有自模的特性，例如在图19、图20可以发现，随 Re 值提高时，所出现两个脉动高峰逐渐接近，这种特性给理论研究带来了困难。不过紊流参数在尾迹后对称性仍被保持着，使得今后研究工作局限于在半圆柱内进行。

最后，在图21中列出方向关联系数 $R_{\omega'_x \omega'_y} = \dfrac{\overline{\omega'_x \omega'_y}}{\sqrt{\overline{\omega'^2_x}} \sqrt{\overline{\omega'^2_y}}}$ 及 $R_{\omega'_x \omega'_z} = \dfrac{\overline{\omega'_x \omega'_z}}{\sqrt{\overline{\omega'^2_x}} \sqrt{\overline{\omega'^2_z}}}$ 的典型变化曲线。如果和图8比较，我们可以发现和自由射流的情况有显著不同，在圆柱尾迹

处经常出现 $R_{\omega'_x\omega'_y} > R_{\omega'_x\omega'_z}$，并且在尾迹边界处具有最小值。从物理意义来看这是很明显的，因为正是这里紊流脉动和混合最为强烈。在尾迹中心处方向关联系数不可能达到最小，亦即气流不可能是"各向同性"的，因为这里有强烈的环流使得横向脉动大于纵向（见图 18）。同样的，方向关联系数的最小值并不直接位于圆柱后，而在距圆柱不远处，约相应气流脉动最大值的截面，之后 $R_{\omega'_x\omega'_y}$ 逐渐升高到射流所特有数值，即圆柱的"激化"作用逐渐消失（图 22）。根据上述试验结果，至少可以得出下列两个一般性结论作为进一步分析燃烧稳定问题的参考：

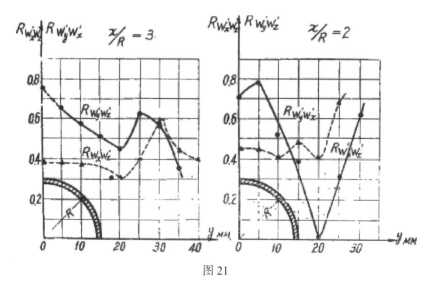

图 21

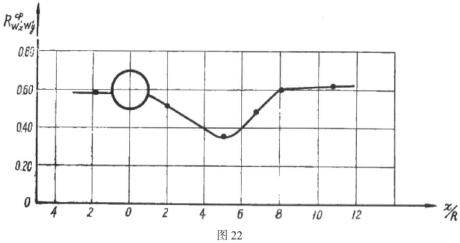

图 22

第一、作为非流线型物体的圆柱是有效的紊流"发生器"，它能激发出比初始值大 4～6 倍的紊流脉动值，同时使方向关联系数相应降低几倍，做成极为有利的紊流混合条件，另一方面在尾迹处气流速度很低，甚至是负值，这样就给火焰的稳定及燃料的燃烬创造了前提。

第二、紊流脉动最大值和方向关联系数的最小值并不直接在非流线型物体后，而是和它有一定距离，即有一个较大的紊流"活化区"。

§3. 射流在管簇后尾迹的紊流结构

在研究绕过圆柱尾迹后，对于管簇的研究有理由主要限于各圆柱间的相互作用及管距对气流结构的影响。为简单起见，下面我们将只提出一些比较典型的结果。

1) 试验证明在管簇每个圆柱后的紊流结构和单圆柱后基本相同，但值得惊异的是在管簇后紊流参数是基本自模的，这在图23中得到很好的证明（总结紊流强度时我们应用公式 $\varepsilon = \dfrac{W'}{W_{cp}}$）在管簇后出现自模的特性我们认为正是各圆柱相互作用的结果，因为各圆柱后尾迹的变化和单圆柱不同，被一定管距所限制，而不能自由膨胀或收缩。同时由图23中看到紊流参数的分布好象由一排圆柱后尾迹组成，似乎管簇第一排（和气流方向相对应算起）对紊流结构作用不大，事实并非如此，管簇是错列的，在第一排管后就已经把气流"活化"起来，如果只有一排管的话，则"活化"气流经过一定距离后就会衰减，第二排管的出现使被第一排管分成两股的射流重新汇合，并加以压缩，因而在第一排管各中心线位置处出现了最大的脉动速度值（单圆柱试验时这里刚好为最小值）。这样至少可以认为第一排管对提高紊流脉动起很大作用，而第二排管则起着加速紊化，并且成为确定尾迹尺寸的主要因素。

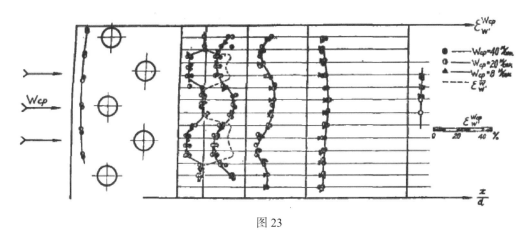

图 23

2) 管簇中各圆柱相互作用第二个结果是各旋涡区互相重叠，横向脉动得到强烈发展，试验中发现，在 $\dfrac{x}{d} \leqslant 14$ 范围内横向脉动 W'_y 几乎到处略大于纵向 W'_x，而在管簇前的射流中则到处 $W'_x > W'_y$，可见管簇具有良好的紊流混合及扰动作用（图24、图25）。并且在 $Re=10000\sim 20000$ 时，脉动的最大值约位于 $\dfrac{x}{d}=4.5$ 截面上，但当雷诺数降至4000时和单圆柱的情况相反，即脉动最大值的截面不但没有后移，反而前移至 $\dfrac{x}{d}=2\sim 3$ 的截面上，这明显的是第一排管作用结果。同时这也相应地引起了方向关联系数最小值截面的前移（图25）。

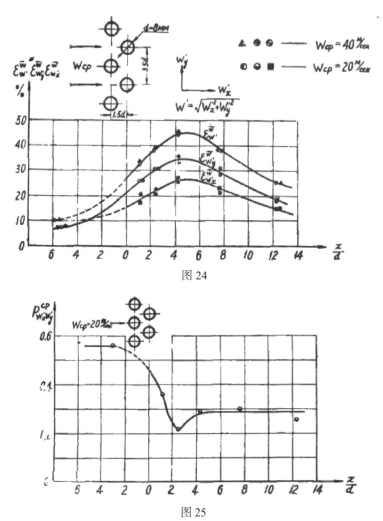

图 24

图 25

3) 横向管距对气流结构影响。从加强紊流混合的角度出发最有利的横向管距应保证有较小的阻力，能激化足够大的紊流脉动，这两个要求有一定矛盾，管距增大阻力会降低，但管距不应大于"活化区"尺寸两倍（"活化区"尺寸也就是尾迹范围），作为例子在下表中列出在 $\frac{x}{d}=1.8\sim2$ 截面上活化区尺寸 $\frac{h}{d}$ 随气流速度的变化（在单圆柱后的试验数据）：

W_{cp} 米/秒	5	8	20	35	40
h/d	2.3	2.3	2.0	1.5	1.3

因此可以预料，随着管距的减小，沿纵向最大脉动的范围必然逐渐扩大，例如当把

横向管距由 $\frac{s_1}{d}$=3.5 降至 2 时，发现在截面 $\frac{x}{d}$=1.8 后紊流脉动已经足够均匀，并且很快达到最大值（图 26），此时紊流强度的最大值并不随 s_1/d 的减小而变化，一般保持在 40%～45%，但最大值范围则扩展到由 $\frac{x}{d}$=0 至 $\frac{x}{d}$=10，同时管距的减小也促使横向及纵向脉动的差距接近，这都是气流混合加强的特征。

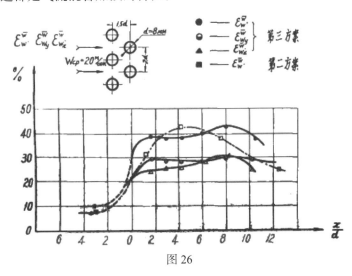

图 26

4）为什么流过非流线型物体紊流脉动及混合都会加强呢？可以从能量的观点加以解释。紊流脉动的形成或加强，都必须耗费部份能量，即流动阻力必须增加。耗费于紊流脉动的能量可以用下面公式来估计：

$$\Delta F_g = \frac{\rho}{2}W'^2 \bigg/ \frac{\rho}{2}W_{cp}^2 \%$$

即脉动动能占气流总动能的百分数。在图 27 中示出所得试验结果，为了比较起见，在图中同时列出 Е.М.Минский 在管内和扩散喷嘴内测量所得的数据（文献 5）。可以发现，在管簇前即自由射流核心区内耗费于紊流脉动的能量接近于管内流动的数值，但在管簇后比它们高出几倍，因而阻力损失也就增大几倍，只有当 $\frac{x}{d}$>11 时，管簇影响才消失。

以上试验结果亦可应用于其他非流线型物体（如燃烧室的稳定器之类）（文献 11、12）。Westenberg 曾经对圆柱、正三角形及 V 型槽道后紊流结构进行过一元脉动的测量（应用扩散法），发现稳定器的几何形状对紊流结构没有影响，作者认为绕非流线型物体后的旋涡频率及其他紊流特性主要取决于非流线型物体的特性尺寸，就这意义上来说，对圆柱及管簇后尾迹作详细的试验研究是有普遍的意义的。

上述试验和分析只是初步的不成熟的工作小结，至于应用上述结果来讨论火焰的稳定及燃料的燃烬已超出本文的范围，而且在今后必须作进一步的燃烧试验。

参 考 文 献

(1) J.O.Hinze　Turbulence　1959
(2) P_Ai Shih—I　Fluid Dynamics of Jets　1955
　　1960年有俄译本
(3) Г.Н.Абрамович　Теория Турбулентных струй　1960
(4) Г.С.Антонова. А) Труды Совещания по прикладной газовой динамике. 1959
　　Б) Диссертация　1957
(5) Е.М.Минский　Турбулентность руслового потока.　1952
(6) В.П.Солнцев.　在论文集 "Стабилизация пламени и развитие процесса сгорания в турбулентном потоке."　1961
(7) Л.Г.Лойцянский.　А) ПММ Т.XVII. вып.1.　1953
　　Б) Труды ЛПИ Энергомашиностроение №.176. 1955
(8) Л.Н.Ляховский.　在论文集 "Теория и Практика сжигания газа"　1958
(9) Л.Н.Уханова.　Труды ЦАГИ "Промышленная аэродинамика" вып.23. 1962
(10) 岑可法　气流紊流结构的研究方法　见本期浙大学报.
(11) A.Westenberg　Journal of Chemical Physics　1954　№5.
(12) А.Г.Прудников　Изв.АН СССР ОТН　1958　№7.
(13) Л.Г.Лойцянский　Механика жидкости и газов　1957
(14) 岑可法　旋风燃烧室内气流紊流结构的研究　见本期浙江大学学报

原文刊于浙江大学学报, 1963, 2: 63-84

旋风燃烧室内气流紊流结构的研究

岑可法

摘 要

本文对旋风炉内强旋转气流的紊流结构进行了理论和试验的研究，得出了紊流运动雷诺方程在强旋转气流下的近似解，并由此作出旋风炉内半自由射流的模型及紊流运动自模的结论。对所提出的物理模型、假定及理论计算进行了试验校核，得出了令人满意的结果。发现旋风炉内的紊流脉动绝对值比一般煤粉炉高近十倍，因而燃料在旋风炉内得到强烈的混合和燃烧。文中对不同几何结构的旋风炉进行了试验，发现出口喷嘴对旋转气流的紊流运动起着决定性的作用。本文最后从紊流运动观点出发，分析了燃料在旋风炉内的燃烧过程。

一、问题的提出

随着燃烧技术的发展，燃烧热强度的不断提高，炉内温度水平不断地增长，使得燃烧过程逐渐由动力区转入扩散区，燃料燃烧速度此时主要取决于物理因素，即氧气和可燃混合物的混合速度，很明显，后面这一过程受着气流紊流脉动结构所限制，要使燃烧过程能够得到继续的强化，首要的任务已经不再是提高过程的温度水平，而是提高气流的紊流脉动程度。但是气流的紊流机理及其对燃烧过程的影响是极端复杂的，各国学者均进行大量研究，可是由于气流紊流理论本身还不够完善，而其试验测量技术又比较困难，因此即使对一些比较简单的问题如均相可燃混合物在各相同性的紊流结构下燃烧，自由射流的紊流结构及其燃烧理论等等还没有得到彻底的、精确的解决，更不用说过程发生在形状复杂、具有三元流动特性的那些燃烧室内的紊流燃烧问题了。

目前所积累的理论及试验资料证明，紊流燃烧速度 U_T 至少与气流脉动速度 W' 有着如下的函数关系：

$$U_T = K \cdot (W')^m \cdot U_H^n \quad \text{米/秒} \tag{1a}$$

这里，U_H 为火焰正常法线扩张速度； K、m、n 为试验常数。

总结试验证明，对均相可燃物在常压下燃烧可取：（文献 1）

$k=2.5\sim5.3$, $\quad m=0.8$, $\quad n=0.2$。

对雾化液体燃料的燃烧：（文献 2）

$k=3.3$, $\quad m=1$, $\quad n=0$。

对粉状固体燃料(文献 3)可近似取为：

$k=1.56 \quad m=1 \quad n=0 \quad$ (当挥发物含量很少时)

及 $k=4.25$ $m=1$ $n=0$ （当挥发物含量很多时）

尽管这些试验的准确度不是很高，但至少可以得出这样一个结论，即燃料紊流燃烧速度几乎与气流脉动速度成正比。

此外理论和试验研究还证明，燃烧速度还与紊流标尺、气流脉动频率等有关，至于数量上的关系，目前还未有统一的结果。

这里我们很明显可以看出，要研究在燃烧室内的紊流燃烧过程实质上可以分为两步，首先研究在该燃烧室内气流的紊流结构，然后再探讨这些紊流结构对燃烧过程有怎样的影响。本文就是着重在第一步的研究——旋转气流(具体在旋风炉内)的气流紊流结构理论及试验研究，而且由于容量的限制及理论试验技术方面的困难，这里我们只讨论气流各脉动分量，各紊流强度，方向关联系数，Prandtl 的位移径，紊流交换(摩擦)系数，局部地探讨了平均脉动频率，紊流标尺，脉动散射能等问题。紊流参数的测量是采用热电风计法，由于旋转气流是一个三元脉动问题，因此其测量是极端困难的。作为研究初步采用了我们所拟制的测量方法，只能用于二元气流，但可满足三元脉动问题，当然今后必须推广到三元气流中。(测量方法在文献 4 中有叙述)。

二、在旋风炉内旋转气流紊流结构近似理论

首先应当指出，在讨论气流的紊流结构时不能不同时讨论气流的平均参数。这里我们的任务在于借半经验紊流理论之助求出旋风炉内旋转气流紊流及平均参数的近似分析解。

在紊流工况下流体运动可用雷诺方程(紊流运动方程)来描述。对于旋转气流来说，把雷诺方程写在圆柱坐标上将会对今后的分析更为方便：

$$\left.\begin{aligned}
& \frac{\partial w_r}{\partial t} + w_r \frac{\partial w_r}{\partial r} + \frac{w_\phi}{r}\frac{\partial w_r}{\partial \phi} + w_z \frac{\partial w_r}{\partial z} - \frac{w_\phi^2}{r} = -\frac{1}{\rho}\frac{\partial P}{\partial r} \\
& +\frac{A}{\rho}\left[\frac{\partial^2 w_r}{\partial r^2} + \frac{1}{r^2}\frac{\partial^2 w_r}{\partial \phi^2} + \frac{\partial^2 w_r}{\partial z^2} - \frac{2}{r^2}\frac{\partial w_\phi}{\partial \phi} - \frac{w_r}{r^2} + \frac{1}{r}\frac{\partial w_r}{\partial r}\right] \\
& +\frac{2}{\rho}\frac{\partial A}{\partial r}\frac{\partial w_r}{\partial r} + \frac{\partial A}{\rho r \partial \phi}\left[\frac{1}{r}\frac{\partial w_r}{\partial \phi} + \frac{\partial w_\phi}{\partial r} - \frac{w_\phi}{r}\right] + \frac{\partial A}{\rho \partial z}\left[\frac{\partial w_z}{\partial r} + \frac{\partial w_r}{\partial z}\right]; \\
& \frac{\partial w_\phi}{\partial t} + w_r \frac{\partial w_\phi}{\partial r} + \frac{w_\phi}{r}\frac{\partial w_\phi}{\partial \phi} + w_z \frac{\partial w_\phi}{\partial z} + \frac{w_r \cdot w_\phi}{r} = -\frac{1}{\rho r}\frac{\partial P}{\partial \phi} \\
& +\frac{A}{\rho}\left[\frac{\partial^2 w_\phi}{\partial r^2} + \frac{1}{r^2}\frac{\partial^2 w_\phi}{\partial \phi^2} + \frac{\partial^2 w_\phi}{\partial z^2} + \frac{1}{r}\frac{\partial w_\phi}{\partial r} + \frac{2}{r^2}\frac{\partial w_r}{\partial \phi} - \frac{w_\phi}{r^2}\right] \\
& +\frac{\partial A}{\rho \partial r}\left[\frac{1}{r}\frac{\partial w_r}{\partial \phi} + \frac{\partial w_\phi}{\partial r} - \frac{w_\phi}{r}\right] + 2\frac{\partial A}{\rho r \partial \phi}\left[\frac{\partial w_\phi}{r \partial \phi} + \frac{w_r}{r}\right] + \frac{\partial A}{\rho \partial z}\left[\frac{\partial w_\phi}{\partial z} + \frac{1}{r}\frac{\partial w_z}{\partial \phi}\right]; \\
& \frac{\partial w_z}{\partial t} + w_r \frac{\partial w_z}{\partial r} + \frac{w_\phi}{r}\frac{\partial w_z}{\partial \phi} + w_z \frac{\partial w_z}{\partial z} = -\frac{1}{\rho}\frac{\partial P}{\partial z} + \frac{A}{\rho}\left[\frac{\partial^2 w_z}{\partial r^2} + \frac{1}{r^2}\frac{\partial^2 w_z}{\partial \phi^2} + \right. \\
& \left.\frac{\partial^2 w_z}{\partial z^2} + \frac{1}{r}\frac{\partial w_z}{\partial r}\right] + \frac{\partial A}{\rho \partial r}\left[\frac{\partial w_z}{\partial r} + \frac{\partial w_r}{\partial z}\right] + \frac{\partial A}{\rho r \partial \phi}\left[\frac{\partial w_\phi}{\partial z} + \frac{\partial w_z}{r \partial \phi}\right] + 2\frac{\partial A}{\rho \partial z}\cdot\frac{\partial w_z}{\partial z}
\end{aligned}\right\} \quad (1)$$

实质上，雷诺方程是包含有各紊流应力项，而上列方程式组已经用半经验的紊流理论加以代换而得出，即取紊流应力与流体平均变形速度之间具有线性的关系。此时，各紊流应力可写成下式：

$$\tau_{rr} = -\rho \overline{w'_r w'_r} = 2A\left(\frac{\partial w_r}{\partial r}\right); \qquad \tau_{\phi\phi} = -\rho \overline{w'_\phi w'_\phi} = 2A\left(\frac{1}{r}\frac{\partial w_\phi}{\partial \phi} + \frac{w_r}{r}\right);$$

$$\tau_{zz} = -\rho \overline{w'_z w'_z} = 2A\left(\frac{\partial w_z}{\partial z}\right); \qquad \tau_{r\phi} = -\rho \overline{w'_r w'_\phi} = A\left(\frac{1}{r}\frac{\partial w_r}{\partial \phi} + \frac{\partial w_\phi}{\partial r} - \frac{w_\phi}{r}\right); \qquad (2)$$

$$\tau_{rz} = -\rho \overline{w'_r w'_z} = A\left(\frac{\partial w_r}{\partial z} + \frac{\partial w_z}{\partial r}\right); \qquad \tau_{\phi z} = -\rho \overline{w'_\phi w'_z} = A\left(\frac{\partial w_\phi}{\partial z} + \frac{1}{r}\frac{\partial w_z}{\partial \phi}\right)$$

我们知道，在解决紊流运动的理论问题中，一般可采用下列三个方向：

1) 把各脉动速度变成平均速度及位移径的函数，之后用经验公式来假定位移径与坐标的关系（假定一般是根据所提出的流动物理模型并和试验比较后来确定），因而使雷诺方程得以封闭。这种方法虽然简单，但对复杂的流动工况很难加以正确的假设。

2) 把各脉动速度变成平均速度及紊流黏度系数的函数，并且假定紊流黏度系数是个常数，因而雷诺方程得以封闭。这样很多复杂的问题，只要在层流工况中得到解决，就可以推广到紊流工况来，此时只不过把所得解的结果中包含分子黏度系数项转为紊流黏度系数项。但这种方法的根本缺点是：第一，在复杂的流动中紊流黏度系数往往是个变数，因而把它变成常数解出的方程式可能会有很大的误差，或者违反出发的物理模型。第二，紊流黏度系数绝对值的大小仍是个未知数。因而所得的解如果没有试验帮助仍然不能应用。

3) 如果按照上法把紊流黏度系数的概念引入以封闭雷诺方程，但它并非常数，而认为是坐标的函数。其函数的形式或者根据物理意义和试验结果加以确定，或者就把它列为未知量，变成要求解的参数之一。这种方法的实现当然比较困难和复杂，但最符合物理意义和实际情况。实质上把解决紊流运动的困难均集中在紊流黏度（交换）系数上。我们这里就企图利用这种方法近似地解决旋转气流在旋风室内的紊流运动规律。因此先把各脉动速度与紊流黏度系数（变数）联系起来再行求解。

此外在紊流工况中，对不可压缩流体其连续方程仍可表示为：

$$\frac{1}{r}\frac{\partial}{\partial r}(r \cdot w_t) + \frac{\partial w\phi}{r\partial \phi} + \frac{\partial wz}{\partial z} = 0 \qquad (3)$$

要准确解出上述三个方程式组目前是不可能的，对于在旋风炉内的旋转气流来说，我们提出如下的物理模型：气流由旋风炉壁切向喷入，一面受着炉壁面的限制，被迫作旋转运动，另一面则自由地向炉室中心膨胀。因此，可以认为这是半自由旋转紊流射流，射流的核心，位于炉室周缘，具体来说就是在"准等势区内"，射流的边界层则一直伸延到炉室中心，也就是处于"准固体区内"。试验证明在射流的核心区，紊流参数并不保持为常量，只不过其变化并不很大(关于自由射流的气流紊流结构问题，我们当另文叙述)，但在射流边界层上，紊流参数急剧地变化。这是由旋风室外吸入不旋转的气流，进行剧烈的紊流交换使之旋转的结果，因此可以预料，这里的紊流摩擦及能量损耗达到最大。由此可见，在解决旋风炉内旋转射流时应该考虑到这些特点，即第一，把炉内气流结构分为两部分：射流核心和射流边界层或"准等势区"和"准固体区"；第二，既然是半自由旋转射流，则在炉室壁面处所有紊流参数应等于零。这就是选择边界条件时根据之一，下面我们将用试验证明，这样的物理"模型"是完全正确的。

根据上述概念，我们提出了下列几个假定：

1) 在旋风炉内气流流动是准稳定的。即

$$\frac{\partial w_\phi}{\partial t} = \frac{\partial w_r}{\partial t} = \frac{\partial w_z}{\partial t} = 0$$

2) 分子黏度系数 μ 比起紊流黏度系数 A 小得多，并可以忽略。同时紊流黏度系数是个标量。即 $\mu + A \approx A$。

3) 气流作对称的旋转运动。即

$$\frac{\partial w_\phi}{\partial \phi} = \frac{\partial w_r}{\partial \phi} = \frac{\partial w_z}{\partial \phi} = 0 \ ; \qquad \frac{\partial A}{\partial \phi} = 0 \ ; \qquad \frac{\partial P}{\partial \phi} = 0$$

4) 各种参数沿径向的变化比沿轴向的变化大得多。即 $\frac{\partial}{\partial z} << \frac{\partial}{\partial r}$ ；

因此各参数沿轴向变化可以用不同的系数来考虑，而不引入复杂的雷诺方程式之中。这样，采取了上述假定之后，雷诺方程和连续方程可简化成：

$$w_r \frac{\partial w_\phi}{\partial r} + \frac{w_r w_\phi}{r} = \frac{A}{\rho}\left(\frac{\partial^2 w_\phi}{\partial r^2} + \frac{1}{r}\frac{\partial w_\phi}{\partial r} - \frac{w_\phi}{r^2}\right) + \frac{1}{\rho}\frac{\partial A}{\partial r}\left(\frac{\partial w_\phi}{\partial r} - \frac{w_\phi}{r}\right) \tag{4}$$

$$w_r \frac{\partial w_r}{\partial r} - \frac{w_\phi^2}{r} = \frac{A}{\rho}\left(\frac{\partial^2 w_r}{\partial r^2} + \frac{1}{r}\frac{\partial w_r}{\partial r} - \frac{w_r}{r^2}\right) + \frac{2}{\rho}\frac{\partial A}{\partial r}\cdot\frac{\partial w_r}{\partial r} - \frac{1}{\rho}\frac{\partial P}{\partial r} \tag{5}$$

$$w_r \frac{\partial w_z}{\partial r} = \frac{A}{\rho}\left(\frac{\partial^2 w_z}{\partial r^2} + \frac{1}{r}\frac{\partial w_z}{\partial r}\right) + \frac{1}{\rho}\frac{\partial A}{\partial r}\frac{\partial w_z}{\partial r} \tag{6}$$

$$\frac{\partial w_r}{\partial r} + \frac{w_r}{r} = 0 \tag{7}$$

这里总共有四个方程式但需要求解下列五个未知数，即：

$$W_\phi = f_1(r,z) \ ; \quad W_z = f_2(r,z) \ ; \quad W_r = f_3(r,z) \ ; \quad P = f_4(r,z) \ ; \quad A = f_5(r,z)$$

而方程式系(2)这里不能被利用，它是留作将来近似求解紊流参数的基本方程，因此为了封闭方程式系必须再引入一个已知方程式。作为初步近似，我们认为在旋风炉内切向速度的分布是已知的，并服从下列规律：

$$W_\phi = k_1 r^{-n} \tag{8}$$

这里，k_1 为由入口速度边界条件决定的常数：$k_1 = W_\phi^{BX} \cdot r_u^n$；$n$ 为试验常数。

通常在准固体区内　　$n = n_1 \leqslant -1$

在准等势区内　　　　$n = n_2$　　$0 < n_2 \leqslant 1$

等号的情况是属于无摩擦理想流体的情况。

Е.А.НахаllеТян（文献5）及 JI.JI.КалИшеВСКий（文献6）证明，无论在冷模或燃烧时，公式（8）都是正确的，并且 n_2 值一般在 0.7 至 1 之间变化，我们的试验也证明了这一点，并且得出指数 n 是和轴向坐标有关（见图1），即

$$n = f_6(z) \tag{9}$$

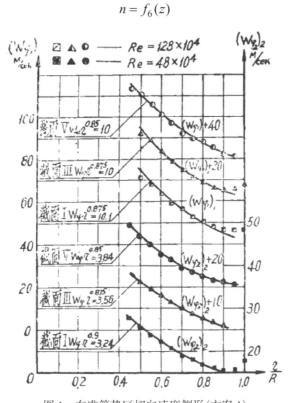

图 1　在准等势区切向速度侧形（方案 1）

在我们的旋风炉冷模中，当 $Re = 48 \times 10^4 - 128 \times 10^4$ 范围时，值在 0.8 至 0.9 内变化。

现在要解出方程式系（4）、（5）、（6）、（7）、（8）是可能的，但首要的任务在于检查所作的四个假定是否正确。下面我们要给出充分的证明：

1）任何的紊流运动都应该认为是不稳定的，但它并不是毫无规则，混乱地运动，而是服从一定的统计规律。既然我们这里只讨论各紊流参数按时间的平方平均开方值，那么完全有理由可以认为过程是准稳定的，采用测量紊流参数平方平均开方值的真空管热

电偶试验证明,在不同时间间隔内测量空间同一点的紊流参数得出的结果都是重复的。

2)过去都认为,除了在壁面附近以外,分子黏度系数 μ 比起紊流黏度系数 A 小得多,因此在工程上 μ 值的影响实际上可以不加考虑。根据我们的试验对旋转气流在准等势区内 A 比 μ 值大一千倍,在准固体区内则大几万至几十万倍(见图5),因此 μ 值影响更不予以考虑!至于紊流黏度系数 A 是个标量的假定,由于测量上极大的困难,目前还没有足够可靠的试验证明。但是这个假定被各国学者顺利地应用于解决一系列重大的工程技术问题,并得出和试验相符的结果(例如文献7、8)。

3)关于在旋风炉内旋转气流对称性问题,对于平均空气动力参数来说已为很多试验所证实,我们在冷模中的测量也重复了这样的结果。重要的问题在于在旋转气流中紊流参数是否也是轴对称的。为此,进行了专门检查性试验,发现紊流摩擦应力 $\tau_{\omega'_\phi \omega'_z}$,紊流强度 ε'_ω 及各脉动分速度在测量准确范围内均可近似认为是轴对称的,作为例子我们只列出紊流摩擦应力的试验曲线(见图2)。由式(2)可知,紊流黏度系数 A 只与 $\tau_{\omega'_\phi \omega'_z}$,$W_\phi$,$W_z$,$\dfrac{\partial W_\phi}{\partial r}$ 等等有关。如果这些参数都是轴对称的,那么我们应有理由认为 A 也是轴对称的,$\dfrac{\partial A}{\partial \phi}$ 应等于零。

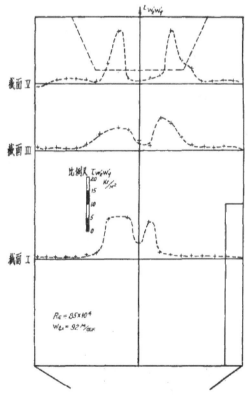

图 2　紊流摩擦应力场($Re=85\times 10^4$)

4)关于气流参数沿轴向变化不大的假定应谨慎对待。如果旋风炉是无限长而且气流在内部旋转运动没有摩擦的影响，那么沿轴向各参数是不会变化的，但是在实际条件下由于摩擦的存在及旋风炉出口喷嘴的作用，各参数沿轴向应有一定变化，平均空气动力参数沿轴向变化不大的结论，过去已经得到证实（见文献5、6及我们的试验）。测量证明，紊流参数的变化也是不大的（见图8及14），因此和各参数沿径向变化比较起来，沿轴向变化是可以忽略，但是完全抛去各参数沿轴向变化是不符合过程的物理模型，这也是ВулисЛ.А.和УстименкоБ.П.所建议旋风炉内气流平均参数计算方法缺点之一（文献9），他们假定，切向速度，径向速度及静压力分布沿炉室长度没有变化。我们在求解这问题时利用了公式（9），即所有参数沿轴向的变化通过指数 n 来考虑，由于摩擦的存在，这个指数离开气流入口处愈远其值下降愈多。例如当 $Re=48\times10^4$ 时，在炉室截面 I，$n_2=0.9$，在截面 III 时 $n_2=0.875$，发展到截面 V 时 n_2 已降低至 0.85（见图1）。

上述假定被试验证明后，就有可能把偏微分方程式系转换成常微分方程式系，也应当指出，我们方法不足之处在于：第一，未能试验证明紊流黏度系数是个标量。第二，由于数学上的困难把各参数沿轴向变化只用变的指数 n 来考虑。但是，目前还未发现有关旋转气流紊流结构试验和理论详细研究，我们作为初步的近似，采用被试验证实的假定应该是容许的，不过要用详细的试验加以检查和证明所提出的近似理论。

下面求解各参数，首先积分方程（7），并考虑到存在有摩擦及出口喷嘴作用的情况，得出径向速度的解为：

$$W_r = k_2^{1-M} \tag{10}$$

M 为试验常数。在准固体区内 $M=M_1$； $M_1<0$。

                在准等势区内 $M=M_2$； $0<M_2\leq 1$。

对于理想流体 $M_2=1$。式（10）即变为物理旋涡的理论公式。

把式（8）和式（10）代入式（4），简化后即可得出求紊流交换系数的微分方程式：

$$\frac{dA}{dr} - \frac{A}{r}(n-1) + \rho k_2 \left(\frac{1-n}{1+n}\right)\frac{1}{r^M} = 0 \tag{11}$$

此式的通解为：

$$A = cr^{n-1} - \rho k_2 \frac{(1-n)r^{1-M}}{(1+n)(2-n-M)} \tag{12}$$

在准等势区内：$n=n_2$；$k_1 = k_1^{\Pi}$；$M=M_2$，$k_2 = k_2^{\Pi}$，$C=C_2$；$A=A_{\text{ПОТ}}$

边界条件为：在旋风室壁面处紊流黏度系数应等于零。

即 $r=r_u$；$A_{\text{ПОТ}}=0$。

因而积分常数应等于 $C_2 = \rho K_2^{\Pi} \frac{(1-n_2)r_u^{n-n_2-M_2}}{(1+n_2)(2-n_2-M_2)} \tag{13}$

由此可得出准势流区中紊流黏度系数的最终公式：

$$A_{\text{ПОТ}} = \rho K_2^{\text{П}} \frac{(1-n_2)r^{1-M_2}}{(1+n_2)(2-n_2-M_2)}\left[\left(\frac{r_u}{r}\right)^{2-M_2-n_2}-1\right] \quad (14)$$

在准固体区内：$n=n_1$；$k_1=k_1^{\text{I}}$；$M=M_1$；$k_2=k_2^{\text{I}}$，$C=C_1$；$A=A_{K.T.}$

边界条件为：当 $r=r_M$；$A_{K.T.}=A_{\text{ПОТ}}$。

这里 r_M 是准等势区和固体区交界处的半径，也近似相应切向速度最大值所处的半径。由此求得积分常数。

$$C_1 = \rho K_2^{\text{I}} \frac{(1-n_1)r_M^{2-n_1-M_1}}{(1+n_1)(2-n_1-M_1)} + \rho K_2^{\text{II}} \frac{(1-n_2)r_M^{2-M_2-n_2}}{(1+n_2)(2-n_2-M_2)}\left[\left(\frac{r_u}{r_M}\right)^{2-n_2-M_2}-1\right] \quad (15)$$

及

$$A_{K.T.} = \rho r^{1-M_1}\left\{K_2^{\text{I}}\frac{(1-n_1)}{(1+n_1)(2-n_1-M_1)}\left[\left(\frac{r_M}{r}\right)^{2-n_1-M_1}-1\right]\right.$$
$$\left.+K_2^{\text{II}}\frac{(1-n_2)r_M^{M_1-M_2}}{(1+n_2)(2-n_2-M_2)}\left(\frac{r_M}{r}\right)^{2-M_1-n_1}\left[\left(\frac{r_u}{r_M}\right)^{2-n_2-M_2}-1\right]\right\} \quad (16)$$

把式(8)、(10)及(12)代入式(5)，经过不复杂的变换，可得出求静压力的微分方程式：

$$\frac{\mathrm{d}P}{\mathrm{d}r} = \rho M K_2^2 r^{-2M-1} + \rho K_1^2 r^{-2n-1} + CK_2\left[(M^2-1)\right.$$
$$\left.-2(n-1)M\right]r^{n-M-3} + \rho K_2^2\frac{(1-n)(1-M)(2M+1)}{(1+n)(2-n-M)}r^{-1-2M} \quad (17)$$

积分可得

$$P = -\frac{1}{2}\rho k_2^2 r^{-2M} - \frac{1}{2^n}\rho k_1^2 r^{-2n} + \frac{CK_2\left[(M^2-1)-2(n-1)M\right]}{n-M-2}r^{n-M-2}$$
$$-\rho k_2^2\frac{(1-n)(1-M)(2M+1)}{(1+n)(2-n-M)\cdot 2M}r^{-2M} + C_3 \quad (18)$$

在准等势区内：

$k_1=k_1^{\text{II}}$；$k_2=k_2^{\text{II}}$；$n=n_2$；$M=M_2$；$C=C_2$；$C_3=C_3^{\text{II}}$ 及 $P=P_{\text{ПОТ}}$。

边界条件为：当 $r=r_M$ 时，静压力等于旋风室外的相对压力，如果旋风室外为大气压，则相对静压等于零，即 $P=P_{\text{ПОТ}}=0$。此时

$$C_3^{\mathrm{II}} = \frac{1}{2}\rho\left(K_3^{\mathrm{II}}\right)^2 r_M^{-2M_2} + \frac{1}{2n_2}\rho\left(K_1^{\mathrm{II}}\right)^2 r_M^{-2n_2}$$
$$-\frac{C_2 K_2^{\mathrm{II}}\left[\left(M_2^2-1\right)-2(n_2-1)M_2\right]}{n_2-M_2-2} r_M^{n_2-M_2-2} \quad (19)$$
$$+\rho\left(K_2^{\mathrm{II}}\right)^2 \frac{(1-n_2)(1-M_2)(2M_2+1)}{(1+n_2)(2-n_2-M_2)(2M_2)} r_M^{-2M_2}$$

及

$$P_{\text{ПОТ}} = \frac{1}{2n_2}\rho\left(K_1^{\mathrm{II}}\right)^2 \left(r_M^{-2n_2}-r^{-2n_2}\right) + \frac{1}{2}\rho\left(K_2^{\mathrm{II}}\right)^2\left(r_M^{-2M_2}-r^{-2M_2}\right)$$
$$-\frac{C_2 K_2^{\mathrm{II}}\left[\left(M_2^2-1\right)-2(n_2-1)M_2\right]}{n_2-M_2-2}\left(r_M^{n_2-M_2-2}-r^{n_2-M_2-2}\right) \quad (20)$$
$$+\rho\left(K_2^{\mathrm{II}}\right)^2 \frac{(1-n_2)(1-M_2)(2M_2+1)}{(1+n_2)(2-n_2-M_2)(2M_2)}\left(r_M^{-2M_2}-r^{-2M_2}\right)$$

在准固体区内：

$$k_1 = k_1^{\mathrm{I}};\quad k_2 = k_2^{\mathrm{I}};\quad n=n_1;\quad M=M_1;\quad C=C_1;\quad C_3=C_3^{\mathrm{I}} \text{ 及 } P=P_{K.T.} \text{。}$$

边界条件为：当 $r=r_M$ 时，$P=P_{K.T.}=0$。代入可得：

$$C_3^{\mathrm{I}} = \frac{1}{2}\rho\left(K_2^{\mathrm{I}}\right)^2 r_M^{-2M_1} + \frac{1}{2n_1}\rho\left(K_1^{\mathrm{I}}\right)^2 r_M^{-2n_1}$$
$$-\frac{C_1 K_2^{\mathrm{I}}\left[\left(M^2-1\right)-2(n_1-1)M_1\right]}{n_1-M_1-2} r_M^{n_1-M_1-2} \quad (21)$$
$$+\rho\left(K_2^{\mathrm{I}}\right)^3 \frac{(1-n_1)(1-M_1)(2M_1+1)}{(1+n_1)(2-M_1-n_1)(2M_1)} r_M^{-2M_1}$$

及

$$P_{K.T.} = \frac{\rho}{2n_1}\left(K_1^{\mathrm{I}}\right)^2\left(r_M^{-2n_1}-r^{-2n_1}\right) + \frac{\rho}{2}\left(K_2^{\mathrm{I}}\right)^2\left(r_M^{-2M_1}-r^{-2M_1}\right)$$
$$-\frac{C_1 K_2^{\mathrm{I}}\left[\left(M_1^2-1\right)-2(n_1-1)M_1\right]}{n_1-M_1-2}\left(r_M^{n_1-M_1-2}-r^{n_1-M_1-2}\right) \quad (22)$$
$$+\rho\left(K_2^{\mathrm{I}}\right)^2 \frac{(1-n_1)(1-M_1)(2M_1+1)}{(1+n_1)(2-n_1-M_1)(2M_1)}\left(r_M^{-2M_1}-r^{-2M_1}\right)$$

讨论：

1) 对理想流体无摩擦的旋转运动，同时径向速度比起切向速度小到可以忽略。此时，$n_1=-1$，$n_2=1$，可得出下列简单的静压力表达式：

$$P_{K.T.} = -\frac{\rho\left(K_1^J\right)^2}{2}\left(r_M^2 - r^2\right)$$
$$P_{\Pi OT} = \frac{\rho}{2}\left(K_1^{II}\right)^2\left(\frac{1}{r_M^2} - \frac{1}{r^2}\right) \quad (23)$$

这就是"物理旋涡"的静压力理论表达式。

2) 公式(20)及(22)是相当复杂的,给实用上带来极端不便,计算证明,旋转气流径向速度对静压力的影响不大,在实际计算中,采用下列的简单式已足够准确:

$$P_{K.T.} = \frac{\rho}{2n_1}\left(K_1^I\right)^2\left(r_M^{-2n_1} - r^{-2n_1}\right)$$
$$P_{\Pi OT} = \frac{\rho}{2n_2}\left(K_1^{II}\right)^2\left(\frac{1}{r_M^{2n_2}} - \frac{1}{r^{2n_2}}\right) \quad (24)$$

3) 公式准确度的讨论。所得出式(20)与(22)是比较全面的考虑了各种因素对用静压力的影响,即使是应用式(24),所得结果也是足够准确的,这里,把按式(24)计算所得结果(图 3 中连续曲线)和 Л.А.ВуЛис 及 Б.Ⅱ.Устименко 的试验相比较(见文献 9)。证明

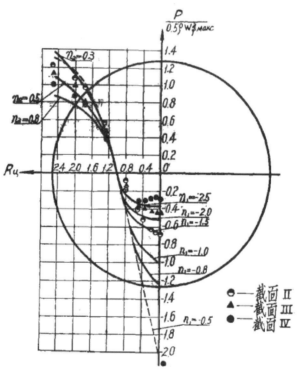

图 3 计算结果(实线)和 Б.Ⅱ.Устименко 等的试验比较
(虚线为 Л.А.ВуЛис 等的理论)。

试验基本上是和理论相符合的。(Б.Ⅱ.Устименко 冷模所作试验得出 n_2 值较低)。为了

比较起见，图3中同时列出 Л.А.Вулис 及 Б.Ⅱ.Устименко 的理论曲线（虚线），应当指出，特别是在准固体区范围内，我们公式的优越性显得更为突出，此时 Вулис Л.А.等理论公式和我们的公式当 n_1=–0.5 时相符合，但试验证明，n_1 经常小于–1。同时 Вулис 的压力公式也没有考虑到沿炉室轴向变化的问题，但在炉室中心部分，这个变化是足够大的。

下面将讨论轴向速度求解问题，把式(8)、(10)及(12)代入式(6)化简后得：

$$\frac{d^2\omega_z}{dr^2} = \frac{d\omega_z}{dr}\left[\frac{\rho K_2 r^{r-1}\dfrac{4-2M-n-n^2}{c(1+n)(2-n-M)} - nr^{c+M-3}}{r^{n+M-2} - \rho K_2 \dfrac{(1-n)}{c(1+n)(2-M-n)}}\right] \tag{25}$$

略去中间的求解过程，最后得出通解为：

$$\omega_z = C_4\left[\frac{1}{1-n}r^{1-n} + \frac{2\rho K_2(1-n)}{C(1+n)(2-n-M)(3-M-5n+Mn+2n^2)}r^{\frac{3-M-5n+Mn+2n^2}{1-n}}\right.$$
$$\left. + \frac{(3-n)(1-n)\rho_2 K_2^2}{C^2(1+n)^2(2-n-M)^2(5+2Mn+3n^2-8n-2M)}r^{\frac{5+2Mn+3n^2-8n-2M}{1-n}} + \cdots\right] + C_5 \tag{26}$$

在准等势区时：

$n=n_2$； $M=M_2$； $k_1=k_1^{\mathrm{II}}$； $k_2=k_2^{\mathrm{II}}$； $C=C_2$； $C_4=C_4^{\mathrm{II}}$； $C_5=C_5^{\mathrm{II}}$； $W_Z=W_{ZПOT}$。

其边界条件为：1) 当 $r=r_u$， $W_{ZПOT}=0$。

2) 当 $r=r_M$， $\int_{r_M}^{r_u} 2\pi W_{ZПOT} r\,dr = Q_0$。

Q_0 为进入旋风炉的空气量。

在准固体区内：

$n=n_1$； $M=M_1$； $k_1=k_1^{\mathrm{I}}$； $k_2=k_2^{\mathrm{I}}$； $C=C_1$； $C_4=C_4^{\mathrm{I}}$； $C_5=C_5^{\mathrm{I}}$； $W_Z=W_{ZK.T.}$。

边界条件为：

1) 当 $r=r_M$， $W_{ZПOT}=W_{ZK.T.}$。

2) 当 $r=r_{o\delta p}$， $W_{ZK.T.}=0$。

这里 $r_{o\delta p}$ 相应于正流和逆流交界处的半径，可按下式求出：

$$\int_0^{r_{o\delta p}} 2\pi W_{ZK.T.} r\,dr = \int_{r_{o\delta p}}^{r_w} 2\pi W_{ZK.T.} r\,dr.$$

这里意义在于：中心逆流在截面 $\pi r_{o\delta p}^2$ 被吸入然后由环形截面 $\pi\left(r_w^2 - r_{o\delta p}^2\right)$ 被排出，根

据 E.A.Haxaцeтян 试验证明，可以取 $r_w = r_M$（文献 5）。

因此按上述条件分别可以求出四个积分常数 C_4^I；C_5^I；C_4^{II}；C_5^{II}。但是由于目前在边界条件中还无法考虑出口喷嘴对轴向速度分布的影响。因此所得出的解在周缘部分显然不符合有出口喷嘴时的情况。计算证明，由式(26)所得的轴向速度沿径向分布在性质上和没有出口喷嘴的旋转气流分布情况相符。Л.A.Вулис 等也同样得出这样类型的曲线（文献 9）。

紊流参数值的近似估计

解出紊流黏度系数 A 后，就有可能利用方程式系(2)来近似估计各紊流参数。这样估计的准确程度一方面取决于求 A 值的准确程度，另一方面更主要的取决于半经试验紊流理论的准确程度。把所解得结果代入方程式系(2)可得：

$$\tau_{rr} = -\rho\overline{\omega'_r\omega'_r} = 2A\left(\frac{\partial \omega_r}{\partial r}\right) = 2AK_2 M r^{-M-1} \tag{27}$$

$$\tau_{\phi\phi} = -\rho\overline{\omega'_\phi\omega'_\phi} = 2A\left(\frac{\omega_r}{r} + \frac{1}{r}\frac{\partial \omega_\phi}{\partial \phi}\right) \approx 2AK_2 r^{-M-1} \tag{28}$$

$$\tau_{r\phi} = -\rho\overline{\omega'_r\omega'_\phi} = A\left(\frac{1}{r}\frac{\partial \omega_r}{\partial \phi} + \frac{\partial \omega_\phi}{\partial r} - \frac{\omega_\phi}{r}\right) \approx -A(1+n)K_1 r^{-1-n} \tag{29}$$

$$\tau_{rz} = -\rho\overline{\omega'_r\omega'_z} = A\left(\frac{\partial \omega_r}{\partial z} + \frac{\partial \omega_z}{\partial r}\right) \tag{30}$$

目前，由于测量上的困难，旋转气流径向速度分布还没有足够可靠的试验数据，因此式(27)、(28)只能用于估计径向和切向脉动分速的大小，比较其数量级可见：

$$\sqrt{\overline{\omega'^2_r}} = \sqrt{M} \cdot \sqrt{\overline{\omega'^2_\phi}} \tag{31}$$

因为在射流核心区内，$M \leqslant 1$，即切向脉动速度大于径向脉动速度，但是和平均速度不同，它们之间之差并不是很大，对射流核心区的测量结果证实了这点。

利用方向关联系数值来简化式(29)，可得

$$\rho\overline{\omega'\omega'_\phi} = \rho\sqrt{\overline{\omega'^2_r}} \cdot \sqrt{\overline{\omega'^2_\phi}} \cdot R_{\omega'\omega'_\phi} = \rho\sqrt{\overline{\omega'^2_\phi}} \cdot \sqrt{\overline{\omega'^2_\phi}} \cdot \sqrt{M} \cdot R_{\omega'_\phi\omega'_r}$$
$$= A(1+n)K_1 r^{-1-n}$$

解之可得切向脉动速度为：

$$\sqrt{\overline{\omega_\phi'^2}} = \left[\frac{A(1+n)K_1}{\rho\sqrt{M}\cdot R_{\omega_\phi'\omega_r'}r^{1+n}}\right]^{\frac{1}{2}} \tag{32}$$

及切向紊流强度值为：

$$\varepsilon_{\omega_\phi'} = \frac{\sqrt{\overline{\omega_\phi'^2}}}{\overline{\omega_\phi}} = \left[\frac{A(1+n)}{\rho\sqrt{M}\cdot R_{\omega_r'\omega_\phi'}K_1 r^{-n}}\right]^{\frac{1}{2}} \tag{33}$$

这里唯一的紊流参数是方向关联系数 $R_{\omega_r'\omega_\phi'}$。但我们的试验证明在旋风炉内其值近似为常量，约等于 0.6。

另外，由式(2)可得。

$$\tau_{rz} = -\rho\overline{\omega_r'\omega_z'} = -\rho R_{\omega_r'\omega_z'}\sqrt{\overline{\omega_r'^2}}\cdot\sqrt{\overline{\omega_z'^2}} \approx A\frac{\partial\omega_z}{\partial r}$$

$$\tau_{r\phi} = -\rho\overline{\omega_r'\omega_\phi'} = -\rho R_{\omega_r'\omega_\phi'}\sqrt{\overline{\omega_r'^2}}\sqrt{\overline{\omega_\phi'^2}} \approx A\left(\frac{\partial\omega_\phi}{\partial r} - \frac{\omega_\phi}{r}\right)$$

$$= A\left(1+\frac{1}{n}\right)\frac{\partial\omega_\phi}{\partial r}$$

由此可得：
$$\sqrt{\overline{\omega_z'^2}} = \frac{R_{\omega_r'\omega_\phi'}}{R_{\omega_r'\omega_z'}}\cdot\frac{1}{\left(1+\dfrac{1}{n}\right)}\cdot\frac{\dfrac{\partial\omega_z}{\partial r}}{\dfrac{\partial\omega_\phi}{\partial r}}\cdot\sqrt{\overline{\omega_\phi'^2}} \tag{34}$$

这里我们没有把式(34)展开，因为上面已经说明，按式(26)所求得的轴向速度和具有出口喷嘴的旋风炉有一定距离。因此，式(34)只能用作估计轴向脉动值的大小。

最后，我们可以利用上述结果求出 Prandtl 的位移径 l，根据 Prandtl 的紊流理论可知，对于三元气流紊流黏度系数和位移径之间有着如下的联系（文献 10）：

$$A = \rho l^2\sqrt{J} \tag{35}$$

即

$$l = \left(\frac{A}{\rho\sqrt{J}}\right)^{\frac{1}{2}} \tag{36}$$

这里，J 为气流平均变形速度张量的不变量。

$$J = 2\left\{\left(\frac{\partial \omega_r}{\partial r}\right)^2 + \left(\frac{1}{r}\frac{\partial \omega_\phi}{\partial \phi} + \frac{\omega_r}{r}\right)^2 + \left(\frac{\partial \omega_z}{\partial z}\right)^2\right\}$$

$$+ \left\{\left(\frac{1}{r}\frac{\partial \omega_z}{\partial \phi} + \frac{\partial \omega_\phi}{\partial z}\right)^2 + \left(\frac{\partial \omega_r}{\partial z} + \frac{\partial \omega_z}{\partial r}\right)^2 + \left(\frac{1}{r}\frac{\partial \omega_r}{\partial \phi} + \frac{\partial \omega_\phi}{\partial r} - \frac{\omega_\phi}{r}\right)^2\right\} \quad (37)$$

按照我们的假设条件可简化为：

$$J = 2\left[\left(\frac{\partial \omega_r}{\partial r}\right)^2 + \left(\frac{\omega_r}{r}\right)^2\right] + \left(\frac{\partial \omega_z}{\partial r}\right)^2 + \left(\frac{\partial \omega_\phi}{\partial r} - \frac{\omega_\phi}{r}\right)^2 \quad (38)$$

综合上述可见，如果我们能够由边界条件和试验求出系数 k_1，k_2，n 及 M，则有可能求出或近似估计出下列气流参数：

空气动力平均参数：$\overline{W_\phi}$，$\overline{W_r}$，$\overline{W_z}$，$\overline{P}$ 及紊流参数：A，$\sqrt{\omega_\phi'^2}$，$\sqrt{\omega_r'^2}$，$\sqrt{\omega_z'^2}$，$\varepsilon_{\omega_\phi'}$ 及 l。

应当指出，上述四个试验系数的求解比紊流参数的测定容易得多。这也是所提出紊流参数近似计算法的最大优点，当然上述十个参数的理论公式，特别是计算紊流参数的理论公式，必须经过详细试验检查和证明。因为这些公式都具有半经验半理论的性质，在进行试验检查以前，可以先由所得出的理论公式作出一些定性的结论。

近似解结果讨论

1) 紊流黏度系数并不是常量。它不但和空间坐标有关[$A=f(r, z)$]，而且还取决于雷诺准则。这里由式(12)可知：

$$C \sim K_2, \quad 而 \quad K_2 \sim W_r \sim W_\phi \sim W_\phi^{BX} \sim Re$$

故 $A=f(C, K_2)=f(Re)$

并且随着雷诺数增加时，A 值也增加，因为：

$$Re\uparrow \to W_\phi^{BX}\uparrow \to W_\phi\uparrow \to W_r\uparrow \to K_2\uparrow \to C\uparrow \to A\uparrow$$

值得指出的是在管内流动时亦得出同样的结论。在图 4 中列出三种不同雷诺数及 n_2 值下的理论曲线并和试验点相比较，证明：第一，上述论断是正确的。第二，理论和试验基本上是符合的。第三，紊流黏度系数随 Re 值变化甚大。

2) 紊流黏度系数与紊流脉动的关系。

紊流黏度(交换)系数愈大，紊流脉动及紊流混合就愈强烈，而能量的损耗也就愈大。这里可以由下面几点事实来证明：

A) 在旋转射流核心区比起边界层紊流交换及脉动弱得多，因此紊流黏度系数也小几十倍。图 5 列出理论和试验的证明。

Б) 由图 1 可知旋转射流入口后,在向前发展过程中,由于摩擦及紊流交换的影响,n_2 值不断降低,因而 A 值也就不断地增加。

В) 紊流黏度系数最大之处和散射能损失最大值是遥遥对应的(见图 16)。这点以后还要详细讨论。

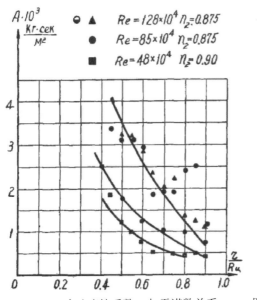

图 4　紊流摩擦系数 A 与雷诺数关系

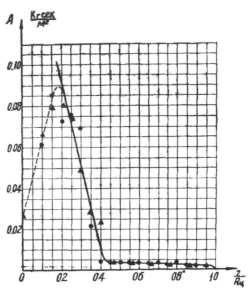

图 5　在旋转半自由射流的核心及边界层上 A 值的变化(连续黑线为理论计算曲线)$Re=188\times10^4$

3) 对于理想流体按"物理旋涡"的系统流动时,那么在势流区 $n_2=+1$,$M_2=+1$,由式(14)可得 $A_{ПОТ}=0$。而在固体旋转区内 $n_1=-1$,$M_1=-1$,由式(16)可得 $A_{К.Т.}=\infty$。亦即在势流区内流动没有摩擦,也没有脉动。因此紊流黏度系数应为零。在固体区内,流体象固体一样旋转,那么,性质上和固体相似,具有无穷大的黏性。可见,所得出的理论公式即使在极限情况下也是符合物理意义的,对实际黏性流体的旋转运动中,例如按照我们试验,在准等势区 n_2 在 0.8～0.9 内变化亦即相应于紊流黏度系数 A 在 0(在壁面)至 0.006 公斤·秒/米^2 范围内变化。这比分子黏度系数 $\mu=1.77\times10^{-6}$ 公斤·秒/米^2 大几千倍,而在准固体区则大几万至几十万倍(见图 5)。

4) 三个方向脉动值大小的比较。对于平均速度来说,毫无疑问,切向旋转速度比轴向速度大得多,轴向速度又比径向速度大得多。但是,对于气流脉动值则要改变这些概念。由式(31)可知,径向脉动分速是小于切向的,但小得不多,至于轴向和切向脉动分速的大小,应按式(34)进行比较。公式第一项表示不同方向关联系数是否相同,由于测量上的困难,在旋风室内我们只测量了 φ、z 方向的关联系数,但既然旋风室内具有射流的"物理模型",因此为了方便,我们在自由射流中进行了校验,试验证明,不同方向的关联系数值是近似相等的(见图 6),并且和雷诺数无关(见图 7),不管对自由射流还是对旋风炉的旋转射流关联系数值约为 0.6(见图 15)。

在射流核心(准等势区内),由于存在有周缘逆流,因此一般 $\frac{\partial W_z}{\partial r} > \frac{\partial W_\phi}{\partial r}$,可见 $\sqrt{\overline{W_z'^2}} > \sqrt{\overline{W_\phi'^2}}$。在射流边界层上,$\left(1+\frac{1}{n_2}\right)$ 值很小,因为 $n_2 \leqslant -1.0$。故尽管此时切向速度梯度 $\frac{\partial W_\phi}{\partial r}$ 很大,$\sqrt{\overline{W_\phi'^2}}$ 值也比 $\sqrt{\overline{W_z'^2}}$ 值为小。只有在远离出口喷嘴的截面,此时中心逆流基本消失,梯度 $\frac{\partial W_z}{\partial r}$ 变得很小,才有可能出现 $\sqrt{\overline{W_\phi'^2}} > \sqrt{\overline{W_z'^2}}$ 的情况,试验完全证实了这样的论断(见图 8)。但一般说来,三个脉动速度值之间的差别并不是很大的。

5) 理论及试验证明位移径 l 值在旋转射流核心近似保持为常量(见图 9)在某种程度上证明了 Prandtl 的假定,即在射流的横截面上位移径是不变化的。而与 Л.А.Вулис 等的假设 $l = \alpha r$ 不符(文献 9)。因为这样的假设无疑承认当 $r = r_u$ 时 $l = \alpha r_u \neq 0$ 这是与半自由旋转射流的物理模型不符。另外,计算和试验证明,在旋转射流的位移径比起同样管径下流体强迫运动的位移径约小五倍,这也说明旋转射流的紊流混合程度比直线流动时大得多。

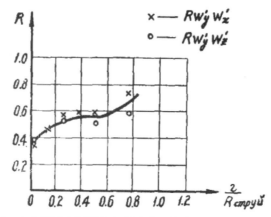

图 6　自由射流内不同方向的关联系数曲线($W_{Bыx}$=20 米/秒)

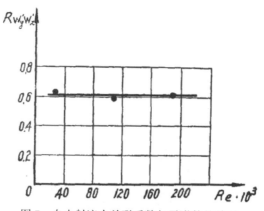

图 7　自由射流中关联系数与雷诺数的关系

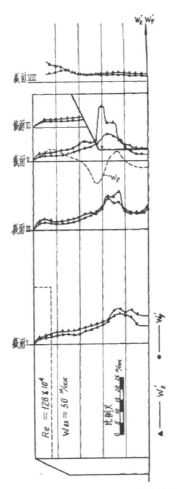

图 8　切向及轴向脉动速度分布曲线

6) 计算及试验完全证明了所假定半自由旋转紊流射流的物理模型是正确的。在射流核心及边界层内平均空气动力参数及紊流参数均按照半自由射流规律变化。

7) 理论计算和试验比较。

上面所提出的近似计算公式,其准确度与要由试验求出的某些常数值有关,图4、8及9就是应用由平均空气动力参数所求得的试验常数值(K_1、K_2、n、M)代入上述近似计算式而得出的各紊流参数变化规律同时也列出和试验的比较。虽然这些图形已经能说明近似理论和试验是符合的,但是它毕竟是对个别工况而言。因此我们认为比较合理和全面的是在同一图形上把各种工况所得试验结果与理论计算相比较。这就有可能较彻底地证明理论计算和试验结果是否处在同一数量级上,试验证明在所有工况下 n_2 均在 0.8～0.9 范围内变动。因此比较的实质在于各工况的试验点是否都落在 n_2=0.8 和 0.9 两条极限理论曲线范围内。在图 10、11、12 分别示出切向紊流强度,紊流黏度系数及位移径的比较结果,可以见到大部分试验点都落在极限理论曲线范围内,只有在靠近壁面附近处,理论值比试验点略低,这可能由于在近似理论中没有考虑分子摩擦系数的结果。我们知道,愈近壁面,分子摩擦所起的作用愈来愈大。另一方面,气流是由炉室切向喷入,喷入射流的紊流脉动值和自由射流相近，因此使旋风室壁面附近紊流参数保持有自由射流核心所具有的数值。此外在图 13 中作出另外一个有趣的比较，即对不同结构的旋风室，当雷诺数及 n 值相同时，试验点均落在一条理论曲线的周围，所有这些都说明所给出的近似计算公式是足够可靠的。

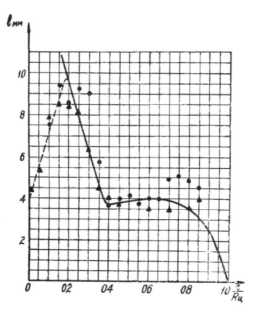

图 9　位移径沿炉室径向的变化（$Re=128\times10^4$）

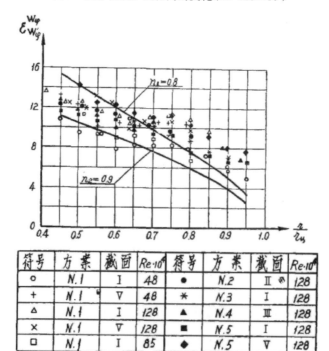

图 10　切向紊流强度理论试验比较图

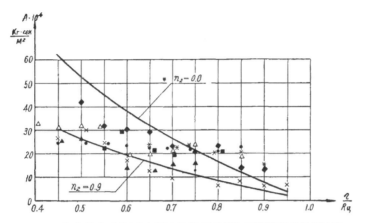

图 11　紊流黏度系数试验理论比较图(试验点符号意义见图 10)

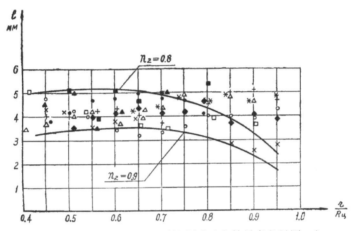

图 12　位移径试验理论比较图(试验点符号意义见图 10)

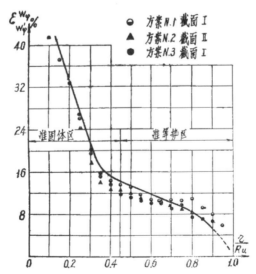

图 13　不同结构方案切向紊流强度值试验理论比较
$Re=128\times10^4$, $n_z=0.875$。

三、半自由旋转紊流射流自模化问题的研究

在旋风室中旋转气流平均参数的自模化问题，在文献 5 中已有叙述，而且也得到普遍公认，问题就是今后还要进一步积累试验资料求出平均参数自模化最低的雷诺准则范围。目前资料证明平均参数自模化的试验只在 $Re^*=50000\text{-}350000$ 范围内适用。（注意：这里雷诺数是按式 $Re^* = \dfrac{d_{PX} \cdot W_{PX}}{\upsilon_{PX}}$ 计算，作为定型尺寸取旋风室入口喷嘴当量直径）。但是单纯考虑平均参数的自模是不够全面的。因为在紊流工况下，自模问题还包括各种紊流参数，它们直接影响到能量损失过程，物质交换混合过程，传热过程等等，这些都是对燃烧过程起着极重要的作用，所以如果能够解决紊流参数自模的问题及其界限，不但可把试验结果推广到几何相似的模型中，而且对生产实践及理论研究均有极重大的意义。不过关于紊流脉动的相似理论今天还发展得很不完善，还不能彻底提出整套的相似准则，因为紊流理论本身还没有研究清楚的缘故。这里，我们先提出这样的假定，即以雷诺方程是能够真正描述气流紊流运动为出发点，今后再补充一些雷诺方程不能包括的紊流特性而出现的准则。

把式(2)代入式(1)，即可得出包含有平均速度及脉动速度的雷诺方程，按照相似理论的法则加以整理(详细整理过程读者可参看文献11)可以得出一系列准则，其中表示气流平均运动的准则有：

$$H_0 = \frac{\overline{w} \cdot t}{l} \ ; \quad Fr = \frac{gl}{\overline{W}^2} \ ; \quad Re = \frac{\overline{w} \cdot l}{\upsilon} \ ; \quad Eu = \frac{\rho}{\rho \overline{W}^2} \tag{39}$$

表示紊流脉动的准则有：

$$\frac{\overline{w_\phi'^2}}{\overline{w}^2} = \text{idem} \ ; \quad \frac{\overline{w_r'^2}}{\overline{w}^2} = \text{idem} \ ; \quad \frac{\overline{w_z'^2}}{\overline{w}^2} = \text{idem} \ ;$$

$$\frac{\overline{w_\phi' w_r'}}{\overline{w}^2} = \text{idem} \ ; \quad \frac{\overline{w_\phi' w_z'}}{\overline{w}^2} = \text{idem} \ ; \quad \frac{\overline{w_r' w_z'}}{\overline{w}^2} = \text{idem} \tag{40}$$

但是正如 М.А.Великанов 所指出(文献 12)，速度的脉动不仅由其振幅所表征，而且与脉动频率有关，而雷诺方程是没有考虑到脉动频率的影响，因此在相似准则中必须再引入 Струхaлb 准则(或称为几何相似准则)：

$$sh = \frac{Hf}{\overline{w}} \quad \text{或} \quad sh = \frac{H}{L} \tag{41}$$

因为紊流标尺和平均频率可用下式联系起来(文献 13)。

$$f = k \frac{w}{L} \tag{42}$$

这里，H 为气流几何尺寸；f 为气流脉动频率；L 为紊流标尺；k 为试验常数。

可见，在紊流工况下，要求两个现象流体动力相似除了式(39)准则相等外，还必需保证式(40)及(41)所列出的紊流脉动准则相等。这就使得模拟紊流工况下的物理现象极为困难。因此在一般工程模拟中或者是不考虑脉动准则相等的法则，或者认为脉动准则是自模的，的确在一般工况所遇到的问题中(如在管内流动等等)。紊流参数都基本上是自模的，但对一些比较特殊的问题，不能任意推广，必须经过试验检查并求出其自模的雷诺数范围。下面对半自由旋转紊流射流的自模化问题进行讨论。如果过程是准稳定的，而且不考虑重力的影响，那么定型准则只有雷诺准则，当黏性的影响小到可以忽略时，即欧拉准则和其他紊流脉动准则与雷诺数无关时，流体紊流运动进入自模区，模拟时只需要遵守几何相似条件和单值性条件，气流平均参数的自模已为试验所证实(文献 5)。我们的任务在于从理论上和试验上探求各紊流参数是否自模。

1) 紊流强度的自模化问题。

由公式(33)可知：$\varepsilon_{w'_\phi} = \dfrac{\sqrt{\overline{w'^2_\phi}}}{\overline{w}} = f\left(\dfrac{A}{K_1}\right)$

上面已经证明过紊流黏度系数 $A \sim K_2$，而 $K_2 \sim W_r \sim W_\varphi \sim W_\phi^{BX} \sim K_1 \sim Re$，系数 K_1、K_2 都与 Re 成同样规律的变化，因此

$$\varepsilon_{w'_\phi} = f_0\left(\dfrac{K_2}{K_1}\right) \neq f_1(Re)$$

亦即切向紊流强度是自模的。

此外，由公式(31)及(34)可见，当平均参数自模时，可以得出

$$\dfrac{\overline{w'^2_r}}{\overline{w}^2} \sim M \dfrac{\overline{w'^2_\phi}}{\overline{w}^2} \neq f_2(Re)$$

$$\dfrac{\overline{w'^2_z}}{\overline{w}^2} \sim N \dfrac{\overline{w'^2_\phi}}{\overline{w}^2} \neq f_3(Re)$$

试验结果完全证实了上述的推论(见图 14)，试验是在 $Re=48\times10^4 \sim 128\times10^4$ 范围内进行的。这里在计算雷诺数时作为定型尺寸采用旋风炉的直径，由图可见，不同雷诺数的试验点，都很好地落在同一曲线上，只有在炉室中央部分，由于平均速度小，脉动速度大，测量准确度降低而导致试验点有某些分散。

2) 方向关联系数的自模化问题。

由式(40)可得：

$$\dfrac{\overline{w'_\phi w'_z}}{\overline{w}^2} = R_{w'_\phi w'_z} \dfrac{\sqrt{\overline{\omega'^2_\phi}} \cdot \sqrt{\overline{\omega'^2_z}}}{\overline{w}^2} \neq f_4(Re)$$

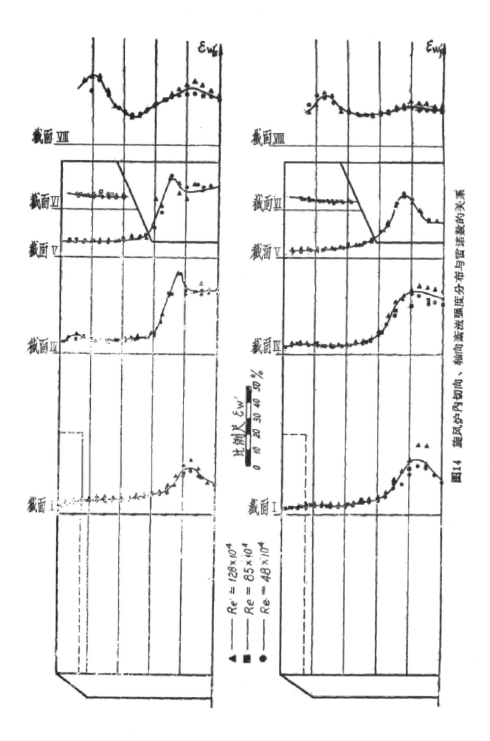

图14 旋风炉内切向、轴向紊流强度分布与雷诺数的关系

$$\frac{\overline{w'_\phi \cdot w'_r}}{\overline{w^2}} = R_{w'_\phi w'_r} \frac{\sqrt{\overline{w'^2_\phi}} \cdot \sqrt{\overline{w'^2_r}}}{\overline{w^2}} \neq f_5(Re)$$

$$\frac{\overline{w'_z \cdot w'_r}}{\overline{w^2}} = R_{w'_z w'_r} \frac{\sqrt{\overline{w'^2_z}} \cdot \sqrt{\overline{w'^2_r}}}{\overline{w^2}} \neq f_6(Re)$$

如果关联系数是自模的话，则上面不等式是成立的，即紊流摩擦应力的相对值与雷诺数无关。图 15 中列出了轴向和切向关联系数测量结果，证明 $R_{w'_\phi w'_z}$ 与 Re 值无关。这里试验点比较分散，因关联系数测量难以高度准确的缘故，至于其他方向的关联系数自模问题，由于测量上的困难在旋风炉内没有进行测定。但上面已经证明，旋风炉内具有射流的模型。试验证明，对于射流来说各方向的关联系数是近似相等的（见图 6），只要一个方向的关联系数是自模的，则其他方向关联系数，也应是自模的。

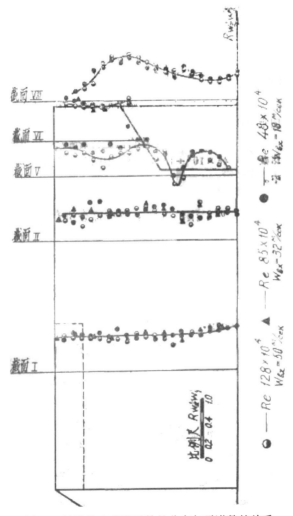

图 15　旋风炉内关联系数的分布与雷诺数的关系

3) 紊流标尺自模问题。

由公式(41)可知，如果紊流标尺 L 能够自模的话，则准则 Sh 亦与雷诺数无关。根据紊流理论，就定义来说，位移径和紊流标尺的数值应该是重合的(文献 2)。我们的试验也证明，这两值是近似相等的，实际上由公式(42)根据测量出来的气流平均脉动频率可以求出 L 值。另外在同样条件下由理论计算和试验亦可得出位移径 l 值。两个数值的比较列于下表。

	截面V				截面VII				截面VI		
r/r_u	0.9	0.9	0.7	0.6	0.9	0.8	0.7	0.6	0.95	0.8	0.6
L_{MM}	6.0	6.76	7.0	8.0	5.9	6.1	6.9	8.5	5.2	6.8	8.9
l_{MM}	4.01	4.71	4.14	4.15	5.2	5.65	6.15	4.74	5.3	5.32	4.16

L 值经常比位移径值高，这是因为在式(42)中试验系数 k 我们只有射流的数据 (0.0267～0.0273)，对于旋转射流可能有更低的数值。

由公式(36)可知 $l = f\left(n, \dfrac{A}{\sqrt{J}}\right)$

$$A \sim K_2, \quad \sqrt{J} \sim \varphi(K_1, K_2) \sim K_1; \quad n \neq f(Re)$$

故

$$L \approx l = f\left(\dfrac{K_2}{K_1}, n\right) \neq f_7(Re)$$

这个论点亦为图 12 的试验近似证明，图中示出不同雷诺数及结构方案下，位移径基本上固定在某范围内不变。

最后可以得出结论，半自由旋转紊流射流的平均及紊流参数是自模的(至少在试验的雷诺数 $Re=48\times 10^4 \sim 128\times 10^4$ 范围内)。因此所得试验数据可以推广到几何相似的其他炉室中。

四、紊流摩擦、紊流混合所产生能量损失问题的讨论

在紊流运动中，如果不考虑重力的影响，则可以写出下列的能量平衡式：

$$P_{cyM_\bullet} = P_g + P_{cr} + P_{TpeH}^{cT_\bullet} + P_{TpeH}^{BH_\bullet} \tag{43}$$

这里 $P_{cyM_\bullet}$ 为总压力；P_{cr} 为静压力；$P_{TpeH}^{cT_\bullet}$ 为气流和壁面摩擦所产生的压力损耗；$P_{TpeH}^{BH_\bullet}$ 为在气流内部由于分子及紊流摩擦所产生的压力损耗；P_g 为动压力。其按时间的平均值为：

$$\overline{P_g} = \dfrac{\pi\rho}{2}\int \overline{\tilde{w}^2} r\mathrm{d}r = \dfrac{\pi\rho}{2}\int \overline{\left(w^2 + 2ww' + w'^2\right)} r\mathrm{d}r$$

$$= \dfrac{\pi\rho}{2}\int \overline{w}^2 r\mathrm{d}r + \dfrac{\pi\rho}{2}\int \overline{w'^2} r\mathrm{d}r = P_g^{cp} + P_g^{ny\pi} \tag{44}$$

由此可见，采用皮托管之类测量速度时所得的平均动压力为平均速度及脉动速度所产生的动压力之和，因此当速度较小，气流脉动较大时，用气力法测量速度是会出现较大误差，如果仪表是惯性很大的，误差最大值可达

$$\Delta = \frac{\varepsilon_{w'}^2}{100}\% \tag{45}$$

在能量平衡式中，过去一般都只算在壁面边界层内的摩擦损失，而在边界层外却认为是无摩擦的运动(或者认为是等势流动)。这对一般工程上气流脉动不是很强的问题是正确的，而且试验也证实了这点，但是对一些比较复杂的问题，例如旋转气流在旋风炉内运动能量损失问题，则要重新考虑，我们认为，任何能量的损耗，至少要分为两部分。一部分是流体和壁面间的摩擦损耗，另一部分是分子和气流脉动的能量损耗，后者或称为散射能。在试验过程中我们发现，在旋风炉内能量损耗和出口喷咀直径有极大的关系，随着出口喷嘴的减小，能量损耗急剧地增加，这个现象，单纯用流体和壁面摩擦损失的理论不能加以完满解释。因为第一，随着出口喷嘴直径的变化，旋风室内壁面面积变化很小。第二，试验证明，改变出口喷嘴直径并没有导致壁面附近平均速度有重大的变化(我们的数据及文献 5，6)。可见，流体与壁面摩擦损失并不是产生能量损耗的唯一原因，但是另一方面仔细测量证明，出口喷嘴的存在使气流紊流脉动强度大为增强(见图 23)。这不能不使我们联想到实际流体的黏性对分子运动及紊流脉动的阻滞问题。很清楚，流体运动时黏性的存在起着双重的作用。第一，它把气流动量由一层传往另外一层。在旋转气流中，在其中心吸入大量外界本来是静止的不旋转的，不脉动的气流。由于黏性传递作用，使它和周缘旋转气流强烈混合，并迫使其作似固体的旋转运动。第二，在传递过程中使部分的运动机械能转变成为热能，这是一个不可逆的能量损失，也即是散射能，可以想象，气流脉动及紊流混合愈强烈，散射能的损失就愈大。下面我们要近似估计在旋风炉内旋转气流所产生散射能的大小。根据紊流理论，平均散射能值可由下式决定：

$$\overline{D} = \overline{D_\mathrm{T}} + \overline{D_\mathrm{c}} = \mu\left[2\overline{\left(\frac{\partial w'_x}{\partial x}\right)^2} + 2\overline{\left(\frac{\partial w'_z}{\partial z}\right)^2} + 2\overline{\left(\frac{\partial w'_y}{\partial y}\right)^2}\right.$$

$$+ \overline{\left(\frac{\partial w'_x}{\partial y}\right)^2} + \overline{\left(\frac{\partial w'_y}{\partial x}\right)^2} + \overline{\left(\frac{\partial w'_x}{\partial z}\right)^2} + \overline{\left(\frac{\partial w'_z}{\partial x}\right)^2} + \overline{\left(\frac{\partial w'_y}{\partial z}\right)^2} + \overline{\left(\frac{\partial w'_z}{\partial y}\right)^2}$$

$$\left. + 2\overline{\frac{\partial w'_x}{\partial y} \cdot \frac{\partial w'_y}{\partial x}} + 2\overline{\frac{\partial w'_x}{\partial z} \cdot \frac{\partial w'_z}{\partial x}} + 2\overline{\frac{\partial w'_y}{\partial z} \cdot \frac{\partial w'_z}{\partial y}}\right] + \mu\left(\frac{\partial w}{\partial y}\right)^2 \tag{46}$$

这里，$\overline{D_\mathrm{c}} = \mu\left(\frac{\partial w}{\partial y}\right)^2$ 为由平均速度所决定的散射能值；$\overline{D_\mathrm{T}}$ 为由脉动速度所决定的散射能值。

在紊流工况中，上式是极端复杂的，在目前还不可能得到准确的理论解答，而且即使用试验求解也是很困难的。为了简化和解决一些实际问题，Е.М.Минский 引入了略去

"三级矩"的假定(文献13)。作者试验证明,"三级矩"(即 $\overline{W_x'^3}$、$\overline{W_z'^3}$、$\overline{W_x' \cdot W_y'^2}$、$\overline{W_x' \cdot W_z'^2}$、$\overline{P' \cdot W_x'}$、$\overline{P' \cdot W_y'}$)数值上是不大的,这样可以把式(46)简化成下面形式:

$$\overline{D_\text{T}} \approx -\rho \overline{w_x' w_y'} \frac{\mathrm{d}w}{\mathrm{d}y} + \frac{\mu}{2} \frac{\mathrm{d}^2}{\mathrm{d}y^2}(\overline{w_x'^2} + \overline{w_y'^2} + \overline{w_z'^2}) + \mu \frac{\mathrm{d}^2 \overline{w_y'^2}}{\mathrm{d}y^2} \text{(公斤/米}^2\text{秒)} \tag{47}$$

对旋风室来说,单位截面上能量损失为

$$P_{Tpe\text{H}}^{BH\bullet} = \overline{D_\text{T}} \frac{V_u}{Q_{\text{CyM}\bullet}} \tag{48}$$

这里,V_u 为旋风室容积,$Q_{\text{CyM}\bullet}$ 为通过炉室的总空气量。

$$Q_{\text{CyM}\bullet} = Q_{\text{BX}} + Q_{\text{OBP}} = (1.2 \sim 1.4) Q_{\text{BX}} \tag{49}$$

Q_{BX} 为进入炉室空气量;Q_{OBP} 为平均逆流空气量。

现在我们有可能试验估计散射能的数值,不过应当指出的是式(47)是近似的,而一般紊流参数测量准确度又并不很高,因此虽然进行了各种结构的测量,但我们认为比较合理的是比较两种极限工况的散射能损失。那就是带有通常出口喷嘴和没有出口喷嘴(圆柱)两种方案。因为这两种方案能量损失相差很远,因而由于理论公式和测量不准确所引起的误差不致对所作定性结论有很大影响。对于这两种方案散射能沿炉室半径变化总结在图16中,试验清楚地表明,在带有出口喷嘴的旋风室内,在旋转射流核心区散射能损失不大,大部分能量损失集中在射流边界层上,在那里散射能数值比核心区高几十倍。这是因为在边界层内有着强烈的紊流混合的结果。对于没有出口喷嘴的旋风室(方案 N)有着另外的特性,散射能损失主要集中在周缘,在炉室中心部分气流几乎是不扰动的,因而能量损失也就很小。如果把各截面散射能分布按面积积分,则可得出平均散射能值沿长度的变化(图17),发现旋转气流散射能损失离开射流入口愈远有降低的趋势。如果把得出的平均散射能值代入式(48)内,则可以计算得由于紊流脉动和混合所引起的压力损失,计算结果表明,在有出口喷嘴的旋风室这个损失达 270 毫米水柱,对没有出口喷嘴的旋风室,则只有 85 毫米水柱。这些数据不是非常准确的,但至少可以给我们作出如下的定性结论:

1)旋风室内旋转气流能量损失应该由两部分组成计算,即与壁面摩擦及气流内部脉动损失,两者均不能忽略。

2)流体和壁面的摩擦损失主要取决于壁面尺寸,状态及壁面附近气流速度的大小。亦即主要由炉室直径和长度决定。而散射能损失主要由紊流摩擦($\tau = -\rho \overline{W_\phi' W_z'}$ 等)所决定,亦即与紊流混合有关。所以当出口喷嘴直径增大时,紊流混合在旋风室内有所恶化,因而散射能损失也降低。总之这两部分能量损失的大小比例与炉室具体结构有关。例如当 $\dfrac{d_c}{D_u} = 0.44$,$Re = 128 \times 10^4$ 时,由于紊流脉动、紊流混合所引起的压力损失占旋风炉压力总损失的 60%。

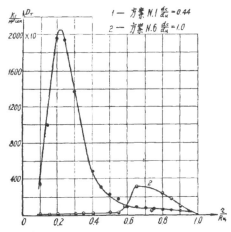 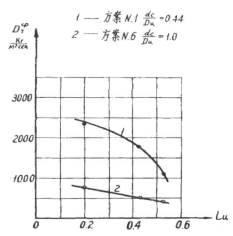

图 16　散射能沿炉室半径变化 ($Re=128\times 10^4$)　　图 17　平均散射能值沿炉室长度变化

3) 旋风炉出口喷嘴不仅是个"紊流强化机构"和"质点网罗机构",而且也是能量损失主要源泉之一。没有了能量损失也就不可能创造出强烈的脉动和混合,使燃烧过程得以强化。因此,没有出口喷嘴的旋风燃烧室,即使燃用气体燃料也不可能得到高的燃烧热强度。 虽然此时能量损失会降低 3~4 倍,但也无法采用。因此我们宁愿损失一些能量来博取燃烧过程的极大强化,而且还应当指出,在旋风炉内散射能的损失并非是完全有害的损失,因为它转化成热能加热了气流,同时也是紊流混合测量的尺度。

由于紊流脉动、混合所产生散射能损失的原因及其属性这样复杂的问题有待今后各国学者进行详细的理论和试验研究,这里我们只能给出一些粗糙的估计和近似的物理解释。

五、旋风燃烧室的几何结构对气流紊流机理的影响

燃烧试验证明改变旋风室的几何参数对燃烧过程都会产生直接的影响,这里只能讨论几种对燃烧影响较大的参数,而且由于试验容量和篇幅的限制,下面主要作结论式的说明。

1) 在旋风室内紊流运动的一般特性。

采用热电风计的研究办法能够同时记录瞬时的及按时间平方平均的脉动速度值。在图 18 中示出典型的瞬时脉动速度示波图,中间黑线是平均速度水准线,脉动速度就在平均速度基准上上下摆动,值得注意的是按照统计规律 $\overline{W'}=0$ 。图 18 中上图是一在旋转射流核心内典型脉动曲线,下图是一在边界层内,在脉动曲线下同时示出时间指标。每个尖端的间隔为 0.02 秒,可以清楚看出,脉动的大小可以用平方平均的脉动振幅值来代表,但是还需要寻求平均脉动频率值。其典型分布可以在图 19 中找到。在旋转射流核心区发现脉动频率和其他紊流参数一样基本上是不变的,脉动频率最大值处于边界层开始不久之处,相应于脉动速度最大的地方。对照图 8 脉动振幅值可以清楚看到,在边界层内脉动会逐渐减少,其原因是因为向弱旋转,微扰动的中心逆流区传递过程中脉动能量不断被吸收的缘故,无论脉动速度或紊流强度其最大值位于周缘逆流和中心正流区。因为这

里刚处于射流边界层开始不久处同时又具有最大的速度梯度(经过平均轴向速度等于零的表面)。试验证明,方向关联系数 $R_{w'_\phi w'_z}$ 和紊流脉动曲线具有恰恰相反的性质(图 15)。这从关联系数物理意义本身也可以得到正确的解释。我们知道,当关联系数 $R_{w'_\phi w'_z}=0$ 时,就说明在这点上纵向和横向脉动并无任何联系。因此在示波图中纵、横向瞬时脉动值均匀地充满在一个圆形面积内,即速度的脉动在各个方面具有相同的或然率。此时的紊流气流称为各向同性。相反,如果 $R_{w'_\phi w'_z}\ne 0$,则瞬时脉动速度的试验点只充满在椭圆的面

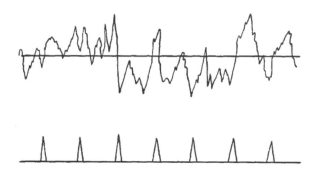

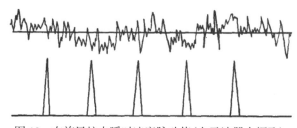

图 18 在旋风炉内瞬时速度脉动值(在示波器中摄取)

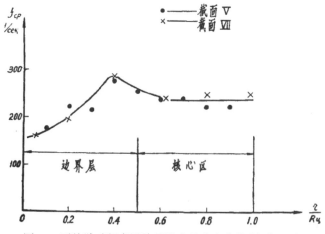

图 19 平均脉动频率沿旋风炉半径分布曲线(方案 N.5)

积,说明在某个方向出现的或然率比另一方面来得大,在极限工况下,$R_{w_\phi' w_z'} \to 1$时,脉动只会出现在某一个方向,而试验点则只集中在一条直线上,由此可见,从强化燃烧及混合过程的观点出发。在同一气流脉动水平下关联系数愈小愈好,因为此时各个方向脉动的或然率最大时,使得氧气和可燃物在该点的混合达到最强烈,遗憾的是目前关于关联系数对燃烧过程影响的理论,试验研究还进行得很少。因此,这里我们只能提出一些间接的判断,旋风炉中,燃烧最强烈是位于出口喷嘴附近周缘逆流和中心正流的交界面上。这里不但有着最大的紊流脉动强度而且方向关联系数的数值最小(图15及22)。另外在风洞中试验证明,在出口处加上管簇或其他非流线形物体后,方向关联系数值急剧地减小,这就说明由于管簇使各个方向的紊流混合加强的缘故。(关于这方面的试验研究,我们将另文叙述)。下面将简单讨论一下旋风炉各主要几何参数对紊流机理的影响,具体的试验方案示于图20内。

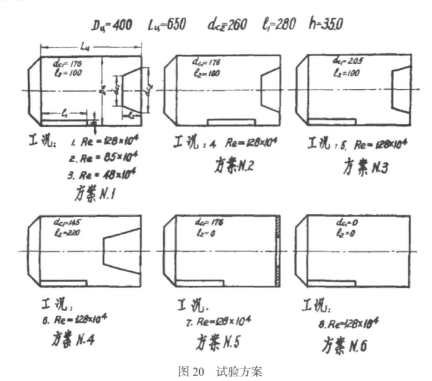

图20 试验方案

2) 旋风室出口喷嘴的作用。

旋风炉出口喷嘴不单只是个灰渣质点网罗机构,能量损耗源泉之一,同时也是一个良好的"紊流发生器",由于它的存在把半自由旋转射流分为核心区和边界层区。强迫气流逆转,因而产生强烈的紊流混合,试验证明,不论在核心区或边界层上,平均紊流强度值在出口喷嘴附近达到最大(图21),关联系数值降至最小(图22)。用下面的事实可以得到更明显的证明:在同样进口速度,雷诺数下,当旋风室没有出口喷嘴时,其平均脉动速度水平比起有出口喷嘴时低2~3倍(见图23)。

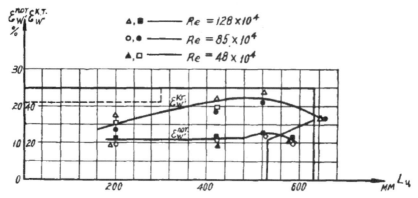

图 21　平均紊流强度沿旋风室长度的变化(方案 1)

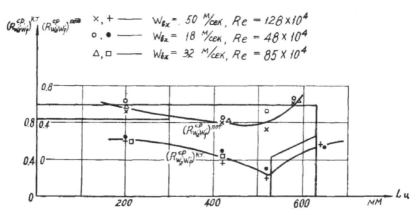

图 22　平均方向关联系数值沿旋风室长度的变化(方案 1)

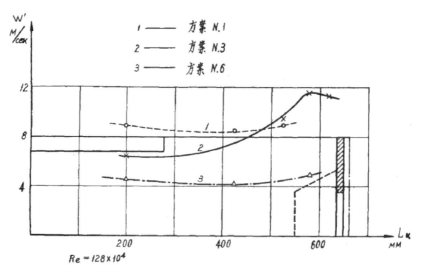

图 23　各种结构方案下平均脉动速度的变化($Re=128\times 10^4$)

3) 改变旋风室出口喷嘴直径的影响。

这里可以归结成下面几点：

A) 出口喷嘴直径减少时，旋转射流核心区增大，而边界层区缩减，中心逆流区强度增大,阻力增加。

Б) 出口喷嘴内环室的大小就决定了射流核心区的大小,当出口喷嘴直径减小时,环室增大,因而射流核心也就扩张。在图24中可以清楚地看到,改变出口喷嘴直径,在核心区内紊流强度仍然维持同一水平。而在边界层内则略有不同。

B) 随着出口直径的减少,旋风炉内总的紊流脉动增强,当然此时能量消耗亦增加。试验证明,紊流脉动的增强主要出现在边界层区内。因为当边界层区范围缩减时,射流核心和边界层之间紊流摩擦和混合亦大为增加,与此相适应的方向关联系数数值亦减小,但在旋转射流核心区,紊流脉动水平是基本上保持不变的,这点在图25中得到证实。

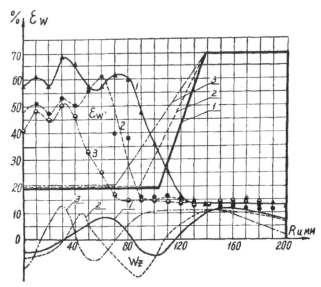

图24 紊流强度变化与出口直径关系($Re=128\times10^4$)

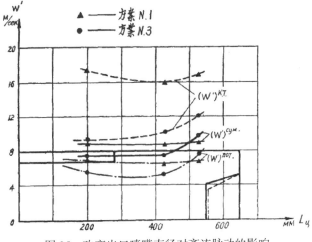

图25 改变出口喷嘴直径对紊流脉动的影响

Γ) 出口喷嘴直径过大,虽然阻力损失甚少,但紊流混合及脉动较差,直接影响到燃烧过程的强化,但出口喷嘴直径过小,则能量损失太大,而且旋风室燃烧过程并不单由混合过程决定。可见这里应有一个最有利的出口喷嘴直径值。这只能在燃烧试验中求出，但亦可以进行近似的计算比较。

4) 出口喷嘴形状的影响。

从紊流混合的观点看，倒圆锥形出口喷嘴并不一定是最理想的结构形式。因为它的阻力损失较大，制造又比较复杂。因此对一些不需要设置专门"陷阱"网罗质点的高温工艺过程，或者当燃用液体或气体燃料时，可以考虑采用结构较简便，阻力损失不大的出口孔板结构。此时孔板的作用变成主要是创造高度的紊流脉动水平，使氧气和可燃物得到充分良好的混合，试验证明采用孔板后，气流在旋风炉内的流动结构是没有重大改变。各个特性区仍然存在。测量证明，采用出口孔板的结构后紊流脉动值并不见得比一般倒锥形出口喷嘴结构为低(见图26)。相反，在出口附近还有某些提高。但关联系数值却比有出口喷嘴的为高(图27)。这是完全可以理解的，因为采用出口孔板后气流紊流混

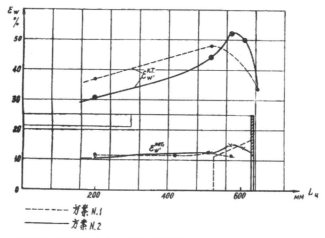

图 26 不同结构方案紊流强度沿长度变化($Re=128\times10^4$)

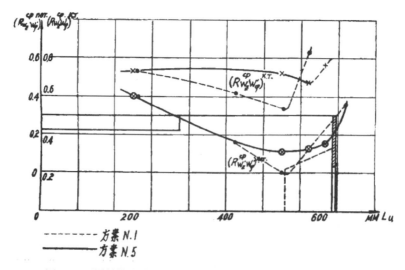

图 27 不同结构方案方向关联系数沿长度变化($Re=128\times10^4$)

合程度有所降低，因而压力损失也较小。这个结论也为 E.A.Haxaпeтян 的燃烧试验所证明。她在带有出口孔板的旋风炉试验台中进行燃用重油燃烧试验。其热应力可达 $14\sim15\times10^6$ 大卡/米^2 时，但这还不是极限。

5) 二次风入口喷嘴位置的影响。

B.И.Xвостов 在旋风炉内进行液体燃料燃烧试验研究时发现，当入口喷嘴向出口方向移动时，燃烧过程得到改善。同时在炉室首部烟气环流区得到强烈的发展。这些现象引起了我们的兴趣。因为可以猜想，二次风入口喷嘴位置的移动，一定会引起炉内气流紊流结构的改变，仔细测量证明，当入口喷嘴由炉室首部向出口处移动时，会出现两个扰动中心：炉室首部的和出口喷嘴附近的(图 28)，首部扰动中心的脉动程度甚至会比后部脉动中心为大。可见燃烧过程之所以得到改善是因为：

A) 两个扰动中心的形成使得有可能做成两个气化区，因而分别进入此两区内的空气量减少，同时在气化区内由于燃料和燃烧产物强烈的紊流混合，使得温度水平急剧提高，改善了燃料气化过程。

Б) 气化和燃烧的"活泼"过程由局限于在出口喷嘴附近而扩展到整个旋风室容积内。

B) 气化可燃物由两气化区流出进入旋转射流边界层，在强烈的气流脉动作用下得到高速的燃烧。同时，两个扰动中心的存在使得进入炉内的冷空气和燃料得到迅速的加热，使燃烧过程稳定。

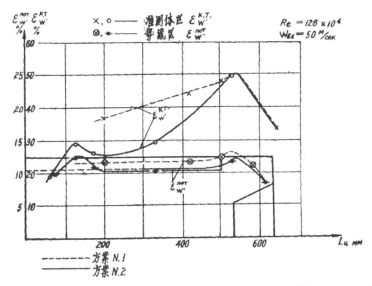

图 28　入口喷嘴移动时平均气流紊流参数沿炉室长度的变化($Re=128\times10^4$)

六、从气流紊流结构来讨论旋风燃烧过程某些特点

燃料在旋风炉内燃烧过程是极其复杂的，而且目前还没有研究清楚，因此各国学者都只能根据一些片断材料来推想旋风燃烧过程。我们认为苏联 Г.ф.Kноppe 教授所提出的旋风燃烧物理模型是比较全面的，但是他没有考虑固体燃料质点在旋风炉壁黏性灰渣膜

附近运动的特殊规律性以及其对气化过程的影响。在这方面陈运铣教授提出了壁面附近燃料"沸腾层"的理论，着实地改正了和发展了 Г.ф.K$_{Hoppe}$ 所提出的物理模型(见本期浙大学报)。 我们这里把任务限制在从气流紊流结构出发来讨论燃料在旋风炉内空气燃烧和气化过程。在旋风炉内的高温高速燃烧过程实际上可以分为两个不同的阶段：气化阶段和气化可燃物燃烧阶段。任何一个阶段组织不够完善，都会导致燃烧过程的恶化。因此，首先提出的一个迫切问题是：在旋风燃烧中究竟哪一个阶段限制和阻碍了燃烧强度和热应力的继续提高，这使得我们今后控制和改善燃烧过程中能够"有的放矢"。

1) 旋风燃烧中燃料气化阶段是限制燃烧强度和热应力的主要因素。

燃烧室极限热负荷一般可以由下列三种原则决定。第一极限热负荷是取决于气流的空气动力危机。亦即在燃烧室内某一截面烟气流速达到音速时所相应的极限热负荷。对一般动力式炉子，这个极限是不现实的。因为目前应用的热负荷只为第一极限热负荷 3%～10%。第二极限热负荷是由化学反应动力学危机所决定。亦即此时热负荷受到化学反应速度的限制(与温度、浓度、压力等有关)。在高温的旋风燃烧过程中这个极限实际上也是达不到的。比较现实的是第三极限热负荷，此时热应力主要取决于混合危机。这相应在燃烧的中间和扩散区域。即物理因素如紊流脉动和混合等起着决定性的作用。这里我们单写出第三极限热负荷的最终理论公式。对截面热应力：

$$\left(\frac{Q}{F}\right)_3^{Пред} = \frac{3600\dfrac{H \cdot K}{\alpha \cdot L_{МИН}} \int_F \varepsilon_w' W_{0 \cdot CP} \mathrm{d}f}{F} \tag{50}$$

及容积热应力

$$\left(\frac{Q}{V}\right)_3^{Пред} = \frac{3600\dfrac{H \cdot K}{\alpha \cdot L_{МИН} \cdot I_T} \int_F \varepsilon_w' W_{0 \cdot CP} \mathrm{d}f}{F} \tag{51}$$

这里，α 为过量空气系数；$L_{МИН}$ 为燃烧理论空气量；H 为燃料发热量；K 为公式(1a)中的试验系数。

同时用 μ_3 表示第三极限截面热应力与目前燃烧技术所达到的极限热应力之比。根据对旋风炉气流紊流结构研究结果代入上式进行计算，并和实际热负荷进行比较，得出下表：(注意：计算所得是平均数据)

	$(Q/F)_{实际}$	$(Q/V)_{实际}$	$(Q/F)^{Пред}$	$(Q/V)^{Пред}$	μ_3
	10^6 大卡/(米^2·时)	10^6 大卡/(米^3·时)	10^6 大卡/(米^2·时)	10^6 大卡/(米^3·时)	%
燃用固体燃料旋风炉	12～15	1～12	18～49	6～49	27～67
燃用重油的旋风炉	12～20	1～12	32	9～29	38～63

由表上可以清楚地看出由脉动强度决定的火焰紊流扩张速度计算所得的第三极限热负荷远比目前容许实际采用的为高。同时应当指出的是：气化产物的燃烧是处在周缘逆流区和中心正流区的交界面附近。即刚好处于旋转射流边界层开始部分，那里有着极高的脉动强度。计算证明，这里的截面热强度可以达到 40÷80×10⁶ 大卡/米² · 时，比截面平均值高两倍。比目前所能达到的截面热应力高 4～6 倍。完全可以相信，由于旋风炉是个强烈的"紊流发生器"。因此限制燃烧强度和热应力的并非紊流混合，而是燃料气化过程的强度。因而对于组织旋风燃烧过程首先就要尽量设法强化和合理安排燃料气化过程，否则就会出现结焦、堵渣，飞灰损失增大种种不良现象。

2) 气流紊流结构与强化燃料气化阶段的关系。

为了强化气化阶段，除了良好的空气动力运载特性外，还必须有高的温度水平和燃烧产物与气化燃料强烈的混合。在气化区内(炉室周沿)，由于氧气的缺乏，主要进行吸热的还原反应，即使有氧化反应也只是按 $2C+O_2=2CO$ 方式进行。再加上大量冷的二次风喷入炉室边缘，因此要提高气化区的温度水平成为组织旋风燃烧主要任务之一。首先气化区的加热可以依靠燃烧区的辐射热，但是很多学者对煤粉炉的研究证明，辐射换热不能作为使新鲜煤粉气流加热到即使是着火温度的主要热源。在旋风炉气化区有类似的情况。因为第一，质点尺寸不大。因而其角系数也很小，同时在沸腾层上质点浓度大，互相掩盖。第二，在气化区内主要充满双原子气体 N_2、O_2、CO、H_2 等，对辐射线它们都是透过体。因而在壁面附近的烟气流也难以得到辐射加热。因此我们认为气化区之所以能得到很高的温度水平主要由于燃烧区的高温燃烧产物向气化区回流，同时在气化区内得到充分的紊流混合。Кацнельсон Б.Д. Палеев И.И.等专门试验证明，紊流脉动对于燃料质点和气流之间的热质交换有很大的影响（文献 14）。并服从下列关系式：

$$Nu \sim 2.8\left(\varepsilon_{W'}^{0.5} \cdot Re^{0.5}\right)C(W')^{0.5} \tag{52}$$

即高温烟气对燃料质点的加热，以及气化产物从质点表面和气化区的排离与气流脉动速度的平方根成比例，在旋风炉内气流脉动速度比煤粉炉大十倍。因此如果空气动力结构组织得合理，则在旋风炉内气化过程应比煤粉炉内强化三至四倍。作为最好的证明是冷的二次风气流喷入旋风室后温度立即提高 3～4 倍，而且沿旋风室截面测量温度分布曲线是比较均匀的，证明在炉内紊流混合非常良好。

关于燃料在旋风室内的气化，燃烧过程及强化气化过程的各种因素等等。读者可参阅陈运铣教授的著作(文献 15)。这里，我们只从紊流结构的角度进行一些简略的讨论。

主 要 符 号

r_u—旋风室半径。　　　　　　r_M—相应于准势流区与准固体区交界的半径。

W，W'——平均及脉动矢量速度。(以后各脉动速度应理解为平方平均开方值)

$\overline{W_\phi}$，W'_ϕ——平均及脉动切向速度。

$\overline{W_z}$，W'_z——平均及脉动轴向速度。

$\overline{W_r}$，W'_r——平均及脉动径向速度。

ε'_w，ε'_{w_ϕ}，ε'_{w_z}，ε'_{w_r}——各为总的，切向，轴向，径向紊流强度。

$R_{W'_n W'_m}$——n、m 各方向的方向关联系数。

$\tau_{w'_n w'_m}$——紊流摩擦力(或紊流切应力)。

l——位移径。　　　　　　L——紊流标尺。

f——气流脉动频率。

D_T——由气流脉动所引起的散射能。

A——紊流黏度系数(或紊流交换系数)。

$$B^{CyM} = \frac{1}{\pi r_\Pi^2} \int_0^{r_u} 2\pi B r \mathrm{d}r \quad \text{——参数 } B \text{ 沿旋风室半径的平均值}$$

$$\left.\begin{aligned} B^{\Pi Or} &= \frac{1}{\pi\left(r_u^2 - r_M^2\right)} \int_{r_M}^{r_u} 2\pi B r \mathrm{d}r \\ B^{\Pi \bullet r} &= \frac{1}{\pi r_M^2} \int_0^{r_M} 2\pi B r \mathrm{d}r \end{aligned}\right\} \text{参数 } B \text{ 在准等势区和难固体区内的平均值}$$

参 考 文 献

1. Власов К.П. Иноземцев Н.Н. Влияние начальных параметров на турбулентную скорость распространения пламени однородной топливовоздушной смеси. М.1959.
2. Лебедев Б.П. 等, Камеры сгорания авиационных ГТД. Том.1. 1957. М.
3. Боровченко Е.А. "Труды ин-та энергетики АН БССР" вып.Ⅲ.1957.
   вып. Ⅳ.1958.
4. 岑可法 气流紊流参数的研究方法（热电风计法） 见本期浙江大学学报。
5. Нахапетян Е.А. 副博士论文. 1952.
6. Калишевский Л.Л. 副博士论文. 1958.
7. Розовский И.Л. Движение воды на повороте открытого русла. Киев. 1957
8. Маккавеев В.М. 等, Гидравлика М. 1940.
9. Вулис Л.А. Устименко Б.П. "Теплоэнергетика" 1956. N.4
10. С. Гольдштейн Современное состояние гидроаэродинамики вязкой жидкости Т.1 1948.
11. Леви И.И. Моделирование гидравлических явлений. М.1960.
12. Великанов М.А. Динамика русловых потоков. Т.1 1954
13. Минский Е.М. Турбулентность русловых потоков. 1951
14. Кацнельсон Б.Д. Палеев И.И.
    в сб. "Третье Всесоюзное совещание по теории горения" Т.Ⅱ 1960. М.
15. 陈运铣. 旋风炉内固相质点的分布规律及炉内燃烧过程机理的商榷.
    见本期浙江大学学报。

气流紊流结构的研究方法

(热电风计的应用)

岑可法

摘 要

本文总结了热电风计的发展近况，阐述了应用热电风计测量平均及紊流脉动参数的物理基础，提出了测量两元气流平均参数及三元紊流脉动参数的方法，拟定了应用 V 型敏感元件线路系统，并对部分元件的制作、调整、校验作出了适当的介绍，对 V 型敏感元件进行了试验，得出其对流放热的准则公式。文章中对所拟定的方法和测量仪表系统的准确度进行了较详细的分析，并与其他方法作了比较，证明所拟定的方法是准确的，操作和试验数据的总结是比较简单的。最后文中提出了实际应用的例子。

一、引 言

气流的紊流结构对燃烧过程，传热过程，气体流动过程均有极大的影响，而在气体力学本身，气流的紊流参数正是重要的研究对象，目前，由于紊流理论还发展得非常不完善，因此，长期以来人们找寻各种各样的研究方法，来测量紊流参数，其中比较主要的有：

球体测量法(文献 1)基于阻力系数在临界雷诺数下有急剧降低的现象。微形放大镜摄测法(文献 2)基于用放大镜来跟踪摄影加入流体内固体质点的运动规律。热扩散法(文献 3)基于用热电偶测量热源后面热扩散的规律，然后换算成气流紊流规律。辉光放电法(文献 4)，基于利用两极间辉光放电来测量气流紊流度。微形风速计法(文献 5)基于记录风速计叶片不稳定的转速。惰性气体扩散法(文献 6、7)，基于研究某种惰性气体在气流中扩散的规律，然后转换成紊流规律，及最后我们准备详细讨论的热电风计法，在上述方法中，应用得最广，最有成效的是热电风计法，因为借它之助原则上可完全测量各种紊流参数(这里指气流平均速度，各脉动分速，坐标关联系数，方向关联系数，温度脉动值，温度速度关联系数各紊流摩擦力，紊流标尺，气流脉动频谱等)，在这点上，其他方法是无可比拟的；因为它们大多数只能测量一元气流脉动分速。热电风计的作用原理是这样的：微小的热电丝在电加热下达到高温，但是它的放热规律取决于气流速度及其紊流结构，因此测量放热情况就能求出各紊流参数，这里稍稍回顾一下热电风计的发展史，将会有利于我们对它的理解及掌握：早在 1912 年 Morris 及 Bordini 就提出了用冷却被电流加热金属丝的方法来测气流平均速度，1914 年 King R.O.提出了这个散热计算的理论公式，这个公式直至今天，还有一定指导意义，之后热电风计按两个方向发展，第一个方向是如何准确测量平均速度，特别在极低速和高速情况下，主要课题要使测量的电参数和气流速度变化尽可能成线性关系，（实际上在下面我们将要证明一般为 3~4 次方关系）于是 Huguenard 建议用补偿得出线性刻度的方法，（在§3 将会详述）而 Davis 及 В. П. Чеббышев 则建议补偿气流温度变化的方法，这样使得在不同的气流温度下，仍然能用一条校正曲线去研究平均气流，第二个方向是热电风计的主要发展方向用它来测量各种紊流参数，在 1926 年 Burgers 提出了计算热电风计热惯性的理论，在 1929 至 1932 年 H.L.Dryden 及 A.M.Kuethe 建议采用补偿线路来克服电热丝的热惯性，并开始企图用复杂的热电丝感应元件来测量紊流尺度，在苏联，从 1938 年起 Ю. Г.Захаров 及 Е.М.Минский 就开始总结、发展热电风计的理论及实践，对其准确度进行了详细分析，同年，Simmons 等开始应用热电风计来研究紊流气流的脉动频谱，后来 Е.С.Семенов 把这些方法运用到内燃机的研究中，可见热电风计的发展已有几十年的历史，但是由于测量仪表本身的复杂性及测量要求的高度精确性（因为不可能直接在脉动气流中进行校正），使得在今天还存在有很多理论和实践的问题，同时也限制了它作为标准仪表成批生产，在苏联，进行试验性生产的热电风计（只适用于一元气的测量）有全苏电工研究所的产品（牌号为 ЭТАМ-3А、ЭТАМ-5А 等）至于要研究复杂的紊流结构，必须由研究者本人亲自设计制作，这也是阻碍广泛应用热电风计原因之一。

热电风计之所以能得到广泛应用，是因为它具有下列主要优点：

1）能够测量三元流动情况下的各种紊流参数，而采用其他方法目前即使在理论上还不可能。

2）能够非常准确测量低速度的气流运动，这在实际研究中经常出现，如旋风炉的核心区，边界层的研究，旋转射流或绕过非流线型体的回流区等等，研究证明，采用热电风计可准确测量到 0.03 米/秒这样小的速度。

3）感应元件体积很小，一般热电丝只有几十微米直径，这点对研究气流紊流结构有重大意义，因为一般应用的感应元件或多或少都会破坏气流的紊流结构，这在紊流边界层的研究中，特别重要（文献 8）。

4）热电风计具有高度的敏感性及相对不很大的热惯性，因此其热惯性的补偿较易，同时，出口的电信号比较容易测量，可以直接换算成紊流参数。

5）感应元件可以伸入测量对象的任何地方，而测量操作可以远离研究对象（其距离可达几十米），以便保护测量仪表不经受振动，灰尘侵蚀，提高测量的准确度。但是热电风计亦有其本身的缺陷：

1）比较严重的是热电丝在长期工作过程中会"变老"，即热电丝特性不够稳定，这是由于丝被局部氧化，受介质及固体微粒的侵蚀磨损，表面被染污，丝直径减少，其中最严重的是在反复高温加热冷却过程中，丝的电阻会变得愈来愈小，因此在感应元件制作完毕后，必须进行"退火"热处理（在§4 中详述）。

2）不可能判断气流的方向，热电风计是能够测量三元气流以及决定矢量速度，但是当它决定矢量速度时不能判断速度矢量指向热电风计还是离开它，因为电阻丝在气流中放热状况基本上与气流方向无关，故在测量气流方向时必须预先估计气流大概朝哪个方向吹。

3）感应元件制作困难（热电丝很易被烧毁，被气流中小质点打坏，因此经常制作是不可避免的),仪表复杂,操作不便,对二次仪表每个环节都有较高的要求。

下面我们将会讨论再设计,操作过程中如何减少和克服这些缺点。

二、气流平均速度的测量

§1. 测量气流平均速度的物理基础

如果有某一金属丝,其直径为 d,质量为 m,比热容量为 δ,电阻为 R_{T_H},被电流 i 所加热,则此金属丝温度随时间的变化等于被电流加热及在气流中被对流散热作用的总和,可以写出下列的热平衡式:

$$m\delta \frac{dT_H}{d\tau} = 0.24 i^2 R_{T_H} - \alpha(T_H - T_0) \tag{1}$$

这里,α 为对流放热系数;T_0 为气流的温度;T_H 为电阻丝温度

如果我们测量的是气流平均速度,并假设在测量过程中气流速度不变,在达到热平衡状态下,式(1)变成:

$$0.24 i^2 R_{T_H} = \alpha(T_H - T_0) \tag{2}$$

对于像热电丝这样细的圆柱,其对流放热情况具有一定独特之处,King 第一个得出了 α 的理论表达式(文献9):

$$\alpha = A + B\sqrt{W} \tag{3}$$

式中,$A = \lambda_f \cdot l \left[1 + 1.14 \times 10^{-4} (T_H - T_0) \right]$;

$B = \sqrt{2\pi \lambda_f C_{vf} \rho_f d} \left[1 + 8 \times 10^{-5} (T_H - T_0) \right]$。

这里,W 为气流速度,注脚 f 代表与气流有关参数,H 代表与金属丝有关参数。King 曾经用试验证明了自己的理论,因此长期以来为人们所普遍采用。

不久前 CepreeBO.A 总结了一些试验,对 King 式提出了修正(文献10),如果把它写成式(2)的样子,则可得

$$i^2 R = 1.35 \lambda_H (T_H - T_0) + 2.2 \lambda_f \left(\frac{d}{\lambda_f} \right)^{0.4} W^{0.4} (T_H - T_0)$$

但是由于热电丝是有限长的,而且联接热电丝两脚对放热也有一定影响,因此实用上采用下式较为恰当:

$$i^2 R = (A + BW^n)(T_H - T_0) \tag{4}$$

式中,常数 A、B 及 n 对具体感应元件由试验决定。

另一方面,热电丝电阻随温度变化的规律性为

$$R = R_0\left(1 + \alpha_\rho T_H\right) \tag{5}$$

R_0 为在与气流温度 T_0 相等时热电丝的电阻，α_ρ 为热电丝电阻温度系数，把(5)代入(4)并经过不复杂的变化可得：

$$i^2 R - \left(A + BW^n\right)\frac{R - R_0}{R_0} = 0 \tag{6}$$

由这个式子出发可以得出两种测量平均速度的原则。

1) 恒电流法，令通过热电丝的电流 i 为定值，则此时速度的变化使放热产生变化，引起热电丝电阻变化，故式(6)可改写成

$$\frac{R \cdot R_0}{R - R_0} = A_1 + B_1 W^n \tag{7}$$

这里，$A_1 = \dfrac{A}{i^2}$，$B_1 = \dfrac{B}{i^2}$

一般测量平均速度时采用惠斯登电桥，热电风计热电丝作为电桥中某一臂，如果采用恒电流法，则在热电丝这一电桥臂中加入可变电阻 R'_H（图1），当气流速度有变化时，丝电阻 R_H 亦产生变化，如果通过电流不变时，则电桥就不平衡，为此，必须要调节 R'_H 使电桥重新达到平衡状态，由式(7)可知，当气流速度增加时丝电阻减少，其典型的校正曲线示于图2。

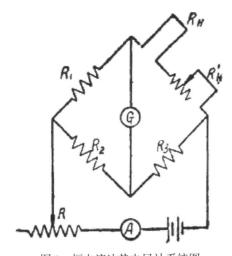

图1 恒电流法热电风计系统图

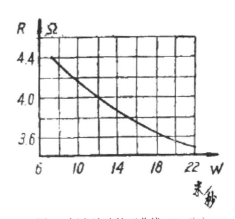

图2 恒电流法校正曲线 $W=f(R)$

2) 恒电阻法，利用热电丝电阻只与温度有关的特性，令热电丝的温度一定，则当气流速度变化时，电阻丝的热平衡受到破坏，改变通过热电丝的电流使平衡重新恢复，故此时由(6)式可改写成：

$$i^2 = A^2 + B^2 W^n \tag{8}$$

这里，$A^2 = A\dfrac{R-R_0}{RR_0}$，$B^2 = B\dfrac{R-R_0}{RR_0}$。

实现恒电阻法的电桥线路原理图示于图 3。当气流速度增大时，电阻丝被冷却丝温降低，因而电阻也降低，电桥平衡被破坏，此时应调节可变电阻 R 使通过热电丝电流增加，电阻增大，重新达到平衡位置。

在实用上，恒电阻法获得比较广泛的应用，因为第一，在电桥中保持恒电阻比保持恒电流来得容易，第二，恒电阻法能测量的速度变化范围较大，图 4 示出我们利用恒电阻法所得出的典型校正曲线。

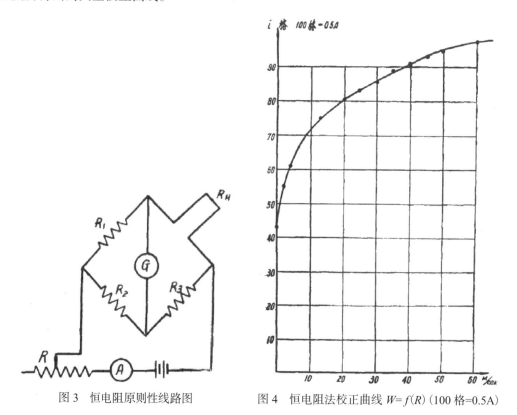

图 3　恒电阻原则性线路图　　图 4　恒电阻法校正曲线 $W=f(R)$（100 格=0.5A）

为了要实现上述测量原则，必须首先要解决下面几个难题。

1) 气流的紊流脉动对热电丝的传热影响问题

由于我们的测量主要处在紊流工况中，在原则上都很清楚，气流的紊流强度，对对流换热是一定有影响的，但是一般传热学课本中推荐的准则方程式只包含

$$Nu = CR_e^n \tag{9}$$

这是不够全面的，只能在和试验条件相同的紊流强度下才能应用，因为雷诺准则是用气流平均速度计算的，当紊流强度不同时，同一平均速度下脉动速度可以变化很大，脉动速度高时，对流传热增强，低时则减弱，由于实际试验基本上是在同一紊流强度下得出(例如在管内，对稳定流动工况紊流强度是自模的)，故试验点都基本上能落在同一

$Nu = CR_e^n$ 曲线上,但是,如果我们研究的对象正是气流不同的紊流度,则式(9)能否应用必须经过严格的检查,否则热电风计理论就丧失了物理基础。

早在1935年 Л.Г.Лойчянский 等就提出利用对流放热系数随紊流强度不同而变化的现象来测量气流的紊流脉动值(文献11)。他们发现,在不同的紊流强度下仍然可以应用式(9),只不过此时系数 A 及指数 n 值随紊流强度 ε 变化而变化,利用卡路里量热计原理,根据测量出 Nu 及 Re 准则,即可求出 ε 值(见图5)具体数值列于下表:(在 $Re=30\times10^3 \sim 300\times10^3$ 范围内有效)。

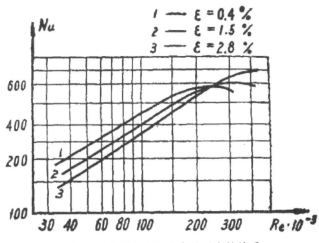

图5 放热强度与气流紊流强度的关系

表1

ε %	0.4	1.1	1.5	1.8	2.8
A	0.216	0.182	0.17	0.155	0.177
n	0.62	0.641	0.65	0.662	0.655

应指出,这些数据是在气流外绕球体($d=70 \sim 150$ mm)情况下得出,后来 М.В.Кирпечев 及 Л.С.Эйгенсон 发展了 Л.Г.Лойцянский 的方法(文献12),采用了圆柱进行准确测量,提出了计算气流紊流强度的公式:

$$\varepsilon = Nu/Nu_0 \qquad \%$$

Nu_0 为当 $\varepsilon=1\%$ 时的努谢尔特准则,Nu 为在该 ε 下试验所得数值,如果紊流脉动对热电丝也是这样影响的话,那么我们就不可能应用热电风计去测量气流的平均和紊流参数了。目前紊流对换热的影响还未研究得很清楚,但是对直径只有几十微米热电丝的对流传热来说,至少可以发现下列两点事实:

A) 在一般气流速度及热电丝直径下,雷诺数不会大于100,因此上述的数据及紊流强度的影响不能机械地搬到小雷诺数情况下,正如 М.В.Кириечев 等后来的试验指出,在较小的雷诺数下,对流放热系数 $\alpha = CW^n$ 曲线随紊流强度变化几乎是平衡移动的,亦

即指数 n 值是不随紊流强度变化的。

Б) 关于在小雷诺数下脉动速度对对流放热影响的问题只能提出 Martinelli 及 Boelter 的试验(文献 13) 他们用圆柱放在静止的水中以一定频率(f)及振幅(A)进行摆动，研究摆动的频率及振幅对对流放热的影响试验结果示于图 6，在横坐标上雷诺数是按下式计算的：

$$Re = \frac{w'd}{v} = \frac{\sqrt{2}\pi f A d}{v}$$

亦即这是按脉动速度计算的雷诺数。由试验可知，当脉动雷诺数小于 900 时，努谢尔特准则与气流脉动状况无关，亦即说明，对于热电丝的对流换热情况来说可以不考虑气流紊流强度的影响。

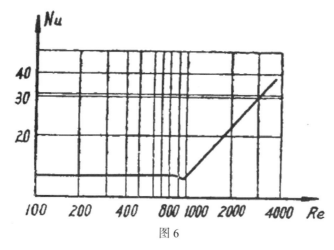

图 6

2) 对于长度有限直径很小的圆柱(热电丝)对流放热规律性问题：

这个问题之所以值得研究是因为我们不可能借用无限长大圆柱所得的试验数据而引起的，如果认为式(9)对热电丝的对流换热仍然是可用的，那么当气流速度为 W_1 时的热平衡式可写成：

$$0.24 i_1^2 R = \alpha_1 F(T - T_0)$$

如果我们应用恒电阻法的测量原则，当气流速度增至 W_2 时，为了保持热平衡，必须适当增加电流至 i_2，即 $0.24 i_2^2 R = \alpha_2 F(T - T_0)$

由式(9)可得出 $\dfrac{\alpha_1}{\alpha_2} = \left(\dfrac{W_1}{W_2}\right)^n$，代入上两式即可求出电流与速度之间的联系：

$\dfrac{i_2^2}{i_1^2} = \left(\dfrac{W_2}{W_1}\right)^n$，这样我们可以由试验得出 n 值：

$$n = 2\frac{\lg \frac{i_2}{i_1}}{\lg \frac{w_2}{w_1}} \tag{10}$$

利用这样的方法我们对单线和 V 型热电丝进行了专门试验,采用的是钨丝,长度为 4.5 毫米,直径为 19 微米,试验时保证加热温度 $T_0=20℃$,$T=170℃$ 即热电丝电阻为常量,即为 1.23 欧姆,试验时速度在 2~30 米/秒范围内变化,相应的雷诺数为 $Re=1~38$,所得出指数 n 的均值为 0.32,在表 2 中列出各个作者试验得出不同 n 值的比较,比较证明对于直径较小的热电丝其指数 n 值比圆柱体来得小,但即使对同一雷诺数范围下的热电丝,各个作者所得数值也不相同,这是因为热电丝的结构以及引出热电丝两脚直径均不同之故。由此可见,对不同类型和结构的热电丝,最好还是由试验直接决定 n 值。

表 2

作者	研究对象	雷诺数范围	指量 n 值	来源
King	热电丝		0.5	文献 14
М.А.Михеев	圆柱	5~80	0.4	文献 15
Hilpert	圆柱	4~40	0.385	文献 16
Hilpert	圆柱	1~4	0.33	文献 16
С.С.Кутателадзе	圆柱	4~50	0.41	文献 17
С.С.Кутателадзе	圆柱	0.1~4	0.305	文献 17
Ю.Г.Захаров	热电丝	2~39	0.357	文献 16
П.В.Чебышев	热电丝	4~63	0.336	文献 18
我们的数据	热电丝	1~38	0.32	文献 19

应当指出,如果采用类似 King 式的总结方法(即式(3)),将会得出较准确的结果,此时

$$\alpha = A + BW^n \tag{11}$$

如果采用恒电流法,代入式(8),可改写成

$$w = \left(\frac{1}{B_2}i^2 - \frac{A_2}{B_2}\right)^{\frac{1}{n}} = \left(Ci^2 - D\right)^{\frac{1}{n}} \tag{12}$$

可见,这里有三个未知数,必须在试验中用三个不同气流速度及相应三个不同电流值,代入式(12)中求出,试验证明 C 值与热电丝加热程度($T-T_0$)及热电丝特性有关,D 及 n 值则仅取决于气流特性,根据 П.В.Чебышев 对钨丝的 36 次测量得出(文献 18):

$n_{CP}=0.397±5.6\%$；$C=22±20\%$；$D=0.913±12\%$，但是我们在实用中没有采用这个方法，因为第一，采用这样的 α 表达式固然可以较精确，但是在以后推广到测量紊流参数时带来很多不便，使方程式复杂化，第二，方程式(11)本身是比较准确，但用试验求它的系数时却很难得出高度的准确性。П.В.Чебышев 的数据是在比较精确的试验室条件下得出的，C、D、n 值的分散性仍然过大，但在同一条件下，П.В.Чебышев 按式(9)进行总结时，对 72 次测量中，得出 $n_{CP}=0.336±0.9\%$（见表 2），即对气流及热电丝的敏感性较小。

3) 气流温度变化的影响及热电丝的电阻温度系数 α_ρ 随温度升高而变化的问题

在推演过程中，我们曾经利用了式(5) $R_H = R_0(1+\alpha_\rho T)$，亦即设 α_ρ 为一常量，但实际上对很多材料发现随温度的增高 α_ρ 亦有所增加，因为以后我们以钨丝为测量元件，故这里以钨丝为例加以讨论，如果以 $T_0=20℃$ 时 $\alpha_\rho = \alpha_{\rho_0}$，当温度升至 $T℃$ 时 α_ρ 变至 α_{ρ_T}，设电阻温度系数变化为 $\Delta\alpha_\rho = \alpha_{\rho_T}/\alpha_{\rho_0}$，其数据列于表 3（文献 18）：

表 3

$T=$	100℃	120℃	140℃	170℃	220℃	270℃
$\Delta\alpha_\rho =$	11105	11205	11305	11405	11605	1179

但实际应用中，我们是采取 (T_H-T_0) 温度下的平均 α_ρ 值，因此电阻温度系数的变化当然比表 3 所得数据小得多，并且可近似地用两段折线来代表（图 7）。一般 $T_{折}=130\sim 160℃$，当 $T_H<T_{折}$，电阻温度系数 $\alpha_\rho=$const，当 $T_H>T_{折}$ 电阻温度系数变成 $\alpha'_\rho=$ const，而且 $\alpha'_\rho>\alpha_\rho$。现在进行研究当气流温度变化时，采用恒电阻法的公式应有些怎样的变化：

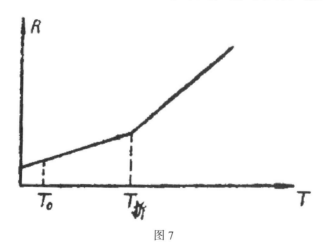

图 7

设当气流温度为 T_0 时，所选取的维持丝恒电阻温度为 $T_H\leq T_{折}$，则按式(1)在热平衡状态下可得

$$0.24i^2 R_{T_H} = \alpha(T_H - T_0) \tag{13}$$

当气流温度升至 T_1 时，为了保持恒电阻所需温度升至 $T'_H > T_{折}$，但在平衡下气流速度不变，为了以后总结方便，令通过电流也不变化，下面求满足这个条件要求电阻变化的规律。

$$0.24 i^2 R_{T'_H} = \alpha (T'_H - T_1) \tag{14}$$

由此得出
$$\frac{R_{T'_H}}{R_{T_H}} = \frac{(T'_H - T_1)}{(T_H - T_0)} = \frac{\Delta T_1}{\Delta T_0} \tag{15}$$

$T = T'_H > T_{折}$ 时热电丝的电阻按式(5)原理可写成

$$R_{T'_H} = R_{T_0} \left[1 + \alpha_\rho (T_H - T_0) + \alpha'_\rho (T'_H - T_H) \right] \tag{16}$$

由(16)式可解出 T'_H 值（考虑到 $\Delta T_0 = T_H - T_0$）：

$$T'_H = T_H + \frac{R_{T'_H} \cdot R_{T_H}}{R_{T_H} \cdot R_{T_0} \alpha'_\rho} - \frac{1 + \alpha_\rho \Delta T_0}{\alpha'_\rho}$$

把(15)结果代入上式： $T'_H = T_H + \frac{\Delta T_1}{\Delta T_0} \cdot \frac{R_{T_H}}{R_{T_0} \alpha'_\rho} - \frac{1 + \alpha_\rho \Delta T_0}{\alpha'_\rho}$

展开上式，同时 R_{T_H} 用式(5)代入，设 $\Delta T = T_1 - T_0$，则经过不复杂的运算可得出最终式：

$$\Delta T_1 = T'_H - T_1 = \Delta T_0 \left(1 + \frac{\alpha'_\rho \Delta T}{1 - (\alpha'_\rho - \alpha_\rho) \Delta T_0} \right) = \Delta T_0 (1 + A_\rho \Delta T) \tag{17}$$

这里，
$$A_\rho = \frac{\alpha'_\rho}{1 - (\alpha'_\rho - \alpha_\rho) \Delta T_0} \tag{18}$$

称为热电丝电阻增长系数。
利用(15)可得出

$$R_{T'_H} = R_{T_H} (1 + A_\rho \Delta T) \tag{19}$$

现在可以理解这些式子的物理意义了；如果认为气流基准温度为 T_0，要求保证恒电阻的温度为 T_H，当气流温度升至 T_1，由于电阻系数出现了曲折点，为了保持恒电阻 T'_H 不能只按比例增加，而其增加值应比没有曲折线出现时来得大，即 $\Delta T_1 > \Delta T_0$，同时 $R_{T'_H} > R_{T_H}$。这说明当气流温度变化时"恒电阻法"就有点名不符实了：原因是这样的，当气流速度不变时，按照式(8)，气流温度增加时 R_0 就会增加，若只保证热电丝电阻不变，则系数 A_2、B_2 不能再保持为常量，亦即电流 i 也会随气流温度升高而变化，这就要求在每种气流温度下给出一条校正曲线，在运行过程中造成极大的不便(图8)。

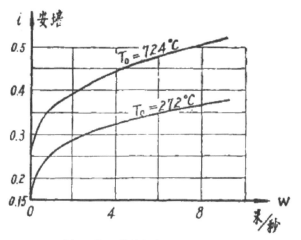

图 8 校正曲线与气流温度的关系

但是，如果我们令电流不随气流温度变化，则必须满足方程式(15)，亦即当气流温度由 T_0 变至 T_1，丝的温度增至 T_H'，而 $T_H' \neq T_H + (T_1 - T_0)$，很明显如电阻温升线没有转折点，则 $\alpha_\rho' = \alpha_\rho$，由式(17)

$$T_H' = T_H + (T_1 - T_0) + \alpha_\rho (T_H - T_0)(T_1 - T_0)$$

相应此时丝的电阻增至

$$R_{T_H'} = R_{T_H} \left[1 + \alpha_\rho (T_1 - T_0) \right] \tag{20}$$

如果再考虑电阻随温升有曲折点，则在式(19)及(20)中以 A_ρ 代换 α_ρ 即可，因此，在气温一样，气流速度不同时，仍然保持"恒电阻法"的称号，但当气温变化时，为了保持加热电流不变，我们就要适当地改变电热丝的加热温度及其电阻，物理意义上即把不同的气流温度下换算成标准的气流温度 T_0 及标准的热电丝加热程度 ΔT_0 情况下，以便有共同的标准，共同的校正曲线，П.В.Чебышев 取 $T_0=20℃$，$\Delta T_0=150℃$（对气流温度不高于 120℃情况下）并把这种方法称为"取决电阻法"（文献 20），总结大量试验在 20～100℃范围内，钨丝的平均电阻温度系数 $\overline{\alpha_\rho}=0.004085\,1/℃$，当 $\Delta T_0=150℃$ 时平均值 $\overline{A_\rho}=0.0049\,1/℃$，至于在电桥中如何实现标准加热 $\Delta T_0=150℃$ 及气流温度不同时如何进行自动补偿等问题将在§5 内讨论。

4) 气流速度很小时的测量问题

在论述热电风计优点时曾说过其优点之一是能准确测量很小的气流速度，其低限取决于两个因素：A) 当速度愈低时，自由对流换热起的作用愈来愈大，此时热电丝所处的位置（与水平面的角度），加热程度，形状，热容等等均有关系，即使保持上述各因素相同，试验证明在极低速时放热系数会比速度为零时的自由放热系数还低，作为例子可以列举 E.Ower 的试验（文献 21）(图 9)。当 $W<0.02$ 米/秒时 $(\alpha - \alpha_0)/\alpha_0 < 0$，E.Ower 把这现象解释作微弱的空气流动会破坏自由放热，因此在这段范围内热电风计原则上不能应

用，实用上一般可应用到 0.03 米/秒。

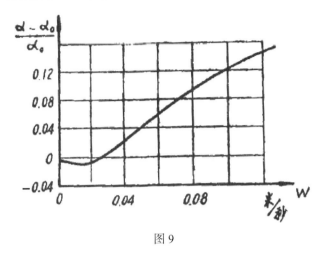

图 9

Б) 取决于校正的准确度，因为热电风计归根到底要在风洞中校正，在低速时校正是很困难的，(这取决于标准皮托管的敏感度)，我们认为，对于小速度的校正最好在具有缩放形喷嘴的风洞中进行，在最大截面处放入要校正的热电风计，而在最窄收缩截面用皮托管测量速度，根据不同截面比换算成校正截面的速度，无论如何，采用热电风计测量小速度是有较大优越性，由图 4 亦可见，速度愈小时，热电风计的灵敏度就愈高。

§2. 决定气流方向及大小的方法

对于最简单的一元气流，只要把热电丝放在与气流方向相垂直的平面上，便能够进行准确测量，但是对测量二元及三元气流，一般可采用下列方法。

1) 气力阴影法

由热电丝元件结构可知(图 10)，当气流方向和热电丝垂直时，散热最大，但当气流方向和热电丝平行时，由于支杆直径比热电丝大几十倍，因此支杆后的"气力阴影"笼罩着热电丝，使其散热达到最小情况，(如图 14 中四个最低点)。因此，我们在操作中可

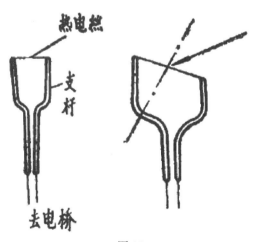

图 10

以旋转热电丝元件,一直到维持电流平衡(用恒电阻法)所需通过加热电流最小为止,为了提高灵敏度及判明气流的方向,П.В.Чебышев 建议使用倾斜式热电丝元件(图 10),这种方法在实用上很不方便,为了决定气流方向要作很多次测量,同时更严重的问题在经常会使热电丝烧毁,因为如果热电丝在与气流相垂直的方向下达到热平衡(此时散热最大)突然转到支杆的气力阴影范围内,试验证明,此时放热系数突然降低几倍,如加热电流不变,热电丝温度急剧上升,直至氧化或烧毁为止。而且这种方法只能测量平面气流,故实用上不予推荐。

2)变换感应元件位置法,这里采用的还是简单热电丝元件,在传热学理论中早已知道气流方向(冲角)对对流热交换有很大的影响,可以想象,在不同的气流与热电丝交角(冲角)φ下,电桥平衡所需的加热电流量均不相同,为了弄清气流方向和热电丝对流热交换间的关系,我们进行了专门试验,其目的有二,第一,找出气流方位角对平均速度测量影响的规律性,从而进一步利用这规律性去决定气流方向,第二,研究雷诺数变化时对这个规律性的影响,即检查它的通用性问题,试验是以 19 微米直径,长 4.5 毫米的钨丝热电元件在风洞中进行的,标准加热温度 $\Delta T_0 = 150℃$,速度在 10～30 米/秒范围内变化,试验结果示于图 11 中,试验证明,速度方向对热电丝的散热作用是有一定规律性,与雷诺数无关,并且很好服从于经验公式:

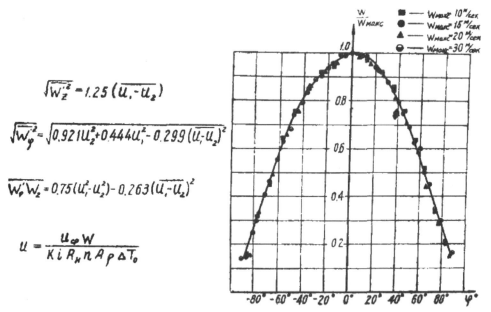

图 11 气流方向与热电丝元件的交角 φ 对测量的影响
(连续曲线是按公式 $W/W_{\max}=0.16+0.84\cos\varphi$ 画出)

$$\frac{W}{W_{\text{Max}}} = a + b\cos\varphi \qquad a+b=1 \tag{21}$$

在我们具体的单丝感应元件结构中 $a=0.16$,$b=0.84$,

在文献 14 中提出的经验公式为 $\dfrac{W}{W_{Max}} = \cos\varphi$，我们认为这个公式未能很好考虑支杆的影响，对不同结构的热电丝元件应有不同的 a、b 值。

知道了气流方向与丝散热的规律后，我们就有可能用不同的角度测量出两个读数，由此决定气流方向及大小，这个方法的主要缺点在于操作复杂，试验总结困难，准确度低，测量时间长，对不很稳定的气流则不适用。

3) 采用 V 型热电丝感应元件法。

用单热电丝测量平面气流有着严重的缺点，因此一般采用 V 型感应元件，但由于结构的复杂，使用的不多，图 12 示出在文献(9)中发现使用的例子，V 型感应元件每一根热电丝顺次安放在电桥的两侧臂中，其他两臂的电阻 $R_1=R_2$，因此，当两根热电丝加热程度及电阻完全一样时，V 型尖端和气流方向相对时(相差 180°)，则电桥应处于平衡状态，因此为了找寻气流方向，我们应旋转感应元件至检流计 G 读数等于零为止，但是在制造过程中往往很难保证两热电丝长短及电阻完全一样，因此加入 R' 电阻作为调节校正之用，为了决定气流平均速度值，此时必须把另一热电丝切断，改接进某一标准电阻，然后，按单热电丝求速度值法则进行测量。采用这样的系统主要缺点在于：A)操作复杂，气流方向及速度值分开求是不合理的，耗费时间多，Б)在加热电流一定下，旋转 V 型感应元件，热电丝容易烧毁(理由如第 2 点中所述，B)不可能通用地用来决定紊流各种参数。Γ)作者推荐 V 型元件的夹角为 8°～10°。我们的试验证明这样小的夹角下仪表的灵敏度并不是最大，而且制造及操作过程对测量准确度影响很大。

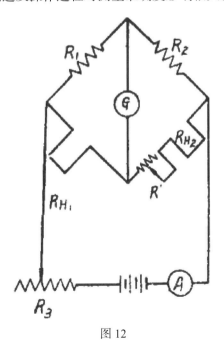

图 12

为此，我们提出采用分开电桥式的测量方案，在拟制这样的方案时考虑到下列问题：
A) 只需用一次操作即能求出气流方向及速度值。

Б) 求气流方向时不需要旋转 V 型感应元件。

В) 考虑能同时测取各种紊流参数。

Г) 要保证足够高的仪表灵敏度及测量准确度。

为了保证上述要求,我们选择了如图 13 所示的电力线路,这里,每一根热电丝所感应到的气流速度值分别送到电桥 1 及 2 中,同时图 13 也示出测量紊流参数的部分线路,首先,必须考虑 V 型元件两热电线的夹角问题,根据 R.C. Pankhurst 的推荐(文献 1)以及我们的试验证明,夹角为 120°时,仪表对气流方向的敏感度最大,因此,当我们把 V 型元件放在气流中,则两热电丝同时感受速度的作用,其大小取决于气流与热电丝的方位角,为了检查 V 型感应元件与气流方位角 φ 的关系,以及式(21)在此情况下是否能应用,在风洞中进行了专门试验,所采用的 V 型元件热电丝直径为 19 微米,长 4.5 毫米,焊于三根引出的支杆上,在风洞中旋转 V 型元件 360°发现 V 型元件的行为基本上与单丝重合(图 14),式(21)仍可应用,只不过此时系数 a、b 有所变更,总结多次试验证明,对 V 型元件采用下式较为准确:

$$\frac{W}{W_{\text{Max}}} = 0.2 + 0.8\cos\varphi \tag{22}$$

由图 14 可见,当热电丝夹角为 120°,而且两丝电阻相等时,若 V 型元件和气流方向相对称,则两丝所感受速度量是一样的,各热电丝达到最高峰点,即为该热电丝与气流方向相垂直之平面,由图中可见,两丝的最高点刚好相隔 120°,证明元件的制作是精确的,在 90°~270°范围内发现曲线变成不对称,这是因为 V 型元件在反方向(即 V 型

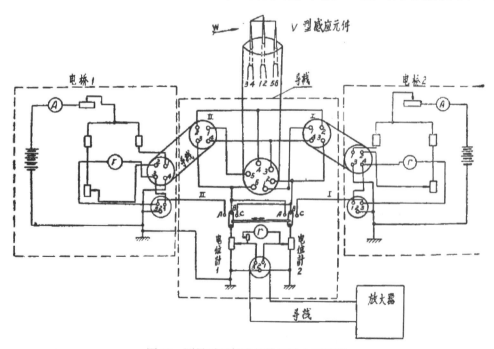

图 13 测量平面气流的热风计电系统图

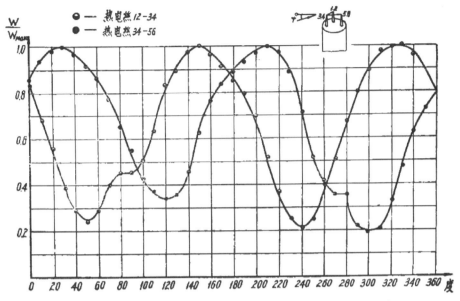

图 14 V 型热电丝元件在风洞中的校正特性(当气流速度 W_{Max}=57 米/秒)

尖端与气流方向相差大于 180°时)工作的结果，而在实际测量中，这段范围是不能应用的，当 φ=90°和 270°，发现曲线有突然变化，这是由于支杆"气力阴影"的影响，因此在测量时我们只容许在下列范围内进行：φ=0°～90°，φ=250°～360°。

现在研究一下如何能借一次测量即可决定气流方向及大小，由式(21)可知对 V 型元件的 A 热电丝：

$$W_A = (a + b\cos\varphi) \cdot W \tag{23}$$

对 B 丝：
$$W_B = \left[a + b\cos(60° - \varphi)\right]W \tag{24}$$

相比可得

$$\frac{W_A}{W_B} = \frac{a + b\cos\varphi}{a + b\cos(60° - \varphi)} \tag{25}$$

在两个独立电桥中，不管气流与 V 型元件成什么方向，W_A、W_B 总可单独测量出。解(25)式即可求出气流方向相对于 V 型元件的 φ 角，我们建议式(25)用图解法较为方便，图 15 中示出图解法的诺莫图，由图中决定了 φ 角后，即可按下式求出速度值：

$$w = \frac{w_A}{a + b\cos\varphi} \tag{26}$$

$$W_x = W \cdot \sin\psi \tag{27}$$

$$W_y = W \cdot \cos\psi \tag{28}$$

这里 ψ 为矢量速度与坐标 oy 所成的夹角,对于成 120°的 V 型元件,$\psi = 30° - \varphi$,由此可见,当 $\varphi > 30°$ 时 ψ 变成负值,即分速 W_x 也变成负值,而分速 W_y 的方向(正的或是负的)正如前述理由不能由热电风计决定,但 W_y 是正或负值很易在实际工作中予以估计。

为了决定空间气流的方向及大小,文献 9 建议采用三角棱形的立体结构,而采用类似于图 12 的线路系统,只不过此时在电极的一个臂中借转换开关之助可以测量棱形结构中任意两热电丝的值。

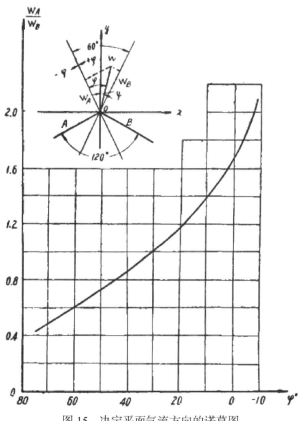

图 15　决定平面气流方向的诺莫图

首先在 oxy 平面中旋转棱形感应元件,直至灵敏电流计读数等于零为止,此时即表示在平面中找出空间气流矢量在该平面的方向,然后利用转换开关使热电丝另外两臂加入工作,此时在 oxz 平面内旋转棱形感应元件至电桥平衡为止,又可求出矢量速度在 oxz 平面的方向,由这两个方向即可决定空间气流方位,然后按照求平面气流大小相同的步骤,分别求出速度的大小值,非常明显,这个方法即使在原理上也很复杂,而实际上要使棱形元件在 oxy 及 oxz 两垂直平面能同时旋转是很难实现,这里值得建议采用类似我们提出的求平面气流的方法,把它推广到空间气流中。

例如采用如图 16 的棱形感应元件,有 A、B、C 三根热电丝,则可组成类似于(23)、(24)式的三个方程,这样由图解法便可求出气流空间方位角,再借类似于式(26)、(27)、(28)之助,便可求得气流大小值,应当指出,采用这种方法最好在三个独立电桥中完成,但亦可借转换开关之助，能在两个甚至一个电桥中实现，至于利用这种感应元件如何求取各种紊流参数问题，将在§12中详细讨论。

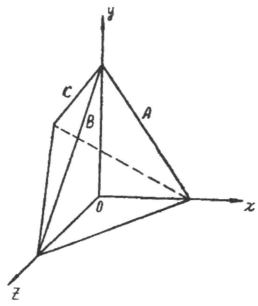

图 16　棱形感应元件

§3. 影响测量平均速度准确性的几个问题

这里我们只能对几个主要的影响因素来讨论。

1) 测量灵敏度随气流速度增加而减小的问题：

例如采用"恒电阻法"的测量系统，由式(8)对速度的微分可得

$$\frac{\mathrm{d}i}{\mathrm{d}W} = \frac{A_2}{2n\sqrt{A_2+B_2W^n}W^{1-n}} \tag{29}$$

由此可见，速度愈高时，$\frac{\mathrm{d}i}{\mathrm{d}W}$ 愈小，即仪表的灵敏度降低，试验也证明了这点，如图 4 所示，当气流速度很小时，$\frac{\mathrm{d}i}{\mathrm{d}W}$ 很大，但当速度高于 50 米/秒时，校正曲线的倾斜度愈来愈小，对"恒电流法"微分式(7)也得类似结果(见图 2)，因此为了增高测量准确度，通常希望校正曲线线性化，为此，可以在测量线路中引入与校正曲线相反的非线性元件，常用的办法有：

A) 电热压降法，在测量电桥的外线路上，例如可在图 3 中电流计旁串联一白金丝电阻，然后在这电阻旁装入测量电流通过时所引起的电压降，当电流升高时，白金丝被加热温度升高，因而在其两端出现的电压降升高，当白金丝直径为 0.1mm 长 75mm 电压降 V 与丝电阻的关系服从下式(见文献 21)：

$$R=0.607V+1 \text{ 欧姆}$$

设通过电桥的电流为 i，则 $i = \dfrac{V}{R} = \dfrac{V}{0.607V+1}$

把这个方程画在图 17 上见到，电压降的增长比电流增长快得多，(甚至比电流平方增长还快)。由式(8)可见电流与速度成 4~5 次方关系，采用这种非线性元件后，则热电风计读数与速度变成 1~2 次方关系，准确度因而得到很大提高。

可以采用实际的校正曲线作为例子(图 18)，当用普通的电流表测量时，$W=0$ 时 $i=40$ 格，$W=23$ 米/秒时，$i=70$ 格(每 100 格相当于 0.5 安培)加入电热元件后，校正曲线变直，$W=0$ 时，$V=14$ 格，$W=23$ 米/秒时，$V=96$ 格(采用电流电压计测量)(文献 16)。可见，这样的补偿线路是能解决这一问题。

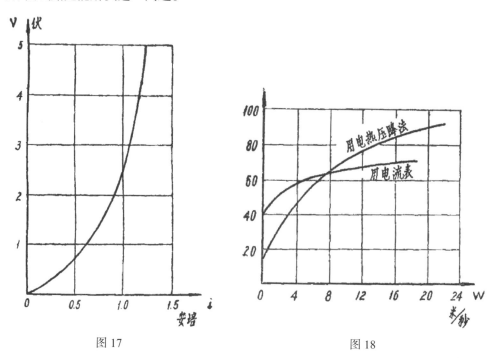

图 17　　　　　　　　　　图 18

Б)真空管补偿法。众所周知，真空管阳极电流与加在阴极的电压的关系刚好和校正曲线的形状(图 4)相反，(例如见图 19)因此，可以在电桥外电路加入一可变电阻，作为电子管阴极加热的电位计，这样，我们可以选择某一电阻，使电子管电流电压曲线刚好与速度电流校正曲线相反，这样可以得出气流速度—阳极电流新的校正曲线，在选择适当时，这曲线可以线性化，Ю.Г.Захаров 的试验就证明了这点(文献 22)所有试验点都落在同一直线上(图 20)详细的线路图读者可在文献 22 及 9 中找出。

2) 热电丝电阻准确测量问题

按照电桥系统，热电丝是作为电桥的一臂，但是一般热电丝电阻只有 1~1.5 欧姆，如果要测量气流脉动参数，则在脉动作用下电阻只会变化百分之几甚至千分之几欧姆，另一方面，热电丝是经过支杆，导线等与电桥联接，这个长度取决于测量仪表与所研究对象的距离，因此，这些联接电阻经常随插头松紧，室内温度，导线长短及通过电流量等变化，严重影响了测量准确度，有时这些电阻的变化甚至比起由于气流脉动引起热电丝电阻变化来得大，为了消除这些可能出现的误差，在设计电桥线路中特采用了双导线系统(见图13)，V型感应元件下每个支杆都用两根导线连接，为了更好地说明其作用，我们把系统简化成图21，热电丝和支杆是紧密焊接在一起，因此其电阻不受外界因素影响，在支杆后各焊上导线两根1、2或3、4，导线直接接到电源上，导线3、4各与大电阻 R_3=2G 欧及 R_2= 100 欧相接(以我们采用的电桥为例)，因此导线电阻的变化只附加在这些大电阻身上，因而也就影响很小，而导线直接与灵敏电流计相接，电阻的变化只会稍微影响检流计的灵敏度，可见，采用了这样的系统后，热电丝便可真正地成为电桥的一臂，与引出导线的长短，状态均无关，当然这样的结构在热电元件的制造工艺上产生一定困难。

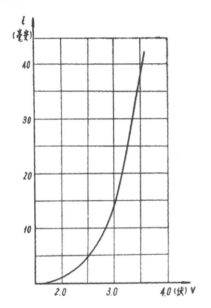
图 19　电子管阳极电流的阴极电压关系

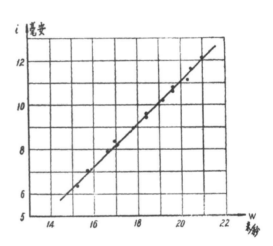
图 20　采用电子管补偿后的热电风计校正曲线

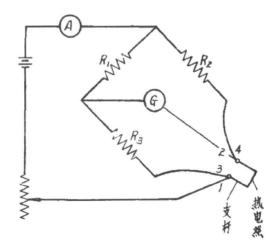
图 21　热电丝双导线连接系统

3) **介质压力变化对测量平均速度的影响**

在测量过程中,往往压力会产生变化(一般仪表校正是在大气压下进行),例如在旋气炉壁面附近压力会比大气压高几百毫米水柱,但在中心则变成负压,因此压力变化对热电风计测量的影响是值得研究的。实际上由式(4)及(5)可得出气流速度值:

$$W = c\left[\frac{i^2 R_{T_0}\left(1+\alpha_\rho \Delta T_0\right)}{\Delta T_0}\right]^{\frac{1}{n}} \tag{30}$$

这里和式(4)、(5)不同的是我们已规定了标准气流温度 $T_0=20℃$,因此丝的标准电阻 R_{T_0} 是从 20℃算起,这时在电阻公式(5)中不能单用丝的温度 T_H,而应该用标准加热温差:

$$\Delta T_0 = T_H - T_0 \tag{31}$$

但实际上, $C = D\dfrac{\mu}{\rho\dfrac{\lambda}{n}}$ (32)

D 为与压力无关的其他常数;μ 为气体动黏度;λ 为气体导热系数;ρ 为气体密度。对于空气,压力变化 1%,μ 只变化 0.00163%,λ 也几乎不随压力变化,剩下只有密度 ρ 是与压力成正比的。故(32)式可改写成

$$C = \frac{D'}{P} \qquad P—气体压力,$$

这样可以把式(30)改写成:

$$W = \frac{P_{HOP}}{P} \cdot D'\left[\frac{i^2 R_{T_0}\left(1+\alpha_\rho \Delta T_0\right)}{\Delta T_0}\right]^{\frac{1}{n}} \tag{33}$$

P_{HOP} 为仪表在风洞校正时的静压力;P 为测量该点时的气流静压。

对于在旋风炉测量来说,一般 $\left(\dfrac{P_{HOP}}{P}\right)_{Max} = 0.95 - 0.97$,即使不引入修正最大误差也不会大于 3%-5%。

§4. **测量感应元件**

下面将讨论热电丝感应元件结构及制造几个问题:

1) **热电丝的材料**:作为热电丝材料一般要求满足下列条件:

A)具有高度热及化学稳定性,这里主要指在反复加温,冷却过程中丝的电阻不应有变化,在高温下丝材料不应与测量介质起化学反应或氧化。

Б)有足够的"柔"性,易于制成只有几微米的丝。

В)有足够高的机械强度,这点在测量高速度时更为重要,热电丝感应元件的制作及

校正是比较复杂的，如果经常断裂或烧毁则在实际工作中不可能被采用。

Γ)有足够高的电阻温度系数，使测量能准确，仪表灵敏度高。

Д)材料易于焊接，易于获得。

要完全满足这些要求的材料，实际上是很难找到的，起初白金丝被广泛应用，白金丝对满足 А.Б.条件是比较好的机械强度也不错，而且能够借特殊的方法——"镀银法"（метод Волластона）可得到极细的直径，现有数据达 0.5 微米，（详细制作过程读者可参考文献 14），因此对测量紊流参数及极低速度(如边界层)极为有利，但是白金丝的电阻温度系数较低，而且难于处理及获得，因此当测量速度较高，感应元件体积不要求太小，测量介质温度不很高的情况下，一般都愿意采用钨丝，钨丝的机械强度比白金丝高得多，在低温下有足够的热及化学稳定性，电阻温度系数也很大，易于获得，易于焊接，其主要缺点是：直径很细的钨丝极难制造，目前数据只能成功地制造出 3.8 微米的钨丝，同时在介质温度高于 400℃时即开始氧化。对于一般在冷模型内的研究来说，采用钨丝是比较现实和合理的。

2)感应元件中热电丝的直径及长度选择原则

A)首先要满足研究对象的需要，研究对象的大小及要求就限制了热电丝的直径及长短。

Б)原则上热电丝不应太长，否则一方面测量出来的并非是该点的速度，而是某一段距离内的平均值，特别是当速度梯度很大的情况下这是不容许的，另一方面，热电丝愈长，其强度愈弱，而且在气流作用下会产生变形(弯曲)，一般在气流中热电丝变有三种负荷，空气动力负荷，随气流速度升高，对丝的压力增大，其二是气流中的灰尘微粒对热电丝的突然打击负荷，这是最易使丝断裂的因素，最后是在气流脉动作用下，支杆及丝产生振动所带来的附加负荷，这是当支架太细，气流脉动强度很大时可能出现的情况。可见，原则上是要求热电丝短些，但是热电丝太短又会出现两个问题，第一，支杆的直径比丝直径大几十倍，当然太短时，支杆对测量的影响就很大，主要表现在支杆的散热作用，支杆的气力阴影作用。第二，热电丝愈短，其电阻愈小，对测量仪表及线路的要求就愈苛刻，同时电阻小时要热电丝加热成高温所需电流很大，(有时要达几个安培)，这都使测量处于不利的条件下。

B)热电丝的直径原则上愈小愈好，一方面可以在相同长度内增加电阻，计算证明，丝直径减小一倍，要保持同样加热温度所需电流减少一倍多，此外，丝的直径愈小，其热惯性也就愈小，这对测量紊流参数有极重要的意义，实际上，丝的直径及长度的选择是互相关联的，丝长度愈短，则要求丝直径愈小，两者不能孤立地选择。

对于测量燃烧室内的平均及紊流参数，我们采用了 4.5mm 长 19 微米直径的钨丝，这是否最合理的尺寸还需进一步试验研究，但由此元件所得的试验结果是令人满意的。

3)热电丝感应元件的焊接和退火

由于热电丝本身具有不大的电阻，因此要求热电丝与支杆之间有很高的机械强度及很小的接触电阻，唯一可靠的方法就是进行接触电焊，对于钨丝来说，用镍支杆能够得到良好的焊接强度，由于热电丝直径很细，支杆尖端处直径也不过 0.2~0.5mm。因此，焊接是在特殊制造的焊接台中完成的(详细结构见文献 19)焊接工作必须在有放大率为

10～50倍的放大镜下完成，原则性线路图示于图22，变压器容量约600瓦特，在副线圈给出电压可在 1～5 伏内变动(一般借变阻器精细调节)焊接时先把热电元件放在焊接台特殊夹具中夹牢，主要目的在于使支杆的距离不变，以免焊完后热电丝出现松弛或拉过紧的现象，之后以一定负荷（几克）把热电丝放在支杆顶端上，焊笔不能直接与热电丝接触，一方面由于丝是圆柱形，焊笔不能很好把热电丝压牢在丝焊上，另一方面直接接触很易使热电丝氧化和烧毁，因此应预先准备好0.1mm宽的薄镍板，压在热电丝上，然后手拿焊接笔紧压在镍板上，脚踏开关使电流接通 0.5～1秒，焊接即完成。焊接所需电压由试验得出，根据具体条件而定，焊接后如果情况良好，两支杆对热电丝的拉力约为80～100克，这样，在高速气流下热电丝仍然能保持为直线，热电丝元件焊好后，为了在工作过程中不会变"老"，(随加热冷却过程电阻减少)，一般必须对热电丝进行退火，我们采用的是脉动方法退火，首先以直流电向电容 C (4 微法拉) 充电，然后借快速转换开关 k 转接至退火路线，电容向热电丝放电，如果充电电压适当，则在电容放电期间可以看见热电丝被加热至赤红色(时间约 0.5 秒) 退火次数取决于钨丝电阻比退火前电阻降低10%～13%为止，此时将会发现，即使继续退火，电阻将不再会降低，因此在工作过程中钨丝也不会"变老"。图23示出对单线型及 V 型热电丝元件的退火线路图，在退火时，必须借电位计之助很好地选择退火电压，一般应从较小电压开始，如发现热电丝并不发红，则可逐渐增加电压，值得注意的是电压不能太高，否则热电丝立即烧毁，根据我们试验结果，一般电压约在75～85伏内变化，这和热电丝长短，粗细均有关。

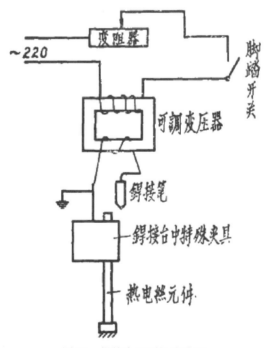

图22　焊接台原则性线路图

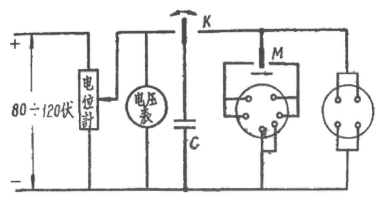

图 23　热电丝元件退火线路图

4) 采用热电元件测量两相气流的探讨

在有固体或液态质点的气流中，采用普通皮托气力测管的办法遇到很大困难，测孔被堵塞问题难于解决，在目前来说，还没有发现更好的其他测量方法，因此这里我们讨论一下采用热电风计的原理是否可能解决两相测量问题，当然在两相气流中，应用直径很小的热电丝是不现实的，热电丝直径增加时为了获得一定的电阻值必须增加其长度，对于一元气流我们认为Ю. Гзахаров 所提出球形热电元件（文献 16）完全可用于测量两相气流，用 5mm 直径的电胶木圆球刻有螺旋形槽，并绕上直径为 40 微米的白金丝（约 16 圈），总长度为 120 毫米，电阻达 10 欧姆，这样可用很简单的电桥系统而得出较高的准确度，同时热电丝被固定在电胶木球内就不再受上述三种负荷的影响，因而在两相气流中测量是很安全的。元件之所以采用球形是考虑到要求测量读数与气流方向无关。为了要测量平面或空间气流我们建议采用 V 型和棱形测量元件，其原理和§2 所述相同，只不过此时用较粗的热电丝，并把它固定在 V 型和棱型的电木槽内。

另一个方案是可以考虑采用半导体热敏电阻作为测量两相气流的感应元件，热电风计以半导体热敏电阻为感应元件来测量气流平均速度在近年来已逐渐被重视，其理论基础及具体线路读者可参阅文献 23、24，不过采用半导体热敏元件目前还未得到完全成功，特别是在高速气流下仪表灵敏度较低，但是这种感应元件应用于两相气流中优点是很明显的，因此值得今后进一步研究。

§5. 测量电桥的特殊要求

1) 电桥各臂电阻选择原则，除了热电丝作为电桥一臂其电阻由感应元件结构已经决定外，其他三臂电阻均应按下列原则选择。首先电流表一般不希望装置在电桥某一臂内，而安置在供给电桥电源线路上，因此为了使电流表测出电流值接近于通过热电丝的电流值，同时为了减少总的电能的消耗，电阻 R_2、R_3 应选择尽可能大，而 R_1 则应较小（符号见图 3），在我们所使用的电桥中，R_2、R_3 各为 2000 欧姆（可调节），R_1 只有 100 欧姆，但应注意，R_1 也不应太小，否则当气流脉动时，热电丝电阻的变化引起通过电流也产生变化，此外为了准确而较快测量出热电丝的电阻值，各可调电阻不宜采用滑线式，应该采用跳跃式，电桥的供电电压取决于测量气流速度及电阻 R_1 值大小，当选择 R_1=100 欧姆时，测量速度要求达 150 米/秒，则电源电压应为 110 伏，同时电源应采用绝对稳定的直流电，因为我们还要利用电桥测量脉动参数，因此如果电源本身有脉动，在放大器中放大几千至几万倍，则会严重影响测量准确度。试验证明用直流电机的直接供电是不能应用的，直流电机输出交流分量的脉动值大大超过了气流脉动所引起电参数的波动，因此要求加入高质量的整流稳定器使整个系统电容或电感增大，滞后惯性增加，可见最好使用电池。在测量中要求电流高达0.5安培（当气流速度为100~150米/秒时）这就要求大量高压电池并联，使用上不经济，但准确度最高，因此我们选用了后一种供应电源方式。

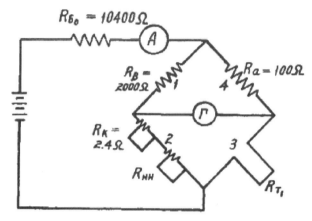

图 24　位置 1 时的电桥线路图

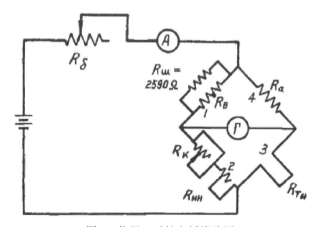

图 25　位置 4 时的电桥线路图

2) 当测量气流温度变化时电桥的自动补偿线路。

在§1 第三点时我们曾经讨论过当气流温度变化时，热电丝的加热度 ΔT_1 及电阻 R'_{T_H} 均应按式(17)及(19)变化，然后才可能应用一根公用的校正曲线，为了能自动实现式(17)、(19)的过程，П. B. чебошеB 建议了独特的电桥线路(图 24)，这样的电桥能保证 A) 仪表读数和所研究气流温度无关，Б) 能换算成并保证热电丝标准加热 ΔT_0 为 150℃。下面我们将讨论如何能达到这种要求。

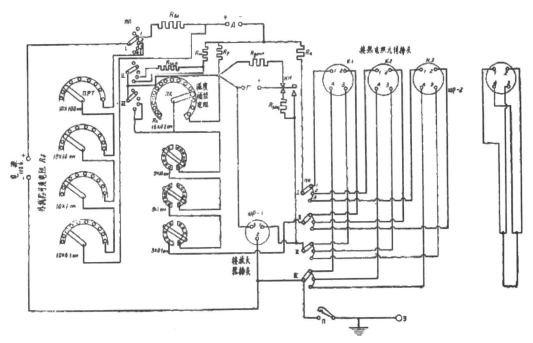

图 26 自动补偿电桥线路图(ЭTAM-3A)

在测量前,我们先把电阻 R_K 放至 2.4 欧姆,然后按图 24 线路把开关 ПП 转放至位置 1(即如图 24 线路简图),此时整个电桥外接大电阻 $R_{Бo}$,热电丝通过电流量极微,保证热电丝不被加热高于测量气流温度 0.5℃的工况,调节电阻 R_{HH} 使电桥平衡,由相应电阻关系比即可求出在此气流温度下相应热电丝的电阻 R_{T1} 及气流温度 T_1,若发现 T_1 大于标准气流温度 T_0,那么为了要维持相应于气流温度为 T_0,热电丝标准加热温为 $\Delta T_0 = T_H - T_0$ 的工况而必须的热电丝加热温度 T_H',此时落于电阻温度曲线的折线段内(图 27),实际上,按上面推导,这些数值可计算得出:

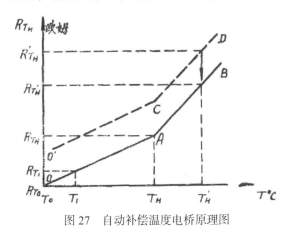

图 27 自动补偿温度电桥原理图

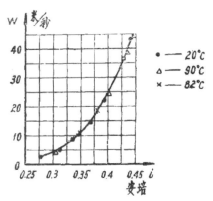

图 28 各种气流温度下的校正曲线

$$\Delta T_1 = T_H' - T_1 = (T_H - T_0)\left[1 + A\rho(T_1 - T_0)\right] \tag{17}$$

及
$$R_{T'_H} = R_{T_H}\left[1 + A\rho(T_1 - T_0)\right] \tag{19}$$

故我们在操作时可预先在电桥 1 臂上(图 26)选择好某一电阻,使得热电丝(3 臂)电阻增至 $R_{T'_H}$ 时,电桥才能达到平衡状态,这样就能自动保证热电丝加热至 T'_H。现在观察一下自动实现这些补偿的电桥线路,一般测量可分两个步骤进行,先把开关 IIII 转到位置 1(图 24),由于串联了大电阻 $R_{бo}$,故目的在于求出所研究气流的温度 T_1 及相应电阻 R_{T1}。R_{T1} 的测量是借助于调节 2 臂上 R_{HH} 电阻值求得,但是我们在测量前预先在 2 臂上给出一个已知电阻 R_k,目的为了能保证当 $T_1 > T_0$ 时 $T'_H > T_H$ 落于电阻温度曲线折线段内,例如当 1、4 臂电阻之比 $\frac{R_B}{R_a} = \frac{2000}{100} = 20$,即电桥平衡时 2 臂中的读数亦相应于热电丝电阻值 R_{TH} 的 20 倍。测量的第二个步骤(平均速度的测量)把开关 IIII 转换至位置 4(图 24),此时电桥线路简图可由图 26 中清楚看出,和位置 1(图 25)主要不同在于:A)大电阻 $R_{бo}$ 改接成可变电阻 $R_δ$,为的是能调节增大通过热电丝的电流,Б)在 1 臂中在原来电阻 R_B 上并联上另一电阻 R_{III},此时 1、4 电臂电阻比由原来 $\frac{R_B}{R_a} = 20$ 降至 $\frac{R_{III} \cdot R_B}{R_{III} + R_B}\Big/R_a = 11.285$,如果电桥 2 臂完全没有变动的话,则为了要使电桥平衡,则热电丝由位置 1(图 25)的 R_{T1} 值必须增长至某一 $R_{T'_H}$ 值,如果气流温度 T_1 比标准温度 $T_0=20℃$ 低,亦即热电丝所需加热温度 T'_H 亦小于 T_H 值,经过这个转换 $R_{T'_H}$ 值必然落在与 OA 线平行的 O'C 线上(图 27),当气流温度 $T_1 > T_0$ 时,则热电线所需加热温度 T'_H 必然落在大于 T_H 的折线段上,因此 $R_{T'_H}$ 可以在与 AB 线平衡的 CD 线段内找出,电阻曲线之所以会平行移动的原因是由于把原来平衡的电桥各臂电阻比例改变后,那么即使是气流温度为 T_0,热电丝温度亦为 T_0,为了使电桥平衡,热电丝电阻已经不能等于 R_{T0} 值,而应该增至某一 $R_{T'_0}$ 值,但是电阻随温度变化规律是不会变动的,因此调整电臂比例后 C'CD 线仍与 OAB 线行。B)应当指出 C'CD 是一条假想线,因为在调节电桥平衡时,热电热丝是处于一个不稳定过程,即随着加热电流增加,温度升高,其电阻也不断增大,但是丝电阻与温度关系只能在一条 OAB 线上变化,因此在位置 4 时,电桥 2 臂上电阻 R_k 被短路,为了要使电桥平衡,等值于在电桥 1、3 臂上要同样减少这样多电阻,因此热电丝电阻由 $R_{T'_H}$ 减至我们所要求的 $R_{T'_H}$ 值,当然由 1 位置热电丝电阻由 R_{T1} 值转到 4 位置变成 $R_{T'_H}$ 是一个动平衡过程,在图中是难于具体表示,图 27 只不过是用准稳定方法作物理意义上解释,理论上如果在设计电桥时预先精确计算好所需要的 R_k 及 $R_{ш}$ 电阻值,则这样的温度补偿是准确的,但实际上由于温度电阻曲线不是一段 OAB 的折线而是一条折线,故这样的补偿在某种程度上也是近似的,П.В.ЧеБышев 的试验证明(文献 20),采用了这种补偿线路后不同气流温度下试验点仍落在同一条速度电流校正曲线上(图 28)。

三、气流紊流参数的测量

§6. 研究气流紊流结构所需测量的紊流参数量

在讨论如何测量紊流参数之前，必须弄清是理论上要求我们测量些什么量，讨论各个紊流参数的物理意义不是本文的任务，但是我们应列出每个紊流参数所包含的必须测量的物理量。

1) 在紊流理论中人为地把某点气流真正速度分为平均速度 $\overline{W}$ 及脉动速度 W' 之和，即

$$W = \overline{W} + W' \tag{34}$$

气流是杂乱的，但是服从一定统计规律在脉动，因此首先要求把瞬时脉动值 W' 随时间变化的规律测量出来，但是由统计规律可知 $\overline{W'}=0$，因此要表达气流脉动强度（或称紊流强度）一方面不能用瞬时脉动值，因为它是一个不稳定参数，但另一方面也不能用各瞬时脉动的平均值，因此一般使用各瞬时脉动值的平方平均开方值，对于三元气流，即 $\sqrt{\overline{W_x'^2}}$，$\sqrt{\overline{W_y'^2}}$ 及 $\sqrt{\overline{W_z'^2}}$。测量出这三个数值后即有可能得出各紊流强度。

$$\varepsilon_x = \sqrt{\overline{\omega_x'^2}}\Big/\overline{w} \;;\qquad \varepsilon_y = \sqrt{\overline{\omega_y'^2}}\Big/\overline{w} \;;\qquad \varepsilon_z = \sqrt{\overline{\omega_z'^2}}\Big/\overline{w} \tag{35}$$

及各脉动压力：

$$P_x' = \frac{\rho}{2}\overline{\omega_x'^2} \;;\qquad P_y' = \frac{\rho}{2}\overline{\omega_y'^2} \;;\qquad P_z' = \frac{\rho}{2}\overline{\omega_z'^2} \tag{36}$$

2) 第二个必须测量的参数是在同一空点上各不同方向瞬时脉动速度乘积的平均值：$\overline{W_x' \cdot W_y'}$，$\overline{W_x' \cdot W_z'}$ 及 $\overline{W_y' \cdot W_z'}$。由此可求得紊流摩擦力（切向应力）：

$$\tau_{xy}' = -\rho\overline{\omega_x' \cdot \omega_y'} \;;\quad \tau_{yz}' = -\rho\overline{\omega_y' \cdot \omega_z'} \;;\quad \tau_{xz}' = -\rho\overline{\omega_x' \cdot \omega_z'} \;; \tag{37}$$

方向关联系数：

$$R_{xy} = \overline{\omega_x' \cdot \omega_y'}\Big/\sqrt{\overline{\omega_x'^2}} \cdot \sqrt{\overline{\omega_y'^2}} \;;\quad R_{yz} = \overline{\omega_y' \cdot \omega_z'}\Big/\sqrt{\overline{\omega_y'^2}} \cdot \sqrt{\overline{\omega_z'^2}} \;;$$

$$R_{xz} = \overline{\omega_x' \cdot \omega_z'}\Big/\sqrt{\overline{\omega_x'^2}} \cdot \sqrt{\overline{\omega_z'^2}} \tag{38}$$

紊流黏度系数（或称紊流交换系数）A：

$$\tau_{xy}' = -\rho\overline{W_x'W_y'} = A\frac{\mathrm{d}w}{\mathrm{d}y} \tag{39}$$

位移径 l（按 Prandtl 定义）：

$$l = \sqrt{\frac{A}{\rho J}} \tag{40}$$

及由脉动速度所引起的散射能损失 $\overline{D_T}$，如果按 Е.М.Минский 的假设略去第三级脉动矩（文献 5），则可写成：

$$\overline{D_T} = -\rho \overline{\omega'_x \cdot \omega'_y} \frac{d\overline{w}}{dy} + \mu \frac{1}{2} \frac{d^2}{dy^2} \left(\overline{\omega'^2_x} + \overline{\omega'^2_y} + \overline{\omega'^2_z} \right) + \mu \frac{d^2 \overline{\omega'^2_y}}{dy^2} \tag{41}$$

3) 必须测量在同一方向不同位置的两点瞬时脉动速度乘积的平均值，即 $\overline{W'_{x_1} \cdot W'_{x_2}}$，$\overline{W'_{y_1} W'_{y_2}}$ 及 $\overline{W'_{z_1} W'_{z_2}}$。由此可得出：

空间关联系数

$$R_{x_{1,2}} = \overline{\omega'_{x_1} \cdot \omega'_{x_2}} \Big/ \sqrt{\overline{\omega'^2_{x_1}}} \cdot \sqrt{\overline{\omega'^2_{x_2}}} \;;\quad R_{y_{1,2}} = \overline{\omega'_{y_1} \cdot \omega'_{y_2}} \Big/ \sqrt{\overline{\omega'^2_{y_1}}} \cdot \sqrt{\overline{\omega'^2_{y_2}}} \;;$$

$$R_{z_{1,2}} = \overline{\omega'_{z_1} \cdot \omega'_{z_2}} \Big/ \sqrt{\overline{\omega'^2_{z_1}}} \cdot \sqrt{\overline{\omega'^2_{z_2}}} \tag{42}$$

紊流标尺

$$L_x = \int_0^\infty R_{x_{1,2}} dx \;;\quad L_y = \int_0^\infty R_{y_{1,2}} dy \;;\quad L_z = \int_0^\infty R_{z_{1,2}} dz \tag{43}$$

4) 脉动频谱的测定，这里要测出脉动频谱的分布规律及气流平均脉动频率值 $\overline{\omega_{cP}}$，由此亦可得出紊流标尺：

$$L = k \frac{\overline{\omega}}{\omega_{cP}} \tag{44}$$

k 为试验常数

5) 瞬时温度脉动值与速度脉动值的乘积 $\overline{T'W'_x}$、$\overline{T'W'_y}$、$\overline{T'W'_z}$，由此可得出：

温度速度关联系数

$$R_{\omega'_x T'} = \overline{T'\omega'_x} \Big/ \sqrt{\overline{T'^2}} \cdot \sqrt{\overline{\omega'^2_x}} \;;\quad R_{\omega'_y T'} = \overline{\omega'_x T'} \Big/ \sqrt{\overline{T'^2}} \cdot \sqrt{\overline{\omega'^2_y}} \;;$$

$$R_{\omega'_z T'} = \overline{\omega'_z T'} \Big/ \sqrt{\overline{T'^2}} \cdot \sqrt{\overline{\omega'^2_z}} \tag{45}$$

应当指出，除了上述参数外还有按拉格朗日法的同一点不同时间的脉动速度乘积 $\overline{W'_{\tau_0} \cdot W'_{\tau_1}}$，由此可求出

时间关联系数

$$R_\tau = \overline{\omega'_{\tau_0} \cdot \omega'_{\tau_1}} \Big/ \sqrt{\overline{\omega'^2_{\tau_0}}} \cdot \sqrt{\overline{\omega'^2_{\tau_1}}}$$

及拉格朗日的紊流标尺 $L_1 = W' \int_0^\infty R_\tau d\tau$

但目前测量技术中还无法测得不同时间内各脉动值之间关联关系，此外理论上还要求我们测量第三级脉动矩如 $\overline{W'^2_x \cdot W'_y}$、$\overline{W'_x \cdot W'^2_z}$、$\overline{W'_x \cdot T'^2}$、$\overline{T' \cdot W'^2_y}$、…等这在目前条件下都是很难实现的，我们也难于讨论。总之，各紊流参数的测定在目前来说还未发展得

足够成熟，特别在准确度方面还距离要求很远，因为我们无法"创造"出各种标准的紊流参数值来校正各种测量仪表，实用上我们只能间接地进行推算。

§7. 测量紊流参数的物理基础

1）在没有热惯性的情况下速度脉动引起热电风计中电参数的变化规律

如果电桥对平均速度所引起电参数的变化已处于平衡状态，那么由于微小的速度脉动必然引起热电丝对流放热瞬间增加或减少，亦即热电丝的温度，因而及其电阻也就瞬间减少或增加（和对流放热相反），所以反映在热电丝支杆出口处所测量出的电压亦产生相应的脉动，其数学表达式为：

$$W = \overline{W} \pm W' \quad 引起 \quad T_H = \overline{T_H} \mp T_H' \quad 及 \quad R_{T_H} = \overline{R_{T_H}} \mp R_{T_H}'$$

及电压变化 $V = \overline{V} + V'$

由于设计电桥时已经取 R_a 比 R_{T_H} 大约 100 倍（见图 25），那么 R_{T_H} 因气流脉动所引起的微弱变化并不会使通过热电丝的电流产生变化，那么在热电丝支杆测得的电压应为：

$$V = iR_{T_H} = iF(W)$$

$F(W)$ 为电阻与气流速度的函数关系式（如式（7））。

当速度增长 ΔW 时，电阻是减少的，故 $F(W) > F(W + \Delta W)$。

此时反映在丝支杆的电势差为：

$$\Delta V = i\left[F(W) - F(W + \Delta W)\right] = i\left[-\Delta W V'(W) + \cdots\cdots\right]$$

当速度变化不大时，可近似认为热电丝电阻及电势差的变化和气流的变化成正比，即

$$\Delta V = k \cdot \Delta W \tag{46}$$

例如当气流按正弦曲线周期脉动：

$$W = \overline{W} + W' \sin \omega \tau \tag{47}$$

式中，$\omega = 2\pi f$，f 为脉动频率。

那么，电压也作相应的脉动：$V = \overline{V} + kW' \sin \omega \tau = \overline{V} + V' \sin \omega t \tag{48}$

同理，电阻的脉动值也可写成：$R_{T_H} = \overline{R_{T_H}} + R_{T_H}' \sin \omega \tau \tag{49}$

其物理意义是这样的，当在平均速度上速度脉动时，脉动速度所引起的电阻变化本来按复杂的函数 $F(W)$ 变化（图 29），但在脉动不大时，我们近似取 AB 段为直线，因而得出（48）、（49）式，当然，这样的假设在脉动很大时是不准确的。

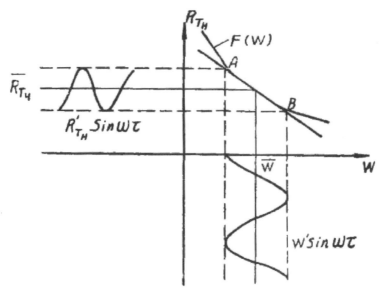

图 29 速度脉动和电阻脉动的关系

2) 在考虑有热惯性的情况下速度脉动所引起热电风计中电参数的变化规律

在实际过程中，由于热电丝有热惯性式(48)会变成

$$\left.\begin{array}{l} V_1 = \overline{V} + V_1' \sin(\omega\tau + \varphi) \\ R_{T_{H_1}} = \overline{R_{T_H}} + R_{T_{H_1}}' \sin(\omega\tau + \varphi) \end{array}\right\} \quad (50)$$

这里 $V_1' < V'$；$R_{T_{H_1}}' < R_{T_H}'$，即脉动振幅减小，方位角落后 φ 角，这是由于当气流速度脉动很快时，热电丝来不及立即冷却和加热。我们的任务在于找出因热惯性使电参数振幅落后多少，方位角滞后多少。由于这是不稳定过程，应该利用方程式(1)来解决，把式(1)改写成：

$$4.2m\delta \frac{dT_H}{d\tau} = i^2 R_{T_H} - (A + BW^n)(T_H - T_0) \quad (51)$$

为方便计，今后 R_T 表示在温度 $T°C$ 下电阻丝的瞬间真正电阻值，$\overline{R_T}$ 则表示在热平衡状态下平均电阻值。

由式(5)可得
$$\frac{R_{T_H}}{R_{T_0}} = \frac{1 + \alpha_\rho T_H}{1 + \alpha_\rho T_0}$$

即
$$T_H - T_0 = \frac{R_{T_H} - R_{T_0}}{\alpha_\rho R_{T_0}}(1 + \alpha_\rho T_0)$$

微分之可得：
$$\frac{dT_H}{d\tau} = \frac{1}{R_{T_0}\alpha_\rho}(1 + \alpha_\rho T_0)\frac{dR_{T_H}}{d\tau}$$

代入(51)式：

$$\frac{4.2m\delta(1+\alpha_\rho T_0)}{R_{T_0}\alpha_\rho}\cdot\frac{\mathrm{d}R_{T_\mathrm{H}}}{\mathrm{d}\tau}=i^2 R_{T_\mathrm{H}}-(A+BW^n)\frac{R_{T_\mathrm{H}}-R_{T_0}}{\alpha_\rho R_{T_0}}(1+\alpha_\rho T_0) \tag{52}$$

设 R_H 代表气流速度为常量时的真正电阻，如果热电丝是没有热惯性的话则测量出来的电阻 R_{T_H} 值在任何时间内均与 R_H 值相等，故当速度为常量时 $R_{T_H}=R_\mathrm{H}$，$\frac{\mathrm{d}R_{T_H}}{\mathrm{d}\tau}=0$，代入(52)：

$$i^2 R_\mathrm{H}-(A+BW^n)\frac{R_\mathrm{H}-R_{T_0}}{\alpha_\rho R_{T_0}}(1+\alpha_\rho T_0)=0 \tag{53}$$

由(53)式可以求出 $(A+BW^n)$ 值，实质上式(53)和(7)式相同，故物理意义上等于用平均速度的水平来求热电丝的热惯性，下面可以看出，这只能作为一个近似，把 $(A+BW^n)$ 值求出后，代入(52)，经整理后可得：

$$4.2m\delta\frac{(1+\alpha_\rho T_0)}{R_{T_0}\alpha_\rho}\frac{\mathrm{d}R_{T_\mathrm{H}}}{\mathrm{d}\tau}=i^2 R_{T_\mathrm{H}}-i^2 R_\mathrm{H}\cdot\frac{R_{T_\mathrm{H}}-R_{T_0}}{R_\mathrm{H}-R_{T_0}}$$

或

$$\frac{4.2m\delta(1+\alpha_\rho T_0)}{i^2 R_{T_0}\alpha_\rho}\frac{\mathrm{d}R_{T_\mathrm{H}}}{\mathrm{d}\tau}=\frac{R_\mathrm{H}-R_{T_\mathrm{H}}}{R_\mathrm{H}-R_{T_0}}$$

如果设

$$M=\frac{4.2m\delta(1+\alpha_\rho T_0)(\overline{R_{T_\mathrm{H}}}-R_{T_0})}{i^2 R_{T_0}\alpha_\rho} \tag{54}$$

这里 $\overline{R_{T_H}}$ 及 $\overline{T_\mathrm{H}}$ 各为热电丝的平均电阻及温度。

则可得：

$$M=\frac{R_\mathrm{H}-R_{T_0}}{\overline{R_{T_\mathrm{H}}}-R_{T_0}}\cdot\frac{\mathrm{d}R_{T_\mathrm{H}}}{\mathrm{d}\tau}+R_{T_\mathrm{H}}=R_\mathrm{H} \tag{55}$$

但是 $\overline{R_{T_\mathrm{H}}}=R_{T_0}\left[1+\alpha_\rho(\overline{T_\mathrm{H}}-T_0)\right]$

即 $\dfrac{\overline{R_{T_\mathrm{H}}}-R_{T_0}}{\alpha_\rho R_{T_0}}=\overline{T_\mathrm{H}}-T_0$ 代入式(54)可得

$$M=\frac{4.2m\delta(\overline{T_\mathrm{H}}-T_0)}{i^2 R_{T_0}} \tag{54'}$$

为了要求解方程式(55)，把由于有热惯性存在热电丝的实际电阻 R_{T_H} 求出必须要知道气流脉动的规律性及引起电阻脉动的规律性，这里我们分两种情况来讨论：

A) 当气流脉动按正弦规律变化，由式(49)可知也引起电阻作相应变化。

即 $R_H = \overline{R_{T_H}} + R'_{T_H} \sin\omega\tau$

代入式(55)得

$$M\left[1 + \frac{R'_{T_H}}{\overline{R_{T_H}} - R_{T_0}}\sin\omega\tau\right]\frac{dR_{T_H}}{d\tau} + R_{T_H} = \overline{R_{T_H}} + R'_{T_H}\sin\omega\tau \tag{56}$$

如果热电丝被加热至比气流温度高很多,那么电阻的脉动值 R'_{T_H} 比起 $\left(\overline{R_{T_H}} - R_{T_0}\right)$ 小得多,可以略去,此时微分方程变为:

$$M\frac{dR_{T_H}}{d\tau} + R_{T_H} = \overline{R_{T_H}} + R'_{T_H}\sin\omega\tau$$

解之得:

$$R_{T_H} = e^{-\frac{\tau}{M}}\left[C + \int\left(\frac{\overline{R_{T_H}} + R'_{T_H}\sin\omega\tau}{M}\right)e^{\frac{\tau}{M}}d\tau\right]$$

$$= C \cdot e^{-\frac{\tau}{M}} + \overline{R_{T_H}} + \frac{R'_{T_H}}{M} \cdot e^{-\frac{\tau}{M}}\int e^{\frac{\tau}{M}}\sin\omega\tau d\tau$$

边界条件:当气流不脉动,电阻也不脉动,即 $R_{T_H} = \overline{R_{T_H}}$
代入可见,积分常数 $C=0$,

故

$$R_{T_H} = \overline{R_{T_H}} + \frac{R'_{T_H}}{M} \cdot e^{-\frac{\tau}{M}} \cdot \frac{e^{\frac{\tau}{M}}}{\left(\frac{1}{M}\right)^2 + \omega^2}\left(\frac{1}{M}\sin\omega\tau + \omega\cos\omega\tau\right)$$

$$= \overline{R_{T_H}} + \frac{R'_{T_H}}{1 + M^2\omega^2}(\sin\omega\tau + M\omega\cdot\cos\omega\tau)$$

如果设 $M_\omega = \text{tg}\varphi$,则 $\cos\varphi = \dfrac{1}{\sqrt{1+\omega^2 M^2}}$。

代入最后得:

$$R_{T_H} = \overline{R_{T_H}} + \frac{R'_{T_H}}{\sqrt{1 + M^2\omega^2}}(\sin\omega\tau + \varphi) \tag{57}$$

比较式(50)和(57)可以发现,热电丝热惯性的存在对于平均电阻的测量是丝毫没有影响,但对脉动值的测量则影响很大。

$$R'_{T_{H1}} = \frac{R'_{T_H}}{\sqrt{1 + M^2\omega^2}} \tag{58}$$

亦即脉动振幅减少了 $\sqrt{1+M^2\omega^2}$ 倍,由于 M 值具有时间的因次,因此称为热电丝的时间常数。由式(54)及(54')可见,M 值主要取决于热电丝的物性、截面、长度、热电丝被加热程度(相对于气流)及该点气流速度(亦即所需加热电流 i),因而对结构已定的热电丝,又采取图 24 的电桥线路,则 M 值主要与气流速度有关。ω 为气流脉动角速度,当气流脉动频率愈高时,测量得电阻脉动值愈小,即热惯性影响愈大。

Б) 当气流脉动按任意周期函数变化时(文献 9)

实际上,气流的脉动形式是无规则的,假设它是按正弦波脉动只是一个近似,但是如果 R_H 按任意周期函数变化,式(55)是很难解出,当脉动值不大的情况下,实际上可以假设:

$$R_H - R_{T_0} \approx \overline{R_{T_H}} - R_{T_0} \tag{59'}$$

则式(55)变成:

$$M \frac{dR_{T_H}}{d\tau} = R_H - R_{T_H} \tag{59}$$

如果把无规则的但周期性变化的电阻脉动分解成傅里叶级数:

即
$$R_H - \overline{R_{T_H}} = \sum_{n=1}^{\infty}(a_n \sin n\omega\tau + b_n \cos n\omega\tau) \tag{60}$$

此时式(59)的解亦类似:

$$R_H - \overline{R_{T_H}} = \sum_{n=1}^{\infty}(c_n \sin n\omega\tau + d_n \cos n\omega\tau) \tag{61}$$

c_n 及 d_n 为待决定的系数。

微分式(61)并一起把(61)、(60)式值代入(59)式,在式左边,右边比较同类项系数,即可求出:

$$c_n = \frac{a_n + M_n b_n \omega}{1 + M^2 n^2 \omega^2}$$

$$d_n = \frac{b_n - M_n a_n \omega}{1 + M^2 n^2 \omega^2}$$

代入(61)可得:

$$R_{T_H} - \overline{R_{T_H}} = \sum_{n=1}^{\infty}\left(\frac{a_n + M_n b_n \omega}{1 + M^2 n^2 \omega^2}\sin n\omega\tau + \frac{b_n - M_n a_n \omega}{1 + M^2 n^2 \omega^2}\cos n\omega\tau\right)$$

同样设 $Mn\omega = \text{tg}\varphi$,可得:

$$R_{T_H} - \overline{R_{T_H}} = \sum_{n=1}^{\infty}\left[\frac{a_n}{\sqrt{1+M^2n^2\omega^2}}\sin(n\omega\tau-\varphi) + \frac{b_n}{\sqrt{1+M^2n^2\omega^2}}\cos(n\omega\tau-\varphi)\right] \quad (62)$$

这就是由于热惯性使电阻脉动值减小倍数的一般式。

现在可以具体估计一下由于热电丝热惯性使测量所得脉动电阻值及电势差值比真正值小多少。由式(54')并考虑到

$$m = l \cdot A \cdot \delta , \quad R_{T_0} = \frac{\rho_0 l}{A} \quad (63)$$

可得

$$M = \frac{4.2 \cdot \delta \cdot A^2 \cdot S\left(\overline{T_H} - T_0\right)}{i^2 \rho_0} \quad (64)$$

这里 l、A 为热电丝的长度及横截面积；δ 为材料密度；ρ_0 为在 T℃下材料的比电阻。

H.L.Dryden 和 A.M.Kuethe 对直径为 20 微米的白金丝热惯性进行了详细计算（文献 22），计算时采取 $\overline{T_H}$ =500℃，δ =21.37 克/CM³，$S = 0.037 \frac{\text{米}\cdot\text{卡}}{\text{克}℃}$，$\rho_0$ =0.000012 欧姆厘米，计算结果详细列于表 4。

表 4

f	1	5	10	20	50	100	200
$\frac{V_1'}{V'}$	0.995	0.884	0.687	0.402	0.186	0.094	0.043
φ	6°	28°	47°	66°	79°	85°	89°

表中 f 为脉动频率，$\frac{V_1'}{V'}$ 有热惯性及没有热惯性时热电丝元件电势差之比。根据我们试验，一般在自由射流中，在旋转气流中，在旋风炉内气流脉动的平均频率为 f =100～200 赫，由表 4 可见，如果不进行热惯性的补偿，紊流脉动参数的测量是完全没有意义的。尽管把热电丝直径减小，热电丝过热温度降低 M 值会显著减小，但这些并不能避免热惯性的歪曲，因此下面我们将详细讨论如何能够准确补偿热电丝热惯性，这是决定能否采用热电风计测量紊流参数的关键。

§8. 热电丝热惯性的补偿方法

1) 热电丝时间常数 M 值的理论和试验求法

为了要准确补偿丝的热惯性，首先就要准确知道特定结构的热电丝感应元件在各种气流速度下 M 的真正值，一般有两种方法求 M 值：

A) 理论计算法，理论计算一般按(64)式进行，一般 M 值除与热电丝元件具体结构有关外，还与气流速度有关，但是为了避免找出速度和电流的函数关系，一般只求 M 值与

电流的关系，因为在测量脉动参数之前，必须使电桥得到平衡，亦即先求出平均参数值，这样电流大小是已知的，因为我们用钨丝热电元件，故亦以钨丝的时间常数为例：对钨来说，密度 $\delta=19.35$ 克/CM³，热容 $S-0.032\dfrac{\text{米}\cdot\text{卡}}{\text{克}\,℃}$ 比电阻 $\rho_0=0.0000055$ 欧姆厘米，由于采用了图 24 的电桥系统保证标准加热度 $(T_H-T_0)=150℃$，计算结果画成诺莫图 30（按 П.В.ЧеБышев），图中列出不同直径时的情况，对我们使用的直径 $d=19$ 微米的钨丝，在 0～150 米/秒速度下，M 值在 0.015 至 0.001 秒内变动。

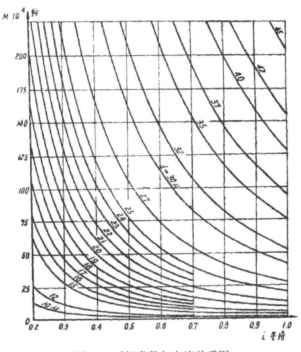

图 30　时间常数与电流关系图

B）用试验方法决定 M 值，公式(64)在理论上是准确的，但进行计算时有一定困难，首先所使用材料的物性 δ、s、ρ_0 很难准确决定，它受使用材料的纯度影响很大，同时热电丝的直径是如此之小，因此要准确决定也是很困难的，况且热电丝直径由于制造不当沿长度可能有变化，故按理说，最可靠的办法是由试验直接求出所使用具体热电元件的 M 值，但是要准确试验求出 M 值也是很困难的。С.И.Кречмер 曾经提出了用不稳定过程试验求 M 法（文献 25），实际上为了简化试验利用式(59)：

$$\frac{\mathrm{d}R_{T_H}}{\mathrm{d}\tau}+\frac{R_{T_H}}{M}-\frac{R_H}{M}=0 \tag{65}$$

如果我们的试验是在气流速度一定的情况下进行，而且 R_H 取该平均气流下的平均电阻，并考虑到如下的起始条件：$T=T_0$（气流温度），$R_{T_H}=R_{T_0}$，积分(65)式得：
$$R_{T_H} = (R_{T_0} - R_H)e^{-\frac{\tau}{M}} + R_H \tag{66}$$

当 $\tau \to \infty$，$R_{T_H} = R_H$，即表示在平衡状态下热电丝的电阻和在该气流速度下电桥平衡所要求的电阻相等，故当未达到平衡时，电阻的瞬时变化值为 $\Delta R = R_{T_H} - R_H$。如果我们测量的是在热电丝支杆下的电势差，而通过热电丝的电流在整个不稳定过程中保持不变的话，则电势差随时间的变化按式(66)可写成：

$$V = i(R_{T_0} - R_H)e^{-\frac{\tau}{M}} \tag{67}$$

因此，如果我们能用试验方法求出不同时间内两个 V 值，则由此可决定时间常数：

$$M = \tau_2 - \tau_1 \bigg/ \ln\frac{V_1}{V_2} \tag{68}$$

下面讨论如何在试验中实现这个要求：

在电桥系统装置有高速转换开关 K，在电桥对角线接上电子示波器(见图 31)，示波器中出现的曲线用拍照方法录取，在摄影前，先调整好电桥系统，使其在该试验风速下达到平衡，然后用开关 K 接入 R_k 电阻，使热电丝冷却到和气流温度相同，此时其电阻为 R_{T_0}，当开关 K 突然和热电丝电臂接通，电流开始加热热电丝，加热规律按式(66)变化，并同时摄下电势差和时间的关系曲线，按(68)式即可求出 M 值，С.И.Кречмер 用白金丝直径为 20 微米，长 25 毫米，热电丝加热温度为 121℃、238℃、324℃，风速在 0~16 米/秒下进行试验，发现试验所得的 M 值比按式(64)的计算值高 1.5~1.8 倍，我们认为，很难判断这是由于试验或者是理论计算的误差，因为第一，采用式(65)本身是近似的，第二，用摄影方法求 $V = f(\tau)$ 关系是很难准确的，如果在电桥对角线接电磁示波器或许会得到较高的准确度，因为电磁示波器计算时间可准确到 1/500 秒以上。

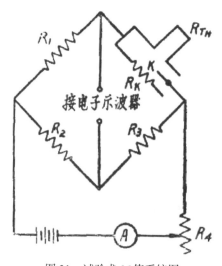

图 31 试验求 M 值系统图

2)热电丝热惯性的补偿方法

当我们准确知道热电元件的时间常数 M 后,下一步问题就是如何去准确补偿它。由于气流脉动所引起的电参数变化非常微弱,一般约为 0.001 至 0.05 伏特,因此必须经过强大的放大器放大,在放大过程中往往就加入特殊的电路系统进行手动的或自动的热惯性补偿,由此而来的补偿方法可以分为两种类型。

A)在放大器内加入特殊的线路,使得放大器出口处的放大系数与入口信号的频率(或角系数 ω)和放大器内可变阻抗比 N 服从下面规律:

$$K = K_0\sqrt{1+\omega^2 N^2} \tag{69}$$

按式(58),如果入口信号不是电阻脉动,而是通过热电丝时电势差的脉动,那么在放大器出口处将会得到:

$$V'_{II} = \frac{V'_1}{\sqrt{1+M^2\omega^2}} \cdot K_0\sqrt{1+\omega^2 N^2} \tag{70}$$

显然易见,如果我们选择阻抗比 N 的数值刚好与时间常数相等,即

$$N=M \tag{71}$$

那么
$$V'_{II} = K_0 V'_1 \tag{72}$$

即在放大器出口处得出被放大了 k_0 倍的真正脉动电压值 V'_{II}(相应于被放大了的真正速度脉动值),采用这种方法补偿的物理实质是让热电丝的温度随气流脉动而变化,从而测量由这个变化所引起的电阻和电压的变化。

自然,目前问题在于研究采用什么样的特殊线路系统能使放大系数按式(69)所示规律变化,以及在使用过程中如何来满足条件式(71)。

a.可以采用电感电阻的补偿系统。为此在放大器的某一级内在阳极线路上加入自感 L 及可变电阻 R_k(图32),在设计这个线路时要求满足:

(i)只有在加入特殊补偿线路这一放大级中与输入信号频率有关,而其他放大级则应与频率无关。

(ii)在选择补偿线路阻抗时,令 $\omega L \ll R_i$ 及 $R_K + R_L \ll R_i$,这里 R_i 表示电子管内阻,如果把图32画成等值电路(图33),可以进行如下的计算:

$$V'_{II} = \frac{k_0 V'_1 \sqrt{(R_L+R_K)^2+\omega^2 L^2}}{\sqrt{(R_i+R_L+R_K)^2+\omega^2 L^2}}$$

$$= \frac{R_L+R_K}{R_i} \cdot \frac{\sqrt{1+\omega^2 N^2}}{\sqrt{\left(1+\frac{R_L+R_K}{R_i}\right)^2+\left(\frac{\omega L}{R_i}\right)^2}} k_0 V'_1$$

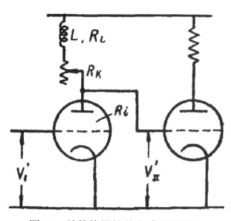
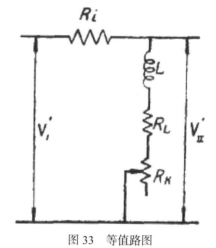

图 32 补偿热惯性的电感电阻系统　　　图 33 等值路图

如果设 $m = \dfrac{R_L + R_K}{R_i}$，

则
$$V'_{II} = k_0 V'_1 \sqrt{1+\omega^2 N^2} \cdot \dfrac{m}{\sqrt{1+m^2}} \tag{73}$$

这里
$$N = \dfrac{L}{R_L + R_K} \tag{74}$$

当满足上述第二点要求时，上式可以简化为

$$V'_{II} = k'_0 \sqrt{1+\omega^2 N^2}\, V'_1 \tag{69'}$$

这就是我们所要求的放大系数变化规律式(69)，而且 N 值的大小完可以由可变电阻 R_K 自由调节。

但要注意的是引入补偿路线后放大系数变成：

$$k'_0 = \dfrac{k_0 m}{\sqrt{1+m^2}} \approx k_0 m \tag{75}$$

因为 m 值很小（一般电子管内阻很大），故使得放大器中放大系数大为降低，要保持同样的放大系数就要增加放大器的级数，此外，采用这样的补偿系统，当输入信号频率为零时，其放大系数随时间常数 N 的调节而改变，使在运行工作中带来了麻烦。

Б.可以采用电容电阻的补偿系统，采用电容电阻系统可以消除上述放大系数随 N 的调节而变化的缺点，按照上面的计算原理，同样可以得出：

$$V_{II}' = \frac{V_1' \cdot R_1 \sqrt{1+\omega^2 C^2 R_K^2}}{\sqrt{(R_1+R_K)^2 + \omega^2 C^2 R_1^2 R_K^2}} = V_1' \frac{R_1}{R_1+R_K} \cdot \frac{\sqrt{1+\omega^2 N^2}}{\sqrt{1+\omega^2 N^2 \left(\frac{R_1}{R_1+R_K}\right)^2}}$$

这里，$N=CR$。 (74')

在设计线路时选择 $R_K \gg R_1$，故 $\frac{R_1}{R_K+R_1}$ 很小。

上式可变成： $$V_{II}' = \frac{R_1}{R_K+R_1}\sqrt{1+\omega^2 N^2}\, V_1'$$

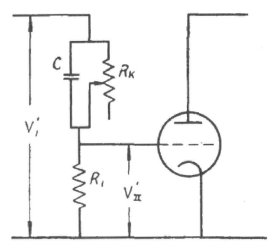

图 34　电容电阻补偿热惯性线路图

这就满足了(69)式的要求，但同样加入此补偿系统后使输入信号减弱了 $\frac{R_1}{R_K+R_1}$ 倍，这个系统固然简单，但是放大器中，任何寄生电容都会对它产生一定影响，这也使对放大器的设计提出了苛刻的要求。

在实用上，还可以找出其他比较复杂的方案，但大多以这两种原理为基础(见文献14)。

B) 采用反馈线路，在放大器中设有反馈系统，热电丝此时也作为自动控制元件之一，此时补偿原理和 A 法有原则上的不同，利用反馈控制线路自动保证了即使在脉动气流下热电丝的温度不变，对于自动调节系统需要测量的是热电丝电阻，例如当气流脉动大于平均速度值，散热加强，热电丝温度，因而也连及其电阻降低，电阻变化的信号被引入放大器中(图 35)，放大器根据这个信号增大输给电桥的电流，直至电桥平衡时才停止工作，亦即供给电桥电流的增减自动跟踪气流的脉动。因此采用反馈方法最大的优点在于自动补偿了热电丝的热惯性，这对气流脉动很大时是有利的。因为第一，当脉动很大时不能满足图 29 所示电参数和气流脉动的线性变化，因而 M 值也是在脉动过程中也是变化的，要知道在§7 所述的补偿理论中 M 值是用平均速度来求出，采用了反馈系统后，如果能自动瞬间跟踪的话，则不管 M 值如何变化均能正确地补偿。第二，能把运行操作手续及时间减至最小。可见从原理来说采用反馈线路是今后发展的方向，但在目前还存在一些困难，其中最主要的有，热电丝电阻很小，因此要求加热电流很大，这就使放大器结构复杂，其次最主要的是气流脉动频率较大，因此要瞬间自动跟踪补偿就比较困难，因此如果反馈系统不够完善，补偿有一定滞后时，就会带来很大的误差，（其误差会比 A 法大得多），故目前使用还不够普遍，详细线路可参阅文献 14、26、27 及 28。

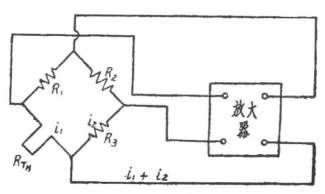

图 35　具有反馈线路的热惯性补偿系统

§9. 有关影响放大器准确工作的几个问题

1) 避免输入信号中直流分量加入放大器的方法

放大器的用途在于把气流脉动所引起的微弱电参数交变分量放大，但热电丝所发出的信号是由平均及脉动速度共同作用下产生的，因此如果直接把这个信号输入放大器，则使放大器栅极经常受有较大的栅偏压作用，更严重的是气流平均速度可能各处不同的，因而作用到栅极上的栅偏压也经常变化引起测量上极端不便，故在实用上一般只容许交变分量输入放大器，为此可以采用下列方法：

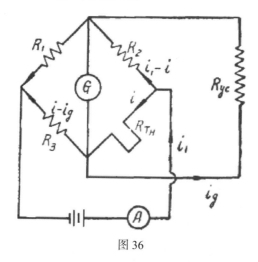

图 36

A) 采用电桥对角线引出输入放大器的信号，如果我们由热电丝支杆脚 AB 直接引入放大器，则经常会有一平均直流电压（$\overline{V}=i R_{T_H}$）作用在放大器上，但是如果我们从电桥对角线 C、D 处引入(图 36)，那么在测量平均速度时，电桥应处于平衡状态。C、D 处直流电势应相等，因此由 C、D 处供给放大器只是交变分量，但此时放大器所感受的不单是热电丝电阻的变化 R'_{T_H}，而且与电桥及放大器各电阻比例有关，由图 36 按克希荷夫定律可得，并考虑到在脉动气流下热电丝电阻按下式变化：

$$R_{T_H} = \overline{R}'_{T_H} + R'_{T_H}$$

则　　$i\left(\overline{R}_{T_H} + R'_{T_H}\right) + i_g \cdot R_{yc} - (i_1 - i)R_2 = 0$

及　　$(i - i_g)R_3 - (i_1 - i + i_g)R_1 - i_g R_{yc} = 0$

解之可得：

$$i_g = \frac{(R_3 + R_1)i_1 R_2 - \left(\overline{R}_{T_H} + R'_{T_H} + R_2\right)i_1 R_1}{(R_3 + R_1)R_{yc} + (R_3 + R_1 + R_{yc})\left(\overline{R}_{T_H} + R'_{T_H} + R_2\right)}$$

在电桥平衡时 $\overline{R}_{T_H} \cdot R_1 = R_2 \cdot R_3$，同时按照§5 所述原理，设计电桥时选取电阻 R_1 及 R_2 值较大，因此 $i \approx i_1$。

可得：$i_g = i \cdot R'_{T_H} \cdot \dfrac{R_1}{R_{yc}\left(\overline{R}_{T_H} + R_{T_H} + R_3 + R_2 + R_1\right) + (R_3 + R_1)\left(\overline{R}_{T_H} + R'_{T_H} + R_2\right)}$

输入放大器的电压为：

$$V'_1 = i_g \cdot R_{yc} = i \cdot R'_{T_H} \cdot \frac{1}{1 + \dfrac{R_2}{R_1} + \dfrac{R_2}{R_{yc}}} \tag{76}$$

这里我们略去比值 $\dfrac{R_3}{R_1}$、$\dfrac{\overline{R}_{T_H} + R'_{T_H}}{R_1}$ 及 $\dfrac{\overline{R}_{T_H} + R'_{T_H}}{R_{yc}}$

R_{yc} 为放大器内阻。

因为 R_1、R_2 及 R_{yc}，对一定结构的电桥，放大器均为定值，可见从电桥对角线引出的脉动电压值与热电丝的脉动电阻值成正比。

6)在我们应用测量平面气流的线路中(图 13)，需要两个电桥，除了用电桥对角线引出信号外，还考虑到由两个电桥共同输出信号时因 V 型元件两热电丝电阻不一定均等而可能出现有电势差，故在放大器输出线路中加入电位计 1 及 2，测量时调整电位计，使两电桥电势均等(由毫伏计监视)

B)在放大器信号入口处，装有电容器(容量为 1 微法拉)，再进一步隔绝直流分量输入的可能性。

2)放大器中 N 值求法。

如果采用了反馈的补偿线路，则不需要预先求出 N 值，但是我们在实际工作中采用了电感电阻的补偿系统(图 32)，在运行中必须要满足式(71) $N=M$，M 值可以在运行中迅速由图 30 得出，而 N 值理论上可按(74)式预先求出，为了要满足式(71)，必须经常调节可变电阻 R_k，考虑到放大器中电路的复杂，因此推荐 N 值预先试验决定，放大器的放大倍数规律是按式(69)变化，当调节某给定 R_k 值时，通入放大器频率各为 f_1 及 f_2（相应于角系数 ω_1 及 ω_2）但其电势相同的信号，按式(69)分别可得出两个不同的放大系数：

$$K_1 = K_0\sqrt{1+\omega_1^2 N^2} \text{ 及 } K_2 = K_0\sqrt{1+\omega_2^2 N^2}$$

解上两式即可得出与频率无关的放大系数 K_0 值及 N 值。

$$N = \frac{1}{\omega_1}\sqrt{\frac{(K_2/K_1)-1}{\left(\frac{\omega_2}{\omega_1}\right)^2 - \left(\frac{K_2}{K_1}\right)^2}} \tag{77}$$

$$K_0 = K_1 \bigg/ \sqrt{\frac{\left(\frac{\omega_2}{\omega_1}\right)^2 - 1}{\left(\frac{\omega_2}{\omega_1}\right)^2 - \left(\frac{K_2}{K_1}\right)^2}} \tag{78}$$

我们所使用的放大器共有五个放大级（П. В. ЧеБышев 所设计），在第一级中有 25 格可变电阻（符号用 ПТ₁ 代表），用来调节出口放大系数，在第二级中引入电感电阻的补偿线路，R_k 可变电阻也同样分为 25 格（用符号 ПТ₂ 代表），在试验放大器时用音频发生器并经过精密的电位计输入频率为 30 及 100 周波、电压为 0.0001 至 0.05 伏的信号，按上两式即可求出 N 及 K_0 值，其结果示于图 37，试验是在 ПТ₁=20 格时做出的，由试验可知，所设计的放大器能够满足于补偿时间常数在 0.0015 至 0.006 秒范围内，图中在 ПТ₂=18～21 格内出现了"平线"段，说明可变电阻 R_k 在该范围内被短路。

3）热惯性没有被准确补偿时所引起误差的讨论。

热电丝热惯性之所以没有被准确补偿，主要是由两个原因引起的：第一，求热电丝的时间常数 M 值时有误差，第二，在使用过程中很难完全满足式(71)的条件，因为 M 值是随速度变化的，要使放大器特殊电路中所得 N 值与之相等就必须要求 N 值能平滑地变化，这只有在采用滑动电阻才有可能，但是采用滑动电阻实用上是不现实的，它不能按上式精确地求出 N 及 K_0 值，因而在运行中也无从选择 N 值，故电阻 R_1（图 32）一般都推荐使用分组的可变电阻，如我们所使用的放大器中分为 25 组（ПТ₂）。这就会产生这样的问题，如果所要求的 N 值刚好落在两个分组之间，那么就会产生"过度补偿"和"补偿不足"的现象，这些问题都迫使我们要研究热惯性没有被准确补偿时可能引起的误差。设热电丝真正的时间常数为 M，而我们计算所得或补偿的是 N 值：

$$N = M + x$$

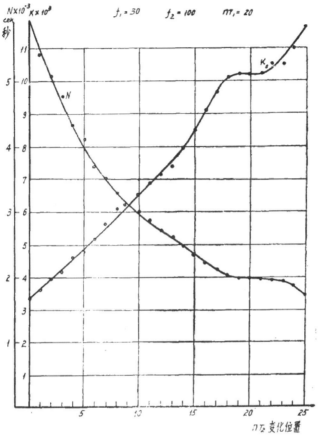

图 37 放大器特性图

那么在放大器出口处得出的电压 V' 与准确补偿时真正电压 V_H 的关系为：

$$V' = V_H' \frac{\sqrt{1+\omega^2 M^2}}{\sqrt{1+\omega^2 N^2}}$$

代入可得：

$$V_H' = V'\sqrt{\frac{1+\omega^2(M+x)^2}{1+\omega^2 M^2}} = V'\left[1+\frac{\omega^2(2Mx+x^2)}{1+\omega^2 M^2}\right]^{\frac{1}{2}}$$

$$= V'\left[1+\frac{2\dfrac{x}{M}+\left(\dfrac{x}{M}\right)^2}{\dfrac{1}{\omega^2 M^2}+1}\right]^{\frac{1}{2}}$$

把上式展开为级数并略去 $\dfrac{x}{M}$ 的高级小项，而 x 值用 $N-M$ 代，则可得：

$$V'_H = V'\left[1 + \frac{N-M}{M}\delta\right]$$

这里 $\delta = \dfrac{1}{1+\dfrac{1}{M^2\omega^2}}$ 与输入信号频率有关，很明显，当 $\omega=0$，$\delta=0$，$\omega=\infty$，$\delta=1$。

但实际上脉动频率的角系数是在 ω_1 及 ω_2 范围内变动，相应于这个范围内 δ 的平均值为：

$$\overline{\delta} = \frac{1}{\omega_2-\omega_1}\int_{\omega_1}^{\omega_2} \frac{1}{\dfrac{1}{\omega^2 M^2}+1}\,d\omega$$

$$= 1 - \frac{1}{M(\omega_2-\omega_1)}(arctgM\omega_2 - arctgM\omega_1) = 1 - \frac{\zeta}{M}$$

这里
$$\zeta = \frac{arctgM\omega_2 - arctgM\omega_1}{\omega_2 - \omega_1}$$

代入上式可得：
$$V'_H = V'\frac{N}{M}\left[1 - \frac{N-M}{M}\cdot\frac{\zeta}{M}\right] \tagged{79}$$

一般气流紊流脉动的平均频率在 $10\sim1000$ 赫范围内变动，根据 Ю.Г.Захаров 的计算（文献 29）ζ 值变化如表 5：

表 5

M(秒)	2×10^{-3}	3×10^{-3}	4×10^{-3}	5×10^{-3}	6×10^{-3}
ζ	0.435×10^{-3}	0.441×10^{-3}	0.438×10^{-3}	0.435×10^{-3}	0.428×10^{-3}

可见 ζ 值变化不大，并可近似取其平均值为 0.435×10^{-3}。如果补偿误差达 $\dfrac{N-M}{M}=10\%$，则

$M=$	2×10^{-3}	3×10^{-3}	4×10^{-3}	5×10^{-3}	6×10^{-3}
$\dfrac{N-M}{M}\cdot\dfrac{\zeta}{M}=$	0.0218	0.0145	0.0109	0.0087	0.0072

亦即（79）式中括号内右面一项可以完全略去，此时可得：

$$V'_H = V'\frac{N}{M} \tagged{80}$$

此式就是用来估计不准确补偿所引起的误差及设计补偿线路时的根据，在我们所使用的放大器中，R_K（即 ПТ$_2$）共分为 25 级（电阻由 21.5 欧至 200 欧内变化），试验证明（见图 37）当气流速度 $W\leqslant 100$ 米/秒时，分级调节所产生的"补偿过度"或"补偿不足"误

差最大值不超过 3%。

4) 气流脉动频率对测量准确度的影响。

在紊流气流中，气流脉动的能量是按一定频谱分布的，而且分布范围也比较广，一般主要集中在 10～2000 赫范围内，如果放大器准确按照式(69)所示规律设计，则无论气流脉动频率如何变化均不会影响放大器的正常工作，但是在放大器出口处，必须装置有测量出口电参数的仪表(出口测量一般为电压或电流)，对这些仪表有两个主要的要求：

A) 测量读数与被测参数的频率、交变波形均无关。

Б) 能够直接测出被测参数的平方平均开方值，因为由紊流理论可知 $\overline{W'}=0$，因此要求我们测量出 $\sqrt{\overline{W'^2}}$。可见，若放大器出口处的参数是电压 V'_{II}，用一般有惯性的仪表进行测量时，则只能得出 $\overline{V'_{II}} \approx 0$。

在实用上，完全满足上述要求的仪表只有静电电压计及真空管热电偶，但前者一般用来测量高压，对于我们要求的规格(测量值在 100～0 伏间)很少生产，因此我们选择了后者作为测量放大器出口脉动参数的主要仪表。真空管热电偶是由极细的加热丝及热电偶组成，为了避免对流散热损失，加热丝及热电偶都放在真空管内，当有电流通过加热丝时，发出热量，使热电偶出口端产生电动势，可由毫伏计(或电位计)准确测量，根据焦耳——楞次定律，同时考虑到加热丝及热电偶有较大的热惯性，因此毫伏计读数可表成下式：

$$E = \frac{1}{\tau}\int_0^\tau K_0 i^2 \cdot d\tau = K_\omega \sqrt{\overline{\omega'^2}} \tag{81}$$

亦即可得出脉动参数的平方平均开方值，由于加热丝与加热电参数的频率及波形均无关，因此，用来测量脉动参数是最适宜的，为了检查真空管热电偶在不同频率下的工作情况，我们曾进行了专门的试验，在放大器入口端利用音频发生器及精密电位计输入频率为 30、100、500、1000 及 2000 赫电压为 0.0001 至 0.05 伏的信号，放大器有两个出口端：一个为电压出口端，用真空管伏特计进行测量，另一为电流出口端，用真空管热电偶进行测量，试验结果示于图 38。

这里，纵坐标是真空管热电偶的读数，横坐标是放大器第一输出端真空管伏特计的读数。在进行专门试验时，采用真空管伏特计是准确的，真空管伏特计读数与输入信号频率无关，但与输入波形有关，因此它只能对测量正弦波形的脉动有效，而音频发生器出口波形是正弦波。试验证明，输入信号频率变化近百倍，真空管热电偶读数仍然不变，所有试验点很好地落在同一条圆滑曲线上，因此今后我们将利用此特性曲线，用来换算出各气流脉动参数值。

§10. 测量紊流参数的热电风计直接校正问题

测量紊流参数和测量平均速度不同在于在实用上很难找出一个标准的"紊流参数"作为校正测量仪表的规范，而一般紊流参数的获得通常借助于理论的换算，这就使得对测量仪表提出种种苛刻的要求，同时也是目前测量紊流参数准确度不高的主要原因。多年以来，很多研究者均企图直接校正热电风计，亦即直接得出气流紊动参数与放大器输出电参数的关系，其中值得提出的是用振动法进行直接校正，其原理简述如下：用一马达，带动一半径为 r 的偏心轮，在偏心轮上固定一长度为 l 的联杆，联杆末端套一滑块在滑槽内运动，热电感应元件就是固定在滑块上，当马达旋转时，滑块带动感应元件作往复振动运动，现在我们来考虑一下滑块（即感应元件）的运动轨迹：

如果把各运动零件简化成图 39 所示。

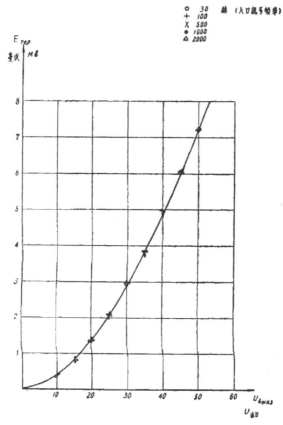

图 38　真空管热电偶特性曲线

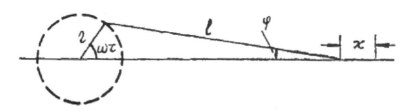

图 39　在校正装置上，感应元件运动图

则显然可见：　　$l\cos\varphi + r\cos\omega\tau = l + x$

及　　　　　　　$l\sin\varphi = r\sin\omega\tau$

解上两式得出

$$x = r\cos\omega\tau - l\left[1 - \sqrt{1-\left(\frac{r}{l}\sin\omega\tau\right)^2}\right]$$

略去 $\frac{v}{d}$ 的高次微小项可得

$$x = r\cos\omega\tau - \frac{r^2}{2l}\sin^2\omega\tau \tag{82}$$

因此如果 $\frac{r}{l}$ 比值足够小时，则运动规律很接近正弦规律，例如文献 22 所设计的校正试验台，取 r=0.5CM，l=10CM，因此，比值 l^2/d^2 =0.01，由此附加的非正弦脉动值约为 2%，在图 40 中并列出上述作者在校正试验台所得结果。

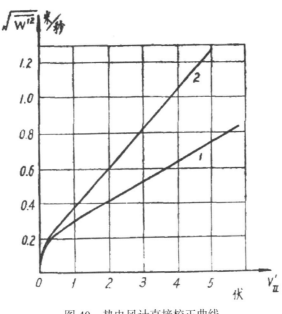

图 40　热电风计直接校正曲线

我们认为，这种方法在原理上是准确的，但在实践过程中，难以得出满意的结果，主要原因如下：

A) 校验是在静止气流中进行的，但当感应元件往复运动时，如果介质还是完全静止的，那当然会感受全部脉动值，问题在于热电感应元件是有一定尺寸的，当作高速往复运动时(要求频率变动范围在 0～2000 周波)，由于摩阻作用不可能不带动附近空气运动，要明确区分空气脉动和感应元件脉动是极端困难的，因而大大影响了校正的准确程度。

Б) 当热电丝感应元件作高速往复运动时，元件支杆本身必然亦产生附加的振动，测量结果证明(文献 22)，这个自身振动对校正的准确度影响很大，Ю. Г.Захаров 用放大镜进行了详细的观察，发现当不考虑元件自身振动时得出校正曲线 1(图 40)，但当考虑了自身振动时，则得出曲线 2，曲线 1、2 之间差别是如此之大足以说明自身振动的影响不可忽略，但是这些在校正过程中都是很难准确知道的。

故一直到目前为止，可以说还未能找出非常妥善的直接校正方法，这就不能不迫使我们尽量提高每个测量环节的准确度及得出更精确的气流脉动参数与电脉动参数联系的理论公式。

§11. 一元气流紊流脉动参数的决定

对于一元气流的脉动参数一般均用单热电丝决定，不过应当指出，在紊流运动中一元气流是指平均气流而言，即此时 $W_y = W_z = 0$，对于脉动速度来说永远都具有"三元"气流的特性，即 W_x'、W_y'、$W_z' \neq 0$，因此即使在一元气流中，要测量出 W_x'、W_y'、W_z' 三个方向的脉动值，并非轻而易举的事情，现在先从最简单的情况研究起，把单热电丝的感应元件放在与平均气流流动方向相垂直的平面上，在一元平均气流及三元脉动速度作用下，热电丝感受到下列瞬间速度的作用：

$$W = \sqrt{\left(\overline{W_x} + W_x'\right)^2 + W_y'^2 + kW_z'^2} \tag{83}$$

系数 $k<1$，因为热电丝和 $\overline{W_x}$ 方向相垂直，因而也和 W_y' 相垂直，而在 z 向则与热电丝平行，由公式(21)及图11可知，此时相应于 $\varphi=90°$，因而 $k \approx a^2 < 1$，即单热电丝不能全部感受 z 向的速度脉动值。但另一方面，由放大器出口处用真空管热电偶测量所得的是平方平均值：

$$\overline{W'^2} = \overline{\left(W - \overline{W}\right)^2} = \overline{W^2} - 2\overline{W}\,\overline{W} + \left(\overline{W}\right)^2 = \overline{W^2} - \left(\overline{W}\right)^2 \tag{84}$$

因为 $\qquad 2\overline{W}\,\overline{W} = 2\overline{\left(\overline{W} + W'\right)\overline{W}} = 2\overline{\overline{W} \cdot \overline{W}} + 2\overline{\overline{W} \cdot W'} = 2\left(\overline{W}\right)^2$

把式(83)展开：

$$W = \overline{W_x}\left[\left(1 + \frac{W_x'}{\overline{W_x}}\right)^2 + \frac{W_y'^2}{\left(\overline{W_x}\right)^2} + k\frac{W_z'^2}{\left(\overline{W_x}\right)^2}\right]^{\frac{1}{2}}$$

$$= \overline{W_x}\left[1 + \frac{1}{2}\left(\frac{2W_x'}{\left(\overline{W_x}\right)^2} + \frac{W_x'^2}{\left(\overline{W_x}\right)^2} + \frac{W_y'^2}{\left(\overline{W_x}\right)^2} + k\frac{W_z'^2}{\left(\overline{W_x}\right)^2}\right)\frac{1}{8}\left(\frac{4W_x'^2}{\left(\overline{W_x}\right)^2} + \cdots\right) + \cdots\right]$$

$$= \overline{W_x}\left[1 + \frac{W_x'}{\overline{W_x}} + \frac{1}{2}\left(\frac{W_y'^2 + kW_z'^2}{\left(\overline{W_x}\right)^2}\right) + \cdots\right]$$

由此可见，瞬间速度的平均值

$$\overline{W} = \overline{W}_x \left[1 + \frac{1}{2}\left(\frac{\overline{W_y'^2} + k\overline{W_z'^2}}{\left(\overline{W}_x\right)^2} \right) + \cdots \right]$$

另一方面由式(83)可知，瞬间速度的平方平均值等于：

$$\overline{W^2} = \overline{\left(\overline{W}_x\right)^2 + 2\overline{W}_x \cdot W_x' + W_x'^2 + W_y'^2 + kW_z'^2}$$

$$= \left(\overline{W}_x\right)^2 \left(1 + \frac{\overline{W_x'^2}}{\left(\overline{W}_x\right)^2} + \frac{\overline{W_y'^2}}{\left(\overline{W}_x\right)^2} + \frac{k\overline{W_z'^2}}{\left(\overline{W}_x\right)^2} \right)$$

由式(84)就可以得出单热电丝所能测出的脉动速度值为：

$$\overline{W'^2} = \overline{W^2} - \left(\overline{W}\right)^2 = \left(\overline{W}_x\right)^2 \left[\left(1 + \frac{\overline{W_x'^2} + \overline{W_y'^2} + \overline{kW_z'^2}}{\left(\overline{W}_x\right)^2} \right) - \left(1 + \frac{\overline{W_y'^2} + \overline{kW_z'^2}}{\left(\overline{W}_x\right)^2} + \cdots \right) \right] = \overline{W_x'^2} \quad (85)$$

上式中，在计算 $\left(\overline{W}\right)^2$ 项时略去了 $\frac{W_y'}{\overline{W}_x}$ 等比值四次方以上各项，因此，当热电丝与一元气流方向（x 方向）垂直时，可直接测出 x 向的脉动值，至于其他方向脉动分量的求法，我们将在下一节中提及。

§12. 二元及三元气流紊流脉动参数的决定

目前对于二元及三元气流中紊流脉动参数的测量问题在一般文献中只有原则上提及，至于如何具体实现以及二元，三元气流脉动参数的试验数据均很少发现，但在工程实用中，二元及三元气流是最经常遇到的，这种情况迫使我们对此问题进行比较详细的讨论。

1) Г.С.Антонова 提出的单热电丝旋转法

Г.С.Антонова 曾对于二元气流紊动参数进行过详细测量(文献30)，她是利用单热电丝感应元件在气流中旋转三次(与气流方向形成三个不同的方位角)得出三个衡量，然后经过转换求得平面气流的脉动值。

如果把单丝热电元件放在 xoy 平面内，在该平面内可以任意旋转三次，与 x 轴方向形成某一 $\beta°$ 角，(见图41)。取平面 N 垂直于热电丝，并称之为主要平面，因为热电丝在此方向上感受脉动作用最大，那么按照图41的投形原理，热电丝在主要平面内所感受的二元平均气流及三元脉动气流的量为：

$$W_N = \sqrt{\left[\left(\overline{W}_x + W_x'\right)\cos\beta \pm \left(\overline{W}_y + W_y'\right)\sin\beta \right]^2 + W_z'^2} \quad (85)$$

这里正负号取决于热电丝与 x 轴位置。

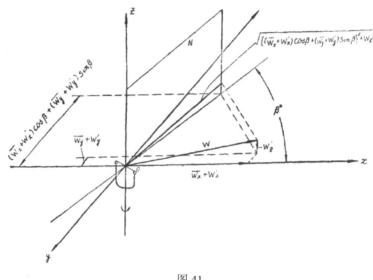

图 41

可见在主要平面 N 上所感受的三元脉动量为：

$$W'_N = W_N - \overline{W_N} = \sqrt{\left[\left(\overline{W}_x + W'_x\right)\cos\beta \pm \left(\overline{W}_y + W'_y\right)\sin\beta\right]^2 + W'^2_z}$$
$$- \overline{\sqrt{\left[\left(\overline{W}_x + W'_x\right)\cos\beta \pm \left(\overline{W}_y + W'_y\right)\sin\beta\right]^2 + W'^2_z}}$$

把上式平方然后平均之可得：

$$\overline{W'^2_N} = \overline{\left(W'_x\cos\beta \pm W_y\sin\beta\right)^2} + \overline{\left(\overline{W}_x\cos\beta \pm W_y\sin\beta\right)} + W'^2_z$$
$$- \overline{\sqrt{\left\{\left[\left(\overline{W}_x + W'_x\right)\cos\beta \pm \left(\overline{W}_y + W'_y\right)\sin\beta\right]^2 + W'^2_z\right\}^2}} \tag{86}$$

方程式是很复杂的，为了得出近似解答，Г.С.Антонова 就非常粗糙的假设：

$$\overline{W'^2_N} = \overline{\left(W'_x\cos\beta \pm W'_y\sin\beta\right)^2} \tag{87}$$

亦即抛弃了式(86)右边除第一项外以后各项，为了证明(87)式是比较近似的，她进行了下列计算,计算时设各脉动分速为平均速度的 7%同时各脉动分速以同一频率按正弦规律变化，计算证明在 A) $\overline{W_x} = \overline{W_y}$ 及 $\beta=20°$，Б) $\overline{W_x} = \overline{W_y} = 0$，$\beta=20°$情况下抛掉(86)式后面各项所引起的误差为 15%~20%。

有了式(87)这样简单的形式后，就可能 xoy 平面旋转热电丝三个不同的角度 $\beta = \beta_1$；$\beta_2 = 0$ 及 $\beta_3 = -\beta_1$，这样可以组成三个方程：

$$\left.\begin{array}{l}\overline{\left(W_{N}^{\prime 2}\right)_{1}}=\overline{\left(W_{x}^{\prime}\cos\beta_{1}+W_{y}^{\prime}\sin\beta_{1}\right)^{2}}\\ \overline{\left(W_{N}^{\prime 2}\right)_{2}}=\overline{W_{x}^{\prime 2}}\\ \overline{\left(W_{N}^{\prime 2}\right)_{3}}=\overline{\left(W_{x}^{\prime}\cos\beta_{1}-W_{y}^{\prime}\sin\beta_{1}\right)^{2}}\end{array}\right\} \quad (88)$$

如果测量得在 β_1、β_2、β_3 角度下的 $\overline{\left(W_{N}^{\prime 2}\right)_{1}}$、$\overline{\left(W_{N}^{\prime 2}\right)_{2}}$ 及 $\overline{\left(W_{N}^{\prime 2}\right)_{3}}$ 值后，解方程式(88)，即可求出 $\overline{W_{x}^{\prime 2}}$、$\overline{W_{y}^{\prime 2}}$ 及 $\overline{W_{x}^{\prime}\cdot W_{y}^{\prime}}$ 值。

我们认为 Г.С.Антонова 提出的方法是不够完善的，而且有很多缺点，其中值得指出的是：

А) 没有提出决定二元平均气流的方法，而在测量脉动速度时又必须预先测量出二元平均气流的方向及大小，使测量复杂而费时。

Б) 在导出公式(85)时，没有考虑某一脉动分速与热电丝平行时，热电丝不能全部感受这个脉动分速的作用。

В) 最终式(87)是极端近似的，而且把(86)式后面几项抛弃是没有根据的，证明误差只有15%～20%的计算例子也是不够全面，因而式(87)的准确度无法估计，即使设理论上误差只有15%～20%，再加上热惯性补偿不足或过度，气流脉动值和放大器出口电参数脉动值之间理论联系不够准确等等误差，则由此法测量得的数据是难以判断所研究过程的实质。

Г) 在运行中要转动热电感应元件三个角度，一方面要花费很多时间，另一方面一点的瞬时脉动值随时间变化的，因此在不同时间内对同一点测量脉动值对热电丝主要平面的投影所得出的结果可能与该点真正值有所不同，因此对测量脉动参数，最好能在同一瞬间(或同一时间间隔内)完成。

Д) 采用此法最多能测量二元脉动气流。

2) V 型热电元件测量法。

为了克服上述缺点，我们提出了用 V 型热电感应元件测量二元平均气流及三元脉动速度(文献19)，V 型元件测量二元平均气流的方法在§2中已有详述，下面我们将着重讨论一下如何测量三元脉动速度。

由紊流运动理论可知，平均气流和脉动气流的方向往往并不重合，因此这里不但要决定脉动速度的大小，而且要决定其方向。

V 型元件由热电丝 A、B 组成其夹角为 40°，A 丝的主要平面为 OC，B 丝的主要平面为 OD，(见图42) V 型元件满在 xoy 平面内，脉动速度 W' 与 x 轴成 α 角作用在 V 型元件上，先把 W' 分成 W_x' 及 W_y'，使后由这些脉动分速分别投影在 A、B 丝的主要平面内，在列出方程式之前，先提请注意的是在我们所采取的测量系统中，平均参数的测量在电桥中完成，而脉动参数的测量则在放大器及后面的测量仪表中完成(见图43)，因此平均及脉动参数的测量是互相分开互不影响(见§9第一点所述原理)。A、B 热电丝所感受到气流脉动的作用在放大器出口处可以得出三个信号(见图42)：

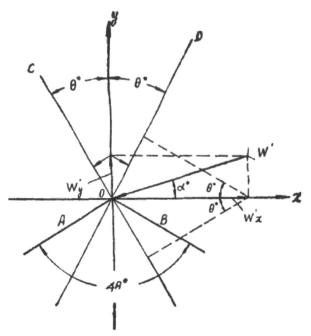

图 42　测量气流脉动参数 V 型感应元件工作图

由 A 热电丝：

$$\sqrt{\overline{W_A'^2}} = \sqrt{\overline{\left[a\sqrt{W_x'^2 + W_y'^2} + bW_y'\cos\theta - bW_x'\sin\theta\right]^2}} \tag{89}$$

由 B 热电丝：

$$\sqrt{\overline{W_B'^2}} = \sqrt{\overline{\left[a\sqrt{W_x'^2 + W_y'^2} + bW_y'\cos\theta + bW_x'\sin\theta\right]^2}} \tag{90}$$

由 A、B 二丝所感受的信号差值：

$$\sqrt{\overline{\left(W_B' - W_A'\right)^2}} = \sqrt{\overline{\left(2bW_x'\sin\theta\right)^2}} \tag{91}$$

在引入这些方程式时没有考虑到 W_z' 对两热电丝的作用，关于 W_z' 值对测量的影响，我们在后面将有详细的讨论。

假设

$$\sqrt{W_x'^2 + W_y'^2} = H\left(W_x' + W_y'\right) \tag{92}$$

这里 H 由 $\dfrac{W_y'}{W_x'}$ 比值所决定的系数。把式(92)代入式(89)及(90)，并取其平方值可得：

$$\overline{W_A'^2} = a^2H^2\overline{W_x'^2} + a^2H^2\overline{W_y'^2} + b^2\overline{W_x'^2}\sin^2\theta + b^2\overline{W_y'^2}\cos^2\theta$$
$$+ 2a^2H^2\overline{W_x'\cdot W_y'} + 2abH\overline{W_y'^2}\cos\theta + 2abH\overline{W_x'W_y'}\cos\theta$$
$$- 2abH\overline{W_x'\cdot W_y'}\sin\theta - 2abH\overline{W_x'^2}\sin\theta - 2b^2\overline{W_x'\cdot W_y'}\sin\theta\cos\theta \tag{93}$$

$$\overline{W_B'^2} = a^2H^2\overline{W_x'^2} + a^2H^2\overline{W_y'^2} + b^2\overline{W_x'^2}\sin^2\theta + b^2\overline{W_y'^2}\cos^2\theta$$
$$+ 2a^2H^2\overline{W_x'W_y'} + 2abH\overline{W_y'^2}\cos\theta + 2abH\overline{W_x'W_y'}\cos\theta$$
$$+ 2abH\overline{W_x'W_y'}\sin\theta + 2abH\overline{W_x'^2}\sin\theta + 2b^2\overline{W_x'\cdot W_y'}\cos\theta\sin\theta \tag{94}$$

及
$$\overline{(W_B' - W_A')^2} = 4b^2\overline{W_x'^2}\cdot\sin^2\theta \tag{95}$$

解上面三个方程式，经过不复杂的变换我们可以得出平面脉动参数的最终解：

$$\sqrt{\overline{W_x'^2}} = \left[\frac{\overline{(W_B' - W_A')^2}}{(2b\sin\theta)^2}\right]^{\frac{1}{2}} \tag{96}$$

$$\sqrt{\overline{W_y'^2}} = \left[\frac{\left(\overline{W_A'^2} + \overline{W_B'^2}\right)4L^2\sin^2\theta - \overline{(W_B' - W_A')^2}\left(2b^2\sin^2\theta - 2a^2H^2\right) - \left(\overline{W_B'^2} - \overline{W_A'^2}\right)4abH\sin\theta}{\left(2a^2H^2 + 2b^2\cos^2\theta + 4abH\cos\theta\right)4b^2\sin^2\theta}\right]^{\frac{1}{2}}$$
$$\tag{97}$$

$$\overline{W_x'\cdot W_y'} = \frac{\left(\overline{W_B'^2} - \overline{W_A'^2}\right)b\sin\theta - aH\overline{(W_B' - W_A')^2}}{4b^2\sin^2\theta(aH + b\cos\theta)} \tag{98}$$

对于我们所使用的 V 型感应元件其夹角为 120°，即 $\theta=30°$，同时假设脉动速度均具有同一数量级，即 $\sqrt{\overline{W_x'^2}} \approx \sqrt{\overline{W_y'^2}}$，此时 $H=0.7$，并根据公式(22)得 $a=0.2$, $b=0.8$ 代入上三式即可得出总结试验时常用的公式：

$$\sqrt{\overline{W_x'^2}} = 1.25\sqrt{\overline{(W_B' - W_A')^2}} \tag{99}$$

$$\sqrt{\overline{W_y'^2}} = \sqrt{0.921\overline{W_A'^2} + 0.444\overline{W_B'^2} - 0.299\overline{(W_B' - W_A')^2}} \tag{100}$$

$$\overline{W_x'\cdot W_y'} = 0.75\left(\overline{W_B'^2} - \overline{W_A'^2}\right) - 0.263\overline{(W_B' - W_A')^2} \tag{101}$$

上述三个参数是当V型感应元件在xoy平面上时所测量得出，如果我们把元件放在yoz平面内，即可求出$\sqrt{W_y'^2}$、$\sqrt{W_z'^2}$及$\overline{W_y' \cdot W_z'}$值，由此可见采用这种测量方法能够决定三元脉动参数，并且，被我们顺利地应用到自由射流，绕非流线型物体及旋转气流的测量中，下面，我们讨论一下这个方法的准确度问题：

A）我们的假设脉动速度具有同一数量级，可取$H=0.7$的准确性。

上面已经说明H值由比值$\sqrt{W_y'^2}/\sqrt{W_x'^2}$决定的，实际上，准确的$H$值可以由连续近似法求得，即先任意设出一$H_1$值求出$\sqrt{W_y'^2}$及$\sqrt{W_x'^2}$第一次近似值后，再反求第二次近似$H_2$值，再代入式(96)、(97)中算出第二次近似的$\sqrt{W_y'^2}$及$\sqrt{W_x'^2}$值，直至得到准确的$H$值为止，但这样总结试验的方法是非常麻烦的，因此我们仍取$H=0.7$，并计算出由此而可能产生的误差，计算时取$\sqrt{W_x'^2}=20$米/秒，$a=0.12$，$b=0.88$，$\varphi=30°$，计算是按公式(89)、(90)、(91)、(97)及(98)完成的，计算结果列于表6，表中H_H表示在比值$\sqrt{W_y'^2}/\sqrt{W_x'^2}$下真正的$H$值，$\Delta\sqrt{W_y'^2}$及$\Delta\overline{W_x' \cdot W_y'}$为采用$H=0.7$所产生的误差，而由式(96)可知，$\sqrt{W_x'^2}$与$H$无关，故不产生误差。

表6

$\sqrt{W_y'^2}/\sqrt{W_x'^2}$	2	1.8	1.6	1.4	1.2	1.0	0.8	0.6	0.4	0.2
H_H	0.745	0.735	0.725	0.716	0.71	0.7	0.712	0.73	0.768	0.85
$\Delta\sqrt{W_y'^2}$ %	1.51	1.48	1.34	1.10	1.1	0	1.3	1.3	2.83	5.5
$\Delta\overline{W_x' \cdot W_y'}$	1.21	0.6	0.33	0.26	0.35	0	0.33	1.23	3.7	15.1

我们的试验证明，紊流脉动比值$\sqrt{W_y'^2}/\sqrt{W_x'^2}$经常在$1.6\sim0.6$范围内变动，因此，选择$H$为常数(0.7)来总结试验所引起的误差不大于1.5%。

Б）第三脉动分速对测量准确度的影响问题

在列出方程式(89)、(90)及(91)时，没有考虑脉动分速W_z'对热电丝的作用，我们认为W_z'的影响不大，而且可以略去，这里和Г.С.Антонова的单丝测量不同，W_z'以同样力量作用在V型感应元件的两根丝上，因此，如果我们把两根丝的出口电参数相减，此时W_z'对两丝的作用刚好大小相等，方向相反而自动被消去，故严格来说，公式(96)和(98)是准确的。因为它们只包含有电参数的差值信号$\overline{(W_B'-W_A')^2}$，而公式(97)则是近似的，因为此式包含有电参数的独立项，W_z'的影响无法去除，故由式(97)所求出的$\sqrt{W_y'^2}$值比真正的$\sqrt{W_y'^2}$值高。为了检查式(97)的准确程度，我们进行了专门的试验，试验是在自由射流中完成的，首先把 V 型元件放在 xoy 平面上，按式(96)，(98)准确决定 $\sqrt{\overline{W_x'^2}}$ 及 $\overline{W_x' \cdot W_y'}$ 值，由式(97)近似决定 $\left(\sqrt{\overline{W_y'^2}}\right)_\Pi$ 值，之后，把 V 型元件改放至 yoz 平面，此时可以准确决定 $\left(\sqrt{\overline{W_y'^2}}\right)_И$ 及 $\overline{W_y' \cdot W_z'}$ 值，近似决定 $\sqrt{\overline{W_z'^2}}$ 值，然后比较 $\left(\sqrt{\overline{W_y'^2}}\right)_\Pi$ 及 $\left(\sqrt{\overline{W_y'^2}}\right)_И$ 之差，即可估计出式(97)的近似程度，试验是在自由射流初始段上进行，因为考虑到这里平均气流速度能保持为常值(20 米/秒)，试验结果列于表 7，表中 $\Delta\sqrt{\overline{W_y'^2}}$ 即为公式(97)的误差程度，r/R —射流横断面位置。

表 7

r/R	0	0.25	0.5	0.75
$\left(\sqrt{\overline{W_y'^2}}\right)_\Pi$ 米/秒	0.948	1.105	1.23	1.346
$\left(\sqrt{\overline{W_y'^2}}\right)_И$ 米/秒	1.03	1.13	1.3	1.49
$\Delta\sqrt{\overline{W_y'^2}}$ %	8.0	2.2	5.4	3.3

虽然测量是在不同时间下进行的(对于同一点，不同平面一次测量是无法完成的)，但试验证明，当测量时间不是很短，而真空管热电偶又有足够大的惯性时，试验结果是能够重复的，例如在自由射流和旋转气流中，在同一工况下我们在不同时间进行测量所得结果是不变的(在准稳定情况下)，因此，可以认为，这样的试验能近似得出公式(97)的误差值，由表 7 中可见，式(97)正如上面所述一样给出了比真正值较高的数据，其"提高部份"就是脉动分速 W_z' 的作用结果，但试验证明 W_z' 的作用而产生的误差平均约为 5%，这在工程中是可以容许的，当然亦可以在不同平面旋转两次的办法进行测量以得出准确的各脉动分量，但会使操作过程极端复杂化，而且研究对象也未必容许在不同平面旋转两次。和 Г.С.Антонова 法比较，我们的方法显示出下列的优越性：

А) 平面气流平均及脉动参数在一次测量中即可全部获得，在不等温气流中，也可以同时把该点气流温度求出。

Б) 理论公式准确度高，只有决定 $\sqrt{\overline{W_y'^2}}$ 值才出现 5%误差，同时能够求出三元各脉动值，特别是在紊流参数很难直接校正的情况下，理论公式的精确度就显得特别重要。

В) 测量手续比较简单，V 型感应元件可固定在某一位置不加变动，把操作中可能出现的误差减至最小，因为操作过程中要准确转动某一角度是不容易的。

下面讨论一下测量如何具体实现的问题，整个仪表装置线路系统示于图 43，整个测量系统可以分为三个主要部份：敏感元件、测量气流平均参数的设备、测量气流脉动参数的设备，作为感应元件正如上述用 V 型互成 120° 的钨热电丝直径为 19 微米，长 4.5 毫米测量平均参数的电桥设备已在§5 中述及，测量紊流参数主要在能补偿热惯性的放大器中完成，$\sqrt{\overline{W_A'^2}}$、$\sqrt{\overline{W_B'^2}}$、$\sqrt{\overline{(W_B'-W_A')^2}}$ 三个信号的获得主要在信号加减盒中完成，（见图 43），脉动信号由两电桥对角线取出，使后输入信号加减盒中，在它内部有一转动开关，能够分别接通 A、B 或 C 线路（见图 13），当放大器和 A 线路接通时，可测量出信号 $\sqrt{\overline{W_A'^2}}$，与 C 线路接通时得信号 $\sqrt{\overline{W_B'^2}}$，在 B 线路上则可得 $\sqrt{\overline{(W_A'-W_B')^2}}$ 值，由于放大器具有较高的放大系数，同时所输入的脉动信号又足够小，因此外界的"噪音"（外界电磁场的影响）对放大器的工作影响很大，试验证明各连接线路若不采用有保护网的导线时，放大器工作极不稳定，甚至"噪音"量比输入脉动信号还大，使测量无法进行，因此所有导线均必须有金属网仔细保护，放大器中有两个调节电阻，第一个用来调节放大系数，第二个是用来调节补偿热惯性的时间常数，放大器中有两个出口，在第一个出口接上电子示波器和真空管伏特计，前者是用来监示 V 型感应元件及放大器工作是否正常；测量出来的是气流脉动值还是线路中的"噪音"，如果测量系统突然受到外界电磁场影响，"噪音"很大，则在电子示波器中立刻出现很不规则的电压变化，测量应马上停止，进行检查校正。真空管伏特计是用来选择适当的放大系数，放大系数过大时，一方面放大器

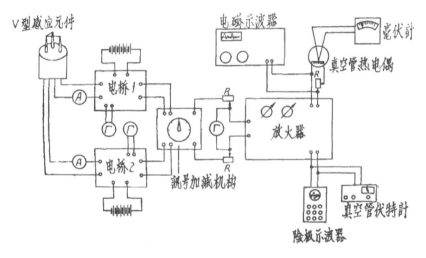

图 43　测量仪表系统图

处于非线性段工作是不容许的，另一方面会把真空管热电偶烧毁。在放大器第二个出口接上电磁（振子）示波器及真空管热电偶与毫伏计，前者是用来记录气流脉动的瞬间值，由此可以测出脉动平均频率，后者是得出脉动参数的平方平均开方值，真空管热电偶是一个精密的热电转换元件，很容易烧毁，因此在运行中必须细心保护，特别要注意的是外线路及放大器中的突然"噪音"以及选择放大系数不能过大（一般在 0~60 伏间）。

现在迫切的问题在于找出气流脉动参数与电参数之间的联系，否则测量出的数据无法总结，如果认为气流脉动与电参数脉动成线性关系是正确的话，（见图 29，及§7 第一点）并考虑到热电丝热惯性的作用，根据公式（19）可得热电丝在具有速度及温度脉动气流

中真正的电阻（文献 31）：

$$R_{T_H} = \overline{R_{T_H}} \sin\varphi\tau = \overline{R_T}\left\{1 + A_\rho \Delta T_0 \left[\frac{T'}{\Delta T_1} \frac{1}{\sqrt{1+\Omega^2 M^2}}\sin(\Omega\tau - \xi)\right.\right.$$
$$\left.\left. - \frac{\alpha_1}{\alpha}\frac{1}{\sqrt{1+\omega^2 M^2}}\sin(\omega\tau - \beta)\right]\right\} \tag{102}$$

这里 $\overline{R_T}$ 为被测气流温度下热电丝的电阻；ΔT_0 为热电丝基准加热温度，在测量中，我们固定取 $\Delta T_0 = 150°C$；T' 及 Ω——温度脉动的振幅及角系数；α'——由气流脉动所引起对流传热系数的脉动值；ξ 及 β 为由于热惯性作用，热电丝感应温度及速度脉动时的滞后相角。如果我们研究的是等温过程，即没有温度的脉动，同时对流传热系数的脉动可与气流速度脉动值作出这样的联系：

$$Nu = cR_e^n$$

即

$$\frac{\alpha d}{\lambda} = c\frac{w^n d^n}{\nu^n}$$

当传热系数变化 $d\alpha$ 时，相应于气流速度变化 dw 值。

即

$$\frac{d\alpha \cdot d}{\lambda} = cdw \cdot n\frac{w^{n-1} d^n}{\nu^n}$$

和上式相比可得

$$\frac{d\alpha}{\alpha} = n\frac{dw}{w}$$

若气流脉动不是很大时可看作：

$$\frac{d\alpha}{\overline{\alpha}} = n\frac{dw}{\overline{w}} \tag{103}$$

把 (103) 式代入 (102)，并考虑到热惯性在放大器中得到准确补偿，对于等温气流电阻的脉动值可以写为：

$$R'_{T_H} = A_\rho \Delta T_0 \overline{R_T} \cdot K_0 \cdot n \frac{\sqrt{w'^2}}{\overline{w}}$$

按 §5 所述原理通过电桥电流不因气流脉动而有所变化，则在放大器出口能够测出的脉动电压值为：

$$V'_H = iR'_{T_H} = iA_\rho \Delta T_0 \overline{R_T} \cdot K_0 \cdot n\frac{\sqrt{w'^2}}{\overline{w}}$$

或

$$\sqrt{w'^2} = \frac{V'_H \cdot \overline{w}}{A_\rho R_T K_0 \cdot n \cdot i\Delta T_0} \tag{104}$$

这里 k_0 为放大器的放大倍数。

因此，由式(104)即可从试验得出公式(96)、(97)、(98)所要求的 W'_A、W'_B 及 $(W'_B - W'_A)$ 值，至于具体总结试验的步骤及过程，我们将示于附录表中。

3) 关于三元平均气流三元脉动速度原理的讨论。

利用上述 V 型元件测量二元气流原理并加以发展，完全可用于三元气流的测量，只不过此时所采用的感应元件正如§2 中讨论一样应为棱形(见图 16)，由棱形感应元件 A.B.C 三热电丝根据公式(21)可列出三个方程：

$$\overline{W_A} = \overline{W}(a + b\cos\varphi)$$

$$\overline{W_B} = \overline{W}\left[a + b\cos(60° - \varphi)\right]$$

$$\overline{W_C} = \overline{W}(a + b\cos\alpha)$$

φ 及 α 为平均矢量气流的方位角。

在电桥中测量出 $\overline{W_A}$，$\overline{W_B}$，$\overline{W_C}$ 平均值后，解方程式组求出三个未知数：φ、α 及 $\overline{W}$。此外，由三个热电丝应该可以得出 6 个脉动信号，即

$\sqrt{\overline{W'^2_A}}$、$\sqrt{\overline{W'^2_B}}$、$\sqrt{\overline{W'^2_C}}$、$\sqrt{\overline{(W'_B - W'_A)^2}}$、$\sqrt{\overline{(W'_C - W'_B)^2}}$ 及 $\sqrt{\overline{(W'_A - W'_C)^2}}$，这样可以组成类似式(89)、(90)、(91)等 6 个方程式组，解之即可得出所求的六个脉动参数：

W'_x、W'_y、W'_z、$\overline{W'_x \cdot W'_y}$、$\overline{W'_y \cdot W'_z}$ 及 $\overline{W'_x \cdot W'_z}$，这样在理论上不会有任何困难，在测量上只需一次即能完成，只不过此时线路系统更为复杂。

§13. 其他紊流参数的测量问题

1) 空间关联系数及紊流标尺的测量

由式(42)可知要求出空间关联系数必须要知道同一方向不同座标点间脉动值的流计联系，如 $\overline{W'_{x1} \cdot W'_{x2}}$ 等等。测量可以用两个独立的单热电丝感应元件完成，如我们把两热电丝放在与气流相垂直的方向，按照§11 所述原理，则这两根丝只感受纵向脉动分速的作用，如果我们把这两根热丝出口信号相加或相减，则可得出：

$$W'_A = \overline{(W'_{x1} + W'_{x2})^2} \text{ 及 } W'_B = \overline{(W'_{x1} - W'_{x2})^2}$$

进行简单运算： $W'_A - W'_B = 2\overline{W'_{x1} \cdot W'_{x2}}$，$\overline{W'_A + W'_B} = \overline{(W'^2_{x1} + W'^2_{x2})}$

两式相除得

$$\frac{W'_A - W'_B}{W'_A + W'_B} = \frac{2\overline{W'_{x1} \cdot W'_{x2}}}{\overline{W'^2_{x1}} + \overline{W'^2_{x2}}} \tag{105}$$

如果脉动值随坐标位置变化不大，可近似取 $\overline{W'^2_{x1}} \approx \overline{W'^2_{x2}} = \overline{W'^2_x}$

即

$$R_{x_{1,2}} = \frac{W'_A - W'_B}{W'_A + W'_B} = \frac{\overline{W'_{x1} \cdot W'_{x2}}}{\sqrt{\overline{W'^2_{x1}}} \cdot \sqrt{\overline{W'^2_{x2}}}} = \frac{\overline{W'_{x1} \cdot W'_{x2}}}{\overline{W'^2_x}} \tag{106}$$

严格来说，上述公式只能在热电丝长度为零时才正确，对于热电丝长度为 l 值时，则此时测量出不是在一点的数值，而是在 $y_1 - y_2 = l$ 一段长度内的数值，因此必须引入修正，修正的方法第一次为 H.L.Dryden 及 G.Schubauer 等提出的（例如可在文献 29 中找到），修正系数值取决于空间关联系数的曲线形状及气流的紊流标尺值 L，作者所得出的结果是：

对于平方平均值修正系数为 k_1：

$$k_1^2 = \left(\overline{W'^2_x}\right)_{\text{ИЗМ}} / \left(\overline{W'^2_x}\right)_\text{И} = \text{用 } l \text{ 长热电丝测出值/真正值}$$

对于紊流标尺的修正系数为 k_2：

$$k_2 = (L)_{\text{ИЗМ}} / (L)_\text{И} = \frac{1}{L} \int_0^\infty R'_{x_{1,2}} \mathrm{d}x$$

$(L)_{\text{ИЗМ}}$ 及 $(L)_\text{И}$ 为用 l 长热电丝测出的及气流真正的紊流标尺。

根据试验结果，作者所算出的修正值列于表 8。

表 8

l/L	0	0.4	0.8	1.2	1.6	2.0	2.4	2.8	3.2
k_1	1.0	1.067	1.133	1.198	1.263	1.327	1.39	1.451	1.512
k_2	1.0	1.105	1.182	1.241	1.289	1.327	1.359	1.384	1406

或可写成下列形式：

$$k_1 = \frac{l}{L} \Big/ \sqrt{2\left(e^{-\frac{l}{L}} - 1 + \frac{l}{L}\right)}$$

和

$$k_2 = 1 + 0.3\frac{l}{L} - 0.05\left(\frac{l}{L}\right)^2 \qquad (\text{当 } \frac{l}{L} \leqslant 1 \text{ 时})$$

值得指出的是当热惯性补偿不够时影响空间关联系数测量准确度甚大。此时测出值比真正值来得高，若两根热电丝由于制造上的困难，电阻不一定相等。（例如长度不相同等），但这对测量准确度影响较小，例如由于两根丝电阻不同，等值于测量出 W'_{x1} 比 W'_{x2} 大 k 倍。则：

$$W'_A = \overline{(kW'_{x1} + W'_{x2})^2} = \left(k^2 \overline{W'^2_{x1}} + \overline{W'^2_{x2}} + 2k\overline{W'_{x1}W'_{x2}}\right)$$

$$W'_B = \overline{(kW'_{x1} - W'_{x2})^2} = \left(k^2 \overline{W'^2_{x1}} + \overline{W'^2_{x2}} - 2k\overline{W'_{x1}W'_{x2}}\right)$$

于是
$$\frac{W'_A - W'_B}{W'_A + W'_B} = \frac{2k}{k^2 + 1} \frac{\overline{W'_{x1} \cdot W'_{x2}}}{\overline{W'^2_x}}$$

即使偏差达 100%，$k=0.8$，$\frac{2k}{k^2+1} = 0.976$ 即误差仅为 2.4%。

详细的线路结构可在文献 1 中找出。为了得出紊流标尺 L，必须先求得空间关联系数与坐标的关系。因此，这两个独立的单线元件要在不同的距离进行测量，一直到两点之间互相没有统计联系为止（此时空间关联系数等于零）。因此要得出某一点的紊流标尺值必须进行大量的试验。特别困难的是：当改变两热电丝距离时，两点间紊流强度及平均速度均不相等。这些问题在测量理论和测量方法上都还未能得到解决。В.П.Солнцев 对自由射流中的空间关联系数及紊流标尺进行过详细的测定（文献 32）。试验证明，所得数据的准确度是不够的。因为严格来说上述方法只是在一元流动，沿长度脉动强度不变的情况下才是准确的，至于对二元及三元气流，目前在理论上还未找出很完满的方法，更不用说测量系统了。

2) 脉动频谱的测定

脉动频谱的决定理论上可以由示波器得出的脉动波形图进行分析，按展开傅里叶级数办法，把它分成一次及高次谱波，但是在实用上这是非常麻烦和复杂的事情。一般均采用频谱分析仪进行分析。频谱分析仪是由一系列性质不同的滤波器组成。每个滤波器只能容许频率为一定的脉动值通过。滤波器的组数愈多，频谱的分析愈准确。例如 E.C.Семенов 采用了五组滤波器（文献 33），而 В.П.Солнцев 则采用 16 组之多（文献 32），在这些文献中还给出了详细的线路图，由于频谱分析仪的结构是极端复杂的，而且准确度也不是非常高。因此 E.M.Минский 建议采用求平均脉动频率的方法（文献 5），他认为平均频率的数值近似等于由脉动示波图中脉动曲线通过零点的数目来决定。因为这实质上等于傅里叶级数一次波谱的频率，而一次波谱是脉动的主要组成部分，求出平均脉动频率后就有可能按下式换算出紊流标尺值（文献 5、32）：

$$L = k \cdot \frac{\overline{W}}{\omega_{OP}} \tag{107}$$

这里 k 为试验常数。

3) 温度脉动和速度脉动之间的联系的测定。

在不等温过程的流动中，要研究紊流脉动不可避免要同时出现温度及速度的脉动，用热电风计测量时能够区分开温度及速度的脉动以及它们之间的联系。当同时存在有温度及速度的脉动时，气流脉动与电参数脉动之间的联系应使用公式(102)。

此时温度与速度脉动的共同结果可写成：

$$\sqrt{\overline{\left(\frac{T'}{\Delta T_1} - n\frac{W'}{\overline{W}}\right)^2}} = \sqrt{\frac{\overline{T'^2}}{\Delta T_1^2} - 2n\frac{\overline{T'W'}}{\Delta T_1 \cdot \overline{W}} + n^2 \frac{\overline{W'^2}}{\overline{W}^2}} \tag{108}$$

如果把速度、温度关联系数式(45)代入可得:

$$\sqrt{\overline{\left(\frac{T'}{\Delta T_1} - n\frac{W'}{\overline{W}}\right)^2}} = \sqrt{\frac{\overline{T'^2}}{\Delta T_1^2} - 2nR'_W\sqrt{\frac{\overline{T'^2}}{\Delta T_1^2}}\sqrt{\frac{\overline{W'^2}}{\overline{W}^2}} + n^2 \frac{\overline{W'^2}}{\overline{W}^2}} \tag{109}$$

为了把温度和速度脉动分开。Г.C.Антонова 建议对热电丝采用三个不同的加热度(文献 30 及 34), 即微加热工况、加热很大工况及中间加热工况, 因此可得出三个和(109)式类似的方程式, 就可解出三个未知数 $\sqrt{\overline{T'^2}}$、$\sqrt{\overline{W'^2}}$ 及 $R_{W'T'}$。对于平面气流, Г.C.Антонова 建议在加热最大工况下旋转热电丝三个角度。此时可以再得出和式(88)相类似的另外三个方程式, 六个方程式可以解出下列六个未知数: $\sqrt{\overline{W_x'^2}}$、$\sqrt{\overline{W_y'^2}}$、$\sqrt{\overline{T'^2}}$、$\overline{W_x' \cdot W_y'}$、$\overline{W_x' \cdot T'}$、$\overline{W_y' \cdot T'}$。详细的试验结果, 读者可以在文献 34 中找出, 值得指出的是采用这种方法的准确度是不高的(对等温气流我们已在§12 中进行过分析)。计算证明, 在决定 $\sqrt{\overline{W_x'^2}}$ 及 $\sqrt{\overline{T'^2}}$ 时误差为 15%, 决定 $\sqrt{\overline{W_y'^2}}$ 时达 20%。至于决定关联系数的准确度取决于其本身值的大小, 当关联系数较小时(在 0~0.2 范围)误差很大, 简直不能判断两脉动值之间是否存在有统计联系, 当关联系数值在 1 附近时, 误差约为 30%。

总的说来, 上述的方法是不能令人满意的, 今后必须找出更完善的方法。我们认为, 采用 V 型感应元件及三个不同的热电丝加热温度, 同样可以组成六个方程式, 同样可以解出上述六个未知数, 但按 §12 的分析, 采用 V 型元件其理论准确度高得多, 并且实现的可能性也大得多。我们认为这是今后发展方向之一, 是值得进行详细的试验研究的。

§14. 附录

试验数据按§12 中我们所提出的 V 型元件法整理的例子, 试验是在旋风炉中完成的, 进口风速为 50 米/秒。V 型两热电丝电阻分别为: $R_{TH1}=1.19$ 欧及 $R_{TH2}=1.21$ 欧, 这里只在准等势区及准固体区中各选取一点作为例子。

表 9

1	2	3	4	5	6	7	8	9	10
τ	i_1	W_1	i_2	W_2	$\frac{W_1}{W_2}$	φ	$\overline{W}$	W_x	W_y
厘米	安培	米/秒	安培	米/秒	/	度	米/秒	米/秒	米/秒
180	0.466	47.0	0.4585	42.0	1.12	23.0	50.2	49.8	6.12
10	0.38	14.0	0.384	15.0	0.93	34.5	16.3	16.0	-1.3

11	12	13	14	15	16	17	18	19	20
ψ	U_{T1}	$U_{\phi 1}$	ΠT_2	ΠT_2	$k_1 \cdot 10^{-3}$	W'_A	U_{T2}	$U_{\phi 2}$	ΠT_1
度	毫伏	伏	/	/	/	米/秒	毫伏	伏	/
7.0	3.0	30.6	18	9	3.89	2.84	1.5	21.0	18
−4.5	2.2	28.0	10	17	0.56	6.54	1.8	23.3	10

21	22	23	24	25	26	27	28	29	30
ΠT_2	$k_2 \cdot 10^{-3}$	W'_B	U_{T3}	$U_{\phi 3}$	$W'_B - W'_A$	$\sqrt{W'^2_x}$	$\sqrt{W'^2_y}$	$\sqrt{W'^2}$	$\sqrt{W'_x \cdot W'_y}$
/	/	米/秒	毫伏	伏	米/秒	米/秒	米/秒	米/秒	米²/秒²
12	3.35	2.01	3.6	33.8	3.24	4.05	2.47	4.74	6.78
17	0.56	5.68	2.5	27.9	6.8	8.5	6.32	10.6	20.0

31	32	33	34	35	36	37	38	39
ε'_W	ε'_{W_y}	ε'_{W_x}	$R_{W'_x W'_y}$	$\Delta P'_g$	f	J	$A \cdot 10^3$	l
%	%	%	/	%	赫兹	1/秒	公斤·秒/米²	毫米
9.4	4.9	8.1	0.57	0.88	223	712	1.37	4.01
65.0	38.8	52.2	0.37	42.3	175	5400	65.0	9.0

由表中可见，即使使用 V 型元件、平面紊流参数的试验总结也是很复杂的，对于总结气流某一点的紊流参数要求试验出 10 个项目计算出 29 项。如果我们要求测量出各截面的紊流场，则测量点数要达几百至几千点，总结工作量极为可观，故目前有利用自动总结的趋势（利用特殊电路直接用读出紊流参数）。（见文献 33 值）。

引用的文献

1. Пэнкхёрст Р.   Холдер Д.   风洞测量技术   М. 1955.
2. Фейдж Тоуненд   用放大镜研究紊流流动。
   在论文集 "Проблемы турбулентности" 一书内。1936. М.
3. Трубчиков Б.Я.   在风洞内测量紊流的"热方法"。
   ЦАГИ论文集, вып. 372. 1938.
4. Захаров Ю.Г.   ЖТФ. 1939 Т. IX   № 21.
5. Минский Е.М.   "河流的紊流"   1952. М.
6. Прудников А.Г.   用光学—扩散方法测量气流及火焰的紊流度:
   在论文集 "Горение в турбулентном потоке" 内。1959. М.
7. Westenberg A.   用氦扩散法测量火焰的紊流度 Jour. chem. phy. 1954. №5
8. Мхитарян А.И. 等.   ИФЖ.   1961 № 11.
9. Попов С.Г.   Измерение воздушного потока   1947. М.
10. Сергеев О.А.   "热电丝放热计算问题" 在ЛИТМО论文集 "Исследование в области тепловых измерений и приборов" вып. 21. 1957. Л.
11. Лойцянский Л.Г.   Шваб Б.А.   "紊流热标尺"
    Труды ЦАГИ. вып. 239. 1935
12. Кирпечев М.В. Эйгенсон Л.С.   "热力紊流计"   Изв. Энер. ин-та.
    АНСССР. Т. 4. вып. 1. 1936.
13. Martinelli and Boelter   "Heating, Piping and Air Conditioning 1939. N.8.
14. Physical Measurements in Gas Dynamics and Combustion 1955. 1957俄文版.
15. Михеев. М.А.   Основы теплопередачи 1949. М.
16. Захаров Ю.Г.   "用热电风计测量平均及脉动速度" труды ЦАГИ 1946
17. Кутателадзе С.С.   "Справочник по теплопередаче"   М. 1960.
18. Чебышев. П.В.   "为研究电机电器通风而拟制的热电风计"
    Отчет ВЭИ.   1949.
19. 岑可法.   "旋风燃烧室气流紊流结构的理论及试验研究" 论文。1962.
20. Чебышев П.В.   "热电风计" "Вестник электропромышленности"
    1952. N.1
21. E. Ower   The Measurement of Air Flow.   1933.
22. Захаров Ю.Г. Минский Е.М.   "用热电风计研究紊流"
    тех. заметки ЦАГИ. N. 172.   1938

23. Дульнев Г.Н. Сергеев О.А. "关于用牛导体热敏电阻测量气流速度问题" 同文献10论文集内。
24. Захаров. Ю.Г. 在ЦАГИ论文集 "Измерение воздушного потока" 1960. вып. 19.
25. Кречмер С.И. "试验决定热电风计热惯性" ДАН СССР. 1948 T.LXI. N.6. P.997.
26. Повх Н.Л. "Аэродинамический эксперимент в машиностроении". 1959. M.
27. Гольштик М.А. "ИФЖ" 1959 N.10
28. Апполонов Г.Ф. "用直流电具有反馈的热电风计" 在 "Техническая гидромеханика" труды ЛПИ. N.217 1961
29. Захаров Ю.Г. Минский М.Е. Филипов М.С. 在 Труды ЦАГИ N.402. 1939.
30. Антонова Г.С. "不等温自由射流紊流研究" 论文. 1957. M.
31. Чебышев П.В. "ВЭИ的热电风计及应用于不等温气流的研究" 在 "Труды совещания по прикладной газовой динамике" 1959.
32. Солнцев. В.П. "在自由射流核心的紊流参数研究"。 在 "Стабилизация пламени и развитие процесса сгорания в турбулентном потоке" 论文集内. 1961.
33. Семенов Е.С. "在活塞发动机内紊流测量仪表" "Приборы и техника эксперимега" 1958. N.1.
34. Антонова Г.С. 同文献31.的论文集内。

原文刊于浙江大学学报, 1963, 2: 165-225

劣质煤沸腾燃烧过程动力特性和双床并联运行沸腾炉提高燃烧效率的试验研究*

陈运铣　岑可法　张鹤声　张学宏　康齐福

(浙江大学)

摘　要

本文介绍了双床并联运行工业沸腾炉的结构原则，根据热平衡对比试验表明：这种沸腾炉在燃用劣质燃料时和单床沸腾炉相比，可以提高热效率16.6%。对燃用宽筛分劣质煤和燃用窄筛分的飞灰的沸腾床的燃烧过程动力特性作了试验测定，并对其结果进行了分析．从而提出在设计和运行上应该注意的问题和双床沸腾炉能够提高热效率的原因．

一、前　言

在近二十年内发展起来的，把化工方面的流态化原理应用到燃烧领域的沸腾燃烧方式，特别是燃用低热值劣质固体燃料引起了全世界的注意，已成为当今重要科研课题之一．我国燃用劣质煤的工业沸腾锅炉，在过去几年内投产的已近 2000 台，蒸发量超过 10000 吨/时．在利用劣质燃料促进工业生产，起了重要作用．但是目前存在二个严重缺点必须克服：(1)对劣质固体燃料沸腾燃烧的机理认识不足．(2)一般燃用劣质固体燃料沸腾燃烧锅炉的热效率较低，因而使我国沸腾炉的发展受到一定限制．

二、劣质煤沸腾炉双床并联运行提高燃烧效率的试验研究

挥发份少（2—8% 工作质）而灰份多（60—75% 工作质）的低热值（$4.2—10.5 \times 10^6$ J/kg 工作质)劣质固体燃料在一般工业锅炉中难以利用，现在广泛地在沸腾炉中得到燃用．但是这些锅炉的热效率较低，一般只有约 60%．

我国的沸腾炉通常燃用粒度为 0—8mm 的劣质固体燃料，其筛分太宽，在燃烧过程中为了维持激烈的沸腾状态，不得不采用较高的沸腾风速．因此，当粗颗粒燃料处于沸腾状态时，小于 1mm 或 0.5mm 的颗粒很大部分，甚至在床内还来不及着火便被烟气流带出炉外，这是形成飞灰固体不完全燃烧热损失高达 20—35% 的原因．

10 吨/时双床并联运行沸腾炉正是为了尽可能减少飞灰热损失而设计的．两床并列装置，中间用矮耐火砖墙相隔，其结构见图 1．主沸腾床燃用 d=8—0mm 劣质固体燃料，从前墙微正压区送入床内，用较高的沸腾风速运行．副沸腾床燃用的是主床烟气流中夹带粒度 d = 1.2—0mm 的细灰，其中一部分细灰是当烟气在流经副床上部时分离下来，另

---

* 本文曾于 1980 年 4 月在桂林全国第三届工程热物理学术会议上宣读．

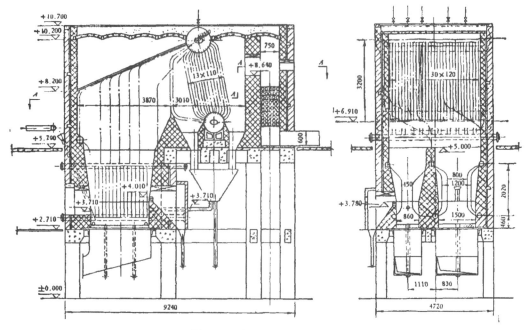

图 1 10 吨/时双床沸腾炉结构简图

一部分则在烟气流经对流管束时分离出来,并在其底部收集下来. 前者由于重力作用沉降到副床之上,后者则用主床风室引来的增压空气通过副床后墙用气力送入副床内. 因为在副床中燃烧的细灰平均尺寸很小,筛分较窄,故用低沸腾风速(约为主床沸腾风速的1/4)运行. 当然还有剩余的一小部分极细飞灰随气流飞出炉外.

为了测定双床并联运行沸腾炉燃烧劣质固体燃料的锅炉蒸发量,热效率以及由于飞灰副床的增设对提高锅炉蒸发量和热效率的作用,进行了二类试验,一是主副两床同时运行;二是主床单独运行(相当于单床沸腾炉)的热平衡多次对比试验. 试验结果表明:

(1) 该炉能长期稳定燃用、煤矸石、劣质煤以及其它层燃炉排出的灰渣等劣质燃料,工作质低位发热量最低为 7.28×10^6 J/kg.

(2) 该炉主床单独运行时和双床并联运行时,锅炉蒸发量平均分别为 9 吨/时和 11.6 吨/时,燃烧效率分别为 65% 和 82.4%,热效率分别为 52.13—55.67% 和 70.20—71.15%,在保持投煤量基本不变情况下,蒸发量提高 2.6 吨/时,燃烧效率提高 17.4%,热效率提高 16.77%,热效率相对提高 31%.

主床单独运行时,被烟气流夹带飞离主床的飞灰,占燃煤总灰分 40% 左右,含炭量约 25%. 飞灰副床投入运行后(双床并联运行)使上述飞灰中 80% 左右在副床中沸腾燃烧后,其含碳量下降到 2.6—4.5% 以下,从而提高了锅炉蒸发量和热效率.

调整沸腾风速更好地适应燃料筛分的要求和整个床面的均匀配风是没有多大困难的,锅炉热效率无疑将得到进一步的提高.

双床沸腾炉的结构较易实现,其所需要的辅机又极简单,因此由于双床结构及其辅机所增加的投资与锅炉整体的投资相比,可略而不计.

三、宽筛分劣质煤与窄筛分飞灰在沸腾炉主床与副床中沸腾燃烧过程动力特性的试验研究

为了了解不同筛分劣质固体燃料沸腾燃烧过程的动力特性,我们对四台不同容量(2—10 吨/时)沸腾炉,在正常运行情况下,对主(或单)床和付副沿各个垂直与水平的不同截面对(1)温度场,(2)烟气各成分的浓度场,(3)沸腾层固体物料所含可燃质的浓度场以及(4)固体物料不同筛分颗粒浓度分布情况,(5)单位容积中固体物料的浓度变化

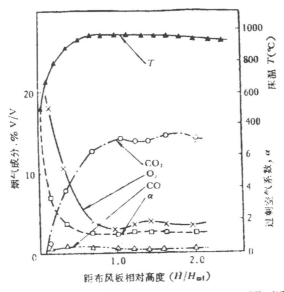

图 2 宽筛分燃煤沸腾床沿垂直高度的烟气成分、过剩空气系数、床温的分布

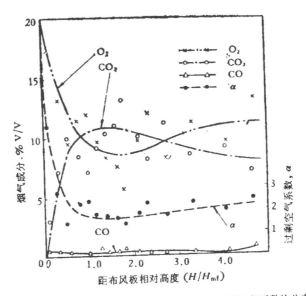

图 3 窄筛分飞灰沸腾床沿垂直高度的烟气成分、过剩空气系数的分布

特性等进行了测定。主（或单）床燃用热值为$(4.8—9.5)10^6$J/kg 颗粒为 8—0mm 宽筛分固体燃料，如，煤矸石、劣质煤，层燃炉渣等，副床燃用为是烟气从主床中携带出来的细灰，其热值约 7.5×10^6J/kg 粒度为 2—0mm. 沿垂直截面测定结果见图 2、3、4 及 5.

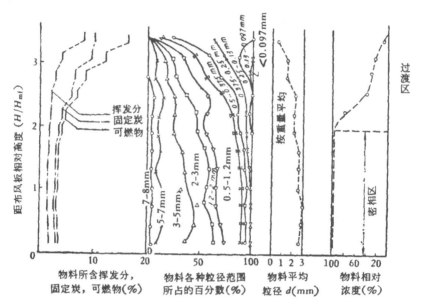

图 4 宽筛分燃煤沸腾床沿垂直高度的物料特性分布情况

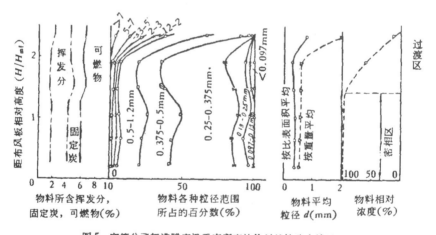

图 5 窄筛分飞灰沸腾床沿垂直高度的物料特性分布情况

从上述测定结果可知：

1. 宽筛分燃煤主床和窄筛分飞灰副床沸腾燃烧过程各动力特性的分布趋向是相似的。

2. 沸腾层中烟气各成分浓度的分布既与层燃炉静止料层中存在氧化区与还原区的情况不同，也与煤粉炉的火焰根部严重缺氧产生大量 CO 的分布情况有别。在沸腾层中基本上只有氧化区，不存在还原区。同时还表明对燃用宽筛分燃料沸腾床在静止料层厚度的对应高度以下的区间内，O_2、α 的减少和 CO_2 的增加从下部到上部，开始极为迅速，之后逐渐减慢，在这个区间的顶部 CO_2 达最小值，说明沸腾层内燃料的燃烧过程基本上在这个区域完成．在静止料层厚度的对应高度以上，O_2、CO_2 及 α 基本上保持不变，燃烧反应已不显著．对于燃用窄筛分的飞灰沸腾床，气体各成分浓度的变化趋势和主（或单）床相似，但其变化范围则扩展到整个床层高度．CO 的分布沿整个沸腾层高度无论对宽筛分煤床还是对窄筛分灰床，基本上变动都很少，一般在 1% 以下．

3. 燃用宽筛分劣质燃料的主（或单）沸腾床中，在床底部的狭窄区间，床温上升甚快．之后逐渐减慢，至静止料层厚度相对应的高度以前达最高值，从此向上至床层表面转而下降．床下部区域由于燃烧放热引起床温上升．所以床温随高度而变化的趋势也表明燃烧过程在床下部进行．尽管在床的上半部燃烧速度很小，而换热面的吸热却很可观，但是由于强烈的沸腾状态导致迅速的质交换和热交换，所以这个区域的温度下降不多．

在副沸腾床中，据过去所测定的床温的变化和主（或单）床相似，但床温的最高值移近布风板．

4. 从沸腾床中固体物料粒度的分布可知，宽筛分燃煤主沸腾床中固体物料上部粒度细，$\bar{d} = 2mm$；下部粒度粗，$\bar{d} = 3mm$；最底层（0—100mm 的高度以内）$d \approx 2—5mm$．存在一定分层现象．在沸腾层密相区与炉膛自由空间之间不存在明显的分界面．在正常运行情况下，沸腾层的膨胀度为 2.5，起伏比为 1.6．

窄筛分飞灰沸腾床中固体物料粒度（假如撇开气力送灰的影响）只在布风板表面有少量粗颗粒沉积，沿整个高度上基本相同，平均粒径约 $\bar{d} = 0.3mm$．沸腾层有比较明显的分界面，在正常运行情况下仍然有较大的起伏比，约为 1.35，料层膨胀度约为 2．

从沸腾层中固体物料粒度分布可知，主床物料粒径基本上在 8—1.2mm 之间，占床中物料总量的 66%，1.2—0.5mm 仅占 24%．而副床中粒径在 1.2—0.5mm 仅占床中物料总量的 14%．基本上为 0.5—0.25mm，占 61.5%．这就充分说明了双床并联运行比单床单独运行在蒸发量和热效率上所以会有显著的提高的原因．

5. 从固体物料的可燃物含量沿整个沸腾炉膛高度的分布可知，在沸腾层内，由于存在激烈的沸腾现象，床上下质交换速度甚高，所以床中物料的可燃物含量基本上相同．但是从沸腾层的上界面到炉膛上部出口，随着水平的提高，物料的绝对浓度越来越小，而物料的可燃物含量却越来越高，特别是宽筛分沸腾床，这种现象尤为显著．烟气从主沸腾床上升时所携带的固体颗粒粗略地可分为两部分：较粗较密的颗粒，由于重力的作用逐渐沉降并重新进入密相区；较细较松的其粒径小于 1.2—0.5mm 的若干部分，停留在密相区中的时间约 1/4 秒或更少，甚至还来不及着火便被烟气流带出沸腾层，还保留原有的可燃物．这也说明为什么飞灰可燃物含量和原煤相接近．因此，副床具有足够数量和很高质量的细灰可供利用，因而有相当可观的热量能够回收．

烟气携带的飞灰飞越炉膛自由空间时由于烟气温度（实际仅 750℃—600℃）氧浓度（实际仅 3.5%）都太低，停留时间太短（实际仅 4—6 秒）而又缺乏强烈的扰动，所以是难以继续燃烧的．

沿水平截面的测定结果见图 6 及 7．从图 6 可见，主床由于在炉前集中进煤，沿整个床的深度形成物料可燃物含量分布的不均匀，进而影响烟气浓度和温度分布的不均匀．

轻则影响燃烧效率,在极端情况下,象我们有时遇到的那样,可能导致熄火或结渣(局部或全部),破坏稳定运行状态. 副床由于气力进料(进灰)较为均匀,所以沿整个床的深度物料可燃物含量,烟气浓度及温度的分布也较为均匀.

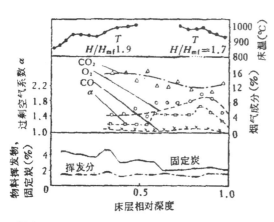

图 6 宽筛分燃煤沸腾床沿床深度的烟气成分、床层温度、固体物料可燃物的分布

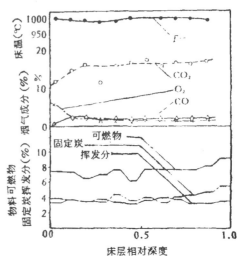

图 7 窄筛分飞灰沸腾床沿床深度的烟气成分、床层温度固体物料、可燃物的分布

四、结 论

(1) 根据长期运行经验和多次热平衡试验表明,双床并联运行工业沸腾燃烧锅炉是燃用低挥发份,很高灰份,低热值的固体燃料的一种合理炉型. 这种锅炉的热效率要比单床沸腾燃烧锅炉提高 16.77%.

(2) 通过沸腾层燃烧过程动力特性的分析,我们了解到沸腾层内燃料的燃烧反应基本上是在沸腾层下部,特别是对于宽筛分沸腾层,是在静止料层高度以下的区域进行. 所以这个区域的沸腾状态和温度水平对沸腾床的运行可靠性影响最大. 相反地,正是在这个区域内,特别是接近底部,物料的粒度较粗较密,而底部的床温又很低. 这种状况对于取得良好的沸腾质量是很不利的. 所以,在设计和运行时,处理好沸腾质量的要求与沸腾条件的矛盾是十分重要的.

本文所叙述的试验均系浙江大学热物理工程系燃烧与传热教研组全体同志共同完成的.

本文所提及的双床并联运行沸腾炉试验研究工作得到湖州化肥厂、杭州锅炉厂和吴兴县夺煤指挥部协作进行.

参 考 文 献

[1] 浙大燃烧理论研究室:劣质石煤在沸腾炉内的燃烧过程及提高燃烧效率的途径,浙江大学学报,2,(1978).
[2] 陈运铣等:劣质煤沸腾燃烧过程的动力特性和双床沸腾炉的试验研究,煤炭学报,2,(1980).
[3] 陈运铣等:提高劣质煤沸腾炉燃烧效率的研究,煤炭科学技术,4,(1980).

EXPERIMENTAL STUDIES OF FLUIDIZED TWIN-BED BOILER FOR LOW GRADE SOLID FUELS

Chen Yun-sien Cen Ke-fa Zhang He-sheng Zhang Xue-hong Kang Qi-fu

(Zhejiang University)

Abstract

The construction principle of the industrial boilers with two fluidized beds operating in parallel is briefly introduced. According to the results of heat balance tests, it is evident that the thermal efficiency of the boiler, when burning low grade solid fuels, may be increased from 53.9 to 70.7% as compared with the single fluidized bed boilers. Dynamic properties of combustion process in fluidized bed boilers when burning low grade solid fuels with broad size distribution and firing fly ash with narrow size distribution were experimentally dettermined, and the results of which were preliminarily investigated. Accordingly, problems in fluidized bed boiler design and operation have been examined, and reasons accounting for the increase of thermal efficiency are also given.

原文刊于工程热物理学报, 1981, 2(4): 366-372

油煤混合燃料的试验研究

岑可法　曹欣玉　袁镇福　陆重庆　洪积瑜

油煤混合燃料（简称COM）就是把一定细度的煤粉和重油均匀混乱的浆状体，它是目前石油短缺的情况下，节约用油，以煤代油的有效措施。我们只要对电站燃油锅炉和燃油工业炉作部分改动，就可利用原来的燃油系统改烧油煤混合燃料，当煤粉浓度为45～50%时可节油约为30%，并可进一步发展到使用油、煤、水混合燃料，直到目前各国正在努力研究的水煤浆燃料，以达到完全以煤代油的目的。

油煤混合燃料的制备，输送和燃烧是一项综合性的交叉技术。对于油煤混合燃料的基本特性、制备、质量监测、贮运、加热和高效地燃烧，有一系列新的课题需要解决。这些课题将涉及到燃料化学，物理化学，胶体化学，化工，材料，流体力学，传热传质学和燃烧学等许多学科领域。为此中国科学院根据国家计委，国家经委和财政部在1979年8月下达了科研任务，在浙江大学建立了油煤混合燃料的制备，输送和燃烧的试验装置，整个试验装置的流程见图1，制备系统最大的制备容量为3吨/时油煤混合燃料，

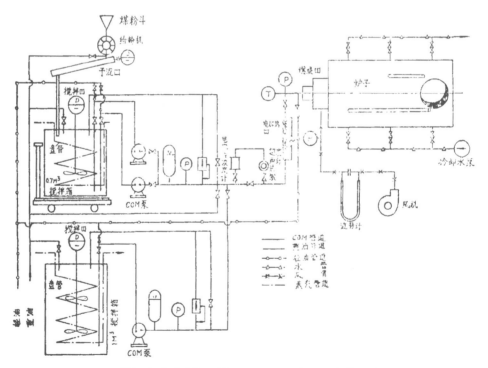

图1　在浙江大学的COM制备和燃烧系统的试验装置流程图

燃烧系统燃烧油煤混合燃料的容量为120公斤/时，试验目的为研究油煤混合燃料的某些基本物理特性和制备、输送及燃烧的特性，本文主要介绍油煤混合燃料的某些基本特性及制备和输送特性的研究成果。

一、油煤混合燃料的流变特性：

由于油煤混合燃料中含有大量的煤粉，使重油的原来流体力学特性有明显改变，而且随煤粉浓度的增加，其流体特性的变化越大。理论上煤粉作为均匀球体，在油中无规则排列的极限重量浓度为73.4%（COM温度为80℃）。但实际上煤粉是分散相，任意形状煤粉颗粒在实验室中能制备的最大重量浓度$C_w = 69\% \sim 70\%$。

为了了解油煤混合燃料的流体属性及其流变特性，采用了EPPreche Rheomat 15型旋转粘度计和直管流动试验装置进行流变特性的研究，试验用平顶山煤粉和大庆渣油及20号机油，在40℃至120℃的温度时，煤粉重量浓度为0～69%，0～2mm的煤粉直径，速度梯度范围$\frac{dw}{dr} = 10 \sim 2000\, sec^{-1}$。试验表明COM是一种拟塑性流体，其剪切应力τ与速度梯度的关系如下：

$$\tau - K\left(-\frac{dw}{dr}\right)^n = K\left(\frac{dw}{dr}\right)^{n-1} \cdot \left(-\frac{dw}{dr}\right) = \mu p\left(-\frac{dw}{dr}\right) \qquad (1)$$

K—均匀系数，n = 流态特性系数

μp—表观粘度，

其试验结果示于图2，两种方法所得的流变特性系数n值示于图3。

对试验油、煤种所得的数据，经计算机回归分析处理所得n、K值的试验关联为：

$$n = n_0 e \times P(-0.5492 C_w^2) \qquad (2)$$

随煤粉浓度C_w变化的均匀系数

$$K = K_0 e \times P\left(\frac{3.113 C_w}{1 - 1.061 C_w}\right) \qquad (3)$$

随COM温度变化的均匀系数

$$K = k_{T_0} e \times P\left[\frac{3.152\left(\frac{T_0 - T}{T_0}\right)}{1 - 0.36\left(\frac{T_0 - T}{T_0}\right)}\right]$$

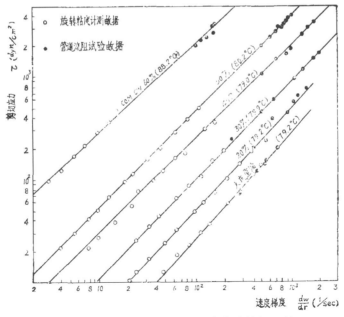

图 2　不同试验方法COM流变特性的比较

式中 n_0, K_0 ——当 $C_w = 0$ 时的数值
k_{T_0} 为 $T_0 = 90℃$ 时的数值

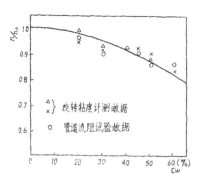

图3 COM的流变指数随煤粉浓度的变化

试验结果表明：当煤粉浓度小于30%时COM的表观粘度 μ_p 增加速度较慢，可近似作为牛顿流体处理，当煤粉浓度大于30%时，μ_p 增加较快，应作拟塑性流体处理。当煤粉浓度大于50%时，粘度剧烈增加，比重油大20～30倍，这时COM的制备、输送和雾化都极为不利，在工程上难于使用。当温度增加时，其表现粘度急剧下降，因此高浓度的COM可用较高的加热温度来降低粘度，以达到安全输送和良好的雾化。当温度小于50℃以后，表观粘度迅速增加，煤粉的沉淀大为减慢，所以COM可以在较低温度下贮存。

二、油煤混合燃料中煤粉颗粒的沉降特性。

油煤混合燃料在制备，加热和输送过程中煤粉颗粒可能产生沉淀，以致阻塞管路和喷咀。因此必须了解COM中煤粉的沉降特性。为此首先对静止状态下COM中煤粒自由沉降规律进行研究。根据COM流变特性试验所得结果，应用资料[1]中的计算公式，对直径为0.1mm的煤粉计算所得结果列于表1。

表 1

煤粉重量浓度 C_w	0%	30%	50%
流态特性系数 n	1.0966	0.9556	0.8613
均匀系数 K(达因秒^n/厘米^2)	0.2647	1.357	8.496
煤粉群终端自由沉降速度 u_t(m/s)	0.1025	7.94×10^{-3}	3.2×10^{-4}
自由沉降雷诺数 R_{et}	0.817×10^{-4}	1×10^{-5}	1.1×10^{-7}
不稳定沉降时间(秒)	6.09×10^{-5}	1.79×10^{-5}	2.57×10^{-6}
不稳定沉降的距离(mm)	2.95×10^{-6}	1.93×10^{-7}	2.19×10^{-7}

我们还采用沉降天平法，灰化法，比重法和电容法进行试验，图4为不同试验方法，不同煤粉浓度的COM沉降过程，图5为不同温度下的COM沉降过程。

由计算和试验表明：(1)其自由沉降雷诺数，$R_{et} \ll 1$，不稳定沉降段很短，在工程上可视为等速沉降。

(2)COM中煤粉浓度愈高，温度愈低，煤粉愈细其沉降显著减慢，因此在工程应用时尽可能提高煤粉浓度，使用较细的煤粉，在制备时温度不宜大于80℃，并可在低温下贮存

三、用机械搅拌制备COM的工艺：

对工业装置制备COM，目前最易于实现的是机械搅拌方法，在本试验装置中成功地采用了螺旋予混器和单、双叶螺旋浆式搅拌器相结合的流程，在连续制备的动态特性的试验过程中，在予混器出口COM的煤粉浓度波动的标准偏差为6.77%，经搅拌器搅拌后，搅拌器出口的浓度波动标准误差降为1.1%。试验证明已能顺利地满足制备，输送雾化和燃烧的要求。

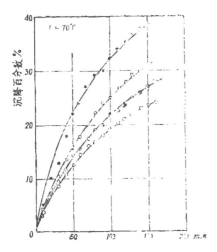

图4 沉降天平法测定的不同煤粉浓度的沉降过程

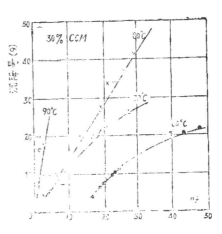

图5 在不同温度下的COM沉降过程

用电容式COM均匀性测量仪及灰化法对不同截面测量COM浓度变化表明，当搅拌器转动雷诺数$Re_m \geqslant 10^4$时，即可制备出浓度均匀的COM。同时还试验了当给粉机突然停止或启动时的动态过程，试验表明，搅拌桶越大，输入的煤粉或重油量的波动，对输出的COM浓度波动影响越小，因此对于一定出力的COM制备系统需要有相当容量的制备桶。

对于螺旋浆式搅拌器的功率消耗可关联成如下的准则方程式：

$$S_P = CR_{em}^n \quad (5)$$

功率数 $S_P = \dfrac{P_A}{\rho N^3 D_m^5}$;

雷诺数 $R_{em} = \dfrac{\rho N D_m^2}{\mu_P}$

P_A——所需的净功率,
N, D_m——叶浆的转数及直径
μ_P, ρ——COM表观粘度及密度
c, n,——试验常数
在$C_w = 0 \sim 60\%$内,仅和叶浆型式有关。

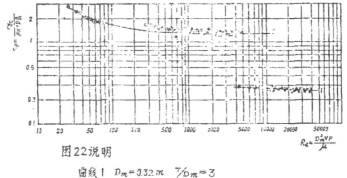

图22说明

曲线1 $D_m = 0.32 m$ $T/D_m = 3$

○:24%; △:36%; ●:45%; ×:重油

曲线2 $D_m = 0.115$ $T/D_m = 3$

×:重油; △:10%; ●:20%; □:30%

▲:40%; ○:50%; ●:60%

图22 搅拌器的无因次功率特性曲线

所得的功率特性曲线见图6，上述所得的结果可供放大设计时参考。

四、COM的流动特性

由于COM是拟塑流体，因而COM在管内流动特性与通常的水和油等牛顿型流体有很大的不同。为此研究了输送系统内的等温和不等温的流动过程。由理论分析和试验结果表明：

(1) COM在管内流动的临界雷诺数为2053～2206，因而在通常的工程输送管道内，COM流动属于层流流动。具体数值见表2。

油煤混合燃料层流流动的临界管经　　表2

煤粉重量浓度		0%	10%	30%	50%
层流临界雷诺数 Rec		2053.2	2115.3	2427.7	2206.2
层流最大管经 Dc	$W_0=1m/s$	36	54.7	88.6	290
	$W_0=0.5m/s$	64.8	113.7	190.3	784.2

(2) 由理论计算可知：COM管内流体动力不稳定段很短，对于一般的工程问题可作为稳定流动处理，但在不等温流动时，热不稳定段则很长，计算时不能忽略。

(3) 在等温和不等温流动工况下，利用COM的流变特性试验数据，可用下式来计算管内流动阻力ΔP及阻力系数ζp

$$\Delta p = 4KL\overline{W}^2 \left(\frac{2+6n}{n}\right)^a d^{\frac{1}{n+1}} \quad (6)$$

$$= \zeta p \cdot \frac{L}{d} \cdot \frac{\rho \overline{W}^2}{2}$$

$$\zeta p = \frac{64}{Re_p} \quad (7)$$

式中：$Re_p = \dfrac{\rho W^{2-n} \cdot d^n}{8^{a-1} \cdot K \left(\dfrac{3n+1}{4n}\right)^a}$

把K，n代入，即可求得任意煤粉浓度及温度下输送COM的流动阻力，这些公式与试验结果是相符的。图7示出了理论阻力系数与等温与不等温流动阻力试验数据的比较，图8示出了不同加热温度对管道流动阻力的影响。由上可得，对于等温和不等温流动，当COM中煤粉浓度$C_w = 0 \sim 50\%$，加与不加添加剂和用机械搅拌或电超声乳化不同制备方法的情况下，利用上式计算，大部分误差均小于10%。

对于不等温流动定性温度采用进出口平均温度和壁温的平均值。应当指出的是当对管道加热，会使流阻有大幅度的降低，由图8可知对于不等温流动每米温升5℃时其阻力下降33.2%。因此只要知道COM的流变特性，即可进行COM的工程流动阻力计算。

五、油煤混合燃料加热特性的研究

由COM的流变特性可知，COM的表观粘度随温度的变化很大，因此在COM的制备、输送和喷咀雾化中要求把COM加热到一定的温度。为此，我们对COM制备过程中盘管加热的过程和在输送过程中夹套管加热进行试验，其试验结果如下：

(1) COM在制备过程中盘管加热COM的研究：试验是在$\phi350$直径的搅拌桶内，用$\phi10\times1$的紫铜管做成$\phi230$直径的盘管，其节距为$S/d=2$，分别试验了COM中煤粉

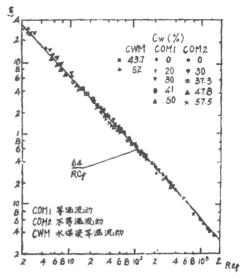

图7　煤浆等温及不等温流动时的阻力系数

浓度为 Cw = 0～53% 等六种不同浓度，测定了其对流放热系数，对试验数据进行回归分析，所得的关联式为：

$$N_d = 3.78\ R_{em}^{0.44}\ P_r^{0.33} \tag{8}$$

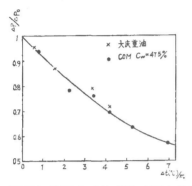

图8 不同加热温度对管道流动阻力的影响

$Nu = \dfrac{\alpha_c T}{\lambda_c}$ 努谢尔特数

$R_{em} = \dfrac{P D_m^2 N}{\mu_p}$ 雷诺数

$P_r = \dfrac{U}{a}$ 伯朗特数

式中：α_c—COM侧放热系数 大卡/米²·时℃

λ_c—COM的导热系数 大卡/米·时℃

T—制备桶的直径 米

D_m—搅拌浆直径 米

μ_p—表观粘度 公斤·秒/米²

a—导温系数 米²/小时

$\upsilon = \mu/\rho$—运动粘性系数 米²/秒

与试验点的标准误差为3.42%，可供工程设计时参考。

(2) 油煤混合燃料在输送过程中用夹管加热的研究：

为保证喷咀良好雾化，高效地燃烧，需把COM加热120℃以上，提出了用蒸汽夹套管边输送边加热的流动方案，这样既可避免在盘管加热器前出现较高的输送阻力，降低能耗，又可防止盘管加热器内众多弯头处容易产生沉淀。为此对夹套管加热COM进行了理论和试验研究。

在流动和传热试验台架上，用微型测温探针和微型测速管为了不同截面的温度场和

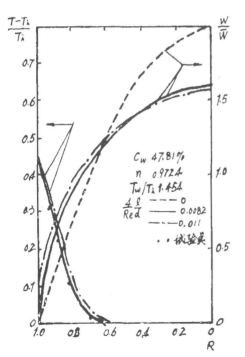

图9a 不等温流动时不同截面上的温度和速度场分布

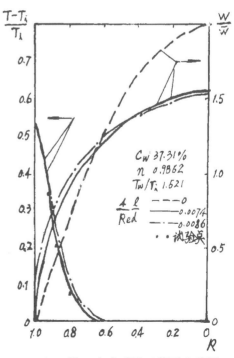

图9b 不等温流动时不同截面上的温度和速度场分布

速度场，测量结果见图9，由图可以发现，在加热时速度场明显变得短而胖，偏离分速度场的抛物线分布规律，由图可见实测和计算能较好的吻合。并对COM管内流动动量、能量和本构方程进行数值求解，所得的计算机程序可供工程放大时使用。数值计算和408组传热准则关联式为：

$$\bar{N}_u = 1.0666 P_r^{0.3989} R_{ep}^{0.3987} (d/L)^{0.3986} \cdot (kf/kw)^{0.1246} \left(\frac{3n+1}{4n}\right)^{0.08085} \quad (9)$$

在缺乏COM的流变参数的情况下，可用表观粘度 μ_p 进行工程估算

$$\bar{N}_u = 1.0709 P_r^{0.3989} R_{ep}^{0.3987} (d/L)^{0.3985} \left(\frac{\mu_f}{\mu_w}\right)^{0.1255} \quad (10)$$

式中：$\bar{N}_u$——平均Nusselt准则 $\bar{N}_u = \frac{\alpha \cdot d}{\lambda}$

L——夹套管加热长度 米

K_f, μ_f——由COM平均温度决定的均匀系数及表观粘度

K_w, μ_w——由壁温决定的均匀系数及表观粘度

理论计算和试验结果对比，大部分误差均小于25%。

为了强化光管管内传热，改善油煤混合燃料的均匀性，防止在运输过程中煤粒的沉降，采用了静态混合器来强化传热，由于静态混合器对介质的分割，内外层流体的交换搅动作用强化了传热，提高了均匀性，使传热提高了2倍，随之阻力增加了4倍，为了不使阻力增加过多，又有良好的传热效果，可将静态混合器间隔一定距离布置。

六、油煤混合燃料的燃烧试验

燃烧试验是在卧式布置带水夹套的筒形炉内进行的[2]，炉子的内径为1060mm，有效长度为5450mm。其燃烧COM的容量为120公斤/小时，燃烧容积热负荷为 250×10^3 大卡/米³时，断面热负荷为 1135×10^3 大卡/米²时炉膛轴向布置二排共35个测孔，用来观察与取样。先后用两种制备方法进行了燃烧试验，一种为机械搅拌加电超声乳化制备方法[3]。此外还进行了加与不加添加剂的燃烧试验[4]，主要结果示于表3。试验表明COM的制备，加热、雾化及燃烧系统运行正常，燃烧稳定，其燃烧效率达97%，能达到以煤代油节约用油的目的

七、下一阶段的工作

上述试验工作，经一年半努力，在1980年12月由中国科学院主持组织浙江大学及科学院系统有关七个研究所在鞍钢发电厂№11炉进行中间试验，锅炉蒸发量为100顿/时，目的是探索COM在电站锅炉的应用前景及其有关的技术关键。目前已完成了COM制备，输送和燃烧系统的设计工作亦已基本完成，预计1982年年底将可完成有关的试验项目。

不同制备方法COM燃烧试验：表3

试验序号	1	2	3
COM制备方法	机械搅拌	机械搅拌+电超声	机械搅拌+电超声+添加剂
COM流量 kg/nr	108	110	112
COM中煤粉浓度%	43.7	43.7	35.8
排烟空气过剩系数α	1.15	1.13	1.08
排烟CO%	0	0	0.052
化学不完全燃烧损失 q_3%	0	0	0.15
机械不完全燃烧损失 q_4%	2.87	2.99	2.07
燃烧效率 η %	97.13	97.01	97.78

引用资料

〔1〕"在浙江大学试验台架上油煤混合燃料的制备和输送的试验研究"。
3th International Symp on COM Orlando U.S.A 1981

〔2〕"COM在筒型炉膛中燃烧过程的试验研究"
3th International Symp on COM Orlando U.S.A.1981

〔3〕"500瓦压电超声聚能乳化换能器"
科学院上海硅酸盐1980年12月

〔4〕"油煤混合燃料添加剂评选试验"
科学院上海有机化学所及山西煤化所1980年12月

原文刊于能源工程, 1982, 1: 51-57

水煤浆在沸腾床内燃烧过程的初步研究*

岑可法　曹欣玉　倪明江　孔水源　洪积瑜
袁镇福　谢名湖　吕德寿　陈运铣
（浙江大学）

摘　要

本文介绍了对于水煤浆在沸腾床内燃烧过程的初步实验研究结果。实验表明水煤浆在沸腾床内的凝聚、结团特性，使得燃烧过程中的飞灰可燃物损失可以被控制得很低，因而其燃烧效率可达很高的水平，高于同煤种的干煤粒。实验还表明采用所提出的异重度床料沸腾燃烧的方案可以成功地消除凝聚、结团造成的大颗粒沉积现象，实现稳定地运行。燃用水煤浆的沸腾床与普通燃煤沸腾床相比还有 NO_x 排放量低和易于脱硫等优点，因此是一项有发展前途的新技术。

引　言

水煤浆燃烧技术作为燃烧工程领域的一项新技术正在日益引起国际上的广泛注意。该项技术的发展不仅会给煤炭的开采、运输和使用带来一些根本性的变革，而且还可能为以煤代油、为洗煤煤浆等低热值工业废液、废渣的能量利用开辟新的发展途径。同时，水煤较易实现低热值煤气化，可以供给燃气轮机实现联合循环。水煤浆又是一种低污染的燃料，其燃烧时 NO_x 排放量较低，用石灰石脱硫可实现较高的脱硫效率。

1981 年以来，我们对水煤浆陆续开展了一些初步的实验研究工作，本文主要讨论其中关于水煤浆在沸腾床内燃烧过程研究的初步结果。

实验设备及方法

整个实验工作是在一个 250mm×250mm 方形沸腾床和一个 ϕ80mm 圆形沸腾床上进行，图 1 为 250mm×250mm 沸腾床的实验系统图。空气由送风机径流量计、空气预热

表　1

元　素　分　析　（分　析　质）　%							低位发热量（大卡/公斤）
C	H	S	O	N	A	W	Q_{DW}
58.77	3.43	1.23	5.48	1.6	26.99	2.50	5447.96

---

* 本文曾于 1982 年 10 月在无锡中国工程热物理学会第三届年会上宣读。
　参加实验工作的还有解海龙、刘关元、葛林甫、曹英武、浦兴国、凌柏林。

器从布风板底部送入沸腾床；燃烧后的烟气从沸腾床上部经减温器，除尘器由引风机排出．水煤浆由机械搅拌制备，然后经齿轮泵由喷嘴从上部送入沸腾床．试验过程中尾部和床内烟气中 CO_2、CO、C_mH_n、SO_2、NO_x 等组分的浓度通过红外烟气分析仪连续取样测定．此外，还配合使用色谱分析仪和奥氏分析仪以测定或校核烟气中 O_2，H_2，CO_2，CO，CH_4 等组分的浓度．为了进一步分析 NO_x 的生成规律还用化学萤光法气体分析仪分别测定了 NO 和 NO_2 的浓度．沸腾床内的温度分布由九点热电偶测定[1][2]．

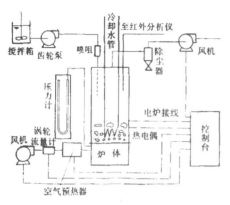

图 1　250 mm × 250 mm 沸腾床系统图

实验使用平均直径为 $126\mu m$ 的平顶山烟煤制备的不同含水量的水煤浆．为了进行比较还在与水煤浆相同的条件下对同煤种的干煤粒进行了实验，实验用平顶山烟煤的分析数据见表 1．

实验结果与讨论

1. 水煤浆液滴在沸腾床内的凝聚、结团现象　水煤浆液滴在沸腾燃烧过程中的一个重要特点就是凝聚结团现象．图 2 中示出了水煤浆液滴在燃烧过程各个阶段的外观显微照片（放大 16 倍）．实验表明在沸腾床内水煤浆液滴在燃烧过程中能形成有一定强度的多孔焦碳球（对煤粉浓度 C_w 为 50% 的水煤浆其水份蒸发造成的空隙度达 0.59）．形成的焦碳球中可燃物的燃烬在表面形成了空隙度更大的灰层，由于表面灰层的空隙度很大（见图 2）使得灰粒之间的联系变得很弱，从而使灰壳显得很疏松．在沸腾床燃烧时疏松的灰层很容易被剧烈运动的床料不断冲刷掉，因此在沸腾床内的水煤浆球团表面始终仅有很薄的一层灰．至于灰层内部的焦碳核心，其强度可一直保持到差不多燃尽为止．

实验表明，水煤浆液滴并非在所有的床温下都能结团．当床温较低时，投入沸腾床内的浆滴大部分不能成球，而是在干燥后散成粉状被吹出沸腾床．床温提高时，虽然能成球但强度不高，因而在床内的磨损很厉害，磨下的煤粉也很快被吹出沸腾床．只有当床温高于一定水平时，投入沸腾床内的水煤浆液滴才能形成具有一定强度的球团，这样的球团在燃烧过程中的磨损和破碎都不大．

以上所述的水煤浆液滴在燃烧过程中形态上的特点对其燃烧过程有重要的意义．燃料球的多孔性使有效反应表面增加，薄灰层使气体传质阻力减小，再加上水蒸汽存在附加的气化反应，使得水煤浆在沸腾床内的燃烧速度大为增加．另一方面，水煤浆在沸腾床内停留时间很长，据估算平均为几十分钟到上百分钟．因此，投入床内的水煤浆球团有足够的时间停留在床内，既不易磨损也不易破碎，燃烧过程中形成的灰不断剥落，就这样逐步燃尽，直到球团直径很小时才有可能被烟气带出沸腾床从而形成很小的飞灰可燃物损失（参见图 2）．

2. 凝聚结团造成的床内沉积现象的克服　水煤浆在沸腾床内的凝聚结团现象是在沸腾床内组织正常燃烧工况的基础，也为提高沸腾床的燃烧效率创造了有利条件．但另一

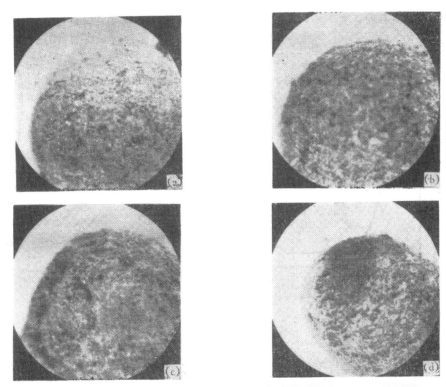

图 2 水煤浆液滴在沸腾床内燃烧过程中的形态

方面,在运行工况组织不好时,大颗粒的水煤浆球团很容易在床内沉积,以至最后破坏沸腾质量使沸腾床无法运行. 所以,煤浆球团在床内的沉积对沸腾床的稳定可靠运行是一个潜在的威胁. 国外的一些文献报道[3][4]都一致指出了解决这个问题的困难性.

在这次实验中,我们后来采用了重度比煤浆球团大的砂作为床料来消除大颗粒的沉积,取得了较好的效果. 床料重度大(2500 公斤/米³)使整个床层的平均重度,即使在床层空隙度为 0.8 时仍大于水份蒸发后煤浆球团的重度,这样整个床层对煤浆球团产生的"水静力浮力"使运行中经常出现的过大凝聚块,即使直径达几十毫米亦不会在床内沉积. 而且实践表明,只要保持适当高的沸腾风速和料层中较低的燃料浓度则由于"浮力效应"造成的煤浆球团在床内的偏折也可减小到很低的程度. 可以认为,本文提出的采用异重度床料沸腾燃烧水煤浆的方案,为解决煤浆球团的沉积问题,保证沸腾床稳定运行开辟了一条新的途径.

3. 沸腾床燃用水煤浆时的燃烧效率 如前所述,由于沸腾床在燃用水煤浆时可实现很低的飞灰可燃物损失,因此其燃烧效率可达到较高的水平. 为了考察运行参数对燃烧过程的影响,我们测定了燃烧效率随床温、过剩空气量和煤浆含水量的变化关系. 作为对比,还在相同的条件下对 0.5—3 mm 的同煤种干煤粒进行了试验.

对燃烧效率影响最大的是温床. 图 3 示出了化学未完全燃烧损失 q_3 随床温的变化规律,q_3 损失主要由 CO 造成. 由图可见,当床温高于 800℃后,q_3 损失迅速降低到很低的数值. 图 4 示出了机械未完全燃烧损失 q_4 随炉温的变化趋势,可以发现,在床温较低时,燃用水煤浆的 q_4 损失很大. 但当床温高于 800℃时,在相同的运行条件下,水煤浆的 q_4

损失开始明显低于筛过的干煤粒.因此,水煤浆的燃烧效率 η_c 在 800℃ 以上时也较煤粒为高(见图5). 在实际的燃煤沸腾床中使用的煤粒一般都含有一定数量的小于 0.5 mm 的细颗粒,而这部分细颗粒最易形成飞灰损失,所以,可以认为燃用水煤浆的沸腾床的燃烧效率将会比实际燃煤沸腾床有更大幅度的提高.

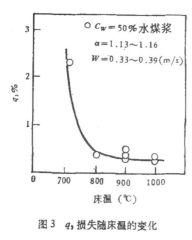

图3 q_3 损失随床温的变化

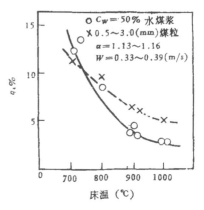

图4 q_4 损失随床温的变化

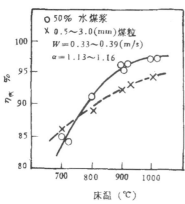

图5 燃烧效率 η_c 随床温的变化

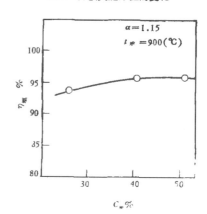

图6 不同煤浆含水量下的燃烧效率

图6给出了煤浆含水量和燃烧效率的关系. 从燃烧及经济性的角度讲,煤浆的含水量越少越好. 目前在管道输送中广泛采用 50% 的含水量. 而洗煤煤浆和其它工业液固废料的含水量则可能很高,例如,洗煤煤浆的含水量有时达 80—90%,为此我们进行了高含水量煤浆的燃烧试验. 试验表明,只要把床温维持在一定水平,即使含水量高达 75% 的水煤浆也能正常地燃烧. 由图6可见,在床温为 900℃ 时其燃烧效率只有很少的降低. 高水份水煤浆燃烧效率的降低主要是源于化学未完全燃烧损失的增加,这是由于水蒸汽量增加以至气化反应强化生成了大量可燃气体的缘故. 只要加强、组织可燃气体的燃烬,燃烧效率可望有进一步的提高.

图7给出了过量空气系数 α 对燃烧效率的影响. 在过剩空气量较小时,随着 α 的增加燃烧效率增加较快,当 α 超过 1.2 以后曲线就变得比较平坦,看来过剩空气系数控制在 1.2 左右是比较合适的. 值得注意的是,在过剩空气量较低时水煤浆的燃烧效率对 α 的变

化不如煤粒那样敏感，由图可见即使过剩空气量为零，水煤浆的燃烧效率仍达92.5%左右。这是由于水蒸汽和焦碳气化反应的存在，使得总的燃烧过程对氧的依赖性不如干煤粒那样大的缘故。

从以上介绍的实验结果可以看出，我们在实验室规模的沸腾床中基本上作到了稳定可靠高效率地燃烧水煤浆。现在正着手进行工业性试验，以期早日使这项技术实用化。

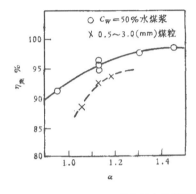

图7 过剩空气系数 α 对燃烧效率的影响

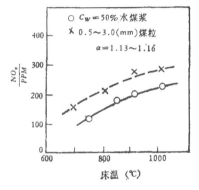

图8 NO_x 排放量与床温的关系

4. NO_x 的排放规律 水煤浆在沸腾床内燃烧时，由于气化反应造成的还原性气氛，抑制了氮氧化物的生成，所以其 NO_x 的排放量较燃煤沸腾床要低。图8中示出了在不同床温下 NO_x 的排放量。由图可见，随着床温的升高，NO_x 的排放量明显增加，这是由于氮的氧化反应强化的缘故。还可以看到，水煤浆的 NO_x 的排放量大约比煤粒低60ppm。显然与燃烧温度很高的火炬燃烧相比将低得更多。

5. 脱硫试验 众所周知，采用沸腾燃烧可以方便地在床内投入脱硫剂进行脱硫。但是在燃煤沸腾床中，由于脱硫剂是以一定粒度的颗粒投入沸腾床的，因此脱硫剂与燃煤的接触不是很密切。而且受传质过程的限制，脱硫剂颗粒只有表面处一定厚度内的成份才是真正有效的。所以一般燃煤沸腾床为取得较高的脱硫率 η_s，往往要用较高的Ca/S比。这次我们对水煤浆进行脱硫试验时是直接将石灰水掺入煤浆，由于脱硫剂在水煤浆内和煤粉得到充分的混合，因而在燃烧过程中脱硫剂得到了充分有效的利用。图9示出了按上述工艺脱硫和按一般燃煤沸腾床脱硫工艺脱硫的对比数据。可以看到，在相同的Ca/S比下，上述工艺的脱硫效果要好得多。

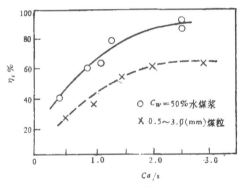

图9 两种脱硫工艺效果的对比

初 步 结 论

1. 在一定的温度水平上，沸腾床内水煤浆液滴能形成具有一定强度的多孔球团，这种球团在沸腾床内燃烧时磨损和破碎都不大，从而使沸腾床燃用水煤浆时的飞灰可燃物损失可以被控制得很低。

2. 本文所提出的用异重度床料沸腾燃烧水煤浆的方案为解决煤浆球团在床内的沉积问题，保证水煤浆沸腾床的可靠运行开辟了新的途径。

3. 在一定的操作条件下，各种水份，甚至水份高达75%的水煤浆都能在沸腾床内稳定可靠地燃烧，其燃烧效率一般高于已筛去细颗粒的干煤粒。

4. 沸腾床在燃用水煤浆时其NO_x排放量比燃煤时有较大幅度的降低。

5. 将CaO直接混入水煤浆内进行脱硫，其脱硫效果明显好于采用普通燃煤沸腾床的传统工艺。

因此，用沸腾床燃用水煤浆是一项有发展前途的新技术，值得进一步深入地开展研究。

参 考 文 献

[1] 倪明江《水煤浆在沸腾床内燃烧过程的研究》浙江大学硕士研究生论文 1981.12.
[2] 孔水源《水煤浆在燃烧过程中有害气体生成的试验研究》浙江大学硕士研究生论文 1981.12.
[3] Randell, A. A. "Disposal of Colliery Tailing by Fluidised Bed Combustion". Fluidization Proc. of the 2nd Eng. Foundation Conf. 1978.
[4] Poersch, W., "Fluidized Bed Combustion of Flotation Tailings" Fluidization Proc. of the 2nd Eng. Foundation Conf. 1978.

EXPERIMENTAL STUDY ON COMBUSTION OF COAL WATER SLURRY IN FLUIDIZED BEDS

Cen Kefa   Cao Xinyu   Ni Mingjiang   Kong Shuiyuan
Hong Jiyu   Yuan Zhenfu   Xie Minghu
Lu Deshou   Chen Yumin

Abstract

This paper reports some experimental results on combustion of coal-water slurry in fluidized beds. The experiments have shown that (i) the agglomerating property of coal slurry make it possible to minimized the elutriation carbon heat loss of the fluidized bed buring coal water slurry. (ii) the proposed idea of using heavy bed material is of effectiveness in eliminating the deposition of big agglomerating lumps in the bed. (iii). under proper operating condition the coal slurry of various water contents can be burnt in the fluidized bed with high combustion efficiency and good stability even though water contents is as high as 75%. (iv) NO_x emission of the fluidized bed burning coal water slurry is lower than that burning dry coal. (v) Sulphur capture effectiveness of mixing CaO directly into coal water slurry is much better than that of using conventional desulfuration method in the fluidized bed. Hence burning coal water slurry in the fluidized bed is a hopeful technology.

原文刊于工程热物理学报, 1983, 4(2): 177-182

水煤浆滴燃烧过程的简化数学模型

岑可法 倪明江 曹欣玉 陆重庆 陈运铣

(浙江大学)

作者以前的工作表明[1]水煤浆滴的燃烧过程可分为水份蒸发、挥发份析出燃烧和焦团燃烧三个有重迭的阶段。本文根据实验数据对这三个阶段进行分析的基础上提出了水煤浆滴燃烧过程的简化数学模型,并与实验结果进行了对照.

一、水煤浆滴燃烧各阶段的速率计算

按照 Splading 液滴蒸发理论计算水煤浆滴的水份蒸发速率和蒸发平衡温度:

$$m_w = (\lambda_m \text{Nu}/2c_{pm}r_p)\ln[1 + c_{pm}(T_g - T_k)/(L - Q_r/4\pi r_p^2 m_w)] \quad (1)$$

$$T_k = T_g - (L/c_{pm} - Q_r/4\pi r_p^2 c_{pm} m_w)[(1-c_{H_2O}^\infty)/(1-c_{H_2O}^s) - 1] \quad (2)$$

结合相应的汽液平衡公式和直径变化规律,容易算出浆滴的水份蒸发时间. 图1表明理论计算和实验结果符合较好,说明对水份较高滴径不大的水煤浆滴可以略去煤粉的影响而按纯水滴计算其水份蒸发速率. 图1中还给出了水煤浆滴在流化床内水份蒸发时间的计算曲线,由于传热传质的强化,其蒸发时间大幅度缩短.

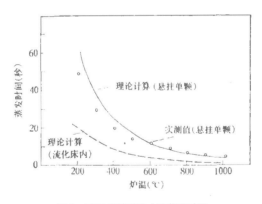

图1. 不同炉温下的水份蒸发时间

挥发份析出过程是复杂的物理化学过程,它既受到挥发份各组分不同的热解动力规律的影响,还受到各种物理因素如颗粒内部传热传质过程的影响. 作为初步近似,本文采用简化的单组分热解动力模型来近似计算水煤浆滴的挥发份析出速率:

$$dV/d\tau = K_{vo}\exp(-E_v/RT_p)(V^f - V) \quad (3)$$

实验测得平顶山烟煤煤浆挥发分析出的有关参数为 $K_{vo} = 12000$ 1/秒, $E_v = 17000$ 卡/摩尔, V^f 值见表I. 图2给出的理论计算与实验结果的比较表明式(3)可用来近似估

本文曾于1983年10月在西安召开的中国工程热物理学会第四届年会上宣读.

计水煤浆滴的挥发份析出速度.

表 1

温度℃	300	400	500	600	700	800	900	1000
V^f%	3.7	11.3	16.7	19.0	20.3	21.0	22.7	24.0

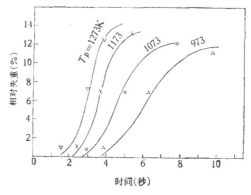

图 2 不同温度下挥发份的析出过程

水煤浆焦团的燃烧反应速率用下式计算:

$$m_c = \varphi \rho_{o_2} c'_{o_2} K_{co} \exp(-E_c/RT_p) \quad (4)$$

对平顶山烟煤煤浆焦团,实验测得其 $K_{co} = 84000$ 米/秒, $E_c = 28000$ 卡/摩尔.

实验表明,氧通过焦团表面灰层的扩散阻力对燃烧速度影响很大. 假定灰层为一均匀球壳,对球壳求解扩散方程,可得焦团未燃核心表面处氧浓度为:

$$c'_{o_2} = c^0_{O_2}/[(1/K_c) + (1/\beta)(r_i/r_p)^2 + (r_p - r_i)r_i/D_i r_p] \quad (5)$$

对平顶山烟煤煤浆,氧通过其灰层的扩散系数 D_i 约为 0.15—$0.18 D_o$.

二、有关问题的进一步讨论

以上讨论了水煤浆燃烧三个主要阶段的计算. 本模型在对整个燃烧过程的计算中没有把各阶段截然分开,因为物理过程进行不平衡造成了各燃烧阶段的互相重迭. 图 3 给出的实验数据说明了挥发份析出和焦团燃烧过程的重迭.

水煤浆滴在燃烧过程中的直径和密度是两个重要的过程参数,一般有:

$$d_p/d_{po} = (G/G_o)^a; \quad \rho_p/\rho_{po} = (G/G_o)^b; \quad 3a + b = 1 \quad (6)$$

图 4 给出了水煤浆滴在不同燃烧方式下的直径变化. 对悬挂浆滴,其焦团燃烧阶段($G/G_o < 50\%$ 时)基本上满足:

$$d_p/d_{po} = 1; \quad \rho_p/\rho_{po} = G/G_o; \quad (a = 0, b = 1) \quad (7)$$

在流化床内燃烧时,颗粒表面灰层不断为剧烈运动的床层所冲刷,故对中等灰分的平顶山烟煤煤浆,基本上满足[2]:

$$d_p/d_{po} = (G/G_o)^{1/3}; \quad \rho_p/\rho_{po} = 1; \quad (a = 1/3, b = 0) \quad (8)$$

高灰分的煤浆,如洗选尾煤,即使在流化床内,由于其灰的生成速度大于磨损速度,其直径仍不遵从式(8)的规律[3],而需用式(6)描述.

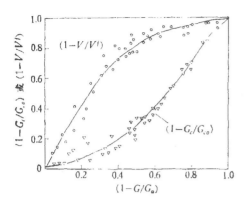
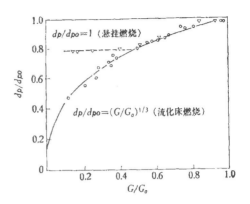

图 3　挥发份析出和焦碳燃烧的重迭　　　　　图 4　水煤浆滴在燃烧过程中的直径变化

关于燃烧各阶段的传热传质计算，对悬浮燃烧用 Ranze-Mashall 方程：

$$Nu = 2 + 0.69 Re^{1/2} Pr^{1/3} \tag{9}$$
$$Sh = 2 + 0.69 Re^{1/2} Sc^{1/3} \tag{10}$$

对流化床燃烧，实验表明用上两式计算结果偏低，故暂用作者得出的初步关联式．

$$Nu = 2hr_p/\lambda_g(1-\varepsilon_b) = 18 + 126.3 Re^{0.98} Pr^{0.33}(r_p/r_m)^{-1.47} \tag{11}$$
$$Sh = 2\beta r_p/D_g(1-\varepsilon_b) = 18 + 126.3 Re^{0.98} Sc^{0.33}(r_p/r_m)^{-1.47} \tag{12}$$

三、综合模型

基于上面的分析，可以写出描述水煤浆滴燃烧过程的微分方程组：

$$dG_w/d\tau = -(2\pi r_p \lambda_m Nu/c_{pm})\ln[1 + c_{pm}(T_g - T_k)/(L - Q_r/G_w)] \tag{13}$$
$$dG_v/d\tau = -K_{vo}G_o \exp(-E_v/RT_p)(V^f - V) \tag{14}$$
$$dG_c/d\tau = -4\pi r_i^2 \varphi \rho_{o_2} K_{co} \exp(-E_c/RT_p) c_{o_2}^i/(1-A) \tag{15}$$
$$dT_p/d\tau = 3[m_c Q_c - h(T_p - T_g) - Q_r - Lm_w]\rho_p c_p r_p \tag{16}$$

初始条件：$\tau = 0$ 时 $G_w = G_{wo}$, $G_v = G_{vo}$, $G_c = G_{co}$, $T_p = T_{po}$

采用实验取得的基础数据，对平顶山烟煤煤浆在流化床内的燃烧过程，用规定精度自动选步长的龙格-库塔法进行了数值求解．图 5 给出了模型计算结果和相应的实测数据，可见计算结果和实测数据符合得较好。

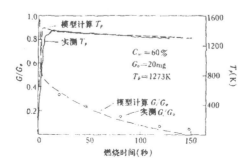

图 5　燃烧过程中重量和温度的变化

四、结论

1. 对水分较高直径不大的水煤浆滴,计算水份蒸发速度时可以略去煤粉影响.
2. 作为初步近似,可用单组分热解动力学公式估计水煤浆滴的挥发份析出过程.
3. 水份蒸发造成的多孔性使水煤浆焦团的燃烧反应活化能较同煤种原煤为低.
4. 所提出的简化计算模型和一些实测数据符合较好,可作为进一步发展的基础.

主 要 符 号

A 焦碳含灰量	G_c 浆滴焦碳重量	r_p 浆滴半径
$c^s_{H_2O}$ 表面水蒸汽浓度	G_0 浆滴初重	T_g 烟气温度
$c^\infty_{H_2O}$ 环境水蒸汽浓度	G_v 浆滴挥发份重量	T_k 蒸发平衡温度
$c^s_{O_2}$ 焦碳核心表面氧浓度	G_w 浆滴水份重量	T_p 浆滴温度
$c^\infty_{O_2}$ 环境氧浓度	h 对流换热系数	V 挥发份析出量
c_p 浆滴比热	k_c 焦碳燃烧反应速度常数	V_f 挥发份最终析出量
c_{pm} 汽气混合物比热	k_{co} 焦碳燃烧反应频率因子	β 传质系数
D_g 气体扩散系数	k_{vo} 挥发份析出综合频率因子	e_b 床层空隙度
D_i 灰层内氧扩散系数	L 汽化潜热	λ_g 气体导热系数
D_0 氧在烟气中扩散系数	m_c 焦碳失重速度	λ_m 汽气混合物导热系数
d_p 浆滴直径	m_w 水份蒸发速度	ρ_{O_2} 氧气密度
d_{po} 浆滴初始直径	Q_c 焦碳热值	ρ_p 浆滴密度
E_c 焦碳燃烧反应活化能	Q_r 辐射传热量	ρ_{po} 浆滴初始密度
E_v 挥发份析出综合活化能	r_i 焦碳核心半径	φ 反应机理因子
G 浆滴重量	r_m 床料半径	

参 考 文 献

[1] 岑可法等: "Experimental Study on Combustion Properties of CWS" Proc. 4th Inter. Symp. on Coal Slurry Combustion U. S. A. III (1982)

[2] 岑可法等; "Experimental Study on Combustion Process of CWS in FBC" Proc. 7th Inter. Conf. on FBC, U. S. A., (1982), pp. 253—263

[3] 倪明江等: "Experimental Study on Fluidized Bed Combustion of Washery Tailings" 国际沸腾燃烧及应用技术学术会议论文集,中国北京,(1983)

SIMPLIFIED BURNING MODEL FOR COAL-WATER SLURRY DROPLETS

Cen Kefa Ni Mingjiang Cao Xinyu Lu Chongqing Chen Yunsien

(Zhejiang University)

Abstract

Based on experimental results, analysis is made for some problems of burning process of CWS droplets such as water evaporation, release and combustion of volatile, combustion of char, heat transfer and mass transfer and so on. Further, a simplified mathematical model for the burning process of CWS droplets is proposed. The computed curves of this model is comparable with some experimental results.

煤及水煤浆燃烧过程中
NO_x 和 SO_2 生成的研究

岑可法　袁镇福　曹欣玉

孔水源　倪明江　陈运铣

(浙江大学)

摘　要

本文系经试验室试验研究，探索煤及水煤浆燃烧过程中 NO_x 产生的原因及加石灰脱硫的效果。试验表明，煤燃烧过程中，NO_x 主要由燃料中氮氧化产生，而热力 NO_x 较低，其量仅为化学平衡值的 $1/12 \rightarrow 1/14$，不到产物中 NO_x 总量的10%。试验还表明，煤中的氧参与此氮的氧化反应。在挥发物析出及其燃烧阶段，NO_x 生成较快，占总生成量的30%，而 SO_2 生成量为总生成量的15—20%，其余在焦碳燃烧阶段生成。随炉温升高，NO_x 生成的增长速度为 3.3ppm/°C。由试验回归得出，生成 NO 的活化能在挥发物析出和燃烧时为 11500 仟卡/摩尔，在焦碳燃烧时为 33500 仟卡/摩尔。薄层脱硫燃烧试验表明，在相同的 Ca/S 比下，水煤浆加入石灰乳的脱硫效率比煤粉加 CaO 粉的为高。

前　言

煤燃烧产生的各种有害气体，特别是氮氧化物 NO_x 和二氧化硫 SO_2 对人类的健康和生态环境构成威胁，一般说来，锅炉烟囱排出的 SO_2 和 NO_x 浓度在几百 ppm 以上时，已足以使人类受害。一台30万瓩锅炉机组，如燃用含硫3%、热值5000仟卡/公斤的煤，其 SO_2 的年排放量高达53000吨/年，而 NO_x 排放量也有同样的数量级。本文的目的是

———1983年1月18日收到。

经过试验室的研究，探索煤及水煤浆燃烧过程中NO_x产生的原因，从而试图预测燃烧不同煤种时NO_x产生的数量。同时还测量了在燃烧各过程中NO_x及SO_2生成的动态特性，以便在各燃烧阶段采用相应措施使之降低。由于水煤浆作为一种新型代用燃料已日益受到世界各国的重视[1]，国内亦开展了研究[2]。本文着重对比煤及水煤浆燃烧过程中NO_x及SO_2的生成规律，探索了水煤浆NO_x排放低的原因。

试验装置和方法

采用热冲击式的方法来试验研究煤和水煤浆在薄层燃烧过程中NO_x，SO_2，CO，C_nH_m等的析出动态过程及规律，其试验系统见图1。电炉装在一台可迅速上下移动的升降装置上，在煤或煤浆薄层中（约1—2克）埋有细丝热电偶，在薄层上部有烟气取样管取样，用自动记录天平测定燃烧过程中的失重[2]，并有脉冲讯号使与温度测量及烟气分析同步，同时记录煤或煤浆的燃烧速度、温升及烟气组成。

采用FP型红外气体分析仪连续测定燃烧过程中CO_2，CO，HC，NO和SO_2的变化，用气敏色谱仪测定H_2，CH_4，CO的含量，用化学萤光法（RS—325L型）测量NO_x（同时可分别测量烟气中NO和NO_x），此外还用碘量法校核SO_2含量。

试验用平顶山煤，其分析数据为：$C^y = 47.79\%$，$A^y = 27.11\%$，$V^t = 23.2\%$，$Q^y_{ow} = 5843$ kcal/kg，$S^y = 2.45\%$，$N^y = 1.45\%$。

煤及煤浆燃烧时NO_x及SO_2的析出规律

1. 影响NO_x及SO_2产生的诸因素

煤燃烧过程中产生的NO_x主要由"热力"氮氧化物及"燃料"氮氧化物组成。所谓"热力"氮氧化物是指空气中氮在高温和催化剂作用下所生成的NO_x，图2中示出了700℃—1000℃下的试验结果，图中同时画有在各温度下反应的平衡浓度。从图可以

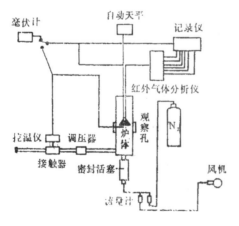

图1 薄层燃烧实验台系统图

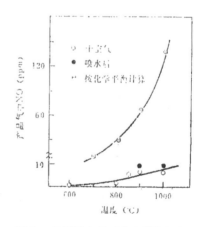

图2 温度和水份对热力NO生成的影响

发现，实测的热力氮氧化物NO_x仅为化学平衡时NO_x浓度的 1/12—1/14，不到煤燃烧时产生的NO_x总量的百分之十，因此热力NO_x不是NO_x的主要来源。

在炉温为900℃、1000℃时，将水喷入热风中（喷水速度为400克/小时），可观察到在水蒸气作用下，离开薄层的气体中热力NO_x没有明显变化，即改燃水煤浆对热力NO_x生成影响不大。

A. 燃料氮对生成NO_x的作用

众所周知，燃料中含有一定数量的氮，但其和氧反应产生NO_x的机理目前还未完全搞清，Be'er等[4]认为煤燃烧时，其中的氮可分为挥发物中氮和焦炭中氮，它们都可产生NO_x。挥发物释出时含有一定量NH_3，按下式与氧进行反应：

$$NH_3 + O_2 \rightarrow NO + \cdots\cdots$$

而焦炭中氮则按下式生成NO：

$$焦炭 N + O_2 \rightarrow NO + \cdots\cdots$$

在炉内空间，NO还会进行一系列还原或氧化反应。

因此NO_x的生成量和炉内气氛、温度水平和混合工况有密切关系。

用平顶山煤粉（$N^y = 1.45\%$）和基本上不含氮的电极炭在炉内进行试验，其结果示于图3。可以明显看出，电极炭的NO_x析出量和热力NO_x的相近，但煤的NO_x析出量大得多，可见NO_x主要由燃料氮产生。对$N^y = 1.45\%$的平顶山煤，当过剩空气系数$\alpha = 1.2$时，理论上由燃料氮全部反应时，在烟气中产生的NO浓度可达2870ppm，因此，并非所有燃料氮均能转变成NO_x。

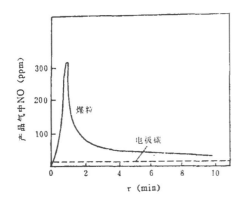

图3 电极碳与煤粒燃烧时NO生成情况
炉温1000℃

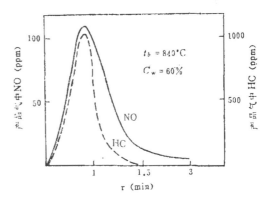

图4 在氮气氛下，水煤浆薄层燃烧时NO、HC的析出情况

B. 在惰性气氛中燃料NO_x的生成

在N_2气氛中，水煤浆薄层燃烧时NO_x的生成规律示于图4，可以明显看出，由于平顶山烟煤含氧$O^y = 3.97\%$，已能生成大量的NO，特别是在挥发物释出和燃烧阶段，产生的NO尤多。在焦炭燃烧阶段，由于反应较慢使产生的NO浓度明显下降。因此，低氧燃烧或"两段燃烧"在焦炭燃烧阶段对控制NO_x形成是有好处的。但是在挥发物释出和燃烧阶段则效果较差，只能设法使产生的NO_x在气相空间还原来降低。同时还表

明，挥发物中形成 NO_x 的机理还应进一步研究，因为此时气相中含氧量极低，未必如 Be'er 所设想的那样，主要由 $NH_3 + O_2 \rightarrow NO + \cdots\cdots$ 生成，是否可能燃料氮就在固相中分解或反应形成 NO_x 。

C. 挥发物释出、燃烧时和焦炭燃烧时产生 NO_x 及 SO_2 的比例

为探明在燃烧过程中形成的 NO_x 及 SO_2 究竟以挥发物的为主还是以焦炭的为主，在不同炉温（700℃—1100℃）下对平顶山煤粉和含煤粉60%的水煤浆分别进行薄层燃烧动态过程的试验，其典型结果示于图5。试验表明，在挥发物释出及其燃烧阶段，NO 和 SO_2 的产生都比较集中，但由于挥发物释出和燃烧的时间较短，仅占燃烬时间的10%，故其绝对量不大。平顶山煤（$V^f = 23.2\%$）挥发物释出和燃烧时生成的 NO_x 约占总 NO_x 产生量的30%左右，SO_2 约占15—20%左右。因此，从数量上讲，焦炭燃烧时产生的有害气体较多，但在挥发物燃烧阶段析出较为集中。

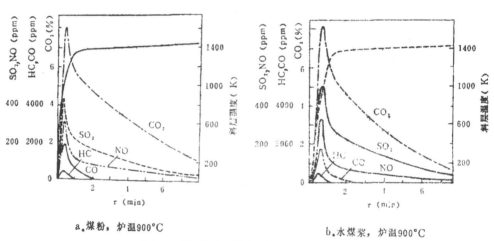

a. 煤粉，炉温900℃　　b. 水煤浆，炉温900℃

图5　薄层燃烧时各气体组份的析出过程

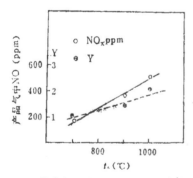

图6　炉温与 NO_x 最大浓度值及最终析出率的关系

$$Y = \frac{NO\text{最终析出量}}{700℃ 下 NO \text{最终析出量}}$$

D. 温度的影响

炉温升高时（700-1000℃）燃料氮产生的 NO 也增加。其增长率约为3.3ppm/℃，见图6，比 Be'er[4] 在燃煤沸腾床所得2.56ppm/℃略高。

2. 煤燃烧过程中 NO_x 释出的动力特性

由于问题较复杂，这里只作一些粗略的分析：从煤的燃烧理论可知，实际上挥发物和焦炭是直到最后同时燃烬的，但可以认为前期以挥发物释出、燃烧为主。根据图4，5所示的试验结果也可以看出，前期以挥发分释出和燃烧生成 NO_x 为主，后期以焦炭燃烧生成 NO_x 为主。这样就能和煤的燃烧理论结合起来考虑，即 NO 的生成速度为：

$$\frac{dC_{NO}}{d\tau} = \left[\frac{d(C_{NO})_V}{d\tau} + \frac{d(C_{NO})_C}{d\tau} \right]$$

按动力学原理，挥发物释出的速度为：

$$-\frac{dV^t}{d\tau} = K_V[\exp(-E_V/RT)](V_T^t - V^t) \quad (kg/sec) \tag{5}$$

式中：V_T^t，V^t 分别为在某一炉温下无穷长时间和 τ 秒内析出的挥发物量，同样可写出由挥发物产生 NO 的速度：

$$\frac{d(C_{NO})_V}{d\tau} = K_{OV}[\exp(-E_{OV}/RT)]\left[1 - \frac{(C_{NO})V}{(C_{NO})V_T}\right](C_{NO})_{V_T} \tag{6}$$

式中：K_{OV}，E_{OV}——挥发物释出和燃烧时生成 NO 的假想频率因子和活化能。

按上式整理平顶山煤粉薄层燃烧的试验结果如图 7 所示，经回归可得 $K_{OV} = 27.9$ 秒$^{-1}$，$E_{OV} = 11500$ 仟卡/摩尔。

另外根据焦炭的燃烧理论：

$$-\frac{dG}{d\tau} = A\varphi_C \rho_{O_2} K_C[\exp(-E_C/RT)]C_{O_2}^n \quad kg/sec \tag{7}$$

式中：　　　　G——焦炭瞬时重量

A、K_C、E_C——碳的反应表面积，频率因子，活化能。

n——反应级数。

φ_C——碳氧比，生成 CO_2 时 $\varphi_C = \frac{12}{32}$

ρ_{O_2}，C_{O_2}——氧气密度和碳表面氧浓度。

由图5可见焦炭燃烧时，氮和氧的反应动态过程与 C 和 O_2 反应生成 CO_2 的动态过程是十分类似的，近似为两根斜率不同的直线，因此 NO 的生成速度亦可近似写成：

$$\frac{d(C_{NO})_C}{d\tau} = A\varphi_{NO}\rho_{O_2}K_{OC}[\exp(-E_{OC}/RT)]C_{O_2}^n \tag{8}$$

式中：　　φ_{NO}——NO_x 生成量与生成 NO 的氧的消耗速度之比。

K_{OC}，E_{OC}——焦炭燃烧生成 NO 的假想频率因子和活化能。

我们在沸腾炉中的试验表明（见另文），当炉温为900℃时，过剩空气系数 $\alpha = 1.15$，用化学荧光法测得 NO_x 为 215ppm，其中 NO 为 210

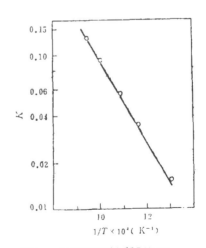

图 7　由挥发份产生NO的K_i与1/T关系图

ppm，NO_2仅5ppm，即只占2.3%，和国外油、气炉测得数据相似[8]，即燃烧过程中主要生成NO，此时$\varphi_{NO}=60/32$。(8)和(7)式相比可得：

$$K_{oc}\exp(-E_{oc}/RT)=\left[\frac{d(C_{NO})_c}{d\tau}\bigg/\frac{dG}{d\tau}\right]\frac{\varphi_c}{\varphi_{NO}}\cdot K_c\exp(-E_c/RT)$$

(9)

等式右边为试验实测值，平顶山煤焦炭的K_c和E_c值经测定为：

$K_c = 84000\text{m/sec}$

$E_c = 28000\text{kcal/mol}$

把燃烧时产生NO及CO_2的试验数据整理如图8所示。代入焦炭的K_c值，经回归可得平顶山煤在焦炭燃烧阶段生成NO假想频率因子及活化能分别为：

$K_{oc} = 553$米/秒

$E_{oc} = 33500$仟卡/摩尔

根据这些反应动力学数据，我们便有可能计算在不同燃烧条件和燃烧方式下，NO_x释出的数量随时间变化的规律。由于条件限制未能对其它典型煤种进行试验，而试验次数也不够多，故上述动力学数据仅供分析问题时参考。

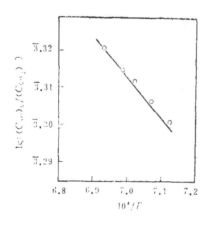

图8 焦炭燃烧生成NO及CO_2的比例

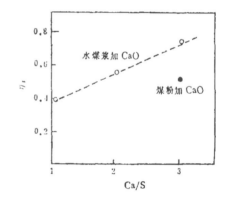

图9 薄层燃烧脱硫率比较

3.燃用煤粉和水煤浆用CaO脱硫时脱硫效率的比较

为了探索加入石灰石的脱硫效果，在本试验炉内用CaO粉（$d<70\mu m$）和平顶山煤粉拌匀，另外按相同的Ca/S比将CaO加入含水量为50%的平顶山煤粉制成的水煤浆内，石灰乳能与水煤浆中的煤粉混合得更为均匀，因而其脱硫效率比粉状CaO加入在煤粉中燃烧时脱硫效果好得多（见图9）。我们在煤及水煤浆的沸腾燃烧中得出同样的试验结果[2]，通常脱硫效率约高20%。

结 论

根据以上初步试验分析，可以得出几点结论：

1. 煤燃烧过程中，NO_x的产生来源于热力氮氧化物和燃料氮氧化物，但热力氮氧化物仅为达到化学平衡时的NO_x量的1/12—1/14，不足煤燃烧时产生的NO_x总量的百分之十，在本试验中90% NO_x为燃料氮所产生。

2. 煤中含有的氧能与煤中含有的氮反应产生大量的NO。

3. 煤燃烧过程中，挥发物释出及燃烧阶段约占燃烬时间10%，但生成的NO_x占总量的30%，SO_2占总量的15~20%，其余在焦炭燃烧阶段生成。随炉温升高（在700℃—1000℃）NO_x成比例地增加，其增长率约为3.3ppm/℃。

4. 用平顶山煤进行试验，得出其NO_x释出反应动力特性为：在挥发物释出阶段，生成NO的假想频率因子$K_{ov}=27.9$ 秒$^{-1}$，活化能$E_{ov}=11500$仟卡/摩尔。在焦炭燃烧阶段，生成NO的假想频率因子为$K_{oc}=553$米/秒；活化能$E_{oc}=33500$仟卡/摩尔，可供分析问题时参考。

5. 水煤浆因能和石灰乳混合均匀，故其脱硫效率比煤粉加CaO粉时为高。

参考文献

[1] Voelker, G. and Foster, C., Proceedings of Fourth International Symposium on Coal Slurry Combustion, vol.1, 1982, PETC, U.S.A.

[2] 岑可法，曹欣玉，洪积瑜，倪明江，孔水源，陈运铣，ibid., vol.3.

[3] Сигал, И.Я., Газовая Промышленность, [2], 24(1969).

[4] Pereira, F.J., Beer, J.M. et al., 15th Symposium (International) on Combustion, The Combustion Institute, Pittsburgh, p.1149, 1974.

INVESTIGATION ON THE FORMATION OF NO_x AND SO_2 DURING COAL AND COAL SLURRY COMBUSTION

Cen Kefa Yuan Zhenfu Cao Xinyu Kong Shuiyuan

Ni Mingjiang Chen Yunxian(Yunsien Chen)

(*Zhejiang University*)

ABSTRACT

This paper presents a study of the formation of NO_x and the effects of desulphuration by limestone in the combustion of coal and coal slurry.

It is shown that the NO_x are formed mainly from the oxidation of nitrogenous components in fuel and the oxygenous components in fuel participate in this reaction.

About 30% of NO_x and 15-20% of SO_2 are formed in the volatilization and in the initial stage of combustion (about 10% of the total combustion time), while the rest in the char combustion stage. The rate of increment of NO_x with oven temperature is 3.3 ppm/°C. The activation energy of NO formation is 11,500 kcal/mol in initial stage and 33,500 kcal/mol in the char combustion stage.

原文刊于燃料化学学报, 1984, 1(12): 33-40

脉冲沸腾床流体动力特性的试验研究

岑可法　康齐福　严建华　蔡安明

(浙江大学)

本文报道了一种新型的沸腾床——脉冲鼓风沸腾床，作者已对这种沸腾床进行了一系列的研究，结果表明：脉冲鼓风沸腾床在强化传热，减少飞灰污染等方面有突出的优点。它可能对现有沸腾炉在提高燃烧效率、减少飞灰污染方面提供有效的途径。

一、背景

沸腾燃烧锅炉由于具有高的燃烧强度、燃料适应性广、床温均匀、混合良好等优点，已得到大量应用。表1是对国内一些沸腾炉的统计数据。可见，飞灰中可燃物损失甚大，燃烧效率低、飞灰污染严重是现有沸腾炉的二大棘手问题。从节能和环境保护角度出发，对现有沸腾炉进行技术改造势在必行。受早期脉冲燃烧发动机的启发，我们提出对脉冲鼓风沸腾床进行研究[1]，目的在于化费较少的改造投资，使燃烧效率增加、飞灰污染降低。

表1　沸腾炉燃用各种煤时的统计数据

煤种	褐煤	石煤	劣质烟煤	无烟煤
C^{fh}, %	10—20	10—25	20—40	30—50
q_4, %	—10	13—26	—20	25—35

二、试验概况

图1是脉冲鼓风沸腾床的系统图，其布风板截面积为 $900 \times 350 mm^2$，风帽开孔率 5.7%，悬浮段截面积为 $1440 \times 350 mm^2$，悬浮段高度1500mm。我们把脉冲鼓风阀安装在近风室的送风道上，用直流电机配以可控硅调速装置在低速段稳定调速。随着阀体的旋转，周期地改变气流的流通面积，达到脉冲鼓风的效果。

三、试验结果

为了研究人为的脉冲鼓风对床内压力脉动的影响，我们采用了电涡流式压力传感器，把模拟信号进行放大，然后记录在模拟磁带记录仪上，并对所记录的信号作了频谱分析。分析结果表示在图2及图3上，纵坐标用分贝数表示。可见，常规沸腾床(不采用脉冲鼓风)内压力脉动信号

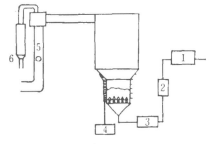

图1　脉冲鼓风沸腾床系统图
1 风机　2 翼形管　3 脉冲鼓风阀　4 多点测压计　5 取样孔　6 除尘器集灰斗

本文曾于1984年10月在天津举行的中国工程热物理学会燃烧学术会议上宣读

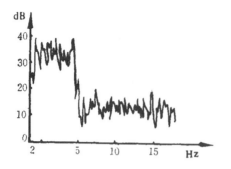 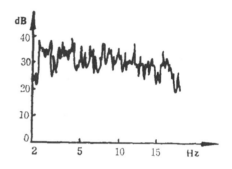

图2 距布风板 230mm 处压力信号的频谱（$f=0$Hz）　图3 距布风板 230mm 处压力信号的频谱（$f=4$Hz）

的能量分布多集中在低频段，而采用脉冲鼓风后，在整个频域信号的能量具有比较平坦的高幅值分布，即表明有更多的谐波分量对床层动力结构起作用。

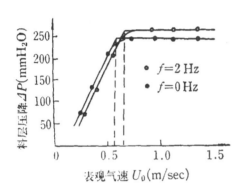

图4 床料为溢流渣的料层压降-流速曲线（静料层高 287mm，平均粒径 $\bar{d}=2.35$mm）　图5 床料为石英砂的料层压降-流速曲线（静料层高 310 mm，平均粒径 $\bar{d}=1.14$mm）

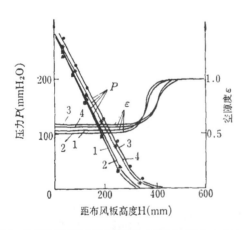 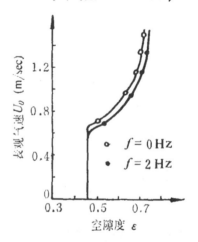

图6 沿床层高度压力及空隙度的变化（静料层高 287 mm，平均粒径 $d=2.38$mm，曲线 1: 流化数 $N=1.1$，$f=2$Hz，曲线 2: $N=1.1$ $f=0$Hz 曲线 3: $N=2$，$f=2$Hz 曲线 4: $N=2, f=0$）

图7 空隙度特性（床料平均粒径 $\bar{d}=2.38$mm）

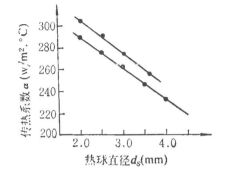

- ● $f=0.4Hz$　○ $f=0Hz$

图 8　不同粒径的电热球传热系数比较

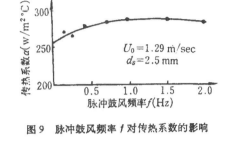

图 9　脉冲鼓风频率 f 对传热系数的影响

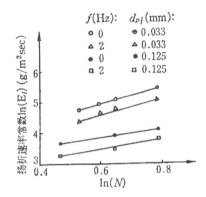

图 10　脉冲与不脉冲送风时扬析量比较

（以溢流渣为床料）

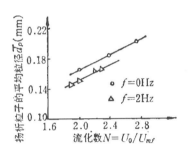

图 11　脉冲与不脉冲送风时扬析粒径比较

（以溢流渣为床料）

通常所说的临界沸腾速度与流体和固体的物性及固体颗粒的大小等因素有关。人为脉冲鼓风后对流化性能有什么影响呢？为此，我们用不同床料对脉冲鼓风的沸腾床作了临界沸腾速度的对比试验，结果见图 4，图 5 所示。试验表明：不论对沸腾炉溢流渣还是对石英砂，采用脉冲鼓风后均能使临界沸腾速度明显降低，这可能是由于脉冲鼓风后，使颗粒间的表面力增大，结果颗粒之间更易于解锁。从频谱图上亦显示脉冲鼓风确对床层的流化特性起作用。

沸腾层内的动力结构与床层的空隙度有关，忽略壁面的摩擦损失及气体的动量变化，固体粒子的宏观运动速度较小，由此引起的压力损失可忽略不计。列出气固混合物的动量方程，忽略高次小量简化后得到：

$$dp/dx = -(1-\varepsilon) \cdot \rho_p$$

式中 ρ_p 为颗粒的表观比重，ε 为空隙度，式 2 表明：通过测定压力 p 沿床高分布可推算出床层的空隙度分布。各个工况的测定和计算结果见图 6，可以发现，距离布风板一段高度内（还未达到沸腾层表面），空隙度几乎为一常数。在这段高度内气泡稳定长大，并有少量聚并，而气泡的大量聚并和破裂多发生在该区域之上。图 7 表明了采用脉冲鼓风后，沸腾床稳定段的空隙度增大。两相流理论告诉我们：颗粒群的空隙度越大，其阻力系数就越小,前述的料层压降-流速曲线也验证了这一理论,采用脉冲鼓风后,床层的压降约降低10%。

曾采用电热球粒子模拟沸腾炉内的煤粒作了气固传热系数的测定。图8是试验结果,显然,脉冲鼓风强化了气固传热过程,这就是说煤粒达到着火所需的时间缩短了。从传热角度讲,看来存在一个最佳的脉冲鼓风频率,其值与风速、粒子尺寸等有关。

有关未燃烬细粒子扬析的研究结果已作了不少的报道[2],我们对脉冲鼓风沸腾床的扬析特性进行了一系列的试验,结果见图10,图11。可见,脉冲鼓风后,被扬析颗粒的平均粒径减小,扬析量减少。这一方面是由于脉冲鼓风后使细粒子在床内停留时间增加了,另一方面是由于脉冲鼓风后,气泡小且分布均匀,这在我们的连续录像中已经发现。Kraft等人[3]的试验亦表明:良好的流化即小气泡其扬析率要比大气泡时小。

四、结论

采用脉冲鼓风后,临界沸腾速度降低、床层阻力下降,折合成相同鼓风量情况下,脉冲鼓风沸腾床的扬析率降低约10%,气固间传热系数提高约10%,初步研究表明:脉冲鼓风沸腾床是一种气泡小、压降低,扬析量少,传热强化,设备简单的新型沸腾床,它可望对提高沸腾炉燃烧效率,降低飞灰污染起积极作用。

参 考 文 献

[1] 岑可法:"关于开展脉动流化燃烧探索的建议",浙江大学研究报告(1984年1月)。
[2] 康齐福等:"宽筛分燃煤沸腾炉的扬析和夹带"工程热物理学报,**6**,1,(1985)
[3] Kraft, W. et al.: In "Fluidization" Othmer, D. F. (ed.) p. 194, Reinhold Publishing Corporation New York, (1956).

EXPERIMENTAL RESEARCHES ON FLOW DYNAMICS BEHAVIOR OF THE PULSATING FLUIDIZED BED COMBUSTOR

Cen Kefa   Kang Qifu   Yan Jianhua   Cai Anming

(Zhejiang University)

Abstract

This paper presents a new type of FBC——Pulsating Fluidized Bed Combustor. Authors have carried out a series of experimental researches in the cold model of the Pulsating FBC. Experimental results show that it will provide an effective way of **increasing** combustion efficiency and preventing the pollution of fly ash.

原文刊于工程热物理学报, 1985, 6(3): 287-290

应用脉冲鼓风来提高沸腾炉燃烧效率的研究
（基础试验部分）

岑可法 康齐福 严建华 蔡安明

提 要

针对流化床的燃烧效率低和飞灰污染严重特点，本文提出了一种新型的流化床燃烧方法：脉冲流化床燃烧。作者对脉冲流化床的各种冷态特性进行了一系列较为系统的研究，结果表明，采用脉冲流化床以后，出现了一系列常规流化床所不具备的优点：临界流化速度小，床层阻力下降，扬析量减少床内颗粒的传热系数提高。研究结果显示，采用脉冲流化床燃烧方式有可能提高燃烧效率，减少飞灰污染和降低风机能耗。

一、研究脉冲流化床的工程意义

煤的流态化燃烧方式具有高的燃烧强度，燃料适应性广，床层混合良好，床温均匀等优点，已在国内外得到推广应用。目前流化床燃烧的关键问题之一是燃烧效率较低，国内现有的沸腾燃烧锅炉大多燃用宽筛分煤粒($0\sim 8mm$)为使其中的大颗粒煤良好流化，不可避免地会导致细颗粒带离炉膛，由于细颗粒在炉内停留时间甚短，这些未燃烬的细颗粒带离炉膛是造成沸腾燃烧锅炉燃烧效率不高的主要原因；沸腾炉另一个关键问题是飞灰排放量大。烧劣质煤时该问题尤其突出。过去几年里，已提出许多方法来解决上述问题[10][11]如飞灰回燃，提高悬浮段温度水平，双床沸腾锅炉[2]，悬浮段设置旋风燃烬室及采用播煤二次风等，并取得了一些结果，但结果还不甚理想。据估计国内现有约二千多台不同容量沸腾炉，其中大部分沸腾炉的燃烧效率较低，从节能和环境保护角度出发，对这些沸腾锅炉进行技术改造势在必行，若化较少的改造费用，使燃烧效率得以提高些，飞灰污染减少些，这是我们提出研究脉冲流态化燃烧的目的所在[1]。作为本研究的初步，我们在浙江大学改装后的脉冲流化床上，作了脉冲流化床流体动力特性的试验研究并考察了脉冲流化床的扬析特性和床内颗粒的传热特性，本文将介绍有关的试验结果。

二、试验概况

脉冲流化床的结构如图1所示：

本文曾在中国工程热物理学会燃烧学学术会议上宣读，1984，天津

试验所用的脉动阀是特别设计的,其可调参数是脉冲频率及脉冲幅度,本文所涉及的气速除非特别说明,一般均指积分气速,扬析试验方法见资料[8],我们采用热球模拟单颗粒研究了脉冲流化床内颗粒的传热特性,采用尾部管道的等速取样及除尘器集灰研究了扬析特性,采用HB—1型电涡流压力传感器,把经过放大器的压力脉冲信号送入磁带机内记录,然后在频谱分析仪上进行了分析。

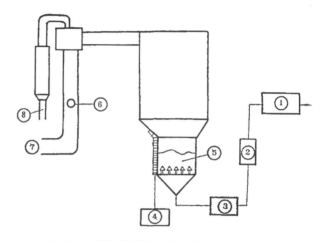

①风 机 ②流量测量 ③脉冲阀 ④多点测压计
⑤取样孔 ⑦引风 ⑧集灰斗

图 1 脉冲流化床结构示意图

表 1 脉冲流化床的主要结构参数

布风板截面 长 宽 比 mm/mm	布风板风 帽开孔率 %	悬浮段截面 长 宽 比 mm/mm	悬浮段高度 mm	出口中心离 布风板距离 mm
900/350	5.7	1440/350	1500mm	2900

三、气流脉动对固体颗粒行为的影响

为说明脉动气流对床内固体粒子运动的影响,我们先来考察脉动气流对单一园球运动的影响。在Stokes流中,脉动气流对一园球的作用可用下式表明[9]:

$$\frac{|V'|}{|W'|} = \frac{\rho_p - \rho_g}{\rho_p + 0.5\rho_g} \frac{1}{\sqrt{1 + \frac{1}{(2\pi f \tau)^2}}} \quad (1)$$

其中:τ 为松驰时间:

$$\tau = \frac{(\rho_p + 0.5\rho_g)d_p^2}{18\mu} \quad (2)$$

式中,$|W'|$ 为气流脉动幅度的绝对值,$|V'|$ 为气流与固粒之间相对速度的脉动幅度的绝对值。式(1)说明了脉动气流对颗粒运动的影响程度。显然,颗粒相对速度的脉动值正比于气流本身的脉动幅度,脉冲气流的频率越低,颗粒直经越小(即松驰时间小),粒子就容易跟随气流脉动。表(2)示出了频率对不同粒径砂子的 $|V'|/|W'|$ 及 τ 的影响。其中取 $\rho_p = 2500kg/m^3$,$\rho_g = 1.165kg/m^3$,$\nu = 16 \times 10^{-6} m^2/s$。

表2 脉动气流的频率不同粒径颗粒下的 τ 和 $|V'|/|W'|$

$\tau(Sec)$ $|V'|/|W'|$ f(Hz) \ d_p	0.1mm	0.25mm	0.5mm	0.75mm	1.0mm	1.25mm
	0.07470	0.4667	1.8668	4.2003	7.4671	11.6674
0.2	0.0934	0.5059	0.9199	0.9825	0.9944	0.9977
0.6	0.27107	0.8694	0.9901	0.9980	0.9994	0.9997
1.0	0.4249	0.9465	0.9964	0.9993	0.9998	0.9999
1.4	0.5492	0.9216	0.9982	0.9996	0.9999	1.0000
2.0	0.6844	0.9858	0.9964	0.9998	0.9999	1.0000
4.0	0.8826	0.9964	0.9998	1.0000	1.0000	1.0000

从表(2)可见，频率 f 越高，气流对颗粒的相对冲刷速度越大，这对加强热质传递，强化燃烧很有利。在流化床中，如气流的脉动频率过高，风速的脉动容易被风室及床层所吸收，不能充分体现脉动鼓风的效果。另一方面，在相同的风速脉动幅值的条件下频率增大会使床层在同一时刻内存在多个脉动气流滞留在床内，使气体与颗粒间的接触时间增大，在一定程度上会影响热质传递过程。但是，频率过低也会影响效果。过低的频率必然导致床内低风速区的持续时间相应延长；各持续时间内的气流速度波动不大，这时实际上已转变成鼓风量为 $\overline{W}+|W'|$ 及 $\overline{W}-|W'|$ 交替进行的普通流化床。因此频率应适中，使得在脉动风速为 $\overline{W}+|W'|$ 时，气流对床内粒子既有一定的相对冲刷速度又能带动颗粒扰动，而当 $\overline{W}-|W'|$ 时，经一定夹带后的粒子开始有所下落，与气流逆向冲刷，上述过程在流化床内表现为周期性的床膨胀、崩塌，使气流与固体颗粒间的相对冲刷要比常规流化床来得强烈，从而加强了热质传递。

适中的气流脉动幅度及频率下，床层周期性的膨胀、崩塌，床内压力波的波动使床内的气泡的长大可能受到抑制，有利于减少扬析量。

四、脉冲流化床的脉动特性

采用了人为的脉冲鼓风后，究竟对床层内压力脉动有何影响，为此我们采用了电涡流压力传感器，经过信号放大，并记录在磁带机上，进行了一系列的频谱分析。

1. 无沸腾床料时，经过布风板后压力脉动的衰减情况见图2和图3，结果表明：压力脉动幅值经过布风板后略有减小，这主要是布风板的阻尼作用之故。

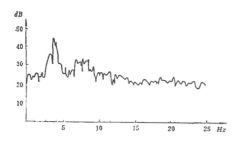 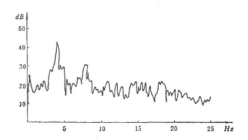

图2 风室内脉动频谱(f=5Hz,无床料)　　图3 距布风板230mm处脉动频谱(f=5Hz,无床料)

2. 脉冲频率对床层内压力脉动频谱的影响见图4和图5。

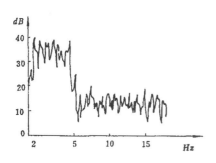 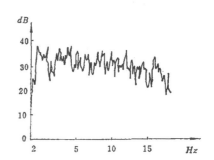

图4 距布板230mm处脉动频谱　　　　图5 距布风板230mm处脉动比频谱
　　(f=0Hz,有床料)　　　　　　　　　　(f=4Hz,有床料)

图4和图5表明，人为脉冲送风对床层内压力脉动确有影响，与不脉冲送风比较，高幅值的频域明显扩展，即有更多的谐波分量对床层起作用，反映着床内扰动剧烈，有利于热质交换。

五、脉冲流化床的临界流化特性

临界流化速度是沸腾炉运行中的重要参数，一般实际沸腾炉的流化数 $N=2\sim3$ 时即可保持良好的流化性能，如果能以较低的空塔速度运行而不致使流化性能恶化，那将是我们所期望的。为此我们用较有代表性的河砂和沸腾炉的溢流渣作了临界流化速度的测定，用传统的压降—流速曲线来整理，见图6和图7及表(3)。

试验表明：不论是对溢流煤渣还是对河砂采用脉冲鼓风后能使临界流化速度明显降低。从频谱曲线中看出，采用脉冲鼓风后，有更多的谐波分量对床层流化特性起作用，这个结果告诉我们采用脉冲流化床后，维持良好的流化性能所需的鼓风量可更小。

从前述的压降流速曲线中不难发现：脉动流化床流化时的料层阻力明显比常规流化床的料层阻力来得低，约降低11%。这对减少风机能耗来说极为有利。

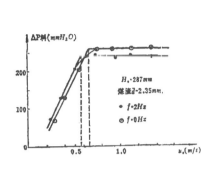 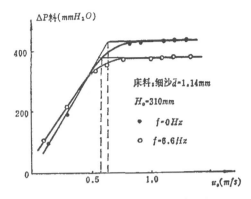

图 6　床料为溢流渣的压降—流速曲线　　　图 7　床料为细砂的压降—流速曲线

表 3　脉冲送风与不脉冲送风临界速度比较

	不脉冲送风	脉冲送风频率 2(Hz)
溢流煤渣	0.68 m/sec	0.58 m/sec
	不脉冲送风	脉冲送风频率 6.6Hz
河　砂	0.66 m/sec	0.59 m/sec

六、脉冲流化床中颗粒与床层的传热

经脉冲后，床内的颗粒与床层的传热有明显的强化，传热系数比普通流化床增大约10%。这里采用热球法来观察床内粒子与床层的传热，热球内装有电热元件，球心引入一细小热电偶以检测球温，用高温粘接剂制成一球形整体。整个热球除了球面有一层很薄的粘结剂之外，几乎都为加热丝所填充，且使用的热球直经较小（$d_s = 1.5 \sim 4.0mm$），故可认为球体温度是均匀一致的，热球的传热系数可从下式得到：

$$\alpha = \frac{I(V-Ir)}{A_s(T_s-T_b)} \qquad (3)$$

其中 T_b 可取为室温。

图8示出了脉冲与非脉冲情况下的传热系数与空塔速度的变化关系。热球的平均高度在离布风板上方210mm处。传热试验中床料是平均粒径为1.14mm的窄筛分砂料，静止高度350mm。

无论是固定床阶段还是流化床阶段，脉动后的传热系数都要比普流化床的要大。但两者随风速度变化的趋势基本是一样的。在固定床阶段，α随空塔速度u_0的增大而略有上升，过了临界流化速度之后。α均迅速增大至最大值α_{max}。

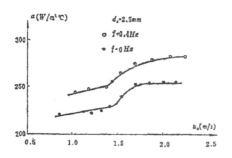
图8 传热系数随风速的变化、脉动与不脉动的比较

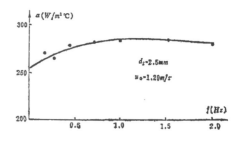
图9 频率f对α的影响

图9反映了脉动频率对传热系数的影响。脉动频率在低频区对传热系数影响较大，在 $f=0\sim1$ 赫兹范围内，α值随f明显增大，脉动频率太低，床层的崩溃时间长，传热系数增加不大，当频率略为加大一些时，从前面的分析可知，在整个周期之内气流对于床内颗粒的平均冲刷加剧，使传热性能变好。似乎存在着一个最佳的传热状态。

图10把脉动与不脉动时的各种粒径（热球粒径）的传热系数作了比较。虽然大颗粒的松弛时间长，但由于小粒子随脉动气流的冲击之后，通过颗粒之间的碰撞和摩擦，在某种程度上对大粒子也起了扰动作用，使大粒子因冲刷也提高了传热系数。这种大小颗粒之间的互相冲刷对流化床的传热无疑是起了重要作用。

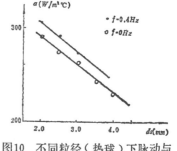
图10 不同粒径（热球）下脉动与不脉动的传热对比

脉冲流化床强调的是利用气流的脉动能量来加强床料的扰动及气固之间的相对冲刷以达到强化传热的目的，这正是我们的出发点之一，脉冲流化床中的传热特点有待于进一步的研究。

七、脉冲流化床的扬析特性

扬析是指从由各种颗粒直径的混合物形成的床层中选择地夹带出细粉颗粒的过程[6]。粒度为 d_p 的固体颗粒从一混合物中扬析出来的速率由下式定义：

$$-\frac{1}{A_t}\frac{dW(d_p)}{dt}=E_{i\infty}\frac{W(d_p)}{W} \tag{4}$$

式中：A_t 为布风板面积，$E_{i\infty}$ 为扬析速率常数，$E_{i\infty}$ 越大说明颗粒的扬析量越大。

关于常规流化床的扬析特性已进行了一些研究，并亦提出了一些准则关联式[8][7]，一般认为扬析速率常数可表示如下函数：

$$E_{i\infty}=f[d_{pi},u_{ti},u_0,\rho_g,(\rho_s-\rho_g),g,N,\mu] \tag{5}$$

我们认为脉冲送风频率对扬析亦有影响，这里我们仅讨论脉冲频率为 $2Hz$ 的扬析特性，并按照如下无因次准则关联式进行回归分析：

$$\frac{E_{i\infty}}{\rho_g u_0} = A_0 \left[\frac{u_0^2(N-1)}{gd_{pi}} \right]^{A_1} \left[\frac{gd_{pi}^3(\rho_s - \rho_g)}{\nu^2 \rho_g} \right]^{A_2} \quad (6)$$

即

$$\frac{E_{i\infty}}{\rho_g u_0} = A_0 \left[\frac{N-1}{F_r} \right]^{A_1} A_r^{A_2} \quad (7)$$

式中流化数 N 反映了床层宏观气流对颗粒群的影响作用，Fr 准则数反映了单颗粒子的惯性力和重力之间的影响因素，Ar 准则数则反映着颗粒的浮力，惯性力与粘性力之间的影响因素。对宽筛分床料的试验点进行了二元线性回归，得到了下述关联式：

$$\frac{E_{i\infty}}{\rho_g u_0} = 1.29 \times 10^{-3} \left[\frac{N-1}{F_r} \right]^{0.4513} [A_r]^{-0.1324} \quad (8)$$

F 检验表明，上关联式在 $\alpha = 0.01$ 水平显著，这说明回归关联式是合理的。

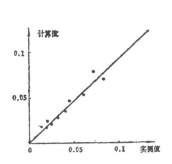

图11 扬析准则数计算值与实测值比较 (f = 2Hz)

结果表明：相同流化数下，采用脉冲送风后扬析量比常规流化床低23%左右，并且扬析粒径更小些。我们认为采用脉冲送风后扬析量减少是由于细粒子在炉内停留时间增加了。

我们曾用连续摄影方法对气泡尺寸进行过粗略的估计，发现采用脉冲送风后，气泡比常规流化床小10%。我们流试验表明：良好流化，即小气泡，其扬析率试有大气泡时要小，kraft etal[6] 的试验也表明了这一点，这可能是脉冲流化床扬析量较少的原因之一。

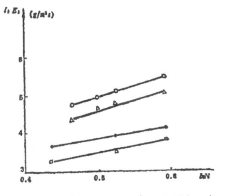

图12 脉动与不脉动送风时扬析量比较

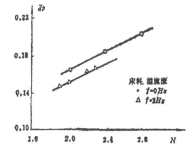

床料：溢流渣 ○ f = 0Hz △ f = 2Hz
图13 脉动与不脉动送风时扬析粒径比较

八、结　论

(一) 采用脉冲流化床后使低频脉动频谱范围扩展，即有更多的谐波分量对床层起作用。

(二)在同样床料、同样静止床高的前提下,脉冲流化床与常规流化床比较,临界风速较低,降低了10%,床层阻力降低11%,混合良好。

(三)脉冲流化床的扬析率比常规的低23%左右($f=2Hz$),并可用以下回归方程表达:

$$\frac{E_{i\infty}}{\rho_g u_0} = 1.29 \times 10^{-3} \left(\frac{N-1}{F_r}\right)^{0.4513} A_r^{-0.1324} \qquad (f=2Hz)$$

(四)在相同条件下,用粗砂粒的脉冲流化床的颗粒与床层之间的传热系数比常规流化床提高约10%。

(五)初步基础试验表明,在相同流化数下,脉冲流化床是小气泡、低压降、扬析损失小、传热强化、设备简单的新型流化床。目前我们正在进行热态燃烧试验。

符 号 说 明

符号	说明	单位
A_t	布风板面积	(m^2)
$A_r = g d_{pi}^3 (\rho_s - \rho_g)/(\nu^2 \rho_g)$	阿基米得数	(—)
A_s	热球表面积	(m^2)
d_p, d_{pi}	颗粒直经	(mm)
$\bar{d}$	平均粒径	(mm)
d_s	热球直经	(mm)
$E_{i\infty}$	扬析速率常数	($g/cm^2 \cdot sec$)
$F_r = \dfrac{g d_{pi}}{u_0^2}$	费鲁特准则	(—)
f	频率	(Hz)
g	$9.8 m/s^2$	重力加速度
H_0	静止料层高度	(mm)
H	距布风板高度	(mm)
H_p	膨胀高度	(mm)
H_f	稳定气泡段高度	(mm)
I	加热电流	($A.$安培)
$N = u_0/u_{mf}$	流化数	(—)
Δp	床层压力降	(mmH_2O)
r	热球外引线电阻	(Ω)
T_s	热球温度	(℃)
T_b	床层温度	(℃)
u_0	空塔速度	(m/s)
u_{ti}	终端速度	(m/s)
u_{mf}	临界流化速度	(m/s)
V	加热电压	($V.$伏特)

v'	气固相对速度的脉动值	(m/s)
w	床料重量	(Kg)
$w(d_p)$	粒径为d_p的床料重量	(Kg)
$\overline{W}$	脉动气流的平均速度	(m/s)
W'	脉动气流的脉动幅值	(m/s)
α	颗粒与床层的传热系数	$(w/m^2℃，瓦/米^2·度)$
ε	空隙率	$(-)$
μ	气体的动力粘度	$(Kg/m·s)$
ν	气体的运动粘度	(m^2/s)
$\rho_p、\rho_g$	分别为粒子和气体的密度	(Kg/m^3)

参 考 文 献

[1] 岑可法，"关于展开脉动流化燃烧探索的建议"，浙江大学，1984年1月26日

[2] 陈运铣、岑可法、张鹤声、张学宏、康齐福，劣质煤沸腾燃烧过程动力特性和双床并联运行沸腾炉提高燃烧效率的试验研究，工程热物理学报，1981.3(2)(4)

[3] 浙江省计量所等，"HB—1型电涡流式薄膜微压计研制报告"，1984年。

[4] HN. 赛罗米亚特尼科夫等，"沸腾层过程"（中译本），1965年。

[5] 国井大藏，O.列文斯比尔编，"流态化工程"（中译本），石油化学工业出版会，1977年

[6] Kraft, W. etal In "Fluidization" othmer, D.F. (ed) P194, Reinhold Publishing Coyperation, New York 1956.

[7] Wen, C.Y. & Hashinger, R.F., AICHE.J.6, 220, 1960.

[8] 康齐福等，"宽筛分燃煤沸腾炉的扬析和夹带"，工程热物理学会第四届年会报告，1983.10.

[9] 岑可法，"燃烧理论"上册，浙江大学，1981年。

[10] Поднмоь. В. Н.: Прмкладнле мсследоьанил Вмбрамионочо чоренмя Казань 1979.

[11] Рассудоь Н.С., О Грименени Импульсной Лодачи Воздуха в Топках с кипящим слоем Теплоэнергетека, 1983, (1) pp62-64.

Investigation on Increasing the Combusion Efficiency of FBC by Means of Pulsating Gas Flow

Cen Kefa, Kang Qifu, Yan Jianhua, Chai Anming

ABSTRACT

Considering lower combustion efficiency and serious pollution of the fly ash from the ordinary FBC (Fluidized Bed Combustor), this paper reports a new method of fluidized bed combustion (Pulsing Fluidized Bed Combustion). Authors have carried out a series of expeirmental researches on the flow dynamics and heat transfer behaviors in the cold model of the pulsing fluidized bed combustor. Experimental results show that the pulsing fluidized bed has following advantages which the ordinary fluidized bed does not possess: (1) minimum fluidizing velocity is lower than that of ordinary F.B. (2) pressure-drop at fluidization stage is lower than that of conventional F.B. (3) the quantity of particle etutriated is less than that of conventional F.B. (4) heat transfer coefficient between the bed materials and single particle is about 10% higher than that of the conventional F.B. We think that perhaps pulsing fluidized bed combustion is an effective way to increase the combustion efficiency and prevent the pollution of fly ash.

原文刊于浙江大学学报, 1985, 4(19): 37-46

煤粉颗粒在气流中的受力分析及其运动轨迹的研究

岑可法 樊建人

提 要

本文分析了煤粉颗粒在气流中所受的力，用Lagrangian方法建立了煤粉颗粒运动方程式，并进行了数值求解，分析了各个力对颗粒运动轨迹的影响。

一、前言

在燃烧技术实践中，经常碰到大量的气固两相流动问题，煤粉颗粒在制粉管道或炉内的输送或燃烧是其中最典型的例子。煤粉在管内和炉内的运动规律十分复杂。第一、煤粉颗粒是分散相，有大有小，其运动规律各异，煤粉之间及煤粉与管壁的相互碰撞对运动带来较大的影响。第二、管内的流动工况通常为湍流。因此气流脉动对煤粉颗粒运动的影响及煤粉的存在对气流的影响均非常复杂。第三、由于惯性不同，气流与煤粉颗粒之间存在着相对速度气流曳引阻力使煤粉加速运动，同时由于流动中存在着压力梯度、速度梯度及煤粉颗粒形状不对称，颗粒之间及管壁的相互碰撞，因此煤粉颗粒出现高速旋转而产生升力效应。第四、在不等温流动过程中会产生热致迁移现象。第五、煤粉输运过程中颗粒的碰碎、炉内着火，燃烧过程中煤粉的燃烧还会发生变质量运动问题。由于上述这些因素，因此要全面的分析和计算煤粉颗粒的管内输送和炉内运动问题是困难的。过去的分析和计算大都是考虑气流曳引阻力对颗粒运动的影响，而略去了其他因素。本文的目的是试图较全面的分析煤粉颗粒在气流中可能受到的力及这些力对煤粉颗粒运动规律的影响，并对部分问题进行了试验验证，并用跟踪颗粒运动的方法在Lagrangian坐标系中建立了运动方程,通过数值求解得出了各个力对煤粉颗粒运动轨迹的影响情况。

二、煤粉颗粒的受力分析及Lagrangian型运动方程式

在Lagrangian坐标系中，根据牛顿第二定律，可以建立起以下颗粒运动方程：

$$m_p \frac{d\vec{v_p}}{dt} = \Sigma \vec{F} \qquad (1)$$

$\Sigma \vec{F}$ 为颗粒所受的合力

煤粉颗粒在气流中所受到的力如下：

1) 颗粒所受的气流曳引阻力 $\vec{F}_r$

$$\vec{F}_r = (\rho_a A_p C_D/2)|\vec{v}_a - \vec{v}_p|(\vec{v}_a - \vec{v}_p) \tag{2}$$

2) 压力梯度力 $\vec{F}_p$

$$\vec{F}_p = -V_p \mathrm{grad} p \tag{3}$$

对于单个颗粒（或浓度很小的悬浮系统）由于小颗粒的存在不影响流体的流动，对流体相来说，作为一种近似可以认为：

$$\rho_a \frac{d\vec{v}_a}{dt} = -\mathrm{grad} p \tag{4}$$

则

$$\vec{F}_p = V_p \rho_a \frac{d\vec{v}_a}{dt} \tag{5}$$

3) 颗粒由于自转而具有的升力 $\vec{F}_e$（Magnus效应[1]）

$$\vec{F}_e = \frac{1}{8}\pi\rho_a d_p^3 (\vec{v}_a - \vec{v}_p) \times \vec{\omega} \tag{6}$$

考虑到实际上由于煤粉颗粒并非球形等因素，引入试验系数 R 来修正。即：

$$\vec{F}_e = \frac{k}{8}\pi\rho_a d_p^3 (\vec{v} - \vec{v}_p) \times \vec{\omega} \tag{7}$$

4) 由于速度梯度引起的Saffman升力 $\vec{F}_s$：

$$\vec{F}_s = 1.61(\mu_a \rho_a)^{\frac{1}{2}} d_p^2 (v_a - v_p)|dv_a/dy|^{\frac{1}{2}} \tag{8}$$

5) 虚假质量效应 $\vec{F}_{vm}$

$$\vec{F}_{vm} = \frac{1}{2}\rho_a V_p (d\vec{v}_a/dt - d\vec{v}_p/dt) \tag{9}$$

6) Basset力 $\vec{F}_B$：

$$\vec{F}_B = \frac{3}{2}d_p^2(\pi\rho_a\mu_a)^{\frac{1}{2}} \int_{t_0}^{t} \left(\frac{d\vec{v}_a}{dt} - \frac{d\vec{v}_p}{dt}\right) \frac{d\tau}{(t-\tau)^{1/2}} \tag{10}$$

7) 温差热致迁移力 $\vec{F}_t$
J.R.Brock 的计算公式为[2]：

$$\vec{F}_t = -\frac{9}{2}\pi \frac{\mu_g^2}{\rho_g T_g} d_p \left(\frac{1}{1+3c_m \frac{2l}{d_p}} \right) \left(\frac{\frac{k_g}{k_p} + c_t \frac{2l}{d_p}}{1 + 2\frac{k_g}{k_p} + 2c_t \frac{2l}{d_p}} \right) \mathrm{grad}\, T \qquad (11)$$

热致迁移力的理论计算公式还有苏联Derjagin所推导的[3]：

$$\vec{F}_t = -3\pi \frac{\mu_g}{\rho_g T_g} \cdot \frac{d_p}{2} \left[\frac{8k_g + k_p + 2c_t(2l/d_p)k_p}{2k_g + k_p + 2c_t(2l/d_p)k_p} \right] \mathrm{grad}\, T \qquad (12)$$

8）颗粒自身重力 $\vec{F}_g$ 和受的浮力 $\vec{F}_f$：

$$\vec{F}_g = \rho_p V_p \vec{g} \qquad (13)$$

$$\vec{F}_f = -\rho_g V_p \vec{g} \qquad (14)$$

9）不均匀燃烧作用力 $\vec{F}_c$ 及颗粒互相碰撞力 $\vec{F}_i$

10）静电力 $\vec{F}_e$

下面我们来具体的分析上述力中对煤粉管道输送及炉内运动较重要的几个力。

阻力是作用于煤粉颗粒上的最大的力，颗粒的加速运动主要是由气流曳引阻力所决定。阻力的计算主要取决于阻力系数。有关阻力系数的问题已有许多实验研究和理论计算公式。我们曾对稳定气流绕球形颗粒运动进行了数值求解[4]，得到了一个形式简便的阻力系数关联式：

$$C_D = \frac{19.65}{Re^{0.333}} \qquad (1 < Re \leqslant 100) \qquad (15)$$

然而在工程实践中，特别是对煤粉的燃烧和在管内输送等问题，流动通常为湍流，所以稳定流动的研究数据很难被应用于实际问题。根据我们[8,9]及Laurence对管内及自由射流湍流脉动频谱的试验研究可得，管内和自由射流中湍流脉动的主频率分别为100Hz和200Hz。通过分别计算各个不同脉动频率的气流绕煤粉颗粒的运动，然后根据脉动频谱进行加权叠加的方法，我们求出了湍流脉动情况下颗粒运动的阻力系数[10]。图1示出了

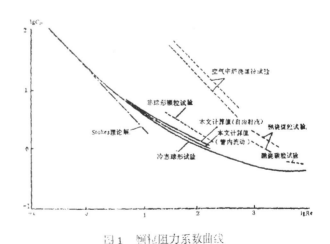

图1 颗粒阻力系数曲线

各种情况下煤粉颗粒阻力系数的实验结果和本文计算结果的对比。

对数值计算结果进行了回归，提出了湍流脉动工况下颗粒阻力系数的关联式：[10]

$$C_D = \frac{24}{Re}\left(1 + 0.19 Re^{0.62}\right)\left[1 + 0.095\left(\frac{f_b Re}{\sigma}\right)^{0.287}\right] \quad (16)$$

无论在管道或炉膛内，煤粉颗粒都会边运动边高速旋转。我们曾用闪频摄影的原理，分别对煤粒、铝片、石英球等进行了在射流中旋转速度的测量[7]，所得旋转速度的数量级为每秒一千转左右。在表1中列出了煤粉和石英球的典型试验结果。

表1 煤粒和石英球与有机玻璃及"质点壁"碰撞前后的转速

颗粒 直径	在气流中的平均转速 转/秒	在碰撞后的平均旋转速度（转/秒）
煤粒 ~2mm	$1\times10^2 \sim 1\times10^3$	
石英球 2.5mm	0.46×10^3	3.96×10^3（与有机玻璃板）
石英球 10mm	0.21×10^3	1.45×10^3（与有机玻璃板）
石英球 2.5mm	0.11×10^3	5.38×10^3（与相同的石英球壁相碰）

煤粉颗粒在运动中产生旋转的原因可能有：

i) 流场中有速度梯度，使冲刷煤粒的力不均匀。

ii) 煤粒形状不规则，使各点所受形状和摩擦阻力不一样。

iii) 煤粒间相互碰撞、摩擦或与管壁的碰撞及摩擦。

不均匀燃烧作用力 $\vec{F_c}$ 的产生原因是：燃料颗粒被喷入炉内后，从水份的蒸发、挥发物的释出，至着火温度后气化和燃烧。由于水份、挥发物释出时向外喷射流和由于煤粉表面温度高于周围烟气温度使燃烧产物向外流动的膨胀流，以及燃烧过程中出现的Stefan流。再加上煤粉颗粒是非球形的，颗粒表面燃烧的不均匀，这样就出现了一些推动煤粉颗粒的附加力，这个力是十分复杂的。

最后我们来分析一下颗粒互相碰撞的影响当煤粉颗粒的浓度较大时，颗粒之间及颗粒与管壁的相互碰撞及摩擦的机会很大，这给气固两相运动带来较大的影响，颗粒由于碰撞而引起的动量损失及在颗粒之间的碰撞过程中的能量的传递等因素都很复杂，目前仍无法定量分析。浙江大学燃烧教研室曾用闪频摄影法测定了固体质点与固定壁及"质点壁"碰撞和摩擦过程中的摩擦系数 f，法向速度恢复系数 e 和综合系数 $(1+e)\dfrac{M}{m+M}$。

用理论力学中动量守恒和动量矩守恒定律推导出碰撞后的法向速度恢复系数为：

$$e = \frac{v_{r2} - u_{r2}}{u_{r1} - v_{r1}} \quad (17)$$

v_{r1}，v_{r2} 为碰撞质点碰撞前后的法向速度，u_{r1}，u_{r2} 为被碰撞的壁或"质点壁"在碰撞前后的法向速度。

$$\text{摩擦系数 } f = \frac{v_{\psi1} - v_{\psi2}}{v_{r1} - v_{r2}} \quad (v_\psi \text{是切向速度}) \quad (18)$$

及综合系数 $(1+e)\dfrac{M}{m+M} = \dfrac{v_{r1}-v_{r2}}{v_{r1}}$ （19）

m、M 分别为碰撞质点和被撞的壁或"质点壁"的质量。

各种碰撞工况下的法向速度恢复系数 e，综合系数 $(1+e)\dfrac{M}{m+M}$ 及摩擦系数 f 的实验数据见表2。关于质点颗粒和质点颗粒碰撞的情况我们还正在研究之中。

表2 各种碰撞工况下的法向速度恢复系数 e、综合系数 $(1+e)\dfrac{M}{m+M}$ 和摩擦系数 f

工况	2.5mm石英珠与光滑有机玻璃板		2.5mm石英珠与不带粘性的2.5mm石英珠"质点壁"						2.5mm石英珠与预先带粘性的石英珠"质点壁"附在"质点壁"上的粘性流体粘度~1000泊	
			"渣膜"粘度44~1200泊 "渣膜"厚度4mm		"渣膜"厚度~5000泊 "渣膜"厚度4mm		"渣膜"厚度8mm			
项目	e	f	$(1+e)\dfrac{M}{m+M}$	f	$(1+e)\dfrac{M}{m+M}$	f	$(1+e)\dfrac{M}{m+M}$	f	$(1+e)\dfrac{M}{m+M}$	f
实验值	0.7806	0.1184	1.187	0.381	1.115	0.280	1.182	0.280	1.145	0.291

为了便于对方程(1)进行计算，我们现在推导单个煤粉颗粒在二维平面流动中运动的方程式，同时作如下假定：

i) 气流速度场已知，为周期性脉动气流，颗粒的存在不影响气流的速度场。

ii) 只沿Y方向存在速度梯度，和温度梯度。

iii) 升力的方向沿Y轴的45°角。

iv) 浮力忽略不计。

v) 为了便于公式的推导，我们人为定义
$f = C_D Re/24$，则阻力表达式为

$$\vec{F}_r = 3\pi \mu_a d_p f \cdot (\vec{v}_a - \vec{v}_p)$$ （20）

在上面这些假定条件下，根据方程(1)可分别建立沿X轴和Y轴方向的运动方程。

$$\frac{1}{6}\pi d_p^3 \rho_p \frac{du_p}{dt} = 3\pi\mu_a d_p f \cdot (u_a - u_p) + \frac{1}{6}\pi d_p \rho_a \frac{du_a}{dt} + \frac{1}{2}\cdot\frac{1}{6}\pi d_p^3 \rho_a \left(\frac{du_a}{dt}\right.$$

$$\left. - \frac{du_p}{dt}\right) + \frac{3}{2}d_p^2 (\pi\rho_a\mu_a)^{\frac{1}{2}} \int_{t_0}^{t} \left(\frac{du_a}{dt} - \frac{du_p}{dt}\right)\frac{d\tau}{(t-\tau)^{1/2}}$$ （21）

$$\frac{1}{6}\pi d_p^3 \rho_p \frac{dw_p}{dt} = 3\pi\mu_a d_p f (w_a - w_p) + \frac{1}{6}\pi d_p^3 \rho_a \frac{dw_s}{dt} + \frac{3}{2}d^2_p (\pi\rho_a\mu_a)^{\frac{1}{2}}$$

$$\int_{t_0}^{t} \left(\frac{dw_s}{dt} - \frac{dw_p}{dt}\right)\frac{d\tau}{(t-\tau)^{1/2}} + \frac{1}{2}\cdot\frac{1}{6}\pi d^3_p \rho_a \left(\frac{dw_s}{dt} - \frac{dw_p}{dt}\right)$$

$$-\frac{1}{6}\pi d^3{}_p\rho_p g + \frac{1}{8}\pi\rho_a d^3{}_p(u_a-u_p)\omega + 1.61(\mu_a\rho_a)^{\frac{1}{2}}d^2{}_p(u_a-u_p)$$

$$\left(\frac{du_a}{dy}\right)^{\frac{1}{2}} - \frac{9}{2}\pi\frac{\mu_a}{\rho_a T_a}d_p\left(\frac{1}{1+3c_m\frac{2l}{d_p}}\right)$$

$$\left(\frac{\frac{k_a}{k_p}+c_t{}^2\frac{2l}{d_p}}{1+2\frac{k_a}{k_p}+2c_t\frac{2l}{d_p}}\right)\frac{dT}{dy} \tag{22}$$

u_p、w_p和u_a、w_a分别为颗粒和气体的X及Y方向的分速度方程(21)和(22)经整理简化后得下列形式：

$$\frac{d}{dt}(u_p-u_a) = -\frac{1}{\tau}(u_p-u_a) + A \tag{23}$$

$$\frac{d}{dt}(w_p-w_a) = -\frac{1}{\tau}(w_p-w_a) + B \tag{24}$$

其中 $\tau = (\rho_p + \frac{1}{2}\rho_a)d_p{}^2/18\mu f$。

方程(23)和(24)可用四阶Runge–Kutta方法积分求解。而颗粒的位置为：

$$x_p = x_{p0} + (u_p+u_{p0})\Delta t/2 \tag{25}$$

$$y_p = y_{p0} + (w_p+w_{p0})\Delta t/2 \tag{26}$$

反复应用上列公式就可以确定颗粒的整个轨迹。

下面我们以一个典型的例子来求解方程(23)、(24)、(25)、(26)并分析上述各种力的数量级关系。

例：来流为周期性脉动的平直流，气流平均速度为20m/s，脉动频率 $f=100$Hz，脉动振幅 $A=10\%$，颗粒旋转速度$\omega=1000$转/秒，Y方向的速度梯度为 $\frac{dw}{dy}=10^1/s$，温度梯度 $\frac{dT}{dy}=100℃/m$，我们对煤粉颗粒直径分别为1μ、10μ、100μ三种情况进行计算结果见表3

由表3可见，在各种不同的煤粉颗粒条件下，虚假质量效应F_{mv}、压力梯度力F_p及Saffman升力F_s的数量级极微，因而一般可以忽略不计。温差热致迁移力在煤粉颗粒直径较大时（$d_p \geqslant 10\mu$）其数量级和重力相比也甚微，只有当颗粒直径很小时（一般$d_p<5\mu$）后，其数量级和重力相等或超过重力。由此可见，热致迁移力对细小颗粒的作用将是很大的。在炉膛内，由于炉内气流和炉壁存在着很大的温差，因而致使细微灰粒垂直于气流速度而沉向受热面。而旋转升力的数量级在各种煤粉颗粒直径下均和重力相等，它起着平衡重力的作用，使煤粉能在管道中安全输送而不致于沉降。在湍流脉动运动中，颗粒不但受粘性阻力和虚假质

表3 各种力的数量级关系

各种力的称名	dp	1μ	10μ	100μ
气流曳引阻力	F_r	$\cdot 59\times 10^{-12}$	$\cdot 15\times 10^{-9}$	$\cdot 82\times 10^{-7}$
压力梯度力	F_p	$\cdot 15\times 10^{-15}$	$\cdot 15\times 10^{-12}$	$\cdot 15\times 10^{-9}$
旋转升力	F_t	$\cdot 82\times 10^{-14}$	$\cdot 82\times 10^{-11}$	$\cdot 82\times 10^{-8}$
Saffman升力	F_s	$\cdot 26\times 10^{-15}$	$\cdot 28\times 10^{-12}$	$\cdot 33\times 10^{-9}$
虚假质量力	F_{vm}	$\cdot 53\times 10^{-16}$	$\cdot 64\times 10^{-13}$	$\cdot 72\times 10^{-10}$
Basset力	F_B	$\cdot 58\times 10^{-12}$	$\cdot 62\times 10^{-10}$	$\cdot 76\times 10^{-8}$
热致迁移力	F_t	$\cdot 19\times 10^{-14}$	$\cdot 20\times 10^{-13}$	$\cdot 20\times 10^{-12}$
重力	F_g	$\cdot 77\times 10^{-14}$	$\cdot 77\times 10^{-11}$	$\cdot 77\times 10^{-8}$

量效应的作用，而且还受一个瞬时的流动阻力，即Basset力，它依赖于气流和颗粒相对加速度的发展过程。因此在湍流运动中，它是非常重要的力。气流曳引阻力是作用于煤粉颗粒上的最大力，起着加速颗粒的作用。

三、各种力对煤粉颗粒轨迹的影响

下面我们将分别讨论几种重要的力对煤粉颗粒轨迹的影响情况，计算是在考虑各种力作用下进行的（个别未考虑的力在说明中注出）。

(1) 气流的脉动对煤粉颗粒运动轨迹的影响

根据我们的试验，在管内和射流核心区，平均湍流强度为10%，平均气流脉动频率为100Hz，在旋转气流中为400Hz[8,9]。因此我们取气流平均速度 $\overline{W_0}=20m/s$，相对振幅 $A=10\%$，

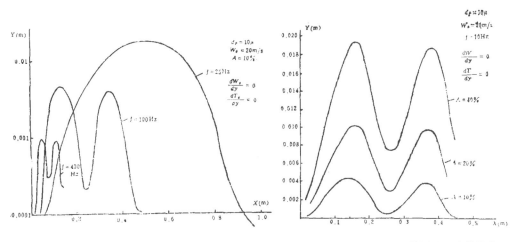

图2 不同脉动频率对颗粒轨迹的影响　　图3 不同脉动振幅对颗粒运动轨迹的影响

颗粒直径 $d_p=10\mu$ 的条件，通过方程(23)、(24)、(25)、(26)，计算不同频率 $f=25\text{Hz}$、100Hz、400Hz 三种情况的颗粒运动轨迹。此处来流为平直流 $\frac{dw}{yd}=0$，温度梯度 $\frac{dT}{dy}=0$，计算结果见图2，由图2可见，随着脉动频率的增加，颗粒在前进相同距离中的上下波动次数越多，但波动的幅度减小。在 $f=25\text{Hz}$ 时，其 $F_B=0.55\times10^{-10}$，而当 $f=100\text{Hz}$ 和 400Hz 时，F_B 将增大到 0.821×10^{-10} 和 0.132×10^{-9}。图3是不同的脉动相对振幅 $A=10\%$、20%、40% 对颗粒轨迹影响的计算结果。

计算结果表明，脉动的相对振幅越大，颗粒随气流上下波动的范围就越大。而图4则计算不同的颗粒直径的情况，由图可见，当颗粒直径很小时 $d_p=1\mu m$，煤粉完全随气流飘扬脉动，随着颗粒直径的增大，脉动运动受到重力的作用而衰减，当颗粒直径较大时 $d_p=100\mu m$ 基本上随气流脉动很小，在重力作用下而慢慢沉降。

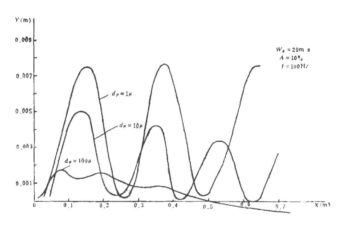

图4　相同的脉动频率和振幅下不同直径颗粒运动轨迹

(2) 颗粒旋转对运动轨迹的影响

图5所示是在不同煤粉颗粒自转速度下运动轨迹的计算结果。

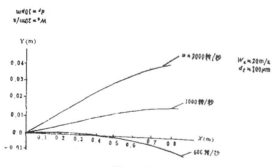 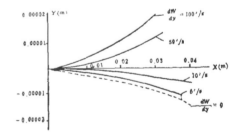

图5　不同自转速度下颗粒的运动轨迹　　　　图6　不同速度梯度下颗粒的运动轨迹

由图可见，当ω=500转/秒时，颗粒在重力的作用下，很快的下沉了，而在ω=1000转/秒，和ω=2000转/秒的情况下，颗粒由于受到很大的升力，在颗粒加速段内，其数量级将大于重力故颗粒将不下降。从计算结果可见，煤粉能在管内不沉降的输送，旋转升力是其中重要的因素之一。

(3) 速度梯度对煤粉颗粒运动轨迹的影响

图6是在不同速度梯度情况下，通过求解方程(23)～27)得到的$d_p=10\mu$的颗粒运动轨迹。

由图6可见，在速度梯度不是很大的时候Saffman升力对颗粒轨迹的影响并不太大，只有当$\frac{dw}{dy}$很大时，（一般是在速度边界层中）Saffman升力才会很大。如设管道直径$D=300mm$管内速度按勃拉修斯1/7次方规律分布，主流的速度为20m/s时，边界层中的速度梯度的平均值约为2529 1/s。这里我们也可以得到这样一个结论，由于在边界层内，速度梯度很大，因而当颗粒沉降到边界层附近时，被Saffman升力所托起而不沉降到管壁，但一般在主流区没有这样大的Saffman升力，所以在主流区Saffman升力不是阻碍颗粒沉降的力。

(4) 温度梯度对煤粉颗粒运动轨迹的影响

图7所示的是温差热致迁移力对不同直径的煤粉颗粒运动轨迹的影响。计算结果和实际情况是一致的，即直径细小的煤粉将垂直于气流方向较快的沉向受热面。其原因可从表3看出。

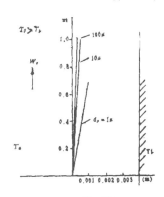

图7 温度梯度对颗粒轨迹的影响

(5) 各种力作用下煤粉运动规律的计算结果

图8所示的是用上节例子的条件，通过求解方程(23)～(27)而得到的不同颗粒运动中加速段的轨迹。由图可见，对煤粉颗粒运动轨迹产生影响的力主要是气流曳引阻力，Basset

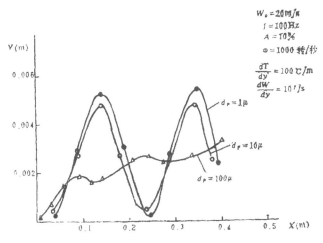

图8 不同直径的颗粒在各种力作用下的运动轨迹

力，旋转升力及颗粒重力。在边界层中，由于颗粒浓度增大，阻力系数增大，阻力也增加。并且$\frac{dw}{dy}$的增大，使Saffman升力增大。

四、结　论

根据理论计算和试验分析，对煤粉在管内和炉内流动时除考虑惯性力和重力外，还要计及下列各因素的影响：

1. 阻力的影响的研究表明气流脉动影响是显著的。
2. 当颗粒高速旋转时（$\omega \geqslant 1000$转/秒）旋转升力将很大，它起着阻碍颗粒沉降的作用，升力可用式（6）计算。
3. 在脉动湍流运动中，Basset将是很重要的力，应该考虑。
4. 炉内水冷壁和火焰中心有很大的温差，因此会产生较大的热致迁移力，使得<5μm的粉粒，垂直于主流气流流动方向沉降于水冷壁上，使水冷壁的吸热能力很快下降，有时形成积灰结渣。
5. 在管底附近，由于$\frac{dw}{dy}$较大，煤粉浓度较大，脉动速度w'指向管心，旋转也会加快，在一定的安全输送速度下使煤粉不沉积于管底。
6. 本文所提出的煤粉多相流的受力方程可供今后工程计算之用。使计算结果更符合实际。

气固多相流是十分复杂的，这里只讨论了部份力作用问题，今后的方向是应用公式（22）、（23）结合流场数值计算以求解工程中的实际流动问题。

参 考 文 献

[1] 岑可法，《燃烧理论》 浙江省电力试验研究所翻印，1982年1月。
[2] Brock.J.R: "On the theory of thermal Force acting on aerosol particles". Jour.Colloid Sci, 1962, Vol.17 pp768—788.
[3] Derjagin.B.V, "Theory of thermopheresis of large aerosol particles." Journel Colloid Sci. 1965 Vol.20 pp555—580.
[4] 樊建人、岑可法，"在脉动气流中球形颗粒的运动阻力计算" 第一届全国计算力学青年研讨会论文。（1986年）。
[5] 樊建人、岑可法，"煤粉颗粒群在湍流气流中运动时阻力系数的数值计算" 待发表（1986年）
[6] LeClair.B.p, "Viscous flow through Particle Assemblages at intemediate Reyrolds numbers" I, &.E.C.Fundamentals, 1968, Vol. 7, No.4。
[7] 刘华信，"固体燃料质点在旋风炉内运动规律性的探讨" 浙江大学电机系燃烧学专业研究生毕业论文 1965.5月
[8] 岑可法，"自由射流、绕圆柱及管簇流动湍流结构的研究"，浙江大学学报，1963，No.3。
[9] 岑可法，"旋风燃烧室内气流湍流结构的研究" 浙江大学学报，1963，No.3。
[10] 樊建人、岑可法, Effecis of Turbulent Fluctuation and Frequency Spectrum on the Drag Coefficient of aSpherical Particle in Gas--Solid Flows, Proc. Inter. Symposium on Multiphase Flows, 1987。

The Analysis of the Forces Acting on Coal Particles and the Trajectories in the Gas Flows

Cen Kefa   Fan Jianren

ABSTRACT

The forces acting on coal particle in gas flow are analyzed. The particle movement equation is described by a Lagrangian approach and numerical calculated Effects of the forces on particle trajectories are also analyzed

原文刊于浙江大学学报(自然科学版), 1987, 6(21): 6-16

洗选煤泥沸腾燃烧技术的研究

岑可法　黄国权　倪明江*

(浙江大学热物理工程系)

洗选煤泥是从选煤厂排出的浆状物料。由于它水份大，灰份高，热值低，持水性强，要加以利用非常困难，因此难以销售，大量堆积。这不仅严重的影响环境卫生，而且还造成大量能源损失。在当前原煤入洗率仅为18%的情况下，每年约排放600万吨煤泥，相当于200万吨原煤。为此国家科委和煤炭部与浙江大学，煤科院煤化所，永荣矿务局电厂签订了一项利用选煤厂洗选煤泥的攻关合同。

沸腾燃烧洗选煤泥的研究工作是在浙江大学燃烧实验室进行，在取得了充分的基础数据及设计依据之后在永荣矿务局电厂进行了10吨/时洗选煤泥沸腾锅炉的中间试验，并于1986年2月通过国家鉴定。

一、洗选煤泥的物理特性

试验采用永荣矿务局选煤厂的洗选煤泥，分析数据及颗粒筛分特性见表1和表2。试验发现这种煤泥属于涨拟性非牛顿型流体，由于煤泥中的固相物绝大部分小于0.5毫米，煤泥的持水性又强而难以脱水，入炉时煤泥往往呈泥状或浆状。

表1　永荣煤泥分析数据（分析基）

	项目	数据		项目	数据
工业分析	水份%	2.80	元素分析	碳%	36.14
	灰份%	51.65		氢%	2.53
	挥发份%	16.41		氧%	5.49
	固定份%	29.14		氮%	0.67
	低位发热量大卡/公斤	3460		硫%	0.72

表2　永荣煤泥筛分特性

粒度范围mm	0.056～0.071	0.071～0.112	0.112～0.154	0.154～0.2
重量份额%	0.25	0.25	35.37	1.27
粒度范围mm	0.2～0.355	0.355～0.8	0.8～0.9	>0.9
重量份额%	38.17	23.41	0.89	0.38

二、洗选煤泥沸腾燃烧工艺特点

由表2可知煤泥颗粒都小于0.8毫米，绝大多数小于0.5毫米且有相当一部份粒度仅为几个微米。如果投入流化床的煤泥在干燥后还原成细粉，在通常的流化风速下必然会立即被吹出床层从而形成巨大的扬析损失，甚至会无法组织正常的燃烧工况。若降低流化风速去迁就细粉不被吹出，又会使流化床的断面热强度低到无法接受的程度。通过试验研究发现在一定

本文1986年12月26日收到。

* 浙大杨家林、项　黔、骆仲泱及永荣矿务局周　涛、北京煤化所王润清、姚　舟参加研究工作。

的工况条件下,浆状或糊状的煤泥投入流化床后并不还原成细粉,而是形成具有一定强度和耐磨性的凝聚团,这种凝聚结团现象将是组织煤泥沸腾燃烧的基础,并为提高断面热强度及燃烧效率创迼了条件。但另一方面留在流化床中的凝聚团往往使床料粒子呈不断增长的趋势,形成的大块凝聚团极易在床内沉积从而对流化床的稳定运行构成威胁。针对煤泥凝聚团对沸腾燃烧影响的二重性,我们提出了如下的煤泥沸腾燃烧新工艺:

1. 采用成型给料,利用煤泥的凝聚结团特点,使煤泥在流化床内形成粒度较大的凝聚团以减少扬析损失。
2. 采用大比重物料组成浮力流化床,防止大颗粒凝聚团在流化床沉积。
3. 采用无溢流运行,以减少重物料的损耗并延长凝聚团在床内的停留时间,提高燃尽度。

三、10吨/时煤泥沸腾炉结构特性

根据所提出的工艺流程,我们对10吨/时煤泥沸腾炉进行了改装设计,如图1所示。其主要参数见表3。该炉布风板面积为4.16m²,床内除了布置少量竖埋管外,主要受热面则为顺列布置倾斜埋管。为保证扬析夹带燃料的燃尽,悬浮段的水冷受热面几乎全部用耐火材料复盖。考虑到燃用煤泥时飞灰量大而粒细,故采用了文丘里水膜除尘器。

表3 10吨/时煤泥沸腾炉参数

项目	蒸汽和给水			运行条件				床层设计参数			
	蒸发量 吨/时	蒸汽压力 kg/cm²	蒸汽温度 ℃	给水温度 ℃	床温	运行风速 m/s	投料量 吨/时	控水料入炉份 %	床截面 m²	风帽数 只	开孔率 %	埋管面积 m²
数据	10	9	300	105	900	3.0	3.3	25	2.6×1.6	1161	2.6	11.31

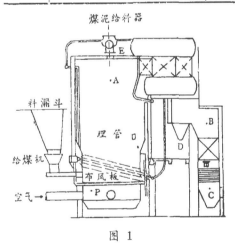

图1

四、10吨/时煤泥沸腾炉燃烧系统

10吨/时洗选煤泥沸腾炉的燃烧系统,如图2所示,它由炉前预处理系统,上料系统及给料系统三部分组成。

炉前预处理系统主要包括双轴拌泥机和气(电)动振动筛。煤泥在入炉前经过预处理系统去除杂质并将水份调整到所希望的范围内,然后送入小车提升到给料平台倒入煤泥给料机煤泥斗,也可通过专门研制的煤泥泵经直径为159毫米长为21000毫米的煤泥输送管道送到给料点。采用管道输送具有系统简单运行调节方便等优点。还可以完全避免煤泥强粘性带来的麻烦。给料系统位于炉子顶部,煤泥可以通过煤泥给料机送入沸腾层,也可以通过泵送管道挤压成型后送入沸腾层。

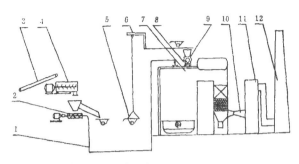

图2 10吨/时煤泥炉燃烧系统图

1.煤泥输送管道　2.煤泥泵
3.上料皮带　　　4.搅拌机
5.煤泥提升车　　6.提升架
7.锅炉　　　　　8.落料口
9.煤泥给料机　　10.文丘里喷水管
11.水膜除尘器　　12.烟囱

五、10吨/时煤泥炉的运行

1. 点火启动　采用"沸腾态点火"方式，即在沸腾状态下将底料加热到引火煤的着火温度，然后过渡到燃烧煤泥。在正常情况下整个点火启动过程大约只需30～40分钟。

2. 燃烧效率　试验时的运行参数为：入炉煤泥水份25～29%，低位发热量2070～2309大卡/公斤，床温850～950℃，沸腾风速0.70～0.85标米/秒，锅炉蒸发量9～12吨/时，投料量3.4～3.8吨/时，平均燃烧效率为95.4%。远大于劣质煤常规沸腾燃烧的燃烧效率，与国外同类工作相比也是令人满意的（见表4）。而我们采用的沸腾风速较高，其难度更大。

表4 燃烧效率比较

资料来源	燃烧效率%	沸腾风速标米/秒	备注
英国CRE	77～89	0.23～0.28	试验台
澳大利亚CSIRO	82～94	0.29～0.47	与洗煤粗料混烧
西德Babcock—BSH公司	58～85	缺	试验台
日本佳友公司	96	0.63	10吨/时
浙大燃烧实验室	95	0.65～0.72	试验台
永荣电厂	95.4	0.70～0.85	10吨/时

3. 运行稳定性　煤泥沸腾炉的运行稳定性问题是当前国际上煤泥沸腾燃烧技术发展中所面临的关键问题之一。本工艺运行是否稳定，主要反映在如下三个方面：1）采用成型给料后形成的大块煤泥凝聚团是否会在床内沉积；2）采用不溢流不放冷渣的运行方式后料层高度会不会呈现不断增长的趋势；3）布置一个给料口，入炉煤泥是否会在给料口处堆积。

沿床高及沿深度和宽度方向测得的床温运行记录稳定均匀，表明所采用的工艺能较好地避免大块凝聚团的沉积，给煤口处亦没有煤泥堆积，保证了良好的沸腾质量。

在运行初期几个小时内由于床内灰渣的积累，料层压降稍有增加，此后料层压降趋于稳定，在运行条件维持不变或变化很小的情况下料层压降的稳定表明料层高度也是稳定的。

4. 烟气中污染物的排放　试验测得烟尘排放量为65.87公斤/时，SO_2排放量为16.49～22.92公斤/时，NO_x排放量为7.97公斤/时。由此可见，洗选煤泥沸腾燃烧象常规沸腾燃烧一样可以有效地控制烟气中的污染物排放量。

原文刊于科技通报，1987，3(4)：18-20

水煤浆燃烧技术的研究

岑可法 谢名湖 吕德寿
曹欣玉 黄镇宇 张铁均
(浙江大学热物理工程系)

水煤浆作为新型低污染燃料而引起人们的广泛重视。它的主要优点是污染低,输送方便,储存不沉淀,能代替燃料油,能雾化燃烧,易于实现煤气化等。所谓水煤浆是把煤加入一定量的水和少量添加剂磨制而成的流体燃料。

国外从70年代末开展对水煤浆制备和燃烧的研究。浙江大学1981年开始进行水煤浆燃烧技术的试验室研究[1],1983年国家科委和煤炭部把水煤浆的制备和燃烧列为国家"六·五"期间重点攻关项目,浙江大学负责水煤浆燃料技术,1986年1月此项技术通过了国家鉴定。

一、水煤浆的着火特性

水煤浆含有30%的水,它的着火特性与原煤的着火特性有明显的差别。

1.着火温度 根据试验,水煤浆的热力着火温度比同煤种煤粉的着火温度一般要低150～200℃。这是因为水煤浆滴在水分蒸发后形成了一个多孔球,大大增加了煤浆的反应表面积。按照热力着火理论,煤反应的比表面积大,氧气与煤表面接触机会多,就能在较低的温度下发生较剧烈的反应,实现着火。

2.着火热 水煤浆中的水分在着火前要蒸发。试验表明:蒸发时间占煤浆总燃烧时间的8%。这就延长了煤浆的着火时间。大量的水分蒸发要吸收热量,含水30%的水煤着火热约为煤粉着火热的1.7倍。同时水分蒸发会降低炉内的温度水平,使煤浆的着火过程延迟。

总的说来,煤浆的着火较之煤粉的着火要困难一些。因此必须研制合适的新型水煤浆燃烧器,以保证着火区附近有强大的热源可供使用。同时要对炉膛进行合理的技术改造,采取适当的措施,如反吹射流、涂抹卫燃带、增加前置燃烧室等以适应水煤浆的着火和燃烧。

二、水煤浆的雾化与火炬燃烧

欲使水煤浆能象油一样燃烧,就需要研制适合水煤浆的雾化喷嘴,使煤浆能雾化成极细的雾滴。雾化颗粒度愈细,水煤浆的着火就愈容易,燃烧效率也愈高。但和油相比,水煤浆的雾化是很困难的,同时磨损大。

我们研制的低压旋流型雾化喷嘴能满足上述要求,并已在北京造纸一厂20t/h燃油锅炉上应用[2]。图1是低压旋流型雾化喷嘴的结构,该喷嘴的使用参数:容量为700—1500kg/h,

---
本文1986年12月26日收到.

煤浆压力为0.25—0.50MPa，气耗率为0.12—0.15kg气/kg浆，雾化颗粒S.M.D约为80μm，雾化介质压力：采用压缩空气时为0.4—0.5MPa，蒸汽时为0.7—0.8MPa。该喷嘴压力低，流阻低，煤浆通道较大且呈流线型，不易堵塞，雾化良好（见图2）。

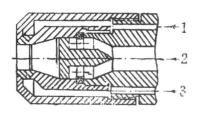

1.气道 2.浆道 3.气道
图1 旋流型喷嘴

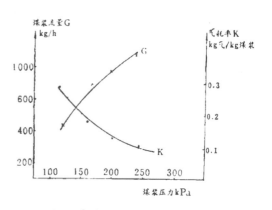

图2 喷嘴流量气耗特性曲线

喷嘴的易磨部份采用耐磨材料如碳化钨、陶瓷等镶嵌。

水煤浆稳定的着火与燃烧，除了需要上述喷嘴外，还得有合适的调风器。图3是我们研制的可调叶片旋流调风器[5]。一次风采用固定叶轮稳焰器。二次风采用可调切向叶片，其旋流强度可在0～8之间较大范围内调节。二次风出口有一可调套筒，可方便地改变二次风切向速度。

该调风器已在工业上应用，它具有回流烟气多，性能好，调节方便，适应性强等优点。

三、水煤浆的工业性试验

我们从1984年5月开始在北京造纸一厂20t/h燃油锅炉上进行改烧水煤浆工业性试验[4]。水煤浆着火燃烧特点有别于油及煤粉，对原燃油锅炉作了相应的改造。

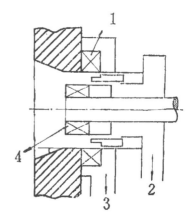

1.可调切向叶片旋流器；2.一次风
3.二次风；4.轴向叶片旋流器
图3 调风器

图4是该锅炉改造后的示意图。由于水煤浆里含有一些灰，所以在燃烧过程中会有部分灰沉降在炉底，长时间运行灰会越积越厚，最终会破坏锅炉的正常运行。我们研制了采用炉底半有限射流连续清灰方法，将原来欲沉积在炉底的灰扬起并将灰中未燃烬的碳再一次燃烧，还可提高燃烧效率。

为了增加水煤浆雾滴在炉内的停留时间和扰动，锅炉改装时，装上射流风，能提高燃烬率。

水煤浆在20t/h工业燃油锅炉上的试验从1984—1985年分三个阶段进行，共烧煤浆2820

吨，累计运行826小时，进行28次燃烧及热工测试。试验结果表明：水煤浆着火可靠，能长期稳定燃烧，操作方便，可在50～100%额定负荷运行，调节范围比烧煤粉大。烧抚顺煤浆时，燃烧效率96.38%，SO_2及NO_x排放量比烧煤粉低20～30%，水煤浆可以取代油在工业上应用。

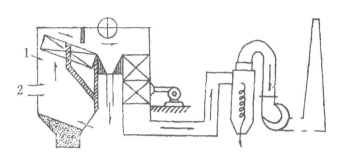

1. 上三次风   2. 燃烧器

图4 20t/h烧油锅炉改烧水煤浆原理系统图

四、前置复合燃烧装置的研究

燃油锅炉改烧水煤浆后，锅炉出力将会降低20—40%。为了达到不降负荷或少降负荷，同时也为了获得更稳定的着火，实现低污染燃烧、部份清灰的目的，我们研制了前置燃烧装置（图5）[3]。该装置由预燃室、燃烧室、反吹射流、水平烟道、沸腾床和燃烬室组成，其中燃烬室是为了模拟锅炉本体的。

该燃烧装置由于采用反吹射流，能将大量高温烟气回到燃烧器出口，用冷空气作为一、二次风实现稳定着火。测量结果表明在预燃室内煤浆已能稳定着火与燃烧，距燃烧器出口175mm处，截面平均温度也有1000℃，在燃烧室内温度比较均匀，达1500℃左右。立式前置燃烧装置的燃烧热负荷较高，所以它的体积很小，温度较高，燃烬率超过40%。同时该装置可采用分段送风，使燃烧室出口处NO_x的排放量仅100ppm左右，实现了低NO_x污染燃烧。

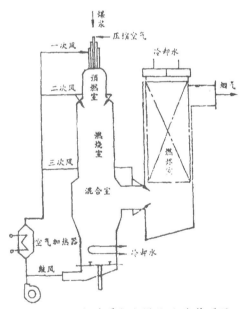

图5 立式前置复合燃烧试验装置图

混合室的下部为一沸腾炉，利用从顶上下来的煤浆转弯时的惯性作用，较大的粒子甩往沸腾炉内燃烧，燃烬后从下部出灰口放出，这样既可使大粒子在炉内有一定的停留时间，增加煤炭的燃烬程度，又可以达到清除部份灰的目的，减轻锅炉本体清灰的难度和减少锅炉磨损。

五、水煤浆的工业应用前景

水煤浆作为一种新型低污染代油燃料，它有其独特的优点，在工业中已逐步获得应用，其前景非常乐观。

1. 在燃油设备中作为代油燃料。
2. 便于实现管道长距离大容量输送。
3. 改造城市的供煤系统。
4. 水煤浆是一种低污染燃料，用水煤浆作燃料，可以减轻大气污染。
5. 实现蒸汽——燃气联合循环。
6. 利用水煤浆气化生产工业煤气。

综上所述，水煤浆是一种新型的、低污染的、可取代油的流态燃料，它将给煤炭的开采、运输、燃烧、气化等带来有希望的变革，为煤炭的加工利用开辟新的途径。目前，水煤浆燃烧技术已在北京造纸一厂20t/h燃油锅炉上试验成功。相信在不久的将来，将会在更多的领域中得到愈来愈广泛的应用。

参 考 文 献

[1] 岑可法等：Experimental study on combustion of coal-water slurry, 第四届国际煤浆会议论文集, 3 (1982), 6 (美国).

[2] 谢名湖等：Experimental study of the characterislies of the swirl-type nozzle for coal slurries, 第四届国际煤浆会议论文集, 4 (1982), 7 (美国).

[3] 曹欣玉等：The combustion of coal-water slurry droplets in tunnel furnace, 第二届欧洲煤浆会议论文集, 1985, 129—146, (伦敦).

[4] 谢名湖等：Experimental study on coal-water slurry combustion on a 20t/h oil-fired boiler, 第七届国际煤浆会议论文集, 1985, 761—773, (美国).

[5] 吕德寿等：Combustion experiment of coal-water slurry in a tunnel furnace, 第六届国际煤浆会议论文集, 1985, 731—749, (美国).

原文刊于科技通报, 1988, 4 (1): 10-13

A NUMERICAL MODEL FOR THE TURBULENT FLUCTUATION AND DIFFUSION OF GAS-PARTICLE FLOWS AND ITS APPLICATION IN THE FREEBOARD OF A FLUIDIZED BED

CEN KEFA and FAN JIANREN

Department of Thermophysics Engineering, Zhejiang University, Hangzhou, PRC

ABSTRACT

A numerical model for the turbulent fluctuation and diffusion of gas-particle flows is presented. This model is based on the idea of treating a turbulent gas flow field as a set of $k-\varepsilon$ equations, and of modeling the turbulent fluctuation velocity of gas flow as a random Fourier series based on the fluctuation frequency and spectrum. The particle properties (trajectory and velocity) are described by a Lagrangian approach. Hence this model is known as the fluctuation-spectrum-random-trajectory (FSRT) model. Finally, particle movements in the freeboard of a fluidized bed and in a turbulent gas-particle-laden jet are analyzed to illustrate the applicability of the model.

1. INTRODUCTION

Up to now mathematical models describing gas-particle flow fall into two categories, namely, continuous medium models and particle dispersion trajectory models. The former include the no-slip model [1], slip models [2,3] and slip-diffusion models [4] while the latter include the particle-source-in-cell (PSIC) model developed by Crowe [5]. The relative advantages and disadvantages of the continuous medium models and the particle dispersion trajectory models for modeling gas-particle flow can be summarized as follows: The no-slip model can deal with the particle phase very simply with little numerical effort, but it has a large error because it does not consider velocity slip and termperature slip between the two phases. Although the slip model considers the diffusion of particles and the resistance caused by the slip between the particle phase and the gas phase, yet it does not take into account the slip of the mean velocity between the two phases, so we cannot get an accurate description of particle movement. The slip-diffusion model considers not only the mean velocity of slip between the two phases caused by the different initial momentum, but also the turbulent diffusion of the particle phase. However, numerical instabilities, numerical diffusion and the large computer storage requirements for calculations involving multiple particle sizes are inherent difficulties. It has also been found that prohibitively small time steps must be used when strong gas-particle momentum coupling is encountered. Besides, the continuous medium models have a common disadvantage. Since Fick's law is used to describe the turbulent diffusion of the particle, an effective diffusion coefficient must be chosen for which no reliable information is currently available [6,7]. The PSIC model considers the movement and resistance of particles along its own trajectory. It is based on treating the particles as sources of mass, momentum and energy to the gaseous phase. One interesting feature of the PSIC model is the absence of numerical diffusion of the particle cloud. There are two approachs by which particle dispersion due to turbulence can be modeled. One approach to modeling particle dispersion is to regard the particles as a gaseous species and use Fick's law, namely, use an effective diffusion velocity (or diffusion "force") in the particle motion equation which is dependent on the particle concentration gradient [8]. This approach has the same problem as the continuous model; that is, the choice of an appropriate value for the effective diffusion coefficient. Another approach to modeling the dispersion process is to treat the turbulent flow as a random field. This is known as the stochastic model, proposed by Gosman and Ioannides [9]. Following this approach the motion of the particles is tracked as they interact with a succession of turbulent eddies, each of which is assumed to have constant flow properties. Velocity fluctuations of the gas phase were assumed to be isotropic with a Gaussian probability density distribution having a standard deviation of $(2k/3)^{1/2}$ and the local distribution is randomly sampled. Application of this model has resulted in reasonably good comparison with experimental data for near-isotropic homogeneous field. However, because the turbulent movement of gas flow is a result of a random superposition of three fluctuation elements (i.e. different frequency, different amplitudes and different directions), it is too much simplified that the constant flow properties in a turbulent eddy and a standard deviation of $(2k/3)^{1/2}$ were assumed. In the authors' previous investigation [10], it was evident that under the condition of different fluctuation frequencies, the number of trajectory fluctuation times of the same size particles increases as the fluctuation frequency increases, but the trajectory fluctuation amplitude decreases. If the relative fluctuation amplitude of gas increases, the particle fluctuation amplitude gets larger. We can see that the detailed turbulent behavior has important effects on particle turbulent fluctuation and turbulent diffusion. On the other hand, large or small particles have the different turbulent diffusion effects in the turbulent flow. Generally the turbulent diffusion of small particles is greater than that of large ones [11].

In order to consider the turbulent diffusion of particle dispersion simply and completely, this paper proposes a new mathematical model (fluctuation-spectrum-random-trajectory model) to describe gas-particle flows, which is bases on the PSIC model. In this model, the k-ε two equation turbulent model is used to solve the gas flow field. The gas fluctuation velocity is modeled a random Fourier series based on the fluctuation frequency and spectrum of turbulence.

The Lagrangian approach is employed to give the movement details of the various size particles along their trajectories. The particle velocity and the concentration field are obtained by a volume-average method. At the same time, the coupling of mass, momentum and energy between the two phases is considered in this model. The FSRT model fully considers the effect of turbulent fluctuation and turbulent spectrum on different size particles. The motion of the particle phase is more consistent with a realistic situation under turbulent gas flow. Finally, the particle movement behavior in the free board of a fluidized bed and in a gas-particle turbulent jet are analyzed. The numerical results tallies well with experimental results.

2. FLUCTUATION-SPECTRUM-RANDOM-TRAJECTORY MODEL

2.1 The Gas-Phase Equation

The basic equations governing the gas-particle flow field for the computational model described above are derived using the principles of conservation of mass, momentum and energy for a control volume. We use a $k-\varepsilon$ turbulent model to simulate gas-phase turbulent flow. The general form of every conservation equation in the gas-phase field is:

$$\frac{\partial}{\partial t}(\rho\phi) + \frac{\partial}{\partial x_j}(\rho V_j \phi) = \frac{\partial}{\partial x_j}(\Gamma_\phi \frac{\partial \phi}{\partial x_j}) + S_\phi + S_{p \cdot \phi} \quad (1)$$

where ϕ denotes any one of the variables of U, V, W, k, ε and h etc. Γ_ϕ is the diffusive coefficient, S_ϕ is the self source term in a gas-phase field and $S_{p \cdot \phi}$ is the source term caused by the particle phase. The particle source terms are defined as Ref. [5]. Detailed meanings of every term in eqn (1) are shown in Ref. [11]. The above equations can be solved by SIMPLER [12].

2.2 Lagrangian Particle Description

When only the particle fluid resistance and the gravitational force are considered, the particle movement equation can be written as

$$m_p \frac{d\vec{V}_p}{dt} = C_D \rho g A_p \frac{\vec{V}_g - \vec{V}_p}{2}(\vec{V}_g - \vec{V}_p) + m_p \vec{g} \quad (2)$$

If the gas flow velocity $\vec{V}_g$ is divided into a time-average part and a fluctuating part then

$$\vec{V}_g = \overline{\vec{V}}_g + \vec{V}_g' \quad (3)$$

and the gas flow turbulent fluctuation velocity V_g' is simulated by a random Fourier Series, namely,

$$U_g' = \sum_{i}^{n} R_1 U_i \cos(i\omega_i t - R_2 \alpha_i^u)$$
$$V_g' = \sum_{i}^{n} R_3 V_i \cos(i\omega_i t - R_4 \alpha_i^v) \quad (4)$$

where R_1, R_2, R_3, R_4 are normal distribution random numbers. U_i and V_i are fluctuation amplitudes based on frequency ω_i which is determined by a turbulent frequency spectrum. The detailed steps are as follows:

First, we use the Boussinewq assumption to determine $\overline{U'^2}$ and $\overline{V'^2}$, where k and ε have been solved in the gas-phase field,

$$\overline{U'}^2 = -2C_1 \frac{\mu_t}{\rho} \left(\frac{du}{dx}\right) + \frac{2}{3}k$$
$$\overline{V'}^2 = -2C_2 \frac{\mu_t}{\rho} \left(\frac{dv}{dy}\right) + \frac{2}{3}k \qquad (5)$$

Then, the energy distribution percentage k_i under frequency ω_i can be obtained through a turbulent fluctuation spectrum

$$U_i^2 = k_i \overline{U'}^2$$
$$V_i^2 = k_i \overline{V'}^2 \qquad (6)$$

By substituting equations (3) and (4) for (2), we have

$$\frac{dU_p}{dt} = \frac{1}{\tau}[U_g + \sum R_1 U_i \cos(i\omega_i t - R_2 \alpha_i^u)] \qquad (7)$$

$$\frac{dV_p}{dt} = \frac{1}{\tau}[V_g + \sum R_3 V_i \cos(i\omega_i t - R_4 \alpha_i^v)] - g$$

where $\tau = \dfrac{\rho d_p^2}{18\mu f}$, $f = \dfrac{C_D}{\left(\dfrac{24}{Re}\right)}$, C_D is the drag coefficient of

particles. The drag coefficients of particles in turbulent flow are very different from those in steady flow. It is pointed out by Uhlerr's experiment [13] that in the range of middle Reynolds numbers (Re ≤ 200) particle drag coefficients in turbulent conditions are larger than the standard ones. It is even more obvious as the turbulent intensity gets stronger. Using the unsteady Navier-Stokes equation, the present authors previously calculated the particle drag coefficients in turbulent fluctuation flow [14]. An empirical relation for drag coefficient is proposed as:

$$C_D = \frac{24}{Re_p}(1 + 0.19 Re_p^{0.62})\left[1 + 0.095\left(\frac{f_a Re_p}{C}\right)^{0.287}\right] \qquad (8)$$

where f_a is average frequency of turbulent fluctuation and C is experimatal constant.

The particle motion equations (9) are solved by the Runge-Kutta integration method. Hence, the particle position is easily determined from:

$$X_p = X_{po} + (U_p + U_{po}) \Delta t/2$$
$$Y_p = Y_{po} + (V_p + V_{po}) \Delta t/2 \qquad (9)$$

In order to intergrate equations (7) and (9), a particle was assumed to interact with an eddy for a time which is the minimum of either the eddy lifetime τ_e or the transit time τ_R required for the particle to cross the eddy [9], namely

$$\Delta t = \min [\tau_e, \tau_R] \qquad (10)$$

where
$$\tau_e = \frac{le}{|u'|} \qquad le = C_\mu^{3/4} \frac{k^{3/2}}{\varepsilon} \qquad (11)$$

and
$$\tau_R = -\tau \ln \left(1 - \frac{le}{\tau |U - U_p|}\right) \qquad (12)$$

where τ is the particle dynamic relaxation time.

2.3 Solving the Particle Velocity Field and the Concentration Field

The number density of particles in each cell is determined from

$$n = \sum_k \sum_j (\eta_{k,j} \Delta \tau)/\Delta V \qquad (13)$$

where $\eta_{k,j}$ is the number of particles of size k at station j. Particle concentration Cv in a cell is

$$Cv = \sum_k \sum_j \left[\frac{1}{6} (\eta_{k,j} \Delta \tau) \pi \, dp_{k,j}^3\right] / \Delta V \qquad (14)$$

where $\Delta \tau$ is the residence time in the cell and ΔV is the cell volume. Furthermore, the particle velocity in the cell can be obtained by the average of the momentum

$$\vec{V}_p = \sum_k \sum_j [(\eta_{k,j} \Delta \tau) \vec{V}_{pk,j}] / \sum_k \sum_j (\eta_{k,j} \Delta \tau) \qquad (15)$$

2.4 Solution Procedure

The sequence of operations for this model is shown as follows:

a) Solving the gas flow field using a numerical scheme developed by Patankar and Spalding [12], using two equations (k-ε) to account for turbulence and to determine the intensity of turbulence and the dissipation rate.
b) Calculating the gas fluctuation velocity based on k and ε.
c) Modifying the gas fluctuation velocity with the random Fourier series and calculating the particle trajectory.
d) Calculating the particle velocity field and the concentration field

by the volume average method.
e) Regarding the particle phase as the source of the conveying gaseous phase. The source terms are incorporated into the gas flow equations, providing the influence of the particles on the gas velocity field.
f) Recalculating the gas flow field and continuing this process until the flow field fails to change with repeated iteration.

3. APPLICATION TO THE FREEBOARD OF FLUIDIZED BEDS AND GAS-PARTICLE TURBULENT JETS

Using the FSRT model, we studied the particle movement properties in the freeboard of fluidized beds and in turbulent gas-particle-laden jets. Measurements have been made to determine the particle velocity and trajectory in the bubble eruption area and the freeboard [15]. The initial velocity distributions of particles and gas flow in the bubble eruption area are shown in Fig. 1. Taking the above distributions as boundary conditions, we calculated the trajectories and velocities of various size particles in the freeboard. Fig.2 show the measured trajectories and velocities of the particle which is at the top of bubble compared with on calculations. From this figure we find that after bubble eruption the particle is thrown away from the bed surface with some initial velocity. After a period of parabolic trajectory it falls back to the bed surface. Fig. 3 gives the measurements of the particle velocity compared with the numerical results which agree well with experimental data. The comparison between numerical results and experimental data for the particle trajectory and velocity in the bubble flank is shown in Fig. 4 and Fig. 5.

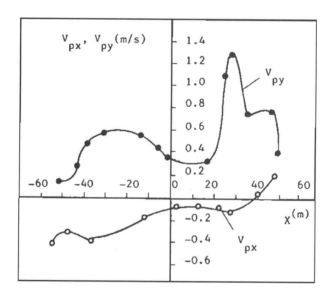

Figure 1. The experimental results of initial velocity distributions of particles in the bubble eruption area [15].

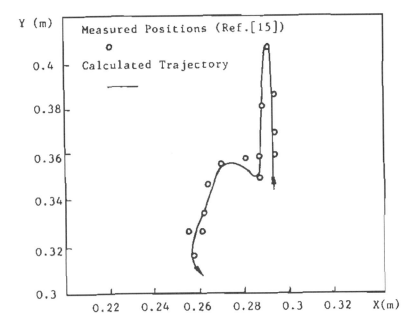

Figure 2. Comparison of numerical predictions with experimental data for trajectory of the particle at the top of bubble.

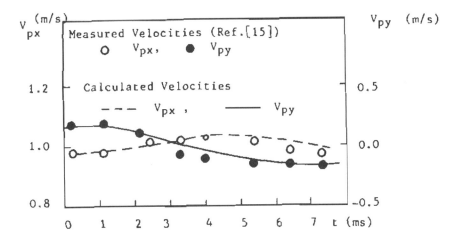

Figure 3. Comparison of numerical predictions with experimental data for velocity of the particle at the top of bubble.

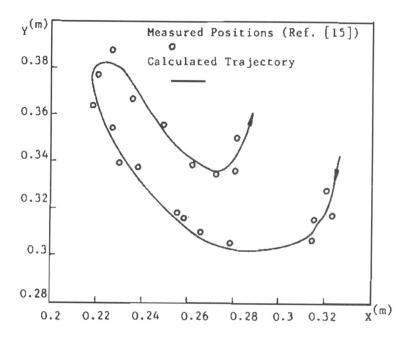

Figure 4. Comparison of numerical predictions with experimental data for trajectory of the particle in the bubble flank.

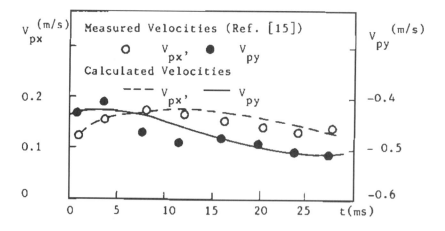

Figure 5. Comparison of numerical predictions with experimental data for velocity of the particle in the bubble flank.

In addition, the theory was tested for the flow of a turbulent axisymmetric gaseous jet laden with spherical solid particles of non-umiform size. The predictions are compared with the data of Levy et al [16]. The experimental specification is shown in Table 1.

Table 1.   Experimental Specification

Code	Particle Size (μm)	Air Flow Rate (kg/sec)	Sand Flow Rate (kg/sec)
A	180 - 250	3.62×10^{-3}	4.13×10^{-3}
B	180 - 250	3.62×10^{-3}	8.45×10^{-3}
C	Clean Gas	4.29×10^{-3}	-

The radial profiles of time-mean axial gas velocity are shown in Fig.6. The particle axial velocity distributions and the concentration distribution of the particle-phase is given in Fig. 7. From these figures the numerical results and experimental data can be seen to be in good agreement.

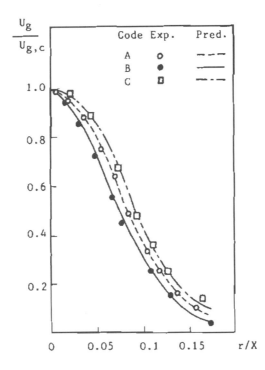

Figure 6. Predicted and measured radial profiles of time-mean axial gas velocity (at X/D = 20).

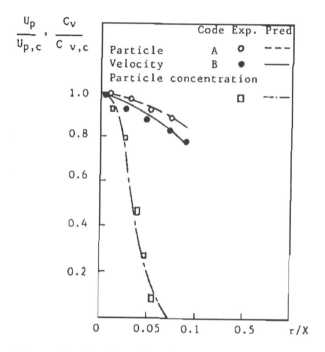

Figure 7. Predicted and measured radial profiles of time-mean axial particle velocity and particle concentration distribution.

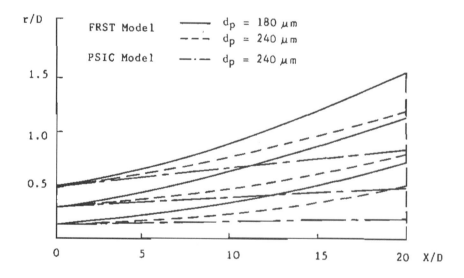

Figure 8. Predicted particle trajectories. Comparison of the FSRT model and PSIC model.

Fig. 8 compares the results of the FSRT model and PSIC model which does not consider turbulent diffusion. In Fig. 8, we see the there is large difference between particle trajectories with turbulent diffusion and without turbulent diffusion. It is evident that the diffusion of small particles is stronger than large ones.

4. CONCLUSION

1. The FSRT model proposed in this paper considers the effects of the fluctuation frequency spectrum of turbulent gas flow on the motion of the particle phase. It also considers the influence of particle size on turbulent diffusion. Thus, it is more consistent with a realistic situation.
2. Although the Lagrangian approach is used to calculate the trajectory and velocity of particles; for convenience in the industrial application, it is changed into the particle velocity profile and concentration profile.
3. This model has been successfully used to calculate the particle motion in the freeboard of a fluidized bed and in gas-particle free jets. It gives good agreement with experimental results.

ACKNOWLEDGEMENT

The authors acknowledge the support of National Science Foundation of P.R. China as well as the Thermoscience and Engineering Department Computer Center.

NOMENCLATURE

A_p	= partick projected area	[m^2]
C, C_1, C_2, C_μ	= empirical constants	[-]
C_D	= drag coefficient	[-]
C_V	= particle concentration	[m^3/m^3]
d_p	= particle diameter	[m]
f_a	= average turbulent fluctuation frequency	[Hz]
g	= gravitational acceleration	[m/s^2]
K_i	= energy distribution percentage	[%]
k	= turbulent kinetic energy	[m^2/s^2]
l_e	= local Lagrangian length scale	[m]
m_p	= particle mass	[kg]
n	= number density of particles	[$1/m^3$]

t	=	time	[s]
$\vec{V}$	=	velocity vector	[m/s]
U	=	X – component of velocity	[m/s]
V	=	Y – component of velocity	[m/s]
ΔV	=	cell volume	[m³]
X	=	coordinate direction	[m]
Y	=	coordinate direction	[m]

Greek Letters

Γ	=	effective turbulent kinematic viscosity	[m²/s]
ε	=	dissipation rate of k	[m²/s³]
η	=	particles per trajectory	[1/s]
μ	=	viscosity	[kg/ms]
ρ	=	density	[kg/m³]
τ	=	particle dynamic relaxation time	[s]
τ_e	=	eddy lifetime	[s]
τ_R	=	transit time	[s]
ϕ	=	arbitrary variable	
ω_i	=	fluctuation frequency	[Hz]

Superscripts

′	fluctuating property
–	time – mean property
u	X – component
v	Y – component

Subscripts

g	gas phase
p	particle phase

REFERENCES

1. D.B. Spalding, Mathematical Models of Continuous Combustion, Emissions From Continuous Combustion Systems, Plenum press, 1972.

2. S.L. Soo, Fluid Dynamics of Multiphase Systems, Blaisdell Publishing Co. Waltham, Massachusetts, 1967.

3. D.A. Drew, Averaged Field Equations for Two-Phase Media, Studies in Applied Mathematics, Vol. 1, No. 2, pp. 133 - 136, 1971.

4. D.B. Spalding, Basic Two-Phase Flow Modeling in Reator Safety and Performance, EPRI Workshop Proceedings, 2, 1980.

5. C.T. Crowe, M.P. Sharma, and D.E. Stock, The Particle - Source - in - Cell (PSI - CELL) Model for Gas - Droplet Flow, ASME J. Fluids Engineering, Vol. 99, No. 2, pp. 325 -332, 1977.

6. C.T. Growe, REVIEW -- Numerical Models for Dilute Gas - Particle Flows, ASME J. Fluids Engineering, Vol. 104, No. 3, pp. 297 -303, 1982.

7. C.T. Growe, Gas - Particle Flow, in L.D. Smoot, and D,T. Pratt (ed.), Pulverized - Coal Combustion and Gasification, Plenum Press, 1979.

8. P.J. Smith, T.H. Fletcher, and L.D. Smoot, Two - Dimensional Model for Pulverized Flow Combustion and Gasification, Proc. 18th Symposium (International) on Combustion, The Combustion Institute, p. 1285, 1981.

9. A.D. Gosman, and E.Ioannides, Aspects of Computer Simulation of Liquid-Fueled Combustors, AIAA paper 81 - 0323, 19th Aerospace Science Meeting, St. Louis, 1981.

10. Cen Kefa, and Fan Jianren, The Analysis of the Forces Acting on the Particles and the Trajectory of Particles in Turbulent Flow, J. of Zhejiang University, Vol. 21, No. 6, pp. 1 - 11, 1987.

11. L.D. Smoot, and P. J. Smith, Coal Combustion and Gasification, Plenum Press, 1985.

12. S.V. Patankar, Numerical Heat Transfer and Fluid Flow, McGraw Hill, 1980.

13. R. Clift, J.R. Grace, M.E. Weber, Bubbles, Drops and Particles, Academic Press, 1978.

14. Fan Jianren, and Cen Kefa, Effects of Turbulent Fluctuation and Frequency Spectrum on the Drag Coefficient of a Spherical Particle in Gas - Solid Flow, Proc. of the International Symposium on Multiphase Flows, pp. 152 - 156, Hangzhou, P.R. China, 1987.

15. Luo Weihong, Fan Jianren, and Cen Kefa, Research on Bubble and Particle Motion and Eruption Model in Fluidized Bed, Proc. of the 9th Inter. Conf. on Fluidized Bed Combustion, p. 1198, Boston, American, 1987.

16. Y. Levy, and F.C. Lockwood, Velocity Measurements in a Particle Laden Turbulent Free Jet, Combustion and Flame, Vol. 40, pp. 333-339, 1981.

气固两相湍流脉动和扩散的数学模型及其在流化床悬浮段的应用

Kefa Cen(岑可法) and Jianren Fan(樊建人)

摘要： 本文提出了一个数学模型来描述气固两相湍流流动中的脉动和扩散现象。该模型基于 k-ε 方程描述气相湍流流场，同时利用基于脉动频谱的随机傅里叶级数来描述气流的湍流脉动速度。颗粒的运动特性(轨迹和速度)则采用拉格朗日方法进行描述。因此，该模型被称为"脉动频谱-随机颗粒轨迹(FSRT)"模型。最后，文章分析了该模型在流化床悬浮段和气固两相湍流射流中的应用。

引用格式： Cen K F, Fan J R. A numerical model for the turbulent fluctuation and diffusion of gas-particle flows and its application in the freeboard of a fluidized bed[J]. Particulate Science and Technology, 1988, 6(1): 81-93.

原文刊于 Particulate Science and Technology, 1988, 6: 81-93

循环流化床内颗粒运动的预测与测量

岑可法　樊建人　骆仲泱　严建华　倪明江

(浙江大学热物理工程系)

摘　要

本文提出了一种二维气固流动的数值计算模型,它考虑了循环流化床中颗粒的脉动和颗粒间碰撞等因素。该模型将颗粒运动的脉动考虑成一种基于气流脉动频谱的随机Fourier级数,称为脉动频谱随机轨道模型(FSRT)。这一模型可用于预测循环床中颗粒相的速度、颗粒轨迹和颗粒相浓度。对循环床中颗粒相速度和浓度进行了测量,并与FSRT模型的计算结果进行对比,同时对循环床中颗粒运动的一些行为也进行了预测。

关键词:预测,FSRT模型,气固两相流测量,颗粒运行,循环流化床

一、前　言

循环流化床燃烧技术(CFBC)相对于其它形式的流化床有许多优点,在世界上得到了迅速的发展。尽管国际上已进行了十余年大规模的工业开发,人们对循环流化床的认识仍不完备,比如(a)循环床内的流体动力学过程,(b)循环床燃烧与传热特性,(c)硫的析出与固硫特性等等。鉴于此,首先应对流体动力学过程进行研究,因为它对上述其它方面有很大影响。在对循环流化床的流体动力学研究中,许多人借鉴化工领域的快速流态化研究的成果,但这些成果不一定能完全适用于CFB。为了详细探讨CFB中固体颗粒运动的机理,我们将注意力集中于循环床中颗粒的轨迹、沿床高的压力分布、固体颗粒浓度分布和循环床的特征速度场分布。与此同时,运用我们提出的频谱随机轨道模型[1],对循环床中颗粒的运动进行了数值预测。本文对颗粒的湍流扩散和颗粒间碰撞的影响都从试验结果和理论计算结果两方面进行了探索。

二、数值模拟

将循环流化床中的气固流动考虑为轴对称充分发展的湍流,诸相间无化学反应。

颗粒浓度对气流的影响也可忽略不计。循环床某截面上的时均流速利用SIMPLE方法求出。同时,气流脉动速度的均方根由Laufer公式给出[6]:

1989年1月26日收到,1989年8月28日收到修改稿

(1) z方向

$$\frac{\sqrt{\overline{U'^2_g}}}{U^*} = -3.05(1-\frac{r}{R})^4 + 5.18(1-\frac{r}{R})^3 - 1.54(1-\frac{r}{R})^2$$

$$-1.98(1-\frac{r}{R}) + 2.11 \tag{1}$$

(2) r方向

$$\frac{\sqrt{\overline{U'^2_g}}}{U^*} = -18.89(1-\frac{r}{R})^4 + 43.51(1-\frac{r}{R})^3 - 33.43(1-\frac{r}{R})^2$$

$$+9.17(1-\frac{r}{R}) + 0.35 \tag{2}$$

为了考虑由流体流动造成的颗粒湍流扩散，我们采用了脉动频谱随机轨道模型[1]来描述颗粒的运动。在这个模型中，颗粒沿轨道的运动是在拉氏坐标系中进行计算的。

当颗粒与颗粒间没有碰撞时，只考虑颗粒的阻力和重力，颗粒运动方程可表述为：

$$\frac{dU_p}{dt} = \frac{1}{\tau_a}(U_a - U_p) - g \tag{3}$$

$$\frac{dV_p}{dt} = \frac{1}{\tau_a}(V_a - V_p) \tag{4}$$

其中颗粒的驰豫时间可采用下式计算：

$$\tau_a = \frac{1}{f_D} \cdot \frac{\rho_p d_p^2}{18\mu} \tag{4-a}$$

系数：

$$f_D = \frac{C_{DT}}{C_{DS}} \tag{4-b}$$

考虑到气流脉动时的曳力系数可由下式求得[5]：

$$C_{D,T} = \frac{24}{Re_p}(1 + 0.19Re^{0.82})[1 + 0.095(\frac{f_b Re_p}{C})^{0.28}] \tag{5}$$

为了修正颗粒的扩散，瞬时气速为脉动气速和平均气速的叠加：

$$\begin{cases} U_a = \overline{U_a} + U'_g \\ V_a = \overline{V_a} + V'_g \end{cases} \tag{6}$$

气相湍流脉动速度 U'_g 和 V'_g 用傅立叶级数来模拟，即：

$$\begin{cases} U'_g = \Sigma R_1 U_{i,m}\cos(iw_i t - R_2 \alpha_i^U) \\ V'_g = \Sigma R_3 V_{i,m}\cos(iw_i t - R_4 \alpha_i^V) \end{cases} \tag{7}$$

上式中z方向及r方向上频率为w_i时的脉动振幅U_{im},V_{im}可由下式计算。

$$\begin{cases} U_{im}^2 = \varepsilon_i \overline{U_g'^2} \\ V_{im}^2 = \varepsilon_i \overline{V_g'^2} \end{cases} \qquad (8)$$

频率w_i下的能量分布率ε_i可由气相湍流脉动的频谱给出。本文中采用了实测值进行计算。将(6)、(7)式代入(4)式,可得一组新的拉氏颗粒运动方程。

$$\frac{dU_p}{dt} = \frac{1}{\tau_a}[\overline{u_g} + \Sigma R_1 U_{im}\cos(iw_i t - R_2 a_i^u) - Up] - g \qquad (9)$$

$$\frac{dV_p}{dt} = \frac{1}{\tau_a}[\Sigma R_3 V_{im}\cos(iw_i t - R_4 a_i^V) - Vp] \qquad (10)$$

颗粒的位置定义为:

$$\frac{dz_p}{dt} = U_p$$

$$\frac{dr_p}{dt} = V_p$$

求解方程(9)、(10)即可得出颗粒的运动速度和轨迹。

从底部进入循环床的颗粒的初始位置及速度(由下标"0"表示)可记为($t=0$时):

$$r_p = r_{p0}, \quad z_p = z_{p0}, \quad u_p = u_{p0}, \quad V_p = 0 \qquad (11)$$

计算的第一个步骤是网格的划分,建立气相边界条件并提供必需的参数数据。第二步就是选取某种离散的颗粒尺寸分布和颗粒入口位置,例如,沿流场的均匀入口分布可以近似为一组有限数目的入口位置,在每个入口处(j网格)有x_j质量份额的颗粒进入床内。当然,随着颗粒尺寸和入口位置划分数目的增加,计算精度会提高,但计算时间也会增加,必须在两者之间进行权衡。此外的工作就是确定沿颗粒轨迹的颗粒数流率。这样,在j网格单位时间内进入流场的K尺寸组颗粒数目为:

$$n_{k,j} = \sigma_K \cdot x_j \cdot \dot{M}_p / m_K \qquad (12)$$

这里σ_K是K组粒子的质量份额。$\dot{M}_p$是单位时间颗粒总质量流率,而m_K是K组颗粒的质量。对于始于j网格的尺寸为K组的颗粒,沿整个轨迹上其n_K是不变的。

每一个计算网格中颗粒的浓度C_V为:

$$C_V = \sum_K \sum_i \left[\frac{\pi}{6}(n_{K,j}\Delta\tau_r)d_{pK,j}^3\right] / \Delta V_c \qquad (13)$$

网格中颗粒速度可由下式获得:

$$\begin{cases} U_p = \sum_K \sum_j [(n_{K,j}\Delta\tau_r)U_{pK,j}] / (\sum_K \sum_j n_{K,j}) \\ V_p = \sum_K \sum_j [(n_{K,j}\Delta\tau_r)V_{pK,j}] / (\sum_K \sum_j n_{K,j}) \end{cases} \qquad (14)$$

在本模型中,我们还考虑了颗粒间碰撞的影响,颗粒碰撞的自由程为:

$$l_p = \frac{1}{\sqrt{2} n_p 4\pi r_p^2} = \frac{r_p}{3\sqrt{2} C_V} \qquad (15)$$

我们假定,各粒度的颗粒碰撞后的参数按其重量份额分配。设碰撞前的颗粒速度为U_{p1},V_{p1}而碰撞后为U_{p2},V_{p2},则:

$$U_{p2}=U_{p1}-(1+e)\frac{M_{p2}}{m_{p1}+M_{p2}}(U_{p1}-V_{p1}) \quad (16)$$

其中m_{p1}是碰撞颗粒的质量,而M_{p2}为另一颗粒的质量,e是从实验数据获得的恢复系数[2]。

三、试验装置和物料

本文试验所用的冷态循环床示于图1,这种直径200mm,高4000mm的床是由玻璃制成的。送入CFB的空气使颗粒流化并将部分颗粒夹带至分离器,分离器出口的气体经布袋除尘器用引风机排入大气,分离器将大部分颗粒分离下来。被分离器捕获的颗粒通过L阀返回床中,完成颗粒的循环。

沿床的中心线安装了四只压力测孔,它们与四个压力变送器相连接,然后送至计算机以计算其统计特性,如相关函数,频谱及特征速度[4]。颗粒浓度分布由两种方法测量。其一是等速取样管,另一是一个具有快速切断功能的颗粒取样管[3]。

表1列出了试验和理论计算中所用到的床料物性参数。

四、试验及计算结果分析

图2给出了循环床中各种粒度的粒子的典型运动轨迹。由图可知,小颗粒会随气流上升,同时会由于颗粒碰撞而向下运动。然而,其总趋势是上升的。因为重力和碰撞的

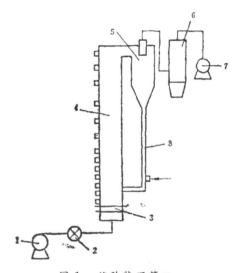

图1 试验装置简图
1—送风机; 2—流量计; 3—布风板;
4—循环床; 5—分离器; 6—布袋除尘器;
7—引风机; 8—阀;

影响,大颗粒的运动情况和小颗粒不同。一些大颗粒在上升到一定高度后就趋于下降,特别是在贴壁处。气流速度下降时,大颗粒将下降。这一现象在试验中可以观察到。

在循环流化床中由于在整个床内固体颗粒浓度较高、颗粒之间的相互作用强烈以及固体颗粒可能会形成颗粒团,固体颗粒运动的变化十分复杂。图2中仅考虑了最简单的碰撞情况,就可以看出其运动规律较为复杂。在循环床中固体颗粒的实际测量更为复杂。图3示出了固体颗粒速度的计算结果和利用文献[7]所述的"微机控制的变色闪频高速摄影系统"测得的初步结果。从图中可以看出二者在总趋势上是一致的。但这还有待于进一步的研究。

表 1 试验及计算中用到的床料物性参数

工况	平均粒径(mm)	重量份额(%)	颗粒密度(kg/m³)	表观速度(m/s)
工况 A 试验 及计算	1.50	3.25	1023 塑料球	9.3
	1.75	20.71		
	2.60	24.86		
	3.50	34.10		
	4.20	17.09		
工况 B 计算	0.133	1.7	2630 砂	9.1
	0.214	12.06		
	0.3	40.00		
	0.425	41.40		
	0.75	4.84		
工况 C 计算	0.055	4.68	1700 煤	9.1
	0.275	37.04		
	0.675	24.70		
	1.52	20.58		
	4.0	13.00		

图 4 给出了计算所得的不同粒径的固体颗粒速度随床高的变化规律。从图中可以看出，小颗粒的速度要比大颗粒大得多，但总的滑移速度都很大，特别是对于大颗粒尤为如此。图 5 给出了计算所得的三种不同颗粒的按颗粒重量加权的颗粒平均速度分布。因为 B 颗粒的平均粒径最小，故 B 颗粒的平均速度最大。

图 6 给出了试验所得的循环床内无量纲压力沿床高的分布。这与一般的观察相符，在循环流化床中存在着下部的密相区和上部的稀相区，但二者之间界限不十分明显，目前采用本模型尚未对压力分布进行计算，拟在以后的研究中进行计算。

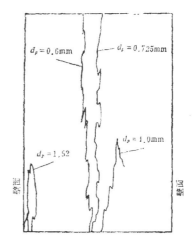

图 2　循环床中各种粒径下典型的颗粒轨迹

图 7 给出了计算和实测的沿床高方向的固体颗粒浓度分布。颗粒的浓度沿床高的变化十分强烈，床底部的颗粒浓度远大于顶部。从图中可以看出实测和计算所得的固体颗粒浓度基本相符。

图 8 给出了三种不同的颗粒沿床高的颗粒浓度分布。从图中可以看出，在一定的气流速度下，循环床中固体颗粒粒径较大时浓度沿高度方向的变化很大，颗粒集中于底部，而粒径

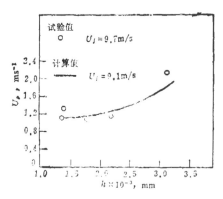

图3　计算和实验所得的固体颗粒速度

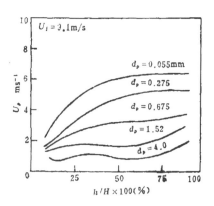

图4　流化速度 $U_a=9.1$ m/s 时循环床中典型的颗粒速度

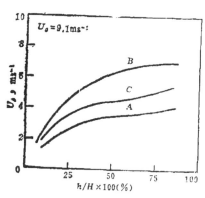

图5　在 $U=9.1$ ms^{-1} 下计算所得的不同工况下的颗粒速度

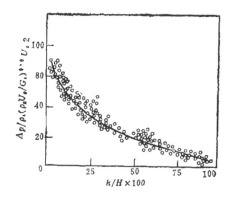

图6　试验所得的典型压力分布曲线

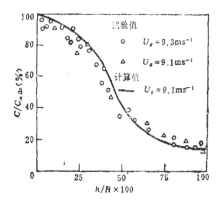

图7　沿床高的颗粒浓度分布理论值与试验值的对比

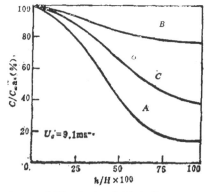

图8　在 $U=9.1$ m/s 时各种工况下的颗粒浓度分布

较小时，固体颗粒浓度变化比较平缓（如B类粒子）。

从上述结果可以看出，计算结果与实验基本相符，但尚需进行进一步的研究

五、结 论

1. 脉动频谱随机轨道模型考虑了湍流气流的脉动频率和频谱对颗粒相运动的影响，并考虑了不同尺寸的颗粒对湍流扩散的影响，因而本模型更接近工程实际；

2. 由于循环流化床内颗粒浓度较大，颗粒与颗粒之间的相互作用较重要，故我们在分子运动论的基础上提出了颗粒运动自由程随机碰撞模型（FPRI），该模型能较好地预测颗粒之间的碰撞；

3. 试验和理论计算均表明，由于颗粒碰撞等因素的影响，循环流化床中颗粒速度比气流要小得多，特别是在床底部滑移速度很大。

4. 循环流化床中浓度场分布取决于粒径分布、气相速度和颗粒密度；

5. 在正常的流化床气流速度范围内，大颗粒的浓度沿高度变化迅速，大颗粒密集于床下部，小颗粒浓度沿床高变化则较小。

致 谢

本文内容受国家自然科学基金资助，在此表示感谢。

符 号 说 明

C ——式(5)中经验常数

C_V ——颗粒相浓度，kg/m^3

C_{DS} ——Stokes曳力系数

$C_{D,T}$ ——湍流脉动的曳力系数

d_p ——颗粒直径，m

e ——方程(16)中的恢复系数

f_b ——脉动主频率，Hz

f_D ——式(4-b)中的系数

g ——重力加速度，m/s^2

l_p ——颗粒碰撞自由程，m

m_{p1}, m_{p2} ——碰撞和被碰撞颗粒的质量

K ——k组分的颗粒质量，kg

$\dot{M}_p$ ——单位时间内颗粒的总质量流量，$kg/m^2 \cdot s$

n_p ——颗粒的数密度

R, r ——半径，m

R_1, R_2, R_3, R_4 ——正态分布的随机数

t ——时间，s

ΔV_c ——网格体积

U_g, V_g ——z及r方向的气流速度，m/s

U_{im}, V_{im} ——z及r方向频率为ω_i时的脉动振幅

U_p, V_p ——z及r方向的颗粒速度，m/s

$\overline{U_g'^2}, \overline{V_g'^2}$ ——z及r方向脉动速度的平均值，m^2/s^2

U_* ——摩擦速度

α_i^U, α_i^V ——脉动初相角

ε_i ——ω_i频率下的能量分布率

μ ——动力粘度，$kg/m \cdot s$

σ_K ——K组分粒子质量份额

τ_g ——颗粒动态驰豫时间，s

τ_r ——网格中的停留时间

ω_i ——脉动频率

ρ_g, ρ_s ——气体和颗粒的密度，kg/m^3

参 考 文 献

1. 岑可法、樊建人 "A Numerical Model for Turbulent Fluctuation and Diffusion of Gas—Particle Flow and Application in the Freeboard of Fluidized Bed" Particulate Science and Technology, an International Journal 6, [1], 81 (1988)
2. 岑可法、樊建人 "煤粉颗粒在气流中的受力分析及其运动轨迹的研究" 浙江大学学报 21, [6], 1 (1987)
3. 骆仲泱、岑可法、倪明江 "循环流化床流体动力特性的试验研究" 浙江大学学报 21, [6], 84 (1987)
4. 岑可法、严建华, "Research on the Frequency Spectrum of Pressure Fluctuation in a Circulating Fluidized Bed" Circulating Fluidized Bed Technology II edited by P. Basu and J.F. Large, Pergamon Press p213
5. 樊建人、岑可法 "Effects of Turbulent Fluctuation and Frequency Spectrum on the Drag Coefficient of a Spherical Particle in Gas—Solid Flow", Proc of Inter. Symp on Multiphase Flows, Hangzhou
6. Lanfer NACA, Tech. Note No 2954, 417 (1953)
7. 罗卫红、樊建人、岑可法 浙江大学学报 21, [6] 32 (1987)

THE PREDICTION AND MEASUREMENT OF PARTICLE BEHAVIOR IN CIRCULATING FLUIDIZED BEDS

Cen Kefa, Fan Jianren, Luo Zhongyang, Yan Jianhua, Ni Mingjiang

(Dept. of Thermoscience and Technology, Zhejiang University)

ABSTRACT

A two-dimensional gas-particle flow numerical model is presented which accounts for particle fluctuation and particle-particle collisions in the circulating fluidized beds. This model treats fluctuation of particle movement as a random Fourier series based on gas flow fluctuation spectrum and hence is known as the fluctuation-spectrum-random-trajetory (FSRT) model. The model is used to predict the velocity, particle trajectory and concentration of the particle phase in circulating fluidized bed. Measurements of particle velocity weand the concentration of the particle phase in circulating fluidized bed were carried out and the predictions of the FSRT model are compared with the experimental results. Some aspects of the behavior of the particle movement in circulating fluidized beds are also predicted.

Key words: Prediction; FSRT Model; Measurement; Particle Behavior; Circulating Fluidized Bed

原文刊于化学反应工程与工艺, 1988, 5(4): 24-31

低倍率中温分离型循环流化床锅炉的设计

岑可法　骆仲泱　倪明江
李绚天　程乐鸣　方梦祥

(浙江大学)

摘　要

本文主要介绍了浙江大学热能工程研究所在循环流化床锅炉设计方面的部分工作，特别是该所发展的低倍率中温分离型循环流化床锅炉的设计思想以及主要部件的设计计算、方法，同时也介绍了这些部件的一些试验结果。

主题词： 循环流化床　锅炉　设计

1 概述

循环流化床锅炉具有清洁，高效的优点。近年来在国内外获得了迅速的发展。我们认为在开发我国循环流化床锅炉时，必须注意不但开发出燃用优质煤的循环流化床锅炉，还应开发出适合中国国情的劣质煤循环流化床锅炉。

在综述了国内外循环流化床型式的基础上，浙江大学热能所提出了自己的循环流化床锅炉类型，即低循环倍率、中温分离，采用下出气型式的循环流化床锅炉，目前已和杭州锅炉厂合作设计了 35t/h 循环流化床锅炉。本文主要介绍锅炉总体布置及参数选择方面的一些考虑以及具体设计时设计方法。

2 循环流化床锅炉的总体布置及主要设计参数

2.1 分离器的位置

循环流化床中分离器的位置是一个十分重要的参数，它直接影响着整个循环流化床的结构布置

循环流化床锅炉的分离装置在国内外大致有如下几种型式：

①高温旋风分离器，内层有防磨层和绝热层，或者采用水冷腔室结构；

②惯性分离器如迷宫式惯性分离器；

③中温分离，即在烟气进入分离器之前，先经过部分辐射和对流受热面，使进入分离器的烟温降至 600°C 以下，这样分离器在设计制造方面都比较容易。

高温分离器的特点是过热器烟气中颗粒浓度低于中温分离,对高循环倍率特别有利,但燃用高灰分燃料时的磨损问题尚未解决,而且旋风分离器体积也较大。由于受旋风分离器最大尺寸的限制,大容量循环流化床锅炉必须配用多个旋风分离器,同时高温分离器的热惯性较大(水冷式除外),因而启动时间较长,散热损失也较大。

中温分离与高温分离相比,亦有其独特的优点:①由于烟气入口温度较低(我们考虑低于600℃),从而使烟气体积减小,因而旋风分离器尺寸可以减小,提高分离器效率,温度降低时,烟气粘度降低,也可以提高分离器效率。②由于分离器温度降低,可以采用较薄的保温层,这样可以缩短锅炉启停炉的时间,在保温相同的情况下,还可以降低锅炉的散热损失。③从运行安全性来考虑,要求分离器内不发生燃烧,采用中温分离,由于分离器温度较低,内部不可能发生燃烧。④采用中温分离对材料的耐高温要求降低,可降低成本,这对于中小型锅炉特别有利。⑤循环流化床锅炉在运行时应当防止床层超温,采用高温分离时,由于分离下来的颗粒本身温度较接近床层温度,这样对抑制床温升高效果很小,在设计中采用中温分离,分离下来的冷物料可以有效地平衡床温,所以不致发生结渣现象,并具有调节负荷的作用。

采用中温分离唯一的缺点是不采取措施时,会增加前面受热面如过热器的磨损。在燃用发热量较低的劣质烟煤、石煤、泥煤及矸石时,一般采用较大的颗粒直径和较小的循环倍率,这样在流过过热器部分的烟气中所含颗粒的绝对量比采用高循环倍率时要低许多,而且我们还可以采用严密的防磨措施,使过热器的磨损控制在可以接受的范围。

基于上述理由,我们在35t/h石煤,煤矸石或煤泥循环流化床锅炉中采用中温分离方案,这一方案目前面临的问题主要是:分离器前受热面必须采取更严密的防磨措施,以保证运行的安全。

2.2 循环物料量

在循环流化床中,随循环量增加而返料量也增加,这样燃烧效率及脱硫效率会有所提高。如在床内布置有受热面则返料量增加,传热系数亦会增加,但返料量增加会使床层阻力及风机压头增加,故必须进行综合的考虑,目前燃用优质煤种时循环倍率为50左右,但返料量对燃烧及脱硫的影响有待进一步研究。

在国内外目前采用的循环流化床锅炉中采用较高的物料量,根据我们的研究这是没有必要的。因为在锅炉中只有小颗粒才参与循环,大颗粒并不参与。如我们设计的煤矸石、石煤循环流化床锅炉倍率为2.36,实际上小颗粒的循环倍率为12。表1示出了在总体循环倍率为2.36时,各档颗粒的循环次数及燃尽所需的循环次数。

表 1 燃尽所需的循环次数与实际循环次数对比

颗粒直径(mm)	0.1	0.5	1.0	2.0	>2.0
最大燃尽时间(s)	0.68	8.9	23.1	50.1	床内循环
运行风速下所需最大循环次数	0	3.6	7.2	16	床内循环
细颗粒实际循环次数	1	6.0	12.0	27	

2.3 受热面布置和床温控制

在循环流化床中燃料产生的热量一部分由高温烟气带至尾部受热面,但由于高温烟气不可能带走全部热量,目前有下列几种主要的方法布置受热面吸热:①在炉膛内布置水冷壁或隔墙,如Ahlstrom的循环流化床锅炉就采用了这种型式;②在炉膛内布置部分受热面,在固体物料循环回路上再布置流化床换热器,如Lurgi/CE和Battelle/Riley等均采用了此种型式,凡采用这两种型式的许多锅炉,通过运行实践,证明都是可行的,但这两种型式的床温控制形式是不同的。方案一主要是靠调节返料量来调节床内固体颗粒浓度来改变水冷壁的换热系数，从而改变床内的吸热量来改变床温。而方案二仅需调节进入流化床换热器和直接返回燃烧室的固体物料的比例即可调节床温，比较灵活，而且布置了外部流化床换热器，使燃烧与传热分离，可使两者达到最佳状态，将再热器布置在流化床换热器中，调节汽温比较灵活，甚至无需喷水减温。

根据美国 Battelle 的研究，采用外部流化床换热器的循环流化床在放大时优点特别明显，因为无外部换热器的循环流化床中床温主要由下列因素决定：①燃料的释热率；②燃烧室中烟气带走热；③燃烧室中的受热面布置，主要是水冷壁；④烟气或固体颗粒对受热面的传热系数；⑤烟气或固体颗粒与燃烧室的温度差。上述各项中①与②随锅炉容量的增大而上升，但③却不然，锅炉容量增大，总输入热量与受热面的比例反而下降，这样会使燃烧室温度升高，此可以采用下述方法解决：①增加燃烧室高度；②改变燃烧室截面，改变长宽比；③使用分隔墙水冷壁，此可以用于 100～150MW 容量锅炉；④在炉顶采用悬吊式过热器，但这会带来磨损问题；⑤采用扩展受热面，如肋等，但也有磨损和其他影响；⑥改变流化风速或固体颗粒循环量，从而改变床内固体颗粒浓度而改变换热系数，但这样会改变床内压降，从而会导致控制问题。上述①、②、③还受到经济性的影响，尽管采用上述方法的组合，采用单个燃烧室的蒸发量为 200～250t/h，而采用外部换热器则由于在燃烧室内无需布置受热面而去控制床温，放大就没有限制。

当然采用外部换热器有其优点，但在小容量锅炉中，优点不甚明显，该方案最大的弱点是增加了设备，而且 Nucla 电站的循环流化床锅炉不采用外部换热器，采用了双炉膛结构容量达到 420t/h，从锅炉运行的可靠性考虑我们不推荐采用外置式换热器。

对于燃用劣质燃料的锅炉，由于采用的燃料粒度较大，所以燃料有较大的份额会在下部的密相区内燃烧，这样在密相区内必须吸走足够的热量，这可采用两个方法来达到，其一是采用较大的循环倍率，利用固体颗粒流来带走热量；其二是采用在密相区内设置较多的受热面的方法来达到。

对于劣质燃料的循环流化床锅炉，从燃烧的角度讲是不需要较高的循环倍率的，从经济及安全运行的角度来讲，如果采用较高的循环倍率，就必须有大量的灰粒在炉内循环，则会增加不必要的能耗，且使受热面的磨损增加，影响锅炉的安全运行。所以我们设想采用较低的循环倍率，且在密相区设置部分埋管。

2.4 分离器的结构型式

在循环流化床锅炉的设计中，采用的分离器种类型式较多，而且还在出现新的结构，大体上讲，有旋风分离器和惯性力分离器两大类，惯性力分离器具有下述特点：①由于不采用高温旋风分离器，则不需要布置很厚的保温层，分离器四周可比较容易地布置受热面，使结构比较紧凑，启停炉也比较容易。②分离器不受单个最大尺寸的限制，且可以使锅炉受热面的设置保持传统的紧凑型式，有利于锅炉的大型化。③分离器阻力相对较小，有利于降低能耗。但惯性力分离器的最大问题是分离效率较低，它对于较小颗粒的分离效率很低，很难保证这部分颗粒的燃尽，特别是在采用脱硫剂时，尤应注意这个问题，目前这种型式采用不多。

在循环流化床锅炉中大部分采用旋风分离器的型式，一般使用在循环流化床锅炉的旋风分离器大都采用上部进气、顶部排气型式，烟气所携带的颗粒由于离心力而向分离器壁面移动下滑，而烟气则经过转折后从顶部的排气管引出。这种结构中，烟气流向与颗粒走向相反，因而已分离的颗粒二次夹带较少，分离效率也较高，不过烟气阻力也较大。对于这种型式的分离器，人们已做过大量研究和流场模化，技术比较成熟。但是，在将它用循环流化床锅炉时，由于烟气从顶部排出，故与尾部烟道烟气向下的流向不能协调，此外，旋风分离器将被迫独立悬置在炉膛和尾部受热面之间，这样势必增加了锅炉深度方向的尺寸，占地面积增大，还必须增设钢架来支承分离器。特别是在采用中、高温分离的情况下，整个锅炉的外形显得庞大（图1）。

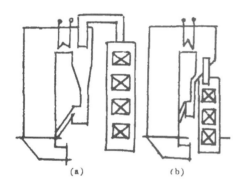

图 1 不同分离器型式的锅炉布置简图

设计中采用下排气结构，烟气从下部引入，颗粒也从下部一个斜锥向下流入回送机构。由于旋风分离器内部烟气流程趋于简单，并且与尾部烟道气流方向一致，因此省去一些不必要的转弯，分离器阻力也较低，约为上排气时的75%。但是由于颗粒与烟气走向一致，故夹带较多，分离效率有所降低。在上排气中分离效率可达95～99%，但在同一容量锅炉中采用下排气的旋风分离器时，分离效率降为93～97%，总的说来，排气位置对分离效率的影响不太大。在旋风分离器中决定分离效率的是颗粒所承受的离心力与其重力之比。因此，通过适当提高烟气入口的速度，减小分离器直筒段直径（中温分离时可以做到），就可以达到所要求的分离效率。况且对于循环床分离器效率不一定要求很高，对于小于0.1mm的颗粒实际上能够一次通过炉膛燃尽而无须分离。

正如上述，中温分离时由于烟气体积较小，旋风分离器直径和高度都较小，并且由于循环流化床炉膛一般远高于普通锅炉，因此，在采用分离器下排气方案时就可以把分离器直接布置于尾部烟道上方，下面仍有足够空间布置省煤器和空预器受热面。这时，锅炉结构显得特别紧凑，支架也可以简化。

这种下排气的旋风分离器在化工冶金方面已经采用。其问题在于分离效率可能稍低于顶部排气的旋风分离器。另外，由于下部排气管直径较小，在经较短的扩容段后，可能会使省煤器上部冲刷不均。在某些场合下，这种布置还可能使锅炉高度被迫增加，因此需要进行多方面的权衡。但总的来讲，由于这种布置方式紧凑、结构新颖、运行可靠，将有利于循环流化床锅炉的大型化。

2.5 负荷变化时床温的控制

在锅炉启动或停炉时，循环倍率为零，即采用鼓泡床运行状态，变负荷时的循环倍率也与正常时不同，因此如何在不同工况下保持床温稳定，防止底部超温结渣是运行可靠性的重要方面。

在我们设计的锅炉中，额定负荷下循环倍率为2.36，因为中温分离下来的颗粒温度只有600℃，相对床层来说，它是冷物料，因此返回物料量较多时，床温会有所降低，根据运行情况，在与正常床温偏离不大时，可以通过调节返料量来控制床温。譬如，在降负荷时，可采用较低的循环倍率，在尖峰负荷时可以提高循环倍率，返料量的变化即可用于控制床温，又可帮助调节负荷。在变负荷时，如果增加一次风率，可以增加底部气流速度，使更多颗粒进入炉膛燃烧，该设计中，额定负荷下，一次风率为0.8，二次风率为0.2，而在启动时，二次风率为0，所有空气都作为一次风送入，床温稳定在900～950℃。在紧急情况下，如果常规方式不能使床温降低，应采用适当的应急降温手段，在该设计中增设了床层紧急喷水降温装置，将水直接喷入床层以保证温度不超过许可的范围。

2.6 燃料颗粒度

一般认为对于高灰的燃料宜采用小一些的燃料颗粒尺寸，如 Ahlstrom 公司认为对于生物燃烧宜采用 30～50mm 的颗粒尺寸，低灰煤种颗粒最大尺寸宜小于 10～20mm，对于高灰燃烧宜采用小于 2～12mm 的颗粒，但各个公司都有自己采用的颗粒尺寸的范围，鲁奇公司的循环流化床采用的燃料颗粒尺寸也取决于煤种，但比 Ahlstram 公司采用的要小一些，一般在 10mm 以下，对于高灰煤种鲁奇公司认为最佳细度为 150～250μm，对于低灰煤种为小于 10mm，而 Riley/Battelle 一般采用小于 25mm 的颗粒，最大值与基本床料及具体使用的煤种有关，应视具体情况而定。

对于燃用优质煤的循环流化床锅炉，一般认为可以采用较大的颗粒度，如 0～8mm 或 0～10mm，这是比较合适的。但对于燃用高灰分煤的循环流化床锅炉，鲁奇公司认为最佳细度为 150～250μm，但如采用这样的颗粒度，就需要把大量的灰磨制成粉，磨煤电耗极大。我国在燃用劣质煤燃烧方面有着丰富的经验可以借鉴，所以我们认为结合鼓泡流化床的经验，可以采用较大的颗粒度，与优质煤的炉子基本相同，但在设计时包括受热面设置及循环倍率的选取上有所不同，这样燃料的处理系统也比较简单，运行也比较有保证。

2.7 受热面的磨损及防止

在循环流化床锅炉中，固体颗粒对受热面的磨损是一个十分重要的问题。目前国外投运的循环流化床锅炉，有些由于设计或运行上的问题，出现了非常严重的磨损，大大地降低了循环流化床锅炉的运行安全性，有的甚至运行个把月就产生爆管。因为在高循环物料量的循环流化床锅炉中，固体颗粒浓度很高，如果不采取措施，会带来严重的磨损。

我们已对循环流化床中受热面的磨损进行了大量的实验和数值试验，结果可参见义献[16]，这里主要介绍设计时防磨的措施。

防止磨损总的原则是：控制烟气中颗粒携带量，因为磨损与颗粒浓度成正比，同时采用低的烟速。试验表明，磨损量与烟速的三次方以上成正比。这意味着烟速降低一倍，磨损量至少可降为原来的八分之一。在此基础上再采用一定的防磨构件，对于不同的受热面我们可采用不同防磨措施。

①对于底部的埋管，除了使底部气流速度保持适当值外，在埋管四周可焊接防磨销钉，它不仅具有防磨作用，而且还作为传热肋片，增大了埋管受热面积。

②对流过热器的防磨，由于采用中温分离，所以这一带的烟气中携带的颗粒量很大，为此可选取较低的平均烟速（4～6.5m/s），并采用顺列布置，与此同时，对流过热器前几排管壁上迎着烟气方向两侧焊两块 2～3mm 厚，高约 10～20mm 的防磨片，这样，颗粒冲刷时就部分被捕获并沿管壁流下。理论与实践都表明，横向冲刷圆管时，磨损最严重的部位不是管子的正面，而是两侧某一角度外，在这一部位焊防磨片效果是很好的，而且可以分离部分颗粒。此外，蛇形管两端部分管子弯头可加防磨罩。

③省煤器，这一区域烟速更高，但是在经过分离器以后，颗粒浓度只有前级的 5% 左右，除了弯头处可以设防磨套以外，由于这一带管子较密，阻力较大，而在靠炉墙外可能存在较大间隔，形成烟气走廊，这一区域磨损将特别严重、为了平衡各处的阻力，可以在贴墙处另加一根空管或圆钢使烟气通流截面各处管子间隔一致，消除烟气走廊，即可有效地防止磨损。

另外，由于分离器直接布置在省煤器上部，而且两者之间距离较短，如果不采取措施产生较大的不均匀冲刷，从而产生较大的局部磨损，所以在设计时在最前面几排可设置假管，管子内部不通工质，另外，在前面几排的省煤器中采用鳍片管，以减轻管子的磨损。

在设计中我们保证了受热面的磨损低于常规锅炉的受热面磨损，在此基础上再采取防磨措施，所以在受热面的磨损问题上锅炉是安全可靠的。

④空气预热器：空气预热器的磨损与常规锅炉差不多，但为了保险起见，我们采用了长达250mm的防磨套管，可以有效地防止空预器进口处的磨损。

3 循环流化床锅炉的设计

循环流化床锅炉主要由下列部件组成：燃烧室、分离器、固体颗粒循环系统和尾部受热面等，其中尾部受热的设计与传统的锅炉相同，恕不赘述。这里主要讨论燃烧室分离装置及物料回送装置的结构设计。

循环床炉膛的设计是最主要的步骤，由此确定燃烧效率和脱硫效率。循环床燃烧室的设计与传统的煤粉炉有较大的不同，床内温度较低(900~950℃)，远低于灰渣的熔点，而且炉内混合强烈，温度比较均匀。设计时的基本参数为燃烧室高度、横截面积、受热面的布置以及下部涂耐火层的高度。

3.1 燃烧室的设计

①循环床燃烧室的横截面积

燃烧室的横截面积可根据锅炉总的热容量和炉子的截面热强度或流化风速确定。它的截面热强度和流化风速这两个参数在燃料量确定时，两者是等同的，选定一个参数另一个也随之确定，它的截面热强度可在3~6 MW/m的范围内选择，此时对应的流化风速大约为6~10m/s的范围。

流化风速的确定一般是选定而不是计算，选择合适的流化风速时必须综合考虑下述因素：燃烧室高度、受热面积、受热面的磨损以及风机压头。流化风速较高，可使炉膛更加紧凑，但同时会加快受热面的磨损，最主要是为了保证煤粒在炉内的一定的停留时间和保证一定的床内吸收量就势必要增加炉膛高度，这样会增加钢材消耗量和增加风机压头，可能会抵消由于炉膛紧凑而带来的好处，设计时必须综合考虑。表2为国外循环流化床锅炉选择的运行风速。

表 2 循环流化床锅炉制造厂家提出的运行风速范围

厂家	Lurgi	Riley/Battle	Ahlstrom
运行风速(m/s)	5~9	10左右	3~10

②燃烧室的高度

在循环流化床中，固体颗粒可以被分成三类：在运行风速下不可能被夹带的大颗粒，在运行风速下可能被夹带但在分离器内被分离的小颗粒，能被夹带但不能被分离的细粉。如果燃烧室高度过低以致细粉不能被燃烬的就会造成较大的损失，所以在燃烧室设计时，燃烧室的高度应以细粉在一次通过时能够燃烬为第一准则。

在循环流化床锅炉中，大约有一半的热量在固体颗粒循环回路上被吸收、燃烧室设计的第二准则是有足够的位置以布置所需的受热面。第三准则是燃烧室高度应使气体有足够的停留时间完成燃烧和脱硫反应。根据第一准则，燃烧室的高度可以采用下式计算：

$$H_{CB} = \tau_0 / U_S \tag{1}$$

上式中 τ_0 是细粉的燃尽时间，U_S 是细粉的上升速度。在循环流化床中，固体颗粒的上升速度不能采用单颗上升的计算公式进行计算，其原因有三：①由于固体颗粒浓度很高，所以固体颗粒之间的相互作用不容忽视，颗粒之间的碰撞使固体颗粒的速度变化比较复杂；②固体颗粒形成了颗粒团，颗粒速度的变化更为复杂；③气流在循环床内流动的紊流度较大。

在上述公式中固体颗粒在床内的上升速度可以采用数值计算方法，比较简单的计算方法可采用 Ravisankar 和 Smith[7] 提出的单颗粒在循环床内上升滑动速度的计算公式：

$$\frac{U_r}{U_t} = A(1-\varepsilon)^B \quad (2)$$

上式中：

$$A = 93.67 (Re_p)^{-0.994} \left(\frac{dp}{dt}\right)^{1.014} \cdot$$

$$\left(\frac{\rho_s}{\rho_g}\right)^{0.706} \quad (3)$$

$$B = 1.075 (Re_p)^{-0.445} \left(\frac{dp}{dt}\right)^{0.476} \cdot$$

$$\left(\frac{\rho_s}{\rho_g}\right)^{0.313} \quad (4)$$

$$Re = d_p \cdot U_t \cdot \rho_g / \mu_g \quad (5)$$

颗粒真实速度为：

$$U_s = U_g - U_r \quad (6)$$

我们可以由公式(1)~(6)计算出满足上述燃烧室高度第一准则的循环流化床最小的燃烧室高度，式中 dt 为床直径。

如果在固体颗粒循环回路中除炉膛内部外不设置其它的受热面，则炉膛高度计算中必须考虑第二准则，反之，我们可以设置其它种类的受热面如外置式流化床换热器等来代替炉膛受热面，所以第二准则不是基本的。

对于第三准则到目前为止由于脱硫及燃烧所需的气体停留时间不能给出，所以比较难确定，尚需进一步的研究。

3.2 旋风分离器的设计

在进行循环流化床旋风分离器的设计时，我们必须注意的是：通常的旋风分离器一般工作在颗粒浓度较低的环境下，所追求的是高效率，而循环床旋风分离器工作的环境为高固体颗粒浓度的环境。在高浓度旋风分离器中，由于固体颗粒形成颗粒群，以及颗粒之间相互作用的增强。分离器的设计与传统的高效分离器是有区别的，下面主要讨论高浓度旋风分离器的设计参数。

①进口管的设计

进口管的面积，形状与位置都直接影响到气流在旋风分离器中的运动，因此对其设计应予以足够的重视。

进口气速越大，分离效率就越高，但当气速过高时，不仅不能提高效率，甚至反而会降低效率，因为气速过高，气流的湍动程度增大，二次夹带严重，分离效率反而下降，而且速度高，压力损失也大，会增加耗电量，加速磨损，所以循环流化床旋风分离器的入口风速应得适中，一般可取为 12~18m/s，不应超过 20m/s，具体可根据综合考虑分离和防磨等要求决定。

进口管一般制成矩形，其宽度的比例要适当，长而窄的进口管与管壁有更大的接触面，此时宽度 B_o 较小，临界颗粒尺寸也较小，分离效率越高，但如果 B_o 小于环隙宽度 $(D_o - D_e)/2$ 也是无益的，一般可取矩形，进口管高与宽之比为：$H_o/B_o = 2$，此时 H_o 取 $0.75 D_o$。

进口的形式一般可取图2所示的三种形式，根据分析由于蜗壳进口处理量大，且压降较小，一般以180°的蜗壳为多，但由于在循环流化床中为了结构布置的方便，采用切向进口具有较大的优点，进口管的位置一般与顶盖相平。

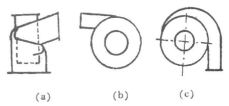

图2 旋风分离器进口方式示意图
(a)螺旋面进口；(b)切向进口；(c)渐开线蜗壳进口

②排气管

排气管的直径和插入筒体的深度对压力和分离效率有着显著的影响、排气管的直径越小，分离效率就越高，但压降的增加也越快，综合考虑分离效率和压降的影响，排气管直径一般可采用 $D_e=(0.5\sim0.75)D_c$，重视分离效率时可取较小值，重视压力损失时可取较大值。

排气管插入过长，会造成排气管与锥体底部的距离过小，从而旋流的有效旋转圈数减小，同时也增加了二次夹带的机率，使分离效率降低，而且还会增加表面摩擦，增大压力降，但插入管过短或不插入筒体，则会造成正常旋流核心的弯曲和破坏，使其处于不稳定状态，亦容易造成短路而降低效率，综合考虑上述影响因素，插入深度范围可取为 $(0.7\sim0.8)D_c$，对于下出气分离器甚至可取 D_c。

③筒体直径及长度

旋风分离器筒体直径越小，离心力就越大，分离效率也越高。但筒直径不能太小，否则出口管离筒壁太近，较大的颗粒容易反弹到中心气流中被带走，一般筒体直径应按处理气流量而确定。

分离器筒体长度增加，可使进入筒体的颗粒停留时间增长，有利于分离，而且也能使尚未到达排气管的颗粒有更多的机会从旋流核心中分离出来，减少二次夹带，从而提高分离效率，但筒体长度还应综合考虑到投资、场地以及与返料机构配合等因素，一般可取 $L_c=1.5\sim2.5D_c$。

④旋风分离器筒锥体

旋风分离器中的自由旋转的轴通常是有一个偏心度，根据实验观察偏心度大约为 $D_c/4$，为了防止由于颗粒重新卷入核心旋流，造成二次夹带，要求锥顶排料管直径 J_c 不得小于 $1/4D_c$，一般考虑顺利排灰，J_c 取 $0.5D_c$。圆锥体的长度 Z_c 一般取为 $2.5D_c$，如果采用圆筒形式的旋风分离器，此值还可

再小些，在决定排料管直径时应与返料机构的立管直径相一致。

对于下排气型分离器基本上可按上述参数考虑。

在循环流化床锅炉中存在的一个问题是在变负荷时的分离器效率问题，因为分离器效率与进口风速有很大的关系，如锅炉负荷从 100% 变至 50% 时，如果保证过剩空气不变，风量应降低一半，进口风速也降至一半，此时分离器效率极低，为了解决此问题，我们认为在分离器的入口处装设可动挡板，在一定风量下可调节分离器入口风速。我们已在试验台上对此进行了试验。试验结果表明效果较好。

3.3 返料机构的设计方法

①L 阀的设计

Know Lton 和 Hirsan(1981) 对 L 阀提出了初步的设计方法，根据压力平衡关系，推出了立管的最低高度计算式：

$$(H_{RS})_{min}=(\Delta P_{LV}+\Delta P_{CB}+\Delta P_{SP})/(\Delta P/L)_{min} \quad (7)$$

根据最小流化关系可得：

$$(\Delta P/L)=\rho_b \cdot g \quad (8)$$

而 L 阀的压力降 ΔP_{LV} 可根据下式得出：

$$\Delta P_{LV}=\Delta P_{LV1}+\Delta P_{LV2} \quad (9)$$

其中 ΔP_{LV1} 和 ΔP_{LV2} 可分别为密相输送时水平管及弯管段的压降，循环流化床的压降，可由如下的经验公式计算，即：

$$\frac{\Delta P_{CB}}{\rho_s \cdot U_f}=1.54\left(\frac{G_s}{A} \quad \frac{g \cdot \Delta H_{CB}}{U_f \cdot \rho_s}\right)^{1.02}$$

$$\cdot \exp\left(-2.61\frac{\Delta H_{CBO}}{H_{CB}}\right) \quad (10)$$

ΔP_{SP} 为分离装置的阻力损失，可根据具体情况而定，从上述公式可求出 H_{min}，但考虑到一定的富裕，则实际的 H_{RS} 为 $(H_{RS})_{min}$ 的 $1.5\sim2.0$ 倍。

已知返料量后返料管直径 D_{LV} 的确定

可采用公式求出，送风口位置的确定稍为复杂一些，根据初步的试验，一般可选择它位于L阀水平段中心线以上$(2\sim4)D_{LV}$处为宜。

L阀充气风机的风压主要应克服循环床后部的压头和L阀的阻力损失，一般可采用下式计算：

$$(P_{LF})_{min}=P_{CB}+(\Delta P_{LV})_{max}+\Delta P_{fr} \quad (11)$$

其中：

$$(\Delta P_{LV})_{max}=2.5(\Delta P_{LV1}+\Delta P_{LV2}) \quad (12)$$

P_{CB}可由炉内零压点的位置由公式(10)求得，ΔP_{fr}是由空气通过充气孔的节流损失，可比较方便地求得。实际选用风机时考虑到一定的富裕量，可取为：

$$P_{LF}=(1.2\sim1.5)(P_{LF})min \quad (13)$$

风机风量的确定主要取决于输送比，在目前无其它数据的情况下可近似按输送比为100选取，再考虑到一定的备用系数，则：

$$V_g=(1.2\sim1.5)G_s/100\cdot\rho_b \quad (14)$$

L阀水平段的长度不光反映在ΔP_{LV1}的压力损失上，而且在很大程度上影响L阀的正常运行，L阀水平段过短会导致物料自流，起不到控制的密封作用，过长会大大影响ΔP_{LV}，引起操作不稳定，使L阀的启动及调节特性变差，要保证L阀能有效关闭时，最短的水平段长度可由下式计算得到：

$$(L_{LV})min=2D_{LV}\cdot ctg\gamma_s \quad (15)$$

L阀水平段的最长长度的影响因素很多，但在一般情况下主要是受立管的高度的限制。已知立管的高度后，我们可以由式(8)和立管上部的静压求出最大的立管密封压力。由(12)和水平段出口处的静压，可求出一定的水平段长度下的充气口压力。此时要求最大的立管密封压力大于充气口压力，否则会在启动时出现反吹现象，计算时必须注意上述比较时使用的是绝对压力。

实际取值时使水平段长度大于$(L_{LV})_{min}$的情况下越小越好，且要满足小于上述的最大水平段长度，具体应由实际布置而定。

②流化床返料机构的设计

流化床返料机构的设计原则上适用于密封罐式和回路密封器式，与上述的L阀设计有类似之处。

从图3可以看出，沿着循环床的固体物料回路，下述压力平衡关系式成立：

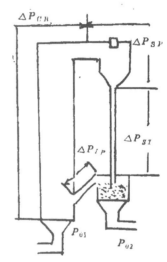

图 3 使用流化床返料机构的循环床

$$\Delta P_{IP}+\Delta P_{CB}+\Delta P_{SP}=\Delta P_{RS} \quad (16)$$

在上述各项中，ΔP_{SP}和P_{CB}上面已予以讨论。ΔP_{IP}在此予以讨论。采用流化床返料机构的优点之一是可以采用主床一次风作为返料风，如果一次空气和返料风采用同一风源，在图4所示的回路中，下述压力平衡式成立：

$$P_{O1}-\Delta P_{DS1}-\Delta P_{H1}+\rho_bgH_3-\Delta P_{H3}$$
$$=P_{O2}-\Delta P_{DS2}-\rho_bg\Delta H_2 \quad (17)$$

若设定两布风板阻力相等，可以得到：

$$P_{O1}-\Delta P_{H1}+\rho_bgH_3-\Delta P_{H3}=P_{O2}$$
$$-\rho_bg\Delta H_2 \quad (18)$$

在极限情况下，倾斜管不满管，故$\Delta P_{rp}=\rho_bg\Delta H_3=\Delta P_{H3}=0$。在该情况下，

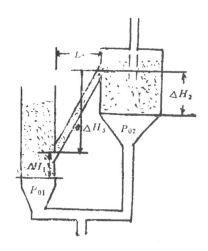

图 4 流化床返料机构简图

我们必须防止循环床主床的气流通过再循环管路旁路,所以:
$$\Delta P_{H1} \geqslant \rho_b \Delta H_2 \quad (19)$$

该等式只有当返料机构的布风板高于主床布风板时才能成立。ΔP_{H1} 可以由式(10)计算。

根据上述分析,当 $\Delta P_{rs} = \rho_b g \Delta H_3 - \Delta P_{H3} = 0$ 时,可以得到:
$$\Delta P_{rs} = \rho_b g H_{rs} = \Delta P_{CB} + \Delta P_{SP} \quad (20)$$

所以:
$$(H_{rs})_{min} = (\Delta P_{CB} + \Delta P_{SP})/\rho_b g \quad (21)$$

立管的高度可由下式计算得到:
$$H_{rs} = (1.5 \sim 2)(H_{rs})_{min} \quad (22)$$

立管的直径是另一个重要的参数,在立管中的流动为重力流动,立管中物料的高度会自动变化使压力降满足等式(20),当立管比较光滑没有阻挡物等时,固体颗粒流速可以取得较高,由于对这种流动的研究较少,目前尚很难给出最佳流速的计算公式,根据我们初步的研究,立管中固体颗粒流量可用下式估算:
$$V_s = 1.6\sqrt{D_{rs}} \quad (23)$$

同时,D_{rs} 必须大于 $(10 \sim 15) d_p$。

流化床返料机构中的流化床部分的设计必须考虑固体物料的储存和合理的流化空气流量与压力,流化床高可取 300~500mm,床截面积可取为 $(8 \sim 12) A_{st}$。如果返料机构与外置式换热器一体,则流化床的设计可主要考虑传热。

在返料机构中还应引起重视的另一问题是物料的再燃,在床内必须避免死区,可以在床内设置挡板或采用不均匀布风等,这样的型式基本上就是上述的密闭罐式和回路密封器式。

流化空气的流量和压力的确定比较简单。空气压力等于一次风压,流量刚好使物料流化,最难确定的参数是连接主床和返料机构流化床的倾斜管的倾角。如果倾角太小,会造成固体颗粒流量过小,而倾角太大,会造成空管,从而造成运行不稳定,根据试验结果,倾角可以比堆积角小 5°~10°,即:
$$\theta = \theta_r - (5° \sim 10°) \quad (24)$$

为了使固体颗粒流动均匀,防止空管和自锁的发生,倾斜管参数应满足下列等式:
$$D_{IP} \text{ctg}(\theta_r - \theta) < \sqrt{H_3^2 + L_3^2} \leqslant 2D_{IP} \cdot \text{ctg}(\theta_r - \theta) \quad (25)$$

4 结论

国内外发展循环流化床的型式很多,但很难说已达到了一种比较完善的程度,在综合了目前存在的循环流化床锅炉的型式,浙江大学热能所工程研究所提出了一种低倍率中温分离型循环流化床锅炉的设计。

本文主要介绍了这类锅炉的设计思想包括主要的设计参数、循环流化床中燃烧室、分离机构以及返料机构的具体设计。

参 考 文 献

[1] 倪明江、骆仲泱、程乐鸣等. Proc. of 15th Biennial Low-Rank Fuel Symp, p. 211

[2] Basu. P. J. of the Institute of Energy. June, 1987, p. 77

[3] 温龙. 东方锅炉, 1988 (4)
[4] 田子平, 锅炉技术, 1988 (2)
[5] Abdullay I. F. and Parham. D. FBC-Technology for Today Edited by A. M. Manafer. ASME, Press. 1989, p. 279
[6] Ravisankav S. and Smith. T. N. Powder Technology, 1986 (47): 167
[7] Tudg S. E. and Willams. G. C. Atmospheric Fluidized Bed Combustion. Final Report US DOE, No. DE-Ac 21-80 Mc 14536 1987, Jan
[8] 岑可法、倪明江、骆仲泱等、工程热物理年会燃烧 1989 年年会论文集
[9] 骆仲泱、倪明江、岑可法, 浙江大学学报, 21 (6): 84
[10] Chong, Y. O., O'Dea, D. P. et al. Circulating Fluidized Bed Technology II, Edited by P. Basu and J. F. Large. Pergomon Press, p. 493
[11] Hill, M. K., Proc. of 9th Inter. Conf. on FBC, p. 862
[12] Knowlton, T. M. and Hirsan, I. Hydrocarbon Processing, March. 1978, p. 149
[13] Leung. L. S. Power Technology. 1987. (49): 271
[14] Wen, C. Y. and Simons H. P., AIChE J. 5, (2): 263
[15] 骆仲泱、倪明江、张政江等. Proc. of 2nd Inter. Symp. on Multiphase Flow and Heat Transfer. p. 310
[16] 岑可法、骆仲泱、樊建人等. 煤炭加工与综合利用. 1990 (5): 50

原文刊于动力工程, 1991, 11 (5): 1-11

(上接第30页)

过不同方式收集和了解了国外其它核电汽轮机高压缸的设计与运行实践。在此基础上，我们进行核电汽轮机高压缸的优化设计。优化后的核电高压缸重量轻、结构简单合理、能降低应力与改善性能，并可方便于加工与安装，减少制造工作量，降低成本。优化后的高压缸不仅能用于 310MW 核电机组，也可用于 600MW 核电机组，它为进一步发展核电机组提供了极为有利的条件。

参 考 文 献

[1] Kalderon, D., Design of full speed 985MW steam turbine for Daya Bay nuclear power station. GEC Review, 1987(2): 76～85
[2] Pawliger, R. l., Ten years experience of wet steam. Steam turbine developments, 1988, (12): 53～57
[3] 三菱原子能发电用汽轮机设备技术交流会资料. 1987.9
[4] Pressure vessels manual. Westinghouse Electric Corporation
[5] HN 310-54.5 型核电站凝汽式汽轮机产品说明书
[6] ASME 锅炉及压力容器规范第Ⅷ卷 第二册 压力容器——另一规程. 1983年 S1版
[7] 汤粮孙. 沸腾传热和两相流动
[8] Аркадьев, Б. А. Режимы работы турбоустановок АЭС. Энергоатомиздат, 1986
[9] 朱丹书. 核电汽轮机高压缸改进设计的探讨. 上海汽轮机. 1990, (1)

ABSTRACT

Cen Ke-fa (Zhejiang University), Luo Zhong-yang, Ni Ming-jiang, Li Xuantian, Cheng Le-min, Fang Meng-xiang, «Design of A Medium Temperature Separating Type Low Circulating Fluidized Bed Boiler», «Power Engineering», 1991, No.5, pp. 1~11

The paper mainly describes some work oh circulating fluidized bed boiler design done by the Thermal Energy Engineering Research Institute of the Zhejiang University, especially as regards the design philosophy of medium temperature separating type low circulating beds and a method for calculating its main compoennts. In addition, some test results of relevant component are presented. 4 figs and 16 refs.

Zeng Han-cai (Huazhong Polytechnic University), Zhong Wen-ying, Lu Huanyao, Li Yan, Guo Jia, «The Experimental Study on Sulphur Removal by Direct Spraying Calcium into Pulverized Coal Boilers», «Power Engineering», «Power Engineering», 1991, No.5, pp. 12~17

The paper mainly describes research work on activation of desulphurization agents, together with some results of experimental study on direct spraying calcium into pulverized coal boilers. 9 figs, 3 table and 4 refs.

Zheng Yun-zhi (Shanghai Turbine Works), «On Quality Control in Steam Turbine Manufacture», «Power Engineering», 1991, No.5, pp.18~22

The paper describes how steam turbines can be made to conform with quality requirements by stringent control during manufacture and thereby improve operational reliability and economy. Major technical requirements and highlights in the steam turbine manufacture quality control are presented.

Zhu Dan-shu (Shanghai Turbine Works), Lu Qing-zhi, Lu Rong-ji, «Optimization of A Nuclear Turbine's H.P.Cylinder», «Power Engineering», 1991, No.5, pp.23~30, 11

This paper presents the optimization design of a nuclear turbine's high pressure cylinder, including the use of single shell instead of double-shell casing, unsysmetical bleed, integration of carrier rings, improvement and perfection of H.P. outer casing configuration, etc., with a comparative thermal stress analysis. A technical/economic evaluation of the optimized design is presented as a conclusion. 4 figs, 7 tables and 9 refs.

数值试验(CAT)在大型电站锅炉设计及调试中应用的前景
——(Ⅰ)基础理论*

岑可法　樊建人

(能源系)

提　要

本文提出了计算机辅助数值试验(Computer Aid Testing)方法,并且讨论了数值试验方法在大型电站锅炉中应用的可能性和必然性。本文阐述了数值试验的基本原理和方法,对锅炉炉内流动、燃烧、传热和磨损等领域建立了各自的数值试验子模型,为锅炉设计及调试工程应用打下基础。

关键词:数值试验(CAT),电站锅炉,数学模型

0　引　言

　　人们常通过利用一些测量仪器对研究的对象进行测试,从而得到所需的各种数据,这即所谓的仪器试验。然而,在某些工程领域中(例如动力、化工和冶金等)由于所研究的对象常处于高温、高压、甚至有害的工作情况下,加上设备十分庞大,这就给仪器试验带来了较大的困难。另一方面,随着技术的发展,要求工程技术人员设计出新型的设备和装置。为了能使设计出的设备和装置具有最佳的工作效率和良好的性能,这仅仅依靠仪器试验研究不但需要花费很多的时间和大量的人力物力,而且也很难达到所要求的目的。近十年来,随着电子计算机容量和计算速度的提高,以及计算方法的日益完善,对各种复杂问题的计算异军突起,成为一种很有利的方法。运用数值模拟方法,对各种具体的实际问题,在物理模型与数理统计理论和计算设计与分析方法的指导下,进行大量的优化计算,从而得出对工程应用有用的数据,这就是本文所提出的计算机辅助数值试验(简称 CAT 方法)的实质。本文的目的是探讨数值试验方法如何在能源动力工程及锅炉设计和调试中的应用问题。

1　数值试验方法在大型电站锅炉中应用的可能性和必然性

　　长期以来,锅炉的设计计算大都依据实验所得的经验公式,或直接对模型进行仪器试验。

---

* 本项目得到国家教委博士点基金资助
  本文于 1991 年 3 月收到

但由于锅炉运行在高温和高压工况下,故仪器试验困难很大,特别是大型的电站锅炉更是如此。为了更快和更经济地设计大型电站锅炉,以及各种新型的燃烧器,并对所设计的新型锅炉和燃烧器进行特性分析和变工况分析,以改善燃料燃烧效率等,数值试验将成为一种很有效的方法。

数值试验在锅炉中应用的特点和作用可以简单的归纳为以下几点:

1. 在新型锅炉设计时可对各种结构方案进行数值分析和仿真,以求得较为合理的设计方案。例如,对大型电站锅炉的炉膛设计等问题,运用数值试验方法,可以确定较佳的炉膛结构和尺寸参数。

2. 可预测变煤种、变负荷、变空气动力工况时可能出现的问题及解决的方法。例如,当四角切圆燃烧器在低负荷运行时,切圆直径的变化以及四角风速、煤粉浓度不等时可能出现的情况可通过数值试验方法予以预测。

3. 可对现在工业或电站锅炉出现的问题用数值试验的办法找出问题的关键因素和提出解决方法。例如,确定炉内结渣是否是切圆过大,或是煤粉离心力过大所致,以及对流换热面的局部磨损最大处位置等。

4. 可利用数值试验方法编制锅炉的合理运行工况。例如,制定合理的变负荷运行方案和调节过热汽温等,以利于高效率和安全运行。

5. 可指导新型燃烧器、炉膛结构、煤粉空气分配系统的设计。例如,对大速差、反吹风、对冲射流等新型燃烧器的研制。

6. 还可对新型对流换热面的强化传热及防磨的措施(如加翅防磨方法)进行数值试验研究等等。

最后也应该指出,虽然数值试验方法有上述很多优点,但还存在着一定的局限性。首先,数值试验必须和仪器试验作适当的配合,这两种试验方法是相辅相成缺一不可的。没有仪器试验结果所证实的数值试验理论计算是不可靠的;相反,没有理论指导的仪器实验也是盲目的。其次,为了能使数值试验的结果更加接近实际情况,更好地反映问题的本质,要提出尽可能符合工程实际的物理模型和数学模型,正确的和合理的物理模型和数学模型是数值试验方法成功的先决条件。再者,参加数值试验的人员不单要求有丰富的锅炉设计、运行等方面的知识,而且还须在计算方法、多相流体力学、传热传质学,以及燃烧学等方面有较扎实的理论基础,同时能熟练的编制计算机程序和上机操作。另外,为能更好地进行数值试验,有一定容量和一定快的计算速度的计算机是必不可少的。

2 数值试验的基本原理和方法

为了使数值试验更接近工程实际情况,并使其结果能成为工程设计的可靠依据,在进行数值试验时,必须考虑到下述五种可能的数值试验原则:

1. 在实物原型进行部分仪器试验和所拟定的数学模型计算结果对比基本符合后,再利用数理统计的试验设计方法对优化后的工况进行数值试验。例如,对于变工况的运行和变结构尺寸等问题。

2. 当实物原型进行仪器试验有困难时(如高温、高压工况),可在冷态或热态的模型上进行局部仪器试验和数学模型计算结果对比,基本符合后再推广到实物原型上。例如,对于60

万千瓦电站锅炉来说,可先进行冷模数值试验和仪器试验;然后再用数值试验方法推广至60万千瓦原型锅炉工况。

3. 当实际情况无法进行仪器试验时,可将所拟定的数学模型把边界条件简化至和能精确求解的计算结果相对比,证明本模型正确后,才进行实物的数值试验。

4. 在上述情况下,也可把所拟定的结果模型经过边界和初值条件的变换,和其它简单的和已知的局部模型仪器试验数学相对比,这样的作法是纯碎为了验证所拟定的数学模型的正确性,肯定后才进行实物数值试验。

5. 对某些新型的结构或新的模型,由于无任何仪器试验可供借鉴,故在进行数值试验时应力求所拟定的数学模型更严格和更接近工程实际。这种情况在新型、结构的设计方案选择中很有意义。

数值试验方法的框图如图1所示。首先,对具体的工程实际问题应该根据其本质和影响因素提出合理的物理模型,然后建立相应的正确的数学模型。对锅炉炉内空气动力学问题而言,描述其运动规律的数学模型可分为单相气体流动模型、多相流动模型、燃烧模型、传热模型以及一些特殊功能的子模型(如碰撞和磨损模型等)。在进行必要的仪器模型或实物试验以证明上述各子模型的正确性和获得一些必要的数据(如初始条件和边界条件)后,就可利用数理统计理论(如正交实验法和最佳逼近法等)来设计数值试验的工况,然后即可进行大量的计算机计算和数据分析(如

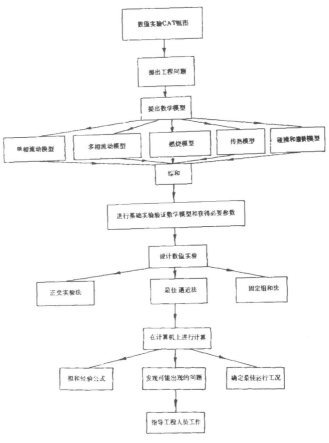

图1 数值试验(CAT)方法框图

确定经验公式,选择最佳运行工况和发现存在的问题等),将上述数值试验编制成专家咨询系统,可为工程技术人员提供设计和运行所需的必要数据。

从数值试验的方框图中可以明显的看出,数值试验方法的核心应该是正确合理的各数学子模型和数值试验工况的设计和组织方法这两大部分,下面我们来分别讨论上述两个关键问题。

3 描述锅炉炉内过程的各子模型

锅炉炉内的实际过程是非常复杂的,它不但包括了气体的流动和煤粉颗粒的运动(多相流动系统),而且还存在煤粉的燃烧和气体组份的燃烧(有化学反应的系统),另外还涉及到热量、质量的传递以及受热面的磨损结渣等方面的问题。因此为了能提出一个较完整的模拟锅炉各部分流动过程的数值试验总体模式,必须对具体的局部的问题进行逐个分解,并提出相应的数值模拟试验子模型。

3.1 单相流动的数值试验子模型

对于不可压缩流体来说,描述其运动的方程可从质量守恒定律、动量守恒定律和能量守恒定律推出,即可用下列张量形式表示:

连续性方程
$$\frac{\partial u_i}{\partial x_i} = 0 \tag{1}$$

动量方程
$$\frac{\partial u_i}{\partial t} + u_j \frac{\partial u_i}{\partial x_j} = -\frac{1}{\rho}\frac{\partial p}{\partial x_i} + \nu \frac{\partial^2 u_i}{\partial x_j \partial x_j} \tag{2}$$

能量和组分方程
$$\frac{\partial \varphi}{\partial t} + u_i \frac{\partial \varphi}{\partial x_i} = \lambda \frac{\partial^2 \varphi}{\partial x_i \partial x_i} + S_\varphi \tag{3}$$

上式中所有参量 u、p、φ 都是瞬时量。为了能进一步的描述单相流体的湍流运动,通常将各瞬时量分解成时均值和脉动值两部分,经这样的变换后,方程(1)~(3) 就可写成下列形式:

$$\frac{\partial \bar{u}_i}{\partial x_i} = 0 \tag{4}$$

$$\frac{\partial \bar{u}_i}{\partial t} + \bar{u}_j \frac{\partial \bar{u}_i}{\partial x_j} = -\frac{1}{\rho}\frac{\partial \bar{p}}{\partial x_i} + \frac{\partial}{\partial x_j}(\nu \frac{\partial \bar{u}_i}{\partial x_j} - \overline{u'_i u'_j}) \tag{5}$$

$$\frac{\partial \bar{\varphi}}{\partial t} + \bar{u}_i \frac{\partial \bar{\varphi}}{\partial x_i} = \frac{\partial}{\partial x_i}(\lambda \frac{\partial \bar{\varphi}}{\partial x_i} - \overline{u'_i \varphi'}) + \bar{S}_\varphi \tag{6}$$

方程(4)~(6)是不封闭的,多了一些脉动相关量(如 $\overline{u'_i u'_j}$ 和 $\overline{u'_i \varphi'}$ 等),为了封闭这组方程,必须有这些相关量的模拟方法,即所谓的湍流模型。目前已有的一些湍流模型有零方程模型、单方程模型、双方程模型以及应力模型等等。工程实际计算中,常采用 $k-\varepsilon$ 双方程湍流模型,它能较简便和较好的反映湍流的特性,其计算结果在大部分流动问题(例如自由射流、有回流的流动以及一些较弱的旋转流动)中都和实验结果吻合较好。$k-\varepsilon$ 双方程模型的形式为:

$$\rho \frac{Dk}{Dt} = \frac{\partial}{\partial x_j}(\frac{\mu_e}{\sigma_k}\frac{\partial k}{\partial x_j}) + k[c_1 \frac{\mu_T}{k}(\frac{\partial \bar{u}_i}{\partial x_j} + \frac{\partial \bar{u}_j}{\partial x_i})\frac{\partial \bar{u}_i}{\partial x_j} - c_2 \rho^2 \frac{k}{\mu_T}] \tag{7}$$

$$\rho \frac{D\varepsilon}{Dt} = \frac{\partial}{\partial x_j}(\frac{\mu_e}{\sigma_\varepsilon}\frac{\partial \varepsilon}{\partial x_j}) + \frac{\varepsilon}{k}[c_1 \mu_T(\frac{\partial \bar{u}_i}{\partial x_j} + \frac{\partial \bar{u}_j}{\partial x_i})\frac{\partial \bar{u}_i}{\partial x_j} - c_2 \rho \varepsilon] \tag{8}$$

其中各常数的取值为 $c_1=1.44$、$c_2=1.94$、$\sigma_k=1.0$、$\sigma_\varepsilon=1.3$。

方程(4)~(8)可写成一个通用的形式,即所谓的通用方程:

$$\frac{\partial (\rho \varphi)}{\partial t} + \text{div}(\rho u \varphi) = \text{div}(\Gamma_\varphi \text{grad} \varphi) + S_\varphi \tag{9}$$

其中 φ 为任一参数量(u、k、ε 等),Γ_φ 为相应的湍流扩散系数,S_φ 为源项。

对于具体的流动来说,在求解方程(9)之前,还必须给定求解的边界条件和初始条件。边界条件和初始条件的确定视流动问题而定,这里值得一提的是固壁附近流场的确定方法。在固壁附近第一层计算网格点 P 上一般采用所谓的壁面函数处理方法,即壁面切应力为:

$$\tau_w = \frac{\bar{u}_p}{y_p} \frac{y_+^+}{u_p^+} \tag{10}$$

其中
$$u_p^+ = \frac{1}{1.4}\ln(Ey_+), \quad y_+ = \rho(k_p C_\mu^{1/2})^{1/2} y_p/\mu$$

而
$$e_p = (C_\mu^{1/2} k_p)^{3/2}/(1.4 y_p) \tag{11}$$

3.2 多相流动的数值试验子模型

到目前为止,描述多相流动的数学模型大致有下列两大类:一类是连续介质模型,另一类是颗粒群轨道模型。属前一类的有:无滑移模型、小滑移模型和滑移扩散模型,而属后一类的有单元内颗粒源模型。这两大类模型各有其优点和不足之处。一般来说,连续介质类模型的气相方程和颗粒相方程相似,程序可通用,但计算需要的计算机内存较大,同时在一些具体的处理上,例如颗粒相的湍流扩散系数和颗粒相的壁面条件等方面仍还存在一些处理上的困难。颗粒群轨道模型与连续介质类模型相比,计算较为简单,在处理颗粒和固壁的碰撞等问题中较为方便,但这种方法对颗粒的湍流扩散运动较难模拟,这方面国内外都有研究者作了不少的努力,提出了多种修正方法。本文作者在这方面也进行了一些探索性的工作,并提出了脉动频谱随机轨道模型(简称 FSRT 模型)。其基本的思想为:在拉格朗日坐标中考察颗粒的运动,其运动方程可写成:

$$m_p \frac{du_p}{dt} = C_D \rho_g A_p \frac{|u_g - u_p|}{2}(u_g - u_p) + F \tag{12}$$

其中 u_p 为颗粒相瞬时速度,u_g 为气相瞬时速度,F 为除阻力外颗粒所受的其它各力之和(例如重力、压差力、Magnus 力、Saffman 力等等)。若我们把气相的瞬时速度 u_g 分解成时均值 $\bar{u}_g$ 和脉动值 u'_g,即

$$u_g = \bar{u}_g + u'_g \tag{13}$$

并用随机的富里叶级数来模拟气相的脉动速度,即

$$u'_{gi} = \sum R_1 u_{gim}\cos(i\omega_t t - R_2 \alpha_i^*) \tag{14}$$

上式中 R_1 和 R_2 正态分布的随机数,e_t 是湍流脉动频率,α_i^* 是脉动初相位,而 u_{gim} 是根据流脉动频谱所确定的脉动幅值。

另外,颗粒的运动轨迹可由颗粒轨迹方程求得,即

$$\frac{dx_p}{dt} = u_p \tag{15}$$

为了得到工程实际所需要的颗粒速度场和浓度场的分布规律,我们采用了体积平均方法来确定它们。具体的方法为:在每个计算单元 ΔV 内,颗粒的数密度 n 为:

$$n = \sum_k \sum_j (\eta_{k,j}\Delta\tau)/\Delta V \tag{16}$$

其中 $\eta_{k,j}$ 为每条轨迹上颗粒的数目流量,$\Delta\tau$ 是颗粒在单元内的停留时间,ΔV 为计算单元体积。而该单元内颗粒体积浓度为

$$C_v = \sum_k \sum_j (\frac{1}{6}\pi d_{pk}^3 \eta_{k,j}\Delta\tau)/\Delta V \tag{17}$$

颗粒相在该单元内的速度则为:

$$U_p = \sum_k \sum_j (\eta_{k,j}\Delta\tau u_{pk,j})/\sum_k \sum_j (\eta_{k,j}\Delta\tau) \tag{18}$$

我们利用脉动频谱随机轨道模型已成功地对多相自由射流、多相同轴射流、循环流化床和鼓泡床内的颗粒相运动，以及颗粒绕管束时对管束的碰撞和磨损问题进行了数值试验，取得了较为满意的结果。

3.3 燃烧过程的数值试验子模型

燃烧是一种伴随着剧烈放热化学反应的流动过程，它包括流体流动、传热、传质和化学反应等分过程，其情况较为复杂。对于单相流体的湍流燃烧过程，国内外有许多研究者进行过这方面的工作，提出了许多数学模型。目前常用的有湍流扩散燃烧和预混燃烧这两种较简单的模型。另外较复杂的模型有几率分布函数的输运方程模型、平均反应率输运方程模型和 ESCIMO 湍流燃烧理论等。

对于煤粉颗粒的燃烧过程国内外也进行了大量的研究。这里我们以单颗粒水煤浆滴的燃烧过程为例，其颗粒的质量守恒方程为：

$$\frac{dm_p}{dt} = \dot{r}_p \quad (\dot{r}_p \text{为质量变化率}) \tag{19}$$

其中 $m_p = m_c + m_h + m_w + m_a$。焦团燃烧反应速度为：

$$\dot{m}_c = \psi \rho_{O_2} C_{O_2,s} K_C \tag{20}$$

挥发分析出速度为：

$$\dot{m}_h = K_{V_e} \exp\left[-\frac{E_V}{RT}\right](V^f - V) \tag{21}$$

水分蒸发速度为：

$$\dot{m}_w = \frac{\lambda_m \text{Nu}}{2 c_{pm} r_p} \ln\left[1 + \frac{c_{pm}(T_g - T_p)}{L - Q_r/(4\pi r_p^2 m_B)}\right] \tag{22}$$

若假定煤浆滴中灰分的含量不变，则

$$\dot{m}_a = 0 \tag{23}$$

式(22)中 Nu 和 Sh 数为：

$$\text{Nu} = \begin{cases} 2 + 0.6/\text{Re}^{0.5} \cdot \text{Pr}^{0.333} \\ 2 + 0.064 [d_p/2(/R_p)]^{-0.16} \cdot \text{Pr}^{0.333} \cdot (\text{Re}/\varepsilon_b)^{1.17} \end{cases} \tag{24}$$

$$\text{Sh} = 18 + 126.3 \text{Re}^{0.98} \text{Sc}^{0.33} \left(\frac{T_g}{T_m}\right)^{-1.47} \tag{25}$$

3.4 传热过程的数值试验子模型

火焰传热数值模拟试验所用的基本方程是能量方程，即

$$(\rho u c_p T) = \nabla(\Gamma_T \nabla T) + S_Q \tag{26}$$

源项 S_Q 中含化学反应释热率 Q_C 和辐射换热率 Q_R。而化学反应释热率 Q_C 可以从燃烧过程的计算中求得，辐射换热率 Q_R 的计算公式为：

$$Q_R = \frac{1}{dx}\left[(q_x^+ - q_x^-) - (q_x^+ + dq_x^+ - q_x^- - dq_x^-)\right]$$

$$= -\left[-(K_a + K_s)q_x^+ + K_a E_b + \frac{K_s}{2}(|q_x^+| + |q_x^-|) \right. \tag{27}$$

$$\left. -(K_a + K_s)q_x^- + K_a E_b + \frac{K_s}{2}(|q_x^+| + |q_x^-|)\right]$$

式中 q_x^+ 和 q_x^- 为 x 方向的半球辐射力，或称辐射热流。

火焰辐射换热数值计算的方法很多，其中主要的有：(1)热流法；(2)区域法；(3)蒙特卡洛法三种。

3.5 含灰气流对受热面磨损过程的数值试验子模型

在锅炉的换热器中，含灰气流对换热管束产生碰撞和冲击磨损。通常锅炉内的换热管束大都是由于灰粒冲击磨损引起破裂，从而导致严重后果的，因此含灰气流对管束磨损的研究具有重要的工程意义。

由于颗粒本身具有惯性，当含灰气流绕固体流动时，颗粒一般无法随气体流线运动，而往往要与壁面进行碰撞和反弹，Tabakoff 等人由实验得到了碰撞-反弹恢复比经验公式如下：

$$\beta_2 = \text{ctg}^{-1}\left[\left(\frac{0.95 + 0.00055\beta_1}{1.0 - 0.02108\beta_1 + 0.0001417\beta_1^2}\right)\text{ctg}\beta_1\right] \tag{28}$$

$$u_{p_2}/u_{p_1} = (1 - 0.02108\beta_1 + 0.0001417\beta_1^2)\sqrt{\frac{1 + \text{ctg}^2\beta_1}{1 + \text{ctg}^2\beta_2}} \tag{29}$$

式中 u_{p_1} 和 u_{p_2} 为颗粒碰撞前后的速度，β_1 和 β_2 分别为碰撞角和反弹角。

Tabakoff 等人还对固粒磨损金属材料（例如锅炉管束所用的钢材）的磨损特性进行了研究，回归出金属磨损量 ε 应满足的关系式：

$$\varepsilon = K_1 f(\beta_1) u_{p_1}^2 \cos^2\beta_1 (1 - R_T^2) + f(u_{in}) \tag{30}$$

其中

$$R_T = 1 - 0.0016 u_{p_1} \sin\beta_1$$
$$f(\beta_1) = [1 + C_b\{K_{12}\sin(90/\beta_0)\beta_1\}]^2$$
$$f(u_{in}) = K_3(u_{p_1}\sin\beta_1)^4$$

这里 β^0 为最大磨损角，u_{p_1} 是碰撞前速度，R_T 是切向恢复系数，K_1、K_{12} 和 K_3 是与金属有关的经验常数。

4 数值试验工况的设计和组织

和一般的仪器实验类似，数值试验工况组织中最常见的是顺序改变某一参量，同时保持其它参量不变。从试验工作量的角度来看，这样的组织方式并不可取，因为尽管数值试验的实施较仪器试验简便，但大规模进行仍很困难，这一点可借助于以数理统计为基础的正交方设计来解决。

在理论上要比较试验取得最高的精度，应在与比较的条件无关的相同条件下进行试验。但在实践中，尤其当试验所需次数较多时，往往有可能将一组实验划分为几个小组，在每个小组内部的变异性很可能小于整个大组里的变异性。只要具备这样的条件，那么把实验划分为若干区组就能提高试验精度。这里应该提出的是虽然任何区组内部的变异性都可能小于整个系统的变异性，但在个别区组内可能还有系统变差。因此，如果对一连串区组的相应相同部位做某种处理，各个区组在部位上的系统变差导致的虚假效应很可能影响到结果。为克服这项弊端，必须在每个区组中将各处理作不同的排列，各处理在任一区组的实际部位以适当的随机方式决定。如果区组数目恰好是处理数目的整数倍，还可以进一步改进这个随机方案，即设计得使每种处理配置于区组的每一部位的次数相等，这样一来，各处理就平衡地分布于各组部位之间了。这种设计方案就是拉丁方设计,正交方是其中的一种。这种设计可以对两三组因素的效应研究提供经济的设计试验方案,只要这因素能假定具有可加性,和假定不受其它因素变化的影响即可。

另外,运用数值试验方法可以来寻找一个最佳运行工况,以使预先给定的某个物理量达到最大或最小化。如果起始的条件与获得最佳结果的条件相距甚远,则可采用最速上升路线的方法,找出比较趋近于最适条件的新的运行工况,当接近最适化时,采用局部探测技术,以更为精确地查明最佳工况。

例如,在构思一项数值试验时,经常规定出(至少近似规定出)那些可能有意义的变量组合相应的变量空间的区域,这种区域叫实验区域。假定只变动一个变量(变动进口风速 x_1 对于结果 y 的效应),如果结果函数 y 和进口风速 x_1 之间的函数关系 $y = f(x_1)$ 是光滑的,则只要适当选择常数 b_0、b_1、b_{11}、b_{111} 等,并在下述级数中取足够多的项,就可以所求的任意近似程度表示这个函数关系

$$y = b_0 + b_1 x_1 + b_{11} x_1^2 + b_{111} x_1^3 + \cdots \tag{31}$$

在上述关系式中是只含一个因素 x_1 的多项式,对于两个变量或两个以上变量的情况,也可得出相应的多项式方程。如果该多项式能精确地表达出域中的函数依赖情况,则该多项式称为回归多项式,系数 b_0、b_1 …… 等称为回归系数。求出回归多项式是数值试验的一个重要目的之一,回归多项式是用于指导实际问题运行和设计等的理论基础,同时它还具有便于工程应用等优点。

用最速上升寻找一个最佳工况的方法如图 2 所示。图中绘出的曲线是等值线,其中 P 表记的初始条件远离最大值。在包含 P 的一个小区域里,实行一组适当的试验组合,估计出局部响应曲面的斜率 b_1 和 b_2,并从这些斜率的相对大小和符号,计算出该平面的最速上升方向或最大斜率。沿此方向前进到一个点 Q,在点 Q 处重新计算斜率,并且重复上述过程,这样就可达到响应值越来越高的点。在上述过程中,回归多项式中的系数 b_0、b_1、…… 等是用它们的最小二乘法估计量来代替的。因为常数的最小二乘法估计量是观测值的正交函数,可以单独计算,即

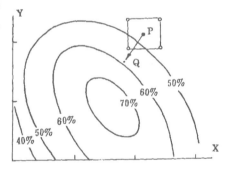

图 2 逼近最佳工况的示意图

$$b = \Sigma yx / \Sigma x^2 \tag{32}$$

在选择数值试验组合时,采用这种统计理论的试验构成方案,使得数值试验工况不用很多但又可以保持一定精度,满足工程需要的目的。

5 结 论

本文所提出的计算机辅助数值试验的新设想如何更好地应用于锅炉设计和调试,目前尚还处于刚刚开始的阶段、我们应用这种方法对锅炉炉内过程的一些问题进行了初步的尝试,结果表明数值试验的结果和仪器试验结果吻合程度还是令人满意的,特别是一些无法测量或很

难测量的情况,数值试验能给我们一个大概的定量和定性数据及趋势,这对大型电站锅炉的设计和调试是很有用的。由此可见,数值试验虽要有待于进一步深入和改进,但它在锅炉设计和调试中无疑将有着很大的发展前景。

The application prospects of computer aid testing (CAT) in furnace design and operation ——(I) basic theory

Cen Kefa　Fan Jianren

(Dept. of Energy Engineering)

Abstract

This paper reports the application posibility and inevitability of the method of Computer Aid Testing (CAT) in furnace of large power stations. The basic theory and method of CAT are discussed, and the mathematical elemental models of fluid flow、combustion、heat transfer and erosion in furnace processes are presented. These work are of great significance for furnace design and operation.

Key words:computer aid testing (CAT), large tangentially — fired furnace, numerical modelling

原文刊于浙江大学学报(自然科学版),1992,1(26):115-123

煤炭洁净综合利用技术的研究与前景

岑可法* 池 涌** 倪明江* 骆仲泱*** 方梦祥***

摘 要 本文主要介绍了浙江大学所提出的煤炭洁净综合利用技术以及在我国的应用前景。浙江大学围绕煤炭洁净综合利用技术所开展的研究工作主要有：(1) 煤泥、煤矸石燃烧及混烧技术；(2) 水煤浆燃烧技术；(3) 循环流化床热电联产技术；(4) 燃煤三联产技术及联合循环发电技术；(5) 溴化锂制冷技术；(6) 发电与水泥联产技术；(7) 流化床燃烧提钒技术；(8) 燃烧脱硫脱硝技术。

关键词 煤，燃烧，气化，综合利用洁净煤
中图分类号 TQ543，TQ517.4，TQ546，TM611.3，TB66

0 引 言

在我国可燃矿产资源中，煤炭占 96%，我国的能源消费结构中也一直以煤炭为主，2000 年全国一次能源消费中煤炭将达到 79%。我国的能源生产和消费以煤为主，这种状况将持续到二十一世纪中叶。目前我国消费的煤炭中有 84% 被直接燃用，我国燃煤技术有两个特点，一是燃烧效率低；二是燃烧产生的大气污染没有得到控制。统计表明，我国每年排入大气的污染物中有 87% 的 SO_2，67% 的 NO_x 来源于燃煤。

以电力工业为例，我国用于发电的煤炭约占全部煤炭的消费量的四分之一，与国外火电技术相比，我国也存在效率低、污染排放严重两大问题。据统计，1980 年美国火电的供电煤耗为 348 克标煤/度电，我国为 448 克标煤/度电；1992 年虽有所降低，但仍高达 419 克标煤/度电，折合供电效率为 29.3%，同时单位发电量的二氧化硫、氮氧化物排放也比发达国家高出数倍。因此，发展高效洁净的发电技术是我国目前急待解决的问题。此外，随着工业和国民经济的发展，对煤化工产品、建材的需求也日益扩大，同时生活水平的提高也要求供热、供冷，因而在高效洁净发电的同时，实现煤的综合利用将为我国的煤炭资源化开辟一种全新的途径。

---

* 教授，博士生导师；** 讲师、博士；*** 副教授，浙江大学能源系，310027，杭州。参加本文工作的还有曹欣玉教授和严建华、姚强副教授。
收稿日期：1994-04-27

浙江大学热能工程研究所自50年代末开始从事煤的燃烧与发电技术以及综合利用方面的研究开发,取得了一系列研究成果。本文介绍了近年来在煤的洁净发电技术与综合利用研究开发方面的成果,以及正在开展的一些工作。

1 煤炭洁净综合利用技术的设想

煤炭洁净综合利用技术路线原理如图1所示。

煤经洗选后分别产出精煤、矸石和煤泥,也可结合洗选过程制备水煤浆,煤炭或水煤浆裂解和气化,气化炉产生的煤气作为工业或民用以及送至蒸汽燃气联合循环装置中的燃气轮机燃烧发电,气化炉产生的残焦和洗选过程产生的煤矸石、煤泥用流化床燃烧产生蒸汽,蒸汽可供汽轮机发电、工业及民用或作为溴化锂制冷和空调。煤的气化和燃烧采用流化床,如此产生的煤灰具有良好的活性,可以生产优质建材,包括水泥、砖瓦等,含钒、铀等品位高的煤还可提取金属钒、铀。

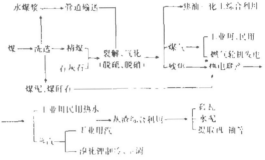

图1 煤炭洁净综合利用技术原理图

从长远考虑,煤不仅是主要能源,而且将取代石油作为原料。从发展战略看,应尽快发展煤的洁净综合利用技术。基于这样的想法,浙江大学提出并一直致力于这方面的研究开发。以大同煤为例的洁净综合利用分析预测如图2所示。

煤炭采用洁净综合利用技术后,脱硫率可达90%,氮氧化物排放低于200ppm.

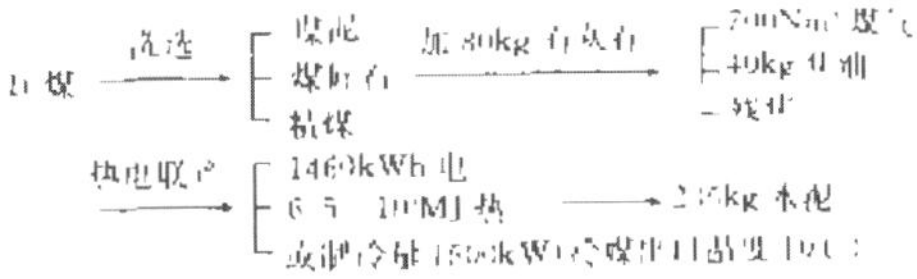

2 浙江大学在煤炭洁净综合利用已开发和正在开发的技术

2.1 煤泥、煤矸石燃烧及混烧技术

图2 大同煤(1吨)的洁净综合利用预测

浙江大学热能工程研究所从1982年开始从事洗煤泥的处理和利用研究。在大量基础研究和半工业性研究的基础上,提出和发展了有效处理和利用洗煤泥的新技术——洗煤泥异重流化床燃烧技术。

浙江大学发展的洗煤泥流化床燃烧新技术的主要内容有:

(1)采用大粒度高位给料,利用洗煤泥的凝聚结团特性使其在流化床内形成粒度较大的凝聚团,以减少燃料的扬析损失,提高燃烧效率。

(2)采用异重流化床技术,以防止大粒度凝聚团在流化床内的沉积,保证稳定的运行。

(3)采用不排渣运行方式,以在料层稳定的前提下减少大重度床料的消耗,并避免燃料的排渣损失,进一步提高燃烧效率。

(4)采用新型高效脱硫技术,将破碎的脱硫剂直接混入洗煤泥,利用洗煤泥的凝聚结团特性使脱硫剂和洗煤泥在床内形成粒度较大的凝聚团,延长脱硫剂在炉内的停留时间,提高脱硫剂的转化率,从而提高脱硫效率。

(5) 采用分段燃烧来控制氮氧化物的排放。

十多年来,我们先后承担了国家"六五"重点攻关项目"洗煤泥流化床燃烧技术",国家"七五"重点攻关项目"35吨/时洗煤泥流化床锅炉"。10吨/时洗煤泥流化床锅炉于1985年在四川永荣矿务局电厂投入运行,燃用干基灰分为52%,干基发热量为14.5MJ/kg,应用基水分为25%的永荣煤泥时,平均燃烧效率超过95%,并于1986年初通过了国家鉴定。35吨/时洗煤泥流化床锅炉于1990年在山东兖州矿务局兴隆庄热电厂投入运行,并于1991年底通过了国家科委组织的鉴定。测试表明该锅炉在额定工况下燃烧效率达到96%,锅炉热效率达到83.5%,居世界领先水平。该洗煤泥流化床锅炉配一台6MW汽轮发电机组,年供电 31.68×10^6 kWh,供热47400MJ,并网发电以来每年可减少矿井电费支出、排污费及烟尘超标费400多万元。

在开发洗煤泥流化床技术的同时,我们还进行了洗煤泥流变特性的研究,洗煤泥管道输送和泵送给料技术的开发,使之成为配套技术。目前正开发75吨/时洗煤泥流化床锅炉。

早在70年代,浙江大学就开始煤矸石燃烧技术的开发,先后研制成功10t/h石煤双床飞灰循环流化床锅炉,35t/h煤矸石循环流化床锅炉等,并已进行75吨/时、130吨/时、220吨/时劣质煤循环流化床锅炉的设计。

在开发单一燃料的流化床锅炉的同时,考虑到国情和用户的需求,浙江大学还十分注重多种燃料的流化床混烧技术的研究和开发,先后进行了洗煤泥和煤矸石、洗煤泥和中煤等的混烧研究。四川永荣矿务局永川电厂的20吨/时煤矸石循环流化床锅炉改造成为煤泥、煤矸石混烧锅炉将于1994年内投入运行,新设计的35吨/时煤泥煤矸石混烧循环流化床发电锅炉计划于1994年底投运。75吨/时混烧锅炉也正在开发之中。

2.2 水煤浆燃烧技术

水煤浆是作为代油燃料从70年代开始得到广泛发展的,随着高浓度水煤浆管道输送技术的发展,水煤浆燃料及燃烧技术将有更强的生命力。

浙江大学从1979年开始从事水煤浆低污染燃烧技术的研究,先后承担了国家重点项目、国家"六五"、"七五"和"八五"的攻关项目。"六五"国家攻关项目"20t/h燃油工业锅炉燃用水煤浆的工业性试验"和国家"七五"攻关项目"60t/h燃油工业锅炉应用水煤浆代油燃烧技术",均通过了国家鉴定,达到了国内首创、国际水平。"六五"和"七五"均获国家攻关重大成果奖。国家"八五"攻关项目"电站锅炉燃用水煤浆技术研究——230t/h燃油锅炉改烧水煤浆"的工程将于明年投产。设计完成了我国第一台35t/h水煤浆专用锅炉,北京造纸一厂两台水煤浆专用锅炉将在年底全部投产使用,白杨河电厂两台35t/h水煤浆锅炉也将于明年投产。北京第三热电厂75t/h燃煤锅炉进行了改烧水煤浆的设计,已经过多次试烧。

目前浙江大学建有1.25MW筒形水煤浆燃烧试验台架,0.3MW立式前置燃烧室试验台架,0.3MW水煤浆喷嘴雾化试验台架,3~8t/h水煤浆喷嘴雾化试验台架,大型四角燃烧锅炉冷态模化试验台架,W型和前墙燃烧锅炉冷态模化试验台架,各种旋流燃烧器,直流燃烧器冷态模化试验台架。并正在建设大型5MW四角燃烧、W型和旋流型综合热态燃烧试验台架。历年来先后进行了水煤浆旋流燃烧器、水煤浆角置式燃烧器的研究,开展了水煤浆液态排渣燃烧技术、可调非对称性引射烟气式水煤浆稳燃技术、水煤浆低污染燃烧技术、水煤浆预燃室燃烧技术的研究,同时进行了燃油炉改燃水煤浆的改炉技术的研究。形成了 4t/h—230t/h 水煤浆锅炉的开发能力。

2.3 循环流化床热电联产技术

循环流化床燃烧技术具有清洁、高效的特点,近年来在国内外得到了迅速发展。但目前的循环流化床锅炉一般燃用优质煤。循环流化床还具有燃料适应性广和能燃用劣质燃料的优点。利用劣质燃料,如煤矸石、石煤等,并实现热电联产是我国能源利用的一个方向,因此循环流化床在煤矸石、石煤方面将是大有可为的。

浙江大学自 70 年代开始进行双床飞灰循环 10t/h 流化床锅炉的研制,成功地在湖州化肥厂连续运行近 10 年,燃用热值为 7.5~8.4MJ/kg 的石煤以及造气炉渣。在此基础上,80 年代又开始了循环流化床锅炉的研究工作,建立了热态和冷态试验台。在这些试验台上研究了循环流化床的流体动力特性,包括鼓泡床过渡时的特性,以及压力、颗粒浓度等沿床层高度的变化,并提出了计算公式;同时还对分离装置进行了研究(包括旋风分离器和惯性分离器),并对两种形式的返料机构进行了详细研究,提出了设计方法。进行了脱硫剂反应特性的研究和高效燃烧脱硫的研究。在试验的基础上,提出了循环流化床内颗粒运动模型,主要由脉动频谱随机轨道模型和颗粒随机脉动模型组成。初步形成了循环床燃烧模型和循环流化床锅炉的设计方法,并在冷、热态试验台上进行了传热试验。

在工业应用方面,主要开发了新型的循环流化床锅炉——采用中温分离(600℃左右)下部排气的旋风分离器结构。这种结构具有分离器尺寸小、分离效率高、启停时间短、运行可靠安全、负荷调节性能好、锅炉结构紧凑、流动阻力低等特点,第一台燃用煤矸石的 35 吨/时煤矸石循环流化床热电联产锅炉已安装于平顶山矿务局,并于 1994 年 1 月投运。另外 220 吨/时的燃煤循环流化床锅炉也已完成方案设计。

2.4 燃煤电热气三联产技术

作为煤洁净综合利用技术的先期应用范例之一,浙江大学开发了电、热、气三联供系统。该系统采用循环流化床热载体技术,集燃料燃烧气化热解于一身。能同时供应电力、蒸汽、民用煤气,既满足了工业发达地区的综合需求,又解决了缺能地区的民用燃料,充分利用余热,提高了能源利用率,减少了环境污染,是一种全新的能源利用模式。整个系统本体是一并联运行双流化床,一床为气化室,另一床为燃烧室,高温物料在两床间循环,气化室为鼓泡流化床,燃烧室为快速流化床。气化室运行温度为 800~850℃,用循环热煤气作为流化介质,以提高煤气热值和增加焦油分解率。燃料首先进气化室热解、气化,产生高热值挥发分和产生部分水煤气,煤气经余热回收、除尘、脱硫、CO 转换后供民用,气化半焦经回送机构送燃料室燃烬,产生热量部分加热循环物料供气化室用热,另一部分产生蒸汽供发电、供热。在炉内采用石灰石脱硫、分段送风等技术降低 SO_2、NO_x 形成和排放。1.5MW 循环流化床气化燃烧联合装置已在实验室建成,经大量试验表明,装置能稳定、连续运行,调节方便,装置燃料转化率大于 90%,煤气热值为 2500~3000kcal/Nm^3,粉尘、SO_2、NO_x 排放达到国家标准。

75 吨/时的电、热、煤气循环流化床三联供系统已设计完成,正在加工制造,将安装于江苏省扬中县。

煤气化燃气蒸汽联合循环发电技术的研究开发是浙江大学目前正在积极从事的一项内

容,根据我们发展的循环流化床煤气化燃烧联合装置,提出了如图3所示的联合循环发电系统。分析计算表明采用先进的燃气轮机技术和先进参数蒸汽循环后,该系统的供电效率可达45%以上。这项技术也可用于现有燃煤电厂的改造,例如对于125MW燃煤电厂,保留原电站的燃煤锅炉、汽轮机和发电机,增加一套南京汽轮机厂生产的MS6001燃气轮机组和浙大发展的循环流化床煤气化燃烧联合装置,可使改造后的电厂供电效率提高6%左右,发电量约增加38MW,同时二氧化硫排放约降低30%,氮氧化物排放约降低20%,达到增加发电容量、提高供电效率、降低烟气污染物排放的目的。

2.5 溴化锂制冷技术

溴化锂水溶液简称溴化锂溶液或溶液,是吸收式制冷机的工质。它是由两种不同沸点(水100℃和溴化锂1265℃)的组元所组成,溶液在发生器G中被热源加热浓缩,产生的冷剂蒸汽进入冷凝器C(见图4),并被冷却水冷凝成冷剂水。该冷剂水经节流,在蒸发器E中喷淋蒸发,吸收冷煤水中的热量,使冷煤水温度降低,达到了制冷目的。在喷淋蒸发时产生的低压冷剂蒸汽进入吸收器,被来自发生器的浓溶液喷淋吸收,生成稀溶液

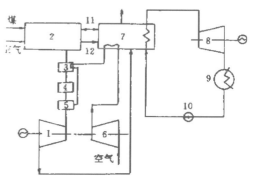

图3 常压循环流化床燃气蒸汽联合循环系统示意图

积聚在吸收器液囊中,再被屏蔽泵P压送到发生器,溶液不断循环,就连续不断地制取冷量。

利用浙江大学所开发的溴化锂制冷技术,与有关厂家合作已开发出系列溴化锂制冷机,规格如下:

(1) 制冷量:230kW(20万千卡/时)至2330kW(200万千卡/时)。

(2) 加热蒸汽:压力为0.25~0.6MPa低压饱和蒸汽。

(3) 冷煤水出口温度:标准机为10℃,同时生产7℃机和13℃机。

2.6 发电与水泥联产技术

煤(尤其是高灰劣质煤)燃烧时会产生大量废渣,若

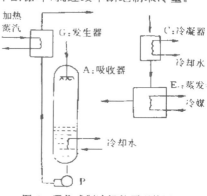

图4 吸收式制冷机的原理简图

不对这些废渣加以处理利用,就会造成环境的二次污染。浙江大学根据多年的研究开发经验,提出了用流化床燃烧来实现发电与水泥联产的技术,在煤的综合利用方面作出了努力。

图5为热电-水泥联产技术的技术工艺图,煤矸石和劣质石灰石按一定比例混合成生燃料投入到床温为950~1000℃的流化床锅炉内进行燃烧,利用煤在燃烧时放出的热量,一部分产汽发电,另一部分使煤的灰分和劣质石灰石经化学反应所形成的"灰渣"变成低标号水泥熟料,再加入适量外加剂,经球磨机粉磨到细度与一般水泥相同,就成为低标号水泥。这样达到一炉两用,既产汽发电又直接生产低标号水泥。

2.7 流化床燃烧提取五氧化二钒技术

一部分煤含有相当数量的稀有金属如钒、铀等，若能结合煤的燃烧转化提取这些稀有金属，将为煤的综合利用开辟另一条途径。

浙江大学从 70 年代开始进行石煤提取五氧化二钒的研究开发工作，先后进行了实验室研究和中间试验，取得了很好的结果。流化床燃烧提取五氧化二钒技术的关键是使煤中的低价钒转化成可溶于水的高价钒，炉内有关的主要反应如下：

$$(V_2O_3)_c + \frac{3}{2}O_2 + 2NaCl \longrightarrow 2NaVO_3 + Cl_2$$

所产生的可溶于水的偏钒酸钠（$NaVO_3$）经水浸和水解沉钒后就可得到五氧化二钒。

年产 10 吨五氧化二钒设备上的中间试验表明能达到以下技术指标：钒的总回收率为 69%～71%，其中焙烧水溶钒转化率 75%～78%；浸出回收率 97%；沉钒回收率 95%。

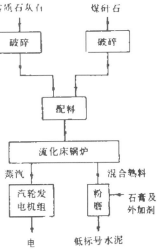

图 5 流化床锅炉热电-水泥联产工艺路线

2.8 燃烧脱硫、脱硝技术

低污染排放技术也一直是浙江大学的研究开发重点，在脱硫、脱硝方面，先后开展了硫析出规律、氮析出规律、脱硫剂的反应特性、高效脱硫技术、注氨降低 NO_x 排放、分段燃烧降低 NO_x 排放等的研究，并在试验研究的基础上，提出了燃烧脱硫以及脱硝的数学模型。

图 6 为流化床燃烧时不同钙/硫比下的脱硫效率，试验结果当钙/硫比为 2 时，脱硫效率可达 80% 以上。图 7 为采用流化床分段燃烧降低氮氧化物排放的结果，试验表明分段燃烧后 NO_x 排放可降低 25%～40%。分段燃烧降低 NO_x 排放的优化研究表明，一次风量空气系数的最佳范围为 0.85～0.95 之间。

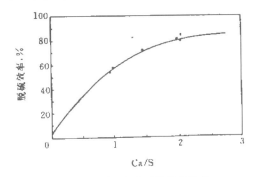

图 6 不同钙/硫比下的脱硫效率

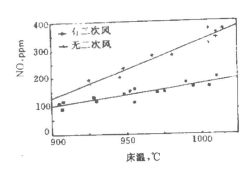

图 7 分段燃烧降低氮氧化物

3 结束语

根据浙江大学多年研究开发经验所提出和发展的煤炭洁净综合利用技术是一种适合国情的高效、洁净的煤炭综合利用和发电新方法,可实现电、热、煤气、制冷、建材、煤化工产品、贵金属等的联合生产,有巨大的经济效益、环保效益和社会效益。

浙江大学在煤炭洁净综合利用技术上已具备了多年的研究开发经验,在煤泥、煤矸石燃烧及混烧技术,水煤浆燃烧技术,循环流化床热电联产技术,燃煤三联产技术,溴化锂制冷技术,发电与水泥联产技术,流化床燃烧提取五氧化二钒技术,燃烧脱硫脱硝技术等方面均已有成熟的工业应用。

(参考文献 18 篇略)

RESEARCHES AND PROSPECTS ON CLEAN UTILIZATION OF COAL

Cen Kefa Chi Yong Ni Mingjiang Luo Zhongyang
and Fang Mengxiang

(*Zhejiang University*, *310027 Hangzhou*)

ABSTRACT In this paper, the clean coal utilization technology which is proposed and developed by Zhejiang University is introduced. The related research and development activities, carried out at Zhejiang University, on clean coal utilization technology are reported. R & D efforts are divided into the following areas: (1) Coal washery sludge and coal washery rejects fluidized bed combustion; (2) Clean combustion of coal water slurry; (3) Coal fired circulating fluidized bed cogeneration; (4) Electricity, heat, and fuel gas tri-generation system and combined cycle power generation; (5) Refrigeration; (6) Production of low grade cement from coal ash; (7) Vanadium extraction from coal; (8) Low pollutants emission combustion of coal.

KEY WORDS coal, combustion, gasification, synthetic utilization

原文刊于煤炭转化, 1994, 17(3): 16-22

新颖的热、电、燃气三联产装置

岑可法　骆仲泱　方梦祥　李绚天　陈　飞　王勤辉　胡国新　倪明江

(浙江大学热能工程研究所　杭州 310027)

摘要：本文介绍一种循环流化床热、电、气三联产新工艺，该工艺是以现有循环流化床技术为基础，集燃烧和气化工艺于一体，能同时产生民用煤气、蒸汽和电力，在小型热态试验台架上进行的一系列试验已成功地证实了该技术的可行性。在此基础上，一台 75t/h 热、电、气三联产装置已完成设计，正在实施之中。

一、前　言

在近几年来，我国能源仍以煤炭为主，而目前煤炭的绝大部分作为燃料直接燃用，现有企业厂矿运行的近 40 万台工业锅炉，其中一半效率低、污染大。另一方面，我国民用煤气供应严重短缺，除少数大城市外，大部分中小城市居民仍用煤做饭取暖，不但生活很不方便，而且环境污染严重，燃料浪费大。解决广大中小城市煤气供应是迫切而又首要的任务。长期以来，煤的气化工艺(包括两段炉、德士古炉等)都追求煤的全气化，为生产中热值煤气，采用细粉给料、高温高压运行、纯氧鼓风等，装置投资成本很高，无法为中小城镇所接受。寻求一种简单、投资省的煤制气工艺，一直是人们追求的目标。

为此，浙江大学热能工程研究所提出了煤干馏和部分气化产生民用煤气，半焦送燃烧炉产汽发电的方案，使煤中成分得到合理利用。该方案中最关键的是气化炉热源和半焦燃烧问题。循环流化床技术发展为解决半焦燃烧和用循环热载体作气化热源提供了可能。该工艺在国家自然科学基金的资助下，经多年的开发，已完成了基础试验和小型装置热态试验研究，证实了方案的可行性，并申请了专利。一台 75t/h 热、电、气三联产装置已完成设计，正进入实施阶段，预计 1995 年投运。本文主要介绍该技术的基本工艺流程、关键技术研究和 75t/h 三联产装置的设计特点和推广应用前景。

二、基本工艺流程及特点

图 1 为本装置的基本工艺流程图，气化室为常压鼓泡床，用水蒸汽和循环煤气作气化剂，运行温度为 750~800℃。燃料经给料机给入气化室，首先受热裂解，析出高热值挥发分，半焦中部分碳和气化剂反应形成水煤气，气化吸热由燃烧室的高温循环物料来提供，气化后半焦随循环物料送入燃烧室燃尽。燃烧室为快速床，空气鼓风，运行温度为 900~950℃。半焦燃烧产生热量，将从气化炉来的低温循环物料加热成高温物料，再送至气化室提供气化吸热和产生水蒸汽。从气化炉出来的高温煤气，经煤气冷却器冷却，净化器净化，除去灰、焦油、水后变成净煤气输出供民用。从燃烧室出来的高温烟气经烟气冷却器冷却，除尘器除尘后，排入大气。由煤气冷却器、烟气冷却器产生的蒸汽，除少量供气化炉用汽外，大部分用于发电、供热，也可制冷。如此实现煤气、热、电三联产。

上述三联产工艺，综合了燃料气化和燃烧过程，工艺独特，其主要特点有：

1. 气化室为一常压鼓泡床，采用自供蒸汽和循环煤气为气化剂，产生的煤气主要由

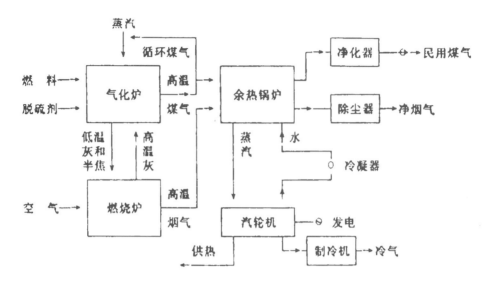

图 1 三联产装置工艺图

干馏气和部分水煤气组成,热值可达 $10.5\sim 12.5 MJ/m^3$ ($2500\sim 3000 kcal/m^3$)甚至更高,CO 含量低,不需特殊的制氧设备和 CO 转换工艺。和同类产生民用煤气的气化炉相比,工艺简单,投资省。

2. 气化后的半焦直接用作锅炉燃料,使燃料中气体和固体成分都得到充分利用。因此,装置的燃料利用率较高。

3. 燃料在热解气化时,燃料中的有机硫和挥发性氮先行脱出,锅炉部分脱硫脱氮比普通 CFB 容易,而且锅炉采用石灰石循环脱硫。因此,装置的 SO_x、NO_x 排放较低。

三、试验设备和试验结果

图 2 为 1MW 热态试验装置的简图,本装置设计投料量为大同煤 120kg/h,实际可达 200kg/h。

我们首先在小型装置上进行了冷态物料循环试验和热态试验,在此基础上,进行了 75t/h 三联产装置设计。以下介绍主要的试验结果和设计思想。

1、物料循环系统

本方案最关键的技术是采用热载体法提供气化热源,因此,物料循环系统是关键,既要保证气化室和燃烧室间有足够的循环物料

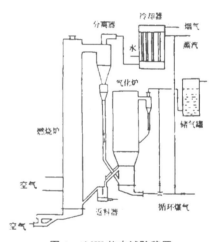

图 2 1MW 热态试验装置

量来提供气化热源,同时要保证各室间不相互串气或串气量很小。为此,我们在小型试验台上进行了大量的物料循环试验,结果表明,循环物料倍率可达 20~30,可满足气化吸热要求。在合理工况下,系统气密性很好,循环密封返料器可良好控制调节循环量。

2、系统运行试验

本系统启动采用热烟气点火方式,试验采用煤种为大同煤,给煤粒度为 0~8mm,平均粒径 2.7mm,其煤质分析示于表 1,试验典型工况为表 2,煤气成分见表 3。

表1 煤质分析(大同煤)

V'	C'_gd	A'	W'	C	H	O	N	S	Q_{DW}(kcal/kg)
25.7	54.49	11.81	8	68.84	11.44	5.6	0.71	0.6	6490

表2 试验典型工况

	床温(℃)	风速(m/s)	给料量(kg/h)	煤气量(标m³/h)	蒸汽量(kg/h)
燃烧室	950	7	—		90
气化室	750	1	150	40	—

表3 煤气成分

H_2%	CO%	CH_4%	C_2H_4%	CO_2%	N_2%	Q(kcal/m³)
45.08	17.41	11.44	2.3	20.1	1.6	3250.6

3、75t/h 三联产装置设计

在小型冷态试验的基础上,我们进行了75t/h热、电、气三联产示范装置的设计。

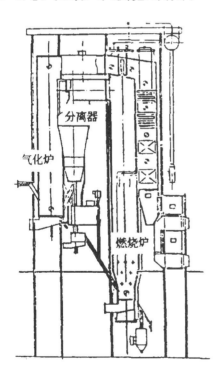

图3 75t/h 循环流化床三联供装置

该设计是由浙江大学、无锡锅炉厂、扬中第二热电厂合作进行的。设计采用了双循环回路结构,既可进行三联产方案运行,也可进行循环流化床锅炉独立运行。这样保证气化炉出故障时,电厂仍能正常发电。整个装置结构如图3,主要设计参数如表4。锅炉炉膛截面为2.45m×5.45m,炉膛高20.30m,采用分段送风,一二次风比为0.6/0.4,二次风口设在密相区上部2m处,在炉膛出口布置两只Φ3m高温旋风分离器,分离高温循环物料,气化炉直径为2.5m,高15m。在旋风分离器下部布置一外置式换热器。在锅炉单独运行时,经分离器分离下的循环物料直接进入外置式换热器,再返回炉膛。在与气化炉联运时,分离器分离下来的循环物料首先进入气化炉,再由气化炉至外置式换热器,返回炉膛。因此,外置式换热器可调节循环物料的温度,从而保证锅炉在单独运行和与气化炉联运时都保持适宜的温度。在锅炉单独运行时,给煤由外置式换热器溢流侧上方给入,与低温灰预混后经返料管进入炉膛密相床层。与气化炉联运时,给煤从气化炉炉前与石灰石一同给入。整个锅炉本体采取了严密的防磨措施,并采用床下热烟气点火。

(下转第30页)

方法清理和冲洗,必须检查清理和冲洗的质量,坚持高标准严要求。

4. 选择微孔粒度均匀、性能稳定、球径较冷却管内径大 1~2mm 的海绵胶球。使用过程中如发现胶球破损或过分发胖,要及时剔除掉,投入新球。也要随时剔除磨损变小的胶球,当球体直径比管内径小 0.1~0.3mm 时,不宜再继续使用。

同批生产的胶球尺寸有大有小,球质也有硬有软,比重也不一样。其密度应与循环水的密度大致相同,这样胶球才能随机地进入冷却管。因此要有专人验收,严把胶球质量关。新胶球投入使用前,应先用循环水浸泡 1~2 小时,并反复用手捏几次,使胶球中的微小孔隙都充满水,避免胶球在出水管中漂浮和滞留。

5. 为保护预防凝汽器铜管腐蚀,延长其使用寿命,要注意铜管保护膜的完整,不要由于胶球清洗而遭损坏,要定期加入硫酸亚铁进行沉膜处理。

6. 在胶球连续清洗过程中,如发现收球率明显降低或运行不正常时,应及时查找原因。如发现网板变形或网板与循环水管壁之间间隙大于 6mm,应及时采取措施整形和调整间隙,使其恢复正常,而不能采取消极措施将其网板固定或焊死。

7. 各级领导重视,建立健全规章制度,科学管理,专人负责。为此我们制订了岗位责任制、运行和日常维护细则,与奖励挂钩的考核办法等,使操作人员有章可循。实践证明,"三分技术,七分管理",只有这样,这项省煤节电的有效措施才能不断进化,日趋完善。

(上接第 19 页)

表 4 主要设计参数

名称	蒸汽量	蒸汽压力	过热蒸汽温度	煤气产量	煤气热值
结果	75 t/h	3.82MPa	450℃	3500m³/h	2500~3000kcal/m³

四、结论

本文介绍了一种循环流化床热、电、燃气三联产新工艺,大量试验研究表明,该装置具有一炉多产、结构简单、高效、低污染、投资省、见效快等特点。技术经济分析表明,建设三联产装置是我国目前中小城市煤气化的一条切实可行的途径。

参考文献(略)

(上接第 20 页)破碎机进料粒度小于 400mm,出料粒度≤80mm,经初级破碎的煤通过皮带进入破碎间,进行下一级破碎。这种两级破碎的布置,既省去一段皮带,节省场地,又可根据粒度情况不同,灵活调节。

3、受煤坑建在地表,防止进水

受煤坑底部基本上与地表面持平,解决了渗水、积水的问题,由桥式抓斗起重机上煤。这样,降低了运输带的倾角,也便于工人维修和维护。

四、结论

燃料是锅炉的粮食,燃料的制备是电厂生产的重要环节。目前还没有非常适应南方湿煤的筛分和破碎设备,所以降低煤的湿度至关重要。本方案所采用的加大干煤棚、增加输煤通廊等方法,只能保证来煤在贮存、运输过程中不增加水分,而干燥、风干则取决于气候条件。若将电厂余热用于煤的烘干,也许是更好的办法。

原文刊于工程热物理学报, 1995, 16(4): 499-502

流化床锅炉床下热烟气点火启动
的理论及试验研究*
——理论模型

岑可法 方建华 倪明江 严建华

（浙江大学）

摘　　要

本文介绍了我们所建立的流化床锅炉床下热烟气点火启动综合数学模型。模型由描述点火启动过程所涉及的流动、传热和燃烧等物理和化学过程的诸多子模型所构成。本文着重讨论了数学模型中的四个子模型：单颗煤粒多相反应和均相反应模型，氧量平衡计算模型，固相物质平衡计算模型和能量平衡计算模型。计算结果和试验结果进行了对比，两者较为吻合。利用本文模型可对不同容量的流化床锅炉点火启动进行模拟计算，为点火装置设计和运行操作提供合理参数。

关键词：单颗煤粒燃烧，氧量平衡，固相物质平衡，能量平衡，计算模型

流化床锅炉，热烟气点火，试验

A Theoretical and Experimental Study on Ignition of Coal-Fired AFBC by Intake Hot Gas I
——Theoretical Model

Cen Kefa　Fang Jianhua　Ni Mingjiang　Yan Jianhua

(Zhejiang University)

Abstract

A Comprehensive mathematical model for ignition of coal-fired AFBC by the intake hot gas is reported in this paper. Of the submodels constituting the comprehensive mathematical model, four

---

* 国家攀登计划资助项目

key ones, i. e. that of single coal particle combustion, mass balance of oxygen, solid matter balance and Energy balance are described. The calculation results have been shown in good agreement with the experimental data. The model can reasonably pridict the effects various operation combustions on ignition process.

Key words: Single coal particle combustion, Mass balance of oxygen, Solid matter balance, Energy balance, Model

符 号 说 明

A_b——床层表面积
C——气体浓度
c_p——比热
d_p——焦碳颗粒直径
d_s——床料平均直径
E——反应活化能
E^*——扬折夹带常数
$\dot{m}_a$——颗粒磨损速率
$P_0(r_1, r_2, \tau)$——投入焦碳粒度分布函数
$P_h(r_1, r_2, \tau)$——床内焦碳粒度分布函数
r_1——未反应核心半径
$\dot{r}_1$——未反应核心收缩速率
r_2——灰壳半径
$\dot{r}_2$——灰壳半径收缩速率
T_b——床层温度
$T_{g,0}$——离开床层的烟气温度
$T_{g,1}$——床层进口烟气温度
T_p——焦碳颗粒温度
T_{zf}——悬浮空间温度
U_0——运行气速
$W(\tau)$——床内焦碳质量
δ——气泡份额
Φ——反应当量系数

f——焦碳燃烧反应当量系数
$F_0(\tau)$——焦碳投入速率
H——床层高度
ΔH——反应热
K_0——反应频率因子
m_b——床料质量
$\dot{m}$——反应速率
Φ_m——单位质量焦碳质量损耗速率
ε_{mf}——最小流化空隙率
ε_b——床层空隙率
ε_{zf}——床层表面和悬浮空间系统黑度
ρ_1——未反应核心密度
ρ_2——灰壳密度
σ_0——波尔兹曼常数

符 号 下 标
CO——一氧化碳
O_2——氧
g——烟气
p——燃料颗粒
R——挥发份燃烧
V——挥发份
C——焦碳

引 言

流化床点火启动是流化床锅炉运行首先遇到的问题,自从流态化燃烧技术问世以来,国内外对流化床点火启动做了大量研究工作,创造了很多点火方法,其中利用热烟气作为流化介质加热床料点燃流化床是一种较为先进的点火启动方法,具有热利用率高,操作简便,易于实现

自控等优点。随着循环流化床燃烧技术的发展,这种点火启动方法用于循环流化床锅炉点火启动更显示出其优越性。对这种点火启动方法国内外虽然已开展不少研究[1],但研究主要依靠实验,理论研究多停留在给煤着火初期阶段[2,3],或从热平衡角度对整个点火启动过程进行定性分析,能够模拟整个点火启动过程的数学模型尚未见发表。利用实验研究流化床热烟气点火启动,对于初步了解和掌握其特性是必不可少的,但由于模化上的困难,实验台中得出的结果难以直接应用于不同容量和受热面布置结构的流化床锅炉的点火启动中。另一方面,由于受到实验条件限制,利用实验研究无法对影响点火启动的诸因素进行逐一深入分析。随着流化床特别是循环流化床锅炉的大型化,点火启动能耗也越来越大,对流化床锅炉点火启动进行参数优化,减少点火启动能耗是很有现实意义的。由于上述原因,对流化床热烟气点火启动的理论研究特别是数学模型方面的研究日益引起重视。本文作者建立了一个能够动态模拟流化床热烟气点火启动整个过程的综合数学模型,计算结果和截面为 $600 \times 600 mm^2$ 的流化床锅炉床下热烟气点火启动试验结果进行对比,两者较为吻合。利用本文所建立的综合数学模型,可对影响流化床锅炉床下热烟气点火启动的各种因素进行模拟计算,对点火启动参数进行优化设计。

1 综合数学模型的基本结构

流化床锅炉床下热烟气点火启动综合数学模型由若干个相对独立的子模型构成,如图1所示。由图1可见,综合数学模型所包含的子模型主要有:预燃室燃烧、风室传热、流化床流体

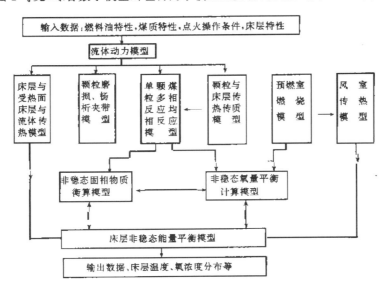

图1 流化床热烟气点火启动数学模型结构框图

动力特性、单颗燃料多相反应均相反应、颗粒磨损、颗粒扬析夹带、颗粒与床层传热传质、床层和受热面床层和烟气传热、固相物质衡算、氧量平衡计算、能量平衡计算等。每个子模型解决一个点火启动过程所涉及到的相对独立的问题。

2 子模型建立

限于篇幅本文着重讨论单颗煤粒多相反应均相反应、氧量平衡计算、固相物质平衡计算和能量平衡计算四个模型。

2.1 单颗燃料多相反应均相反应模型

本模型解决单颗煤粒的挥发份析出和燃烧反应速率计算,焦炭颗粒燃烧反应速率计算。本模型中忽略煤中水分影响。

(1)挥发份析出和燃烧

为简化计算,本模型中采用最简单的单方程模型。

$$\frac{dW_v}{d\tau} = (W_{v,\infty} - W_v)K_v\exp(-E_v/RT_p) \quad (1)$$

式(1)中 K_v 和 E_v 的数值采用傅维标提出的大颗粒挥发份析出通用模型[4]中的实验数据。

$$K_v = 8.087 \times 10^4 \, 1/s \quad E_v = 7.94 \times 10^4 \, J/mol$$

文献[5]指出,对于大颗粒燃料,挥发份析出时间与燃料在床内的循环时间相比是够长的,可以认为床内各点处挥发份析出速度是相同的。在流化床点火启动过程中,煤粒着火初期床层温度较低,挥发份析出速率相对较慢,所以本模型中采用文献[5]的假设,认为床内各点处挥发份析出速率是相同的。由于流化床内紊流运动强烈,计算挥发份燃烧反应时可以不考虑气体间的传质和混合问题,认为挥发份的燃烧反应完全由反应动力学所控制。在上述假设条件下,挥发份燃烧速率为:

$$\dot{m}_v = C_v \cdot C_{O_2} \cdot K_R\exp(-E_R/RT_b) \quad (2)$$

对Ⅱ类烟煤式(2)中 K_R 和 E_R 值分别为 $K_R = 1.375 \times 10^8 \, m^3/kg \cdot s$, $E_R = 5.0244 \times 10^4 \, J/mol$[4]。

(2)单颗焦碳燃烧反应

我国流化床所燃用的煤大多灰份较高,针对这种情况,对单颗焦碳燃烧采用"双收缩表面模型"如图2所示。"双收缩表面模型"假定燃烧发生在焦碳核心表面,燃烧过程核心表面不断向内收缩,核心密度始终不变。外层灰壳外表面由于磨损向内收缩,收缩速率受磨损率控制。两个表面的收缩速率分别为:

$$\dot{r}_1 = \frac{\dot{m}_c}{4\pi(\rho_1 - \rho_2)r_1^2} \quad (3)$$

$$\dot{r}_2 = \frac{\dot{m}_a}{4\pi\rho_2 r_2^2} \quad (4)$$

计算焦碳燃烧速率 $\dot{m}_c$ 时考虑了化学反应动力、气体在灰壳中扩散、灰壳表面对流传质三项阻力,由下列公式计算:

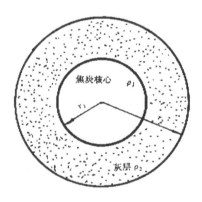

图2 双收缩表面燃烧模型示意图

$$\dot{m}_c = \frac{4\pi r_1^2 \cdot f \cdot C_{O_2}^0}{\frac{1}{K_c} + \frac{1}{\beta}\left(\frac{r_1}{r_2}\right)^2 + \frac{r_2 - r_1}{\varepsilon_e \cdot D_g}\left(\frac{r_1}{r_2}\right)} \quad (5)$$

式(5)中，K_C 为

$$K_C = K_{c_0} \exp(-E_O/RT_P) \quad (6)$$

对流传质系数按文献[5]传热传质类比式计算：

$$\beta = \frac{Sh \cdot D_g}{d_p} \quad (7)$$

$$Sh = 2.0 + 0.064 \left(\frac{Re}{\varepsilon_{mf}}\right)^{1.07} \cdot \left(\frac{r_p}{r_s}\right)^{-0.16} S_c^{0.333} \quad (8)$$

灰层有效扩散孔隙率 ε_e 采用严建华(1989)[7]等的实验关联式：

$$\varepsilon_e = \varepsilon_o \exp[-B(r_2 - r_1)] \quad (9)$$

对于大同煤 $\varepsilon_0 = 0.236, B = 0.60 \times 10^{-3} 1/m$。

式(4)中磨损速率 $\dot{m}_a$ 由颗粒磨损子模型计算。

焦碳燃烧一次反应产物一般认为既有CO又有CO_2，可用下式表示：

$$C + \xi O_2 \rightarrow 2(1-\xi)CO + (2\xi-1)CO_2 \quad (10)$$

其中

$$\xi = [0.5(CO/CO_2) + 2]/[(CO/CO_2) + 1] \quad (11)$$

CO 和 CO_2 的比值按 Arthur[11] 公式计算：

$$CO/CO_2 = 2500\exp(-6239/T_P) \quad (12)$$

一次反应产生的CO进入床层燃烧，同挥发份燃烧反应计算一样，认为CO的反应完全由反应动力学所控制，其反应速率为：

$$\dot{m}_{CO} = C_{CO} \cdot C_{O_2} \cdot K_{gc} \exp(-E_g/RT_b) \quad (13)$$

实际上为了求得各化学反应速度，尚需知道焦碳颗粒和床层温度。床层温度可通过床层非稳态能量平衡模型计算，焦碳颗粒温度可根据焦碳燃烧能量平衡求得。忽略焦碳颗粒内部温度分布，点火启动过程焦碳颗粒温度可按下式计算：

$$m_P c_P \frac{dT_P}{d\tau} = \dot{m}\Delta H_C - 4\pi r_2^2 [\sigma_0 (T_P^4 - T_b^4) + \alpha_P (T_P - T_b)] \quad (14)$$

式(14)两边均为颗粒温度 T_P 的函数，据此可求出颗粒温度。

2.2 氧量平衡计算模型

氧量平衡计算模型是建立在慢速气泡流体动力模型基础上的。根据慢速气泡气体流型特点，假设气体在乳化相和气泡相内都为平推流，氧浓度沿床高的变化等于该段高度内挥发份和焦碳燃烧的氧气消耗量，其质量守恒方程为：

$$\frac{\partial C_{O_2}}{\partial \tau} + U_O \frac{\partial C_{O_2}}{\partial Z} = \frac{\partial}{\partial Z}\left(D_g \frac{\partial C_{O_2}}{\partial Z}\right) - \sum_{r_1=0}^{r_{1max}} \sum_{r_2=0}^{r_{2max}} (1-\delta) n(r_1, r_2) 4\pi r_1^2$$
$$K_{cc} C_{O_2} - \Phi_v C_v C_{O_2} K_{RV} - \Phi_{co} C_{co} C_{O_2} K_g \quad (15)$$

式(15)中 $n(r_1, r_2)$ 为单位体积乳化相中内径为 r_1，外径为 r_2 的焦碳粒子数：

$$n(r_1, r_2) = \frac{W(\tau) P_b(r_1, r_2, \tau) \Delta r_1, \Delta r_2}{A_b H(1-\delta) \frac{4}{3}\pi [(\rho_1 - \rho_2) r_1^3 + r_2^3]} \quad (16)$$

K_{cc}、K_{RV}、K_g 分别为：

$$K_{tc} = \cfrac{1}{\cfrac{1}{K_t} + \cfrac{1}{\beta}\left(\cfrac{r_1}{r_2}\right)^2 + \cfrac{(r_2 - r_1)}{\varepsilon_s \cdot D_g}\left(\cfrac{r_1}{r_2}\right)} \tag{17}$$

$$K_{RV} = K_R \cdot \exp(-E_R/RT_b) \tag{18}$$

$$K_g = K_{g_o}\exp(-E_g/RT_b) \tag{19}$$

式(15)的初始条件和边界条件为：

$$\tau = 0, \quad C_{D_2} = C_{O_{2,1}}, \quad C_V = 0, \quad C_{O_{2,1}} = 0 \tag{20}$$

$$Z = 0, Co_2 = Co_{2,1} \tag{21}$$

式(20)、(21)中 $Co_{2,1}$ 为流化床底部进口气体氧浓度，烟气加热床料时等于热烟气中的氧浓度，关闭点火装置后等于空气中的氧浓度。

2.3 固相物质平衡计算模型

由于燃料中的挥发份在燃料入炉后不久即成为气相物质，本文模型中假定挥发份是由燃料带入流化床内的气相物质单独计算，固相物质平衡计算模型则只考虑挥发份析出后的焦碳。

关于固相物质平衡计算模型一般都是针对稳定燃烧工况建立的，但流化床点火启动是一个非稳态燃烧过程，床内固体燃料质量、粒度分布、燃烧速率等都是随时间变化的，因此必须建立非稳态的固相物质平衡计算模型。由于单颗焦碳燃烧采用"双收缩表面模型"，本文的非稳态固相物质平衡计算模型用未反应焦碳核心半径 r_1 和灰壳外半径 r_2 来表征颗粒的性质。

在任一时刻 τ，对于床内固相可燃物可列出其单位时间总的质量平衡方程

$$\begin{bmatrix}\text{给料带入的}\\\text{颗粒质量}\end{bmatrix} - \begin{bmatrix}\text{扬析夹带带出}\\\text{的颗粒质量}\end{bmatrix} - \begin{bmatrix}\text{由于燃烧反应损}\\\text{失的颗粒质量}\end{bmatrix} = \begin{bmatrix}\text{床内颗粒}\\\text{质量的增量}\end{bmatrix} \tag{22}$$

将各项数学表达式代入式(22)，整理后其显式差分方程为：

$$W(\tau + \Delta\tau) = W(\tau) + \Delta\tau\{F_0(\tau) - \sum_{r_1=0}^{r_{1max}}\sum_{r_2=0}^{r_{2max}}[E^*(r_1,r_2,\tau)P_b(r_1,r_2,\tau)A_b + W(\tau)P_b(r_1,r_2,\tau)\Phi_m(r_1,r_2,\tau)]\Delta r_1 \Delta r_2\} \tag{23}$$

在任一时刻 τ，对粒度在 r_1 到 $r_1 + \Delta r_1$，r_2 到 $r_2 + \Delta r_2$ 之间的焦碳颗粒其单位时间质量平衡方程为

$$\begin{bmatrix}\text{给料带入}\\\text{的颗粒质量}\end{bmatrix} - \begin{bmatrix}\text{扬析夹带带出}\\\text{的颗粒质量}\end{bmatrix} + \begin{bmatrix}\text{由于}r_1\text{收缩从大尺寸}\\\text{档进入该间隔质量}\end{bmatrix}$$
$$- \begin{bmatrix}\text{由于}r_1\text{收缩离开该间}\\\text{隔进入小尺寸档质量}\end{bmatrix} + \begin{bmatrix}\text{由于}r_2\text{收缩从大尺寸}\\\text{档进入该间隔质量}\end{bmatrix}$$
$$- \begin{bmatrix}\text{由于}r_2\text{收缩离开该间}\\\text{隔进入小尺寸档质量}\end{bmatrix} - \begin{bmatrix}\text{由于燃烧反应}\\\text{损失的质量}\end{bmatrix} = \begin{bmatrix}\text{该粒度范围颗}\\\text{粒质量增量}\end{bmatrix} \tag{24}$$

将各项数学表达式代入式(24)，整理后其显式差分方程为

$$P_b(r_1,r_2,\tau + \Delta\tau) = \frac{W(\tau)}{W(\tau + \Delta\tau)}P_b(r_1,r_2,\tau) + \frac{W(\tau)}{W(\tau + \Delta\tau)}[F_0(\tau)P_0(r_1,r_2,\tau)$$
$$- E^*(r_1,r_2,\tau)P_b(r_1,r_2,\tau)A_b + W(\tau)P_b(r_1 + \Delta r_1,r_2,\tau)\dot{r}_1(r_1 + \Delta r_1,r_2)/\Delta r_1$$
$$- W(\tau)P_b(r_1,r_2,\tau)\dot{r}_1(r_1,r_2)/\Delta r_1 + W(\tau)P_b(r_1,r_2 + \Delta r_2,\tau)\dot{r}_2(r_1,r_2 + \Delta r_2)/\Delta r_2$$
$$- W(\tau)P_b(r_1,r_2,\tau)\dot{r}_2(r_1,r_2)/\Delta r_2 + W(\tau)P_b(r_1,r_2,\tau)\Phi_m(r_1,r_2)] \tag{25}$$

方程(23)和(25)的初始条件为

$$\tau = 0 \quad W(\tau) = W_0$$
$$P_b(r_1, r_2, \tau) = P_{b0}(r_1, r_2, 0) \tag{26}$$

式(26)中 $W_0=0$ 表示底料中不含燃料,为纯惰性床料。$W_0>0$ 表示点火前底料中预混有一定的燃料,其粒度分布为 $P_{b0}(r_1,r_2,0)$。

式(23)和(25)中 $\Phi_m(r_1,r_2)$ 为单位质量焦碳质量损耗速率:

$$\Phi_m(r_1, r_2) = \frac{3[(\rho_1 - \rho_2)r_1^2\dot{r}_1 + \rho_2 r_2^2 \dot{r}_2]}{(\rho_1 - \rho_2)r_1^3 + \rho_2 r_2^3} \tag{27}$$

由方程(23)、(25)和(26)可求解任一时刻 τ 床内焦碳颗粒的总质量 $W(\tau)$ 及其粒度分布 $P_b(r_1,r_2,\tau)$。

2.4 床层能量平衡模型

床层能量平衡计算的目的是用来求点火启动过程床温变化。根据能量守恒,床内任一时刻有:

$$\begin{bmatrix}\text{给煤带入}\\\text{的物理热}\end{bmatrix} + \begin{bmatrix}\text{煤燃烧}\\\text{放热量}\end{bmatrix} + \begin{bmatrix}\text{烟气或空气带}\\\text{入床内的热量}\end{bmatrix} - \begin{bmatrix}\text{烟气带出}\\\text{床内的热量}\end{bmatrix}$$
$$- \begin{bmatrix}\text{颗粒扬析夹带}\\\text{带走的热量}\end{bmatrix} - \begin{bmatrix}\text{埋管带走}\\\text{的热量}\end{bmatrix} - \begin{bmatrix}\text{炉墙吸收和}\\\text{散失的热量}\end{bmatrix} - \begin{bmatrix}\text{床层表面和}\\\text{悬浮段换热量}\end{bmatrix}$$
$$= \begin{bmatrix}\text{床层能}\\\text{量的增量}\end{bmatrix} \tag{28}$$

由于流化床热烟气点火启动过程床层流化状态良好,可假定床内温度均匀,将各项表达式代入方程(28)得:

$$[m_b + W(\tau)] \cdot c_{pb} \frac{dT_b}{d\tau} = F_c(\tau) c_{pc} T_0 + \sum_{r_1=0}^{r_{1max}} \sum_{r_2=0}^{r_{2max}} W(\tau) P_b(r_1, r_2, \tau)$$
$$\Delta r_1 \Delta r_2 (\dot{m}_c \Delta H_c + \dot{m}_{co} \Delta H_{co} + \dot{m}_v \Delta H_v) + \dot{m}_g c_{pg} T_{g,i} - \dot{m}_g c_{pg} T_{g,o} -$$
$$\sum_{r_1=0}^{r_{1max}} \sum_{r_2=0}^{r_{2max}} E^*(r_1, r_2, \tau) P_b(r_1, r_2, \tau) \Delta r_1 \Delta r_2 A_b c_{pp} T_p - K_m F_m (T_b - T_m)$$
$$- K_w F_w (T_b - T_w) - \sigma_0 \varepsilon_{xf} A_b (T_b^4 - T_{xf}^4) \tag{29}$$

为封闭方程(29)还需补充烟气和床层的传热方程、烟气和床层的传热方程为:

$$\frac{dT_g(Z,\tau)}{dZ} = -\frac{\alpha a A_b}{\dot{m}_g c_{pg}} [T_g(Z,\tau) - T_b(\tau)] \tag{30}$$

$$T_g(Z,\tau)|_{Z=0} = T_{g,i} \tag{31}$$

式(28)中 a 为床层比表面积:

$$a = \frac{6(1-\varepsilon_b)}{d_s} \tag{32}$$

α 为烟气和床层之间的换热系数,在流化状态下可按文献[8]的实验关联式计算:

$$Nu_{gp} = 0.0247 Re^{1.18} \tag{33}$$

3 数学模型计算

流化床锅炉床下热烟气点火启动数学模型建立了点火启动过程流动、传热和燃烧等一系列控制方程,利用数值计算方法联立求解这些方程,可求出点火启动过程各参数的变化。

4 计算结果

由于本文篇幅所限,以下只给出少量典型的模型计算结果和实验比较,作为模型合理性的初步验证。计算以模拟截面为 $0.6\times0.6 m^2$ 的流化床锅炉燃用大同煤的点火启动为例,详尽的模拟计算和实验结果分析参见"流化床床下热烟气点火启动的理论及试验研究 II. 点火试验与理论计算分析"一文。

图 3 为一典型点火启动过程床层温升变化模拟计算结果。点火启动床料为纯石英砂,床温加热至 500℃ 时开始投煤,投煤量由 50kg/h 逐渐增加大到额定给煤量,升温到 600℃ 时,关闭点火装置。从图中可以看出,床层温升变化可以为三个阶段。第一个阶段为投煤前的纯加热阶段,第二阶段为给煤着火后,床层温度快速上升阶段,第三个阶段为床温接近正常运行工况时的平稳上升阶段。这和实际点火启动过程床层温度变化曲线是一致的。

图 4 为点火启动过程热烟气温度和床层温度模拟计算与试验数据比较。从图中可以看出,

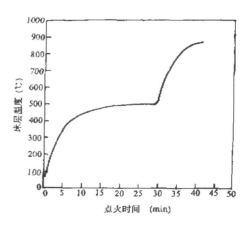

图 3 点火启动过程床层温升曲线

在整个点火启动过程中,热烟气和床层温度的计算结果都和实验数据较为吻合,这表明流化床床下热烟气点火启动综合数学模型能较准确地模拟实际的点火启动过程。图 4 的模拟计算工况中,开始点火时油量为 15kg/h 风量为 700m³/h,3 分钟后油量增大到 30kg/h,风量变为 800m³/h。从图 4 中可以看出,3 分钟时增加油量热烟气温度有一明显的快速升温,但床层由于热惯性较大,没有明显的温度波动。点火装置关闭后烟气变为冷空气,床层底部进气温度迅速下降,但床层温度由于煤的燃烧作用继续上升。

图 5 为上面所述工况床层进出口氧量的计算与实验结果比较,两者也较吻合。从图中还可

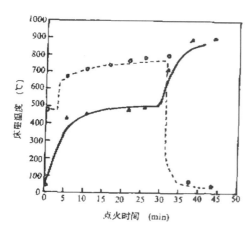

图 4 热烟气和床层温度计算与实验数据比较
——— 床温计算曲线
– – – 风室烟温计算曲线
○ 风室烟温实测值
△ 床温实测值

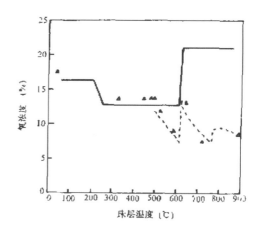

图 5 氧量变化计算与实验数据比较
——— 床层进口氧浓度计算曲线
– – – 床层出口氧浓度计算曲线
△ 床层出口氧浓度实测值

以看出,油量增大时烟气中氧含量降低。在给煤着火前加热阶段,床层进出口氧量相同,给煤着火后床层出口氧量降低,关闭点火装置床层进口氧浓度变为21%,相应地床层出口氧量有所提高,但随着床内煤燃烧强度增大,床层出口氧量又开始下降,直到进入稳定燃烧。

5 结论

本文介绍了流化床锅炉床下热烟气点火启动综合数学模型的基本概况,其主要特点为:

1. 较为全面综合地考虑了点火启动过程的各个环节,能够对整个点火启动过程进行动态数值模拟,为点火装置设计和点火启动操作提供合理参数。

2. 采用模块式结构,为模型在今后进一步发展提供了便利。

3. 针对点火启动过程参数变化特点,建立了非稳态的氧量平衡计算、固相物质平衡计算、能量平衡计算模型。

计算结果表明,点火启动模型能较准确地模拟实际的点火启动过程,可为不同容量的流化床锅炉床下热烟气点火启动提供操作参数。

参 考 文 献

1　王致钧等.常压沸腾炉预燃室点火的试验研究.重庆大学学报,1985,5
2　谢洪勇,王达三.进气升温引燃烧煤流化床的研究.工程热物理学报,1987,8(3)
3　刘柏谦.煤在流化床中的着火过程和流化床锅炉启动过程研究.[学位论文]:浙江大学,1993,12
4　Weibiao Fu, Yanping Zhang et al. A Study on devolatilization of Large Coal Particles". Combustion and Flame. 1987, 70: 253~266
5　倪明江.煤水混合物的流化床燃烧技术.[学位论文]:浙江大学,1986
6　张鹤声,卜成.链条炉内煤层的着火模型及试验.工程热物理学报,1987,8(2)
7　严建华.煤在流化床中燃烧特性的研究.[学位论文]:浙江大学,1990
8　张勖奎等.流化床移动叠置式灰渣冷却装置的研究与设计.[学位论文]:浙江大学,1992
9　Kunii D, et al. Fluidization Engineering. John Wiley & Sons, Inc. (1969)
10　Badzioch S, Hawksley P G W and C T Peter. Ind. Eng. Chem. Process Des, Dev. 1970, 9: 501
11　Arthur J A. Trans, Faraday Soc. 1951, 47: 164

原文刊于燃烧科学与技术, 1995, 1(1): 34-42

Experimental study of a finned tubes impact gas-solid separator for CFB boilers

Kefa Cen*, Xiaodong Li, Yangxin Li, Jianhua Yan, Yueliang Shen, Shaorong Liang, Mingjiang Ni

Institute for Thermal Power Engineering, Zhejiang University, 310027 Hangzhou, People's Republic of China

Received 10 May 1995; accepted 17 October 1996

Abstract

We present a finned tubes impact gas–solid separator with enhancing heat transfer behavior which can be used in circulating fluidized bed boilers (CFBB). Experimental results show that this separator has advantages such as high separation efficiency, low pressure drop and compact structure. Furthermore, it can also enhance heat transfer, and fins welded on the tubes can be cooled at the same time by the work medium in the tubes so that they can be made by common boiler steel. We describe the fluid dynamics characteristics, heat transfer and separation performance of the finned tubes separator through a series of experiments. © 1997 Elsevier Science S.A.

Keywords: Impact separation; Fluid dynamics; Fluidized bed; Boiler

1. Introduction

Worldwide attention is now paid to the development of clean coal combustion technologies. Among these technologies, circulating fluidized bed combustion (CFBC), which has been widely applied, has good characteristics such as satisfactory desulfurization efficiency, combustion efficiency, low NO_x emission, suitability for different types of coal, good performance for control and appropriate capital and operation costs.

As key equipment for circulating fluidized bed boilers (CFBB), gas–solid separators, which help the circulation of the solids in the furnace, have strong effects on the combustion efficiency, the circulation rate, the desulfurization efficiency, and so on. At present, cyclones are the most widely used separators in CFBB. Although used for a long time and having high efficiency, cyclones have disadvantages such as high pressure drop (~2 kPa), serious wear, high thermal inertia, and long start-up and cool-down times. An especially serious problem is that the size of hot cyclones increases rapidly with the capacity of CFBB; for example, the diameter of some commercial cyclones reaches 10 m [1] thus making them almost as large as the furnaces of the boilers. Large cyclones also result in some manufacturing, installation and operation problems. Therefore, it is essential to develop new

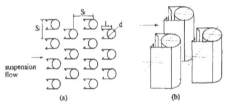

Fig. 1. Targets of the finned tubes impact separator.

types of gas–solid separators that should have high efficiency, low pressure drop and compact structure, and so on.

Compared with cyclones, inertial separators have low pressure drop, compact structure and low capital cost, and their application in CFBB is now paid great attention, for example the U-beams separator of Sweden's Studsvik Corporation [2,3], the slots separator of Germany's Steinmuller Corporation [4], and the slotted-tube impact separator of China's Xi'an Jiaotong University [5]. These separators have shown that the separation efficiency of inertial separators can, to some extent, meet the demand of some CFBB.

The above mentioned inertial separators do not combine separation with heat transfer. This paper puts forward a new type of finned tubes impact gas–solid separator with enhancing heat transfer behavior [6]. Targets of this separator (see Fig. 1) are superheated or convective tubes welded with shaped fins, thus also improving the heat transfer behavior of

* Corresponding author.

the tubes. Flow stagnant regions are created in the slots between pairs of fins. When passing through the targets, the suspension impacts on the targets and repeatedly changes its flow path. Because of inertia, the particles that have a larger momentum than that of the gas can be separated from the main gas stream. When entering the flow stagnant regions, the particles may fall down along the slots to the ash hopper. Moreover, the fins can be cooled by the work medium in the tubes so that fins made of common boiler steel can be used in high temperature conditions. With the support of the tube, it will be difficult to change the shape of the fin when the fin's length is too long along the tube, which is one of the features of the separator.

This paper presents the fluid dynamics characteristics of finned tubes measured by means of a laser Doppler velocimeter (LDV) for a single phase and a particle dynamics analyzer (PDA) for two phases, and the heat transfer characteristics of the fin. Based on this study, cold model experiments for separation and pressure drop performance of the separator have been carried out and some optimum design parameters are obtained.

2. Fluid dynamics characteristics of finned tubes

A laser Doppler velocimeter (LDV) made by American TSI Co. was used to measure the flow structure of a single finned tube. Fig. 2 shows the LDV measurement system. A finned tube is placed horizontally in the measuring part which is a Plexiglas® square-cross-sectional pipe. The flow rate of the gas fed into the pipe is $102 \text{ m}^3 \text{ h}^{-1}$, and the inlet velocity is 4.2 m s^{-1}. By measuring the velocity of the tracer loaded in the gas, the velocity and turbulent intensity of the gas phase are obtained [7]. The coordinate systems of the measurement parts are shown in Fig. 3.

A three-dimensional particle dynamics analyzer equipped with a He : Ne laser with a maximum power of 10 W (Dantec Measurement Technology A/S) was adopted to measure the two-phase fluid dynamics characteristics of a single tube and finned tubes. Fig. 4 shows the PDA measurement system. The particle tracers are white resin. The flow rate of the gas was the same as that of the LDV. The particle concentration ranged from 0.17 to 0.44 kg m^{-3}. The coordinate systems are also shown in Fig. 3.

Fig. 5 shows the velocity distribution of the gas in the x-direction of a single finned tube on several cross sections before the tube (in the figure, x is the distance from the cross section to the center of the tube). As shown in the figure, for $x/d = -1.43$ the streamline begins to curve distinctly. For $x/d = -1.14$, a reflux region appears and then the gas velocity reduces to a very small value and a flow stagnant region is created in the slots between couples of fins from $x/d = -1.0$ to 0.0.

Fig. 6 shows the two-dimensional velocity vector of a particle in a finned tube; the stagnant and reflux regions are obvious. Fig. 7 shows the two-phase x-directional flow and the slip velocity distributions of a single finned tube along the flow direction. From $x/d = -2.0$ to $x/d = -1.0$ before the tube, the slip velocity is somewhat larger than zero because of the particle's inertia at the $-0.5 < y/d < 0.5$ section, and from $x/d = -1.0$ to $x/d = 0.0$, due to the effect of the stagnant regions, the slip velocity and the particle's velocity are nearly zero.

The measurement results illustrate, according to the above flow field, that most of the particles in the separator can be separated from the gas phase from the section $x/d = -1.43$ and enter the region between the two fins. Furthermore, the reflux region makes it possible that the rebounded particles can be dragged back into the region between the two fins.

Fig. 8 shows the gas turbulence intensity distribution of the single finned tube along the tube. At the entrance of the region between the two fins, the turbulence intensity is small, which prevents the particles in that region from escaping from it. In addition, the experimental results show that the largest value of the cross-section-mean turbulence intensity appears in the section $x/d = 3.0$ behind the tube.

The gas–solid turbulence intensity distribution when particles are put into the gas flow of a finned tube is shown in

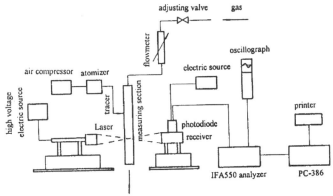

Fig. 2. Schematic diagram of the laser Doppler velocity measurement system.

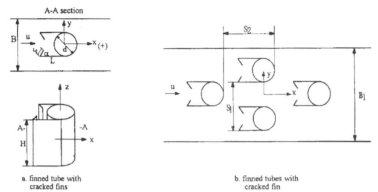

a. finned tube with cracked fins

b. finned tubes with cracked fin

Fig. 3. The coordinate systems of the measurement parts.

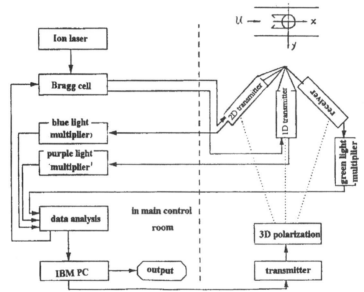

Fig. 4. Schematic diagram of the particle dynamics analyzer measurement system.

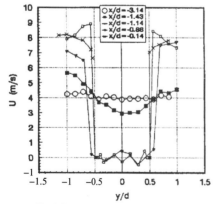

Fig. 5. Before-tube gas velocity distribution.

Fig. 9. From the figure, the gas turbulence intensity before the tube (up to $x/d = -1.0$) is nearly equivalent to that of the single tube flow, but larger than that of the single tube flow from $x/d = -1.0$. Before $x/d = -1.0$, the motion of the particles is uniform and has a small effect on the gas turbulence intensity. However, from $x/d = -1.0$, i.e. from the fins area, the gas turbulence intensity increases due to the particle's violent motion and the impact effects among the particles. Fig. 9 also shows that the particle turbulence intensity is higher than that of the gas phase before $x/d = 0.0$ because of the collision effects of the particles and lower than that of the gas phase after $x/d = 0.0$. From this section, due to the reflux flow and acceleration, the gas turbulence intensity is increased to a high value. Meanwhile the particle turbulence intensity does not follow the variation of the gas phase because the particle's inertia is large and the impact proba-

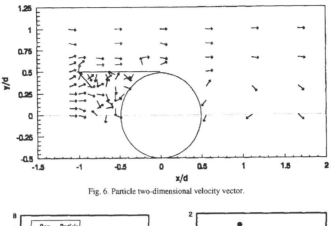

Fig. 6. Particle two-dimensional velocity vector.

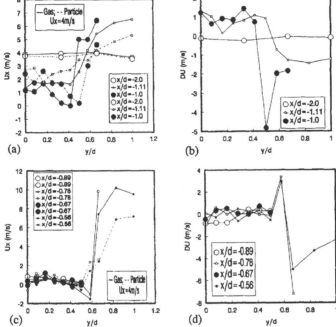

Fig. 7. Gas–solid two-phase flow and slip velocity distribution of a single finned tube. (a) Two-phase velocity distribution before the finned tube. (b) Gas–solid slip velocity distribution before the finned tube. (c) Two-phase velocity distribution between the fin's sections. (d) Gas–solid slip velocity distribution between the fin sections.

bility among particles decreases due to the low particle concentration at the back of the finned tube.

PDA measurement results also show that for two-phase flow the largest value of the cross-section-mean turbulence intensity appears at the section $x/d = 1.0–1.5$ behind the tube and not at $x/d = 3.0$ as for the single phase flow.

The effect of neighboring tubes on the flow field in the finned tubes separator has been studied by PDA. Four finned tubes arranged in three rows are placed vertically in a plexiglas measuring part with a square-cross-sectional area of 80×144 mm^2. The parallel and normal intervals are twice the diameter of the finned tube, 36 mm. The coordinate system is shown in Fig. 3.

Fig. 10 presents the particle velocity distributions in the normal direction along the finned tubes. Due to the effects of neighboring tubes, the particle reflux flow behind the first second row's tubes does not appear.

Fig. 11 shows the particle turbulence intensity distributions of the finned tubes at the front section before the tubes. In the figure, $x/d = -1.11$ is the front section of the second

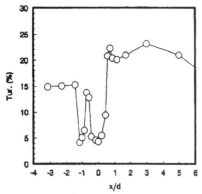

Fig. 8. Gas turbulence intensity distribution of a finned tube.

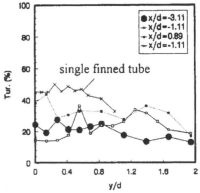

Fig. 11. Particle turbulence intensity distribution along finned tubes.

before the three tube rows increase with the tube row and all of them are lower than that of a single tube flow. This phenomenon is probably due to the flow stagnant and direction-guide effect of the latter tube row.

From the above, the LDV and PDA measurement results of the fluid dynamics characteristics of a finned tube and finned tubes show that a helpful flow structure is created in the finned tubes separator and lays the foundation of the following study.

3. Temperature distribution and heat transfer performance of a finned tube

Under high temperature conditions, the temperature of the fins is higher than that of the tubes. In order to avoid overly high temperatures in the fins, the fins must not be too long, thus relatively cheap steel can be employed for the fins. In this paper, the temperature distribution in a thin fin under hot conditions is calculated to find out a reasonable fin length [7].

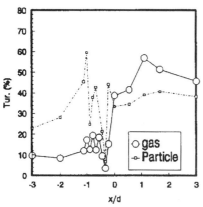

Fig. 9. Gas and particle turbulence intensity distribution of a single finned tube.

Fig. 12 shows the heat flux through a fin (let the height of the fin, Z, be 1). For a cuboid whose length in the X-direction is dX, based on the heat flux balance of the cube and neglecting the radiation for the neighboring fins to simplify the problem, the following differential equation is obtained:

$$\lambda A_f \frac{d^2 t}{dx^2} = \alpha_0 U(t - t_f) + \sigma \epsilon \left[\left(\frac{t+273}{100} \right)^4 - \left(\frac{t_f+273}{100} \right)^4 \right] \quad (1)$$

with the boundary condition:

$$X = 0, \qquad t = t_f$$

$$X = L, \qquad -\lambda \frac{dt}{dx} \bigg|_{X=L} = \alpha_0 (t_L - t_f)$$

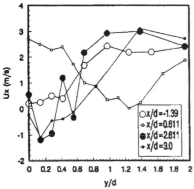

Fig. 10. Particle normal-directional velocity distribution along finned tubes.

tube row, $x/d = -3.11$ is that of the first tube row and $x/d = 0.89$ is that of the third tube row. The dotted line in the figure represents the measurement results of the single tube flow. From the figure, the particle turbulence intensities

The convective coefficient α_0 for the tube surface is calculated from the following equation which is obtained from the heat transfer experimental results [7].

$$Nu_f = 0.0423 Re_f^{0.7397} \quad (2)$$

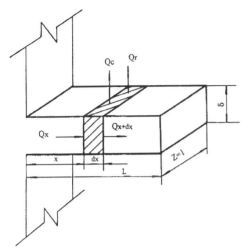

Fig. 12. Heat flux through the fin.

In Eq. (1), the fin emissivity ϵ is chosen to be 0.8. In Eq. (2), the influence of the solid loading on α_0 is not considered because the influence is very complex. In generally, α_0 will increase with solid loading so that it is safe if the influence of solid loading on α_0 is not taken into account.

After applying the finite-difference numerical method to solve Eq. (1), the following results are obtained. Curves (a) and (b) in Fig. 13 are the variations of the fin temperature with the tube wall temperature and the fin length, for which the gas temperature is 850 °C and the gas velocity is 6.6 m s^{-1}. In the calculation, the tube diameter is 35 mm. As can be seen in the figure, when the temperature of the tube wall is lower than 450 °C and the fin length is less than 45 mm, the highest temperature in the fin is lower than 790 °C. Hence heat-resisting steel can be employed.

Curves (c) and (d) in Fig. 13 show the variations of the temperature of the fin top with the temperature of the tube wall and the length of the fin when the gas temperature decreases to 550 °C. The curves indicate that when a finned tube is arranged in the intermediate temperature region, the highest temperature in the fin will be much lower than that shown in curves (a) and (b). Therefore the fin can be made of common steel.

The fins can also enhance the heat transfer behavior of the finned tubes. Initial cold experiments have been carried out to study the heat transfer enhancing capacity of the finned tubes. The results show that when the gas velocity is between 3.3 and 10.0 m s^{-1}, the temperature of the outer wall of the tubes is 81 °C, the gas temperature is 23 °C, the fin length is 35 mm, and the heat transfer rate of a single finned tube is about 60% higher than that of a single bare tube.

4. Separation and pressure drop performance of the separator

4.1. Experimental apparatus and method

Fig. 14 shows a schematic diagram of the experimental apparatus. The size of the separator is $630 \times 360 \times 275$ mm^3 in length, width and height respectively. The solids are introduced into a Venturi tube from a hopper via a screw feeder. There is a long enough straight pipe (about 3.5 m long, 9.5 times the hydraulic diameter of the separator's cross section) between the Venturi tube and the separator to ensure that the velocity of the fed solids is as high as that of the gas. At most nine staggered rows of finned tubes with a diameter of 35 mm (for U-beams, the fin length is 40 mm) can be arranged in longitude and five columns can be arranged in latitude. The solid material is quartz sand whose density and bulk density are 2650 and 1550 kg m^{-3} respectively. The outer mean diameter of the sand is 260 μm. The particle size distribution is shown in Fig. 15.

The gas flow rate is measured by a rotameter. The feed and the separated particles are weighted by a precise balancer; the maximum deviation is 1%. The pressure drop of the separator is measured using a U-type glass column; the maximum deviation is 9.8 Pa.

The particle concentration is calculated from the following equation:

$$C = \frac{G_i}{Q\tau} \tag{3}$$

The separation efficiency can be determined by analyzing any two of the three material streams involved in the separation: the feed, the separated product and the unseparated fine product. The first two factors are adopted to determine the separation efficiency from the following equation:

$$\eta = \frac{G_s}{G_i} \times 100\% \tag{4}$$

It is useful to perform an uncertainty analysis for the separation performance tests. Leaving aside systematic errors, i.e. assuming perfect calibration of the instruments, the ran-

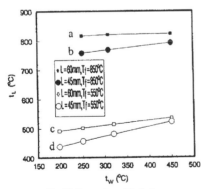

Fig. 13. Temperature of the fin top.

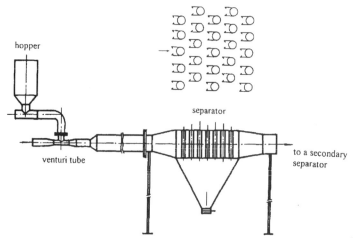

Fig. 14. Schematic diagram of the experimental set-up for the separator.

dom errors in all independent measured variables will be reflected in an error in the efficiency determined. The standard deviation of the efficiency can be calculated from the standard deviations of the feed G_i and the separated particles G_s using the following equation:

$$\sigma_\eta = \sqrt{\left(\frac{1}{G_s}\right)^2 \sigma_j^2 + \left(\frac{G_i}{G_s^2}\right)^2 \sigma_s^2} \qquad (5)$$

In Eq. (5), it is acceptable that G_i and G_s are made equal to 1 kg because all of the G_i and G_s in the test runs are larger than this value. According to error theory, the standard deviations σ_j and σ_s for G_i and G_s respectively are calculated by the maximum deviation, i.e. a third of the maximum deviation. Based on Eq. (5) and the above parameters, the standard deviation of the efficiency is about 0.47%.

4.2. Results and discussion

The following parameters have great effects on the separation efficiency of the separator: normal and parallel tubes interval, fin length, number of the tube row, particle diameter, gas velocity and particle concentration. The results are presented to show the effect of these parameters and the pressure drop performance of the separator.

4.2.1. Effect of fin length

The study of the fluid dynamics characteristics revealed that when the fin is not long enough, the particles may be easily bounced back to the main flow which leads to a low separation efficiency, so the fin length should be long enough. However, due to the temperature limit of the fin material, fins that are too long are not feasible because of the temperature distribution of the fin as shown in Fig. 13. It is recommended to choose $L/d = 1.0$, where L is the fin length and d is the diameter of the tube.

4.2.2. Effect of tubes interval

The variation of tube interval affects the separation remarkably. Let the parallel interval of the tubes be S_1/d, the normal interval of the tubes be S_2/d. Fig. 16 shows the effect of S_1/d on the separation efficiency. As shown in the figure, S_1/d has an optimum value of about 2.0. When S_1/d is too small, the suspension velocity between the tubes increases, thus leading to an increase of the entrainment ability of the main flow. Therefore the separation efficiency decreases with the increase of the re-entrainment. When S_1/d is larger than 2.0, the increase of S_1/d means that the number of the tubes in each row of the separator may decrease, so S_1/d must not be too large. Fig. 17 shows the effect of S_2/d on the separation efficiency; the efficiency decreases with S_2/d. Since the tubes are welded with fins and sufficient flow space between the tubes and the fins is needed, a value of S_2/d that is too small is not advisable. According to the results, $S_2/d = 2.0$ is reasonable.

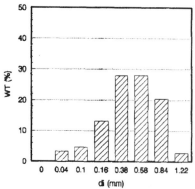

Fig. 15. Quartz sand size distribution.

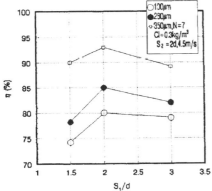

Fig. 16. Effect of parallel interval on separation efficiency.

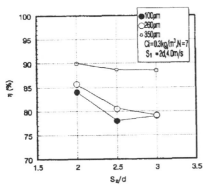

Fig. 17. Effect of normal interval on separation efficiency.

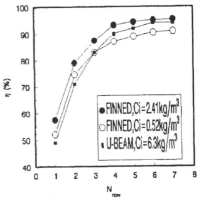

Fig. 18. Effect of number of tube rows on separation efficiency.

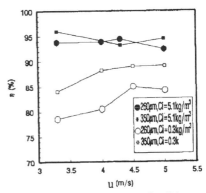

Fig. 19. Effect of gas velocity on separation efficiency.

4.2.3. Effect of the number of tube rows

Let the number of tube rows be N. Fig. 18 shows the effect of N on the separation efficiency. The figure shows that with increasing N the separation efficiency increases, but most of the separated particles are captured by the first four rows. The amount of particles captured by the first four rows reaches up to 90–95% of the total amount captured by seven rows. When the inlet particle concentration increases, this percentage will increase slightly. Therefore four or five rows are enough for a certain efficiency requirement. If there are too many rows, the increase of the separation efficiency will not be significant while the pressure drop may increase rapidly.

In Fig. 18, the result of the U-beams separator based on our experiments is also given. The trend of increasing efficiency with tube rows, and the total efficiency, are nearly the same as that of the finned tubes separator.

4.2.4. Effect of gas velocity

In impingement separators, the particle inertia increases with suspension velocity, thus making the particles more easily separable from the main flow. On the other hand, the entrainment capacity of the main flow, which depends directly on the square root of the particle diameter, increases with the gas velocity. This effect will lead to a decrease of the separation efficiency. Due to these two effects, there is an optimum value of the gas velocity. For the finned tubes impact separator, the gas velocity is based on the cross-section ($A = 360 \times 275$ mm^2) of the separator. Fig. 19 presents the effect of the gas velocity. It can be concluded that at a low inlet particle concentration the separation efficiency reaches its highest value when the velocity is about 4.5 m s^{-1}, while at a high inlet particle concentration the gas velocity has almost no effect on the separation efficiency. Therefore a finned tubes impact separator used in CFBB may have good separation performance even if it operates at low gas velocity. Considering the pressure drop of the separator, a gas velocity of about 4.0 m s^{-1} is appropriate.

4.2.5. Effect of inlet particle concentration and diameter

Much has been written on the effect of the inlet particle concentration on the separation efficiency [8–11]. In most of these publications, it is reported that if the inlet particle concentration is low, the separation efficiency increases with increasing inlet particle concentration. Some researchers have found that if the inlet particle concentration is very high the

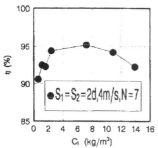

Fig. 20. Effect of inlet particle concentration on separation efficiency.

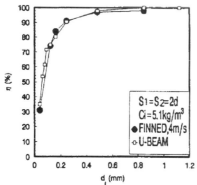

Fig. 21. Effect of inlet particle diameter on separation efficiency.

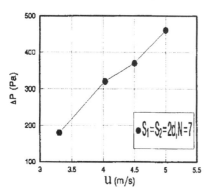

Fig. 22. Effect of gas velocity on pressure drop.

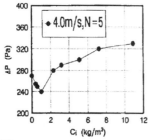

Fig. 23. Effect of inlet particle concentration on pressure drop.

separation efficiency may decrease. In this paper, the variation of the separation efficiency with the inlet concentration ranging from 0 to 14 kg m^{-3} has been studied. Fig. 20 shows the relationship between the separation efficiency and the inlet particle concentration. As shown in the figure, the separation efficiency increases with the inlet concentration before it reaches its highest value (95.4% with an inlet particle concentration of 7.0 kg m^{-3}), beyond which it falls off.

The separation efficiency increases with the particle diameter because the particle inertia increases with its diameter. According to the experimental results shown in Fig. 21 the cut size of this separator is about 60 μm.

According to our experimental results, the grade efficiency of the U-beams separator is also shown in the figure and the cut size is about the same as that of the finned tubes separator.

4.2.6. Pressure drop performance

The pressure drop of the separator is low and increases with the number of tube rows. Therefore it is better not to use too many rows in order to obtain a low pressure drop and a high enough separation efficiency. Fig. 22 shows the effect of the gas velocity on the pressure drop. The pressure drop increases with increasing gas velocity.

Many researchers have shown much interest in the effect of the inlet particle concentration on the pressure drop of gas–solid separators [9,10,12]. Most carried out their experimental study under low particle concentration conditions (maximum value up to several hundreds of g m^{-3}) and found that with the increase of the particle concentration, the pressure drop may decrease. In this paper, the experiments cover a wide range of inlet particle concentrations. The results (see Fig. 23) show that at a low inlet particle concentration the profile of the pressure drop is similar to that reported by other researchers, while at a high inlet particle concentration (higher than 1.0 kg m^{-3}), the pressure drop may increase with the increase of the inlet particle concentration.

However, when the inlet particle concentration is high enough, the increase of the pressure drop will be less.

Based on the experimental results, two empirical correlations are presented. Eq. (6) is the pressure drop equation of the pure gas phase and the other is that of a gas–solid suspension. Compared to the experimental data, the maximum error of these equations is less than 3%.

$$\Delta P_g = 6.34 N^{0.78} \frac{\rho W^2}{2} \quad (6)$$

$$\Delta P_{gs} = \begin{cases} 6.34 N^{0.78} \dfrac{\rho W^2}{2} e^{-0.110 \left(\frac{C_i}{C_{cr}}\right)} & C_i < C_{cr} = 1.0 \text{ kg m}^{-3} \\ 6.63 N^{0.78} \dfrac{\rho W^2}{2} \left(\dfrac{C_i}{C_{cr}}\right)^{0.139} & C_i > C_{cr} = 1.0 \text{ kg m}^{-3} \end{cases}$$

(7)

5. Wear of the separator

Since the separation targets are welded on the tubes, it is more important for this separator to protect the tubes from being worn. However, at low gas velocity (about 4.0 m s^{-1}) and high inlet particle concentration, wear of the tubes will be reduced because the particles must pass through a dense particle region before impacting on the tubes. The site experimental results of Studsvik's U-beams separator have verified the above point of view [2]. To prevent contingent serious wear of the tubes, a heat treatment such as carbonization or boronization can be used for the fins and the front of the first two rows to increase their wear-resisting capacity.

6. Conclusion

The flow structure of the two-phase flow of finned tubes is measured showing the good fluid dynamics characteristics of the finned tubes separator. The temperature in a fin is calculated and initial heat transfer experiments indicate that the fins may lead to a ~60% increase of the heat transfer rate. Based on a series of experiments, the optimum values of the following parameters are presented: fin length, parallel and normal interval between the tubes, suspension velocity, number of tube rows. The effect of the inlet particle concentration and particle diameter on the separation efficiency and pressure drop performance of the separator are shown.

The above results show that the finned tubes impact gas–solid separator presented in this paper has advantages such as high efficiency, low pressure drop and more compact structure, it can enhance heat transfer as well and the fins can be made of common steel or heat-resisting steel contrary to other impact separators such as the U-beams separator. In brief, this separator is very suitable for circulating fluidized bed boilers (CFBB) with a high furnace outlet particle concentration.

7. Nomenclature

A	cross-sectional area of the separator	mm^2
A_f	cross-sectional area of the cuboid (see Eq. (1))	m^2
C_{cr}	inlet particle concentration with the lowest pressure drop	kg m^{-3}
C_i	inlet particle concentration	kg m^{-3}
D_i	inlet particle diameter	mm
d	tube diameter	mm
G_i	particle quantity fed during given time	kg
G_s	particle quantity separated during given time	kg
L	fin length	mm
N	number of tube rows	
Nu_f	Nusselt number (see Eq. (2))	
Q	gas flow rate	$\text{m}^3 \text{ s}^{-1}$
Q_c	convective heat flux (see Fig. 12)	W
Q_r	radiate heat flux (see Fig. 12)	W
Q_x	conductive heat flux at location x (see Fig. 12)	W
Re_f	Reynolds number based on d	
S_1	parallel tubes interval	mm
S_2	normal tubes interval	mm
Tur	turbulence intensity	%
T_w	temperature of tube wall	°C
t	temperature of the cuboid (see Eq. (1))	°C
t_f	temperature of gas	°C
t_L	temperature of fin top	°C
U	inlet velocity (see Fig. 5)	m s^{-1}
U	circumference of the cross section (see Eq. (1))	m
W	gas velocity	m s^{-1}
wt	weight percent of test particles	%
ΔP_g	pressure drop of the separator (pure gas phase)	Pa
ΔP_{gs}	pressure drop of the separator (gas–solid suspension)	Pa
α_0	convection coefficient	$\text{W m}^{-2} \text{ K}^{-1}$
λ	thermal conductivity of the fin	$\text{W m}^{-1} \text{ K}^{-1}$
ϵ	emissivity of the fin	
ρ	gas density	kg m^{-3}
σ	Stefan–Boltzman constant	$\text{W m}^{-2} \text{ K}^{-4}$
σ_η	the standard deviation of the separation efficiency	
σ_j	the standard deviation of the feed particles	
σ_s	the standard deviation of the separated particles	
τ	given time (see Eq. (3))	s
η	separation efficiency	%

References

[1] S.L. Darling, H. Beisswenger and A. Vechsler, The Lurgi/combustion engineering circulating fluidized bed boiler—design and operation, in: Proc. of the 1st International Conference on CFB, 1985, pp. 297–308.

[2] R.F. Johns and R.E. Wascher, Design and construction of a wood-fired circulating fluidized bed boiler, in: Proc. of the 9th International Conference on FBC, 1987, pp. 385–391.

[3] F. Belin and T.J. Flynn, Circulating fluidized bed boilers solids system with in-furnace particle separator, in: Proc. of the 11th International Conference on FBC, 1991, pp. 385–391.

[4] J. Makansi, Special report for fluidized bed boilers, power, March 1991, pp. 15–21.

[5] Zhang Yongzhao, Li Yingtang, Zhang Ximing, Jing Pingan, Duan Jianzhong, Bao Bongwen and Wang Ganghua, Experimental study of a slotted tube impact separator, Chinese J. Power Eng. 6 (1989) 9–13.

[6] Chinese Patent, No. 93.235633.8, February 1994.
[7] Li Xiaodong, Experimental and Theoretical Study of Gas–Solid Separation Technologies for Circulating Fluidized Beds, PhD Thesis, Zhejiang University, 1994.
[8] M. Trefz, E. Muschelknautz, Extended cyclone theory for gas flows with high solids concentration, Chem. Eng. Technol. 16 (1993) 153–160.
[9] L. Svarovsky (Ed.), Handbook of Powder Technology, vol. 3, Elsevier, Amsterdam, 1981.
[10] A.C. Stern, K.J. Caplan and P.D. Bush, Cyclone Dust Collector, API Report, 1955.
[11] W. Licht, Air Pollution Control Engineering—Basic Calculation for Particulate Collection, 1980.
[12] A. Ogawa, Separation of Particles from Air and Gases, CRC Press, 1984.

翅片管惯性气固分离器对循环流化床锅炉运行的试验研究

摘要：提出了一种带翅片管的惯性气固分离器，该分离器具有增强传热性能的特点，可用于循环流化床锅炉。试验结果表明，该分离器具有高分离效率、低压降和结构紧凑等优点。此外，它还可以增强传热效果，翅片焊接在管子上可以通过管子内的工作介质同时冷却，因此可以采用普通锅炉钢制造。通过一系列实验研究认识了翅片管式惯性分离器的流体动力学特性、传热和分离性能。

原文刊于 Chemical Engineering Journal, 1997, 66: 159-169

Experimental studies on municipal solid waste pyrolysis in a laboratory-scale rotary kiln

A.M. Li, X.D. Li, S.Q. Li, Y. Ren, N. Shang, Y. Chi, J.H. Yan, K.F. Cen*

Institute for Thermal Power Engineering, Zhejiang University, Hangzhou 310027, China

Received 2 April 1998

Abstract

A laboratory-scale, externally heated, rotary-kiln pyrolyser was designed and built. Pyrolysis tests were performed. Solid wastes (paper, paperboard, waste plastics including PVC and PE, rubber, vegetal materials, wood, and orange husk) were tested. The effects of heating methods, moisture contents and size of waste on pyrolysis gas yields and compositions, as well as heating values, were evaluated. © 1999 Elsevier Science Ltd. All rights reserved.

1. Introduction

MSW (municipal solid waste) production and accumulation are global environmental issues [1]. Both pyrolysis and incineration have been applied in recent years for the purpose of energy recovery in WTE (waste to energy) plants. Studies have indicated that a suitable thermal pyrolysis method may be used for resolving the disposal problems and effective energy conversion from MSW [2–6]. Properly designed and operated thermal treatment devices do not pose serious threats to the environment [7]. In pyrolysis, emissions of NO_x, SO_2 and heavy metals are much lower than those in incineration [8]. Controls of emissions are easier for pyrolysis than for incineration due to reduced oxygen content, lower temperature and greatly reduced air-flow rate. In pyrolysis, the ability to control the production of pollutants minimizes the need for expensive post-process gas scrubbing. With pyrolysis, reaction parameters may be easily varied to alter the product composition [9]. Pyrolysis is an endothermic process. However, in most cases, the quantities of gener-

---

* Corresponding author. Fax: 0086-0571-7951616; E-mail: itpe@sun.zju.edu.cn

0360-5442/99/$ - see front matter © 1999 Elsevier Science Ltd. All rights reserved.
PII: S0360-5442(98)00095-4

ated gas are more than sufficient to meet thermal process needs by using recycled gas combustion [9].

The use of rotary kilns is advantageous for other pyrolysis technologies. Solid wastes with different shapes, sizes and heating values can be fed into rotary kilns in batches or continually. Rotary kilns have been widely used in incineration and studied by many researchers [10–13]. In incineration, a kiln is often followed by an after-burner to ensure a complete combustion [14–16]. In the 1970s, a field-scale rotary kiln was first used to pyrolyse waste tires at Rocky Flats [17]. Since then, the pyrolysis technology has been greatly improved and widely applied. Most kilns used for pyrolysis are internally heated. The heating value of the produced gas is very low and it must be co-combusted with oil or natural gas in close-by utilities [18]. In order to raise the heating value of the produced gas and make the gas more widely usable, externally heated kilns should be designed [19]. There are very few published reports on pyrolysis of municipal solid waste in externally heated rotary kilns.

An externally heated rotary-kiln pyrolyser was designed and built at Zhejiang University to produce medium heating value gas from solid wastes. In this paper, the characteristics of the gases produced by pyrolysis from various materials under different operating conditions are described.

2. Experimental studies

2.1. Rotary kiln

The rotary-kiln pyrolyser is shown schematically in Fig. 1. The furnace has an internal diameter of 0.205 m and is 0.450 m long. The pyrolyser can process up to 4.5 litres of feedstock in each run and its rotation rate ranges from 0.5 to 10 rpm, in our tests, the kiln rotation rate was adjusted to 3 rpm. Two k-type thermocouples were used to record the temperature–time history in the kiln. One was fixed on the inner surface and the other suspended at the center of the freeboard. The kiln is heated by a 12 kW electric furnace. The temperature of the kiln freeboard may reach 850°C when the temperature of the internal surface of the furnace is at 900°C.

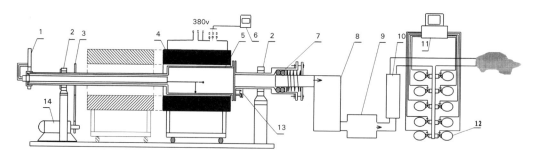

Fig. 1. Schematic of the apparatus; 1 = thermometer, 2 = bearing, 3 = gear transmission, 4 = electric furnace, 5 = rotary kiln, 6 = temperature controller, 7 = seal, 8 = two-step condenser, 9 = filter, 10 = total flowmeter, 11 = computer, 12 = gas sampler, 13 = feed and discharge opening, 14 = adjustable-speed motor.

2.2. Sealing

Sealing of rotary kilns is a difficult technology, especially for a pyrolyser. In our pyrolyser, the internal pressure of the kiln was higher than atmospheric. The hot gas must be carefully contained. A special friction-type seal was designed and successfully applied to contain the hot pyrolysis gas.

2.3. Gas processing

A two-step, directly water-cooled condenser is used to cool the hot pyrolysis gas. Most of the tar and moisture are condensed. The pyrolysis gas passes next through a filter (a glass tube filled with quartz wool). The remaining tar and moisture in the gas are here removed. Before and after each test, the condenser and filter are weighted. Thus, the tar and moisture generated during the test may be calculated. The gas flow rate is recorded using a total flowmeter. A gas sampler is located at the end of the system.

2.4. Gas sampling and analysis

A computer-controlled gas-sampling device is used. The computer controls the sampling procedure by adjusting the electrical magnet valves of the sampling balls. A gas chromatograph is used to analyse the compositions (e.g. H_2, CO, CH_4, CO_2, O_2, N_2, C_2H_4 and C_2H_6) of the gas.

2.5. Raw materials

The primary and ultimate analyses of the wastes are listed in Table 1.

2.6. Heating methods

Two methods are selected to heat the raw materials. During slow heating, the kiln filled with raw materials is kept outside of the furnace until the temperature of the electrical furnace is stable at a defined value. With this process, the kiln is cool at the start of the tests and the temperature of the raw materials increases slowly. During fast heating, the empty kiln is first placed into the electrical furnace and then heated. Raw materials are introduced into the kiln promptly when the temperature of the electrical furnace is stable at a defined value. With this fast procedure, the temperature of the raw materials increases faster than with slow heating method. Fast heating resembles continuous feed operation, while slow heating is similar to a batch-feed operation.

3. Results and discussion

In the fast heating method, the gas release is completed in 7 min except for vegetal and orange husk as shown in Fig. 2. For the pyrolysis of vegetal and orange husk, about 15 min are required due to their high moisture content. The moisture content of vegetal materials is 86.86%, while

Table 1
Primary and ultimate analyses of the raw materials

Materials	Primary analysis					Ultimate analysis				
	M_{ad} (%)	A_{ad} (%)	V_{ad} (%)	FC_{ad} (%)	$Q_{b,ad}$ (J/g)	C_{ad} (%)	H_{ad} (%)	N_{ad} (%)	S_{ad} (%)	O_{ad} (%)
Paper	10.25	2.09	74.45	13.21	32816.5	36.12	5.37	0.09	0.17	45.91
Paper board	9.28	12.76	65.56	12.40	14074.8	38.6	4.90	0.21	0.17	34.08
Wood-chip	14.83	2.94	69.41	12.82	15855.5	40.32	4.68	0.18	0.06	36.99
Cotton cloth	5.14	0.79	86.94	7.13	15443.3	42.5	5.32	0.2	0.13	45.92
Vegetal	86.86	2.51	8.60	2.03	201.1	4.39	0.33	0.57	0.07	5.27
Orange husk	73.93	0.72	21.51	3.84	4759.5	13.42	1.88	0.35	0.02	9.68
PE plastic	0.17	0.06	99.77	0	37575.2	89.28	13.66	0.06	0.02	0
PVC plastic	0.28	14.95	64.88	19.89	15854.6	34.24	3.85	0.17	0.08	46.43*
Rubber	0.45	47.03	39.53	12.99	10001.6	25.22	2.45	0.14	0.19	24.52

*Cl is also contained in this item.

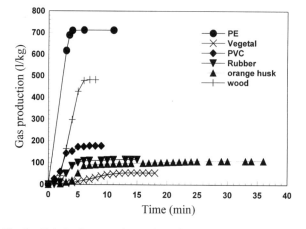

Fig. 2. Relation between the total gas flow and the pyrolysis time.

that of orange husk is 73.93%. Typical pyrolysis results are summarized in Table 2. The heating value of the produced gas ranges from 13 000 to 23 000 kJ/Nm³.

3.1. Temperature

Temperature is a very important factor in the pyrolysis of solid wastes. Although various heating rates were used in many laboratory studies, the heating rate was kept constant during each

Table 2
Pyrolysis product yields at 800°C

Materials	Gas output (Nm³/kg)	Tar and water (wt %)	Semicoke (wt %)
PVC plastic	0.191	0.436	0.359
PE plastic	0.720	0.457	0.141
Rubber	0.126	0.130	0.769
Paper	0.374	0.466	0.098
Paperboard	0.282	0.440	0.248
Wood-chip	0.376	0.379	0.186
Cotton cloth	0.480	0.303	0.159
Vegetal	0.051	0.888	0.045
Orange husk	0.176	0.749	0.031

test. However, in practice, the heating rates of materials are not constant but will vary with time. In our tests, two heating methods are used to simulate continuous and batch-feed operations. Figs. 3 and 4 show typical time–temperature curves. In Fig. 3, cotton cloth is put into the kiln when the temperatures of the kiln freeboard and wall are stable at 800°C. Both the kiln-freeboard and wall temperatures drop quickly and reach the lowest value at about 674°C before they start to increase. It first increases quickly and then more slowly. In Fig. 4, the wood chips are fed into the kiln when the kiln was still cool and the charge and kiln are heated up together, the rate of the temperature increase is now much slower than before, about 10 min are required to reach 600°C. In Figs. 3 and 4, the kiln wall temperature is higher than the freeboard temperature during the entire process. This is a characteristic of externally-heated kilns because heat is transferred from the outside to the inside.

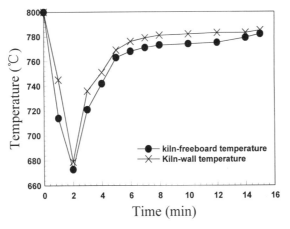

Fig. 3. Variation of the temperatures of the kiln-freeboard and internal wall during cotton-cloth pyrolysis with the rapid heating method.

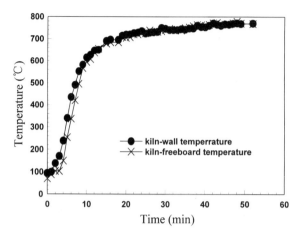

Fig. 4. Variation of the temperatures of the kiln-freeboard and internal wall during wood-chips pyrolysis for the slow heating method.

3.2. Effects of heating methods on the pyrolysis-gas yield

Wood and PVC plastics were pyrolysed with the different heating methods in order to study the effects of heating methods on gas yield and pyrolysis time. Figs. 5 and 6 show comparisons of the results. The quantities of gas obtained with the fast heating method are much greater than those obtained with slow heating. The quantities of the gas obtained in pyrolysis are mainly determined by the residence time of the materials at the high temperature.

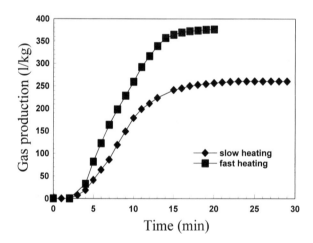

Fig. 5. The relation between total gas flow and pyrolysis time for wood-chips with different heating methods.

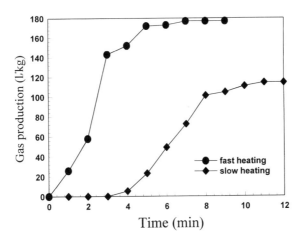

Fig. 6. The relation between total gas flow and pyrolysis time for PVC with different heating methods.

3.3. Gas composition and heating value

Fig. 7 shows that the fraction of H_2 in the pyrolysis gas increases with the temperature because larger molecules are easier to break into small molecules with increasing temperature [5]. As shown in Fig. 7, the large molecule contents of C_2H_4 and C_2H_6 decrease with increasing temperature. The CO content increases when the temperature is below 670°C, then decreases and becomes stabilized at 28–29%. The CH_4 content increases when the temperature is below 750°C and then decreases. The CO_2 content drops steeply when the temperature is below 565°C and then fluctuates

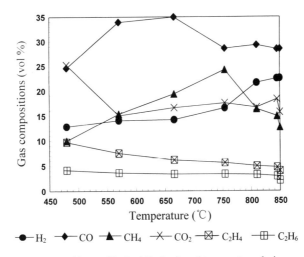

Fig. 7. Changes of gas compositions with the kiln-freeboard temperature during wood-chips pyrolysis.

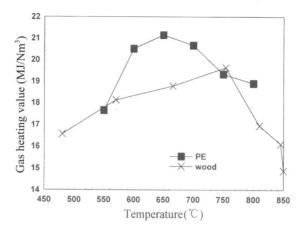

Fig. 8. Changes of the gas heating values with the kiln-freeboard temperatures during wood-chips and PE pyrolyses in the kiln.

from 14.8 to 18%. The change of the gas heating value with the kiln-freeboard temperature is shown in Fig. 8. The trends of heating values of the pyrolysis gases are similar for wood chips and PE plastics. There is a maximum value of 750°C for wood and 660°C for PE plastics.

3.4. The effects of moisture and size of raw material

Different sizes and moisture contents of wood-chips are described and their physical characteristics shown in Table 3.

Fig. 9 shows the effects of the size and moisture of the wood-chips on the total flow of pyrolysis gas. The moisture content has an obvious effect on the pyrolysis time. When the moisture content increases from 5.25 to 14.83%, the pyrolysis time increases from 6 to 12 min. Higher moisture content means that more heat is required for evapovation and thus more time is needed. Compared to the moisture, the size has little effect on the pyrolysis time. It is interesting that 50 mm wood-chips produce more gas per unit mass than 24 mm wood-chips during a specified reaction time. The size distributions for the semicokes produced from pyrolysis are shown in Fig. 10. The weight percentage of the semicoke, which size is smaller than 8 mm, is about 62% for 50 mm wood-chips and it is about 37% for 24 mm wood-chips. Thus, fragmentation for the wood-chips with

Table 3
Physical characteristics of wood-chips

Code	Diameter (mm)	Length (mm)	M_{ad} (%)
1	22	40	14.83
2	24	40	5.25
3	34	40	14.83
4	50	40	5.25

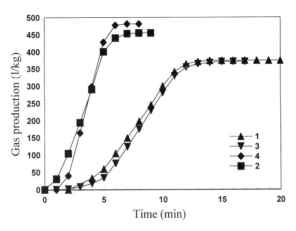

Fig. 9. Effects of size and moisture content of wood-chips on the total flow of pyrolysis gas. (Legend 1, 2, 3, 4 refer to Table 3.)

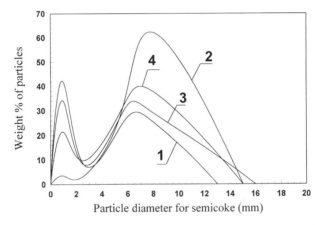

Fig. 10. Particle sizes distributions for semicoke formed from wood-chips. (Legend 1, 2, 3, 4 refer to Table 3.)

larger initial size is more violent than that with smaller initial size. Fragmentation aids rapid heating and promotes the gas yield. The conditions are different for 24 and 34 mm wood-chips, they produce nearly the same quantities of gas and required nearly the same pyrolysis time. It is probably because that their high moisture ($M_{ad} = 14.83\%$) content slows the rate of increase of temperature and alleviates the breakage.

4. Conclusion

Pyrolysis with fast heating method not only needs less reaction time but also produces more gas than that with slow heating method. The compositions and the heating values of the pyrolysis gas vary during the pyrolysis. With the increase of temperature, small molecule compositions increase and big molecule compositions decrease. The heating values first increases and decrease after reaching a maximum value. Compared to the size, the moisture has more effects on gas yields and pyrolysis time. The breakage of raw materials will take place during the pyrolysis. The lower moisture content and the larger original size will enhance the breakage.

References

[1] Penner SS, Wiesenhahn DF, Li CP. Ann Rev Energy 1987;12:415.
[2] Hoffman DA, Fitz RA. Environmental Science and Technology 1968;2:1023.
[3] Buekens AG, Schoeters JG. Conservation and Recycling 1986;9:253.
[4] Aroğuz AZ, Önsan ZI. Chimica Acta Turcica 1987;15:415.
[5] Levie B, Diebold JP, West R. In: Bridgwater AV, Kuester JL, editors. Research on thermochemical biomass conversion. London: Elsevier, 1988:312.
[6] Diebold J, Evans R, Scabill J. Energy from Biomass and Wastes 1990;13:851.
[7] Pershing DW, Lighty JS, Silcox GD, Heap MP, Owens WD. Combust Sci Tech 1993;93:245.
[8] Avenell CS, Griffiths AJ, Syred N. IEEE Clean Power 2000 Conference, London, 1993.
[9] Arenll CS, Sainz CL, Griffiths AJ. Fuel 1996;75:1167.
[10] Lighty JS, Eddings EG, Lingren ER, Xue DX, Pershing DW, Winter RM, Mcciennen WH. Combust Sci Tech 1990;74:31.
[11] Lester TW, Cundy VA, Montestruc AN, Leger CB, Acharya S, Sterling AM. Combust Sci Tech 1990;74:67.
[12] Lemieux PM, Linak WP, Mcsorley JA, Wendt JOL. Combust Sci Tech 1992;85:203.
[13] Cook CA, Cundy VA, Sterling AM, Lu C, Montestruc AN, Leger CB, Jakway AL. Combust Sci Tech 1992;85:217.
[14] Kroll PJ, Chary RC. Environmental Progress 1991;10:45.
[15] Lewis CR, Edwards RE, Santoro MA. Chemical Engineering 1976;83(18):115.
[16] Waterland LR, Fournier DJ, Lee JW, Carroll GJ. Waste Management 1991;11:103.
[17] Ricci LJ. Chemical Engineering 1976;83(16):52.
[18] Schaefer WD. Environmental Science and Technology 1975;19:98.
[19] Heek KH, Strobel BO, Wangle W. Fuel 1994;73:1135.

垃圾在实验室规模回转内热解的试验研究

摘要：设计并建造了一台实验室规模的外加热回转窑热解炉。在这个炉子上进行了热解试验。固体废物(纸、纸板、包括 PVC 和 PE 的废塑料、橡胶、植物材料、木材和橘皮)进行了测试。加热方式、水分含量和废物粒径大小对热解气产率、组分及热值的影响进行了评价。

原文刊于 Energy, 1999, 24: 209-218

循环流化床锅炉炉膛热力计算

程乐鸣，岑可法，倪明江，骆仲泱

（浙江大学热能工程研究所，能源清洁利用与环境工程教育部重点实验室，浙江杭州 310027）

THERMAL CALCULATION OF A CIRCULATING FLUIDIZED BED BOILER FURNACE
CHENG Le-ming, CEN Ke-fa, NI Ming-jiang, LUO Zhong-yang
(Institute for Thermal Power Engineering, Clean Energy and Environment Engineering Key Lab of Ministry Education, Zhejiang University, Hangzhou 310027, China)

ABSTRACT: Different types of circulating fluidized bed (CFB) boilers have been designed and put into industrial operation by different designers during their rapid development. In a design every designer has his own thermal calculation method. However, less information is available in published papers. A simple thermal calculation method for designing a CFB boiler was introduced based on most recent information worldwide, especially research and industrial experience achievements at Zhejiang University on the CFB boiler technologies in recent decade. Determination of the furnace size, heat balance and heat transfer in a furnace is reported. A CFB boiler with a high temperature cyclone is applied as an example and other types of CFB boiler could be designed with modifying the thermal calculation method.

KEY WORDS: circulating fluidized bed boiler; boiler deign; thermal calculation method

摘要： 结合作者在循环流化床锅炉传热和设计理论研究及实践的基础上，提出一种循环流化床锅炉炉膛的热力计算方法，包括循环流化床锅炉炉膛的几何尺寸确定、炉膛热量平衡和炉膛传热计算。考虑循环流化床锅炉炉型不同，其热力计算方法有所不同，该方法针对采用高温分离装置的循环流化床锅炉，提出的计算方法可用于一般高温分离的循环流化床锅炉的设计计算，其余炉型可在此基础上根据具体炉型特点修改使用。

关键词： 循环流化床锅炉；锅炉设计；热力计算

中图分类号： TK212　　**文献标识码：** A

1 引言

循环流化床锅炉燃烧效率高，污染排放低，燃料适应性广，被广泛应用于蒸汽生产中。随着循环流化床锅炉的发展，其容量和规模都在增大。目前美国在建的 300 MWe 循环流化床锅炉即将投入运行，600 MWe 容量的循环流化床锅炉也已在设计中。利用国内技术生产的 35 t/h、75 t/h 循环流化床锅炉有大量运行，目前国内投入运行的最大循环流化床锅炉是高温高压 420 t/h 容量的锅炉，高温高压 450 t/h 循环流化床锅炉也已在建，但运用的是国外技术。

在循环流化床锅炉的开发与发展过程中，各设计单位和锅炉制造厂家开发出各种炉型，针对各自不同的炉型采用各自的热力计算方法，即使是相同的炉型设计方法也可能不同，各有特点。这与煤粉锅炉和鼓泡流化床锅炉在设计过程中有统一的热力计算方法[1]可供参考不同。有关循环流化床锅炉热力计算方法在文献中也少见发表。本文结合作者在循环流化床锅炉传热和设计理论研究及实践的基础上，建立了一种简单的循环流化床锅炉炉膛热力计算方法[2~9]。

与一般沸腾燃烧鼓泡流化床锅炉不同，循环流化床锅炉类型较多，炉型不同，其热力计算方法有所不同。本方法针对采用高温分离装置的循环流化床锅炉，提出的计算方法可用于一般高温分离的循环流化床锅炉的设计计算，其余炉型可在此基础上根据具体炉型特点修改使用。典型的高温分离器型循环流化床锅炉采用高温立式旋风分离器，安置在锅炉炉膛上部烟气出口处。离开炉膛的大部分颗粒，由高温分离器所捕集并通过固体物料再循环系统从靠近炉膛底部的物料回送口送回炉膛。经高温分离器分离后的高温烟气则进入尾部烟道，与布置在尾部烟道中的受热面进行换热后排出。计算中未

基金项目：教育部留学回国人员科研启动基金项目。

考虑添加石灰石的影响,若添加石灰石,则入炉热量、灰浓度和烟气量等有变化,需修正。

2 循环流化床锅炉炉膛几何尺寸的确定

2.1 炉膛横截面积

循环流化床锅炉炉膛一般由膜式水冷壁构成,其传热面积以通过水冷壁管中心面的面积计算。若炉膛由轻型炉墙或敷管炉墙构成,则需考虑角系数的影响。

炉膛尺寸的确定主要包括炉膛密相区和稀相区的长、宽、高以及是否有截面收缩等确定。

炉膛横截面积的确定取决于床层运行风速或截面热负荷的选取。密相区的运行风速类似于鼓泡流化床。一般循环流化床锅炉稀相区运行风速在3～7 m/s,考虑磨损的危险性和为降低风机能耗,可选取运行风速在4～6 m/s左右。运行风速数值与燃料种类也很有关系。截面热负荷的选择与运行风速的选择是相关的,实际上只要燃料和过剩氧量确定,运行风速与截面热负荷中只要一个参数确定后,另一个参数也随之确定。截面热负荷一般可选择在3～4 MW/m² 左右[2]。

2.2 炉膛深度

炉膛横截面积确定后,根据炉膛长宽比确定炉膛的长宽。炉膛的深度一般不超过 8 m,以保证二次风的穿透,长宽比以 1:1 至 2:1 都是合适的。具体在确定炉膛的长、宽时,一般还应考虑尾部受热面的布置,使之相适应[2]。

2.3 炉膛密相区高度

若循环流化床锅炉有二次风,则其密相区与稀相区的分界面取二次风入口高度平面。对于没有二次风或三次风的情况,或负荷变化较大时,若 H_0 为静床料高,则其密相区高度 H_{den} 可通过计算膨胀比 R_{den} 得到[10]

$$H_{den} = R_{den} H_0 \qquad (1)$$

R_{den} 可用下式计算:

当 $\dfrac{U_g}{U_t} > 0.267(d_p \rho_p)^{-0.6}$ 时,

$$R_{den} = 6.01(d_p \rho_p)^{0.3}\left(\dfrac{U_g}{U_t}\right) \qquad (2)$$

当 $\dfrac{U_g}{U_t} < 0.267(d_p \rho_p)^{-0.6}$ 时,

$$R_{den} = 3.1(U_g / U_t)^{0.5} \qquad (3)$$

式中 U_g 为床层运行风速;U_t 为颗粒终端速度;d_p 为颗粒平均粒径;ρ_p 为颗粒密度。

2.4 炉膛高度

循环流化床锅炉炉膛高度是循环流化床设计的一个关键参数。炉膛越高,则锅炉钢架就越高,因而锅炉的造价也会提高。因此,在满足锅炉和炉膛的下述要求下,尽可能地降低炉膛高度。一般地,炉膛高度应满足以下条件:

(1) 保证分离器不能捕集的细粉在炉膛内一次通过时能够燃尽;

(2) 炉膛高度应容纳炉膛能布置全部或大部分蒸发受热面;

(3) 炉膛高度应保证返料机构料腿一侧有足够的静压头,从而使循环流化床锅炉有足够的循环物料在循环回路中流动;

(4) 炉膛高度应保证脱硫所需最短气体停留时间;

(5) 炉膛高度应和循环流化床锅炉的尾部烟道或对流段所需高度相一致;

(6) 炉膛高度应保证锅炉在设计压力下有足够的自然循环。

具体设计时,一般可根据常规循环流化床锅炉的炉膛高度确定一个数值,布置受热面是否足够,然后考虑分离器的切割直径,再根据上述(1)的要求考虑固体颗粒的燃尽和其他要求条件,使之满足上述要求,若条件容许偏高些为好[2]。

3 循环流化床锅炉循环倍率 n

循环流化床锅炉循环倍率是循环物料重量与计算给煤重量的比值,其值的选取比较经验,可参考表1。

表 1 锅炉循环倍率
Tab.1 Circulation ratios of boiler

项目	较高脱硫效率	不考虑脱硫	劣质燃料
循环倍率 n	20～50	10～20	1～10

4 密相区和稀相区的燃烧份额 δ

密相区和稀相区的燃烧份额受燃料粒径、煤种、流化风速、一二次风率、床层温度等诸多因素影响,尤其是煤种的影响较大,如挥发份高易爆的煤在密相区的燃烧份额会降低。在目前缺乏数据的情况下,设计时可以参考有关不同煤种的燃烧特性试验数据取值[2]。一般地,固体颗粒粒径越大,燃烧份额相对增加。如果采用宽筛分燃料,可以采用鼓泡流化床计算标准中推荐的方法并考虑一次风率的影响而求取。

5 炉底排渣量与飞灰量比

炉底排渣量和飞灰量之比受许多因素影响,其中随煤的特性、床内物料粒径、和运行速度的变化较大,其取值相当经验,一般可在 0.2~1。

6 焓温表中炉膛内飞灰焓 I_{fh} 的计算

在计算焓温表炉膛内的飞灰焓时,对于分离器前部分需考虑循环固体颗粒的影响,其飞灰焓 I_{fh} 以下式计算:

$$I_{fh} = (c\vartheta)_h \left[\frac{A_{ar} a_{fh}}{100} \times \frac{100}{100 - C_{fh}} \times \frac{100}{100 - q_4} + n \right] \quad (4)$$

式中 c_h 和 ϑ_h 分别为灰的比热和温度;A_{ar} 为煤收到基灰分;a_{fh} 为飞灰份额;C_{fh} 为炉膛出口飞灰含碳量;q_4 为机械不完全燃烧损失。计算密相区的飞灰焓时,上式中的 C_{fh} 和 q_4 应代以密相区出口飞灰含碳量 C^*_{fh} 和密相区机械不完全燃烧损失 q_{4ft}。

7 密相区和稀相区热量平衡

密相区的入炉热量 Q_l:

$$Q_l = Q_r \frac{\delta(100 - q_3 - q_4) - q_{6lh}}{100 - q_4} + x\alpha_{ft} I_{lk}^0 + n I_{flh} \quad (5)$$

式中 Q_r 为锅炉输入热量;q_3 为化学不完全燃烧损失;q_{6lh} 为炉底排渣损失;x 为一次风率;α_{ft} 为密相区出口处的名义空气过剩系数;I_{lk}^0 为理论冷空气焓,I_{flh} 为分离灰焓。

埋管受热面吸热量 Q_m:

$$Q_m = \varphi B_j \left(Q_l - \frac{100 - q_{4ft}}{100 - q_4} I_{ft}'' \right) \quad (6)$$

式中 φ 为保热系数;B_j 为计算给煤量;I_{ft}'' 为密相区烟气焓。

带入稀相区的热量 Q_{xx}':

$$Q_{xx}' = \frac{100 - q_{4ft}}{100 - q_4} I_{ft}'' + \frac{Q_r}{100 - q_4}(1 - \delta)(100 - q_3 - q_4) + (1-x)\alpha_{ft} I_{lk}^0 + \Delta\alpha_{xx} I_{lk}^0 \quad (7)$$

其中 $\Delta\alpha_{xx}$ 为稀相区漏风系数。

稀相区吸热量 Q_{xx}:

$$Q_{xx} = \varphi B_j (Q_{xx}' - I_{xx}'') \quad (8)$$

式中 I_x'' 为烟气在稀相区的出口焓。

8 循环流化床锅炉炉膛传热计算

8.1 炉膛下部密相区的传热计算

循环流化床锅炉炉膛密相区的流体动力特性属鼓泡流态化,和鼓泡床密相区相似,若循环流化床密相区中布置有埋管受热面,其传热计算可直接参照鼓泡流化床中计算传热系数的方法进行方法。由于一般地循环流化床锅炉床内运行风速比鼓泡流化床高,计算时有关床层空隙率数值的选取应根据情况适当增大[1]。

8.2 炉膛上部稀相区的传热计算

8.2.1 壁面平均传热系数 h

循环流化快速床中,包含含分散固体颗粒(固体颗粒分散相)的连续上升气流和相对密的颗粒团两部分。根据循环流化床的流体动力特性,可以将稀相区横截面分为中心核心区和壁面环形区两部分。在核心区,颗粒在其中由下向上运动,固体颗粒浓度较小;在床体壁面为密相环形区中,固体颗粒汇集成各种不同的密相结构(颗粒团),颗粒团与固体颗粒分散相在其中交替地与床壁面接触,沿传热壁面下滑、离散(图1)。

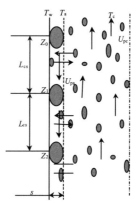

注:T_w—壁面温度;T_a—环形区温度;T_c—核心区温度;s—密相区环形厚度;U_{pa}—壁面环形区内的颗粒速度;U_{pc}—核心区内的颗粒速度

图1 连续上升的固体颗粒分散相和沿壁面下滑的颗粒团

Fig. 1 Illustration showing dispersed phase moving upwards and clusters moving downwards

假定 δ_{cs} 是被颗粒团覆盖的壁面面积的平均百分率,用 h_{conv} 表示对流传热系数,h_r 表示辐射传热系数,则壁面的平均传热系数可表示为 h_{conv} 与 h_r 之和(对于壁面来说忽略气相传热):

$$h = h_{conv} + h_r = \delta_{cs}(h_{cs} + h_{csr}) + (1 - \delta_{cs})(h_d + h_{dr}) \quad (9)$$

式中 h_{cs} 和 h_d 分别为颗粒团与固体颗粒分散相的对流传热系数,h_{csr} 和 h_{dr} 则分别表示颗粒团与固体颗粒分散相的辐射传热系数。

8.2.2 颗粒团覆盖壁面的时均覆盖率 δ_{cs}

在任何时刻,循环流化床锅炉壁面的一部分被颗粒团所覆盖,其余部分则暴露在固体颗粒分散相中(图1)。颗粒团覆盖壁面的时均覆盖率 δ_{cs} 可由下式计算:

$$\delta_{cs} = K\left(\frac{1-\varepsilon_w - Y}{1-\varepsilon_{cs}}\right)^{0.5} \quad (10)$$

参数 K 的取值范围 Basu[11]建议取为0.5。程乐鸣等[7]提出对于循环流化床密、稀相区 K 值取不同数值。对于稀相区,推荐 $K=0.1$;对于密相区,推荐 $K=0.25$。此值仍需考察。壁面空隙率 $\varepsilon_w = \varepsilon^{3.811}$,$\varepsilon$ 为稀相区空隙率,$\varepsilon = 1 - \frac{\rho_{xx}}{\rho_p}$,$\rho_{xx}$ 是稀相区固体颗粒浓度。ε_{cs} 为颗粒团中的空隙率,可取值为临界流态化下的空隙率值;Y 为固体颗粒相中固体颗粒的百分比,可取 $Y=1-\varepsilon$。

8.2.3 循环流化床对流传热系数 h_{conv}

对流传热包括颗粒团与颗粒分散相的对流传热两部分,根据(9),对流传热系数 h_{conv} 以下式表示,

$$h_{conv} = \delta_{cs} h_{cs} + (1-\delta_{cs}) h_d \quad (11)$$

(1) 颗粒团与壁面间对流传热系数 h_{cs}

颗粒团沿着壁面下滑,在与壁面接触一段时间后,颗粒团或者破裂消失或者运动到别处。颗粒团与壁面接触时,其初始温度为床温,这样,颗粒团与壁面间产生非稳态传热。快速床中颗粒团与壁面间的传热热阻主要有两部分,一是与壁面的接触热阻,二是颗粒团本身的平均热阻(图2)。

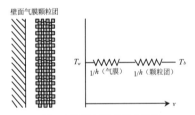

图2 壁面与颗粒团间的传热
Fig. 2 Heat transfer between clusters and a water wall

假定传热只在水平方向由壁面传入颗粒团,而忽略竖直方向的任何传热量,则壁面与颗粒团间的传热系数可用下式表示

$$h_{cs} = \frac{1}{1/h_w + 1/h_e} \quad (12)$$

式中 h_w 为壁面接触传热系数;h_e 为常温壁面向均匀半无限介质的不稳态导热过程中的有效传热系数。

对于锅炉内的情况,连续传热面较长,颗粒团的贴壁时间就会长些。这时与接触热阻相比,颗粒团中的非稳态导热阻较为重要,这就减弱了固体颗粒径对传热系数的影响,对于颗粒团贴壁时间较短的情况,传热限于颗粒群的贴壁层。

1) 常温壁面向均匀半无限介质不稳态导热过程中的有效传热系数 h_e

若颗粒团与传热壁面的接触时间为 t_{cs},则其平均传热系数为

$$h_e = \sqrt{\frac{4K_{cs}\rho_{cs}C_{cs}}{\delta t_{cs}}} \quad (13)$$

式中 K_{cs} 为气-固颗粒团的有效导热系数。即 Mickley 和 Fairbanks[12]根据颗粒团理论推导所得。

上式中,颗粒团与壁面的导热情况取决于其在壁面的停留时间 t_{cs}。贴壁的颗粒团在重力作用下加速下滑,同时受到壁面的阻力与向上气流的曳引力作用。图1中,在壁面传热面的上部 Z_0 位置,形成一空隙率为 ε_{cs},温度为 T_a^0(假定与床温相同)的颗粒团,该颗粒团与壁面接触,沿壁面以 U_{pa} 的速度下滑至 Z_1 位置,在壁面上的特征停留长度为 L_{cs},这样,颗粒团的每一部分在壁面上的停留时间 t_{cs} 就可以如下式计算:

$$t_{cs} = \frac{L_{cs}}{U_{pa}} \quad (14)$$

式中 L_{cs} 根据 Wu 等[13]的试验求出,$L_{cs} = 0.0178\rho_{susp}^{0.596}$,其中 ρ_{susp} 取边壁区固体颗粒浓度,$\rho_{usp} = (1-\varepsilon_w)\rho_p$。$U_{pa}$ 是固体颗粒贴壁下滑速度,可取值为 1.2～2.0 m/s。

气-固颗粒团的有效导热系数 K_{cs},推荐采用 Xavier 和 Davidson[14]提出的下式来计算。

$$K_{cs} = K_g\left(\frac{K_p}{K_g}\right)^{[0.28-0.757\log10\varepsilon_{cs}-0.057\log10(K_p/K_g)]} +$$
$$0.1\rho_g C_g d_p U_{mf} \quad (15)$$

式中 K_p 和 K_g 分别为固体颗粒和气体的导热系数;ρ_g 为气体密度;C_g 是气体比热;U_{mf} 为临界流化速度。

在公式适用范围内,该式与采用其它方法计算的颗粒团的有效导热系数基本一致,而采用该式的优点是该式还考虑了空气密度的影响。

颗粒团密度 $\rho_{cs} = (1-\varepsilon_{cs})\rho_p + \varepsilon_{cs}\rho_g$,
颗粒团比热 $C_{cs} = (1-\varepsilon_{cs})C_p + \varepsilon_{cs}C_g$。

2) 颗粒团与壁面间气膜传热系数 h_w

关于颗粒团与壁面传热系数 h_w,可根据颗粒团

与壁面接触间的相应气体薄层厚度的热阻计算，颗粒团与壁面传热系数可用气体间隙厚度来计算：

$$\frac{1}{h_w} = \frac{d_p/n}{K_g} \quad (16)$$

根据试验，在模型中选取参数n=2.5。

（2）固体颗粒分散相的传热系数h_d

循环流化床锅炉的壁面并不总是与颗粒团接触的。在与两颗粒团接触之间，壁面与床中的上升气流接触，在上升气流中含有分散相的固体颗粒。对流传热系数项中的固体颗粒分散相的传热系数h_d的计算，选用Wen和Miller[15]基于稀相气–固混合物而导出的传热系数计算公式近似计算：

$$h_d = \frac{K_g}{d_p} \frac{C_p}{C_g} \left(\frac{\rho_{dis}}{\rho_p}\right)^{0.3} \left(\frac{U_t}{gd_p}\right)^{0.21} Pr \quad (17)$$

式中 C_p为固体颗粒比热；ρ_{dis}为固体颗粒分散相的密度，其值可经由$Y\rho_p+(1-Y)\rho_g$计算，U_t为固体颗粒的终端速度。

8.2.4 循环流化床辐射传热系数h_r

辐射传热是循环流化床锅炉中传热的一种重要方式，尤其是在高温($>700\ ℃$)和低床密度($<30\ kg/m^3$)的情况下。循环流化床锅炉中的辐射传热包括两部分，一部分主要来自与壁面接触的颗粒团的辐射，另一部分是固体颗粒分散相壁面的辐射。床层向壁面的总辐射系数根据式(9)：

$$h_r = \delta_{cs}h_{csr} + (1-\delta_{cs})h_{dr} \quad (18)$$

式中 h_{cr}为来自与壁面接触的颗粒团的辐射；h_{dr}为固体颗粒分散相向壁面的辐射。

（1）固体颗粒分散相对壁面的辐射传热系数h_{dr}

对于大型循环流化床锅炉，床吸收率e_d可根据下式计算[16]：

$$e_d = \left[\frac{e_p}{(1-e_p)B}\left(\frac{e_p}{(1-e_p)B}+2\right)\right]^{0.5} - \frac{e_p}{(1-e_p)B} \quad (19)$$

式中 e_p为颗粒表面的吸收率。对各相同性漫反射B＝0.5，对漫反射颗粒B=0.667。

固体颗粒分散相的辐射传热系数h_{dr}可以根据下式计算：

$$h_{dr} = \frac{\sigma(T_b^4 - T_s^4)}{(1/e_d + 1/e_s - 1)(T_b - T_s)} \quad (20)$$

式中 e_s为传热表面的吸收率；σ为斯蒂芬–波尔兹曼常数；T_b是稀相区床温；T_s为表面温度。

（2）颗粒团对壁面的辐射系数h_{csr}

颗粒团的辐射系数h_{csr}，可将式(20)中的e_d换成e_{cs}同样进行计算。颗粒团的吸收率e_{cs}可由下式计算：

$$e_{cs} = 0.5(1+e_p) \quad (21)$$

9 结论

本文结合作者在循环流化床锅炉传热和设计理论研究及实践的基础上，针对采用高温分离装置的循环流化床锅炉，提出一种简单的循环流化床锅炉炉膛热力计算方法，可用于一般高温分离的循环流化床锅炉的设计计算，其余炉型可在此基础上根据具体炉型特点修改使用。

参考文献

[1] 工业锅炉技术手册·层状燃烧和沸腾燃烧工业锅炉热力计算方法编写组 (Editorial group of technical handbook of industrial boilers · Thermal calculation method for grate-firing and bubbling fluidized bed industrial boilers). 《工业锅炉技术手册（1）·层状燃烧和沸腾燃烧工业锅炉热力计算方法（报批稿）》(Technical Handbook of Industrial Boilers (1) · Thermal Calculation Method for Grate-firing and Bubbling Fluidized Bed Industrial Boilers) [S]. 1981.

[2] 岑可法, 倪明江, 骆仲泱, 等 (Cen Kefa, Ni Mingjiang, Luo Zhongyang, et al)，《循环流化床锅炉理论设计与运行》(Circulating fluidized bed boilers - Theoretical design and operations) [M]. 北京：中国电力出版社 (Beijing: China Electric Power Press), 1998.

[3] 岑可法, 倪明江, 骆仲泱, 等译 (translated by Cen Kefa, Ni Mingjiang, Luo Zhongyang, et al.)，《循环流化床锅炉的设计与运行》(Circulating fluidized bed boilers – design and operations) [M]. Prabir Basu, Scott A. Fraser, 北京：科学出版社 (Beijing:Science Press), 1994.

[4] 程乐鸣, 骆仲泱, 倪明江, 等 (Cheng Leming, Luo Zhongyang, Ni Mingjiang, et al). 循环流化床中传热综述（试验部分）(A summary of the circulating fluidized bed heat transfer (testing part)) [J]. 动力工程 (Power Engineering), 1998, 18(2): 20-34。

[5] 程乐鸣, 骆仲泱, 倪明江, 等 (Cheng Leming, Luo Zhongyang, Ni Mingjiang et al). 循环流化床中传热综述（数学模型）(A summary of the circulating fluidized bed heat transfer (mathematical model part)) [J]. 动力工程(Power Engineering), 1998, 18(2): 48-53。

[6] Cheng Leming, Cen Kefa, Ni Mingjiang, et al. Heat transfer in circulating fluidized bed and its modeling [A]. Proc. of 13th Inter. Conf. on FBC, ed. by K. J. Heinschel[C], ASME Press, ISBN No.0-7918-1305-3, 1995：487.

[7] 程乐鸣, 骆仲泱, 李绚天, 等 (Cheng Leming, Luo Zhongyang, Li Xuantian, et al). 循环流化床膜式壁传热试验与模型 (Membrane wall heat transfer in a circulating fluidized bed and its modeling) [J]. 工程热物理学报 (Journal of Engineering Thermophysics), 1998, 19(4): 514-518.

[8] 程乐鸣 (Cheng Leming). 大型循环流化床锅炉传热 (Heat transfer in a commercial circulating fluidized bed boiler) [J]. 动力工程 (Power Engineering), 2000, 20(2): 587-591.

[9] 程乐鸣 (Cheng Leming). 循环流化床与压力循环流化床传热研究 (Heat transfer in a circulating fluidized bed and a pressurized circulating

fluidized bed) [D]. 杭州：浙江大学 (Hangzhou: Zhejiang University), 1996.

[10] 清华大学电力学系锅炉教研室 (Boiler Section, Electricity Department, Tsinghua University),《沸腾燃烧锅炉》(Fluidized combustion boiler) [M]. 北京：科学出版社 (Beijing：Science Press), 1972.

[11] Basu P. Heat transfer in high temperature fast fluidized beds [J]. Chem. Eng. Sci., 1990, 45(10)：3123-3136.

[12] Mickley H S, Faiebanks D F. Mechanisms of heat transfer to fluidized beds [J]. AIChE J.,1955,1(3)：374-384.

[13] Wu R L, Grace J R, Lim C J. A model for heat transfer in circulating fluidized beds [J]. Chem. Eng. Sci., 1990, 45(12)：3389-3398。

[14] Xavier A M, Davidson J E Heat transfer in fluidized beds：convective heat transfer in fluidized beds [M]. Fluidization 2nd Edition, Ed. by J. F. Davidson, R. Clift, D. Harrison, Academic Press London, ISBN 0-12-20552-7, 1985：443-450.

[15] Wen C Y, Miller E N. Heat transfer in solid-gas transport lines [J]. Ind.Eng,Chem.,1961,53：51-53.

[16] Brewster M Q. Effective absorptivity and emissivity of particulate slender with application to a fluidized beds [A]. in Circulating Fluidized Bed Technology IV[C], A. Avidan, ed., AIChE, New York, 1986：137-144.

收稿日期：2002-02-04。

作者简介：

程乐鸣（1965-），男，博士，教授，主要从事低污染燃烧与多相流传热理论与技术研究；

倪明江(1935-)，男，博士，教授，中国工程院院士。主要从事低污染燃烧技术，垃圾焚烧技术与生物质气化技术研究；

骆仲泱(1962-)，博士，教授，主要从事低污染燃烧技术与生物质气化技术研究。

（责任编辑　贾瑞君）

原文刊于中国电机工程学报，2002, 22(12)：147-152

Experimental research on solid circulation in a twin fluidized bed system

M. Fang*, C. Yu, Z. Shi, Q. Wang, Z. Luo, K. Cen

Clean Energy and Environment Engineering Key Laboratory of MOE, Institute for Thermal Power Engineering, Zhejiang University, Hangzhou 310027, China

Received 16 July 2002; accepted 27 January 2003

Abstract

A twin fluidized bed solid circulation system, in which two adjacent fluidized beds exchange solids, was developed for coal gasification to produce middle heating value gas. The effects of bed material, operation velocity, and bed structure on the solid circulating rate were tested at a small-scale test facility. Experimental results showed that the solid circulation rate of the system could be adjusted by changing gas velocities of two beds and could attain 30–40 times of the fuel feed rate, which would meet the demands of heat supply to an endothermic process of a gasifier. On the basis of experiments, reasonable operation and design parameters were put forward, which can be used to as a reference for the commercial gasifier design. A mathematical model was erected to calculate the solid circulation rate of the system, and it could predict well the solid flow rate through a horizontal orifice by comparison with experimental data.
© 2003 Elsevier Science B.V. All rights reserved.

Keywords: Coal gasification; Twin fluidized beds; Solid circulation; Mathematical model

1. Background

Multi-fluidized bed solid circulation systems are widely used in the process of catalyst regeneration, coal gasification, coking, thermal cracking, drying and incineration of waste [1–11]. The two main types of solid circulation systems are the external circulation system and the internal circulation system. In the external circulation system, two fluidized beds are collected by two circulation pipes (riser and downcomer), and particles flow in pipes, controlled by means of valving and aeration rates [1]. In the internal circulation system, one vessel is divided into several chambers or compartments by internal walls. There are orifices on the walls and solid can flow through orifices between different chambers. Compared with the external circulation system, this type of the system has a simple structure, low energy consumption and low investment. Rudolph and Judd [2] put forward a vessel with a draft tube for coal gasification process. Snieders et al. [3] made a research on a four-compartment interconnected fluidized bed system, and Chong et al. [4] introduced an adjacent fluidized bed with no gas mixing for char production process.

Presently, most of the small-scale industry gasifiers operate with air/steam aeration and can only produce low heating value gas. The gasifier which produces the middle heating value gas usually operates at high temperature and high pressure with oxygen aeration such as Texaco, U-gas gasification process, and need very high investment and are difficult to be accepted by small-scale enterprises. A new type of gasifier which produced middle heating value gas with air and steam blown, was developed by Zhejiang University, by means of a twin fluidized bed system.

A sketch of the twin fluidized bed system is shown in Fig. 1. The reactor consists of two adjacent fluidized beds, divided by a vertical wall with two orifices at a certain distance. The two fluidized beds operate at different gas velocities and form so-called the fast bed (fluidized vigorously) and the slow bed (aerated slowly). Fig. 2 shows a typical pressure distribution along the bed height of two beds. Due to differences in height and void of two dense beds, pressure gradients are established at the lower and upper orifices. Driven by the pressure gradient, solid particles in the slow bed flow into the fast bed through the lower orifice, moving upward in the fast bed and recycling back into the slow bed through the upper orifice. In this way, solid circulation between the two fluidized beds is formed.

In the process of coal gasification, the slow bed acts as a gasifier and the fast bed acts as a combustor. The gasifier

---

* Corresponding author. Fax: +86-571-87951616.
E-mail address: mxfang@cmee.zju.edu.cn (M. Fang).

Nomenclature

A_o	cross section area of orifice (m^2)
b	orifice width (m)
B	fuel feed rate (kg/s)
C_0, C_1	constants in Eq. (12)
C_D	discharge coefficient
d_b	bubble diameter (m)
d_p	mean particle diameter (mm)
f_b	bubble frequency
g	gravity constant
Gs	solid circulation rate (kg/s)
Δh_j	height of the j zone (m)
H	fluidized bed height (m)
ΔH	distance between two orifices (m)
H_1	bed height at the top of the lower orifice (m)
ΔH_1	height of the lower orifice (m)
$\Delta H_1'$	particle flow height at the lower orifice (m)
ΔH_2	height of the upper orifice (m)
m	constant in Eq. (11)
n_1, n_2	constants in Eq. (12)
Δp	pressure gradient at the orifice (Pa)
Δp_1	pressure gradient at the top of the lower orifice (Pa)
R	solid circulation ratio
u	particle flow velocity (m/s)
u_0	particle flow velocity at the top of the orifice (m/s)
u_{sj}	particle velocity at j zone of the lower orifice (m/s)
U	fluidized gas velocity (m/s)
y	height at a orifice (m)

Greek letters

δ_b	bubble fraction
ε	void fraction
ρ_p	particle density (kg/m^3)
ϕ	shape factor

Subscripts

f	fluidized condition
h	fast bed
l	slow bed
mf	minimum fluidized condition

operates at the temperature of 800–850 °C with steam aeration, while the combustor operates at 900–950 °C with air aeration. Coal is first fed into the gasifier, heated and pyrolysed, and char reacts with steam. The absorption heat of gasification process is provided by high temperature circulating solids from the combustor. The semi-coke from the gasifier recycles into the combustor with circulating solids and burnt out there. The heat produced in the combustor is used for heating-up the circulating solids and steam generation. The raw gas from the gasifier is cooled and purified to be a clean gas in which the main composition is H_2, CO and CH_4 and used for middle heating value gas supply.

The key technology of the scheme is how to maintain large enough and stable solid circulation rate between the two beds to supply enough heat for endothermic process of the gasifier. A series of experiments on the solid circulation rate was carried out at a small-scale test facility. This paper introduced main experimental results.

2. Experiments

A two-dimensional test facility of the twin fluidized bed system is shown in Fig. 3. The reactor was 2 m high and the section of the slow bed was 280 mm × 40 mm while the fast bed was 280 mm × 40 mm. There were two orifices allocated on an internal divided wall at a certain distance with the width of 40 mm. The height and distance of the two orifices can be adjusted by changing the vertical wall. The front and back walls of the reactor used transparent glass, which was used to observe and measure particle flow through the orifice. Two other sides used steel plates, on which 20 small pressure pipes with jam-prevention air aeration devices were allocated along the bed height for pressure measurement. Each of two beds had its own wind-box and air supply.

There are many methods to measure the particles flow rate, such as the light fiber velocity detector [5], radiotracer residence time method [6], heat response measurement [7], and the high-speed video recorder [8]. Here we used a high-speed digital video recorder to record particle flow in the lower orifice through the transparent glass wall of the reactor. Combined with a computer correlation analysis method, particle flow velocity could be determined by replaying video tapes slowly. The experiments showed that particle velocities varied along the height of the orifice and fluctuated with time (see Fig. 8). So the orifice was divided into several zones along the height. Particle velocities at different times were measured and time mean velocities were obtained. The total solid flow rate can be calculated according to the following equation:

$$\text{Gs} = \rho_p (1 - \varepsilon_{mf}) b \sum_{j=1}^{k} u_{sj} \Delta h_j \quad (1)$$

where ρ_p is the particle density (kg/m^3), ε_{mf} the void fraction at the minimum fluidized velocity, b the orifice width (m), u_{sj} the particle time mean velocity at the j zone (m/s), and Δh_j the height of the j zone (m).

The bed materials experiments used were plastic balls with narrow size distribution and industry fluidized bed ash with broad size distribution. The properties were shown in Table 1.

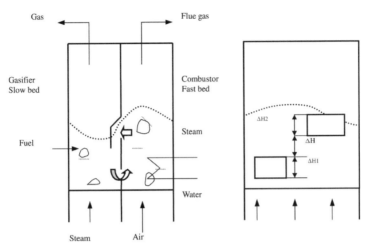

Fig. 1. Scheme of the dual fluidized bed system.

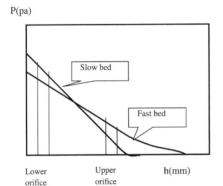

Fig. 2. Pressure distribution of the two beds.

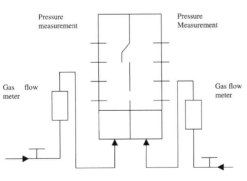

Fig. 3. A small-scale test facility.

Table 1
The properties of bed materials

Bed material	Size range (mm)	ϕ	d_p (mm)	ρ_p (kg/m^3)	U_{mf} (m/s)	ε_{mf}
Plastic balls	3–5	1	4	997	1.05	0.571
Fluidized bed ash	0–8	0.7	2.26	2433	0.71	0.45

3. Results and discussion

The main factors, which affect the solid circulation rate in the twin fluidized bed system, are bed material properties, fluidization velocities of the fast bed and the slow bed U_h and U_l, the size and distance of the orifices.

3.1. Effect of the bed material

Bed material properties such as particle density, diameter, and fluidization characteristics have an obvious effect on solid circulation. Fig. 4 shows the solid flow rate variation with the gas velocity by using plastic balls and fluidized bed ash as a bed material separately. The circulation rate of fluidized bed ash was much higher than that of plastic balls at the same gas velocity and bed structure, because fluidized bed ash has a higher density and thus a higher pressure gradient to drive particle flow at the orifice than plastic balls.

3.2. Effect of gas velocities

The gas velocities of the two beds have a great effect on the solid circulation rate. Experiments showed that the beds had some dead zones or obvious segregation when gas velocity U/U_{mf} was less than 1.2 for plastic balls and

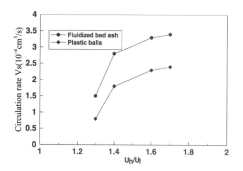

Fig. 4. Effect of the bed material on the solid circulation rate (H_{mf} = 390 mm, ΔH = 170 mm, ΔH_1 = 100 mm, ΔH_2 = 100 mm, U_h/U_{mf} = 2.2).

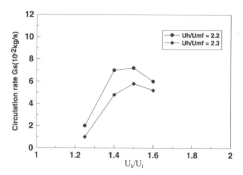

Fig. 6. Effect of the velocity of the slow bed on the solid circulation rate (plastic balls) (H_{mf} = 380 mm, ΔH = 170 mm, ΔH_1 = 50 mm, ΔH_2 = 100 mm).

2 for fluidized bed ash separately. This means fluidization velocities of the two beds must be larger than a certain value.

Fig. 5 shows the effect of the gas velocity of the fast bed U_h on the solid circulation rate, while the gas velocity of the slow bed kept constant. The solid circulation rate increased with U_h sharply and then slowly when U_h attained a certain value. This was due to two opposition influences on the solid circulation rate by U_h. In one aspect, with increase of U_h, the dilute density of the fast bed decreased and the pressure gradient at the orifice increased, and led to the solid flow rate increasing. In another aspect, too high U_h enhanced bubbling near the lower orifice and hindered the solid flow into the fast bed.

Fig. 6 shows the effect of the gas velocity of the slow bed U_l on the solid circulation rate, while U_h was kept constant. When U_l decreased, i.e. U_h/U_l increased, the solid circulation rate increased firstly, and then decreased slowly. This is because the decrease of U_l caused the increase of pressure gradient at the orifice, but has a negative influence on bed material fluidization.

The experiments showed that the gas velocity ratio of the two beds U_h/U_l should be kept in the range of 1.5–1.7 to attain high solid circulation rate.

3.3. Effect of orifice allocation and size

The location of the two orifices had a distinct effect on the solid circulation rate. When the lower orifice was nearer the distribution plate of the bed, the pressure gradient at the orifice became higher and the solid circulation rate increased. Similarly, when the upper orifice was nearer the surface of the dense bed, the pressure gradient and the particle flow rate increased. In this case, the upper orifice is usually allocated near but lower than the static bed height. Fig. 7 shows that the solid circulation rate increased with the vertical distance of two orifices ΔH as predicted before.

The horizontal distance of the two orifices would affect residence time of circulation solid in the two beds. In order to keep enough heating time for solids in the combustor (the fast bed), the two orifices must keep a certain distance in the vertical and horizontal direction.

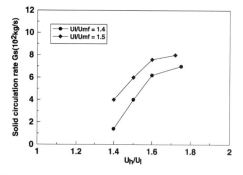

Fig. 5. Effect of the velocity of the fast bed on the solid circulation rate (plastic balls) (H_{mf} = 380 mm, ΔH = 170 mm, ΔH_1 = 50 mm, ΔH_2 = 100 mm).

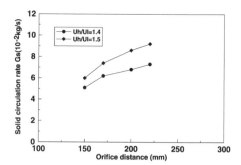

Fig. 7. Effect of the vertical distance of two orifices on the solid circulation rate (plastic balls) (H_{mf} = 400 mm, U_l/U_{mf} = 1.4, ΔH_1 = 50 mm, ΔH_2 = 150 mm).

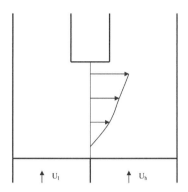

Fig. 8. Particle velocity distribution along the height of the orifice.

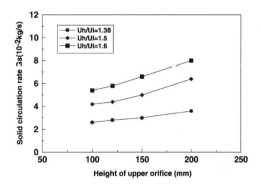

Fig. 10. Effect of the height of the upper orifice on the solid circulation rate (plastic balls) ($H_{mf} = 400$ mm, $U_l/U_{mf} = 1.4$, $\Delta H_1 = 95$ mm, $\Delta H = 170$ mm).

The size of the orifice is an important parameter for design. For a rectangular orifice, the bigger the orifice width, the higher the solid circulation rate. Experiments showed that particle velocity profile through the orifice formed a similar shape of parabola, which attained the maximum at the top of the orifice and decreased to zero at the bottom of the orifice as shown in Fig. 8. Surprisingly, particle flow height was usually less than or equal to the orifice height. So the solid circulation rate first increased with the height of the lower orifice, attained the largest when the orifice height was kept at a certain value, and then decreased as shown in Fig. 9. Fig. 10 shows the effect of the height of the upper orifice on the solid circulation rate. Different from the lower orifice, the solid circulation rate increased with the upper orifice height. This is because solid flow at the upper orifice was more likely to overflow from the low bed to the fast bed.

3.4. Gas leakage rate between the two beds

In the coal gasification process, gas leakage rate is expected to be zero or a little and too much gas leakage will affect the quality of the product gas. Experiments were made to measure the gas leakage rate between the two beds by means of a hydrogen trace method. Hydrogen gas was injected into the slow bed and the fast bed separately. Hydrogen concentrations along the height of the two beds were measured by a gas detector, and hydrogen leakage ratio through the lower and upper orifice were calculated. The gas leakage ratio of the lower orifice was higher than the upper orifice as shown in Fig. 11.

3.5. Reasonable design and operation parameters

Experiments showed that the solid circulation rate could be controlled by changing the two bed velocities and the orifice structure. To keep a large solid circulation rate, the reasonable orifice sizes and gas velocities of the system were put forward on the basis of experiments.

1. The lower orifice should be near the bed bottom and the upper orifice near but under the bed static height. The

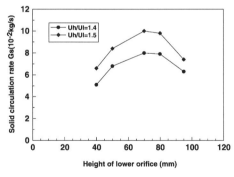

Fig. 9. Effect of the height of the lower orifice on the solid circulation rate (plastic balls) ($H_{mf} = 400$ mm, $U_l/U_{mf} = 1.4$, $\Delta H_2 = 150$ mm, $\Delta H = 170$ mm).

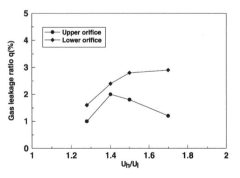

Fig. 11. Gas leakage ratio through the orifice changes with gas velocities (plastic balls) ($H_{mf} = 400$ mm, $U_l/U_{mf} = 1.4$, $\Delta H_1 = 95$ mm, $\Delta H_2 = 150$ mm, $\Delta H = 170$ mm).

horizontal and vertical distance were determined by the particle heating time and the residence time.

2. The height of the lower orifice ΔH_1 and the height of the upper orifice ΔH_2 should be kept in the range of 0.15–0.3H_{mf} and 1–1.5 ΔH_1 separately. The width of the two orifices could be determined according to Eq. (1).

3. The gas velocities of the two beds should be controlled in the following ranges:
   - for plastic balls with narrow size distribution: $U_l/U_{mf} \geqslant 1.4$;
   - for fluidized bed ash with broad size distribution: $U_l/U_{mf} \geqslant 2.2$;
   - $U_h/U_l = 1.4$–1.8.

Experiments showed that the solid circulation rate could attain 1000 kg/h for the fluidized bed ash given the small-scale test facility. To compare with a circulation fluidized bed, a concept of the solid circulation ratio R is introduced here:

$$R = \frac{G_s}{B} \quad (2)$$

where G_s is the solid circulation rate (kg/s), and B the fuel feed rate (kg/s).

According to the calculation, the solid circulation ratio in the system could attain 30–40, which is enough for the heat supply to an endothermic process of a gasifier.

4. Solid circulation rate calculation

Most of the literature [10] on the flow of a particulate material through an orifice is using the well-known relation for the discharge of a liquid through an orifice, derived from the Bernoulli equation:

$$G_s = C_D A_o [2\rho_p (1-\varepsilon_f) \Delta p]^{1/2} \quad (3)$$

where the mass flow through an orifice G_s has been plotted against the pressure gradient of the orifice Δp and other influence factors sum up to the discharge coefficient C_D [2,3,10]. Snieders et al. [3] predicted that G_s increasing with the square-root Δp is not reflected by the experiment data and put forward the following equation:

$$G_s = C_D A_o (\Delta p)^{1.64} \quad (4)$$

According to our experiments, the solid flow is different from the liquid flow. Despite of the influence of the bed pressure, the gas velocity and the bed material also affect the solid flow through the orifice. So a mathematical model to calculate the solid circulation rate in a twin fluidized bed system was erected here on the basis of experiments.

4.1. Pressure difference at the orifice

Experiments showed that there existed a narrow gas channel on the top of the lower orifice, where, driven by the pressure gradient, gases flew quickly and carried particles to move layer by layer under the effect of the friction force, which forms a particle velocity distribution as shown in Fig. 8. So, here the pressure gradient at the top of the lower orifice Δp_1 is a very important parameter, which directly controls the particle flow through the orifice. The Δp_1 can be calculated according to Eq. (5):

$$\Delta p_1 = p_l - p_h = ((1-\varepsilon_{fl})(H_{fl} - H_1)\rho_p g) \\ - ((1-\varepsilon_{fh})(H_{fh} - H_1)\rho_p g) \quad (5)$$

where ε_{fl} and ε_{fh} are the void fractions of the slow bed and the fast bed in the dense zone, H_{fl} and H_{fh} the dense bed height of the slow bed and the fast bed, and H_1 the height from the top of the lower orifice to the bed distributor.

According to the 'two-phase theory', the dense bed consists of two phases, a bubble phase and an emulsion phase at incipient fluidized condition. Thus, void fraction ε_f can be obtained by the following equations:

$$\varepsilon_f = \delta_b + (1-\delta_b)\varepsilon_{mf} \quad (6)$$

$$\delta_b = 0.3 f_b d_b^{0.5} \quad (7)$$

where δ_b is the bubble fraction in the dense bed, f_b the bubble frequency, and d_b the bubble diameter. For coal fired fluidized bed, f_b and d_b can be calculated according to the following empirical equations [11]:

$$f_b = 1.74(U - U_{mf})^{0.725} H_f^{-0.434} \quad (8)$$

$$d_b + 0.9 U_{mf} d_b^{0.5} - 0.862(U - U_{mf})^{0.275} H_f^{0.434} = 0 \quad (9)$$

The height of the dense bed H_f can be got from operation data:

$$\Delta p_f = (1-\varepsilon_f)\rho_p g H_f \quad (10)$$

where Δp_f is the pressure drop of the dense bed.

4.2. Particle flow model

As mentioned earlier, the particle velocity profile through the orifice formed a shape of parabola, which attained the maximum at the top of the orifice and decreased to zero at the bottom of the orifice. So the particle velocity profile can be described by the following equation:

$$\frac{u}{u_0} = \left(\frac{y}{\Delta H_1'}\right)^m \quad (11)$$

where u_0 is the particle flow velocity at the top of the orifice, and $\Delta H_1'$ the particle flow height at the lower orifice. According to the experiment, when $\Delta H_1/H_{mf} = 0.4$, $\Delta H_1' = H_1$ and when $\Delta H_1/H_{mf} > 0.4$, $\Delta H_1' = 0.4 H_{mf}$. A fit to the experimental data resulted in $m = 0.55$ with a correlation coefficient, r^2 of 0.966 as shown in Fig. 12. The u_0 mainly related with the bed material, the fluidization velocity, and

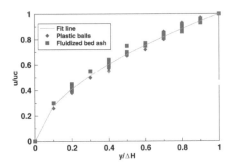

Fig. 12. Particle velocity profile through the orifice.

the pressure gradient. So here u_0 was correlated with Δp_1 and U_1 using the following equation:

$$u_0 = C_0 \left(\frac{U_1}{U_{mf}} - C_1\right)^{n_1} \Delta p_1^{n_2} \quad (12)$$

where C_0 is a factor related to bed material property, and C_1 the minimum fluidization number from the experiment. By fitting with data, u_0 was obtained by the following equations as shown in Figs. 13 and 14. For plastic balls with narrow size distribution:

$$u_0 = 1.604 \times 10^{-6} \left(\frac{U_1}{U_{mf}} - 1.2\right)^{1.256} \Delta p_1^{1.78} \quad (13)$$

For fluidized bed ash with broad size distribution:

$$u_0 = 3.312 \times 10^{-6} \left(\frac{U_1}{U_{mf}} - 2\right)^{1.256} \Delta p_1^{1.78} \quad (14)$$

So particle flow rate Gs:

$$Gs = (1 - \varepsilon_f)\rho_p b \int_0^{\Delta H_1'} u_0 \left(\frac{y}{\Delta H_1'}\right)^m dy$$
$$= (1 - \varepsilon_f)\rho_p b u_0 \frac{\Delta H_1'}{m+1} \quad (15)$$

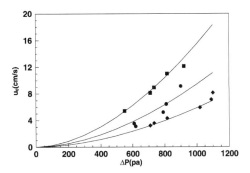

Fig. 13. Plot of solid flow rate vs. pressure gradient (plastic balls) ((■) $U_l/U_{mf} = 1.72$, (●) $U_l/U_{mf} = 1.55$, (◆) $U_l/U_{mf} = 1.44$).

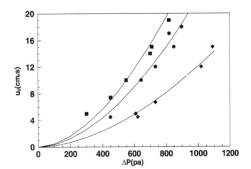

Fig. 14. Plot of solid flow rate vs. pressure gradient (fluidized bed ash) ((■) $U_l/U_{mf} = 2.48$, (●) $U_l/U_{mf} = 2.39$, (◆) $U_l/U_{mf} = 2.28$).

Using Eq. (12):

$$Gs = \frac{C_0}{m+1}(1-\varepsilon_f)\rho_p b \Delta H_1' \left(\frac{U_1}{U_{mf}} - C_1\right)^{n_1} \Delta p_1^{n_2} \quad (16)$$

For plastic balls:

$$Gs = 1.035 \times 10^{-6}(1-\varepsilon_f)\rho_p b$$
$$\times \Delta H_1' \left(\frac{U_1}{U_{mf}} - 1.2\right)^{1.256} \Delta p_1^{1.78} \quad (17)$$

For fluidized bed ash:

$$Gs = 2.136 \times 10^{-6}(1-\varepsilon_f)\rho_p b$$
$$\times \Delta H_1' \left(\frac{U_1}{U_{mf}} - 2\right)^{1.256} \Delta p_1^{1.78} \quad (18)$$

Fig. 15 shows the comparison of calculation results using the given Eqs. (3)–(18) and experimental data and the relative deviation between calculation results and experimental data is less than 15%. So this model can predict solid flow through a horizontal orifice well.

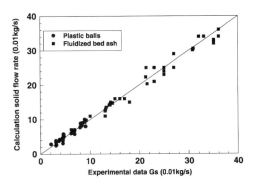

Fig. 15. Comparison of calculation results and experimental data.

5. Conclusions

A twin fluidized bed solid circulation system, in which two adjacent fluidized beds operated in different gas atmospheres and exchanged solids but not gases, was developed to be utilized in coal gasification to produce middle heating value gas. Experiments on a small-scale test facility showed that the main factors, which affected the solid circulation rate, were bed material property, gas velocities of the two beds, orifice size and distance. The solid circulation rate for the fluidized bed ash could attain 1000 kg/h, given the small-scale test facility. On the basis of experiments, the reasonable gas velocity and size of orifice structure were put forward, which could be used as a reference for the coal gasifier design. A mathematical model was erected to calculate particle flow rate through the orifice in the system, which considered the effects of bed material, pressure gradient and gas velocity. The relative deviation between calculation results of the model and experimental data is less than 15%.

Acknowledgements

Supported by the China NKBRSF Project nos. G1999022105 and 2001CB409600.

References

[1] V.R. Burugupalli, et al., Process analysis of a twin fluidized bed biomass gasification system, Ind. Eng. Chem. Res. 27 (1988) 304–312.
[2] V. Rudolph, M.P. Judd, Circulation and slugging in a fluid bed gasifier, in: P. Basi (Ed.), Circulating Fluidized Bed Technology, 1985, pp. 437–441.
[3] F.F. Snieders, et al., The dynamics of large particles in a four-compartment interconnected fluidized bed, Powder Technology 101 (1999) 229–239.
[4] Y.O. Chong, et al., Solids circulation between adjacent fluidized beds without gas mixing, in: Circulating Fluidized Bed Technology, 1985, pp. 397–340.
[5] M. Ishida, et al., Measurement of the velocity and direction of flow of solid particles in a fluidized bed, Powder Technol. 44 (1985) 77–84.
[6] R.D. Abellon, A single radio trace particle method for the determination of solids circulation rate in interconnected fluidized beds, Powder Technology 92 (1997) 52–60.
[7] M. Yamazaki, Solid flux measurement by heat response detector, Kogaku Kogku Ronbunshu 5 (1979) 155–159.
[8] M. Fang, Mechanism research on circulation fluidized bed gas steam cogeneration, Ph.D. Thesis, Zhejiang University, 1991.
[9] B. Bhattacharya, et al., Solid circulation in a compartmented gas fluidized bed, Powder Technology 101 (1999) 191–204.
[10] B.H. Song, et al., Circulation of solids and gas bypassing in an internally circulating fluidized bed with a draft tube, Chem. Eng. J. 68 (1997) 115–122.
[11] A.M.C. Janse, et al., A novel interconnected fluidized bed for the combined flash pyrolysis of biomass and combustion of char, Chem. Eng. J. 76 (2000) 77–86.

双流化床系统中固体循环的实验研究

摘要：本文提出了一种双流化床固体循环系统用于煤气化以生产中热值气体，该系统由两个相邻的流化床构成，通过固体物料交换。在一个小型试验装置上测试了床料、操作速度和床结构对固体循环速率的影响。实验结果表明，该系统的固体循环速率可通过调整两个流化床的气速而得到调节，最大可达到燃料进料率的30～40倍，可满足气化炉吸热过程的热量供给需求。基于实验，提出了合理的操作和设计参数。建立了一个数学模型来计算系统的固体循环速率，通过与实验数据的对比，该模型能较好地预测水平孔口处的固体流量。本研究结果可为商用气化炉的设计提供了依据和设计参考。

原文刊于 Chemical Engineering Journal, 2003, 7(94)：171-178

中国能源与环境可持续发展问题的探讨(一)

岑可法,邱坤赞,朱燕群

(浙江大学,浙江杭州 310027)

摘 要:概述了我国能源现状和发展趋势,分析了我国环境污染现状和控制,以及所造成的危害,指出了只有能源与环境协调发展,才能实现国家经济可持续发展的必要性。重点介绍了几种污染控制技术、新能源开发及新能源利用系统。

关键词:能源;环境;污染控制;新能源开发

中图分类号:TK01,X5 **文献标识码**:A **文章编号**:1671-086X(2004)05-0245-06

On the Possibility of Continuous Development of Energy Sources as well as of the Environment in China

CEN Ke-fa, QIU Kun-zan, ZHU Yan-qun

(Zhejiang University, Hangzhou 310027, China)

Abstract: A review of the status quo of energy sources and their development trend in China is being presented, with an analysis of prevailing pollution and a prognosis of the future, together with an account of the harm already done. It is moreover pointed out that only by a harmonious co-development of energy sources and environment can the required continuous economic development of the country be effectuated. In the main, a few techniques for containing pollution and some new systems for the exploitation and utilization of new energy sources are being introduced.

Keywords: energy source; environment; pollution containment; exploitation of new energy sources

21世纪人类所面临的重大挑战是实现经济和社会的可持续发展,而保证优质可靠的能源供应是现代社会实现可持续发展的必要条件之一。我国经济的持续快速发展,对能源的需求增长必然造成对资源短缺和环境的压力。近年来出现全国性的电力紧张,使能源生产倍受关注,2003年国家调整了电力行业十五规划,加紧电力改革与电力建设,但是电力紧张局面的全面缓解预计要到2006～2007年。同时能源生产带来的环境问题也日益加剧,只有能源与环境协调发展,才能实现国家经济可持续发展的目标。

1 我国能源现状及趋势

我国能源需求不断增长,资源短缺现象逐渐崭露。

2003年我国能源消费总量16.78亿t标准煤,比上年增长10.1%,其中,原油消费量2.52亿t,增长12%;全年原油进口量9112.63万t,同比增长31.29%,进口原油占国内原油消费量的比重达到36.1%;出口原油813.33万t,同比增长12.84%。

2003年全国生产原煤16.67亿t,其中发电用煤炭的消费已经达到8.5亿t,占全部煤炭消费

收稿日期:2004-07-28
作者简介:岑可法(1935-),男,中国工程院院士,中国动力工程学会副理事长,浙江大学机械与能源学院院长,浙江大学热能工程研究所所长,主要从事能源与环保研究工作。

的50%以上；原油产量1.69亿t；天然气产量达341.28亿m³[1]。

我国是一个以煤为主要一次能源的国家，目前煤占能源消费的70%以上。我国煤94%集中在大别山～秦岭～昆仑山一线以北地区。我国水能资源虽居世界第一位，理论蕴藏量为6.76亿kW，按发电量计可开发的水电资源约占世界总量的15%。但大部分集中在西南地区，其次在中南地区，而东部地区水能资源蕴藏量只占全国的7%；石油、天然气资源主要分布在西部地区，而东部沿海经济发达地区是负荷中心却缺乏能源资源。我国石油的储采比小于15，而世界平均为40左右。2003年我国进口石油9100万t，我国的石油进口依存度从1995年的7.6%，提高到2003年的36%。中国已成为仅次于美国的第二大石油消费国；而由于经济快速增长，国内石油需求仍在不断增长，预计到2020年，中国石油需求量为4.5亿t～6亿t，年均递增12%，石油进口依存度将达到60%以上。

经济快速发展，使能源需求不断增长，特别是电力的需求增长迅速，将对能源利用和环保技术提出更高的要求。

根据党的十六大提出的全面建设小康社会的发展目标，到2020年，国内生产总值将比2000年翻两番，全国电力装机容量需达到10亿kW～13亿kW以上，才能基本满足全社会的用电发展需求。按照我国GDP增长速率，对不同能源电力生产的规划见表1。

如前10年发电量年增长按6%～6.75%计算，后10年按发电量年均4.5～5%预测，相应的装机容量见表2。

电力的大幅度增加，使煤炭需求剧增（见表3），到2020年中国一次能源的需求在25亿t～33亿t标准煤之间，均值为29亿t标准煤，是2000年的2.2倍。我国人均能源可采储量远低于世界平均水平，2000年人均石油开采储量只有2.6t，人均天然气可采储量1074m³，人均煤炭可采储量90t，分别为世界平均值的11.1%、4.3%和55.4%。能源需求的大幅度增加对我国的资源和能源利用技术提出了更大的挑战，也是能源生产

表1 电力与新能源发电组装机容量规划[2]

年度		2000年	2010年	2020年
总装机容量/万kW		32740	58770	95000～100000
水电	装机容量/万kW	7930	15000	25000
	所占比例/%	24	25.5	26
煤电	装机容量/万kW	23700	40000	60000
	所占比例/%	73	68	63
气电	装机容量/万kW	700	2000	5000
	所占比例/%	2	3.4	5
核电	装机容量/万kW	210	1170	3600～4000
	所占比例/%	0.6	2	4
新能源（风、太阳、生物质、地热等）	装机容量/万kW	240	600	1500～2000
	所占比例/%	0.7	1	2

表2 2010-2020年电力装机容量预测[3]

年度		2000年	2003年	2005年	2010年	2020年
总装机容量/万kW		31932	39142	48500	70000～73000	110000～120000
水电	装机容量/万kW	7935	9490	11000	16300	16000～27000
	所占比例/%	24.8	24.2	22.7	23.3	14.5～22.5
煤电	装机容量/万kW	22115	27357	34080	47730～50330	71000～78500
	所占比例/%	69.3	69.9	70.3	68.2～68.9	64.5～65.4
气电	装机容量/万kW	96	120	800	2800～3000	6000～7000
	所占比例/%	3.0	3.1	1.65	4.0～4.1	5.5～5.8
核电	装机容量/万kW	210	619	870	1070	4000
	所占比例/%	0.7	1.6	1.8	1.53	3.6
新能源（风、太阳、生物质、地热等）	装机容量/万kW	36	56	250	600～800	1500～2000
	所占比例/%	1.1	1.4	0.5	0.86～1.1	1.36～1.67

和利用的极大机遇,因此洁净煤技术、低 NO_x 技术及新能源发电等技术将得到大力的发展。

表3 煤炭需求预测

年度	2000年	2005年	2010年	2020年
原煤(亿t)	13	14.9	20	29

2 我国环境现状及趋势

我国小康社会的建设目标对环境的要求提出了巨大的挑战。

2.1 大气主要污染物二氧化碳、二氧化硫和氮氧化物等

中国目前并没有摆脱先污染后治理的老路。仅以大气污染情况为例,中国的二氧化硫和二氧化碳排放量分别居世界第一位和第二位。虽然单位GDP的碳排放量明显下降(1990年至2001年下降了52%),但二氧化碳排放总量却从1980年的3.94亿t碳增加到2001年的8.32亿t碳;燃煤排放的二氧化硫是造成酸雨的主要原因,90年代中期酸雨区面积比80年代扩大了100多万平方公里,年均降水pH低于5.6的区域面积已占全国面积的30%左右。由于较严重的环境污染,造成了高昂的经济成本和环境成本,并对公众健康产生较明显的损害,国内外研究机构的成果显示,大气污染造成的经济损失占GDP的3%～7%。造成大气质量严重污染的主要原因是中国以燃煤为主的能源结构,并且没有对煤炭利用采取有效的环保措施,烟尘和二氧化碳排放量的70%、二氧化硫的90%、氮氧化物的67%来自于燃煤,此外机动车快速增长所带来的污染加剧。

全国环境统计公报(2003年)[4]显示:2003年全国废气中二氧化硫排放量2158.7万t,比上年增加12.0%。其中工业二氧化硫排放量为1791.4万t,占二氧化硫排放总量的83.0%,比上年增加14.7%;生活二氧化硫排放量367.3万t,占二氧化硫排放总量的17.0%,比上年增加0.7%。烟尘排放量1048.7万t,比上年增加3.6%。其中工业烟尘排放量846.2万t,占烟尘排放总量的80.7%,比上年增加5.2%;生活烟尘排放量202.5万t,占烟尘排放总量的19.3%,比上年减少2.9%。工业粉尘排放量1021.0万t,比上年增加8.5%。

2003年,全国工业固体废物产生量10.0亿t,比上年增加6.2%。工业固体废物排放量1940.9万t,比上年减少26.3%。全国危险废物产生量1171.0万t,比上年增加17.1%。危险废物排放量为0.2万t,比上年减少83.5%。

2003年我国经济快速增长,钢铁、电解铝、水泥等9种重要工业原材料生产量大幅增长,电力、煤炭等能源供不应求。高能耗、高污染行业的快速发展,对环境造成重大压力。各主要污染物排放量,特别是废气中工业二氧化硫、烟尘和粉尘,一改近几年逐年下降的趋势,呈现较大幅度的反弹。

在表4、表5中对大气主要污染物进行了预测,从环境容量看,根据估算,要使二氧化硫排放量处在生态系统能承受的降解能力之内,全国最多能容纳1620万t左右;氮氧化物的环境容量也不会高于1880万t,这些标准应是"环境小康"的最低要求。但是,表4的相关数据表明,目前已经存在着环境"透支"。

表4 国内的污染现状[4](单位:万t)

年度	2000	2001	2002	2003
CO_2	88900	83200		
SO_2	1995.1	1947.8	1926.6	2158.7
NO_x		1988		
烟尘	1165.4	1069.8	1012.7	1048.7

表5 火电厂 SO_2 排放总量预测(单位:万t)

年度	2000	2010	2020
无控制状态SO_2排放量(火电)		890	1450	2250
SO_2总量控制不变,需年减排			560	1360
减排后SO_2平均排放因子(g/kg煤)		15.1	9.3	6.0

1997年12月1日《京都议定书》,到2010年发达国家的六种温室气体排放总量应比1990年时减少5.2%,其中:美国消减7%;东欧消减5%～8%;日本消减6%;爱尔兰增加10%;欧盟消减8%;澳大利亚增加8%;加拿大消减6%;挪威增加1%;发展中国家自愿参加减排。

目前的温室效应显示的影响和破坏力越来越大,二氧化碳的减排还存在很多经济和技术上的

难题，迄今为止美国尚未签署协议，但各国正在大力开展减排二氧化碳的新技术。2003年二氧化硫、氮氧化物产生量也远远超过环境容量所承受的范围，如果满足目前的控制水平，两类污染物需削减量分别达到538.7万t、108万t，到2020年煤炭的需求量是现在的2倍，二氧化碳、二氧化硫和氮氧化物的产生量也将成倍增加。国际上要求中国限排温室气体的国际压力将越来越大，那时中国将难以回避温室气体排放增长限制的承诺。随着温室气体的限排，二氧化碳的边际削减成本将趋于上升。可以明确地说，以现有技术、现有的用能结构、现有的政策来利用40亿t～50亿t当量煤的化石燃料，是绝对不可持续的。因此，未来中国能源的发展将受到来自全球环境的压力。从国际经验和中国的潜力看，在保持经济增长和能源发展的同时，明显减少环境污染，满足小康社会对环境质量的要求，不是不可能，但面临着十分严峻的挑战。

根据满足小康社会对环境质量的要求，我国大气污染物总量在2010～2020年要控制达到如下目标[5]：

SO_2排放总量控制目标：2010年1600万t，2020年1300万t；

烟粉尘排放总量控制目标：2010年1600万t，2020年1000万t；

NO_x排放总量控制目标（以NO_2计）：2010年1800万t，2020年1600万t。

按此目标对我国污染控制市场进行预测，到2020年，新装600MW火电机组约700台；新装机组均装置脱硫、脱硝设备。目前污染控制投资成本：SO_2 = 250元/kW装机容量；NO_x：200元/kW装机容量。因此我国在污染控制上的投入将大大增加，污染控制技术产业也将大有可为。

2.2 我国的其他污染物重金属汞、氟、氯化氢等

长期以来，不同于燃煤的首要污染物SO_2和NO_x，燃煤造成的微量元素污染问题一直没有引起人们的足够重视。煤中的有毒痕量元素重金属及其化合物的释放会对包括大气、水以及土壤在内的生态环境产生污染，它们不为微生物降解，可以在人体内沉淀，并转化为毒性很大的金属有机化合物，对人类的身体健康产生直接或间接的危害。能源利用过程对环境的污染还有重金属、有机污染物，如二恶英和多环芳烃等，还有一些无机污染物，如氟(F)、氯化氢(HCl)等，这些污染物也是环境污染的重要因素，虽然有些污染物排放量很少，甚至微量，但是对人类和环境的毒害也相当大。表6列举了部分重金属对人类的危害。

汞(Hg) 近年来国际上对重金属汞污染的研究最为热门，造成汞环境污染的来源主要是天然释放和人为两方面。从局部污染来看，人为来源是相当重要的。据统计，全球每年向大气中排放总量约为5000t，其中4000t是人为的结果。汞的人为来源与以下方面有关：汞矿和其他金属的冶炼，氯碱工业和电器工业中应用以及矿物燃料的燃

表6 重金属污染对人体的危害

元素	危害
汞 Hg	损害人体细胞内的酶系统蛋白质的巯基，可引起中枢神经系统的障癌，汞属于蓄积性毒物，汞毒可分为金属汞、无机汞和有机汞三种，其中甲基汞毒性最大，破坏了肝脏细胞的解毒作用，中断了肝脏的解毒过程，损害了肝脏合成蛋白质的功能和其它功能导致细胞坏死，肾功能衰竭中枢神经中各处均可产生神经细胞变性，脱落，发生感觉障碍。
铅 Pb	以气溶胶即铅尘和铅烟形式存在，通过消化道、呼吸道或皮肤进入人体的造血系统、神经系统和肾脏，造成心肌损伤。
镉 Cd	镉的氧化物毒性较大，以肾脏损害最为明显，容易引起肺气肿，肾病和骨痛病。
铬 Cr	六价铬毒性最大，吸入过多，会导致鼻出血、鼻膜炎。
铜 Cu	引起肝硬化。
砷 As	砷与人体细胞酶系统的结合，抑制基酶的活性，引起代谢障碍；饮用含砷量高的水可引起皮肤角质化，皮肤癌、肝癌、肾癌、肺癌。
锌 Zn	人体不可缺少的微量元素，如过量会导致刺激胃肠道，重者引起中毒。
硒 Se	破坏一系列酶系统，对肝、胃、骨髓和中枢神经系统有破坏作用。

烧。以美国为例，美国每年汞的排放量占全世界向大气排放汞总量的3%，大约158t左右，其中份额最大的来源于燃烧行业占87%，10%来源于制造工业，1%来源于其它方面。燃烧行业中，燃煤汞排放所占比例最大，达到33%，垃圾焚烧炉年排放汞量约占19%；工业锅炉汞排放比例约为18%，医疗垃圾焚烧约占10%[6]。

1995年全国煤炭中平均汞含量为0.22mg/kg，与煤中无机硫有关，电力生产供应业向大气排放72.86t/年，占各行业燃煤汞排放量的34.1%，排在第二位。利用文献资料中的统计数据计算了中国燃煤电站年汞排放量，2000年我国向大气中排放汞为60.34t，排入灰渣和洗煤废渣的汞为18.88t[7]。2003年我国原煤产量17.28亿t，电煤消耗量约为8.26亿t，2004年将比2003年增加9000万t以上，达到9亿t，按此估算全国燃煤电厂年排放汞量达到100t以上。

美国对于燃煤电站汞排放的标准也已经着手制订，预计2004年12月15日之前将会正式颁布燃煤电站锅炉汞排放标准，2007年底前电站锅炉都将必须执行汞排放标准，届时各燃煤电站为达到汞排放标准将会纷纷加大对燃煤汞污染的研究力度。目前，Wisconsin州已经初步拟定了汞控制的计划，3年后减排30%，10年后减排50%，15年后减排90%。我国于2000年3月公布了生活垃圾焚烧污染控制标准(GWKB3-2000)中对汞的排放标准为0.2mg/m³(标准状态下)(11%O₂)。对燃煤电厂我国目前还没有相应的标准。

氟(F) 氟化物是大气污染控制物种，煤燃烧是大气氟污染的最大污染源之一。我国每年因燃煤可向大气中排放15万t氟化物。

燃煤大气氟污染物为气态HF、SiF_4。

对植物的危害：使植物中毒、影响发育、减产，毒性是SO_2的20倍～100倍。对桑叶损害造成蚕死亡。

对人和动物的危害：氟中毒、斑釉齿症、氟骨症。

对锅炉危害：尾部结露后形成氢氟酸，对设备有严重腐蚀作用。

燃煤引起的大气氟污染已给我国的农牧业造成了较大的损失。

我国不同煤种氟含量的分布特征见表7。

表7 我国各煤种含氟量[8]

煤种	氟含量范围/$\mu g \cdot g^{-1}$	平均值/$\mu g \cdot g^{-1}$	样品数量
褐煤	151～615	241	7
烟煤	17～696	173	81
无烟煤	61～1800	308	28
石煤	193～3313	1058	31
煤矸石	259～1956	794	33

18个省从褐煤到石煤共180个煤样的氟含量。结果表明：(1)不同煤种的氟含量都存在相当的差异。(2)烟煤氟含量最低，石煤氟含量最高。(3)应当特别重视石煤和煤矸石利用时氟污染的危害。

氯化氢(HCl) 常温下HCl为无色气体，有刺激性气味，极易溶于水而形成盐酸。能腐蚀皮肤和粘膜，致使声音嘶哑，鼻粘膜溃疡，咳嗽直至咯血，浓度大于1000×10^{-6}时出现肺水肿以至死亡。导致叶子褪绿，进而出现变黄、棕、红至黑色的坏死现象。在坏死区域的边缘可能出现白色或乳白色，在叶子的上表面出现斑块或斑点和穿孔。对燃煤锅炉及垃圾焚烧设备的危害：

(1)会造成炉膛受热面的高温腐蚀损毁和层部受热面的低温腐蚀。

(2)在锅炉中，温度在426℃以上时HCl气体可能会使管壁发生高温腐蚀。而在露点温度以下时，则会导致低温腐蚀。

(3)氯化氢还易形成露点腐蚀，溶解在设备壁面的凝结水珠中，生成强酸，使金属溶解腐蚀，在壁面生成大大小小的凹坑，严重部件甚至穿透。

图1为我国煤种中氯的含量，石煤和煤矸石的氯含量较低，并且比较接近，燃烧后氯化氢的平均排放浓度约为125×10^{-6}。烟煤的平均氯含量最高，约为246×10^{-6}，褐煤次之[9]。

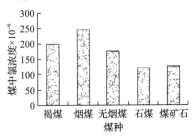

图1 我国煤种中氯含量

对燃煤电厂我国目前还没有相应的标准，但对垃圾焚烧炉有排放标准，表 8 为各国垃圾焚烧炉 HCl 的排放标准，我国的限值大大高于欧美等国的标准。

表 8　各国垃圾焚烧炉中 HCl 排放标准($O_2 \leqslant 11\%$)

	德国	奥地利	瑞士	荷兰	美国	中国(GB16297-1996)
$HCl/mg \cdot m^{-3}$	10	15	30	10	15	75

（待续）

原文刊于发电设备，2004, 5: 245-250

Modulations on turbulent characteristics by dispersed particles in gas-solid jets

BY KUN LUO, JIANREN FAN AND KEFA CEN

State Key Laboratory of Clean Energy Utilization and Institute for Thermal Power Engineering, Zhejiang University, Hangzhou 310027, People's Republic of China

(fanjr@zju.edu.cn)

A direct numerical simulation technique combined with a two-way coupling method was developed to study a gas–solid turbulent jet with a moderately high Reynolds number. The flow was weakly compressible and spatially developing. A high-resolution solver was performed for the gas phase flow-field and the Lagrangian method was used to trace particles. The modulations on flow structures and other turbulent characteristics by particles at different Stokes numbers were investigated. It is found that the particles at Stokes numbers of 0.01 and 50 can advance the development of the large-scale vortex structures and make the turbulence intensity profiles wider and lower, but the particles at a Stokes number of 1 delay the evolution of the large-scale vortex structures and decrease the turbulence intensities. The jet velocity half-width and the decay of the streamwise mean velocity in the jet centreline are reduced by all particles, in which particles at a Stokes number of 0.01 result in a larger reduction of the velocity half-width and particles at a Stokes number of 1 lead to a larger reduction of the streamwise mean velocity decay. All particles decrease the vorticity thickness, but increase the fluid momentum thickness. In addition, the two-way coupled particle distribution is more uniform than that of the one-way coupled case.

Keywords: direct numerical simulation; gas–particle jet; two-way coupling; turbulence modulation

1. Introduction

Gas–particle two-phase turbulent jets are usually found in many engineering applications, such as jet propulsion, pulverized coal combustion and environmental control systems. In these applications, the dispersion of particles becomes a controlling factor in the efficiency and stability of the engineering processes. So the ability to predict and control particle dispersion in jets is of great significance in obtaining efficient engineering applications. In addition, the interactions between gas phase and particle phase in the jets have been a hot research topic for many years.

Chung & Troutt (1988) used a discrete vortex element approach to simulate the particle dispersion in an axisymmetric jet. They reported that the particle dispersion extent depends strongly on the ratio of particle aerodynamic response time to the characteristic time of the jet. Hardalupas *et al.* (1989) studied the velocity and particle flux characteristics of turbulent particle-laden jet and suggested that the effect of the mass mixture ratio on the particle concentration distribution is small. Yuu *et al.* (1996) directly simulated a gas–particle turbulent free jet with a low Reynolds number. Their computational results of two-phase characteristics are in agreement with their experimental data. Glaze & Frankel (2000) investigated the effect of dispersion characteristics on particle temperature in an idealized non-premixed reacting jet. It was found that the particle temperature behaviour is a strong function of the spatial dispersion behaviour, and the spatial dispersion is characterized by the particle Stokes number and the injection location in both reacting and non-reacting jets.

Most of the above studies are based on one-way coupling, which neglects the effects on turbulence by dispersed particles. However, moderate mass loadings of particles can distinctly modulate turbulence. Parthasarathy (1995) studied the spatial and temporal stabilities of the circular particle-laden jet. It was observed that the presence of particles decreases the amplification rate at all frequencies, decreases the wave amplification but increases the wave velocity. Recently, Despirito & Wang (2001) examined the linear instability of a viscous, two-way coupled particle-laden jet. They demonstrated that the addition of particles can destabilize the flow at smaller particle Stokes numbers, but the particles at Stokes numbers on the order of 1 correspond to the maximum flow stability.

However, the modulations on the turbulent characteristics such as the coherent structures and statistical variables by particles at different Stokes numbers are still unclear. Furthermore, the previous studies are all based on the hypothesis that the gas phase is incompressible. Whereas in some gas–solid two-phase turbulent jets the fluid is compressible and the momentum transferral takes place between the phases, so the method of two-way coupling which includes the modulations on compressible turbulence by different particles has to be employed.

To the author's knowledge, there is no direct numerical simulation (DNS) of the compressible, gas–solid turbulent flows with moderate high Reynolds numbers by using two-way coupling method. This paper will demonstrate DNS of a weakly compressible, spatially developing and two-way coupled two-phase turbulent jet with a Reynolds number of 4500. The main objective of this study is to reveal the modulations on turbulent characteristics by the dispersed solid particles at different Stokes numbers in the particle-laden turbulent plane jet and provide references for the development of gas–solid two-phase turbulent models and the application in related industries.

2. Mathematic description

(a) *Computational domain and boundary conditions*

Figure 1 shows the computational domain and flow configuration of a gas–solid two-phase turbulent plane jet. The high-speed stream velocity is U_1 and the co-flow stream velocity is U_2. The mean convective Mach number, $Ma_c = (U_1 - U_2)/(c_1 + c_2) = 0.15$. The ratio of the nozzle width d to the initial momentum thickness δ_0 is set to be 20. The initial flow Reynolds number, Re, based on the nozzle width and the velocity difference between two streams, is 4500.

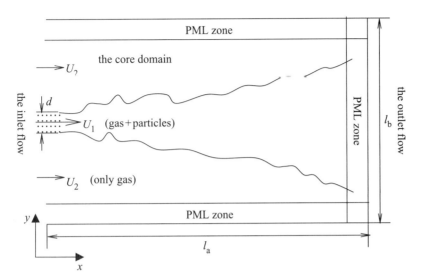

Figure 1. Computational domain and flow configuration of the two-dimensional particle-laden jet.

It is well known that for the simulation of turbulent flows in an open, non-periodic environment system the proper artificial boundary conditions at the computational domain boundaries are crucial to the simulation results. However, the accurate and efficient boundary conditions for the three-dimensional Navier–Stokes equations have not been found yet. In practical simulations, it seemed that combining the one-dimensional non-reflecting boundary conditions (Thompson 1987) and the perfectly matched layer (PML) buffer zone method could give reasonable results (Adams 1998). In the present simulation, to get higher resolution results, the boundary conditions are treated carefully. At the inflow boundary, the typical top-hat inflow profile for streamwise velocity and the viscous boundary conditions (Poinsot & Lele 1992) are adopted. At the outflow boundary, non-reflecting boundary conditions, sponger layers based on Hu (1996, 2001) and Freund *et al.* (2000), viscous boundary conditions (Poinsot & Lele 1992) and the pressure correction (Rudy & Strikwerda 1980) are used. At the sidewall boundaries, non-reflecting boundary conditions, sponger layers based on Hu (1996, 2001) and Freund *et al.* (2000) and viscous boundary conditions (Poinsot & Lele 1992) are utilized.

The whole computational domain is divided into a core domain and three PML buffer zones. In the core domain, a uniform grid of $\Delta x = \Delta y = (1/15)d$ is used. But in the buffer zones the stretching grid with a 5% stretching ratio is employed. A total 257×315 computational grid system is utilized with the whole domain dimension $l_a = 17.656d$, $l_b = 22.112d$. Many experimental results have shown that the plane jet is largely two-dimensional near the nozzle within 20 widths of the nozzle. So, the present study is of significance in both theory and practice.

(b) *Governing equations for gas phase*

In this simulation, the gas phase fluid is regarded as an ideal, Newtonian gas. As for the dispersed phase, the diameter of the largest particle is smaller than the smallest grid scale. So the momentum coupling effect of a particle on the fluid can be approximated by a point force. For this two-way coupled flow, the non-dimensional governing equations in the core domain can be expressed as follows.

– Continuum equation:

$$\frac{\partial \rho}{\partial t} + \frac{\partial (\rho u_j)}{\partial x_j} = 0. \tag{2.1}$$

– Momentum equations:

$$\frac{\partial (\rho u_i)}{\partial t} + \frac{\partial (\rho u_i u_j)}{\partial x_j} = -\frac{\partial p}{\partial x_i} + \frac{1}{Re}\frac{\partial \tau_{ij}}{\partial x_j} - f_i. \tag{2.2}$$

– Energy equation:

$$\frac{\partial p}{\partial t} + u_k \frac{\partial p}{\partial x_k} + \gamma p \frac{\partial u_k}{\partial x_k} = \frac{\gamma}{\Pr Re}\frac{\partial}{\partial x_k}\left(\kappa \frac{\partial T}{\partial x_k}\right) + \frac{\gamma - 1}{Re}\phi. \tag{2.3}$$

– The state equation of an ideal gas:

$$p = \rho R T. \tag{2.4}$$

In those the shear stress tensor

$$\tau_{ij} = \mu\left(\frac{\partial u_i}{\partial x_j} + \frac{\partial u_j}{\partial x_i}\right) - \frac{2}{3}\mu \frac{\partial u_j}{\partial x_j}\delta_{ij}$$

and the viscous dissipation $\phi = \tau_{ij}(\partial u_i / \partial x_j)$. The inter-phase momentum coupling term f_i is the net force exerted per unit volume of fluid by N particles, and can be computed from

$$f_i = \frac{1}{V_i}\sum_{m=1}^{N} f_{im} = \frac{1}{V_i}\sum_{m=1}^{N}\left(3\pi\mu d_p f(u_i - v_i) + 1.61\nu^{1/2}\rho_p d_p^2 (u_i - v_i)\left|\frac{d u_i}{d y}\right|^{1/2}\right), \tag{2.5}$$

where V_i is the pseudo integration control volume, $V_i = \Delta x \times \Delta y \times d_p$ (Xu & Yu 1997). f_{im} is the force acting on the mth particle in the x_i direction. The other symbols are explained in §2c. Because one can only obtain f_{im} at the position of a particle, the volume-weighted linear partitioning scheme is used to partition the term to the surrounding grid points. Other methods for handling the momentum coupling term can be found in the work of Boivin et al. (1998). For one-way coupling cases, $f_i = 0$.

In three PML buffer zones, the exponential damping terms are added to the above standard governing equation to improve the numerical accuracy near the computational boundaries. This numerical algorithm has been validated by comparison with the linear stability theory.

(c) Governing equations for dispersed phase

In the particle dispersion simulation, Lagrangian method is used. At first, several assumptions about the particles are made as follows.

(i) All particles are rigid spheres with identical diameter d_p and density ρ_p.
(ii) The ratio of the material density of particles to fluid approximates to 2000.
(iii) Regarding the flow as dilute two-phase flows, and the particle–particle interactions are neglected.
(iv) Before the particles are ejected into the flow-field, they distribute evenly at the nozzle and their velocity equal to the local gas phase velocity.

Because the larger velocity gradient in the free shear flow could exist at certain local areas, the Saffman lift force (Saffman 1965) is considered in the present study; it was usually was neglected in the previous simulations (Chein & Chung 1988; Tang *et al.* 1992; Fan *et al.* 2001). In this work, we consider the main forces acting on a sphere particle are the Stokes drag force, the Saffman lift force and the gravity force. Thus the governing equation of the particle can be expressed as follows:

$$m_p \frac{dv_i}{dt} = \frac{\pi d_p^2}{8} C_D \rho |u_i - v_i|(u_i - v_i) + 1.61 \nu^{1/2} \rho_g d_p^2 (u_i - v_i) \left|\frac{du_i}{dy}\right|^{1/2} + m_p g. \quad (2.6)$$

Then the non-dimensional governing equation of a particle becomes

$$\frac{dv_i}{dt} = \left(\frac{f}{St} + \frac{9.66 \nu^{1/2} l_r \rho_g \left|\frac{du_i}{dy}\right|^{1/2}}{\pi d_p \rho_p u_r}\right)(u_i - v_i) + Fr, \quad (2.7)$$

where v_i is particle velocity and u_i is fluid velocity at the particle position. f is the modification factor for the Stokes drag coefficient, which can be described by $f = 1 + 0.15 Re_p^{0.687}$ for $Re_p \leq 1000$. The particle Reynolds number $Re_p = |u_i - v_i| d_p / \nu$; here d_p is the particle diameter and ν is the kinematical viscosity of the fluid. In this simulation, the Reynolds numbers for all particles are less than 1. St is the particle Stokes number, defined as $St = (\rho_p d_p^2 / 18\mu)/(l_r/u_r)$. ρ_p is the particle material density, μ is the fluid dynamics viscosity, l_r is the characteristic length scale and u_r is the characteristic velocity scale. $Fr = gl_r/u_r^2$. The velocity and position of a particle can be obtained by integrating equation (2.7). Since there is only the velocity of fluid at every grid point, we use the fourth-order Lagrange interpolating polynomial to get the fluid velocity at the position of the particle.

(d) *Numerical scheme*

For any DNS of turbulence, the general requirements are that the numerical scheme can provide the higher accuracy both in space and time as well as can be solved efficiently.

To solve the fluid governing equations, the fourth-order compact finite difference schemes of Lele (1992) are chosen to discretize the spatial derivatives in the interior mesh nodes, and the third-order compact finite difference schemes of Carpenter *et al.* (1993) are used at the boundary nodes. All these derivate evaluations are done in the computational space on the uniform meshes. For the non-uniform meshes, these derivatives are transformed to the physical space by utilizing the Jacobian of the grid transformation to get the non-uniform derivative schemes. The five-stage fourth-order Runge–Kutta scheme is adopted

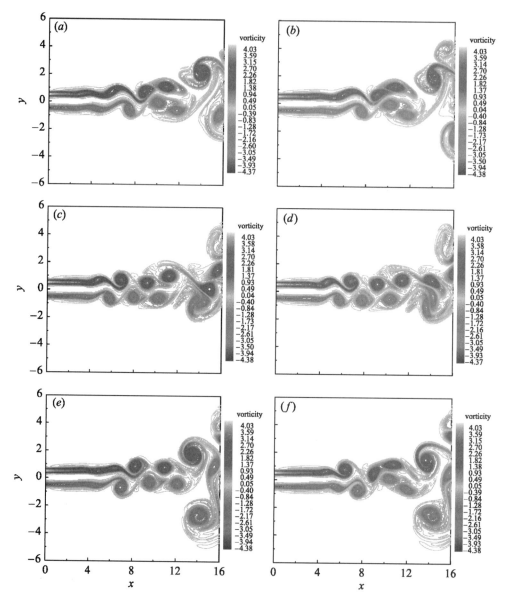

Figure 2. Modulations on the large-scale vortex structures by particles at different Stokes numbers under the same mass loading of 0.3. (a) One-way coupling, $St=0.01$, $t=152.02$. (b) Two-way coupling, $St=0.01$, $t=152.02$. (c) One-way coupling, $St=1$, $t=168.02$. (d) Two-way coupling, $St=1$, $t=168.02$. (e) One-way coupling, $St=50$, $t=252.02$. (f) Two-way coupling, $St=50$, $t=252.02$.

to integrate the Euler terms in the above governing equations with sufficient numerical accuracy and lower memory requirement. However, due to the larger jet Reynolds number, the viscous, coupling and conduction terms are much smaller than the Euler terms. As a result, they are integrated by using the first-order Euler integration scheme to reduce the computational work by 20% and maintain the considerable accuracy. In addition, the non-uniform fourth-order compact filter is utilized after getting the flow solutions at each time step to eliminate the high wave number errors.

In order to reveal the typical modulations on turbulence by different dispersed solid particles, the particles at Stokes numbers of 0.01, 1 and 50 are traced by using the two-way coupling method and compared with one-way coupling cases. The concept of the computational particles is introduced to change the particle mass loading and reduce computational work. A computational particle represents a group of particles with the same velocity and position (Elghobashi 1994; Fan *et al.* 2003). The use of the computational particle will lead to numerical errors, but the errors are slight and can be neglected when the real particle number replaced by a computational particle is not very large. For each case, 121 computational particles are ejected into the flow-field at the inflow plane every five time steps. The non-dimensional time step $\Delta t = 0.02$, and a total of about 840 000 computational particles are traced for each case.

3. Results and discussion

(a) Modulations on large-scale vortex structures

It is well known that the flow-field coherent structures in the jet play very important roles in the particle dispersion. In reverse, the particles can modulate the flow-field coherent structures under moderate mass loadings. Figure 2 shows the modulations on the large-scale vortex structures by particles at different Stokes numbers. The particle's mass loading Z_m is the same and equal to 0.3 for each case. It is clear that the modulation degrees on large-scale vortex structures by particles at different Stokes numbers are different. At the earliest non-dimensional time, $t = 152.02$, the modulation by particles at a Stokes number of 0.01 becomes obvious, because the number of these particles interacting with the vortex structures is largest and the effects are prominent under the same mass loading. These particles enhance the development of the large-scale vortex structures (figure 2a,b) and destabilize the flow, which is coincident with the previous linear instability conclusion (Despirito & Wang 2001). For particles at a Stokes number of 50, the modulation on flow structures becomes evident at the later non-dimensional time $t = 252.02$. These particles also speed the evolution of the large-scale vortex structures (figure 2e,f). However, for particles at a Stokes number of 1, the modulation can be seen at the moderately non-dimensional time $t = 168.02$, and the particles delay the growth of the large-scale vortex structures (figure 2c,d). It means that the particles at Stokes number of 1 stabilize the flow due to the local-focusing dispersion pattern, which also agrees with the previous linear instability conclusion. As the particle's Stokes number increases, it takes a longer time for the modulation on the large-scale vortex structures to become obvious. The reason is that the number of particles interacted with vortex becomes smaller for the particle at larger Stokes number under the same particle mass loading. As a result, it needs a longer time to have obvious effects on the large-scale vortex structures. In addition, the large-scale vortex structures are slightly compressed along the lateral direction by the presence of the particles. With increasing time, the modulations on the large-scale vortex structures by different particles become more and more distinct.

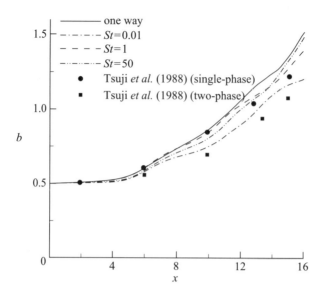

Figure 3. Modulations on the jet velocity half-width by particles at different Stokes numbers under the same mass loading of 0.3.

(b) *Modulations on mean velocity*

Modulations on the jet half-width based on mean velocity by particles at different Stokes numbers under the same mass loading of 0.3 are shown in figure 3. The jet velocity half-width, b, is the distance from the jet centreline to the point where the mean streamwise velocity excess is half of the centreline velocity excess. It is obvious that all particles decrease the jet velocity half-width because the momentum transfers from the gas phase to the particles. This qualitative trend is in good agreement with the previous experiment data (Tsuji et al. 1988) though a quantitative difference exists. The results show that the jet is made narrow by all particles, and the particles at a Stokes number of 0.01 cause the lowest spreading of the jet due to their larger effects on the jet under the same mass loading.

The developments of the streamwise mean velocity in the jet centreline for all cases are shown in figure 4. Here $\Delta U_0 = U_1 - U_2$ is the mean velocity excess at the inflow position. $\Delta U_c = U - U_2$ is the centreline velocity excess. The particle mass loading is same as 0.3. The presence of particles reduces the decay of the streamwise mean velocity because of the mutual momentum transferring between the fluid and the particles. When the fluid velocity is larger than that of the particle, momentum transferral from the fluid to the particle occurs. But when the particle velocity is larger than that of the fluid, the momentum transfers back to the fluid. In the jet centreline, the particle velocity can exceed the fluid velocity easily and the reverse momentum transferring often occurs. So the decay of the streamwise mean velocity in the jet centreline becomes slower due to the effects of the particles, which agrees well with the laser Doppler anemometry measured results (Modarress et al. 1984). In addition, beyond the potential core region of the jet, the reduction of the streamwise mean velocity decay in the jet centreline is higher by particles at a Stokes number of 1 and lower by particles at a Stokes number of 50. This is associated with the different

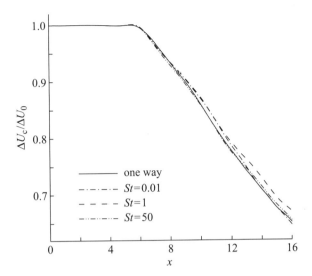

Figure 4. Decay of the streamwise mean velocity in the jet centreline for all cases.

dispersion patterns for particles at different Stokes numbers in the plane jet. Particles at a Stokes number of 1 are thrown out of the large-scale vortex structures at higher velocity and local-focusing dispersion occurs. So the frequency of the reverse momentum transferral is higher. Particles at a Stokes number of 50 disperse slowly along the lateral direction due to their larger mass inertia. As a result, most of the particles distribute near the jet centreline and need a longer time to accelerate to obtain a higher velocity than that of the fluid.

To show the modulations on the mean velocity along the lateral direction by different particles, the distribution of mean velocity profiles at the position of $x=12$ is shown in figure 5. For each case, the mass loading is same as 0.3. The two-way coupled streamwise mean velocity profiles become wider than the one-way coupling case, as shown in figure 5a. For $y>0$, all the particles reduce the streamwise mean velocity near the jet centreline, but for $y>1$, the reverse trend is found. This trend is in excellent agreement with the corresponding experimental data at $x=10$ (Fleckhaus et al. 1987). In that experiment, the particle diameters are 64 and 132 μm. To qualitatively compare our simulated results with that data, we multiply the x-coordinate, y/x in their figure, by 10. However, for $y<0$, the streamwise mean velocity is increased by the particles. This means that little deflection of the jet occurs due to the addition of the particles, which is associated with the action of particle gravitation. As a whole, the modulation on the streamwise mean velocity by the particles at a Stokes number of 0.01 is still the largest under the same mass loading. As to the lateral mean velocity, the particles at a Stokes number of 1 always increase it, but the particles at a Stokes number of 0.01 and 50 decrease it in the region of $y<1$, as shown in figure 5b.

(c) Modulations on turbulence intensities and Reynolds stress

Modulation on the turbulence intensity profiles at the position $x=10$ for each case under the same mass loading of 0.3 is examined. It is found from figure 6

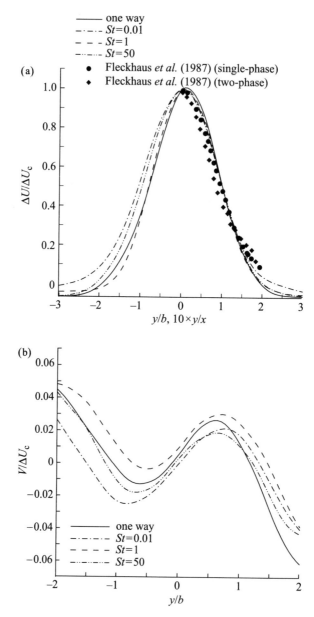

Figure 5. Modulations on the mean velocity profiles by different particles at the position of $x=12$ under the same mass loading of 0.3. (a) Streamwise mean velocity profiles. (b) Lateral mean velocity profiles.

that the modulations on turbulence intensity profiles by different particles are different. The particles at Stokes numbers of 0.01 and 50 reduce the turbulence intensities in the jet centre region but increase them in other regions, which makes the turbulence intensity profiles wider and lower. The particles at a Stokes number of 1 always decrease the streamwise and lateral turbulence intensities, which makes the profiles narrower and lower. The different modulation patterns

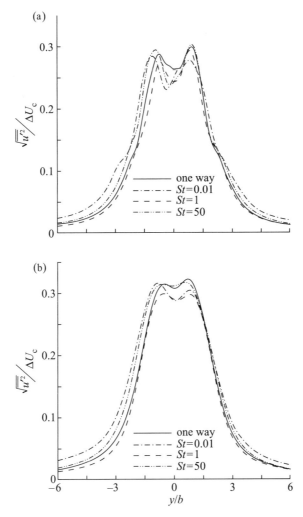

Figure 6. Modulations on the turbulence intensity profiles by different particles at the position of $x=10$ under the same mass loading of 0.3. (a) Streamwise turbulence intensity profiles. (b) Lateral turbulence intensity profiles.

on the turbulence intensity profiles by different particles are also closely related to their different dispersion patterns. Due to the local-focusing dispersion and the higher concentration distribution, the particles at a Stokes number of 1 reduce the turbulence intensities. For the particles at Stokes numbers of 0.01 and 50, the particle concentration is relatively higher in the jet centre region, but it is lower in other regions. So the modulations on the turbulence intensity profiles by these particles are non-uniform.

The modulations on the Reynolds stress profile at $x=10$ for all cases under the same mass loading of 0.3 are depicted in figure 7. The Reynolds stress of two-way coupling cases is decreased by particles at Stokes number of 0.01 and 50 compared with that of the one-way coupling cases. The trend also accords with previous experimental data of pipe jet at the position of $x/d=20$ (Modarress

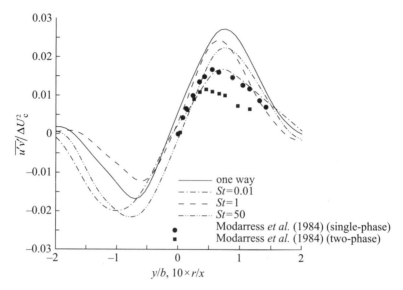

Figure 7. Modulations on the Reynolds stress profiles by different particles at the position of $x=10$ under the same mass loading of 0.3.

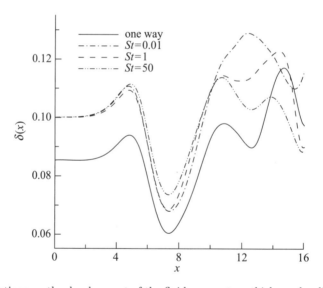

Figure 8. Modulations on the development of the fluid momentum thickness by different particles under the same mass loading of 0.3.

et al. 1984). We also multiply the x-coordinate r/x in their figure by 10 to qualitatively compare our simulated results with their experimental data. Although the trends between simulated and experimental results are very consistent, the quantitative difference is obvious. This is related to the limit of the present two-dimensional plane simulation. Particles at a Stokes number of 1 reduce Reynolds stress for $y<0$, but increase it for most of $y>0$. In addition, the maximum values of the Reynolds stress for all cases are obtained at about $y=0.75$.

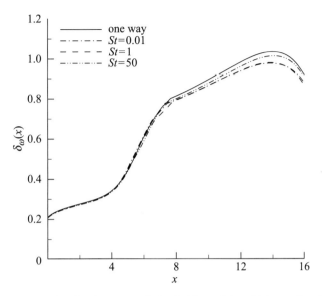

Figure 9. Modulations on the development of the fluid vorticity thickness by different particles under the same mass loading of 0.3.

(d) Modulations on shear layer thickness and turbulent kinetic energy

The momentum thickness is an important parameter for the free shear layer flows. In the present paper, the jet momentum thickness is defined as

$$\delta(x) = \frac{1}{\Delta U_0^2} \int_{-\infty}^{+\infty} (U_1 - \overline{u(y)})(\overline{u(y)} - U_2) \mathrm{d}y. \tag{3.1}$$

Figure 8 shows the modulations on the jet momentum thickness by particles at different Stokes numbers. It is found that the addition of the particles enhances the jet momentum thickness compared with that of the one-way coupling case. Particles at a Stokes number of 50 lead to a larger effect on the development of the momentum thickness in most of the region under the same mass loading of 0.3, but in the region of $10 < x < 14$, the modulations by particles at a Stokes number of 0.01 and 1 become more remarkable.

The vorticity thickness, defined as $\delta_\varpi(x) = \Delta U/(\partial \overline{u(y)}/\partial y)_{\max}$, is also studied to show the largest vortex scale along the lateral direction in the jet flow. As shown in figure 9, all particles decrease the fluid vorticity thickness. It indicates that the particles compress the large-scale vortex structures along the lateral direction. These results agree with the above discussion about the modulations on large-scale vortex structures and jet velocity half-width. The reduction of the vorticity thickness by particles at a larger Stokes number of 50 is not obvious, because these particles disperse less along the lateral direction.

(e) Modulations on particle dispersion

To quantify the modulations on the dispersion of particles at different Stokes numbers in the flow-field, the particle number density is calculated. The particle number density is defined as the number of particles located in each grid at some

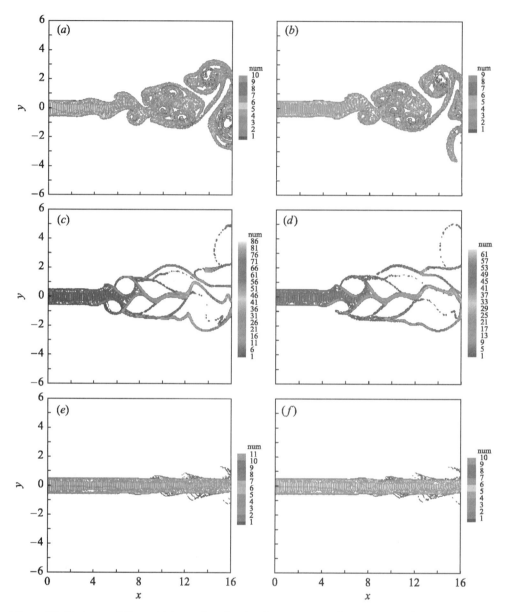

Figure 10. Contours of the particle number density for one-way and two-way coupling cases under the same mass loading of 0.3. (a) One-way coupling, $St=0.01$, $t=152.02$. (b) Two-way coupling, $St=0.01$, $t=152.02$. (c) One-way coupling, $St=1$, $t=168.02$. (d) Two-way coupling, $St=1$, $t=168.02$. (e) One-way coupling, $St=50$, $t=252.02$. (f) Two-way coupling, $St=50$, $t=252.02$.

time to describe the particle concentration distribution. The contours of the particle number density for one-way and two-way coupling cases are shown in figure 10. It should be noted that the scales of the contour plots are different. Obviously, the number densities of particles at Stokes numbers of 0.01 and 50 are lower, suggesting that these two kinds of particles distribute more uniformly in the flow-field due to the much smaller and much larger aerodynamic time, respectively, though the dispersion scope for these two kinds of particles is very different. For an intermediate Stokes number of 1, the particle number density in the outer regions of the large-scale vortex structures is much higher. Particularly, in the common boundary connecting multiple vortex structures, the particle number density becomes extremely high, which means that local-focusing dispersion occurs for these particles. The different dispersion patterns of particles at different Stokes numbers lead to different modulations on the jet. By comparing two-way with one-way coupling cases, it is found that the two-way coupling effect reduces the particle number density in the flow-field and the reduction for particles at a Stokes number of 1 is more remarkable. These results indicate that the particles distribute more evenly in the two-way coupled jets.

4. Conclusions

To investigate the modulations on turbulent characteristics by particles at different Stokes numbers, a two-way coupled, weakly compressible and spatially developing particle-laden turbulent jet was studied by using the DNS technique. The gas phase flow-field was directly solved by using high-resolution algorithm. The particles were traced in the Lagrangian framework. Due to the smaller Mach number, the effect of the compressibility was not discussed. According to the DNS results, the following conclusions can be reached.

(i) The evolving of large-scale vortex structures is accelerated by the particles at Stokes numbers of 0.01 and 50, but delayed by the particles at a Stokes number of 1. Under the same mass loading, the particles at the smaller Stokes number of 0.01 have the largest effects on the large-scale vortex structures.
(ii) The presence of particles reduces the jet velocity half-width and reduces the decay of the streamwise mean velocity in the jet centreline. The decrease in the velocity half-width by the particles at a Stokes number of 0.01 is the largest, and the reduction of the decay of streamwise mean velocity by the particles at a Stokes number of 1 is more obvious.
(iii) A slight deflection of the streamwise mean velocity profiles towards the gravity direction occurs due to the addition of particles. The lateral mean velocity is increased by the particles at a Stokes number of 1, but the modulation is very finite.
(iv) The particles at Stokes numbers of 0.01 and 50 make the turbulence intensity profiles wider and lower. However, the particles at a Stokes number of 1 always decrease the turbulence intensities.
(v) All particles enhance the fluid momentum thickness and reduce the vorticity thickness. Compared with one-way coupling case, the particles distribute more evenly in the flow-field under two-way coupling cases.

The authors gratefully acknowledge the financial support of this research by the Key Program of Natural Science Foundations of China (grant no. 50 236 030). Thanks to the referees for their suggestions and to Miss Annie for her helpful grammar revision to improve the paper.

References

Adams, N. A. 1998 Direct numerical simulation of turbulent compression corner flow. *Theor. Comp. Fluid Dyn.* **12**, 109–129. (doi:10.1007/s001620050102.)

Boivin, M., Simonin, O. & Squires, K. D. 1998 Direct numerical simulation of turbulence modulation by particles in isotropic turbulence. *J. Fluid Mech.* **375**, 235–263. (doi:10.1017/S0022112098002821.)

Carpenter, M. H., Gottlieb, D. & Abarbanel, S. 1993 The stability of numerical boundary treatments for compact high-order finite-difference schemes. *J. Comput. Phys.* **108**, 272–295. (doi:10.1006/jcph.1993.1182.)

Chein, R. & Chung, J. N. 1988 Simulation of particle dispersion in a two-dimensional mixing layer. *AICHE J.* **34**, 946–954. (doi:10.1002/aic.690340607.)

Chung, J. N. & Troutt, T. R. 1988 Simulation of particle dispersion in an axisymmetric jet. *J. Fluid Mech.* **186**, 199–222.

Despirito, J. & Wang, L. P. 2001 Linear instability of two-way coupled particle-laden jet. *Int. J. Multiphase Flow* **27**, 1179–1198. (doi:10.1016/S0301-9322(00)00067-7.)

Elghobashi, S. 1994 On predicting particle-laden turbulent flows. *Appl. Sci. Res.* **52**, 309–329. (doi:10.1007/BF00936835.)

Fan, J. R., Zheng, Y. Q., Yao, J. & Cen, K. F. 2001 Direct simulation of particle dispersion in a three-dimensional temporal mixing layer. *Proc. R. Soc. A* **457**, 2151–2166.

Fan, J. R., Luo, K., Zheng, Y. Q., Jin, H. H. & Cen, K. F. 2003 Modulation on coherent vortex structures by dispersed solid particles in a three-dimensional mixing layer. *Phys. Rev. E* **68**, 036 309. (doi:10.1103/PhysRevE.68.036309.)

Fleckhaus, D., Hishida, K. & Maeda, M. 1987 Effect of laden solid particles on the turbulent flow structures of a round free jet. *Exp. Fluids* **5**, 323–333. (doi:10.1007/BF00277711.)

Freund, J. B., Lele, S. K. & Moin, P. 2000 Numerical simulation of a Mach 1.92 turbulent jet and its sound field. *AIAA J.* **38**, 2023–2031.

Glaze, D. J. & Frankel, S. H. 2000 Effect of dispersion characteristics on particle temperature in an idealized nonpremixed reacting jet. *Int. J. Multiphase Flow* **26**, 609–633. (doi:10.1016/S0301-9322(99)00013-0.)

Hardalupas, Y., Taylor, A. M. P. K. & Whitelow, J. H. 1989 Velocity and particle-flux characteristics of turbulent particle laden jets. *Proc. R. Soc. A* **426**, 31.

Hu, F. Q. 1996 On absorbing boundary conditions for linearized Euler equations by a perfectly matched layer. *J. Comput. Phys.* **129**, 201–219. (doi:10.1006/jcph.1996.0244.)

Hu, F. Q. 2001 A stable, perfectly matched layer for linearized Euler equations in unsplit physical variables. *J. Comput. Phys.* **173**, 455–480. (doi:10.1006/jcph.2001.6887.)

Lele, S. K. 1992 Compact finite difference schemes with spectral-like resolution. *J. Comput. Phys.* **103**, 16–42. (doi:10.1016/0021-9991(92)90324-R.)

Modarress, D., Tan, H. & Elghobashi, S. 1984 Two-component LDA measurement in a two-phase turbulent jet. *AIAA J.* **22**, 624–630.

Parthasarathy, R. N. 1995 Stability of particle-laden round jet to small disturbances. In *Proceedings of the FEDSM'95 San Francisco, CA* (ed. D. C. Wiggert et al.), vol. 228, pp. 427–434. New York: American Society of Mechanical Engineers.

Poinsot, T. J. & Lele, S. K. 1992 Boundary conditions for direct simulations of compressible viscous flows. *J. Comput. Phys.* **101**, 104–129. (doi:10.1016/0021-9991(92)90046-2.)

Rudy, D. H. & Strikwerda, J. C. 1980 A nonreflecting outflow boundary condition for subsonic Navier–Stokes calculations. *J. Comput. Phys.* **36**, 55–70. (doi:10.1016/0021-9991(80)90174-6.)

Saffman, P. G. 1965 The lift on a small sphere in a shear flow. *J. Fluid Mech.* **22**, 385–400.

Tang, L., Wen, F., Yang, Y., Crowe, C. T., Chung, J. N. & Troutt, T. R. 1992 Self-organizing particle dispersion mechanism in free shear flows. *Phys. Fluids A* **4**, 2244–2251. (doi:10.1063/1.858465.)

Thompson, K. W. 1987 Time dependent boundary conditions for hyperbolic systems. *J. Comput. Phys.* **68**, 1–24. (doi:10.1016/0021-9991(87)90041-6.)

Tsuji, Y., Morikawa, Y., Tanaka, T. & Karimine, K. 1988 Measurement of an axisymmetric jet laden with coarse particles. *Int. J. Multiphase Flow* **14**, 565–574. (doi:10.1016/0301-9322(88)90058-4.)

Xu, B. H. & Yu, A. B. 1997 Numerical simulation of the gas–solid flow in a fluidized bed by combining discrete particle method with computation fluid dynamics. *Chem. Eng. Sci.* **52**, 2785–2809. (doi:10.1016/S0009-2509(97)00081-X.)

Yuu, S., Ikeda, K. & Umekage, T. 1996 Flow-field prediction and experimental verification of low Reynolds number gas–particle turbulent jets. *Colloids Surfaces A Physicochem. Eng. Aspects* **109**, 13–27. (doi:10.1016/0927-7757(95)03470-6.)

气固射流中分散颗粒对湍流特性的调制

摘要：本文发展了一种结合双向耦合方法的直接数值模拟方法，以研究具有中等雷诺数的气-固两相湍流射流，该流动是弱可压缩且空间发展的。针对气相流场进行高分辨率求解，并采用拉格朗日法追踪粒子。研究了不同斯托克斯数下颗粒对流动结构和其他湍流特性的调制。结果表明，斯托克斯数为 0.01 和 50 的颗粒可以促进大尺度涡旋结构的发展，使湍流强度分布更宽、更低，而斯托克斯数为 1 的颗粒会延迟大尺度涡旋结构的演变并降低湍流强度。所有颗粒都减少了射流中心线的速度半宽和流向平均速度的衰减，其中斯托克斯数为 0.01 的颗粒导致速度半宽的较大减少，斯托克斯数为 1 的颗粒则导致流向平均速度衰减的较大减少。所有颗粒都减小了涡度厚度，但增加了流体动量厚度。此外，双向耦合的颗粒分布比单向耦合的更均匀。

原文刊于 Proceedings of the Royal Society A-Mathematical Physical and Engineering Sciences, 2005, 461: 3279-3295

高效低污染燃烧及气化技术的最新研究进展

岑可法,程 军,池 涌,周 昊

(浙江大学 能源洁净利用与环境工程教育部重点实验室,杭州 310027)

摘 要:根据参加 2003 日本神户国际动力工程会议的情况,着重评述了化石燃料低 NO_x 燃烧、固体废弃物焚烧、煤及固体废弃物气化等技术领域的最新研究进展。目前烟气脱硝成为中国电厂技术革新的热点问题,炉内高效低 NO_x 燃烧是一种适于我国国情的高效廉价的技术路线。热解和焚烧相结合的固体废弃物利用技术在日本已得到较为广泛的应用,但是它利用蒸汽轮机发电对于未来小型发电系统而言效率较低,而热解和气化相结合的固体废弃物利用技术是利用内燃机或燃气轮机发电,对于日处理废弃物量小于 $200t/d$ 的发电系统其效率较高,因此成为下一代分布式能源利用系统的良好选择。图7参1

关键词:环境工程学;燃烧;气化;评论;煤;废弃物;低 NO_x

中图分类号:TK31 **文献标识码**:A

Recent Progress in Research on Efficient Low Polluting Combustion and Gasification Technology

CHEN Ke-fa, CHENG Jun, CHI Yong, ZHOU Hao

(Clean Energy and Environment Engineering Key Lab of the Ministry of Education,
Zhejiang University, Hangzhou 310027, China)

Abstract: According to proceedings of the International Conference on Power Engineering, held in Kobe, Japan, in 2003, recent technical progress in low NO_x combustion, incineration of solid wastes, gasification of coal and solid wastes are being reviewed in the main. At preset, denitration of flue gas has become an urgent problem in the retrofit of Chinese power plants; highly efficient combustion with low NO_x emission is a cost efficient technology, suitable for Chinese conditions. The technology of waste utilization by combining pyrolysis and incineration has already been popularized in Japan, but its power generation with steam turbines has but an efficiency too low for future small scale power generation systems. Only the technology by means of gas engines or gas turbines proffers a higher efficiency for power generation systems which have a waste handling capacity of below $200t/d$, and may thus become a good option for distributed type of energy utilization. Figs 7 and ref 1.

Key words: environment engineering; combustion; gasification; review; coal; waste; low NO_x

---

收稿日期:2004-11-15
基金项目:国家重点基础研究发展规划项目($G1999022204$);国家高技术研究发展计划项目($2002AA529122$)
作者简介:岑可法(1935-),男,广东南海人,中国工程院院士,浙江大学机械与能源学院院长,教授,博导。主要研究领域为:能源清洁利用与环境工程。

2003年11月在日本神户召开了国际动力工程会议 ICOPE-2003,来自日本、中国、美国等10多个国家的400多名代表参加了会议。大会宣读了3篇特邀报告,日本 Kaneko S 在《21世纪的发电站》特邀报告中指出:在21世纪后半叶,由燃料电池、燃气轮机和蒸汽轮机组成的具有超高效率的三级联合循环发电机组将会得到广泛应用。随着石油和天然气资源的枯竭,太阳能、风能和生物质能等可再生能源将越来越多地应用于分布式电站。美国 Stenzel W 在《美国电力工业的发展前景》特邀报告中指出:过去3年美国在燃气轮机联合循环发电机组方面发展很快。将来美国天然气价格会更贵,而煤炭和核燃料的价格会有所降低。美国新建和现有发电机组下一步污染治理的重点将是汞和超细颗粒。中国岑可法院士在"煤和废弃物燃烧发电过程中污染物的生成及控制"特邀报告中介绍了炉内燃煤脱硫脱硝、煤与垃圾混烧抑制二噁英、吸收剂脱除 HCl、HF 和汞等污染控制技术的最新进展。

大会着重对锅炉、汽轮机和发电技术、燃烧和气化技术、废弃物能源化利用技术、污染控制技术、燃料电池、可再生能源、分布式能源系统等进行了分组讨论。本文就此次会议中了解到的高效低污染燃烧及气化技术的最新研究进展进行评述。

1 化石燃料低 NO_x 燃烧技术

日本日立公司早在1990年和1992年开发出两代低 NO_x 旋流燃烧器,通过在高温富燃料区加速 NO_x 分解从而达到显著降低 NO_x 排放的效果,最近开发出更加高效低污染的第三代低 NO_x 燃烧器(HT-NR3),如图1所示。其特点为:① 对煤粉浓缩器的进出口角度进行优化,在火焰稳定环处加装1块隔板以增大高温烟气的回流区,在强化着火性能的同时增大还原区以减少 NO_x 生成;② 在喷口处增加1块导流板使三次风有效分离,使燃烧火焰更宽更短以降低 NO_x 生成。1997年 HT-NR3 开始应用于芬兰 INKOO 热电站1台265MW锅炉改造等工程,在保持相同飞灰含碳量3%的情况下,比改造前采用传统旋流燃烧器的单级燃烧方式下 NO_x 排放降低约75%;后来应用于日本东京电力公司 Hitachi Naka 1 号机组1台1000 MW 的新建锅炉,飞灰含碳量保持在3%~4%,折算到氧量6%条件下 NO_x 排放浓度低于140 mg/kg。日本三菱重工株式会社在 A-PM 型低 NO_x 燃烧器基础上,开发了一种燃用挥发份低于10%的石油焦的燃烧器,通过优化风速、将燃料与一次风浓淡分离以及增加稳燃结构等,提高了石油焦的着火稳定性并且实现了低 NO_x 排放。首先在炉内燃烧器区域控制过量空气系数小于1,利用 A-PM 型燃烧器喷嘴内部的浓淡分离装置产生外浓内淡的环状火焰,有利于过量空气分布均匀并且延长煤粉的实际停留时间,以减少高温区 NO_x 生成。然后在炉膛上部送入燃尽风使过量空气系数大于1,控制炉膛出口氧量在2%~4%的较低水平,使飞灰含碳量保持在较低水平的同时降低 NO_x 排放。

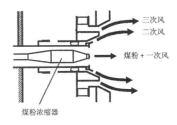

(a) 第二代低 NO_x 燃烧器 HT-NR2

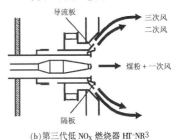

(b) 第三代低 NO_x 燃烧器 HT-NR3

图1 日本日立公司低 NO_x 旋流燃烧器

Fig 1 Hitachi vortical burner for low NO_x combustion

日本中央电力研究院针对水份高于20%的次烟煤燃烧过程研究了分级优化配风以降低飞灰含碳量和 NO_x 排放的技术。采用日本中央电力研究院和石川岛播磨重工株式会社合作开发的 CI-α 型低 NO_x 旋流燃烧器,如图2所示,在燃烧器喷口附近形成高温烟气的回流区,煤粉首先燃烧加速放出 NO_x,其后在宽广的强还原性火焰区内 NO_x 快速分解生成 N_2,最后在炉膛上部喷入燃尽风促使残碳燃尽。研究表明:当燃尽风后移或者燃尽风量增加时,炉膛出口的 NO_x 排放浓度降低,但是当燃尽风喷射位置与燃烧器出口之间的距离达到炉膛当量直径的4.5倍时或当燃尽风率达到30%时,降低 NO_x 排放的效果达到饱和,而此时飞灰含碳量并没有显著增加。对于传统的液态排渣锅炉,燃烧过程一般在凝渣管之前完成,导致 NO_x 排放浓度高达500~600 mg/kg,高于传统的干式排渣煤粉炉 NO_x 排放浓度一般为200 mg/kg 的水平。日本川崎重工株式会社在1台1.8 MW 的 U 型燃烧液态排渣试验炉上研究了低 NO_x 燃烧技术,提出将燃尽风分级的位置从凝渣管上游移到下游，在产生融渣的高温区内造成还原性气氛，从而降低 NOₓ 排放。当燃尽风喷射位置与燃烧器喷口之间的距离为煤粉提供的停留时间约为传统干式排渣煤粉炉的 3 倍时，NOₓ 排放浓度可以降低到 100 mg/kg 左右。

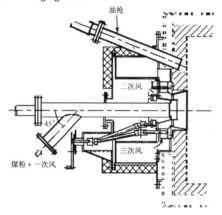

图 2　日本中央电力研究院 CI-α 型低 NOₓ 旋流燃烧器
Fig 2　Japanese CI-α type of vortical burner for low NOₓ combustion

日本川崎重工株式会社研究了直接燃用重油如沥青和奥里油的锅炉降低 NOₓ 排放的技术，提出一个称为 KACC 的先进洁净燃烧系统，即首先在下部炉膛内燃烧器区域铺设耐火泥，造成 1500℃～1600℃的高温和强还原性气氛以减少燃烧过程产生的 NOₓ，当燃烧器区域的过量空气系数控制在 0.65～0.75 时 NOₓ 生成量最低；然后在上部水冷炉膛内送入分级风完成燃尽过程。在 1 台 1.8 MW 试验炉上研究表明：该技术能使 NOₓ 排放浓度降低至 100 mg/kg 以下(折算到氧量 4%)。

若再采用炉内喷射尿素溶液的方法，控制氨氮摩尔比为 0.8～1.3，则 NOₓ 排放浓度能再降低 20～30%。目前该技术已应用于 3 台 100～200 t/h 的重油锅炉，NOₓ 排放浓度达到预期指标 100 mg/kg 左右。日本东京工业大学研究了重油锅炉采用高温空气燃烧降低 NOₓ 排放的技术，将预先加热到 1000℃的高温空气与贫氧的循环烟气混合后送入炉膛，控制整体的过量空气系数低于 1.0，则在一般氧量过剩条件下产生的强发光火焰消失，重油燃烧过程拖长并且充满整个炉膛，炉内温度分布均匀仅为 900℃左右，故 NOₓ 排放量显著降低到 80 mg/kg 以下。高温空气燃烧技术不仅具有很高的燃烧效率，而且能够显著降低 NOₓ、烟尘和二噁英等污染物排放，是不仅适合于天然气，而且适合于石油和煤炭的先进的高效低污染燃烧方式。

2　固体废弃物焚烧技术

据世界能源组织调查显示，世界原油可采储量 138.3 Gt，天然气 2.4 Gt，占世界能源供给 90% 的化石燃料的储量在日益枯竭。已探明世界原油可供再利用 50～80 年，天然气可供再利用 100 年，煤可供再利用 200～300 年。面临能源危机和环境污染日益严峻的形势，将作为可再生资源的固体废弃物进行能源化利用，具有节能和环保的双重效益，已成为国际能源领域的研究热点。日本政府 2003 年立法，

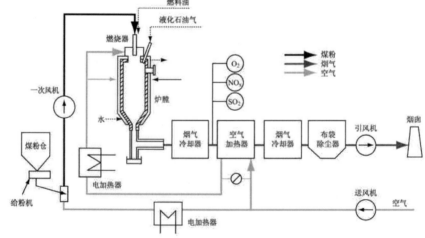

图 3　日本石川岛播磨的 1.2 MW 煤粉炉混烧生物质工业试验装置
Fig 3　IHI 1.2MW pulverized coal and biomass co-fired experimental boiler

强制要求到 2010 年可再生能源发电份额要达到 1.35%，而利用现有电站锅炉中占据绝大多数的煤粉炉混烧生物质就是一种廉价高效的较佳选择。石川岛播磨重工株式会社进行的试验表明：利用直立纺锤型磨煤机进行混合研磨，约有重量比 75%~85% 的生物质可以磨制到与煤粉同样的细度，不过研磨生物质的电耗(120~130 kWh/t)比磨煤电耗(15 kWh/t)要高得多。将锯木屑、木片等生物质以低于 3% 的重量比掺入煤中混合磨制是可行的，混合燃料的粒度只是略有增加，混磨耗电量增加 23%。利用 1 台 1.2MW 煤粉炉进行了混烧生物质(添加量 6%) 的工业性试验(图 3)，表明炉内燃烧情况良好，混烧时 NO_X 排放浓度有所降低，而燃烧效率基本不变。

由于流化床燃料适应性广，故特别适于燃烧生物质、淤泥等低热值燃料，目前其燃烧技术已日趋成熟，故关于其多种污染物的控制问题倍受重视。日本三菱重工株式会社于 2001 年建成了 1 台木片等生物质(热量输入占 57%~100%)与煤混烧的 60 t/h 鼓泡流化床锅炉，发现在悬浮段高温氧化区的 CO 和二噁英浓度降低，而在密相还原区的 NO_X 浓度降低，如何同时控制 CO、二噁英和 NO_X 的低污染排放是一项困难的技术。因此，提出四级送风的技术，系统布置如图 4 所示。将流化床密相区保持为还原性气氛(过量空气系数约为 0.6，床温约为 850℃)以降低 NO_X 排放，然后逐级送入二次风，使烟气温度逐渐升高到 1200℃，以降低 CO 和二噁英的浓度，炉膛出口过量空气系数控制在 1.3(温度降至接近 800℃)，从而实现 CO、二噁英和 NO_X 的最终排放量能够分别低于 100 mg/kg($12\%O_2$)、150 mg/kg($6\%O_2$)和 0.1 ngTEQ/Nm^3($12\%O_2$)的排放要求。利用增压流化床可以实现低热值固体燃料的高效低污染燃烧，日本已有 3 座增压流化床与燃气轮机相结合的联合循环发电机组(85 MW~360 MW)投入商业运行。日本先进工业科技研究院实验室研究表明：利用增压流化床(最大运行压力为 1 MPa)焚烧淤泥是可行的，淤泥具有良好的可燃性，在增压条件下的燃烧速率比常压下要高 3 个数量级，飞灰含碳量低于 0.1%，燃烧效率高达 99.9%。当焚烧湿淤泥时，70% 以上的入炉灰以平均粒径为 150μm 的飞灰形式排放，其余 30% 粒径为 mm 级的灰份保留在床层中。而焚烧干淤泥时约有 80wt% 的入炉灰保留在床层中。虽然淤泥中具有较高的挥发份和氮含量，但它燃烧时 CO 和 NO_X 排放量都非常低(CO<0.1%，NO_X<50 mg/kg)，不过 N_2O 排放量很高(N_2O>200 mg/kg)。由于悬浮段温度升高时 N_2O 排放量降低很快，故当床温运行在较高温度 900℃ 时，N_2O 和 NO_X 排放量都低于 50 mg/kg。

3 煤及固体废弃物气化技术

从能源安全和环境可持续发展的观点来看，整体煤气化联合循环发电系统(IGCC)很可能会成为

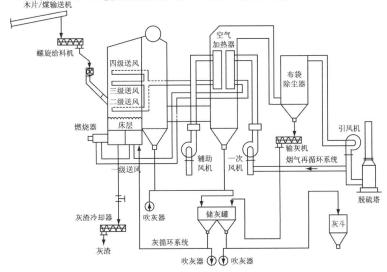

图 4 日本三菱重工混烧木片和煤的 60 t/h 鼓泡流化床锅炉
Fig 4 MHI 60t/h wood chip and coal co-fired bubbing fiuidized bed boiler

下一代燃煤电厂的选择。日本多家电力公司和中央电力研究院已开始进行 250 MW 的 IGCC 示范工程建设,采用了空气喷射的两段夹带流煤气化装置,整座电厂将于 2007 年投入运行。日本中央电力研究院针对该煤气化炉进行了三维的气固反应流动数值模拟,研究了当富含氧气的空气作为气化介质时氧气浓度和蒸汽流速对气化性能的影响。发现当气化介质中氧浓度升高时整体碳转化率升高,原因是不仅燃烧室由于烟气温度升高而导致其出口的碳转化率升高,而且还原室的碳转化率也升高。然而,在相同氧浓度下仅当蒸汽流速增加时整体碳转化率几乎不变,原因是虽然燃烧室由于烟气温度明显降低而导致其出口的碳转化率降低,但是还原室的碳转化率升高。为了在运行稳定情况下尽量提高整体气化效率,建议气化炉最佳运行条件如图 5 所示,即烟气温度为 1850 K~2173 K,氧浓度为 21%~38%,蒸汽流量为 0~20%,则整体碳转化率可达 66%~72%。

利用超临界液体如水或 CO_2 进行碳氢燃料的裂解或气化反应,由于其具有环境良好和热转化效率高的优点而已成为国际上研究的热点。超临界液体具有象液体一样的密度、象气体一样的扩散性以及完全不同于理想气体的可压缩性,因此具有一些独特的性质,如反应和相态的可控性等。当反应温度和压力增大到足够高时,超临界液体、碳氢燃料和气体能够形成均一物相,因此使反应速度大大增加,故该技术工艺是很有发展前途的。日本三菱材料公司研究了在超临界水中进行煤气化的新工艺,在超临界水条件下(汽温 374 ℃ 以上,汽压 22 MPa 以上)利用氧气对包括煤在内的碳氢燃料进行部分氧化,以产生包含 H_2、CO_2、CH_4 和 CO 的气体,其主要成分是 H_2 和 CO_2。以水煤浆燃料(煤粉浓度为 50%~67%)在气化压力 25 MPa 下进行试验,将气化温度控制在 1000 ℃,氧气比例控制在 0.3~0.4(均明显低于常规气化温度和氧气比例),能够使碳转化率高达 98%,而且烟气冷却效率高达 80%。以烟煤为原料进行气化所需的氧气比例要低于次烟煤,而烟煤产生的煤气量、碳转化率以及烟气冷却效率都高于次烟煤。煤中杂质硫和氮在气化过程中转化成 H_2S、N_2 以及少量 NH_3,这些气体容易从煤气中分离。COS 几乎不会产生,因此煤气的净化比较简单。将 CO_2 以液态形式分离是有可能的,在 20 MPa 压力下将煤气冷却到 -28 ℃ 时,液相中回收的 CO_2 含量可达 80%,煤气中的其它成分 H_2、CH_4 和 CO 等也会溶入液相中。

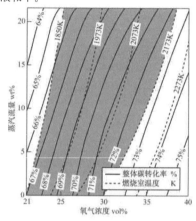

图 5 燃烧室中整体碳转化效率(PPCCE)和烟气温度的分布
Fig 5 Distribution of per pass carbon conversion efficiency and gas temperature in combustor

日本政府 2002 年颁布了长期能源供应计划,要求到 2010 年废弃物发电容量要从当时的 120 MW 增加到 417 MW。日本应用能源研究院组织 6 家有实力的制造厂商引进开发了 3 种先进的废弃物能源化利用技术:① 由于传统的废弃物焚烧炉烟气中 HCl 对过热器管产生强烈的高温腐蚀,故过热蒸汽温度被限制在 300 ℃ 左右而使得发电效率低到 10% 左右。通过改进锅炉设计并且提高过热器材料的性能,能使过热蒸汽温度和压力分别提高到 500 ℃ 和 9.8 MPa,目前已建成 1 台日处理废弃物量达 50 t/h 的示范电厂,发电效率可达 30%。② 热解和燃烧相结合的废弃物利用技术在日本得到了较为广泛的应用,其优点是能够利用自身能量产生高密度的熔融灰、铁和铝等金属具有良好的再生特性、二噁英排放量低以及能够防止灰中重金属浸出等。③ 由于废弃物热解和燃烧技术是利用蒸汽轮机发电,故对于小型发电系统其效率较低。而热解和气化相结合的废弃物利用技术是利用内燃机或燃气轮机发电,对于日处理废弃物量小于 200t/d 的发电系统其效率较高,因此被认为是下一代分布式能源利用系统的良好选择。下面选择其中几个实例进行简要介绍。

由气化炉和燃烧室组成的双床循环气化系统虽然已经得到较为广泛的应用,但是其辐射热损失较多导致其热效率较低。因此,日本 EBARA 株式会社开发了一种内循环流化床气化炉,将 1 个炉膛分为两部分:以蒸汽作为流化气体的气化室和以空气作为流化气体的燃烧室,(图6)。2 个反应室内都有石英沙床料作为流化介质,并且在两室之间进行循环。含有焦碳的气化残渣通过床料循环从气化室转移到燃烧室,在燃烧室内焦碳燃烧释放的热量供给石英床料,这些加热的床料再回到气化室中为气化反应供给热量。由于气化过程得到 H_2/CO 摩尔比较高的气体与焦碳燃烧得到的气体是从两个反应室中分别排出而没有混和,故该内循环流化床气化炉可以利用生物质和废弃物等低热质固体燃料产生高热值煤气。1 台以生物质或废弃塑料为原料的功率为 3 MW 的内循环流化床气化示范装置已于 2002 年建成,气化室和燃烧室温度分别控制在 873 K~1073 K 和 1073 K~1173 K,目前正处于试验改进阶段,初步试验得到气化合成气的热值为 4554 kJ/Nm^3。东芝株式会社 1997 年引进固体废弃物热解气化系统的设计技术,依次设计完成了日处理废弃物为 4.6 t/d 的小型气化系统、10 t/d 的工业性示范系统和 60 t/d 的商业化运行系统,并分别于 1998 年、2000 年和 2001 年投入运行。首先将废弃物送入炉温为 500℃~600℃的回转窑内,加热 60min 左右,有机成分几乎全部热解成干馏气体,然后将其送入温度为 1000℃~1200℃的重整反应器中进一步裂解成短链小分子气体,同时水蒸汽与碳发生的水煤气反应生成 CO 和 H_2。最后将裂解气通入湿式洗涤装置和脱硫装置除去其中的无机酸性污染物、灰尘、重金属和 H_2S 等杂质,得到纯净的可燃气体用于内燃机发电。利用热值为 10265 kJ/kg 的城市固体废弃物和热值为 22204 kJ/kg 的汽车内饰废弃物进行气化试验,得到精华后的合成气热值分别为 4432 kJ/Nm^3 和 4719

kJ/Nm^3,热解得到焦碳的热值分别为 18106 kJ/kg 和 8280 kJ/kg,碳黑的热值分别为 31738 kJ/kg 和 30776 kJ/kg。对于 1 MW 的工业性示范系统和 15 MW 的商业化运行系统,能源转化率分别为 53% 和 81%。另外对于热解得到的焦碳和碳黑,可送入温度为 1300℃~1500℃的反应器中与 O_2 作用气化生成 CO,则 1 MW 和 15 MW 气化系统的能源转化率分别为 36% 和 68%。热解气化后得到各种气固态产物中二噁英的含量均远小于 $1ng\text{-}TEQ/g$ 的排放标准。

气化是生物质等固体废弃物能源化利用的最佳工艺之一,但是如何高效脱除副产物焦油从而提高气化热效率是目前存在的一个难题。由于多孔颗粒具有电容作用,在温度 873 K~1073 K 范围内能够有效地捕集焦油,因此将循环流化床或移动床中的沙粒简单地换成多孔颗粒就有可能实现不含焦油的生物质气化工艺。日本东京工业大学利用硅胶、活性氧化铝和沸石三种多孔颗粒作为固定床的床料,对锯木屑的气化过程试验研究表明:多孔颗粒对于高沸点(>773 K)焦油成分如重油等具有物理吸附作用,对于中沸点(673 K~773 K)焦油成分如萘和芘等具有催化脱氢作用,对于低沸点(<673 K)焦油成分如苯等则无法捕集。多孔颗粒的电容作用受到固体酸性催化的影响很小,然而气化产量和成分则受到固体酸性催化的强烈影响,因此当活性氧化铝和沸石作为气化炉床料时得到的氢气产量分别是硅胶作床料时的 5 倍和 3 倍。

由于建立大规模固体废弃物发电厂经常受到周围居民的反对、收集废弃物需要一定的运输成本以

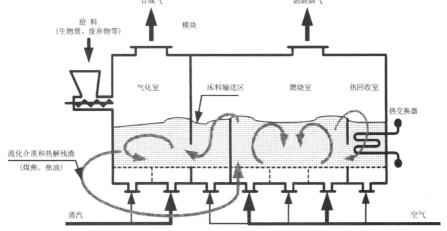

图 6 内循环流化床气化炉的气化和燃烧双室结构

Fig 6 Schematic of an internally circulating fluidized-bed gasification furnace

及余热利用率低等问题,故能否在产生固体废弃物的地点建立新型的处理发电系统成为一个令人感兴趣的课题。日本东京工业大学据此开发了一种称为 STAR-MEET (Steam/Air Reforming type Multi-staged Enthalpy Extraction Technology)的新型固体废弃物小型气化系统,主要包括一个固定床气化炉和一个高温蒸汽/空气重整反应器。利用木片和聚烯烃胶片作为原料的气化试验表明:将高温蒸汽/空气混合物喷射到热解气中能够有效地使焦油成分裂解,重整后得到不含焦油主要含 CO 和 H_2 的洁净气体。对木片而言,当蒸汽直接喷射到热解炉中时,原料消耗量以及重整气中 CO 和 CH_4 的浓度都有所升高;对聚烯烃胶片之类低熔点的原料来说,固定床形式的热解炉更加合适。木片和聚烯烃胶片得到含焦油的气化气的最大热值分别为 5018 kJ/Nm^3 和 11290 kJ/Nm^3,利用高温蒸汽/空气混合物对其重整处理后得到的不含焦油的气化气的最大热值分别为 4386 kJ/Nm^3 和 9659 kJ/Nm^3。通过水蒸气和原料送入该反应系统的氢元素中 22% 转化为气态的 H_2 燃料。日本东京工业大学在实验室研究基础上与石川岛播磨重工株式会社合作建立了一个 200 kg/h 商业化示范电厂(称作 MEET-II)。该系统主要由 1 台融渣气化炉,1 个烟气净化系统,1 个高温空气发生器和 1 台燃气轮机组成。该融渣气化炉的上部是 1 个夹带流反应床,下部是 1 个充填了直径为 50 mm 陶瓷球的固定反应床(图7)。陶瓷球能够蓄热、增加燃料颗粒的停留时间,从而对气化反应具有促进作用。气化炉保持高温,不仅使灰分熔融流入炉底的融渣池,而且使产生的气体中不含焦油。利用粒径为 0.3 mm 的木材粉末进行气化试验,将 1000℃ 左右的高温空气用作气化介质以得到高热值合成气。当空气当量比由 0.32 增加到 0.49 时,碳转化率由 75.3% 增加到 87.9%,烟气冷却效率由 52.9% 增加到 55.6%,但是合成气热值却由 4.3 MJ/Nm^3 减少到 3.7 MJ/Nm^3,原因是空气过量会使气化反应向燃烧反应转化。

4 结 论

中国环保局自 2004 年 1 月 1 日起实施了最新的火电厂大气污染物排放标准,并且在排放浓度达标的前提下要对 NO_X 排放总量进行收费,故脱硝成为中国电厂技术革新的热点问题。目前中国已有少

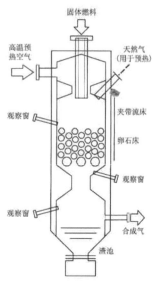

图 7 日本 MEET-II 生物质气化发电系统中的融渣气化炉结构
Fig 7 Schematic of the MEET-II pebble bed slagging gasifier

数电厂采用了脱硝技术,如江苏常熟电厂 300 MW 锅炉采用了天然气再燃脱硝系统,上海外高桥电厂 900 MW 锅炉采用了分级燃烧脱硝系统。因此,与锅炉排烟中低温条件下各种昂贵的催化脱硝系统相比,发展炉内高效低 NO_X 燃烧技术是一种效果较好、成本较低、适于我国国情的有竞争力的技术路线。

对生物质、淤泥、生活垃圾等固体废弃物发展更为先进的焚烧及气化技术继续成为国际能源领域的研究热点,利用超临界流体如水或 CO_2 进行碳氢燃料的裂解或气化反应,由于具有反应速度大、热转化效率高和环境友好的优点也已成为新的研究热点。热解和燃烧相结合的废弃物利用技术在日本已得到较为广泛的应用,但是它利用蒸汽轮机发电对于未来小型发电系统而言效率较低。而热解和气化相结合的废弃物利用技术是利用内燃机或燃气轮机发电,对于日处理废弃物量小于 200 t/d 的发电系统其效率较高,因此成为下一代分布式能源利用系统的良好选择。

参考文献:

[1] Japan Society of Mechanical Engineers [C]. Proceedings of the International Conference on Power Engineering-03, Kobe, Japan, 2003.

Simultaneous removal of NO_x, SO_2 and Hg in nitrogen flow in a narrow reactor by ozone injection: Experimental results

Zhihua Wang*, Junhu Zhou, Yanqun Zhu, Zhengcheng Wen, Jianzhong Liu, Kefa Cen

State Key Laboratory of Clean Energy Utilization, Institute for Thermal Power Engineering, Zhejiang University, Hangzhou 310027, Zhejiang, China

Received 1 August 2006; received in revised form 7 February 2007; accepted 3 April 2007

Abstract

A process capable of removing NO_x, SO_2 and mercury simultaneously was proposed, which utilizes the injection of ozone and assist with a glass made alkaline washing tower. Experiments were conducted in a quartz flow reactor within an electrical heated furnace. Oxidation properties of NO and Hg, removal efficiency of NO and SO_2 behind the washing tower were investigated. Results show that the oxidation efficiency of NO and Hg greatly depends on the amounts of ozone injected. With the increasing amounts of ozone added to the main flow, NO and Hg oxidation efficiency all improved individually. About 85% of NO can be oxidized with 200 ppm of ozone added and 89% of elemental Hg can be oxidized with 250 ppm of ozone added. The optimal temperature for NO oxidation should be lower than 473 K, and the optimal temperature range for mercury was 473 K to 523 K. The appearance of SO_2 has little effect on the NO oxidation process. NO has priority compared to mercury when react with ozone. With the assistance of washing tower, about 97% of NO and nearly 100% of SO_2 can be removed simultaneously with 360 ppm of ozone added.
© 2007 Elsevier B.V. All rights reserved.

Keywords: Ozone injection; Nitrogen oxides; Sulfur dioxide; Mercury

1. Introduction

Various kinds of air pollutants are emitted from power plants, incinerators and boilers during coal combustion. Sulfur dioxide (SO_2) and nitrogen oxides (NO_x) are the most abundant air pollutions in flue gas. Up to now, Wet flue gas desulfurization (WFGD) technologies are still the most effective and widely used methods for the removal of SO_2 in utility boilers[1–3]. For the reduction of NO_x in coal-fired boilers, combustion modification technologies such as low NO_x burners, over fire air, reburning and even choosing low content of nitrogen coals will be first considered [4–6]. After that, the post-combustion technologies including selective catalytic reduction (SCR) and selective non-catalytic reduction (SNCR) will be applied with further demand of NO_x emissions control[7]. SCR is considered as the most effective one for NO_x reduction in the available technologies. The combined application of WFGD and SCR can meet the further limitation by environmental legislation. But the individual treatment strategy will result in expensive investment and operating cost. Therefore, many new technologies are to be updated for high efficiency, low cost, simultaneous removal of SO_2 and NO_x, including the nonthermal plasma, electric beam irradiation and absorption process etc [8–10].

In the nonthermal plasma process, there are two kinds of reactions for NO_x reduction. One side, NO_x can be converted into N_2 directly by N radicals. Another side, NO_x can be oxidized into NO_2, NO_3, N_2O_5 etc. by O, O_3 radicals[11–14]. About 95% of NO_x is NO in the coal combustion flue gas. But NO is insoluble in water and difficult to remove. However, high order nitrogen species such as NO_2, NO_3 and N_2O_5 etc. can reactive with water forming NO_2^- and NO_3^- species which can be efficiently removed in downstream wet scrubber such as WFGD system. Therefore, it's possible for simultaneous removal of NO_x and SO_2 in one washing tower by pre-oxidation of NO.

Besides the most abundant air pollutions emitted from stacks, trace heavy metals are also considered as hazardous air pollutants. As one of these heavy metals, mercury is the first one regulated by the U.S. Environmental Protection Agency (EPA) for the coal-

---

* Corresponding author. Institute for Thermal Power engineering, College of Mechanical and Energy Engineering, Zhejiang University. 310027, Zheda road 38#, Hangzhou, Zhejiang, China. Tel.: +86 571 87953162; fax: +86 571 87951616.

E-mail address: wangzh@zju.edu.cn (Z. Wang).

0378-3820/$ - see front matter © 2007 Elsevier B.V. All rights reserved.
doi:10.1016/j.fuproc.2007.04.001

fired electric power plants. Therefore, all kinds of mercury control technologies were developed and investigated recently including the injection of active carbon, fly ash and various kinds of sorbent [15–17]. The efficiency of methods for removing mercury depends greatly on the species composition of mercury. Mercury in the flue gas usually exists in the form of elemental mercury (Hg^0) and oxidized mercury (Hg^{2+}). It's known that WFGD system can remove nearly 90% of the Hg^{2+} but essentially none of the Hg^0 [18,19]. Research to enhance mercury removal in the wet scrubber focuses on converting Hg^0 into the oxidized form using oxidation reagents and catalyst. Argonne national laboratory (ANL) has developed a chloric acid solution and national energy technology laboratory (NETL) has introduced ultraviolet (UV) light for the oxidation of elemental mercury [20–22]. Published results show that Cl_2, O_3, H_2O_2 and even SCR catalyst can convert Hg^0 into Hg^{2+} efficiently[23–25].

Ozone or disassembly by-products O radical can easily oxidize NO into high order nitrogen species such as NO_2, NO_3 and N_2O_5 etc., which have been found in the nonthermal plasma and electric beam technology. At the same time, O_3, NO_2, NO_3 and O radical can also convert Hg^0 to Hg^{2+}. Unlike the other short life radical species, the life time of O_3 is longer enough under the traditional flue gas temperatures. So, it's likely to discharge only small amounts of pure O_2 or air to produce O_3 highly effectively and then inject it into the flue gas. Ozone can also be considered as O radical carrier in the reaction system. Hence, ozone injection technology is an attractive method which is much more energy-efficient than typical nonthermal plasma and electron beam process directly applied to the exhaust gas. Cannon Technology Inc. in collaboration with BOC Gases has developed a low temperature oxidation (LTO) process for removing NO_x emissions by ozone injection [26]. Yan et al. has conducted the cost-effectiveness analysis of the ozone injection process for NO_x reduction [27]. Young et al. has realized the simultaneous removal of NO_x and SO_2 by ozone injection in a bench scale test facility [28]. But the possibility and property of simultaneous removal of NO_x, SO_2 and Hg by ozone are still unclear. This paper will present detailed investigations on the reactions between O_3 and NO, SO_2, Hg mixture system in nitrogen flow. Combined with glass made alkaline solution washing tower, the multi-pollution removal efficiency will be investigated by the ozone injection technology.

2. Experimental setup

Fig. 1 shows the schematic diagram of the experimental apparatus including the ozone generation, mercury generation, quartz flow reactor, glass made alkaline washing tower and online gas analysis system. Ozone was generated by a dielectric barrier discharge (DBD) device manufactured by Qingdao Guolin Co. with 3.7–4 kV AC voltage and 5 kHz (model CF-G-3-010G). The output concentration of O_3 was monitored by an ozone analyzer continuously (in 2000, USA Co.) which has 0–10000 ppm measuring range and 1 ppm precision.

The simulated flue gas was prepared by N_2 and small amount of concentrated NO gas (0.6% (v/v), balanced with N_2,) and SO_2 gas (1.5% (v/v), balanced with N_2) purchased from New Century Gas Co., China. Oxygen was not absence in main flow in the current test. But the ozone generated from air source, it's inevitable existing oxygen in the reaction system. The flow rate of the gas was all controlled by mass flow controller (MFC, Qixing Huangchuang Co., China). The elemental mercury was generated from mercury osmotic tube (VICI Metronics Co., USA) heated in a thermostatic water bath with 300 mL/min N_2 as carrier gas.

The mixture of NO, SO_2, Hg and N_2 reacted with O_3 in a quartz flow reactor which was located in an electrically heated horizontal furnace. To reduce the impact of non-uniform temperature profile of the furnace, a special delicate flow reactor was designed and made by quartz glass as shown in Fig. 2. It's a three channels homocentric reaction tube. Total length of the reactor is 642 mm with outer-diameter of 20 mm. Heating length of the electric

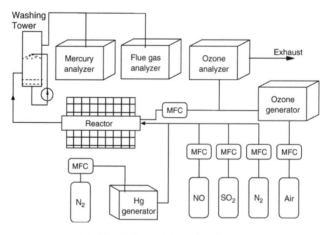

Fig. 1. Schematic diagram of the experimental apparatus.

3. Results and discussion

3.1. Oxidation of NO

In most practical flue gas, NO is the predominant nitrogen species of NO_x. Unlike NO's low solubility, NO_2, NO_3 and N_2O_5 etc. are highly soluble in water, which can be captured in downstream SO_2 removal equipment such as WFGD system. Therefore, the oxidation of NO is the first step of the simultaneously removal process. When O_3 is injected into the center tube, NO in the flue gas can be oxidized into NO_2, NO_3 and N_2O_5 as shown in the following reactions:

$$NO + O_3 = NO_2 + O_2 \tag{1}$$

$$NO_2 + O_3 = NO_3 + O_2 \tag{2}$$

$$NO_2 + NO_3 = N_2O_5 \tag{3}$$

$$NO + O + M = NO_2 + M \tag{4}$$

$$NO_2 + O = NO_3 \tag{5}$$

From reactions (1)–(5), NO_2 is main products when O_3/NO stoichiometric ratio small than 1.0, which are also confirmed in this experiment. The results are shown in Fig. 3 at temperature of 373 K–673 K. For the interference of NO_2 detection module(UV) of CEMs by ozone, the NO_2 concentration is not accurate and can only be considered as nitrogen tracer. The oxidation products at 473 K are shown in Fig. 4. The N_2O, NO_3 and N_2O_5 are minority products[29] and also difficult to measure. Therefore, only NO of NO_x in this experiment is trusted and the oxidation rate is calculated. There is only NO in the original simulated flue gas with N_2 as dilution. As observed, NO can be effectively oxidized but

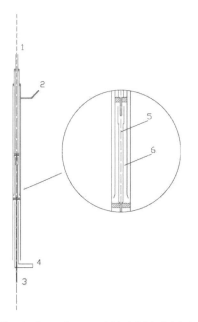

Fig. 2. Schematic of quartz flow reactor. 1.Inlet 1; 2. Inlet 2; 3.Outlet; 4. Air cooling; 5.Quartz flow reactor; 6. Preheat channels.

furnace is about 600 mm. Most part of the reactor was located in center of the furnace. The reactions proceed in the center tube (5 in Fig. 2) with inside diameter of 5 mm and 100 mm in length. The center tube can be considered as isothermal reactor. The NO/SO_2/Hg enriched nitrogen gas flow enter the reactor through inlet 2 and then pass through a reciprocating preheating channel before entering the center tube. O_3 was introduced into the reactor through inlet 1 entering a separate preheating channel. The two well-preheated streams mixed each other instantly just at the nozzle of center tube by high velocity due to the narrowing gas passage.

The washing tower is a glass-made cylinder with 37 mm inner diameter and 150 mm in length. Total volume of the washing tower is about 161 mL and running temperature is about 293 K. The alkaline absorber is 1% (w/w) $Ca(OH)_2$ solutions and 1 L is used in one test. The alkaline solutions are recycled using a pressure pump and sprayed into the tower through an atomizer at a flow rate of 0.5 L/min. The concentration of NO, NO_2, O_2, N_2O and SO_2 were analyzed by continuous emissions monitors (CEMs) (Rosemount Analytical NGA2000, Emerson Process Management Co., Ltd.). Mercury in the gas was monitored by mercury CEMs (Ms-1/dm-6a, Nippon Instrument Co.).

The total flow rate of all the reactants including O_3 and simulated flue gas were fixed to 1000 mL/min. The residence time in the center tube is from 0.049–0.089 s varied with temperature calculated by 33.3 K/T s. The residence time in the washing tower is approximate to 9.7 s. The initial gas concentrations used in the test were NO: 215±10 ppm, SO_2 220±10 ppm, O_3:4000±200 ppm, Hg: 50±0.5 ug/m^3 appeared in different cases.

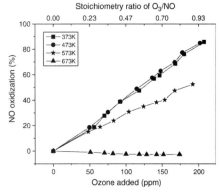

Fig. 3. NO conversion property with ozone added at different temperature.

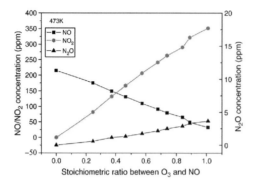

Fig. 4. By-products of NO oxidation by ozone at 473 K.

various with temperature. The two lines of 373 K and 473 K are almost overlapped and nearly 85% of NO can be oxidized when 200 ppm of ozone are added with stoichiometric ratio around 0.97. The oxidation rates are almost linearly increased with the amount of ozone added.

Ozone is a kind of unstable gas which will decompose into O_2 automatically especially at high temperature. Fig. 5 is the thermal decomposition property of enriched ozone gas. The experiment was carried out in a multi-sample glass tube with oil bath as heat source. The initial ozone concentration was 4400±250 ppm, temperature is 298 K–523 K and residence time is 0.2 s–10 s. At room temperature of 298 K, only 0.5% of ozone disappears within the 10 s. With the increasing of temperature, the decomposition rate dramatically increased especially when temperature large than 473 K. As can be seen, more than 80% of ozone decomposed within the first 1 s at 523 K.

The decomposition of ozone will undoubtedly weaken the NO oxidation rate. At the same time, residence time in the reactor will also lessen from 0.089 s to 0.049 s when temperature increased from 373 K to 673 K. The two mentioned reasons cause the decrease of NO conversion at 573 K compared with 373 K and 473 K. As shown in Fig. 3, only 52.5% of NO can be oxidized at 573 K with 192 ppm ozone added with stoichiometric ratio around 0.89. At 673 K, there is almost no effect on the oxidation of NO for the decomposition of ozone. So in future industrial application, the optimal temperature for the ozone injection technology should be lower than 473 K and 0.09 s of residence time is needed. This temperature range just exists behind the air preheater. It's feasible and easy to retrofit for the boilers equipped with WFGD system not to say with a new design.

3.2. Influence of SO_2 on the oxidation of NO

There are always large amounts of SO_2 in the flue gas. And for the simultaneously removal of NO_x and SO_2, it's inevitable existing SO_2 during the NO's oxidation process. Therefore, the influence of SO_2 on the oxidation of NO should be investigated. There are two aspects when SO_2 appearance in the reaction system. One side, SO_2 may be also oxidized into SO_3 by ozone or O radical. With the help of moisture in the flue gas, H_2SO_4 will be formed by SO_3, which will increase the corrosion of the ducts. Another side, the dosage of ozone will increase besides NO's consumption. Although SO_3 have high solubility than SO_2. Because more than 95% of SO_2 can be removed by WFGD system effectively, the oxidation of SO_2 is redundant. The oxidation of SO_2 is undesirable. Experiments were carried out for the existing SO_2 or not in the NO/N_2 reaction system. Results are shown in Fig. 6. As can be seen, the appearance of SO_2 has little impact on the conversion of NO, which means the reactions between O_3 and SO_2 is weakly. The appearance of SO_2 in flue gas will not consume large amount of O_3.

3.3. Oxidation of Hg

It's known that the oxidized mercury Hg^{2+} can be easily trapped in WFGD, but elemental mercury Hg^0 cannot be removed. Although the proportion of mercury in gas phase varies with the combustion condition, coal types and other factors, absorption or oxidation of elemental mercury still are the key steps for improving the mercury removal efficiency in

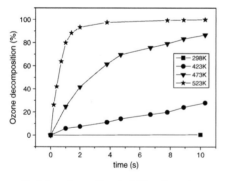

Fig. 5. Ozone decomposition rate at different temperature.

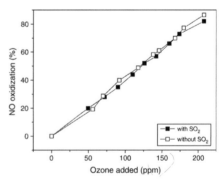

Fig. 6. Influence of SO_2 on the conversion of NO by ozone. Solid symbol denotes the condition with 200 ppm SO_2 in the flue gas; hollow symbol denotes the condition without SO_2, T=373 K.

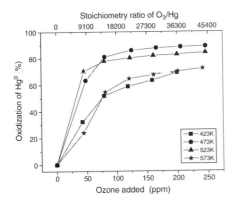

Fig. 7. Mercury oxidation property by ozone at different temperature.

flue gas. Therefore, all kinds of reagents and method were developed to promote the oxidation rate of elemental mercury. Apart from the oxidation of NO, O_3 and O can also efficiently convert Hg^0 into Hg^{2+} as shown in the following reactions:

$$Hg + O_3 = HgO + O_2 \qquad (6)$$

$$Hg + O = HgO \qquad (7)$$

Fig. 7 shows the experiment results at the temperature range of 423 K–573 K. Only elemental mercury and N_2 in the flue gas. As observed, the elemental mercury can be efficiently converted into oxidized mercury with 50–100 ppm ozone added with stoichiometric ratio of O_3/Hg around 8900–17,900. After that, the oxidation rate can only increase slowly with the increasing amounts of ozone. From 423 K to 473 K, there is more than 20% increasing of the mercury oxidation. At temperature of 523 K, there is a little decreasing compared with 473 K but still much more effective than that of 423 K. At 573 K, there is a great decreasing of the oxidation rate. The oxidation rate of elemental mercury first increase with the temperature and then falling down. May be the decomposition of O_3 and HgO counteract the oxidation efficiency at higher temperature. Fig. 8 shows the equilibrium analysis of the HgO thermal decomposition property by FACTsage a software of chemical equilibrium analysis. It can be seen, the solid state of HgO(s) began to decompose into Hg and O_2 at temperature around 700 K. That means the decomposition of HgO will not cause the decreasing of mercury oxidation efficiency under 573 K. The reason still lies on the decomposition of ozone itself.

Nearly 80% of elemental Hg can be oxidized with 80 ppm of ozone added at stoichiometric ratio around 14,300 and about 89% of elemental Hg can be oxidized with 250 ppm of ozone added at stoichiometric ratio around 44,700 at 473 K. The conversion rate is relative low at 423 K and 573 K, but 55% of oxidation rate can still be observed when 80 ppm of ozone is added. The optimal temperature range for mercury oxidation is from 473 K to 523 K, which is a little different compared with NO.

3.4. Oxidation of NO/Hg mixture

As discussed previously, either NO or Hg can be effectively oxidized by ozone or O radical. In the practical flue gas, NO and Hg exist together. Therefore, the competition between NO and mercury with ozone needs to be clarified. At the same time, mercury can also be oxidized by high order nitrogen species such as NO_2, NO_3 etc. shown in the following reactions.

$$NO_2 + Hg = HgO + NO \qquad (8)$$

$$NO_3 + Hg = HgO + NO_2 \qquad (9)$$

Fig. 9 indicates the reactions between ozone and a mixture of NO/Hg at 423 K. It's quite obviously that the oxidation of NO is much easier than the elemental mercury. Nearly 100% of NO can be converted into higher order nitrogen species when stoichiometric ratio large than 1.0. While, oxidation rate of Hg

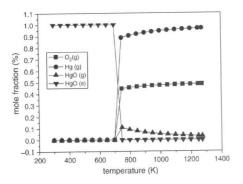

Fig. 8. Equilibrium analysis of the HgO decomposition.

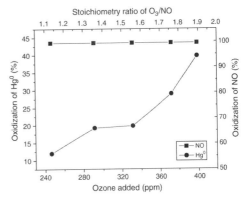

Fig. 9. Reaction competition between NO and Hg^0 with ozone. T=423 K.

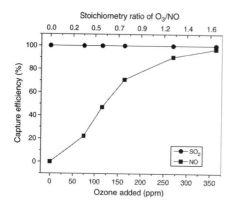

Fig. 10. Simultaneous capture efficiency of NO, SO$_2$ behind the washing tower with ozone injection. T = 423 K.

only increased from 12% to 39.7% with the increasing amounts of ozone. In the reaction competition between NO and mercury with ozone, NO has a priority. Apart from the direct oxidation of NO into NO$_2$, further reactions such as the formation of NO$_3$ will consume some of the excessive ozone, which will counteract some of the Hg oxidation.

3.5. Simultaneous removal of NO and SO$_2$ behind washing tower

With the assistant of alkaline washing tower, we can investigate the simultaneous removal performance of NO, SO$_2$ and Hg in the flue gas by ozone injection technology. Fig. 10 describes the removal efficiency of NO and SO$_2$ behind the washing tower. Few NO$_2$ and N$_2$O found behind the washing tower. SO$_2$ is liable to be captured from gas phase because of its high solubility in water and is seldom affected by ozone. Nearly 100% of SO$_2$ can be absorbed in all the tests in despite of the ozone injected or not. With the injection of ozone, NO can be gradually converted into soluble species and then absorbed in alkaline solutions. About 97% of NO can be removed through washing tower with 360 ppm of ozone added. The absorbed species in the alkaline solutions need further investigation later. The removal efficiency of mercury greatly depends on the species composition. It's known the oxidized mercury can be effectively removed in WFGD and elemental mercury cannot. It will be removed in washing tower if the elemental mercury can be converted into oxidized mercury. Due to the complex influence and trace amount property of mercury, the solubility of oxidized mercury will be investigated in the future.

4. Conclusion

The preliminary tests conducted to date clearly show that the ability of the ozone injection process to oxidize elemental mercury, NO and the ability to capture NO and SO$_2$ simultaneously in a washing tower. More than 80% of elemental mercury can be oxidized with 80 ppm of ozone added. It's known that the oxidized mercury can be removed in wet absorption process, but the absorption efficiency should be investigated in detail in the near future. The temperature range for NO oxidation should be lower than 473 K. The optimal temperature was in the range of 473 K–523 K for the oxidation of mercury. The appearance of SO$_2$ has little effect on the oxidation of NO by ozone and O radical. In the mixture of NO and Hg, ozone will first react with NO. With the assistant of washing tower, about 97% of NO and nearly 100% of SO$_2$ can be removed with 360 ppm of ozone added. Future work including reducing the power consumption and requirements for ozone generation, influence of moisture, CO, CO$_2$, HCl etc., absorption efficiency and species compositions in the washing tower should be investigated in detail.

Acknowledgements

The work was supported by National Natural Science Foundation of China (50476059), the Key Project of Chinese National Programs for Fundamental Research and Development (2006CB200303), the National Natural Science Foundation for Distinguished Young Scholars (Grant No.50525620) and Program for New Century Excellent Talents in University (NCET-04-0533).

References

[1] A.B. Lopez, A.G. Garcia, Combined SO$_2$ and NO$_x$ removal at moderate temperature by a dual bed of potassium-containing coal-pellets and calcium-containing pellets, Fuel Processing Technology 86 (2005) 1745–1759.

[2] J. Cheng, J. Zhou, J. Liu, Z. Zhou, Z. Huang, X. Cao, X. Zhao, K. Cen, Sulfur removal at high temperature during coal combustion in furnaces: a review, Progress in Energy and Combustion Science 29 (2003) 381–405.

[3] T.W. Chien, H. Chu, Removal of SO$_2$ and NO from flue gas by wet scrubbing using an aqueous NaClO$_2$ solution, Journal of Hazardous Materials B 80 (2000) 43–57.

[4] A.A. Patsias, W. Nimmo, B.M. Gibbs, P.T. Williams, Calcium-based sorbents for simultaneous NO$_x$/SO$_x$ reduction in a down-fired furnace, Fuel 84 (2005) 1864–1873.

[5] A. Molina, E.G. Eddings, D.W. Pershing, A.F. Sarofim, Nitric oxide destruction during coal and char oxidation under pulverized-coal combustion conditions, Combustion and Flame 136 (2004) 303–312.

[6] C.K. Man, J.R. Gibbins, J.G. Witkamp, J. Zhang, Coal characterization for NO$_x$ prediction in air-staged combustion of pulverized coals, Fuel 84 (2005) 2190–2195.

[7] J.O.L. Wendt, W.P. Linak, P.W. Groff, R.K. Srivastava, Hybrid SNCR-SCR Technologies for NO$_x$ Control Modeling and Experiment, AIChE Journal 47 (2001) 2603–2617.

[8] J.S. Chang, K. Urashima, Y.X. Tong, W.P. Liu, H.Y. Wei, F.M. Yang, X.J. Liu, Simultaneous removal of NO$_x$ and SO$_2$ from coal boiler flue gases by DC corona discharge ammonia radical shower systems: pilot plant tests, Journal of Electrostatics 57 (2003) 313–323.

[9] M.T. Izquierdo, B. Rubio, C. Mayoral, J.M. Andres, Low cost coal-based carbons for combined SO$_2$ and NO removal from exhaust gas, Fuel 82 (2003) 147–151.

[10] Y.H. Lee, W.S. Jung, Y.R. Choi, J.S. Oh, S.D. Jang, Y.G. Son, M.H. Cho, W. Namkung, Application of pulsed corona induced plasma chemical process to an industrial incinerator, Environmental Science and Technology 37 (2003) 2563–2567.

[11] G.B. Zhao, S.V.B.J. Garikipati, X. Hu, M.D. Argyle, M. Radosz, Effect of oxygen on nonthermal plasma reactions of nitrogen oxides in nitrogen, AIChE Journal 51 (2005) 1800–1812.

[12] X. Hu, J.J. Zhang, S. Mukhnahallipatna, J. Hamann, M.J. Biggs, P. Agarwal, Transformations and destruction of nitrogen oxides NO,NO₂ and N₂O in a pulsed corona discharge reactor, Fuel 82 (2005) 1675–1684.

[13] H. Lin, X. Gao, Z. Luo, K. Cen, Z. Huang, Removal of NO$_x$ with radical injection caused by corona discharge, Fuel 83 (2004) 1349–1355.

[14] H. Lin, X. Gao, Z. Luo, K. Cen, M. Pei, Z. Huang, Removal of NO$_x$ from wet flue gas by corona discharge, Fuel 83 (2004) 1231–1235.

[15] S. Wu, M.A. Uddin, E. Sasaoka, Characteristics of the removal of mercury vapoer in coal derived fuel gas over iron oxide sorbents, Fuel 85 (2006) 213–218.

[16] Z. Luo, C. Hu, J. Zhou, K. Cen, Stability of mercury on three activated carbon sorbents, Fuel Processing Technology 87 (2006) 679–685.

[17] T. Morimoto, S. Wu, M.A. Uddin, E. Sasaoka, Characteristics of the mercury vapor removal from coal combustion flue gas by activated carbon using H₂S, Fuel 84 (2005) 1968–1974.

[18] J.H. Pavlish, E.A. Sondreal, M.D. Mann, E.S. Olson, K.C. Galbreath, D.L. Laudal, S.A. Benson, Status review of mercury control options for coal-fired power plants, Fuel Processing Technology 82 (2003) 89–165.

[19] Y. Zhuang, C.J. Zygarlicke, K.C. Galbreath, J.S. Thompson, M.J. Holmes, J.H. Pavlish, Kinetic transformation of mercury in coal combustion flue gas in a bench-scale entrained-flow reactor, Fuel Processing Technology 85 (2004) 463–472.

[20] W.J. O'Dowd, R.A. Hargis, E.J. Granite, H.W. Pennline, Recent advances in mercury removal technology at the National Energy Technology Laboratory, Fuel Processing Technology 85 (2004) 533–548.

[21] C.R. McLarnon, E.J. Granite, H.W. Pennline, The PCO process for photochemical removal of mercury from flue gas, Fuel Processing Technology 87 (2005) 85–89.

[22] C.D. Livengood, M.H. Mendelsohn, Process for combined control of mercury and nitric oxide, EPRI-DOE-EPA Combined Utility Air Pollutant Control Symposium, August 16–20 1999.

[23] J.G. Calvert, S.E. Lindberg, Mechanisms of mercury removal by O₃ and OH in the atmosphere, Atmospheric Environment 39 (2005) 3355–3367.

[24] X. Liang, P.C. Looy, S. Jayaram, A.A. Berezin, M.S. Mozes, J.S. Chang, Mercury and other trace elements removal characteristics of DC and pulse-energized electrostatic precipitator, IEEE Transactions on industry applications 38 (2002) 69–76.

[25] H. Agarwal, H.G. Stenger, S. Wu, Z. Fan, Effects of H₂O,SO₂ and NO on homogeneous Hg oxidation by Cl₂, Energy & Fuels 20 (2006) 1068–1075.

[26] J.B. Jarvis, A.T. Day, N.J. Suchak, LoTOx™ process flexibility and multi-pollutant control capability, Combined Power Plant Air Pollutant Control Mega Symposium Washington, DC, 19–22(2003).

[27] Y. Fu, U.M. Diwekar, Cost effective environmental control technology for utilities, Advances in Environmental Research 8 (2003) 173–196.

[28] Y.S. Mok, H.J. Lee, Removal of sulfur dioxide and nitrogen oxides by using ozone injection and absorption reduction technology, Fuel Processing Technology 87 (2006) 591–597.

[29] Z. Wang, J. Zhou, J. Fan, K. Cen, Direct Numerical Simulation of ozone injection technology for NO$_x$ control in flue gas, Energy & Fuels 20 (2006) 2432–2438.

在狭窄流反应器中喷射臭氧同时脱除N2中NO$_x$、SO$_2$和Hg的试验研究

本文提出了一种利用臭氧耦合碱性洗涤塔，同时脱除NO$_x$、SO$_2$和Hg的工艺。实验是在电加热炉内的石英流反应器中进行的，考察了NO和Hg的氧化性能以及洗涤塔后NO和SO$_2$的脱除效率。结果表明，臭氧的投入量对NO和Hg的氧化效率有很大影响。随着臭氧投入量的增加，NO和Hg的氧化效率分别提高。当臭氧投入量为200ppm时，约85%的NO可被氧化；当臭氧投入量为250ppm时，可氧化89%的汞元素。NO氧化的最佳温度应低于473K，Hg的最佳温度范围为473～523K。SO$_2$的存在对NO的氧化过程影响不大。与臭氧反应时，NO优先于汞反应。在洗涤塔的辅助下，当臭氧投入量为360ppm时，可同时脱除约97%的NO和近100%的SO$_2$。

原文刊于Fuel Processing Technology, 2007, 88: 817-823

On coherent structures in a three-dimensional transitional plane jet

LUO Kun[†], YAN Jie, FAN JianRen & CEN KeFa

State Key Laboratory of Clean Energy Utilization, Zhejiang University, Hangzhou 310027, China

Direct numerical simulation of coherent structures in the three-dimensional transitional jet with a moderate Reynolds number of 5000 was conducted. The finite volume method was used to discretize the governing equations in space and the low-storage, three-order Runge-Kutta scheme was used for time integration. The comparisons between the statistical results of the flow field and the related experimental data were performed to validate the reliability of the present numerical schemes. The emphasis was placed on the study of the spatial evolution of the three-dimensional coherent vortex structures as well as their interactions. It is found that the evolution of the spanwise vortex structures in three-dimensional space is similar to that in two-dimensional jet. The spanwise vortex structures are subject to three-dimensional instability and induce the formation of the streamwise and lateral vortex structures. Going with the breakup and mixing of the spanwise vortex structures, the streamwise and transverse vortex tubes also fall to pieces and the mixing arranged small-scale structures are formed in the flow field. Finally, the arrangement relationship among the spanwise, the streamwise and the lateral vortex structures was analyzed and their interactions were also discussed.

coherent structures, three-dimensional plane jet, direct numerical simulation, finite volume method, interactions

Due to the particular flow characteristics itself, the fundamental study on the plane jet not only helps to deeply understand the free shear turbulence and provides convenience for evaluating different physical models, but also offers reference for related engineering applications. Therefore, the study on plane jet has always been one of the hot topics in turbulence research for the recent 40 years.

Most of the early studies focused on the statistic and correlation quantities in plane jets by means of experiments[1,2]. With the discovery of turbulent coherent structure and its important effects on turbulent transport, noise and mixing enhancement, some researchers also tried to investigate the development of coherent structures in plane jet[3]. To identify the existence of co-

---

Received November 8, 2006; accepted August 13, 2007
doi: 10.1007/s11431-008-0032-x
[†]Corresponding author (email: zjulk@zju.edu.cn)
Supported by the National Natural Science Foundation of China (Grant No. 50506027)

herent structures in turbulent jet, Crow and Champagne[4] investigated the round jet based on visualization and found that as the Reynolds number increased from order 10^2 to order 10^3, the instability of the jet evolved from a sinusoid to a helix, and finally to a train of axisymmetric waves. Liepmann and Gharib[5] used the technology of flow visualization to study the role of streamwise vortex structures in the near field of the round jet. It was found that the streamwise vortex pairs firstly emerged in the braid region between the spanwise vortical structures and were able to drastically alter the entrainment process in the near field after evolution and amplification. As the flow evolved downstream, the efficiency of the streamwise vorticity in entraining fluid increased relatively to that of the azimuthal vorticity. Thomas and Goldschmidt[6] studied the developing characteristics of the coherent structures in a naturally developing 2D plane jet and found that the flow structures were initially characterized by strong symmetric modes and then by an asymmetric pattern formed beyond the jet potential core.

Wang and his group[7,8] examined in detail the coherent structures in single-phase and two-phase round jets. They observed the typical large-scale structures, such as the vortex ring, the single and double helices, and found that the development of coherent structures was influenced by the external forcing and dominated the particle dispersion. Liu et al.[9] performed direct numerical simulation of compressible axisymmetric jet based on high-order finite difference. It was found that the Kelvin-Helmholtz instability appeared first when the jet lost its stability, and then with the increasing of the nonlinear effects, the secondary instability appeared and the streamwise vortices formed. Recently, Sakai et al.[10,11] used hot wire probe and proper orthogonal decomposition (POD) to study the characteristics of coherent structures in different regions in the plane jet. They used the eigenvalues and eigenfunctions as well as the spatial velocity correlations to characterize the development of coherent structures and found that the characteristics of coherent structures could be identified by the reconstructed two-point spatial velocity correlation. This is a progress compared with previous visualization study of coherent structures.

Although many achievements have already been obtained, the study on coherent structures in plane jet is not so deep as that on plane mixing layer. So far, there has been little in-depth research on the coherent structures in three-dimensional turbulent transitional plane jet. The development of the large-scale structures and the small-scale structures as well as their interactions during turbulence transition, the development of the spanwise and the streamwise vortex structures as well as their relationships and so on still need to be further investigated.

Under this background, the characteristics of coherent structures in three-dimensional turbulent transitional plane jet are investigated in detail in the present study. In order to observe the transitional process of the flow, a moderate Reynolds number of 5000 is chosen. The main objective is to examine the spatial evolution and their interactions of three-dimensional coherent vortex structures in the region with 25 nozzle width d in the down stream of the jet. To validate the reliability of the numerical schemes, the second-order statistics are also calculated and compared with the related experimental data.

1 Governing equations

Assuming the gas phase to be an incompressible Newtonian fluid and the body force to be neglectable, the non-dimensional governing equations for the fluid can be expressed as

Continuum equation: $\dfrac{\partial u_i}{\partial x_i} = 0.$ (1)

Momentum equation: $\dfrac{\partial u_i}{\partial t} + \dfrac{\partial u_i u_j}{\partial x_j} = -\dfrac{1}{\rho}\dfrac{\partial p}{\partial x_i} + \dfrac{1}{Re}\dfrac{\partial}{\partial x_j}\left(\dfrac{\partial u_i}{\partial x_j} + \dfrac{\partial u_j}{\partial x_i}\right).$ (2)

2 Numerical algorithms

To ensure the second-order resolution in space, the above governing equations of the fluid are discretized by the finite volume method[12] with central differences. According to the requirements of direct numerical simulation on turbulence[13], the computational grid scale that is smaller than or in the same order of the Kolmogorov microscale is used to capture the small-scale vortex structures in the flow during the simulation. In time integration, an explicit low-storage, third-order Runge-Kutta scheme[14] is adopt. Considering the requirements of resolution in time, the stability of computation and the statistics of flow field, the non-dimensional time step is set as $\Delta t = 0.02$ and 20 streamwise convection times are simulated.

Initially, the following velocity profile is given in the region with the nozzle width d in the flow-field:

$$u = \dfrac{U_0}{2} + \dfrac{U_0}{2}\tanh\left(\dfrac{y}{2\theta_0}\right), \quad v = w = 0,$$ (3)

where U_0 is the initial streamwise inflow velocity, θ_0 is the initial momentum thickness selected as 0.05 d. u, v and w denote the streamwise, lateral and spanwise velocities, respectively. At the streamwise outflow boundary, Neumann boundary conditions for velocity and pressure are used and the pressure is also corrected. At the lateral boundaries, the pressure is set as zero. In the spanwise direction, the periodic boundary conditions are applied.

For the detailed numerical schemes, grid arrangement and boundary condition, please refer to the previous study[15].

3 Results and discussions

3.1 Statistics of flow field and comparisons with experimental data

To validate the reliability of the numerical schemes, the second-order variables are calculated statistically in the present study and are compared with the previous related experimental data. For convenience of comparison, only the turbulence intensity and shear Reynolds stress profiles in the right half of the self-similar regions are given.

Figure 1 shows the profiles of the streamwise, lateral and spanwise turbulence intensities in the self-similar region of the plane jet and the comparisons with the experimental data. It is found that the predicted streamwise turbulence intensity is in good agreement with the experimental data of Ramaprian et al.[2] and Namer et al.[16], but is obviously smaller than that of Gutmark et al.[1], indicating that the experimental data of Gutmark et al.[1] in the strong-shearing regions of the jet are a little higher. In addition, the lateral turbulence intensity is also in good agreement with the experimental data of Ramaprian et al.[2], but is slightly higher than that of Gutmark et al.[1] in some areas. Whereas the spanwise turbulence intensity is coincident with the experimental data of Gutmark et al.[1]. Figure 2 demonstrates the profiles of shear Reynolds stress profiles in the self-similar regions of the jet and the comparisons with the experimental data. Similarly, it can be seen that the numerical results in the present study agree well with the experimental data of Gutmark et al.[1] and Ramaprian et al.[2]. These good agreements with previous related experimental data show that the numerical algorithms used in the present study are feasible and reliable.

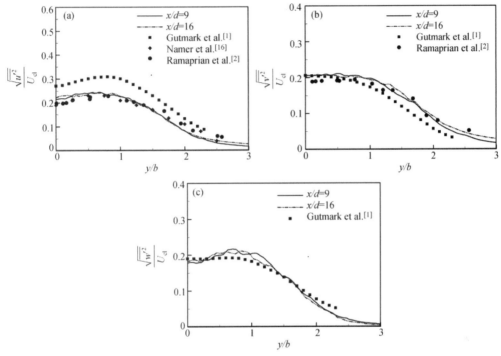

Figure 1 Turbulence intensities in the self-similar region of the plane jet and the comparison with experimental data. (a) Streamwise turbulence intensity; (b) lateral turbulence intensity; (c) spanwise turbulence intensity.

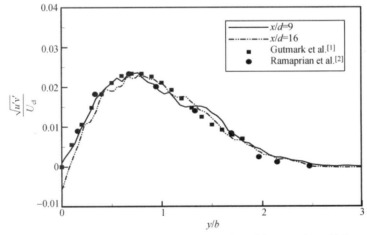

Figure 2 Shear Reynolds stress in the self-similar region of the plane jet and the comparison with the experimental data.

3.2 Evolution of the spanwise vortex structures

To describe the three-dimensional evolution process of the coherent vortex structures in planar jet, the vorticity vector is defined as $\boldsymbol{\omega} = \nabla \times \boldsymbol{u}$. Then the streamwise vorticity $\omega_x = \left(\dfrac{\partial w}{\partial y} - \dfrac{\partial v}{\partial z}\right)$, lateral vorticity $\omega_y = \left(\dfrac{\partial u}{\partial z} - \dfrac{\partial w}{\partial x}\right)$ and spanwise vorticity $\omega_z = \left(\dfrac{\partial v}{\partial x} - \dfrac{\partial u}{\partial y}\right)$ can be obtained.

Figure 3 shows the development of the spanwise vorticity ω_z in the early stage of the jet at different times. It is found that the development of the spanwise vortex structures can be divided into three stages. At the non-dimensional time $t=6$, the superior mode corresponding to the natural frequency is amplified until saturation in the free shear layer, and the instability of Kelvin-Helmholtz occurs in the flow field. This K-H instability leads to the formation of the rolling up of the spanwise vortex structures in the position $x/d=2.0$ in the flow field. The rollers in the upper and lower two shear layers appear at the same time with opposite signs and form the typical dipole structure, as shown in Figure 3(a). They develop symmetrically and move towards down stream with convection velocity of U_0. With the increasing of the interactions between two shear layers, the vortex core of the rolling spanwise structures becomes larger and larger, and the entrainment of the surrounding fluid is stronger and stronger. As a result, the instability occurs again in the down stream of the spanwise vortex core at the time $t=16.4$, which causes the shifting of the spanwise vortex structures from the symmetric mode to the asymmetric mode, as shown in Figure 3(b). Once, the asymmetric mode is triggered, it dominates the subsequent evolution progress of the spanwise vortex structures. These results are consistent with previous study in two-dimensional simulations[17].

During the development of the asymmetric mode, the disturbance is amplified gradually and the two initial spanwise rolling vortex structures further develop by entraining the surrounding fluid which influences the up stream and down stream more and more. At the time $t=29.2$, the spanwise rolling vortex structures form again in the up stream and down stream. In the up stream, the rolling structures develop in opposite directions, but in the down stream, the fluid in the two layers rolls up in the same direction and form the continuous structure of "λ" shape due to the strong stretching there, which is somewhat similar to the Karman vortex street in the wakes, as shown in Figure 3(c). With the further development in time and space, the interactions between the two initial spanwise vortexes become stronger and stronger, resulting in the formation of mixing phenomena at the time $t=32.4$, i. e., the large-scale structures begin to break up and the vorticity reconstructs. Thus, the flow field is full of lots of small-scale vortexes with opposite signs and the jet finally changes to small-scale turbulence, as shown in Figure 3(e).

In order to further investigate the characteristics of spanwise vortex structures in jet, the comparisons among the picture obtained by PIV study of a round jet, the spanwise vortex structures in the middle plane of the flow obtained by the present DNS and the spanwise vortex structures obtained by previous two-dimensional DNS are performed, as shown in Figure 4. It can be found that the three-dimensional simulation is able to capture the small-scale vortex structures and the pairing phenomenon is fewer when compared to the two-dimensional simulation. This indicates that the development of the flow three-dimensionality can delay the pairing of vortex structures, which is in accordance with the observed phenomenon in plane mixing layer[18]. Furthermore, the results obtained by the present three-dimensional simulation are more consistent with the PIV experimental picture compared to those obtained by two-dimensional simulations.

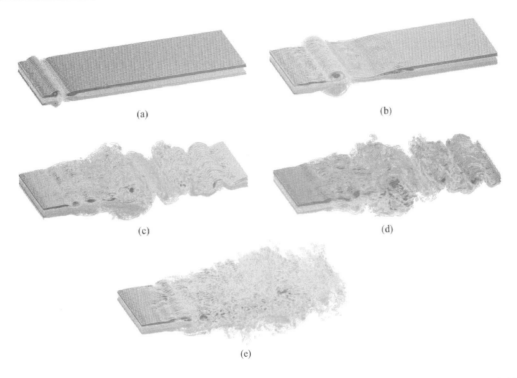

Figure 3 Evolution of the spanwise vortex structures at different time at the early stage of the plane jet. (a) $t=6$; (b) $t=16.4$; (c) $t=29.2$; (d) $t=32.4$; (e) $t=48.4$.

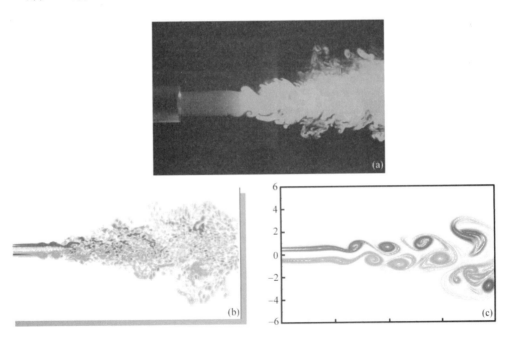

Figure 4 Comparison of the spanwise vortex structures in jets. (a) Picture obtained by PIV in round jet; (b) spanwise vortex structures in the present three-dimensional simulation; (c) spanwise vortex structures in previous two-dimensional simulation.

3.3 Evolution of the streamwise vortex structures

In previous study of the coherent structures in plane mixing layers, it was found that the three-dimensional instability could influence the spanwise vortex structures and induce the formation of the streamwise vortex structures. In this paper, similar phenomena are observed in the simulation of plane jet. Figure 5 presents the evolution process of the three-dimensional streamwise vortex structures. For clearness, only the three-dimensional iso-surface with the values of $\omega_x=\pm 0.059$ is plotted. It is found that the obvious streamwise vortex structures appear until the time $t=8.4$. Besides the small-scale streamwise vortex structures caused by the initial inflow disturbance, the new formed streamwise vortex structures caused by the three-dimensional instability of spanwise vortex structures occur first in the braid region of the spanwise vortex structures. The streamwise vortex structures appear in pairs and rotate counter. Their vortex tubes surround the spanwise rolling structures from the upper and lower sides respectively, moving towards down stream with the development of time, as shown in Figure 5(b). With the growing of the spanwise vortexes, the streamwise vortex structures also develop gradually. The vortex tubes are stretched and extended to the vortex braid region in the front side of the spanwise rolling vortex structures. When the new spanwise rolling vortex structures appear in the flow field, the new streamwise vortex tubes also come into being and extend from one side to the other side to surround the corresponding spanwise vortex, as shown in Figure 5(c). Although the development of each streamwise vortex tube is different, the tubes with opposite signs always take on an alternate and staggered arrangement. When the breaking up and mixing of the spanwise vortex structures happen at the time $t=32.4$, the fracturing of the streamwise vortex tubes also occurs and only few of them are reserved in the outer region of the vortexes. As the time increases, the streamwise vortex structures are all translated from large-scale into small-scale gradually. Finally, the small-scale streamwise turbulent structures with opposite signs and mixing arrangement are observed in the flow field, as shown in Figure 5(e).

3.4 Evolution of the lateral vortex structures

At present, the discussions on the lateral coherent structures are fewer. In fact, the three-dimensional instability originated from the spanwise vortex structures can also induce the formation of the lateral vortex structures. In the present study, the obvious lateral vortex structures are observed in the flow field, as shown in Figure 6. For clearness, only the three-dimensional iso-surface with the values of $\omega_y=\pm 0.13$ is given. At the non-dimensional time $t=8.4$, the new lateral vortex tube structures occur in the flow field. It is found that the new formed lateral vortex structures also appear first in the braid region of the spanwise vortexes and take on an inclined arrangement towards the direction of the normal stress produced by the large-scale spanwise structures. Similar to the streamwise vortex structures, the lateral vortex structures are also stretched gradually due to the strong tension by the spanwise vortex structures. But the difference is that the lateral vortex tubes extending to the upper and lower sides of the spanwise structures fall into pieces so much that there is not integrated lateral vortex tube any more in the flow field accompanying the development of the lateral vortex structures, as shown in Figure 6(c). Some of the fractured lateral vortexes rotate along with the spanwise vortex and enter the core, but others still move towards the direction of the normal stress. When the new spanwise vortex structures appear in the flow field, the new lateral vortexes also form subsequently. But due to the fracture and breakup, the arrangement of the lateral vortex structures is comparatively complex. However, the lateral vortex tubes in front of the initially formed spanwise vortexes are still visible, as demonstrated in Figure 6(c). With the breaking up and mixing of the large-scale spanwise vortex structures, the lateral vortexes are also translated from large-scale into small-scale to form the small-scale turbulence, as shown in Figure 6(d) and (e).

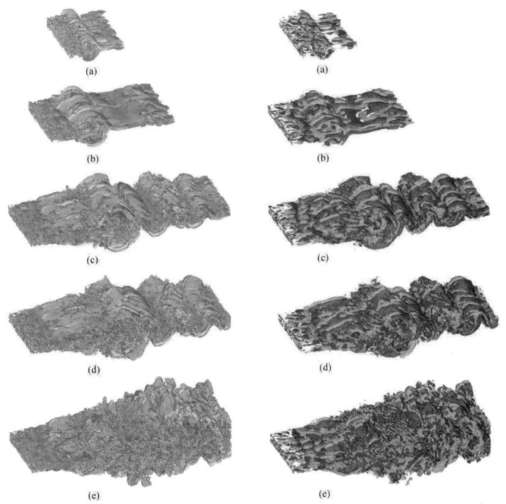

Figure 5 Iso-surface of the streamwise vortices in the early stage of the plane jet. (a) $t=8.4$; (b) $t=16.4$; (c) $t=29.2$; (d) $t=32.4$; (e) $t=50$.

Figure 6 Iso-surface of the lateral vortices in the early stage of the plane jet. (a) $t=8.4$; (b) $t=16.4$; (c) $t=29.2$; (d) $t=32.4$; (e) $t=50$.

3.5 Arrangement of the streamwise and spanwise vortices

Figure 7 shows the arrangement relationship between the three-dimensional iso-surfaces of the spanwise and the streamwise vortex structures in the flow field at the time $t=31.6$. Here, Figure 7(a) is the corresponding oblique view with 45° and Figure 7(b) is the corresponding front view. The red and green segments represent the spanwise vortex structures and the yellow segment represents the streamwise vortex structures. The streamwise vorticity is set as $\omega_x = -1.35$ and the spanwise vorticity is set as $\omega_z = \pm 1.55$. It can be found that the streamwise vortex structures form first in the downstream vortex braid region of the spanwise structures and then are stretched towards the rolling direction of the spanwise vortices. As a result, the streamwise vortices surround the spanwise vortex cores from the upper and lower sides and some of them can even be entrained into the vortex core regions of the spanwise vortices. When multiple spanwise rolling vortex structures appear, the streamwise vortex tubes will extend between the two adjacent spanwise vortex structures up and down. This extension from the top of one spanwise vortex structure to the bottom of the other spanwise vortex structure forms the arrangement of "clipper" shape, as shown in Figure 7(b).

Figure 7 The spanwise and the streamwise vortex structures in the flow field at the time $t=31.6$. (a) The oblique view with 45°; (b) the spanwise front view.

Based on the above results, the topological arrangement of the spanwise and the streamwise vortex structures in the spanwise section can be plotted, as shown in Figure 8. In the figure, the circles represent the spanwise vortex structures and the lines denote the streamwise vortex structures. Figure 9 shows the sketch map for the formation of the streamwise vortex structures. Due to the three-dimensional instability, the early formed spanwise rolling vortex structures lose stability and the normal stress appears in the flow. Thus the vortex tubes are stretched, as shown in Figure 9(b). With the amplification of the stress, the counter-rotating pair of streamwise vortices forms in the end, as shown in Figure 9(c).

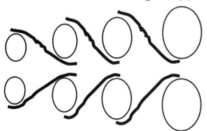 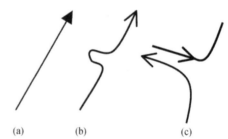

Figure 8 Sketch map for the topological arrangement of the spanwise and the streamwise vortex structures.

Figure 9 Sketch map for the formation of the streamwise vortex structures in z section.

Figure 10 The spanwise and the lateral vortex structures in the flow field at the time $t=31.6$. (a) The oblique view with 45°; (b) the spanwise front view.

3.6 Arrangement of the lateral and spanwise vortices

Figure 10 shows the arrangement relationship of the spanwise and the lateral vortex structures in the flow field at the time $t=31.6$. Figure 10(a) is the corresponding oblique view with 45° and Figure 10(b) is the corresponding spanwise front view. The red and the green segments represent the spanwise vortex structures, and the blue segment represents the lateral vortex structures. The value of the streamwise vorticity is set as $\omega_x = -1.35$ and the values of the spanwise vorticity are set as $\omega_z = \pm 1.55$.

Similar to the streamwise vortex structures, the lateral vortex structures also occur first in the downstream braid region of the spanwise vortices and then are stretched by the spanwise vortex structures and extend towards the direction of the normal stress in the flow field. But the difference is that the lateral vortex structures will fall to pieces beside the spanwise vortex structures during the stretching. As a result, some lateral vortices take on an inclined arrangement and some others are entrained into the vortex core regions of the spanwise vortices. Due to the fracturing, the lateral vortex tubes become thicker and shorter and finally present an interlaced arrangement in the braid regions between the adjacent spanwise vortices.

4 Conclusions

Direct numerical simulation of the three-dimensional spatially-developing transitional jet is performed in this paper. The second-order statistics, such as the turbulence intensities and the shear Reynolds stress are calculated and compared with the previous experimental data. The good agreement between them validates the reliability of the used numerical algorithms. Based on this, the evolution characteristics and their interactions of the coherent structures are specially investigated. It is found that with the emergence of the Kelvin-Helmholtz instability and the three-dimensional instability in the flow field, the spanwise vortices form first and then the streamwise and the lateral vortex structures are induced. Going with the changes of the coherent structures from two-dimensional to three-dimensional and from symmetric mode to asymmetric mode, the large-scale structures begin to break up and the vorticity reconstructs. As a result, a lot of small vortices with opposite signs present in the flow field and gradually switch to small-scale turbulence. The change in development of the three-dimensional spanwise vortices from the symmetric mode to the asymmetric mode in the initial jet is similar to that of the two-dimensional jet. Finally, the arrangement relationships among the spanwise, the streamwise and the lateral vortex structures are explored, and it is found that the streamwise and the lateral vortex structures form initially in the vortex braid regions of the spanwise structures and have strong interactions with the latter.

Thanks are due to Dr. Klein for his help, and to the referees for their insightful comments.

1 Gutmark E, Wygnanski I. The planar turbulent jet. J Fluid Mech, 1976, 73(3): 465—495
2 Ramaprian B R, Chandrasekhara M S. LDA measurements in plane turbulent jets. ASME J Fluids Engrg, 1985, 107: 264—271
3 Goldschmidt V W, Bradshaw P. Flapping of a plane jet. Phys Fluids, 1973, 16(3): 354—355
4 Crow S C, Champagne F H. Orderly structure in jet turbulence. J Fluid Mech, 1971, 77: 397—413
5 Liepmann D, Gharib M. The role of streamwise vorticity in the near-field entrainment of round jets. J Fluid Mech, 1992, 245: 643—668

6 Thomas F O, Goldschmidt V W. Structural characteristics of a developing turbulent planar jet. J Fluid Mech, 1986, 163: 227−256

7 Wang X L, Gu H X, Lin W Y. Coherent structure visualization of round turbulent jet. Tsinghua Sci Technol, 1997, 2(2): 624−627

8 Fan Q L, Zhang H Q, Guo Y C, et al. Experimental studies of two-phase round turbulent jet coherent structures. Tsinghua Sci Technol, 2000, 5(1): 105−108

9 Liu M Y, Ma Y W, Fu D X. Evolution of three-dimensional coherent structures in compressible axisymmetric jet. Sci China Ser G-Phys, Mech, Astron, 2003, 46(4): 348−355

10 Sakai Y, Tanaka N, Kushida T. On the development of coherent structure in a plane jet: Part 1, Characteristics of two-point velocity correlation and analysis of eigenmodes by the KL expansion. JSME Int J Ser B, 2006, 49(1): 115−124

11 Sakai Y, Tanaka N, Kushida T. On the development of coherent structure in a plane jet: Part 2, Investigation of spatio-temporal velocity structure by the KL expansion. JSME Int J Ser B, 2006, 49(3): 714−721

12 Partankar S V. Numerical Heat Transfer and Fluid Flow. Washington D C: Hemisphere Publishing Corp, 1980

13 Moin P, Mahesh K. Direct numerical simulation: A tool in turbulence research. Ann Rev Fluid Mech, 1998, 30: 539−578

14 Williamson J. Low-storage Runge-Kutta schemes. J Comput Phys, 1980, 35: 48−56

15 Luo K, Fan J R, Cen K F. Transitional phenomenon of particle dispersion in gas-solid two-phase flows. Chin Sci Bull, 2007, 52(1): 1−10

16 Namer I, Ötügen M V. Velocity measurements in a plane turbulent air jet at moderate Reynolds numbers. Exp Fluids, 1988, 6:387−399

17 Fan J R, Luo K, Ha M Y, et al. Direct numerical simulation of a near-field particle-laden plane turbulent jet. Phys Rev E, 2004, 70: 026303-1−026303-14

18 Ling W, Chung J N, Troutt T R, et al. Direct numerical simulation of a three-dimensional temporal mixing layer with particle dispersion. J Fluid Mech, 1998, 358: 61−85

三维转捩平面射流的拟序结构研究

摘要：本文对雷诺数 5000 的三维过渡射流中的相干结构进行了直接数值模拟。采用有限体积法对控制方程进行空间离散，采用低存储三阶 Runge-Kutta 方法进行时间积分。为了验证现有数值方法的可靠性，对流场的统计结果与相关实验数据进行了比较。重点研究了三维相干涡结构的空间演化及其相互作用。研究发现，三维空间中展向涡结构的演化与二维射流中存在相似性。展向涡结构受到三维不稳定性的影响，诱导了纵向和横向涡结构的形成。随着展向涡结构的破裂和混合，纵向和横向涡管也分裂，流场中形成了混合排列的小尺度结构。最后，分析了展向、纵向和横向涡结构之间的排列关系，并讨论了它们之间的相互作用。

原文刊于 Science in China Series E: Technologica Sciences, 2008, 51(4): 386-396

An experimental investigation of a natural circulation heat pipe system applied to a parabolic trough solar collector steam generation system

Liang Zhang[a], Wujun Wang[b], Zitao Yu[a,*], Liwu Fan[c], Yacai Hu[a], Yu Ni[d], Jianren Fan[d], Kefa Cen[d]

[a] *Institute of Thermal Science and Power Systems, Department of Energy Engineering, Zhejiang University, Hangzhou 310027, PR China*
[b] *Division of Heat and Power Technology, Department of Energy Technology, Royal Institute of Technology (KTH), S-10044 Stockholm, Sweden*
[c] *Mechanical Engineering Department, Auburn University, Auburn AL 36849-5341, USA*
[d] *State Key Laboratory of Clean Energy Utilization, Zhejiang University, Hangzhou 310027, PR China*

Received 3 June 2011; received in revised form 20 October 2011; accepted 13 November 2011
Available online 18 January 2012

Communicated by: Associate Editor Bibek Bandyopadhyay

Abstract

A U-type natural circulation heat pipe system is designed and applied to a parabolic trough solar collector for generating mid-temperature steam. Thermal performance of the heat pipe system is investigated experimentally. A detailed heat transfer analysis is performed on thermal behaviors of the system, especially the solar collector. The results show that the system can generate mid-temperature steam of a pressure up to 0.75 MPa. The thermal efficiency is found to be 38.52% at discharging pressure of 0.5 MPa during summer time.
© 2012 Elsevier Ltd. All rights reserved.

Keywords: Solar thermal energy; Parabolic trough solar collector; Heat pipe; Natural circulation; Steam generation

1. Introduction

Solar thermal energy is an environmentally friendly and sustainable energy, which likely has the greatest potential for any single renewable energy area. The implementation of solar power plants is a promising option for environmentally compatible electricity supply strategy (Trieb et al., 1997). Advantages and study progress of solar thermal technologies were summarized by Price et al. 2002) and Mills (2004) and Thirugnanasambandam et al. (2010). Although there is flourishing development, solar thermal energy is still uncompetitive to traditional fossil fuels in generating electricity without government policy support. Efficiency improvement and cost reduction are deemed necessary for the development of solar thermal energy.

As a kernel part of solar thermal systems, steam generation plays a critical role in improving system efficiency. In recent years, a more efficient steam generation system, the DISS (Direct Solar Steam) or DSG (Direct Steam Generation) project is proposed, especially in Europe (Rodnguez et al., 1999; Eck and Steinmann, 2001; Eck et al., 2003; Zarza et al., 2002, 2004). In combination with further improvements of the collector field and overall system integration, a 26% reduction in the electricity cost seems to be achievable comparing with SEGS plants in California (Zarza et al., 2002). Muñoz et al. (2009) proposed a conceptual design of a solar boiler form the conventional thermal power plants boiler, with the difference that the heat comes from mirrors that concentrate the solar radiation on wall-type array of

* Corresponding authors. Tel./fax: +86 571 87952378.
E-mail address: yuzitao@zju.edu.cn (Z. Yu).

Nomenclature

A	available aperture area, m²	$\eta_{optical}$	optical efficiency
c_{pf}	specific heat of liquid phase, J/kg K	v_f	specific volume of liquid, m³/kg
Di	inner diameter of receiver, m	φ	surface heat flux, W/m²
G	mass velocity, kg/m² s	η	empirical factor, m³ °C /J
I	direct radiation, W/m²	η_{all}	thermal efficiency
m_{vapor}	production of steam at Δt, kg	ε	dimensionless ratio
P	pressure, Pa		
Q_t	heat received by collector, W	*Subscript*	
R	total thermal resistance, K/kW	A–D	Cross-section A–D of receiver
r	latent heat of vaporization, kJ/kg	*boiler*	boiler
T	temperature, °C	*collector*	collector/receiver
$\bar{T}$	average temperature, °C	*coil*	coil of boiler
$(\Delta T_{SUB})_i$	inlet subcooling, °C	*coil in*	entrance of boiler coil
Δt	time interval, s	*coil out*	exit of boiler coil
z^*	length of the tube to the point of thermal equilibrium, m	*water*	water pipe
		vapor	steam/vapor pipe
θ	incident angle, °		

solar collectors, instead of coming from fuel flames and hot gases. In their analysis the overall efficiency of the conversion from direct solar irradiation energy to electricity is above 20%. In a word, finding a more effective and inexpensive solar steam generation system for solar thermal electricity system is what researchers and engineers are interested right now. Nandi and De (2007) installed a solar thermal system with parabolic concentrators for sweetmeats production. The system had documented a temperature rise up to 200 °C.

Moreover, the dependence of the forced convection system upon a fossil fuel or electricity supply for stimulating the pump makes the self-sufficient, free-convective thermosyphon relatively attractive (Norton and Probert, 1982). Comparing with forced-circulation, free convective (or natural circulation) solar heat pipe presents some advantages, such as simple construction, wide adjustability, easier control and high heat transfer ability at low temperature differences etc. The conception of natural circulation solar heat pipe (or thermosyphon) has been widely applied in flat plate solar water system for a century. Experiments, theoretical analysis and simulation work on characteristics of thermosyphonic flow (Shitzer et al., 1979; Morrison and Ranatunga, 1980), thermal performance (Ezekwe, 1990; Bong et al., 1993; Ismail and Abogderah, 1998; Chun et al. 1999; Yu et al., 2005; Riffat et al. 2005; Rittidech and Wannapakne, 2007; Hussein, 2007; Azad, 2008; Rittidech et al., 2009), system design and optimization (Gupta and Garg, 1968; Ortabasi and Fehlner, 1980; Sodha and Tiwari, 1981; Ribot and McConnell, 1983; Hussein et al., 1999; Mathioulakis and Belessiotis, 2002; Tanaka and Nakatake, 2004; Tanaka et al. 2004; Hobbi and Siddiqui, 2009) have been carried out thoroughly. All these results clearly demonstrated that solar heat pipe performs a better result for solar hot water system. Further, U-tube collector has been successful in domestic hot water system for its convenient to seal and better performance (Morrison et al., 2004; Shah and Furbo, 2007). Moreover, recent researches show that the application of nanoparticles in thermosyphons or heat pipes brings about an important enhancement of thermal characteristics (Noie et al., 2009; Shafahi et al., 2010; Huminic et al., 2011).

Though solar heat pipe system has a good thermal performance, thermal performance of solar heat pipe for generating mid/high temperature steam applied in parabolic trough collector (PTC) is barely discussed. Fadar et al. (2009a, 2009b) studied a solar adsorptive cooling system, using activated carbon–ammonia pair, where the reactor was heated by a parabolic trough collector coupled with a water-stainless steel annular heat pipe. However, no experimental validation was provided.

The objective of the present study was to propose a natural circulation heat pipe system to generate steam of mid-temperature (120–200 °C). In order to avoid large temperature difference in axis as DISS and difficulty of glass–steal seal, we designed and fabricated a novel U-type natural circulation heat pipe system which was combined with a PTC. Thermal performance of the solar steam generation system was investigated.

2. Experimental setup and procedure

The system presented here consisted of a U-type parabolic trough solar receiver, a steam pipe, an ascending pipe, a descending pipe, a water pipe, and an unfired-boiler with associated pipes (see Fig. 1). Water was heated to vapor in the U-type solar receiver (evaporation section). The vapor steam flowed into the boiler coils (condensation section) to exchange heat with water in the boiler. The condensed outflow water from the boiler then flowed back to the U-type receiver through the descending and water pipes. Potential

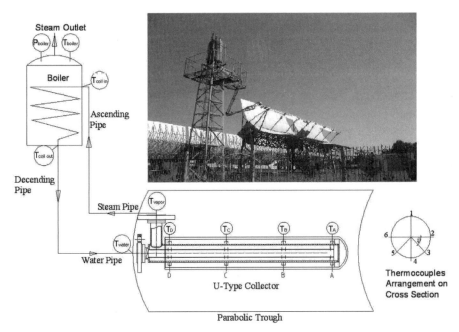

Fig. 1. Diagram of natural convection U-type heat pipe solar steam generation system.

Table 1
The specifications of the solar collector steam generation system.

	Item	Value/type
PTC system	Collector aperture area	78.47 m²
	Collector aperture	5.77 m
	Receiver diameter	70 mm
	Receiver length	2 m
	Receiver number	6
	Receiver coating	Black (Fe_2O_3 & Fe_3O_4)
	Glass tube	Double glazing (100 mm/125 mm)
	Concentration ratio	82
	Working fluid	Water
	Tracking mechanism type	Electronic
	Mode of tracking	Collector axis: N–S horizontal E–W tracking
Boiler	Dimensions	Φ 800 mm × 1902 mm
	Volume	~950L
	Maximum capacity	25.1 kW
	Coil length	10 m

gradient between descending and ascending pipes provided the driving force of circulation, which is called the natural circulation. The boiler was a vertical cylinder pressure vessel as shown in Fig. 1. Pressure regulating valve and liquid level meter were installed for assuring qualified discharging steam and enough feed water. The specifications of the system are tabulated in Table 1.

In order to investigate thermal performance of heat pipe steam generation system, a total of 4 × 6 T-type thermocouples were placed on the receiver (T_A, T_B, T_C, T_D) with a distance of 65 cm in axis direction each. Two thermocouples were placed on the end of steam pipe (T_{vapor}), ascending pipe ($T_{coil\,in}$), descending pipe ($T_{coil\,out}$) and water pipe (T_{water}) in two ends of one diameter, respectively. Note that, the thermocouples were all placed on the tube wall. One T-type thermocouples (T_{boiler}) and a WMB3351 pressure sensor (P_{boiler}) were placed in the boiler. Details are shown in Fig. 1. Thermocouples were placed on the surface of the pipes and covered with insulation. Thermocouples were calibrated to have an accuracy of ±0.1 °C with a standard thermistor. Solar radiation was measured using a TBS-2-2 radiometer with an accuracy of ±2%. The temperature and pressure signals were gathered by a data acquisition board 34970A (Agilent Company).

In this work, field experiments were performed to evaluate characteristics of system natural circulation and reliability for generating mid-temperature steam at three working conditions (0.2 MPa, 0.5 MPa and 0.75 MPa). The filling ratios of working fluid (water) in all three cases were the same, 70%. The radiation and ambient temperature conditions of the three cases are shown in Fig. 2.

3. Results and discussion

3.1. Thermal characteristic of natural circulation of heat pipe in PTC system

Fig. 3a–c shows temperature distributions of natural circulation heat pipe PTC system at three operating conditions at boiler discharging pressure of 0.2 MPa, 0.5 MPa and 0.75 MPa, respectively. As shown in Fig. 3a, four typical phases characterize the system working process: starting (*o–a*), boiler heating (*a–b*), steam discharging (*b–c*) and ending (*c–d*) phase. In the starting phase (*o–a*), average temperature in collector ($\overline{T}_{collector}$), water and steam pipes ($\overline{T}_{vapor}$) is rising rapidly, but the other parts are stable. This means that mass transfer and system circulation have not started yet. With the feature of acute temperature rise and fall in water pipe ($\overline{T}_{water}$) and rapid temperature up in coil in ($\overline{T}_{coil\ in}$) and coil out ($\overline{T}_{coil\ out}$), the boiler heating phase (*a–b*) begins. Accordingly, mass (vapor) and heat transfer from collector to boiler starts and temperature ($\overline{T}_{boiler}$) and pressure (P_{boiler}) in boiler gradually rise. As presented in Fig. 4, a quasi-stable state characterizes the steam discharging step (*b-c*). It is important to notice that temperature vibrating trends of $\overline{T}_{coil\ out}$ and $\overline{T}_{water}$ are opposite when discharging pressure fluctuates (see Fig. 4a and b). This means that heat pipe system is going on a self-equilibrium process. Accordingly, the boiler pressure (P_{boiler}) is gradually recovered. In addition, fluctuations of $\overline{T}_{coil\ out}$ and $\overline{T}_{water}$ are less in the beginning of discharging than that in the self-equilibrium process. This could be an explanation of the breaking of heat transfer equilibrium of the system.

It is important to point out that, in Fig. 3, $\overline{T}_{water}$ and $\overline{T}_{coil\ out}$ are closed and even $\overline{T}_{water} > \overline{T}_{coil\ out}$ especially at 0.2Mpa operating condition. This is because the test point of $\overline{T}_{water}$ which is 0.5 m away from the inlet of the fist U-type receiver as shown in Fig. 1. Actually, $\overline{T}_{water}$ is also inlet temperature of the receiver. Further, the temperature test point is not within the fluid but on the tube wall. Thus, heat conduction from the collector to inlet water results in this consequence. Moreover, comparing with three figures in Fig. 3, it is found that $\overline{T}_{water}$ in Fig. 3a is greatly fluctuated. This becomes clear when we examine $\overline{T}_{coil\ out}$. In the boiler heat phase, $\overline{T}_{coil\ out}$ nearly retained stable in Fig. 3a, while

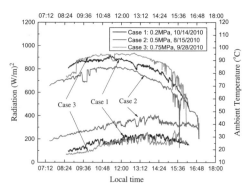

Fig. 2. The radiation and ambient temperature conditions of the three field experiment cases.

in the other two cases, trends of $\overline{T}_{coil\ out}$ are increasing. This indicates that in this case, subcooled heat transfer in coils of boiler is much more serious. This means that heat flux is lower. In another way when heat flux increased, $\overline{T}_{coil\ out}$ rise, then more water in descending pipe flow through $\overline{T}_{water}$ test point which decreases $\overline{T}_{water}$ quickly. Temperature vibrating trends of $\overline{T}_{coil\ out}$ and $\overline{T}_{water}$ shown in Fig. 4a and b confirm this explanation. In the mean time, because of lower heat flux in 0.2 Mpa case, when there is flow fluctuation, the relative fluctuation is larger, and then temperature fluctuation would be more obviously.

3.2. Thermal efficiency and reliability of natural circulation heat pipe PTC system

Comparing results in Fig. 3a–c, a valid point is that this heat pipe PTC system can generate mid-temperature steam up to 0.75 MPa dependable. Table 2 illustrates steam generation process at 0.5 MPa. Thermal efficiency is defined as

$$\eta_{all} = \frac{\sum m_{vapor} r}{\sum AI \Delta t \cos\theta} \quad (1)$$

where m_{vapor}, r, A, I, Δt, and θ are the production of steam, latent heat of vaporization, available aperture area, direct radiation, time interval and incident angle, respectively. Results show that thermal efficiency can reach up to 38.52%.

On the other hand, Table 3 tabulated thermal loss of the heat pipe system in pipes. The results demonstrate that thermal loss in pipes is 47.35% and 52.57%. This suggests that thermal efficiency would have substantial improvement by enhancing thermal insulation of pipes.

3.3. Total thermal resistance of heat pipe system

Considering the three conditions, Fig. 5 displays the total thermal resistance of the heat pipe. Total thermal resistance is defined as

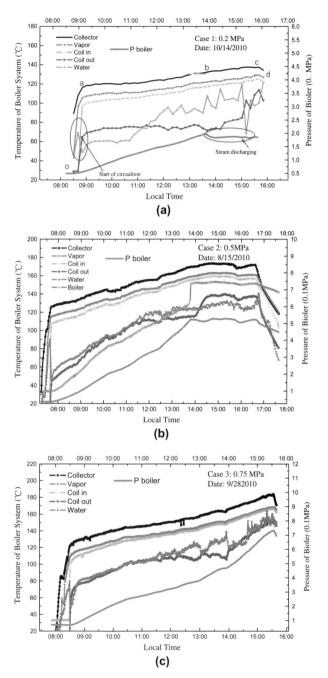

Fig. 3. Temperature distributions of natural circulation heat pipe PTC system at three operating conditions ((a): 0.2 MPa, (b): 0.5 MPa and (c): 0.75 MPa).

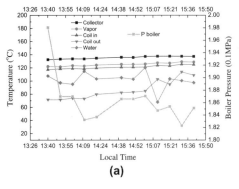

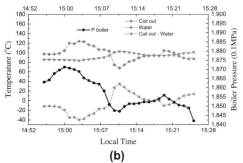

Fig. 4. Thermal characteristics of steam generation system at discharging phase ((a): The whole phase and (b): Characteristics of $T_{\text{coil out}}$ and T_{water}).

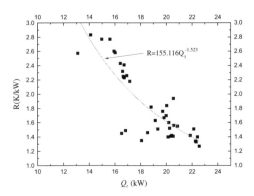

Fig. 5. Total thermal resistance of heat pipe changes with heat power.

$$R = \frac{\overline{T_{collector}} - \overline{T_{coil}}}{Q_t} = \frac{\overline{T_{collector}} - \overline{T_{coil}}}{\eta_{optical} \sum AI} \quad (2)$$

where $\overline{T_{coil}}$, Q_t and $\eta_{optical}$ are average temperature of coil, heat flux received by collector and optical efficiency. As shown in Fig. 5, total thermal resistance decreases reasonably with increase of heat flux. It is important to point out that the total thermal resistance is less than 3.0 K/kW in these conditions. Furthermore, the correlation of thermal resistance and heat flux is fitted as

$$R = 115.116 Q_t^{-1.523}. \quad (3)$$

Table 2
Thermal efficiency result calculating at 0.5 MPa discharging in summer.

Items	Value	Items	Value
Start time	13:50	Steam temperature (°C)	153.5
End time	15:40	Steam pressure (MPa)	0.523
Average direct radiation (W/m²)	649.66	Enthalpy (kJ/kg)	2750.592
Available aperture area (m²)	62.75	Latent heat (kJ/kg)	2103.045
Average Incident angle (°)	44.45	Steam production (kg)	35.18
Thermal efficiency (%)	38.52		

Table 3
Enthalpy drop of the heat pipe system.

Section	Temperature (°C)		Temperature difference (°C)		Enthalpy difference (kJ/kg)	
	0.2 MPa	0.5 MPa	0.2 MPa	0.5 MPa	0.2 MPa	0.5 MPa
Receiver	138.06	173.09				
			9.734	10.15	13.147	10.28
Vapor	128.33	162.94				
			3.855	3.78	5.417	4.12
Coil in	124.47	159.16				
			27.641	21.06	41.813	25.53
Coil out	96.83	138.1				
			11.722	10.28	19.036	13.9
Water	85.11	127.82				
Thermal loss percentage of pipes					47.35%	52.57%

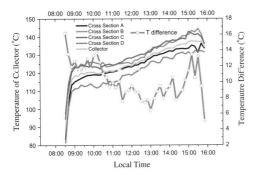

Fig. 6. Temperature characteristics of U-type receiver in axis direction.

3.4. Heat transfer characteristic in collector

To facilitate the understanding of boiling phenomenon in the collector, length of the tube to the point of bulk boiling z^* is defined as (Collier and Throme, 1994)

$$z^* = \frac{Gc_{pf}D_i}{4}\left[\frac{(\Delta T_{SUB})_i}{\phi} + \frac{\eta}{G\varepsilon v_f}\right] \quad (4)$$

where G, v_f, D_i, c_{pf}, $(\Delta T_{SUB})_i$, φ, η and ε are mass flow, specific volume of liquid, inner diameter, specific heat of liquid phase, inlet subcooling, surface heat flux, empirical factor and dimensionless ratio. For cases of 0.2 MPa, $z^* \sim 1.0$ m. This means that bulk boiling happens between Cross section B and Cross section C. Figs. 6 and 7 display temperature distribution characteristics of U-type receiver in axis and cross section. In Fig. 6, it clearly shows that temperature difference of the receiver is less than 16.2 °C and most of the time is around 10 °C. This indicates that this U-type can avoid large temperature difference that happened in DISS (Rodnguez et al., 1999; Eck and Steinmann, 2001; Eck et al., 2003; Zarza et al., 2002, 2004). In detail, average temperature of Cross section C ($\overline{T_C}$) and Cross section B ($\overline{T_B}$) are similar. While temperature of cross section C is lower than Cross section B in the inner annular tube. On the basis of counter flow, this reveals that there is a highest temperature value between B and C, which is in support of bulk boiling happening in the inner annular

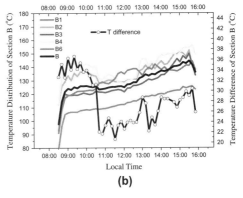

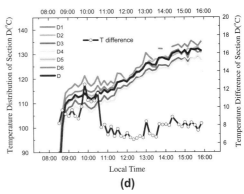

Fig. 7. Temperature characteristics of U-type receiver in cross sections ((a): Cross section A, (b): Cross section B, (c): Cross section C and (d): Cross section D).

tube between B and C according to temperature acceleration speed in subcooled boiling and bulk boiling (Collier and Throme, 1994). Further, in Fig. 7b and c, it is noticed that temperature difference are both nearly 30 °C in Cross section B and C. It is quite possible that steam there is overheated.

However, overheated steam is cooled by feed-in water from Cross section D, in another way, this section can be understood as a preheat zone. It is deserved to notice that temperature difference in Cross section D (see Fig. 7d) is much less comparing with B or C, which means that an optimizing thermal design should be adopted to assure qualified steam leaving from collector for much higher working temperature.

In addition, as shown in Fig. 7, temperature difference in cross sections decreased quickly after about 1.5 h of circulation start. In the discharging phase, temperature difference is increasing especially in Cross section A, which indicates worse temperature uniformity in starting and discharging phases. During both phases, thermal equilibrium is broken and a new equilibrium is needed. What's more, it is observed that fluctuation in Cross section C is more clearly. In the overheat zone, it is easier to observe this phenomena due to less thermal capacity when thermal equilibrium is unstable.

4. Conclusions

A U-type natural circulation heat pipe system applied to a parabolic trough solar collector for generating mid-temperature steam is introduced. Heat transfer performance of the system is investigated experimentally. A detailed heat transfer analysis is performed on thermal behaviors of the system, especially the solar collector. The results show that the system can generate mid-temperature steam of a pressure up to 0.75 MPa. The thermal efficiency is found to be 38.52% at discharging pressure of 0.5 MPa during summer time. The total thermal resistance is correlated as $R = 115.116 Q_t^{-1.523}$. Additionally, optimization on thermal insulation and collector exit design would greatly increase system efficiency and dependability.

Acknowledgements

This work was supported by the National Natural Science Foundation of China (NSFC) and the National High-Technology Research and Development Program of China (863 Program) under Grant Nos. 50706044 and 2007AA05Z254, respectively.

References

Azad, E., 2008. Theoretical and experimental investigation of heat pipe solar collector. Experimental Thermal and Fluid Science 32, 1666–1672.
Bong, T.Y., Ng, K.C., Bao, H., 1993. Thermal performance of a flat-plate heat-pipe collector array. Solar Energy 50, 491–498.
Chun, W., Kang, Y.H., Kwak, H.Y., Lee, Y.S., 1999. An experimental study of the utilization of heat pipes for solar water heaters. Applied Thermal Engineering 19, 807–817.
Collier, J.G., Throme, J.R., 1994. Convective Boiling and Condensation, third ed. Oxford.
Eck, M., Steinmann, W.D., 2001. Direct steam generation in parabolic troughs: first results of the DISS project, proceedings of solar forum 2001. The Power to Choose, Washington D.C, Solar Energy.
Eck, M., Zarza, E., Eickhoff, M., Rheinländer, J., Valenzuela, L., 2003. Applied research concerning the direct steam generation in parabolic troughs. Solar Energy 74, 341–351.
Ezekwe, C.I., 1990. Thermal performance of heat pipe solar energy systems. Solar & Wind Technology 7, 349–354.
Fadar, A.E., Mimet, A., Azzabakh, A., Garca, M.P., Casting, J., 2009a. Study of a new solar adsorption refrigerator powered by a parabolic trough collector. Applied Thermal Engineering. 29, 1267–1270.
Fadar, A.E., Mimet, A., Garca, M.P., 2009b. Study of an adsorption refrigeration system powered by parabolic trough collector and coupled with a heat pipe. Renewable Energy 34, 2271–2279.
Gupta, C.L., Garg, H.P., 1968. System design in solar water heaters with natural circulation. Solar Energy 12, 163–182.
Hobbi, A., Siddiqui, K., 2009. Experimental study on the effect of heat transfer enhancement devices in flat-plate solar collectors. International Journal of Heat and Mass transfer 52, 4650–4658.
Huminic, G., Huminic, A., Morjan, I., Dumitrche, F., 2011. Experimental study of the thermal performance of thermosyphon heat pipe using iron oxide nanoparticles. International Journal of Heat and Mass Transfer 54, 656–661.
Hussein, H.M.S., 2007. Theoretical and experimental investigation of wickless heat pipes flat plate solar collector with cross flow heat exchanger. Energy Conversion & Management 48, 1266–1272.
Hussein, H.M.S., Mohamad, M.A., Belessiotis, V., 1999. Optimization of a wickless heat pipe flat plate solar collector. Energy Conversion & Management 40, 1949–1961.
Ismail, K.A.R., Abogderah, M.M., 1998. Performance of heat pipe solar collector. Journal of Solar Energy Engineering 120, 51–59.
Mathioulakis, E., Belessiotis, V., 2002. A new heat-pipe solar domestic hot water system. Solar Energy 72, 13–20.
Mills, D., 2004. Advances in solar thermal electricity technology. Solar Energy 76, 19–31.
Morrison, G.L., Ranatunga, D.B.J., 1980. Thermosyphon circulation in solar collectors. Solar Energy 24, 191–198.
Morrison, G.L., Budihardjo, I., Behnia, M., 2004. Water-in-glass evacuated tube solar water heaters. Solar Energy 76, 135–140.
Muñoz, J., Abánades, A., Val, J.M.M., 2009. A conceptual design of solar boiler. Solar Energy 83, 1713–1722.
Nandi, P., De, R., 2007. Production of sweetmeat utilizing solar thermal energy: economic and thermal analysis of a case study. Journal of Cleaner Production 15, 373–377.
Noie, H.S., Heris, S.Z., Kahani, M., Nowee, S.M., 2009. Heat transfer enhancement using Al_2O_3/water nanofluid in a two-phase closed thermosyphon. International Journal of Heat and Fluid Flow 30, 700–705.
Norton, B., Probert, S.D., 1982. Natural circulation solar energy simulated systems for heating water. Applied Energy 11, 167–196.
Ortabasi, U., Fehlner, F.P., 1980. Cusp mirror-heat pipe evacuated tubular solar thermal collector. Solar Energy 24, 477–489.
Price, H., Lüpfert, E., Kearney, D., Zarza, E., Cohen, G., Gee, R., 2002. Advances in parabolic trough solar power technology. Journal of Solar Energy Engineering 124, 109–125.
Ribot, J., McConnell, R.D., 1983. Testing and analysis of a heat-pipe solar collector. Journal of Solar Energy Engineering 105, 440–445.
Riffat, S.B., Zhao, X., Doherty, P.S., 2005. Developing a theoretical model to investigated thermal performance of a thin membrane heat-pipe solar collector. Applied Thermal Engineering 25, 899–915.
Rittidech, S., Wannapakne, S., 2007. Experimental study of the performance of a solar collector by closed-end oscillating heat pipe (CEOHP). Applied Thermal Engineering 27, 1978–1985.

Rittidech, S., Donmaung, A., Kumsombut, K., 2009. Experimental study of the performance of a circular tube solar collector with closed-loop oscillating heat-pipe with check valve (CLOHP/ CV). Renewable Energy 34, 2234–2238.

Rodnguez, L.G., Marrero, A., Camacho, C.G., 1999. Application of direct steam generation into a solar parabolic trough collector to multi effect distillation. Desalination 125, 139–145.

Shafahi, M., Bianco, V., Vafai, K., Manca, O., 2010. Thermal performance of flat-shaped heat pipes using nanofluids. International Journal of Heat and Mass Transfer 53, 1438–1445.

Shah, L.J., Furbo, S., 2007. Theoretical flow investigations of an all glass evacuated tubular collector. Solar Energy 81, 822–828.

Shitzer, A., Kalmanoviz, D., Zvirin, Y., Grossman, G., 1979. Experiments with a flat plate solar water heating system in thermosyphonic flow. Solar Energy 22, 27–35.

Sodha, M.S., Tiwari, G.N., 1981. Analysis of natural circulation solar water heating systems. Energy Conversion & Management 21, 283–288.

Tanaka, H., Nakatake, Y., 2004. A vertical multiple-effect diffusion-type solar still coupled with a heat pipe solar collector. Desalination 160, 195–205.

Tanaka, H., Nakatake, Y., Watanabe, K., 2004. Parametric study on a vertical multiple-effect diffusion-type solar still couple with a heat-pipe solar collector. Desalination 171, 243–255.

Thirugnanasambandam, M., Iniyan, S., Goic, R., 2010. A review of solar thermal technologies. Renewable and Sustainable Energy Reviews 14, 312–322.

Trieb, F., Langiniß, O., Klaiß, H., 1997. Solar electricity generation – a comparative view of technologies costs and environmental impact. Solar Energy 59, 89–99.

Yu, Z.T., Hu, Y.C., Hong, R.H., Cen, K.F., 2005. Investigation and analysis on a cellular heat pipe flat solar heater. Heat and Mass Transfer 42, 122–128.

Zarza, E., Valenzuela, L., León, J., Weyers, H.D., Eickhoff, M., Eck, M., Hennecke, K., 2002. The DISS project: direct steam generation in parabolic trough systems operation and maintenance experience and update on project status. Journal of Solar Energy Engineering 124, 126–133.

Zarza, E., Valenzuela, L., León, J., Hennecke, K., Eck, M., Weyers, H.D., Eickhoff, M., 2004. Direct steam generation in parabolic troughs: final results and conclusions of the DISS project. Energy 29, 635–644.

自然循环热管系统应用于抛物线槽式太阳能集热器蒸汽发生系统的实验研究

摘要：本文为产生中温蒸汽设计了一种 U 型自然循环热管系统，并将其应用于抛物线槽式太阳能集热器。通过实验研究，对热管系统特别是太阳能集热器的热行为进行了详细的传热分析。结果表明，该系统可以产生压力高达 0.75MPa 的中温蒸汽。在夏季，当排汽压力为 0.5MPa 时，热效率可达 38.52%。

原文刊于 Solar Energy, 2012, 86: 911-919

Nitrogen oxide absorption and nitrite/nitrate formation in limestone slurry for WFGD system

Chenghang Zheng, Changri Xu, Yongxin Zhang, Jun Zhang, Xiang Gao*, Zhongyang Luo, Kefa Cen

State Key Laboratory of Clean Energy Utilization, Zhejiang University, Hangzhou 310027, China

HIGHLIGHTS

- The NOx absorption and nitrite/nitrate formation in limestone slurry was studied.
- The NOx compositions and SO_2 significantly affects NOx absorption rates.
- The mechanism of NOx absorption and conversion in limestone slurry was proposed.

ARTICLE INFO

Article history:
Received 6 December 2013
Received in revised form 21 April 2014
Accepted 4 May 2014
Available online 24 May 2014

Keywords:
Nitrogen oxide
Absorption
Nitrite
Nitrate
Limestone
WFGD

ABSTRACT

Promoting the nitrogen oxide (NOx) absorption step is the major challenge for simultaneous removal of sulfur dioxide (SO_2) and NOx in wet flue gas desulfurization. This paper studied NOx absorption and absorption product formation in limestone slurry. The effect of $CaCO_3$ concentration, NOx composition, and SO_2 and O_2 concentration was investigated. A proposed reaction pathway of NOx absorption and conversion was developed based on the experimental results. NOx absorption is thought to involve two types of absorption pathways, namely, pure NO_2 absorption and simultaneous NO and NO_2 absorption. A positive correlation was found between $CaCO_3$ concentration and NO_2 absorption rate, and the best molar ratio of NO_2/NOx in gas stream for NO absorption was 0.5. The main NOx absorption products were NO_2^- and NO_3^-. NO_2^- inhibits further NOx absorption and could be partially decomposed with release of NO, while NO_3^- has no obvious effect on NOx absorption. Nitrite can also be converted into nitrate through a reaction with dissolved oxygen. Addition of SO_2 provides another NOx absorption pathway due to the reaction between S(IV) and NOx, which indicates the feasibility of simultaneously removing SO_2 and NOx by wet scrubbing.

© 2014 Elsevier Ltd. All rights reserved.

1. Introduction

As typical pollutants emitted from fuel combustion in stationary sources (e.g., power plants, steel plants, and industrial boilers), sulfur dioxide (SO_2) and nitrogen oxides (NOx) cause many environmental problems such as acid rain, photochemical smog, and ozone depletion. Thus, SO_2 and NOx removal has been the focus of air pollution control in the last decades.

Pollution control methods for a single pollutant have been developed previously; such methods include wet scrubbing and selective catalytic reduction (SCR) for SO_2 and NOx removal, respectively. However, these one-to-one control technologies are costly and require a large amount of space. Unlike conventional controls, control technologies that are capable of simultaneously reducing emissions of multiple pollutants may solve these problems [1]. Limestone-based wet flue gas desulfurization (WFGD), which is one of the most widely applied wet scrubbing processes, is less expensive and has high SO_2 removal efficiency. However, its advantages do not apply to NOx removal, because insoluble NO is the main component of NOx in typical flue gas [2,3].

Recently, simultaneous removal of SO_2 and NOx by using integrated oxidation-absorption process has gained increasing interest [4–6]. In this process, insoluble NO is converted into soluble NO_2 through environmentally friendly oxidation processes (e.g., selective catalytic oxidation [7,8], corona discharge oxidation [9,10], and photocatalytic oxidation [11,12]) and then further removed with SO_2 in the subsequent wet absorption process. However, higher NO oxidation efficiency requires higher operating cost, and the NO cannot be oxidized completely into NO_2. Some researchers have reported that the NOx absorption in solution was affected by the NO_2/NOx ratio in gas phase [13–15]. Thus,

* Corresponding author. Tel.: +86 571 87951335.
  E-mail address: xgao1@zju.edu.cn (X. Gao).

http://dx.doi.org/10.1016/j.apenergy.2014.05.006
0306-2619/© 2014 Elsevier Ltd. All rights reserved.

Nomenclature

N_{NOx}	total absorption rate of NOx, mol/(m² s)
N_{NO2}	absorption rate of NO$_2$, mol/(m² s)
N_{NO}	absorption rate of NO, mol/(m² s)
P_0	total pressure, Pa
V_G	gas volume rate, m³/s
R	gas constant (8.314 J/(mol K))
S	interfacial area, m²
T	temperature, K
P_{NO}	partial pressure of NO in bulk of gas, Pa
P_{NO_2}	partial pressure of NO$_2$ in bulk of gas, Pa
$P_{N_2O_4}$	partial pressure of N$_2$O$_4$ in bulk of gas, Pa
P_I	partial pressure of inerts, Pa
OR	defined as molar ratio of NO$_2$ to (NO + NO$_2$)
DO	dissolved oxygen, mg/L

Subscripts

in	inlet of the gas stream
out	outlet of the gas stream

the compositions of NOx are an important factor. If a certain amount of NO can be absorbed by the solution in the presence of NO$_2$, the total treatment cost for NOx control will be reduced.

To the best of our knowledge, the subsequent absorption process of NOx in CaCO$_3$ slurry, which is the main absorbent of limestone-based WFGD, has not yet been investigated. This paper studies NOx absorption and nitrite/nitrate formation in CaCO$_3$ slurry as well as investigates operational parameters, such as CaCO$_3$ concentration, NOx composition, and SO$_2$ and O$_2$ concentrations. A proposed global reaction pathway of NOx absorption in CaCO$_3$ slurry was developed to depict the experimental findings, which include the fate of nitrite and nitrates, reaction between NOx and S(IV), and gas–liquid phase oxidation of NOx.

2. Experimental setup

The schematic of the experimental system is shown in Fig. 1. NO and NO$_2$ were diluted by N$_2$ to the desired concentrations in the mixing chamber before being fed into the reactor. A certain concentration of SO$_2$ and O$_2$ was employed to investigate their effects on NOx absorption. The gas flow rate was controlled by the mass flow meters (SevenStar Huachuang Co., Ltd., China) with an accuracy of ±0.1 mL/min.

All experiments for NOx absorption were performed in a double-stirred tank reactor. The reactor was cylindrical with a volume of 1250 mL (100 mm in diameter, 160 mm in height). Two rotating axial impellers were installed on two stirring rods, with one nested within another for gas and liquid phase mixing. Each stirring rod was locked to a driven pulley, and each driven pulley was connected to a drive pulley with a belt. The drive and driven belt and pulley system were actuated by a JJ-1A digital tachometer (Jintan instrument plant, China). Various stirring speeds of gas and liquid impellers could be obtained by adjusting the diameters of the drive and driven pulleys. Four baffles (100 mm in height, 10 mm in width) were positioned 90° apart to keep the gas–liquid interface flat during the operation. The reactor was immersed in a water bath to achieve a stable desired temperature. NOx and SO$_2$ concentrations at the outlet of the double-stirred tank reactor were measured continuously by using Testo 350XL [Testo Instrumental Trading (Shanghai) Co., Ltd.]. All the gas concentration information during the experimental procedure was transferred to the data acquisition system for analysis.

All the chemicals, such as calcium carbonate (⩾99.0%, AR), sodium nitrate (⩾98.0%, AR), sodium nitrite (⩾98.0%, AR), and deionized water, were used as received without further purification. Gases, such as NO (0.6%), NO$_2$ (0.5%), SO$_2$ (1.0%), O$_2$ (⩾99.999%), and N$_2$ (⩾99.999%), stored in the steel cylinders were all provided by Hangzhou New Century Gas Co., Ltd.

For all experimental runs, the simulated flue gas continuously flowed through the reactor with a total flow rate of 500 mL/min at atmospheric pressure, and the reactor was filled with 800 mL CaCO$_3$ slurry. According to previous research [16], as the liquid stirring speed was 100–187 rpm and the gas stirring speed was 120–800 rpm, the gas and the liquid phases can be considered completely mixed. In this study, all the experiments were performed at a liquid stirring speed of 180 rpm and a gas stirring speed of 250 rpm. Therefore, the gas and liquid phases in the

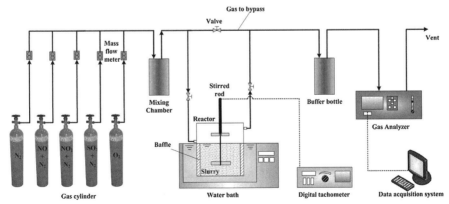

Fig. 1. Schematic of the experimental system.

Table 1
Experimental conditions.

Parameter	Range	Parameter	Range
NO (ppm)	100–700	N_2 (vol%)	Balance
NO_2 (ppm)	100–700	$CaCO_3$ (wt%)	0.1–1.0
SO_2 (ppm)	500–2000	Temperature (°C)	50
O_2 (vol%)	0–15		

double-stirred tank reactor were completely mixed. The absorption rate of NO and NO_2 can be calculated as [17]

$$N_{NO} = \frac{V_G P_0}{RTS}\left[\left(\frac{P_{NO}}{P_I}\right)_{in} - \left(\frac{P_{NO}}{P_I}\right)_{out}\right] \quad (1)$$

$$N_{NO_2} = \frac{V_G P_0}{RTS}\left[\left(\frac{P_{NO_2}}{P_I}\right)_{in} - \left(\frac{P_{NO_2}}{P_I}\right)_{out}\right] \quad (2)$$

Then, the total absorption rates of NOx can be calculated by

$$N_{NOx} = N_{NO} + N_{NO_2} \quad (3)$$

The concentrations of nitrite and nitrate ions in the slurry sample were analyzed by using ion chromatography (Dionex ICS-900). The operating conditions for all the experiments are listed in Table 1.

3. Results and discussion

3.1. Effect of $CaCO_3$ concentration on NOx absorption

The effect of $CaCO_3$ concentration on NO and NO_2 absorption rates was investigated, and the results are shown in Figs. 2 and 3. Fig. 2 shows that the NO concentration difference between reactor inlet and outlet is only about 4–7 ppm, and the calculated NO absorption rate is as low as 2.885×10^{-7} mol/(m^2 s) when the absorption process reaches equilibrium. This result indicates that the effect of $CaCO_3$ concentration on the NO absorption rate is trivial because nitric oxide is water insoluble. However, Fig. 3 shows that the absorption rate of NO_2 increases as the $CaCO_3$ concentration increases from 0 wt% to 0.5 wt% and then becomes nearly constant when the $CaCO_3$ concentration is greater than 0.5 wt%, which is probably due to the mass-transfer limitation of NO_2 into the slurry. The main reactions of the NO_2 absorption in $CaCO_3$ slurry can be summarized as

$$2NO_2(g) \leftrightarrow N_2O_4(g) \quad \Delta_f G^o = -2.8 \text{ kJ/mol} \quad (4)$$

$$2NO_2(g) + H_2O(l) \rightarrow HNO_2(aq, \text{undissociated}) + H^+ + NO_3^-$$
$$\Delta_f G^o = -22.8 \text{ kJ/mol} \quad (5)$$

$$N_2O_4(g) + H_2O(l) \rightarrow HNO_2(aq, \text{undiss}) + H^+ + NO_3^-$$
$$\Delta_f G^o = -20.0 \text{ kJ/mol} \quad (6)$$

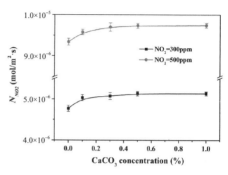

Fig. 3. Effect of $CaCO_3$ concentration on NO_2 absorption rates (T = 50 °C).

$$HNO_2(aq, \text{undiss}) \leftrightarrow H^+ + NO_2^- \quad \Delta_f G^o = +13.8 \text{ kJ/mol} \quad (7)$$

$$HNO_3(aq, \text{undiss}) \leftrightarrow H^+ + NO_3^- \quad \Delta_f G^o = -30.6 \text{ kJ/mol} \quad (8)$$

$$CaCO_3(s) \leftrightarrow Ca^{2+} + CO_3^{2-} \quad \Delta_f G^o = +47.4 \text{ kJ/mol} \quad (9)$$

$$CO_3^{2-} + H^+ \leftrightarrow HCO_3^- \quad \Delta_f G^o = -59.0 \text{ kJ/mol} \quad (10)$$

$$HCO_3^- + H^+ \leftrightarrow H_2O(l) + CO_2(g) \quad \Delta_f G^o = -44.7 \text{ kJ/mol} \quad (11)$$

The above absorption reactions indicate that H^+ depletion by reactions (10) and (11) will accelerate NO_2 hydrolysis and could increase the NO_2 absorption rate. Additionally, the absorbent particles that are suspended in the $CaCO_3$ slurry could provide more reactive surface. Dagaonkar et al. [18] reported that the gas absorption rate in slurry that contains fine particles of absorbent was considerably higher than that in clear solution, and the absorption rate would increase with solid content within a certain range. The NO_2 absorption rate changes minimally as the $CaCO_3$ concentration is greater than 0.5 wt%. It indicates that when the NO_2 absorption in $CaCO_3$ slurry, the reaction may be mainly influenced by gas-film as the $CaCO_3$ concentration is above 0.5 wt%. Therefore, all the following experiments were carried out with a $CaCO_3$ concentration of 1.0 wt% to investigate the effect of different gas phase conditions on NOx absorption rate.

Fig. 3 also shows that the NO_2 absorption rate increases with increased NO_2 inlet concentration under the same $CaCO_3$ concentration and temperature. This increase is due to the fact that the hydrolysis reaction of NO_2 was 1.5 order with respect to NO_2 [15,19]. Thus, a higher NO_2 inlet concentration indicates that a greater absorption rate can be obtained.

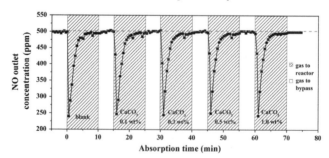

Fig. 2. Effect of $CaCO_3$ concentration on NO absorption (NO inlet concentration = 500 ppm, T = 50 °C).

3.2. Reaction products of NOx absorption

Nitrite and nitrate ions were analyzed to verify the main products of the NOx absorption reaction to better understand the NOx absorption mechanism in $CaCO_3$ slurry. In the current system, gas flow rate and NOx inlet concentration were previously set, and the content of absorbed NOx for a fixed period could be calculated by graphical integration under the curves of NOx absorption rate, i.e., according to Eq. (12).

$$n_{NOx} = \int_0^{t^*} N_{NOx}(t) \cdot S dt \quad (12)$$

where $nNOx$ is the total amount of NOx absorbed in the slurry (mol), $N_{NOx}(t)$ is NOx absorption rate [mol/(m² s)] at time t, and t^* is the total absorption time. Fig. 4a shows the compositions of NOx absorption products in $CaCO_3$ slurry. Both nitrate and nitrite ions in the slurry increase as the absorption proceeded, and the calculated values of total nitrate and nitrite concentration are in good agreement with the experimental values. In addition, the mole percentage of nitrite ions, i.e., the molar ratio of $NO_2^-/(NO_2^- + NO_3^-)$ gradually decreases within absorption time. The products in the slurry should be equimolar nitrate and nitrite according to reactions (5) and (6). However, the results in Fig. 4a shows that the mole percentage of nitrite was only about one-third when the absorption reaches equilibrium. This result can be attributed to the HNO_2 decomposition. Generally, HNO_2 is unstable in liquid phase and can decompose with NO release as [19,20]

$$3HNO_2(aq, undiss) \leftrightarrow 2NO(g) + NO_3^- + H^+ + H_2O(l)$$
$$\Delta_f G^o = -35.2 \text{ kJ/mol} \quad (13)$$

According to reaction (13), HNO_2 decomposition can increase nitrate content and decrease nitrite content. Fig. 4b shows the outlet NOx concentration during the NO_2 absorption period. A certain amount of NO begins to be released after 2 min of NO_2 absorption in $CaCO_3$ slurry, which is similar to the experimental results reported by Zhang et al. [21]. This finding confirms the hypothesized HNO_2 decomposition. Fig. 4b also shows that from the moment of NO release, both the outlet concentration of NO and NO_2 are gradually kept constant, which indicates that NO_2 uptake reaches equilibrium and further NO_2 absorption is inhibited. This result implies that the release of NO does not occur as soon as NO_2 is absorbed in $CaCO_3$ slurry, but is necessary for HNO_2 accumulation. Another possible reason why the mole percentage of nitrite is lower than 50% is liquid phase oxidation of NO_2^- by dissolved oxygen (DO), which will be confirmed later in this paper. The reaction is expressed as

$$2NO_2^- + O_2(aq) \rightarrow 2NO_3^- \quad \Delta_f G^o = -158.2 \text{ kJ/mol} \quad (14)$$

As stated above, HNO_2 accumulation and decomposition can considerably influence the NO_2 absorption rate. Therefore, as the main NO_2 absorption products, the effect of NO_2^- and NO_3^- concentration on NO_2 absorption rate was investigated; the results are shown in Fig. 4c. The initial NO_2^- and NO_3^- concentration in the slurry was adjusted by adding analytically pure sodium nitrite and sodium nitrate, respectively. Fig. 4c shows that the NO_2 absorption rate decreases with increased NO_2^- concentration, whereas it remains nearly constant with increased NO_3^-. This result indicates that NO_2^- inhibits further NO_2 absorption. Thus, promoting NO_2^- depletion in the slurry could enhance the NO_2 absorption level.

3.3. Effect of NOx composition on NOx absorption

Aside from reactions (4)–(11), the following reactions occur when NO and NO_2 coexist in the gas stream:

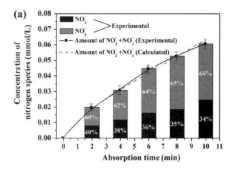
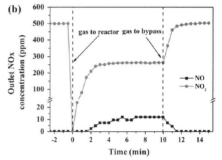
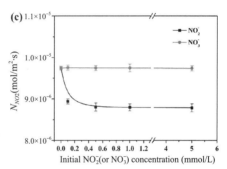

Fig. 4. NO_2 absorption in $CaCO_3$ slurry: (a) experimental and calculated concentration of absorption products; (b) NO and NO_2 concentrations in the outstream during the experimental period; (c) effect of initial NO_2^- (or NO_3^-) concentration on NO_2 absorption rate. ($CaCO_3$ = 1.0 wt%, NO_2 inlet concentration = 500 ppm, T = 50 °C).

$$NO(g) + NO_2(g) \leftrightarrow N_2O_3(g) \quad \Delta_f G^o = +3.5 \text{ kJ/mol} \quad (15)$$

$$N_2O_3(g) + H_2O(l) \leftrightarrow 2HNO_2(aq, undiss) \quad \Delta_f G^o = +2.7 \text{ kJ/mol} \quad (16)$$

Therefore, the existence of NO_2 may promote NO absorption. Fig. 5a shows the effect of OR [defined as molar ratio of NO_2 to (NO + NO_2)] on NOx absorption rates. As shown in Fig. 5a, NO absorption rate increases from 2.885×10^{-7} mol/(m² s) to 1.057×10^{-6} mol/(m² s) as OR increases from 0 to 0.5 and reaches a maximum of 0.5. This result was obtained because N_2O_3, which generated the reaction between NO and NO_2, has a higher solubility than NO. Hence, N_2O_3 formation significantly increases NO

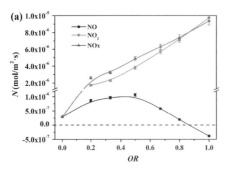

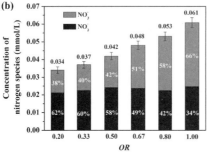

Fig. 5. (a) Effect of NOx compositions in gas stream on NOx absorption rates; (b) nitrite/nitrate concentrations in the slurry for different ORs (total NOx inlet concentration = 500 ppm, T = 50 °C).

absorption rate. However, NO absorption rate decreases with continuous increase of OR and becomes negative, which means that the outlet concentration of NO was higher than that of the inlet when OR is greater than 0.8 because excessive NO_2 at higher OR can release NO, which has been proved with pure NO_2 absorption as shown in Fig. 4b.

Additionally, as OR increases from 0.2 to 1, the absorption rate of NO_2 is significantly increased from 1.720×10^{-6} mol/(m^2 s) to 9.741×10^{-6} mol/(m^2 s). The absorption rate of NO_2 is much higher than that of NO under the same condition. Therefore, the total NOx absorption rate increases with increased OR.

Fig. 5b shows the product compositions in $CaCO_3$ slurry when the NOx absorption process reaches equilibrium. The figure shows that the concentration of nitrite/nitrate ions increases as the OR increases. This increase occurs because NOx is more soluble [14] and reactive when more oxidized NOx species are present in the gas stream. Thus, a higher total of NOx dissolved quantities are obtained. Meanwhile, the mole percentage of nitrite is greater than in the case of pure NO_2 absorption, and it increases as the OR decreases, while nitrite concentration exhibits a slight variation. For example, the mole percentage of nitrite is 62% at the OR of 0.2, and the value is only 34% at the OR of 1.0 while all nitrite concentrations are about 0.02 mmol/L with different ORs. This result is due to the coexistence of NO and NO_2 in the gas stream. The NOx absorption process contains two reaction pathways: reaction (16) and reactions (5) and (6). In addition, in consideration of the combined results in Fig. 5a and b, the total NOx absorption rate and the mole percentage of nitrite in $CaCO_3$ slurry have a negative correlation. Apparently, a higher nitrite proportion in $CaCO_3$ slurry leads to a lower total NOx absorption rate because of the NO_2^- inhibition effect on NOx absorption, as mentioned previously.

3.4. Effect of SO_2 on the absorption of NOx

Fig. 6a shows the effect of SO_2 on the NOx absorption rates as the total NOx inlet concentration is 500 ppm with different ratios of NO and NO_2 at 50 °C. When the SO_2 concentration increases from 0 ppm to 2000 ppm, the NOx absorption rate remains nearly constant at 2.887×10^{-7} mol/(m^2 s) with OR = 0, while the NOx absorption rates both increase with OR of 0.5 and 1. These results suggest that SO_2 introduction does not play a promoting role in pure NO absorption. However, SO_2 introduction can promote the total NOx absorption rate when the inlet NOx is the mixture of NO and NO_2. A reaction between NOx and SO_2 in gas phase is difficult, because the SO_2 that promoted NOx absorption may be related to S(IV) ions, which are the hydrolysis products of SO_2. Fig. 6b shows the effect of S(IV) on the NOx absorption rates when the total NOx inlet concentration is 500 ppm and OR is 0.5. This finding indicates that the S(IV) ions indeed play a positive role in NOx absorption, and the mechanisms can be summarized as [22–24]

$$SO_2(g) + H_2O(l) \leftrightarrow H^+ + HSO_3^- \quad \Delta_f G^o = +9.5 \text{ kJ/mol} \quad (17)$$

$$HSO_3^- \leftrightarrow H^+ + SO_3^{2-} \quad \Delta_f G^o = +41.2 \text{ kJ/mol} \quad (18)$$

$$2NO_2(g) + H_2O(l) + SO_3^{2-} \rightarrow 2NO_2^- + SO_4^{2-} + 2H^+$$
$$\Delta_f G^o = -187.9 \text{ kJ/mol} \quad (19)$$

$$2NO_2(g) + H_2O(l) + HSO_3^- \rightarrow 2NO_2^- + SO_4^{2-} + 3H^+$$
$$\Delta_f G^o = -146.7 \text{ kJ/mol} \quad (20)$$

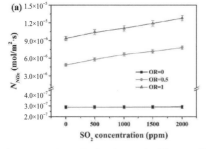
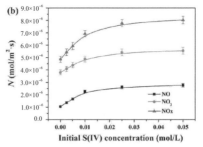

Fig. 6. (a) Effect of SO_2 concentration on NOx absorption rates for different Ors; (b) effect of S(IV) concentration on NOx absorption rates at OR = 0.5 (total NOx inlet concentration = 500 ppm, T = 50 °C).

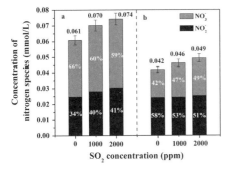

Fig. 7. Effect of SO$_2$ concentration on nitrite/nitrate concentration in the liquid phase as (a) NO$_2$ inlet concentration = 500 ppm; (b) total NOx inlet concentration = 500 ppm and OR = 0.5 (T = 50 °C).

The sulfite and bisulfite ions can react with NO$_2$, which explains why the addition of S(IV) can increase the NOx absorption rate. In addition, as mentioned above, the N$_2$O$_3$ that was formed by the reaction between NO and NO$_2$ can be hydrolyzed into nitrous acid as shown in reaction (16), which tends to react with bisulfate ions and form hydroxylamine-disulfonate (HADS) [25], a precursor to numerous sulfur–nitrogen compounds such as (HO$_3$S)$_2$NOH, HO$_3$-SNO, and (HO$_3$S)$_3$N [26].

$$HNO_2(aq, undiss) + 2HSO_3^- \rightarrow HADS(aq, undiss) + H_2O(l) \quad (21)$$

The nitrate and nitrite ions in CaCO$_3$ slurry with different inlet SO$_2$ concentrations are shown in Fig. 7. The concentration of nitrogen species increases in the presence of SO$_2$. Furthermore, the mole percentage of nitrite ion increases from 34% to 41% as the SO$_2$ concentration increases from 0 ppm to 2000 ppm (see Fig. 7a). This result indicates that reactions (19) and (20), which increase the NO$_2^-$ content, rather than reaction (21), is another pathway of NO$_2$ absorption aside from NO$_2$ hydrolysis in the presence of SO$_2$. However, when NOx is the mixture of NO and NO$_2$ (see Fig. 7b), the mole percentage of nitrite ion decreases from 58% to 51% with increased SO$_2$ concentration, which means that reaction (21) plays a dominant role among the reactions between nitric oxides and S(IV).

3.5. Effect of O$_2$ on NOx absorption

Fig. 8a shows the effect of O$_2$ concentration on NO and NO$_2$ absorption rates. Both NO and NO$_2$ absorption rates in CaCO$_3$ slurry increase with the increase of O$_2$ concentration. For example, the introduction of O$_2$ can markedly increase the NO absorption rate from 2.885×10^{-7} mol/(m^2 s) to 7.694×10^{-7} mol/(m^2 s) at 6% O$_2$ (increased 1.67 times) and 1.346×10^{-6} mol/(m^2 s)v at 15% O$_2$ (increased 3.66 times). The insoluble NO can be partially oxidized into soluble NO$_2$ in the presence of O$_2$ according to an irreversible third-order reaction [27]

$$2NO(g) + O_2(g) \rightarrow 2NO_2(g) \quad \Delta_f G^\circ = -72.6 \text{ kJ/mol} \quad (22)$$

Additionally, as mentioned above, NO$_2$ may be oxidized by the DO in the slurry, which may enhance the NOx absorption rate. To understand the role of oxygen in NOx absorption, the DO content in CaCO$_3$ slurry was measured by using a SevenGo™ SG6 DO meter (METTLER TOLEDO, China) with temperature compensation. First, N$_2$ gas was used to strip off DO in CaCO$_3$ slurry until its concentration was lower than 0.2 mg/L. Then, different simulated flue gases were blown into the reactor for 10 min at T = 50 °C; the experiment

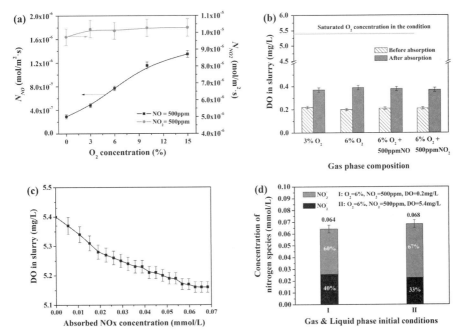

Fig. 8. (a) Effect of O$_2$ concentration on NO and NO$_2$ absorption rates; (b) DO concentration in slurry for different gas phase compositions; (c) DO concentration and the absorbed NOx concentration in slurry; (d) NOx absorption product compositions in CaCO$_3$ slurry for different gas and liquid phase conditions (T = 50 °C).

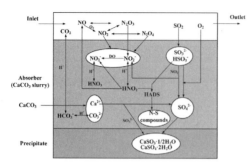

Fig. 9. Proposed absorption pathways for NOx absorption in $CaCO_3$ slurry which with the presence of SO_2 and O_2.

results are shown in Fig. 8b. It shows that when the simulated flue gases make contact with the $CaCO_3$ slurry at the gas–liquid interface, the DO concentration in slurry were increased from about 0.20 mg/L to 0.37 mg/L and 0.39 mg/L as the O_2 concentration in gas stream were 3% and 6%, respectively. It indicates that the mass transfer resistance of O_2 dissolving may lies on liquid phase. Furthermore, under the same O_2 concentration, the DO concentration hardly changed with the introduction of NO or NO_2 in gas stream, which means the O_2 in gas phase tend not to be involved in the liquid phase reaction. The DO concentration in $CaCO_3$ slurry during 500 ppm NO_2 absorption is shown in Fig. 8c. The figure shows that the NOx absorption process is associated with DO consumption, which means that DO is involved in NOx oxidation in liquid phase and that the oxygen mainly originated from the original DO in $CaCO_3$ slurry rather than gas phase O_2. This result explains why the NO_2 absorption rate cannot keep increasing as the gas stream O_2 concentration increases from 3% to 15%.

The analysis of absorption product in $CaCO_3$ slurry with different gas and liquid phase conditions is shown in Fig. 8d. The total concentration of nitrogen species increases as DO concentration increases from 0.2 mg/L to 5.4 mg/L. Meanwhile, the mole percentage of nitrite in $CaCO_3$ slurry decreases with the increase of DO concentration under the same gas phase condition. This finding confirms the existence of the reaction between nitrite and DO in $CaCO_3$ slurry; the reaction product is nitrate.

3.6. Proposed pathways for NOx absorption

The NOx absorption in $CaCO_3$ slurry was thought to involve two absorption pathways, namely, pure NO_2 absorption and simultaneous NO and NO_2 absorption. The absorption model of NOx in $CaCO_3$ slurry with the presence of SO_2 and O_2 is more complex. A global pathway of NOx absorption and nitrite/nitrate formation, which includes a variety of reactions such as reversible, redox, and disproportionation, is developed based on the conclusions derived from experimental results and is shown in Fig. 9.

Main components considered in the gas phase include NO, NO_2, SO_2, and O_2. NO solubility is considerably enhanced by oxidation or formed to N_2O_3 via the reaction between NO and NO_2. The absorption products of NO_2 or N_2O_4 are nitrite and nitrate while that of N_2O_3 is just nitrite. Nitrite can combine with hydrogen ions to produce nitrous acid, which is unstable in liquid phase and partly decomposes, which leads to NO release. The NO_2^- in liquid phase could be oxidized by DO to form NO_3^-. In the presence of SO_2, sulfite and bisulfate formation is in favor of NO_2 absorption, with NO_2 converted into NO_2^- and S(IV) converted into SO_4^{2-}. In addition, the HNO_2 can react with HSO_3^- to form HADS, which is a precursor to numerous sulfur–nitrogen compounds. When the slurry pH value decreases due to the NOx and SO_2 absorption, the dissolution rate of $CaCO_3$ increases, the carbonate transforms into bicarbonate, and the bicarbonate turns into carbon dioxide, which will flee from the liquid phase. In addition, the calcium ions react with S(IV) and SO_4^{2-} to form precipitates such as $CaSO_3 \cdot 1/2H_2O$ and $CaSO_4 \cdot 2H_2O$.

4. Conclusions

A series of experiments were performed to study NOx absorption and nitrite/nitrate formation in $CaCO_3$ slurry. NO_2 absorption rate increased with increased $CaCO_3$ concentration at the range of 0–0.5 wt%, while $CaCO_3$ content did not affect NO absorption. The best NOx composition for NO absorption was at a NO_2/NOx molar ratio of 0.5. The proportion of nitrite and nitrate, which were the main products of NOx absorption, changes with different operating conditions, and a negative correlativity was observed between the total NOx absorption rate and the mole percentage of NO_2^-. The NOx absorption rate was effectively enhanced in the presence of SO_2 or S(IV) ions. The mechanism for this behavior was postulated to be the redox reaction between S(IV) and NO_2 or formation of sulfur–nitrogen compounds. Both gas-phase and liquid-phase oxygen participated in NOx absorption reactions. Results suggest that a global reaction pathway of NOx absorption in $CaCO_3$ slurry was developed, which can explain NOx absorption and conversion in $CaCO_3$ slurry.

Acknowledgments

The authors acknowledge the financial support from the National Science Fund for Distinguished Young Scholars (No. 51125025), the National Key Technology R&D Program in the 12th Five-Year Period (No. 2011BAA04B08) and the Key Innovation Team for Science and Technology of Zhejiang Province (Grant No. 2011R50017).

References

[1] Tavoulareas ES, Jozewicz W. Multipollutant emission control technology options for coal-fired power plants. EPA-600/R-05/034; 2005.
[2] Nelli CH, Rochelle GT. Nitrogen dioxide reaction with alkaline solids. Ind Eng Chem Res 1996;35:999–1005.
[3] Pereira CJ, Amiridis MD. NOx control from stationary sources: overview of regulations. Technol Res Front ACS Symp Ser 1995;587:1–13.
[4] Sun WY, Ding SL, Zeng SS, Su SJ, Jiang WJ. Simultaneous absorption of NOx and SO_2 from flue gas with pyrolusite slurry combined with gas-phase oxidation of NO using ozone. J Hazard Mater 2011;192:124–30.
[5] Wu ZB, Wang HQ, Liu Y, Jiang BQ, Sheng ZY. Study of a photocatalytic oxidation and wet absorption combined process for removal of nitrogen oxides. Chem Eng J 2008;144:221–6.
[6] Barman S, Philip L. Integrated system for the treatment of oxides of nitrogen from flue gases. Environ Sci Technol 2006;40:1035–41.
[7] Bhatia D, McCabe RW, Harold MP, Balakotaiah V. Experimental and kinetic study of NO oxidation on model Pt catalysts. J Catal 2009;266:106–19.
[8] Weiss BM, Iglesia E. Mechanism and site requirements for NO oxidation on Pd catalysts. J Catal 2010;272:74–81.
[9] Kim HJ, Han J, Kawaguchi I, Minami W. Simultaneous removal of NOx and SO_2 by a nonthermal plasma hybrid reactor. Energy Fuel 2007;21:141–4.
[10] Yang JX, Chi XC, Dong LM. Effect of water on sulfur dioxide (SO_2) and nitrogen oxides (NOx) removal from flue gas in a direct current corona discharge reactor. J. Appl. Phys. 2007;101:103304.
[11] Sheng ZY, Wu ZB, Liu Y, Wang HQ. Gas-phase photocatalytic oxidation of NO over palladium modified TiO_2 catalysts. Catal Commun 2008;9:1941–4.
[12] Wu ZB, Sheng ZY, Liu Y, Wang HQ, Tang N, Wang J. Characterization and activity of Pd-modified TiO_2 catalysts for photocatalytic oxidation of NO in gas phase. J Hazard Mater 2009;164:542–8.
[13] Pradhan MP, Joshi JB. Absorption of NOx gases in aqueous NaOH solutions: selectivity and optimization. AIChE J 1999;45:38–50.
[14] Patwardhan JA, Joshi JB. Unified model for NOx absorption in aqueous alkaline and dilute acidic solutions. AIChE J 2003;49:2728–48.
[15] de Paiva JL, Kachan GC. Absorption of nitrogen oxides in aqueous solutions in a structured packing pilot column. Chem Eng Process 2004;43:941–8.
[16] Gao X, Du Z, Ding HL, Wu ZL, Lu H, Luo ZY, et al. Kinetics of NOx absorption into $(NH_4)_2SO_3$ solution in an ammonia-based wet flue gas desulfurization process. Energy Fuel 2010;24:5876–82.

[17] Gao X, Du Z, Ding HL, Wu ZL, Lu H, Luo ZY, et al. Effect of gas–liquid phase compositions on NO2 and NO absorption into ammonium-sulfite and bisulfite solutions. Fuel Process Technol 2011;92:1506–12.
[18] Dagaonkar MV, Beenackers AA, Pangarkar VG. Enhancement of gas-liquid mass transfer by small reactive particles at realistically high mass transfer coefficients: absorption of sulfur dioxide into aqueous slurries of Ca(OH)2 and Mg(OH)2 particles. Chem Eng J 2001;81:203–12.
[19] Thomas D, Vanderschuren J. Analysis and prediction of the liquid phase composition for the absorption of nitrogen oxides into aqueous solutions. Sep Purif Technol 2000;18:37–45.
[20] Decanini E, Nardini G, Paglianti A. Absorption of nitrogen oxides in columns equipped with low-pressure drops structured packings. Ind Eng Chem Res 2000;39:5003–11.
[21] Zhang XW, Tong HL, Zhang H, Chen CH. Nitrogen oxides absorption on calcium hydroxide at low temperature. Ind Eng Chem Res 2008;47:3827–33.
[22] Littlejohn D, Wang T, Chang SG. Oxidation of aqueous sulfite ion by nitrogen dioxide. Environ Sci Technol 1993;27:2162–7.
[23] Shen CH, Gary T. Nitrogen dioxide absorption and sulfite oxidation in aqueous sulfite. Environ Sci Technol 1998;32:1994–2003.
[24] Turšič J, Grgić I, Bizjak M. Influence of NO2 and dissolved iron on the S(IV) oxidation in synthetic aqueous solution. Atmos Environ 2001;35:97–104.
[25] Susianto M, Pétrissans A, Zoulalian A. Experimental study and modelling of mass transfer during simultaneous absorption of SO2 and NO2 with chemical reaction. Chem Eng Process 2005;44:1075–81.
[26] Siddiqi MA, Petersen J, Lucas K. A study of the effect of nitrogen dioxide on the absorption of sulfur dioxide in wet flue gas cleaning processes. Ind Eng Chem Res 2001;40:2116–27.
[27] Macneil JH, Berseth PA, Westwood G, Trogler WC. Aqueous catalytic disproportionation and oxidation of nitric oxide. Environ Sci Technol 1998;32:876–81.

石灰石浆液中的氮氧化物吸收及亚硝酸盐/硝酸盐形成用于湿式烟气脱硫系统

摘要：推进氮氧化物（NO_x）吸收步骤是湿式烟气脱硫过程中同步去除二氧化硫（SO_2）和 NO_x 的主要挑战。本文研究了在石灰石浆液中 NO_x 的吸收和吸收产物的形成，以及 $CaCO_3$ 浓度、NO_x 组成以及 SO_2 和 O_2 浓度的影响，根据实验结果，提出了 NO_x 吸收和转化的反应路径。NO_x 吸收被认为涉及两种吸收途径：纯 NO_2 吸收和同时吸收 NO 和 NO_2。研究发现 $CaCO_3$ 浓度与 NO_2 吸收速率呈正相关，气体流中 NO_2/NO_x 的最佳摩尔比为 0.5；主要的 NO_x 吸收产物是 NO_2^- 和 NO_3^-。NO_2^- 抑制进一步的 NO_x 吸收，可部分分解释放 NO，而 NO_3^- 对 NO_x 吸收没有明显影响。亚硝酸盐也可通过与溶解氧反应转化为硝酸盐。添加 SO_2 可提供另一条 NO_x 吸收途径，即 S(IV) 与 NO_x 之间的反应，这表明湿式吸收同时去除 SO_2 和 NO_x 是可行的。

原文刊于 Applied Energy, 2014, 219: 187-194

Pilot-scale investigation on slurrying, combustion, and slagging characteristics of coal slurry fuel prepared using industrial wasteliquid

Liu Jianzhong*, Wang Ruikun, Xi Jianfei, Zhou Junhu, Cen Kefa

State Key Lab of Clean Energy Utilization, Zhejiang University, Hangzhou 310027, China

HIGHLIGHTS

- Wasteliquid is used as a substitute for clean water to prepare coal slurry.
- The combustion characteristics of CWLS are studied in a pilot scale furnace.
- Wasteliquid enhances the slurryability of the coal and saves the additive agent.
- Wasteliquid improves the ignition and combustion performances of coal slurry.
- The metal ions in the wasteliquid reduce the SO_2 and NO_x emissions.

ARTICLE INFO

Article history:
Received 20 June 2013
Received in revised form 22 October 2013
Accepted 9 November 2013
Available online 1 December 2013

Keywords:
Industrial wasteliquid
Waste disposal
Coal slurry
Combustion
Slagging

ABSTRACT

The large amount of industrial wasteliquid generated during various industrial processes has raised serious environmental issues. A coal–wasteliquid slurry (CWLS) is proposed to dispose such wasteliquids, which are used as a substitute for clean water in the preparation of a coal-based slurry fuel. By the use of this method, a significant amount of clean water is conserved, and the environmental problems caused by wasteliquid discharge are resolved. However, the high content of organic matters, alkaline metal ions, and sulfur and nitro compounds considerably affects the slurrying, combustion, slagging, and pollution emission characteristics of CWLS. In this study, these characteristics are experimentally studied using a pilot-scale furnace. The results reveal that, compared with conventional coal–water slurry (CWS), CWLS exhibits a good performance with respect to slurrying, combustion, and pollution emission, i.e., low viscosity, rapid ignition, high flame temperature, high combustion efficiency, and low pollution emission. CWLS has a relatively low viscosity of 278 and 221 mPa·s and exhibits shear-thinning pseudo-plastic behavior without the use of any additive agent. In contrast, CWS requires the use of an additive agent to achieve good fluidity, and its viscosity is 309 mPa·s. The maximum flame temperature of the two CWLSs (CWLS-A and CWLS-B) is 1309.0 and 1303.1 °C, respectively, and their respective combustion efficiency is 99.61% and 99.42%. The values of both these parameters are greater than those obtained in the case of CWS. However, the alkaline metal ions in the wasteliquid lead to a considerable slagging status. This status improves significantly after turning down the operating load.

© 2013 Elsevier Ltd. All rights reserved.

1. Introduction

Coal–water slurry (CWS), which approximately consists of 65–70% coal, 30–35% moisture, and a little additive, is a coal-based clean fuel that can be used as a substitute for oil [1,2]. It not only exhibits similar performances as oil in terms of pumping, storing, and atomizing, but can also be used as an alternative fuel with high combustion efficiency in power station boilers, industrial boilers, and furnaces. Further, NO_x emissions can be reduced during the combustion of CWS, because the furnace temperature is low compared with that in the case of the combustion of oil and pulverized coal [3].

Along with the development of industries, the emissions of various industrial wasteliquids, sewage sludge, and waste residues have increased rapidly, particularly the emission of concentrated non-biodegradable toxic organic wasteliquids, which are generally discharged into the environment from various industries such as food, fermentation, slaughter, textile, paper, rubber, plastic, cosmetics, pesticide, petrochemical, and domestic sewage. When discharged into the environment, organic wasteliquids cause serious environmental pollution. The disposal of such volumes of organic wasteliquids is a long-standing problem in the field of environmental protection. Therefore, reliable methods are very necessary for the disposal of organic wasteliquids.

* Corresponding author. Tel.: +86 571 87952443 5302; fax: +86 571 87952684.
E-mail address: jzliu@zju.edu.cn (L. Jianzhong).

0306-2619/$ - see front matter © 2013 Elsevier Ltd. All rights reserved.
http://dx.doi.org/10.1016/j.apenergy.2013.11.026

A considerable amount of effort has been devoted to the effective disposal of organic wasteliquid and the removal of toxic matters from wasteliquids. For the disposal of organic wasteliquids, various treatment techniques are available such as adsorption [4], chemical precipitation [5], liquid–liquid extraction [6], ion exchange, coagulation [7,8], reverse osmosis [9], electrolysis, membrane process [10], glow discharge plasma [11], full cell [12], and oxidation [13,14]. However, building a complete set of wasteliquid-disposing technology and matching installation, which meets national standards, needs high investment and running costs. Numerous enterprises are unable or unwilling to invest the significant money to treat their wasteliquid. This is a main reason for serious environmental pollution in China. Therefore, exploring for a low cost way of disposing wasteliquid is of great concern. Coal–wasteliquid slurry (CWLS) is considered an effective, efficient, and economic method for the disposal of wastewater. Wasteliquids are used as a substitute for clean water in the preparation of a coal slurry fuel, thereby saving a considerable amount of clean water. It usually contains a certain amount of lignin, semicellulose, sugar, organic acid, and sulfonic substances, which are active matter or active-group-containing polymers. When the wasteliquid is used for preparing coal slurry, these active functional groups act as surfactants and improve the surface activity and wettability of the coal particles; thus, the coal particles are well dispersed in suspension [15,16]. Therefore, all or part of the chemical additive agent needed for slurrying can be replaced, and the slurrying cost can be reduced [17,18]. During the combustion of CWLS, the caloric value of the organic matter in the wasteliquid is utilized sufficiently. The organic matters can also improve the ignition and burnout performances of the slurry fuel in boilers. At the same time, the emissions of some pollutants such as SO_2 and CO are reduced because of sulfur-retention components of the wasteliquid and the complete combustion of the coal particles [19]. Hence, the CWLS technique provides a low-cost solution for pollution problems caused by wasteliquids, leads to savings of a considerable amount in terms of pollution treatment fees, and provides large volumes of cheap fuel for enterprises. The CWLS technique reduces the environmental hazards of the wasteliquid discharged by various industries, broadens the disposal ways of wasteliquids, and substantially reduces the production costs of the enterprises.

However, wasteliquids contain a high concentration of organic materials, metal ions, and sulfur and nitro compounds, which significantly influence the slurrying, combustion, slagging, and pollution emission properties of CWLS. Thus far, there have been no reports on the pilot-scale combustion and slagging properties of CWLS. Therefore, the ignition characteristics and combustion, slagging, and pollution emission properties of two CWLSs are investigated using a pilot-scale horizontal cylindrical furnace in this study. The obtained results provide a foundation for the disposal of industrial wasteliquids and the industrial application of CWLS and can be used as a reference for the optimal operation of full-scale boilers.

2. Experimental section

2.1. Materials

Two chemical wasteliquids from Beijing Yanshan Petrochemical Company were used for preparing CWLS. The two wasteliquids (wasteliquid A and wasteliquid B) were caustic liquids obtained from two typical processes in petrochemical field, ethylene separation from dry gas and petroleum refining, respectively. Before being discharged, they must be processed for environmental safety. The quality analysis of these two wasteliquids is shown in Table 1.

Table 1
Wasteliquid quality analysis (Unit: mg/L).

Test items	Wasteliquid A	Wasteliquid B
K	37.9	44.3
Na	1.13×10^3	2.66×10^3
Ca	123	154
Mg	<0.02	<0.02
Cu	<0.01	<0.01
Pb	<0.50	<0.50
Ni	<0.01	<0.01
Cr	<0.01	<0.01
Cd	<0.03	<0.03
V	0.096	<0.01
Mn	0.097	0.129
Co	<0.120	<0.120
SO_4^{2-}	2.8×10^3	9.6×10^3
Cl^-	27	9.78
S^{2-}	0.411	0.087
Ammonia nitrogen	24	19.5
Volatile phenol	17.4	7.7
TDS	1.6×10^4	1.5×10^4
COD	1.06×10^3	596
Total nitrogen	42.8	27.8

TDS is the total dry solid content; COD is the chemical oxygen demand.

Three coal slurry samples were prepared using tap water, wasteliquid A, and wasteliquid B, and their respective names are CWS, CWLS-A, and CWLS-B, respectively. The coal used was bituminite from Datong, Shanxi province, China.

Because of the lack of an active agent, tap water could not disperse the coal particles sufficiently, and the coal–water mixture could not flow. A little amount of a chemical additive agent (0.8% of the coal) was added to prepare good-fluidity coal slurry. In contrast, the coal slurry prepared using wasteliquid could flow easily; thus, no additive agent was added during slurrying.

The proximate and ultimate analysis of the three slurry samples is given in Table 2. The solid concentration, equal to $(1 - M_t)$, of the three slurry samples is similar: 63.35% (CWS), 63.27% (CWLS-A), and 63.20% (CWLS-B. Because wasteliquids contain a certain amount of TDS, COD, and S and N components, the corresponding CWLSs show a higher content of FC_d, V_d, and $S_{t,d}$ and N_d elements than CWS. Between the two wasteliquids, wasteliquid A contains a higher content of TDS, COD, and N components; thus CWLS-A shows a higher content of FC_d, V_d, and N_d elements. On the other hand, wasteliquid B contains a higher content of SO_4^{2-}, and thus, CWLS-B shows a higher content of $S_{t,d}$ elements.

The high content of alkaline metal ions such as K, Na, and Ca in the wasteliquids considerably affects the ash fusion temperature and slagging during the combustion of the coal slurry. The ash fusion temperatures of the three slurry samples are shown in Table 3. The ash fusion temperatures of coal slurry samples prepared by using the wasteliquids evidently decrease. As a result, it is considerably easy to melt fly ash particles, and the slagging status is then aggravated. Moreover, CWLSs have a smaller difference between the deformation and fluxion temperatures (FT–DT) than CWS; thus, the fly ash of CWLSs is easier to transform from the deformation to the flow state, i.e., the melted ash could easily adhere to the boiler/furnace heating surfaces.

2.2. Slurrying properties

The rheological properties of the coal slurry were determined using a rotational viscometer (HAAKEVT550, Thermo, USA). The temperature was maintained constant within (20 ± 0.5) °C and was controlled using a water bath. During the measurements, the slurry samples in the viscometer underwent three stages: (1) increasing shear rate stage, in which the shear rate was increased

Table 2
Proximate and ultimate analysis of the slurry samples.

Sample	Proximate analysis (%)				Ultimate analysis (%)					$Q_{net,ar}$ (J/g)
	M_t	A_d	V_d	FC_d	C_d	H_d	N_d	$S_{t,d}$	O_d	
CWS	36.65	6.49	28.18	65.34	80.11	4.44	1.17	0.63	7.17	19,570
CWLS-A	36.73	5.09	29.11	65.80	81.56	4.64	1.43	0.73	6.56	19,876
CWLS-B	36.80	5.41	28.86	65.73	81.47	4.62	1.32	0.92	6.27	19,744

M_t refers to the total moisture; A_d, V_d, and FC_d refer to ash, volatile, and fixed carbon on a dry basis, respectively; ultimate analysis was conducted on a dry basis; $Q_{net,ar}$ refers to net calorific value on an as-received basis.

Table 3
Ash fusion temperatures of the three coal slurry samples (Unit: °C).

Slurry samples	DT	ST	HT	FT
CWS	1233	1258	1268	1290
CWLS-A	1228	1244	1263	1281
CWLS-B	1225	1231	1254	1273

DT denotes the deformation temperature; ST, the softening temperature; HT, the hemispheric temperature and FT, the fluxion temperature.

from 0 to 100 s^{-1}; (2) constant shear rate stage, in which the shear rate was maintained constant at 100 s^{-1} for 5 min; the viscosity data were recorded every 30 s during the 5-min period; and (3) decreasing shear rate stage, in which the shear rate was decreased from 100 to 0 s^{-1}. The rheological properties of coal slurry were referred to as shear stress (τ)-shear rate ($\dot{\gamma}$) and apparent viscosity (η)-shear rate ($\dot{\gamma}$) in Stage1.

Given that apparent viscosity is closely related to the shear rate, the characteristic viscosity (η_c), which is defined as the apparent viscosity of the slurry sample at the shear rate of 100 s^{-1}, was used for evaluating the slurryability of the sludge. In this study, an average of 10 records at 100 s^{-1} was taken as η_c. A lower η_c indicated better slurryability of the sludge.

2.3. Pilot-scale combustion experiment of CWS/CWLS

2.3.1. Pilot-scale combustion system

Fig. 1 shows the pilot-scale combustion system of CWS/CWLS. The test system is composed of four subsystems: slurry fuel supply, compressed air, water cooling and cleaning, and furnace.

The slurry fuel supply subsystem is composed of slurry storing and stirring tanks, slurry pump, filter, flowmeters, and pressure gauges. Using the slurry pump, we can transport the prepared slurry from the storage tanks to the slurry nozzle and then spray it into the furnace. Along with the slurry pump speed, the furnace-front slurry return circuit is set to regulate the slurry flow into the furnace.

The compressed air provided by an air compressor is used for slurry spraying and atomization. The flow and the pressure of the compressed air are measured online by using a flowmeter and a pressure gauge, and the flow is regulated by a valve.

Water is used for cooling the furnace and cleaning the pipes, nozzle, and storage tanks.

The cylindrical furnace is arranged horizontally, with a swirl burner installed at the inlet side. The flue gas flows from the horizontal into the vertical gas duct and is then sent into the stack by a suction fan, after removing the dust by a cyclone. The temperature of the combustion-aid air can reach over 200 °C after the two-stage preheating. The combustion-aid air is first heated by the waste heat of the flue gases and then by an electric heater.

Thirteen measuring points, numbered 1–13, are distributed on the furnace wall along its axis line, where the temperature and gas compositions inside the furnace are measured, and the solid particles are sampled. The temperature profile measured at point 12 from the ignition to the stable combustion is used for investigating the process of the furnace temperature increase. The oxygen content of the flue gas is measured at point 13. The distance of points 1 to 13 from the outlet of the coal slurry nozzle is 200, 400,700, 1000, 1200, 1500, 1900, 2200, 2500, 2800, 3100, 3800, and 4520 mm, respectively. Fig. 2 shows the layout of the measuring points on the furnace.

2.3.2. Test method

A complete combustion test mainly consists of two periods: ignition and stable combustion. At the beginning of combustion test, i.e., ignition period, light diesel oil is first fed and burned in the furnace. When the furnace temperature exceeds a certain value, the CWS/CWLS is fed into the furnace. After the combustion stabilizes, oil injection is reduced gradually until extinction. After the oil sprayer is removed, operating conditions such as the slurry

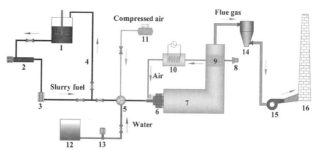

1. Slurry storage tank; 2. Slurry pump; 3. Filter; 4. Slurry return circuit; 5.Three way valve; 6. Slurry burner; 7. Horizontal furnace; 8. Air blower; 9. Air preheater; 10. Air electrical heater; 11. Air compressor; 12 Water storage tank; 13. Water pump; 14. Cyclone; 15. Suction fan; 16. Stack.

Fig. 1. Pilot-scale combustion system of CWS/CWLS.

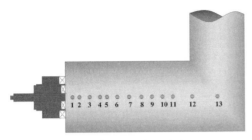

Fig. 2. Layout of measuring points on the furnace.

flow, pressure, and inflow air amount are regulated, and then the combustion is stabilized for approximate 1 h. Then, the combustion enters into a stable period, when the main test begins.

2.3.3. Operating conditions
Besides being operated at the design load (100 kg slurry per hour), all three slurry samples are also operated at a low load, i.e., each slurry sample is operated under two load conditions-high and low. The operating conditions are listed in Table 4.

2.4. Combustion efficiency calculation

The combustion efficiency is a measure of the degree of completeness of combustion. It can be calculated as follows:

$$\eta_r = 1 - (q_3 + q_4).$$

where η_r denotes the combustion efficiency, %; q_3, the heat loss due to the incomplete combustion of the combustible gas, %; and q_4, the heat loss due to the incomplete combustion of the solid, %.

The detailed calculation processes of q_3 and q_4 can be found in literature [20].

2.5. Gas pollutant emission determination

The emission of the gas pollutants was determined using a Fourier transform infrared gas analyzer (Gasmet FTIR Dx4000, Gasmet Technologies Oy Co., Finland).

3. Results and discussions

3.1. Slurrying properties of CWS/CWLS

3.1.1. Apparent viscosity of CWS/CWLS
The apparent viscosities of the three slurry samples are low. The characteristic viscosity of the three slurry samples is 309, 278, and 221 mPa·s for CWS, CWLS-A, and CWLS-B, respectively. Although no chemical additive agent is used, CWLSs show a relatively low viscosity. It is clear that the active ingredients in the wasteliquid play a role in improving the surfactivity and slurryability of coal particles. Therefore, wasteliquids can be used for replacing all or

Table 4
Operating conditions of combustion test.

Slurry sample	Load	Slurry dosage (kg/h)	Test duration (h)
CWS	High	98.79	3.5
	Low	66.53	2.0
CWLS-A	High	98.76	6.5
	Low	66.05	2.5
CWLS-B	High	99.66	4.5
	Low	66.15	3.0

part of the additive agents during the preparation of the coal slurry to achieve slurrying cost savings and a good slurrying performance.

3.1.2. Rheological characteristics
Fig. 3 shows the rheological characteristics of CWS/CWLS. All the three slurry samples exhibit a shear-thinning flow characteristic, i.e., the apparent viscosity decreases with increasing shear rate. This flow characteristic favors the industrial application of CWS/CWLS; on one hand, when the slurry stands still, the viscosity is high, and thus, the solid particles in the slurry do not easily deposit. On the other hand, when the slurry is pumped, the viscosity decreases under the shear force, and the pumping pressure can thus be decreased [21].

Therefore, rheological characteristics must satisfy two demands. The first is the yield stress part, which reflects stability, and the second is the shear stress part, which reflects flow resistance. The rheological characteristics can be suitably described by the Herschel–Bulkley model [22,23].

$$\tau = \tau_y + K \times \dot{\gamma}^n,$$

where τ denotes the shear stress (Pa); τ_y, the yield stress (Pa); and K, the consistency coefficient (Pa·s^n). A higher K indicates a higher slurry viscosity; $\dot{\gamma}$ denotes the shear rate (s^{-1}) and n, the rheological index (dimensionless). The flow patterns of CSS can be determined by the rheological index as follows: $n = 1$, Newtonian fluid; $n > 1$, dilatant fluid; and $n < 1$, pseudoplastic fluid.

The rheological parameters of flow behavior are listed in Table 5. The experimental data fit well with the Herschel-Bulkley model, with correlation coefficients greater than 0.9995. All the values of n are less than 1, suggesting that the CWS/CWLS belong to the category of pseudoplastic fluids. Compared with CWS, CWLSs have a relatively low yield stress and consistency coefficient. This also leads to an improvement in the slurrying performance of coal particles by the active ingredients in the wasteliquid.

3.1.3. Static stability of CWS/CWLS
Solid–liquid separation is a common phenomenon in a two-phase slurry. The static stability of a slurry can be evaluated on the basis of the water separation ratio (WSR), which is defined as the mass ratio of the supernatant to the total water in the slurry sample after it is statically held for some period of time (7 d in this study). A higher WSR value implies more serious solid–liquid separation and worse slurry stability. The WSR results of the three slurry samples are shown in Table 6.

As evident from Table 6, the WSR values of the three slurry samples are very high. Further, a hard sediment is formed. This suggests that serious solid–liquid separation occurs. This is because of the fact that all the three slurry samples have a very low yield stress, which cannot prevent the settling of solid particles. Among them, the two CWLSs show higher WRS values, implying that slurry stability is relatively poor. This is because CWLSs have a lower yield stress than CWS. Further, the chemical additive agent used in preparing CWS has certain stabilizing effects, whereas no additive agent is used in the preparation of CWLSs. The active ingredients in the wasteliquid only play a role in dispersing coal particles. In an actual industrial application, appropriately increasing the coal concentration and adding a little stabilizing agent are effective ways for preventing settlement.

3.2. Combustion characteristics

3.2.1. Process of furnace temperature increase during ignition period
The ignition process of each slurry sample consists of three stages: oil ignition and combustion, co-combustion of oil and CWS/CWLS, and complete CWS/CWLS combustion.

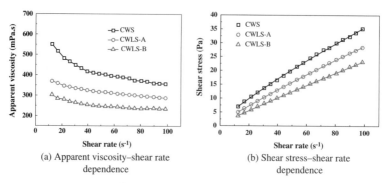

(a) Apparent viscosity–shear rate dependence

(b) Shear stress–shear rate dependence

Fig. 3. Rheological characteristics of CWS/CWLS.

Table 5
Rheological parameters of CSS.

Sample	τ_y	K	n	R^2
CWS	1.01	0.71	0.84	0.9995
CWLS-A	0.58	0.46	0.90	0.9999
CWLS-B	0.51	0.30	0.95	0.9998

Table 6
Static stability of coal slurry—WRS.

Sample	WRS (%)
CWS	17.5
CWLS-A	23.7
CWLS-B	24.0

Fig. 4 shows the profiles of the furnace temperature increase during ignition period at measuring point 12. Fig. 4a shows the case of CWS. The oil is fed and ignited at time zero ($t = 0$), and the initial oil injection rate is 55 L/h. The temperature is recorded from 142 °C. The temperature increases continuously and reaches 500 °C within 2.5 min, and then, CWS is fed at a rate of 98.79 kg/h into the furnace since the CWS and oil co-combust. Because of the injection of the cold slurry and the evaporation of a considerable amount of water, the temperature stops increasing. After a certain period of time, the temperature continues to increase rapidly. Oil injection is reduced to 40 L/h at $t = 19$ min;

the temperature decreases from 1100 °C to 1025 °C and then increases gradually. At $t = 32$ min, the temperature reaches 1090 °C, and oil injection is stopped; hereafter CWS is combusted alone without oil. Because oil injection is stopped, the temperature decreases again. After decreasing to approximately 1000 °C, the temperature begins to rise smoothly, and the combustion eventually stabilizes.

The furnace outlet temperature profiles of the two CWLSs are very similar, and the case of CWLS-B is analyzed as typical case in the study given the paper length limitation. Fig. 4b shows the case of CWLS-B. The operating conditions, including initial oil injection rate, slurry injection rate, and oil reduction amount, are the same as those in the case of CWS. The temperature increase profile is similar to that shown in Fig. 4a. The differences are as follows: (1) CWLS-B has a rapid temperature increase rate. At $t = 17$ min, the furnace temperature increases to 1105 °C, after which oil injection is reduced. The temperature first decreases and then increases. At $t = 26$ min, the temperature reaches 1090 °C; (2) when the coal slurry is fed, the furnace temperature shows a constant period in the case of CWS, while smoothly increasing in the case of CWLS-B; and (3) after oil injection is stopped, the temperature decreases for both CWS and CWLS-B; however, in the case of CWLS-B, the temperature instantly increases again, while in the case of CWS, it increases again very slowly, i.e., CWS needs more time to reach stable combustion than CWLS-B when oil injection is stopped. This is due to the combustible materials in the wasteliquid, which are easily volatilized and

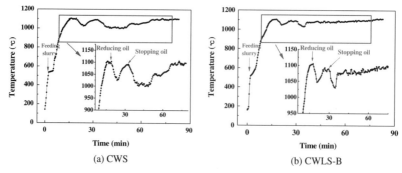

(a) CWS

(b) CWLS-B

Fig. 4. Furnace outlet temperature increase profiles during ignition period.

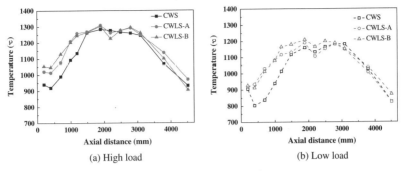

Fig. 5. Axial temperature distribution in the furnace.

ignited. Moreover, the metal ions in the wasteliquid play a catalytic role in promoting the combustion of the volatiles and coal particles.

3.2.2. Axial temperature distribution in the furnace during stable combustion period

The axial temperature distributions in the furnace under both high and low loads are shown in Fig. 5. In the early combustion stage, i.e., the zone close to the burner, the furnace temperatures of the two CWLSs are obviously greater than those of CWS, suggesting that the CWLSs are ignited earlier than CWS. This is because the combustible matter contained in the wasteliquid, similar to the volatiles in coal, can be easily ignited [24,25]. Further, the higher content of metal ions in the wasteliquid can promote the release of volatiles and play a catalytic role in combustion [26–28]. Given that early ignition enhances combustion stability, the use of wasteliquid as an alternative to clean water for the preparation of the slurry fuel improves combustion performance.

The relative low temperature value at 2200 mm (measuring point 8) is likely resulted from the experimental errors due to the measuring method and the external structure drawbacks of furnace. The furnace temperature is measured using an infrared radiation thermometer. A supporting steel column just stands at the measuring point 8, and thus affects the radiation temperature measurement.

As evident in Fig. 5a, within a distance of 200 mm from the burner outlet, the flame temperature exceeds 1000 °C for the two CWLSs, while it is below 950 °C for the CWS. This is because of the rapid burning of the combustible matter and the catalytic combustion effect of the metal ions contained in the wasteliquid. For all three slurry samples, the temperature at the main combustion zone reaches exceeds °C, and the maximum reaches approximately 1300 °C, suggesting that combustion is stable.

When the load is decreased, the furnace temperature has an evident decrease. This is because the slurry dosage is reduced under the low load condition; thus, the heat output decreases. Comparing the high and low load conditions, it is obvious that the maximum temperature occurs at 2000 mm under both high and low load conditions for the two CWLSs, while it shifts downwards from 2000 to approximately 3000 mm for CWS when the load is decreased.

The axial temperature distribution in the furnace from Fig. 5b shows that temperature distribution of CWLS-B is unexpectedly higher than CWLS-A under a condition of low load. This phenomenon can be explained as follows. Generally, the air supplying rate has an increasing effect on the combustion status, i.e. combustion

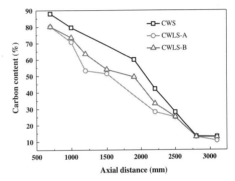

Fig. 6. Carbon content of solid particles in the furnace.

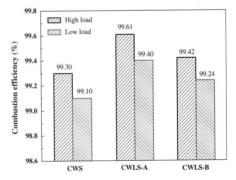

Fig. 7. Combustion efficiency of CWS/CWLS.

temperature, as the load of the boiler decreases. From the oxygen content in the exhaust gas (presented in Table 8 and Section 3.4), at low load condition, the air supplying rates are all beyond the normal range for the three slurry samples, and then higher air supplying rate decreases the combustion temperature instead. By comparing the low load condition of the three slurry samples, the CWLS-B case has the lowest but at the same time also the most appropriate value of air supplying rate, while the CWLS-A case has a higher air supplying rate, and thus its temperature is likely decreased by the superfluous supplying air. As a result, the temperature distribution of CWLS-B is a higher than CWLS-A under low load condition.

Under the low load condition, the temperature in the main combustion zone exceeds 1100 °C for all three slurry samples, and the maximum exceeds 1200 °C, suggesting that combustion is also stable, i.e., there is room for further reduction in the load.

3.2.3. Burnout rate of the particles in the furnace

The solid particles in the furnace are sampled at different points along the furnace axis using a homemade water-cooled sampling device, and their carbon contents are determined. The results are shown in Fig. 6.

Coal particles are burnt while moving along in the furnace; hence, the carbon content of the solid particles decreases with an increase in the axial distance from the burner. In the front zone of the furnace, the carbon content of the particles is lower for the two CWLSs than for CWS. This is because the metal ions in the wasteliquid can act as oxygen carriers to promote the transfer of oxygen and improve the burnout rate of coal [29]. From 2500 mm (measuring point 9), the carbon content of the solid particles begins to get close to each other and decreases to approximately 25% for all three slurry fuels. At the furnace outlet, the carbon content of the solid particles approaches around 10%.

3.2.4. Combustion efficiency

The combustion efficiency is shown in Fig. 7. The combustion efficiencies of the three slurry samples are 99.30% (CWS, High load), 99.10 (CWS, Low load), 99.61% (CWLS-A, High load), 99.40% (CWLS-A, Low load), 99.42% (CWLS-B, High load), and 99.24% (CWLS-B, Low load). All the efficiencies are above 99% for all three slurry samples, suggesting that combustion exhibits good performance. The combustion efficiencies of the slurry samples reach the level of a pulverized-coal-fired boiler.

As evident from the figure, the combustion efficiency is slightly lower under the low load condition. This is because when the load is decreased, the furnace temperature decreases, and the combustion effect of the coal particles is negatively influenced.

3.3. Slagging characteristics

Wasteliquid, which usually contains a high content of alkali metal ions, is used as an alternative to clean water to prepare CWLS. As a result, CWLS contains more alkali metal ions than CWS. Previous studies and experiences have shown that the alkaline mineral

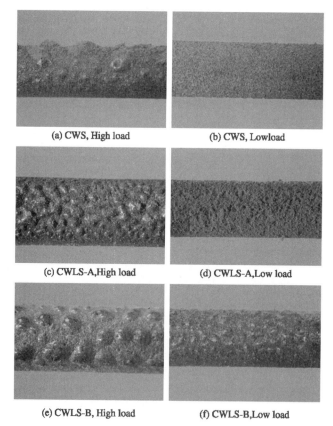

Fig. 8. Slag pictures at a distance of 1900 mm from the burner outlet.

in the fuel is the main incentive for slagging in the boiler and furnace [30,31].

To study the slagging characteristics of CWLS, silicon carbide rods are placed at certain positions in the furnace during combustion. The slagging characteristics are obtained by analyzing the appearance and amount of the slag deposited on the silicon carbide rods.

3.3.1. Appearance of the slag

Fig. 8 shows the slag pictures at a distance of 1900 mm (measuring point 7) from the burner, which is the highest temperature position. The deposition time is 30 min.

From the pictures, we observe that the slagging status is closely related to the slurry sample and load. The slag deposits more heavily when CWLSs are burnt than when CWS is burnt, because the high content of alkaline metal ions in the wasteliquid results in a reduced ash fusion point. When the two CWLSs are burnt under the high load condition, the slag deposits at the highest-temperature position (measuring point 7) have dense textures, and the slag surface exhibits a metallic luster. These molten slags strongly stick to the silicon carbide rods. CWS also shows a similar slagging status. However, compared with that of the CWLSs, the slagging status of CWS is better in that its slag is not melted as significantly as that of CWLSs. When the load is decreased, the slagging status is improved. The slags change from molten states to loose powder clusters in the cases of CWS and CWLS-A. However, CWLS-B still shows molten slags although the slagging status improved.

During high-load combustion, the actual flame temperature at measuring point 7 is 1283.8, 1309.0, and 1303.1 °C for CWS, CWLS-A, and CWLS-B, respectively. All of them are greater than the FTs of slags, which are 1290, 1281, and 1273 °C, respectively (as shown in Table 3). As a result, the fly ashes in the high-temperature zone can be melted; thus, the molten ashes stick to the furnace heating surface once they come into contact with the surface. In contrast, during low-load combustion, the flame temperature at this position decreases to 1159.2, 1189.8, and 1209.1 °C, respectively. All of them are less than the DTs of the slags, which are 1233, 1228, and 1225 °C; respectively, and the slagging statuses are thus improved. The slagging status of CWLS-B is more significant, because this sample has a higher flame temperature than CWS and CWLS-A. The melted ash droplets may not have solidified before reaching the heating surface.

Fig. 9 shows the slag pictures at a distance of 4520 mm (measuring point 13) away from the burner, which is the position of the outlet of the horizontal furnace. The deposition time is also 30 min. The actual flame temperatures of the three slurry samples at point 13 are 931.4, 969.3, and 905.3 °C, respectively, under the high load condition, and their respective values under the low load

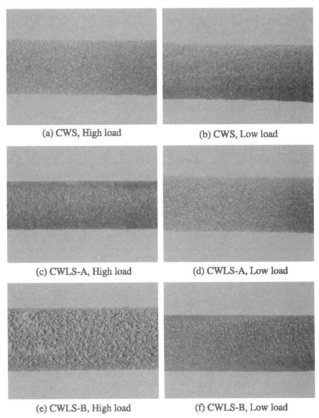

Fig. 9. Slag pictures at a distance of 4520 mm from the burner outlet.

condition are 825.0, 824.5, and 872.7 °C; these temperatures are considerably less than the DTs of the slags. Thus, only a little amount of slag is loosely deposited on the silicon carbide rods. The slags are not sticky and can fall off easily. After the load is decreased, the slags on the silicon carbide rods decrease as well.

3.3.2. Micro-morphology of the slag

Fig. 10 shows the micromorphology of the slag (at point 7) of CWS and CWLS-B. Under the high load condition, the slag of CWS is loose, and the surface is very rough and rugged. The slag shows certain molten signs, indicating that the ash is partially melted during CWS combustion; however, it does not reach the full-scale flow state; hence, an integrated slag block is not formed. On the other hand, the slag of CWLS-B shows significant fusing and sintering signs, dense texture, smooth surface, and internal bubbles. This suggests that the ash of CWLS-B is fully melted and exhibits a flow state, and when the fluid molten ash sticks to the heating surface of the furnace, it congeals and condenses into one whole slag piece. This is attributed to the fact that the large amount of alkaline metal contained in the wasteliquid promotes slagging. In general, a higher content of alkaline metal leads to a stronger sintering strength. The mineral substances containing the alkaline metal have a low ash fusion temperature, and thus are more likely to exhibit a fluid feature. The fluid state ash dramatically accelerates slagging.

When the load is decreased, the slagging status is improved. The slag of CWS changes from partially molten slag having a blocky structure to loosely piled ash granules, and the slag of CWLS-B changes from a dense sintered block to small pieces of molten slag. This is because the furnace temperature is low under the low load condition, and thus, the fusing tendency of the slag is decreased.

3.3.3. Energy dispersion spectrum analysis of the slag

Table 7 shows the energy dispersion spectrum analysis of the slags under the high load operating condition. The main component elements are O, Si, and Al. Compared with CWS, the two CWLSs have a higher content of Na and K, which is the reason for their more serious slagging status. On one hand, Na and K are sublimated and released in a gaseous form in a high-temperature atmosphere [32,33]. The gaseous Na and K coagulate onto the furnace wall when the temperature is low and form the initial slag layer. On the other hand, alkaline metal oxides and salts react with SiO_2 and form a low-melting-point melt-Na_2O–SiO_2 and K_2O–SiO_2-resulting in serious sintering phenomena [34].

3.3.4. Slag growth rate

The slag growth rate is defined as the slag amount deposited on per unit of area in unit time. Fig. 11 shows the slag growth rate at measuring points 7 and 13. The slag growth rate of the two CWLSs is slightly higher than that of CWS, resulting from the lower ash fusion temperature of the two CWLSs.

As evident from the figure, the slag shows a lower growth rate at point 13 than at point 7. This is because the temperature at point 13 is significantly less than that at point 7. The temperature significantly influences the physical properties of fly ash, including the phase, viscosity, and fluidity. At a higher temperature, more fly ash is melted or semi-melted; thus, it can easily stick to the silicon carbide rod once it comes into contact with the rod surface. When the temperature exceeds the FT of the fly ash, the adhered ash finds it difficult to escape from the silicon carbide rod. In contrast, molten fly ash is solidified when the temperature is below the DT of the ash. The solid-phase ash can be blown away easily and finds it difficult to stick to the rod.

Under the low load condition, the flame temperature decreases, and thus, the slag growth rate decreases considerably. This also confirms that the flame temperature significantly affect slagging.

3.4. Gas pollutant emission characteristics

Gas pollutants contents in exhaust gas are continually measured in the stable combustion period, during which the values of pollutants concentration are relatively stable, although they fluctuate along with time. The mean value of the continual measuring process is calculated as the primary data of the gas pollutants

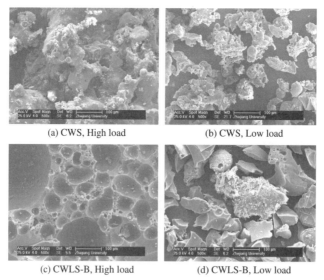

(a) CWS, High load (b) CWS, Low load

(c) CWLS-B, High load (d) CWLS-B, Low load

Fig. 10. Slag microstructures of the three slurry samples (×500 magnification).

Table 7
Energy dispersion spectrum analysis of the slag.

Slurry samples	Components (Wt%)									
	O	Na	Mg	Al	Si	P	K	Ca	Ti	Fe
CWS	44.43	2.1	0.89	13.32	28.49	0.47	0.6	3.34	0.85	3.81
CWLS-A	43.98	3.14	0.72	13.04	29.4	0.52	0.62	3.47	1.05	3.15
CWLS-B	43.25	3.96	0.69	13.15	29.23	0.52	0.63	3.46	0.71	3.64

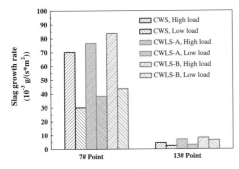

Fig. 11. Growth rate of the slag.

contents. The oxygen content in the exhaust gas is between 8% and 12% because the air leaks into the furnace at the tail. By professional convention, to reach a uniform standard, the oxygen content is converted to 6%, i.e., the excess air coefficient is 1.4, and then the gas pollutant emissions on the uniform standard are obtained according to the following formula: $[G] = [G'] \times (21 - 6)/(21 - [O_2])$, where, $[G]$ is the final value after converting, $[G']$ is the actual value, $[O_2]$ is the actual oxygen content. The converted gas pollutant emissions are presented in Table 8. The CO concentration is very low, suggesting that volatiles combusted well. SO_2 and NO_x are the main gas pollutants. The SO_2 emissions are 761.74, 609.97, and 623.89 mg/N m³ for CWS, CWLS-A, and CWLS-B, respectively. The lower SO_2 emission by the two CWLSs probably results from the sulfur retention or catalytic retention effects of the alkaline metal ions in the wasteliquid [19]. The NO_x emissions of the two CWLSs are also less than those of CWS. The NO_x emissions are 552.48, 529.37, and 507.35 mg/N m³ for CWS, CWLS-A, and CWLS-B, respectively. This is attributed to the fact that the metal ions in the wasteliquid, such as V, Ti, Cu, Fe, and Mn, have catalytic effects on the transformation of NO_x to N_2 [35–37]. Hence, it is concluded that the wasteliquid used for the preparation of CWLS can reduce the emissions of the gas pollutants.

When operated under a low load condition, the generation of gas pollutants is subjected to some restrictions because of the lower furnace temperature. Further, the air supply is relatively sufficient and plays a dilution effect.

4. Conclusion

The slurrying, combustion, slagging, and pollutant emission characteristics of CWLS are studied in a pilot-scale furnace. The following conclusions have been obtained:

(1) Without using an additive agent, CWLSs show a relatively low viscosity and shear-thinning pseudoplastic behavior, suggesting that the active ingredients in the wasteliquid play a role in improving the surfactivity of the coal particles. Therefore, the use of wasteliquid to prepare coal slurry can save all or part of the additive agent, thus achieving slurrying cost savings and a good slurrying performance.

(2) The ignition and combustion performances are improved when the wasteliquid is used for preparing the coal slurry. The maximum flame temperature of CWLS-A and CWLS-B reaches 1309 and 1303 °C, respectively; their respective combustion efficiency is 99.61% and 99.42%. In contrast, the maximum flame temperature of CWS is 1283.8 °C, and the combustion efficiency is 99.30%.

(3) The flame temperatures of the two CWLSs are greater than that of CWS in the early combustion zone, suggesting that the wasteliquid facilitates the release and ignition of the volatiles.

(4) The slagging status is closely related to the slurry sample, load, and flame temperature. The slag deposits more heavily when CWLSs are burnt than CWS. Decreasing the operating load improves the slagging performance.

(5) The SO_2 and NO_x emissions of the two CWLSs are less than those of CWS because of the sulfur and nitrogen retention or catalytic retention effects of the alkaline metal ions in the wasteliquid. The use of the wasteliquid as a substitute for clean water to prepare the coal slurry plays a dual role in environmental protection. One is the role of wasteliquid disposal, and the other is that of the reduction in the gas pollutant emissions after coal slurry combustion.

Table 8
Gas pollutant emissions of the three slurry samples.

Components	CWS		CWLS-A		CWLS-B	
	High load	Low load	High load	Low load	High load	Low load
O_2	9.27	11.73	8.56	11.08	7.99	10.70
CO_2	9.03	8.29	10.20	8.74	11.08	9.21
CO	56.96	18.10	24.63	1.42	37.05	3.01
SO_2	761.74	608.43	609.97	576.77	623.89	589.72
NO_x	552.48	510.35	529.37	489.89	507.35	444.92

The unit of O_2 and CO_2 is %, while that of other components is mg/N m³. The content of O_2 is actually a measured value, while that of other components is the corresponding value when the O_2 content is converted to 6%.

Acknowledgment

The authors are grateful to the National Basic Research Program of China (Grant No. 2010CB227001) for their financial support.

References

[1] Xu Z, Chong L, Wang W, Chen Y, Tu Yn, Zhang R. Coal water mixture preparation technology and application in replacing oil to generate electricity. In: Asia-Pacific power and energy, engineering conference; 2009. p. 2487–91.
[2] Papachristodoulou G, Trass O. Coal slurry fuel technology. Can J Chem Eng 1987;65:177–201.
[3] Cen KF, Yao Q, Cao XY, Zhao X, Zhou JH, Huang ZY, et al. Theory and application of combustion, flow, heat transfer, and gasification of coal slurry. 1st ed. Hangzhou: Zhejiang University Press; 1997 [in Chinese].
[4] Wang T, Liang L, Wang R, Jiang Y, Lin K, Sun J. Magnetic mesoporous carbon for efficient removal of organic pollutants. Adsorption 2012;18:439–44.
[5] Rabii A, Bidhendi GN, Mehrdadi N. Evaluation of lead and COD removal from lead octoate drier effluent by chemical precipitation, coagulation–flocculation, and potassium persulfate oxidation processes. Desalin Water Treat 2012;43:1–7.
[6] Zhang TX, Jiang AP, Harrison JH, Chen SL. Pigment removal in anaerobically digested effluent through polyelectrolyte flocculation and liquid–liquid extraction. J Chem Technol Biotechnol 2012;87:1098–103.
[7] Yuksel E, Gurbulak E, Eyvaz M. Decolorization of a reactive dye solution and treatment of a textile wastewater by electrocoagulation and chemical coagulation: techno-economic comparison. Environ Prog Sustain Energy 2012;31:524–35.
[8] Yang CL, McGarrahan J. Electrochemical coagulation for textile effluent decolorization. J Hazard Mater 2005;127:40–7.
[9] Won CH, Choi J, Chung J. Evaluation of optimal reuse system for hydrofluoric acid wastewater. J Hazard Mater 2012;239:110–7.
[10] Coskun T, Debik E, Demir NM. Operational cost comparison of several pre-treatment techniques for OMW treatment. Clean-Soil Air Water 2012;40:95–9.
[11] Wang XY, Zhou MH, Jin XL. Application of glow discharge plasma for wastewater treatment. Electrochim Acta 2012;83:501–12.
[12] Sevda S, Dominguez-Benetton X, Vanbroekhoven K, De Wever H, Sreekrishnan TR, Pant D. High strength wastewater treatment accompanied by power generation using air cathode microbial fuel cell. Appl Energy 2013;105:194–206.
[13] Chen HL, Yang G, Feng YJ, Shi CL, Xu SR, Cao WP, et al. Biodegradability enhancement of coking wastewater by catalytic wet air oxidation using aminated activated carbon as catalyst. Chem Eng J 2012;198:45–51.
[14] Chen FT, Yu SC, Dong XP, Zhang SS. High-efficient treatment of wastewater contained the carcinogen naphthylamine by electrochemical oxidation with gamma-Al_2O_3 supported MnO_2 and Sb-doped SnO_2 catalyst. J Hazard Mater 2012;227:474–9.
[15] Zhou DW, Chen M, Tao PS. Industrial waste fluids as additive agent of coal water slurry. Coal Chem Ind 1994;1:23–7 [in Chinese].
[16] Zhou M, Kong Q, Pan B, Qiu X, Yang D, Lou H. Evaluation of treated black liquor used as dispersant of concentrated coal-water slurry. Fuel 2010;89:716–23.
[17] Liao YF, Ma XQ. Thermogravimetric analysis of the co-combustion of coal and paper mill sludge. Appl Energy 2010;87:3526–32.
[18] Zhu R, Huang DG, Wu YM, Xu ZB. Burning character and kinetics analysis on coal water slurry with new black liquid. Coal Conv 2007;30:49–52 [in Chinese].
[19] Wolf KJ, Smeda A, Muller M, Hilpert K. Investigations on the influence of additives for SO_2 reduction during high alkaline biomass combustion. Energy Fuel 2005;19:820–4.
[20] Rong LE, Yuan ZF, Liu ZM, Tian ZP. The theory of power plant boiler. 1st ed. Beijing: China Electric Power Press; 1997 [in Chinese].
[21] Kaushal DR, Tomita Y. Solids concentration profiles and pressure drop in pipeline flow of multisized particulate slurries. Int J Multiphas Flow 2002;28:1697–717.
[22] Liu JZ, Wang RK, Gao FY, Zhou JH, Cen KF. Rheology and thixotropic properties of slurry fuel prepared using municipal wastewater sludge and coal. Chem Eng Sci 2012;76:1–8.
[23] Wang RK, Liu JZ, Gao FY, Zhou JH, Cen KF. The slurrying properties of slurry fuels made of petroleum coke and petrochemical sludge. Fuel Process Technol 2012;104:57–66.
[24] Demirbas A. Combustion characteristics of different biomass fuels. Prog Energy Combust Sci 2004;30:219–30.
[25] Williams A, Pourkashanian M, Jones JM. Combustion of pulverised coal and biomass. Prog Energy Combust Sci 2001;27:587–610.
[26] McKee DW, Chatterji D. The catalytic behavior of alkali metal carbonates and oxides in graphite oxidation reactions. Carbon 1975;13:381–90.
[27] Ciambelli P, Corbo P, Gambino M, Palma V, Vaccaro S. Catalytic combustion of carbon particulate. Catal Today 1996;27:99–106.
[28] Fahmi R, Bridgwater AV, Darvell LI, Jones JM, Yates N, Thain S, et al. The effect of alkali metals on combustion and pyrolysis of Lolium and Festuca grasses, switchgrass and willow. Fuel 2007;86:1560–9.
[29] Ma Bg, Xu L, Li XG, Ke K, Wan XF. Catalysis of alkali/alkali-earth metal salt on combustion behavior of high ash coal. Coal Sci Technol 2007;35:69–72 [in Chinese].
[30] Nutalapati D, Gupta R, Moghtaderi B, Wall TF. Assessing slagging and fouling during biomass combustion: a thermodynamic approach allowing for alkali/ash reactions. Fuel Process Technol 2007;88:1044–52.
[31] Wei XL, Lopez C, von Puttkamer T, Schnell U, Unterberger S, Hein KRG. Assessment of chlorine-alkali-mineral interactions during co-combustion of coal and straw. Energy Fuel 2002;16:1095–108.
[32] French RJ, Milne TA. Vapor phase release of alkali species in the combustion of biomass pyrolysis oils. Biomass Bioenergy 1994;7:315–25.
[33] Wei X, Schnell U, Hein KRG. Behaviour of gaseous chlorine and alkali metals during biomass thermal utilisation. Fuel 2005;84:841–8.
[34] Wibberley LJ, Wall TF. Alkali-ash reactions and deposit formation in pulverized-coal-fired boilers: experimental aspects of sodium silicate formation and the formation of deposits. Fuel 1982;61:93–9.
[35] Camarillo MK, Stringfellow WT, Hanlon JS, Watson KA. Investigation of selective catalytic reduction for control of nitrogen oxides in full-scale dairy energy production. Appl Energy 2013;106:328–36.
[36] Zhang RD, Luo N, Yang W, Liu N, Chen BH. Low-temperature selective catalytic reduction of NO with NH_3 using perovskite-type oxides as the novel catalysts. J Mol Catal A – Chem 2013;371:86–93.
[37] Roy S, Hegde MS, Madras G. Catalysis for NOx abatement. Appl Energy 2009;86:2283–97.

工业废液制备水煤浆的成浆特性、燃烧机理及结渣特性中试试验研究

摘要：各种工业过程中产生的大量工业废液会引起严重的环境问题，将这些工业废液替代水来制备废液水煤浆（CWLS）是一种可行的处理办法。采用该方法能够保护大量水资源，解决了废液排放引起的环境问题。然而，废液中高浓度的有机物、碱金属离子、硫和硝基化合物等对CWLS的浆体、燃烧、成渣和污染物排放等特性有显著影响。本文对上述特性在中试炉上进行了实验研究。结果表明，与传统的水煤浆（CWS）相比，CWLS在浆体、燃烧和污染排放方面具有更良好的性能，即低粘度、快速点火、火焰温度高、燃烧效率高、低污染排放等。CWLS具有相对较低的粘度，为278和221mPa·s，并且在不使用任何添加剂的情况下表现出剪切变薄的假塑性行为。相比之下，CWS需要使用一种添加剂才能达到良好的流动性，其粘度为309mPa·s。两种CWLSs（CWLS-A和CWLS-B）的最大火焰温度分别为1309.0℃和1303.1℃，燃烧效率为99.61%和99.42%，

CWLS 的这两个物性参数均大于 CWS。但废液中的碱金属离子导致了 CWLS 相当大的残渣状态，在降低操作负载后该现象则会显著改善。

原文刊于 Applied Energy, 2014, 115: 309-319

基于煤炭分级转化的发电技术前景

岑可法,倪明江,骆仲泱,方梦祥,王勤辉,王智化,岑建孟

(能源清洁利用国家重点实验室(浙江大学),杭州 310027)

摘要: 面对我国一次能源以煤为主、能源需求日益增长以及大气环境持续恶化的现实,构建资源、能源、环境一体化的可持续发展能源系统是我国能源的战略方向。本文介绍了基于"煤炭既是能源又是资源"的理念提出的煤炭转化利用新技术——煤炭分级转化发电技术的路线和特点。从节能减排等方面对该技术的发展前景进行了展望,指出煤炭分级转化的发电技术可提高煤炭发电的综合效益,改变煤炭单一用于发电的产业结构,可形成基于煤炭资源化利用发电的新产业链并缓解我国油气等资源的紧缺状况,对于改变和优化国家煤电产业结构、循环经济和节能减排具有重要意义。

关键词: 分级转化;梯级利用;节能减排;煤炭

中图分类号: TQ536   **文献标识码:** A

The Prospect of Power Generation Technology Based on Coal Staged Conversion

Cen Kefa, Ni Mingjiang, Luo Zhongyang, Fang Mengxiang, Wang Qinhui, Wang Zhihua, Cen Jianmeng

(State Key Laboratory of Clean Energy Utilization (Zhejiang University), Hangzhou 310027, China)

Abstract: In the face of the coal-dominated primary energy structure, increasing energy demand, and serious air pollution, the sustainable development of energy system integrated resources, energy and environment is the direction of China's energy strategy. This paper introduces a new technology of coal conversion and utilization based on coal staged conversion technology on the concept of "coal is not only energy but also resource". From the aspects of energy conservation and emission reduction, the development of this technology is prospected. It is pointed out that this technology can improve the comprehensive efficiency of coal power generation, changes the industrial structure and forms a new industrial chain, and ease the shortage of oil and gas in China. It is of great significance to change and optimize the structure of China coal power industry, cyclic economy and energy conservation and emission reduction.

Key words: staged conversion; staged utilization; energy conservation and emission reduction; coal

一、我国能源和煤炭利用现状和问题

受我国化石能源资源以煤为主的制约,2013 年我国煤炭生产量占一次能源生产总量的 75.6 %,占消费总量的 66 %[1]。

尽管我国大力发展核电、水电以及新能源发电

收稿日期:2015-11-09;修回日期:2015-11-12
作者简介:岑可法,能源清洁利用国家重点实验室(浙江大学),教授,中国工程院院士,主要研究方向为洁净煤技术、能源与环境系统工程等;
E-mail: kfcen@zju.edu.cn
基金项目:中国工程院重大咨询项目"推动能源生产和消费革命战略研究"(2013-ZD-14)
本刊网址:www.enginsci.cn

技术,但火电机组所占比例仍然居高不下。预计到 2020 年年底,火力发电装机容量仍将占 62% 左右,虽然煤炭在一次能源中的比重有所下降,但是其绝对消费量依然保持着较大增幅,预计我国将在 2020 年使用燃煤 4.2×10^9 t,与 2013 年的 3.6×10^9 t 相比,增长 17%,这也是我国和国际主流的一次能源利用不同的显著特征[1]。

目前,我国一半左右的煤炭资源用于发电。随着我国社会和经济的快速发展,电力需求越来越大,未来这一比例将会不断提高。煤炭火力发电产生的颗粒物、CO_2 和 SO_2 等有害物质,与当前我国大部分地区的雾霾、$PM_{2.5}$ 超标和水资源枯竭有密切关系。严重的能源和环境问题决定了在未来一个时期内我国面临的节能减排压力将越来越大,能源短缺及环境污染问题已经成为制约我国社会与经济可持续发展的瓶颈。

我国的能源资源和煤炭利用现状决定了以提高煤炭利用的综合能效、控制煤转化过程中的污染排放、解决短缺能源需求为近中期能源领域的首要任务。

二、基于煤炭分级转化的发电技术路线和特点

煤炭由不同组分组成,各组分具有不同的性质和反应活性,如煤炭中所含挥发分是富氢组分,反应活性很高,而固定碳部分反应活性则相对较差。另外,煤炭组分在燃烧和气化两种反应过程中表现出的反应特性相差较大,一般情况下,燃烧反应要比气化反应容易得多,反应条件也要低得多。

传统的燃煤方式忽视了煤的资源属性,将煤炭完全作为燃料燃烧,导致煤炭综合利用水平和效益不高。煤分级转化技术是基于煤炭各组分具有的不同性质和转化特性,突破传统的利用方式,将煤炭同时作为原料和燃料的热解、气化、燃烧等过程有机结合,将煤炭中容易热解、气化的部分转化为煤气和焦油。所产生的煤气作为后续合成工艺的原料生产具有高附加值的化工产品,所产生的焦油可分馏出各种芳香烃、烷烃、酚类等,也可经加氢制得汽油、柴油等产品;难热解气化的富碳半焦去燃烧提供热电,灰渣进行综合利用,从而在同一系统中获得低成本的煤气、焦油和蒸汽,如图 1 所示。

根据煤种特性、转化途径和目标产物不同,煤炭分级转化技术可以组合不同的热解气化燃烧等煤转化方式。不仅可以通过热解实现煤炭中挥发分提取,而且结合各种生产技术路线的优越性,使生产过程耦合在一起,结合热解气化燃烧过程调节目标产物油、气、电的比例,彼此取长补短,提高煤炭转化效率和利用效率,降低污染排放,实现系统整体效益最优化,从而真正实现煤炭的分级综合利用[2,3]。该技术适用于我国十多亿吨不同品质的煤炭资源,可用于新建工厂和大量旧电厂的改造,从而使煤炭分级转化发电技术有更广阔的应用前景。

三、基于煤炭热解分级转化技术开发

采用温和的热解方法从煤炭中提取液体燃料和

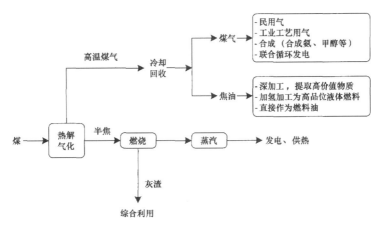

图 1 煤的热解、气化、燃烧分级转化技术

化学品的重要性和必要性已逐渐被认识和接受。日本通产省在《21世纪煤炭技术战略》报告中，特别提到了提高燃料利用率的高增值技术，其中把低温快速热解制取燃气、燃油及高价值化学品作为重要的研究项目。美国能源部也把从煤中提取部分高品位液体燃料和化学品列入《21世纪能源展望》计划中成为一项重要内容。

国内外各研究机构在该领域已开展了较多的研究开发工作，开发了各种不同类型的煤炭热解气化燃烧工艺。国外主要的煤加工技术有德国的鲁奇三段炉（L-S）低温提质工艺、Lurgi-Ruhrgas（L-R）提质技术，苏联的褐煤固体热载体提质（ETCH-175）工艺，美国的温和气化（Encoal）技术、Toscoal工艺、西方提质（Garrett）法、焦-油-能量开发法（COED）、澳大利亚的流化床快速热解工艺（CSIRO）和日本的煤炭快速提质技术。国内的热解工艺目前主要可以分为以获得半焦和焦油为目的和热解半焦燃烧相结合的煤气、焦油和蒸汽联产为目的两大类。其中以获得半焦和焦油为目的的典型技术有大连理工大学开发的褐煤固体热载体干馏多联产（DG）工艺、煤炭科学研究总院北京煤化工研究分院开发的多段回转炉（MRF）提质工艺等，而将煤的热解、气化、燃烧相结合的典型的分级转化技术则有浙江大学的煤炭循环流化床分级转化工艺、国家电力公司北京动力经济研究所、中国科学院工程热物理研究所以及中国科学院山西煤炭化学研究所提出的以移动床热解为基础的固体热载体热电气三联产技术，中国科学院过程工程研究所的基于下行床的多联产工艺和清华大学的基于流化床的多联产工艺。

浙江大学是国内较早开发煤炭分级转化工艺[4-6]的研究单位之一，早在1981年就提出了循环流化床煤热解热电气联产综合利用方案，并建立了一套1 MW热态试验装置，在上面对不同煤种和不同运行参数进行了大量的试验[7,8]，证实了技术和工艺上的可行性。试验结果表明该方案具有燃料利用率高，污染低，煤气热值高，结构简单，投资省等特点，并获得国家发明专利授权（专利号 92100505.2）。利用该技术开发了 12 MW[9,10] 和 25 MW[11] 循环流化床煤炭分级转化装置。

浙江大学所提出的循环流化床煤炭分级转化工艺是将循环流化床锅炉和热解炉紧密结合，在一套系统中实现热、电、煤气和焦油的联合生产。图2为浙江大学开发的煤炭循环流化床分级转化工艺流程：循环流化床锅炉运行温度在850～900 ℃，大量的高温物料被携带出炉膛，经分离机构分离后部分作为热载体进入以再循环煤气为流化介质的流化床热解炉。煤经给料机进入热解炉和作为固体热载体的高温物料混合并加热（运行温度在550～800 ℃），煤在热解炉中经热解产生的粗煤气和细灰颗粒进入热解炉分离机构，经分离后的粗煤气进入煤气净化系统进行净化。除作为热解炉流化介质的部分煤气再循环外，其余煤气则经脱硫等净化工艺后作为净煤气供民用或经变换、合成反应生产相关化工产品。收集下来的焦油可提取高附加值产品或改性变成高品位合成油。煤在热解炉热解产生的半焦、循环物料及煤气分离器所分离下的细灰（灰和半焦）一起被送入循环流化床锅炉燃烧利用，用于加热固体热载体，同时生产的水蒸汽用于发电、

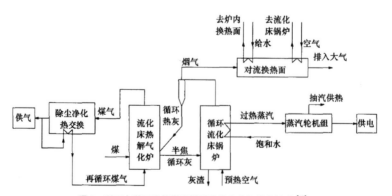

图2 浙江大学开发的煤炭循环流化床分级转化技术[12]

供热及制冷等。

浙江大学所研发的煤炭分级转化工艺具有如下优势。

(1) 工艺简单先进：将循环流化床锅炉和热解气化炉紧密结合，通过简单而先进的工艺在一套系统中实现热、电、煤气和焦油的联合生产。所产煤气品质高，是生产合成氨、甲醇、合成天然气等多种化工产品的优质原料，也可以作为燃气蒸汽联合循环发电的燃料气，所生产的焦油可以在提取高价值化学品的同时加氢制取液体燃料，从而有效地利用了煤中的各种组分，实现了以煤为原料的分级转化梯级利用。

(2) 燃料适应性广：收到基挥发分在 20% 以上的各种褐煤、烟煤都适用于这种工艺。同时煤的颗粒度要求与现有循环流化床锅炉一致，避免了现有煤气化和干馏工艺对煤种和煤粒度有较严格的限制的缺点。

(3) 工艺参数要求低，设备投资低：煤在常压低温无氧条件下热解气化，对反应器及相关设备的材质要求低（常规气化炉操作温度为 1 300～1 700 ℃，压力 2~4 MPa），设备制造成本低，同时热解气化过程不耗氧气和蒸汽，避免了常规气化炉所需的氧制备装置和蒸汽锅炉，大幅度降低气化系统的设备建设成本。

(4) 运行成本低：煤热解单元不需要氧气、蒸汽作为气化剂，系统能量损耗低，与常规气化技术相比，过程热效率大幅度提高，因此运行成本也得到大幅度降低。

(5) 高温半焦直接燃烧利用：原煤热解气化后的半焦直接送锅炉燃烧发电，避免了散热损失，使能源得到充分利用；而锅炉燃用不含水分的半焦，锅炉烟气量大幅度减少，从而降低了引风机的电耗，装置能耗降低，锅炉系统效率也有所提高。避免了以半焦为产品的工艺过程存在的需要半焦冷却的过程，同时所产生的细半焦颗粒存在运输和利用困难的问题。

(6) 易实现大型化：所采用的流化床热解炉具有热灰和入煤混合剧烈，传热传质过程好，温度场均匀的特点，有利于给煤在炉内的热解气化，同时流化床热解炉易于大型化，而且布置上易与循环流化床锅炉匹配，实现与循环流化床锅炉有机集成，从而避免固定床或移动床热解反应器的不易放大和布置的问题。

(7) 煤气产率高，品质好，实现煤气的高值利用：循环流化床分级转化工艺的热解过程以循环灰为热载体，热解所产出的煤气有效组分高，而且所产出的煤气全部用于后续利用，从而保证后续煤气合成工艺的煤气量，避免燃烧热解煤气提供热解热源使得外供煤气量小的问题。

(8) 具有很好的污染物排放控制特性：煤中所含硫大部分在热解气化炉内的热解过程中以 H_2S 形式析出，并与所产生的煤气进入煤气净化系统进行脱硫，仅有少量的硫进入循环流化床燃烧炉以 SO_2 形式释放。同时，与煤直接燃烧后烟气脱硫相比，从煤气中脱除 H_2S 具有较大的优势：①所处理的气体量大大减少，因此脱硫设备的体积、投资及运行成本较小；②目前煤气脱硫的副产品一般是硫磺，其利用价值较大。煤中所含的氮大部分（80% 以上）在热解过程中主要以氮气和氨的形式析出，同时由于循环流化床燃烧过程是中温燃烧，几乎不产生热力 NO_x，因此分级转化工艺可进一步降低循环流化床燃烧炉所产生的烟气中 NO_x 的排放浓度。同样从体积流量较小的煤气中脱出少量的氨是相对比较容易且成本较低的。

2007 年 6 月，浙江大学和淮南矿业（集团）有限责任公司合作完成了 12 MW 循环流化床热电气焦油分级转化工业装置安装，2007 年 8 月完成 72 h 试运行，2008 年上半年完成性能优化试验，2008 年 10 月系统投入试生产运行。12 MW 循环流化床热电气焦油分级转化装置的热态调试运行表明，系统运行稳定，调节方便，运行安全可靠，焦油和煤气的生产稳定，实现了以煤为资源在一个有机集成的系统中生产多种高价值的产品。

2009 年中国国电集团公司小龙潭发电厂、小龙潭矿务局和浙江大学合作以云南小龙潭褐煤为原料，结合中国国电集团公司小龙潭发电厂现有 300 MWe 褐煤循环流化床锅炉的结构和现状，把 300 MWe 褐煤循环流化床锅炉改造为以干燥后褐煤为原料的 300 MWe 循环流化床分级转化装置，目前一期工程已完成试运行及性能参数测试，运行结果表明，系统运行稳定，操作方便，以未干燥褐煤为原料，热解气化炉给煤量达到设计的 40 t·h^{-1}，煤气产率及组分、焦油产率达到设计要求。

四、基于煤炭分级转化的发电技术前景分析

煤炭是我国今后相当长时期内的主要能源。

如何清洁高效地利用发电用煤是我国面临的重大挑战。现有火电厂只将煤炭作为燃料直接燃烧，造成系统效率偏低，污染物控制成本高，且浪费了煤中具有高附加值的油、气和化学品及硫、铝等资源。

以发电为主的煤热解气化燃烧分级转化近零排放污染物灰渣资源化回收技术具有巨大潜力。2013年年底超过 $8.6×10^8$ kW 火电装机，$1.8×10^9$ t 耗煤，90 % 以上为烟煤和褐煤，其所含挥发分可转化为 $2.7×10^{11}$ m³ 合成天然气，相当于我国天然气消费量 1.6 倍多（我国 2013 年天然气消费量为 $1.676×10^{11}$ m³），或 $2.2×10^8$ t 燃油，接近我国石油消费量的一半，与石油进口量相当。由此可见，利用我国电煤所含挥发分采用分级转化为合成天然气，量大且稳定可靠，可作为天然气的重要补充来源，因而提取电煤挥发分替代油气资源前景十分广阔。另外，煤炭的灰分是潜在的建材和矿产资源。以灰分为 25 % 计算，我国燃煤发电排放灰渣作为掺合材料可制取 $1.1×10^9$ t 水泥，或提取 $9×10^7$ t Al_2O_3（约为 2013 年我国 Al_2O_3 产量的 2 倍）。煤炭灰分中含有的锗、镓、铟、钍、钒、钛、铀等贵重金属达到工业品位时，就可提取利用。污染物也是潜在的资源，全国电煤中硫资源若回收利用每年约可生产 $4×10^7$ t 硫酸等产品（相当于 2013 年全国硫酸产量的近一半）。

由此可见，推广应用煤炭分级转化技术适合我国的国情和特色，充分体现煤炭既是能源又是资源的理念，既可对现有近 $8×10^8$ kW 燃煤电厂进行分级利用改造，又可适用于新建电厂，可应用于高效清洁发电、替代工业锅炉燃煤、运输燃料替代和煤化工等领域。

五、结语

根据煤炭既是能源又是资源的理念，在燃煤发电过程中先通过热解提取煤的轻质组分（挥发分）用于生产油、气，提取后的半焦（以碳元素为主）再用于燃烧发电，实现煤炭的分级转化。该技术对于我国清洁高效煤炭发电、油气等资源替代、大幅度节能减排、循环经济等具有重要的战略意义。因此，建议建立煤分级转化技术创新体系，通过出台产业政策促进其推广应用，打造适合我国国情的煤炭利用新模式，从而推动形成煤分级转化战略性新兴产业，来解决我国煤炭的高效、洁净利用问题。

参考文献

[1] 中华人民共和国国家统计局. 中国统计年鉴[M]. 北京: 中国统计出版社, 2014.

[2] 梁晓晔. 煤基多联产系统的全生命周期评价及关键问题研究[D]. 杭州: 浙江大学博士学位论文, 2013.

[3] 郭志航. 褐煤热解分级转化多联产工艺的关键问题研究[D]. 杭州: 浙江大学博士学位论文, 2015.

[4] 岑可法, 骆仲泱, 方梦祥, 等. 新颖的热、电、燃气三联产装置[J]. 能源工程, 1995(1): 17–19.

[5] 方梦祥, 骆仲泱, 王勤辉, 等. 循环流化床热、电、气三联产装置的开发和应用前景分析[J]. 动力工程, 1997, 17(4): 21–27.

[6] 骆仲泱, 方梦祥, 王勤辉, 等. 循环流化床热电气焦油多联产装置及其方法: 200610154581X[P]. 2007.06.13.

[7] 吕小兰. 煤部分气化燃烧集成系统的研究[D]. 杭州: 浙江大学硕士学位论文, 2002.

[8] 刘耀鑫. 循环流化床热电气多联产试验及理论研究[D]. 杭州: 浙江大学硕士学位论文, 2005.

[9] 王勤辉, 骆仲泱, 方梦祥, 等. 12兆瓦热电气多联产装置的开发[J]. 燃料化学学报, 2002, 30(2): 141–146.

[10] 方梦祥, 岑建孟, 石振晶, 等. 75 t/h 循环流化床多联产装置试验研究[J]. 中国电机工程学报, 2010, 30(29): 9–15.

[11] 方梦祥, 岑建孟, 王勤辉, 等. 25 MW 循环流化床热、电、煤气多联产装置[J]. 动力工程, 2007, 27(4): 635–639.

[12] 岑可法, 骆仲泱, 王勤辉, 等. 煤的热电气多联产技术及工程实例[M]. 北京: 化学工业出版社, 2004.

煤炭清洁发电技术进展与前景

岑可法, 倪明江, 高翔, 骆仲泱, 王智化, 郑成航

(能源清洁利用国家重点实验室(浙江大学), 杭州 310027)

摘要: 近年来, 我国大气复合污染问题日益突出, 燃煤是造成大气污染的主要原因之一, 我国电力行业耗煤量约占全国耗煤总量的一半, 实现燃煤电厂烟气污染物高效控制是重中之重。本文介绍了我国煤炭清洁发电实现超低排放的最新进展及未来发展前景。通过理论研究、技术研发及集成应用, 形成了符合我国国情的燃煤烟气污染物超低排放技术路线, 建立了超低排放清洁环保岛, 实现了污染物排放优于天然气机组排放标准限值, 为我国大气污染防治特别是高用能密度区域的污染物减排提供了一条重要出路。研究综述了近年来我国超低排放技术的示范应用情况, 通过费效分析表明超低排放可实现污染物大幅度减排, 具有良好的环境、经济和社会效益。未来, 我国燃煤电厂还将进一步发展烟气污染物深度脱除技术及二氧化碳捕集技术, 最终实现燃煤烟气污染物的近零排放, 为建设全世界最清洁的燃煤电厂奠定坚实的技术基础。

关键词: 煤炭; 高用能密度; 大气污染; 超低排放; 清洁发电

中图分类号: X511 **文献标识码**: A

Progress and Prospects on Clean Coal Technology for Power Generation

Cen Kefa, Ni Mingjiang, Gao Xiang, Luo Zhongyang, Wang Zhihua, Zheng Chenghang

(State Key Laboratory of Clean Energy Utilization (Zhejiang University), Hangzhou 310027)

Abstract: Recently, air pollution has taken Chinese a great concern. The emission control of electricity industry is the key to the air quality improvement since electricity industry accounts for over 50 % of China's coal consumption. This work introduces the progress and prospects of ultra-low emission control technologies based power generation technologies. The ultra-low emission control technologies and integrated systems are developed and optimized by the support of a series of national projects to reach the emission level of natural gas power generation units in using coal-fired units. The application and promotion of these technologies are proven to be an important solution for the air quality improvement, especially in regions with high energy-consumption, since these technologies have shown great potential for deep-cut in pollutant emissions, which will benefit both the society and the environment. In the future, the research will focus on the deep-cut of various pollutants and high-efficiency CO_2 capture to lay a solid foundation for the construction of the world's cleanest coal-fired power plants.

Key words: coal; high energy consumption density; air pollution; ultra-low emissions; clean power generation

收稿日期: 2015-11-09; 修回日期: 2015-11-18
作者简介: 岑可法, 能源清洁利用国家重点实验室(浙江大学), 教授, 中国工程院院士, 主要研究方向为洁净煤技术、能源与环境系统工程等; E-mail: kfcen@sun.zju.edu.cn
基金项目: 中国工程院重大咨询项目 "先进清洁煤燃烧与气化技术"
本刊网址: www.enginsci.cn

一、前言

当前,我国能源开发利用和环境保护面临诸多问题和挑战。一是伴随着我国经济的持续快速增长,我国能源消耗呈快速增长态势。回顾我国能源消费史,改革开放头 20 年,能源消耗翻一番,支撑了国内生产总值(GDP)翻两番;21 世纪头 10 年,能源消耗又翻一番,支撑了国内生产总值翻 1.4 番;若延续这样的增长态势,按 2020 年国内生产总值翻一番测算,能源消费总量将超过 5.5×10^9 tce (tce 为吨标准煤)。二是能源消耗的快速增长使我国大气环境面临十分严峻的挑战。2013 年全国 SO_2、NO_x 和烟(粉)尘排放 $2.043\ 9 \times 10^7$ t、$2.227\ 4 \times 10^7$ t 和 $1.278\ 1 \times 10^7$ t [1],已远超出我国大气环境承载容量,需要进一步加快能源清洁利用,大幅减少大气污染物的排放。三是在大力发展清洁能源过程中(主要包括发展可再生能源或相对清洁的天然气等)存在诸多不足问题。如 2013 年我国发电量为 $5.397\ 6 \times 10^{12}$ kW·h,其中可再生能源发电仅约占 19.5 %(水电 16.9 %,风电、太阳能及其他 2.6 %)[1]。我国一次能源消耗中非化石能源(水能、核能、风能、太阳能等)约占 9.4 %,根据《可再生能源发展"十二五"规划》到 2020 年我国非化石能源占一次能源比例可望达到 15 %。由此可见,我国以煤炭为主体的能源消费结构短期内难以发生重大改变。

为解决燃煤造成的环境污染,全国各地相继推出了以天然气替代燃煤的措施;但我国天然气资源有限,2013 年我国天然气产量为 1.21×10^{11} m³,仅相当于 2.06×10^8 t 煤炭,"煤改气"将会造成严重的天然气供需失衡。因此,2013 年年底国家发展与改革委员会和国家能源局连续发文指出:各地在发展"煤改气"、燃气热电联产等天然气利用项目时,不能"一哄而上",避免供需出现严重失衡。着力抓好煤炭清洁使用,确保已建燃煤发电机组脱硫脱硝设施改造达标并正常投运,实现既改善环境质量又缓解天然气供应压力的目标。

鉴于我国能源生产及消费现状,亟需大力推进煤炭清洁高效利用。其中电力行业是我国主要耗煤行业,与国外发达国家相比,我国煤炭利用整体仍较为分散,如图 1 所示,2013 年中国电力行业耗煤量约占煤炭总消耗量的一半(46 %),远低于 2010 年美国的 92 %、德国的 80 %[1-3]。面对煤炭利用过程中不够集中、不够高效、不够清洁等问题,一方面需要压缩煤炭的比例,另一方面要实现煤炭清洁利用。同时,针对我国煤炭污染集中源和分散源的特点需采取不同的对策。集中源如火电厂,其特点是便于实时监控和污染便于集中治理,可通过烟气污染物超低排放技术的应用,实现污染排放优于天然气发电标准限值;分散源如工业锅炉、民用散煤等,其特点是难以监控和在末端进行经济性可接受的深度治理,可通过改用天然气、生物质等相对清洁的燃料,在减排的同时将分散煤集中于大型燃煤电厂利用,提高煤炭利用效率和治污水平,大幅降低污染物排放水平。

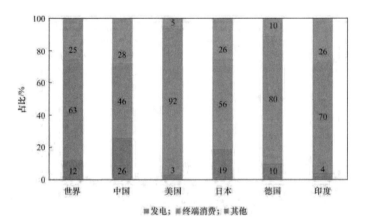

图 1 不同用途煤炭消费结构的国际比较 [1-3]

注:美国、德国、日本数据是 2010 年,中国是 2013 年;"其他"包括供热、制气、煤炭转换、液化、能源部门自用及损失等

另一个需引起重视的问题是：与欧洲、美国、日本等发达地区和国家相比，我国经济发展仍严重依赖于高能耗、高污染的产业；能源结构方面，还存在着空间分布不平衡、消费结构不合理与清洁高效利用水平较低等突出问题。如表1、图2所示，京津冀、长三角、珠三角等重点地区的一次化石能源消费强度约为全国平均值的5.10倍、美国的5.66倍、日本的1.10倍；煤炭消费强度则约为全国平均值的4.92倍，美国的15.7倍、日本的2.74倍；且上述高用能密度区域化石能源消费仍以煤为主。大量化石能源的消费也造成上述地区大气污染物排放强度约为全国平均水平的5倍，更是远高于欧洲、美国、日本等发达地区和国家的水平。因此，针对上述区域能源结构及污染排放问题，一方面需进行产业结构调整，降低煤炭能源消费强度，另一方面需鼓励以燃煤烟气污染物超低排放技术为代表的先进煤炭清洁发电技术在火电行业的推广应用，并进一步拓展实现其他行业烟气污染物的超低排放。超低排放已成为中国现阶段高效清洁集中可持续利用煤炭、保障能源安全的一条重要出路。

二、燃煤电厂烟气污染物超低排放技术

（一）超低排放关键技术发展现状

近年来，国家通过"863"计划、科技支撑计划、自然科学基金、"973"计划等科技项目部署了大量经费用于支持燃煤电厂大气污染物控制理论提升及技术研发工作，在SO_2、NO_x、颗粒物（PM）、汞等污染物控制方面取得了重大突破，为探索建立一套使燃煤电厂主要污染物排放达到国家天然燃气轮机排放限值的多种污染物高效协同脱除技术系统提供了有力保障。

SO_2控制方面，发展了石灰石/石灰-石膏湿法、烟气循环流化床法、海水法等脱硫技术，其中石灰石/石灰-石膏湿法烟气脱硫技术在我国已投运燃煤脱硫机组中占90%以上的份额，其脱硫效率一般可达95%以上。针对当前量大面广的石灰石/石灰-石膏湿法脱硫机组难以满足环保新要求的现状，浙江大学深入研究了湿法烟气脱硫的强化传质与多种污染物协同脱除机理，在此基础上开发了pH值分区控制、筛板塔内构件强化传质、脱硫添加剂等

表1 一次化石能源消费强度对比[1~3]

能源消费	纽约州	华盛顿州	加利福尼亚州	美国	日本	中国	中国重点地区
化石能源消费 /×10^4 tce	9 370	3 844	20 967	280 214	59 157	327 791	105 053
煤炭消费 /×10^4 tce	201	168	123	71 700	16 814	240 812	74 573
化石能源消费强度 /×10^4 tce·km^{-2}	766.2	223.0	519.1	305.9	1 578.6	339.8	1 731.6
煤炭消费强度 /×10^4 tce·km^{-2}	16.4	9.8	3.0	78.3	448.7	249.6	1 229.2

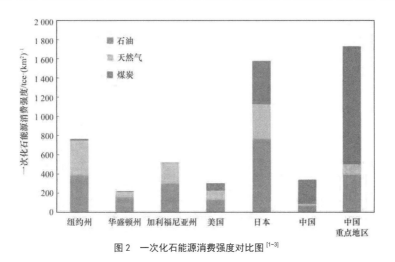

图2 一次化石能源消费强度对比图[1~3]

系列脱硫增效关键技术[4-8]，并在 50~1 000 MW 燃煤机组上实现了示范应用，脱硫效率突破了 99 %，SO_2 排放浓度可低于 20 mg·m^{-3}；同时可协同脱除颗粒物、NO_x、汞等污染物。高效脱硫关键技术也在钢铁烧结机、玻璃炉窑、垃圾焚烧等行业得到了推广应用。

NO_x 控制方面，发展了有低 NO_x 燃烧技术、选择性非催化还原法（SNCR）烟气脱硝技术、选择性催化还原法（SCR）烟气脱硝技术和 SNCR-SCR 耦合脱硝技术等，其中选择性催化还原法烟气脱硝技术在我国已投运燃煤脱硝机组中占 95 % 以上的份额，其脱硝效率一般为 70 %~85 %，最高可达 90 % 以上。针对部分机组 NO_x 排放超标，尤其是低负荷下 NO_x 超标现象严重，大量废烟气脱硝催化剂面临再生等问题，浙江大学通过技术研发形成了具有高脱硝效率、高单质汞 / 二价汞（Hg^0/Hg^{2+}）转化率、低 SO_2/SO_3 转化率、宽温度窗口、高抗磨性能的催化剂配方及其活性恢复方法[8-11]，在含 1 000 MW 等级燃煤机组上也实现了产业化推广应用，排放浓度可低于 50 mg·Nm^{-3}；具有自主知识产权的脱硝催化剂再生改性工艺技术及装备，已成功应用于 300 MW 及 1 000 MW 机组等催化剂再生改性项目，实现 NO_x 高效脱除的协同控制汞等污染物。

颗粒物控制方面，发展了静电除尘、袋式除尘和电袋复合除尘等除尘技术，其中现有近 80 % 的火电机组安装了静电除尘器，而随着袋式除尘器滤袋材料性能的改善及排放标准的严格，袋式除尘器和电袋复合除尘器应用呈上升趋势。为提高颗粒物控制效率，浙江大学近年来还研发了湿式静电除尘、高效凝并、高效供电电源等多种高效除尘关键技术[12-14]。通过在湿法烟气脱硫塔后采用新型湿式静电除尘技术（WESP），形成脱硫塔前除尘、脱硫塔内除尘及脱硫塔后除尘的多级 $PM_{2.5}$ 控制系统，$PM_{2.5}$ 总捕集效率可达到 99 % 以上，烟尘排放浓度小于 5 mg·Nm^{-3}，可实现脱硫塔后污染物控制装备的 SO_3 脱除效率达 70 % 以上[13,14]。目前，浙江大学研发的湿式静电多种污染物协同控制技术已在 300 MW 机组、热电机组等实现了示范应用。

汞等污染物协同控制方面，脱硫塔前一级除尘装备本身可协同控制一部分吸附在颗粒上的 Hg、SO_3 等污染物；而通过对选择性催化还原法脱硝催化剂配方改性及向烟气中添加活性组分，可以将大部分单质汞氧化成二价汞，以利于在后续的脱硫塔内吸收脱除并固定于脱硫副产物中；而脱硫后的湿式静电除尘技术可高效脱除 $PM_{2.5}$ 的同时，协同脱除塔后烟气中携带的 SO_3 酸雾、细小浆液滴、汞等多种污染物，脱汞效率可达 85 % 以上，Hg 排放浓度小于 0.002 mg·Nm^{-3}，SO_3 酸雾去除效率可达 80 % 以上，能有效解决蓝烟 / 黄烟、"石膏雨"以及汞、雾滴排放等污染新问题。

针对单一污染物高效脱除及其他污染物协同控制技术上，通过对 SO_2、NO_x、颗粒物、汞等多种污染物高效脱除与协同控制关键技术的集成开发，形成了能达到天然气燃气轮机排放标准限值要求的燃煤电站超低排放环保岛技术，其系统工艺流程简图见图 3。

目前，燃煤电站超低排放技术正在京津冀鲁、长三角、珠三角等重点区域的燃煤发电机组和热电联供机组上推广应用。如在嘉兴电厂 1 000 MW 燃煤机组上实施烟气清洁化排放改造，采用高效协同脱除技术，对现有的除尘、脱硫、脱硝系统进行提效，实现超低排放，该机组是国内首台达到超低排放的燃煤机组，被国家能源局授予"国家煤电节能减排示范电站"称号；在广东顺德五沙热电 300 MW 燃煤机组上采用选择性催化还原法高效脱硝、筛板强化脱硫除尘一体化、卧式湿式静电除尘等关键技术，实现烟气污染物超低排放；在嘉兴新嘉爱斯热电 220 t·h^{-1} 热电联产锅炉上通过耦合选择性催化还原法脱硝 + 常规除尘 + 高效湿法烟气脱硫 + 湿式静电深度净化等高效协同脱除技术，实现烟气污染物超低排放；上述工程示范通过不同的减排技术路线均使燃煤机组烟气的主要污染物排放浓度达到国家燃气排放标准限值要求。随着燃煤发电机组超低排放示范工程的深入推进，我国煤电行业将取得革命性进步，可望建成世界上最大的清洁高效煤电体系[15]。

（二）超低排放技术经济效益分析

根据《火电工程限额设计参考造价指标（2012 年水平）》[16] 及某 300 MW 发电机组的运行情况调研，分别对燃煤锅炉超低排放发电、燃气锅炉发电及燃气蒸汽联合循环发电成本进行核算，结果如表 2 所示（其中，燃料价格以燃气 3.6 元·Nm^{-3}，煤炭 600 元·t^{-1} 计算）。若燃煤机组进行超低排放改造，发电成本增加约 0.016 分·(kW·h)$^{-1}$，而

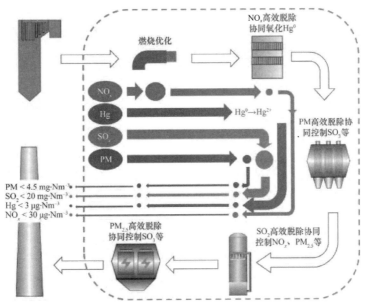

图 3 燃煤电站超低排放环保岛

改造成燃气蒸汽联合循环发电,发电成本增加 0.482 元·(kW·h)$^{-1}$(未考虑机组投资),改造成燃气锅炉发电成本增加 0.636 6 元·(kW·h)$^{-1}$。针对不同容量、不同污染物排放水平的燃煤机组,其超低排放改造的投资成本及运行成本有所差别,超低排放改造增加发电成本 1.5~2.0 分·(kW·h)$^{-1}$;而"煤改气"根据不同地区、燃气价格不同发电成本亦有所不同,发电成本增加 0.3~0.8 元·(kW·h)$^{-1}$。超低排放技术发电与天然气发电相比,具有较好的经济效益。

(三)超低排放技术环境效益分析

2013 年,全国烟(粉)尘、二氧化硫、氮氧化物排放量分别为 1.278 1×10^7 t、2.043 9×10^7 t、2.227 4×10^7 t,其中电力行业烟尘、二氧化硫、氮氧化物年排放量分别为 1.42×10^6 t、7.8×10^6 t、8.34×10^6 t,分别占全国排放量的 11.1 %、38.2 %、37.4 %[1]。

随着超低排放技术的进一步推广,烟气污染物的减排效益逐渐凸显。据中国电力企业联合会初步统计,2014 年火电行业烟尘、SO$_2$、NO$_x$ 排放量为 9.8×10^5 t、6.2×10^6 t、6.2×10^6 t,分别比 2013 年下降约 31.0 %、20.5 %、25.7 %,全面提前完成《节能减排"十二五"规划》规定的电力二氧化硫排放 8×10^6 t,氮氧化物 7.5×10^6 t 的减排目标。经测算,若燃煤烟气污染物超低排放技术在全国燃煤机组推广应用,预计燃煤烟气污染物排放量仅为:颗粒物为 8×10^4 t、SO$_2$ 为 5.3×10^5 t、NO$_x$ 为 7.6×10^5 t,为解决我国燃煤大气污染与能源资源双重约束问题提供了一条重要出路。

三、燃煤清洁发电技术发展趋势及前景

2011 年环保部颁布"史上最严格"的《火电厂大气污染物排放标准(GB 13223—2011)》[17],2013 年环保部颁布《关于执行大气污染物特别排放限值

表 2 不同发电方式的发电成本对比

发电类型	常规燃煤电站发电技术	燃煤电站超低排放技术	燃气-蒸汽联合循环发电技术	燃气锅炉发电技术
发电成本 / 元·(kW·h)$^{-1}$	0.450	0.466	0.932	1.083 6
与常规燃煤电站发电对比增加成本 / 元·(kW·h)$^{-1}$	—	0.016	0.482	0.633 6

的公告》，国家针对燃煤电厂持续采取严格的大气环境管理措施，严格控制大气污染物新增量。2013年9月，国务院出台《大气污染防治行动计划》[18]，2014年10月，国家发改委、环保部、国家能源局联合印发《煤电节能减排升级与改造行动计划（2014—2020年）》（2093号文）[19]，要求"东部地区新建燃煤发电机组大气污染物排放浓度基本达到燃气轮机组排放限值，中部地区新建机组原则上接近或达到燃气轮机组排放限值，鼓励西部地区新建机组接近或达到燃气轮机组排放限值。"2015年3月，李克强总理在政府工作报告中提出要加强煤炭清洁高效利用，推动燃煤电厂超低排放改造；浙江、山东、江苏、广东、山西、河南等地方政府也纷纷出台超低排放相关扶持政策。一系列政策的出台有力地推动了燃煤发电机组超低排放升级改造，据《"十二五"生态环境保护成就报告》统计，目前，我国已完成煤电行业超低排放改造 8.4×10^7 kW，约占全国煤电装机 1/10，正在进行改造的超过 8.1×10^7 kW[15]。

随着我国经济的进一步发展，人均用能水平的不断提高，以及对空气质量改善的需求，未来大气污染物排放要求必将日趋提高。在新能源发展尚不满足我国现阶段经济和社会发展需求时，煤炭清洁发电技术是我国目前能源客观条件下的必然选择。尤其是在人口密集、经济发达的重点地区，更清洁的煤炭发电技术是未来燃煤电厂发展与立足的必由之路。

未来，燃煤电厂将进一步发展燃煤烟气深度净化技术，浙江大学已研发了活性分子多种污染物一体化脱除技术，目前已在 6×10^4 $Nm^3\cdot h^{-1}$ 炭黑尾气上工业应用，实现了烟气 NO_x 由初始浓度 800 $mg\cdot Nm^{-3}$ 降至 10 $mg\cdot Nm^{-3}$，SO_2 由初始浓度 1 000 $mg\cdot Nm^{-3}$ 降至 30 $mg\cdot Nm^{-3}$，脱除效果远低于火电厂污染物国家燃煤排放标准（GB 13223—2011，重点地区 $NO_x<100$ $mg\cdot Nm^{-3}$、$SO_2<50$ $mg\cdot Nm^{-3}$，而且也优于超低排放要求（$NO_x<50$ $mg\cdot Nm^{-3}$，$SO_2<35$ $mg\cdot Nm^{-3}$），为我国燃煤电厂超低排放提供了具有自主知识产权的新技术方案。未来活性分子多种污染物一体化脱除技术将在大型燃煤电厂中实现应用。同时，随着 CO_2 排放控制需求的提高，未来将进一步大力发展廉价 CO_2 捕集技术及 CO_2 利用技术，并在燃煤电厂实现示范应用，真正实现燃煤电厂烟气污染物的近零排放，达到甚至优于天然气发电污染物与温室气体排放水平，为燃煤电厂的绿色清洁发电、经济的可持续发展提供一条有效途径。

四、结语

当前在我国能源资源短缺和节能减排双重约束下，发展清洁煤技术是当前我国重大战略需求，通过利用煤炭清洁发电最新的研究和工程实践，实现煤电产业转型升级，是我国大气污染防治的一条重要可持续发展路线。

燃煤清洁发电技术是当前国际能源环境领域的战略性前沿课题之一，也是研究热点和难点问题之一。针对我国大气污染治理的严峻态势，通过燃煤烟气污染物超低排放的新思路，实现燃煤烟气颗粒物、二氧化硫、氮氧化物、汞及其化合物等多种污染物排放达到或优于燃气机组排放水平，具有良好的经济、环境和社会效益。通过活性分子污染物一体化脱除技术等具备主要烟气污染物排放进一步降低的能力，实现近零排放。未来，污染物的深度脱除及二氧化碳捕集及封存技术将会持续发展，为我国建设全世界最清洁的燃煤发电体系奠定了坚实的基础。

参考文献

[1] 中华人民共和国国家统计局. 中国统计年鉴[M]. 北京: 中国统计出版社, 2014.
[2] BP中国. BP世界能源统计年鉴2013[EB/OL]. [2015-11-08]. http://www.bp.com/zh_cn/China/reporte-and publications/bp_2013.html.
[3] IEA. Energy Balances of OECD Countries 2014 [EB/OL]. [2015-11-08]. http://dx.doi.org/10.1787/energy_bal_oecd-2014-en.
[4] 李存杰, 张军, 张涌新, 等. 基于pH值分区控制的湿法烟气脱硫增效研究[J]. 环境科学学报. doi: 10.13671/j.hjkxxb. 2015.0137
[5] 张军, 张涌新, 郑成航, 等. 复合脱硫添加剂在湿法烟气脱硫系统中的工程应用[J]. 中国环境科学, 2014, 9: 2186–2191.
[6] 许昌日. 燃煤烟气NO_x/SO_2一体化强化吸收试验研究[D]. 杭州. 浙江大学硕士学位论文, 2014.
[7] 王惠挺. 钙基湿法烟气脱硫增效关键技术研究[D]. 杭州: 浙江大学博士学位论文, 2013.
[8] 谢克昌, 等. 中国煤炭清洁高效可持续开发利用战略研究[M]. 北京: 科学出版社, 2014.
[9] 姜烨, 高翔, 吴卫红, 等. 选择性催化还原脱硝催化剂失活研究综述[J]. 中国电机工程学报, 2013, 14:18–31, 13.
[10] 姜烨, 张涌新, 吴卫红, 等. 用于选择性催化还原烟气脱硝的V_2O_5/TiO_2催化剂钾中毒动力学研究[J]. 中国电机工程学报, 2014, 23:3899–3906.

[11] 毛剑宏, 宋浩, 吴卫红, 等. 电站锅炉SCR脱硝系统导流板的设计与优化[J]. 浙江大学学报(工学版),2011,06:1124–1129.

[12] 万益, 黄薇薇, 郑成航, 等. 湿式静电除尘器喷嘴特性[J]. 浙江大学学报 (工学版), 2015, 02: 336–343.

[13] 杨正大, 常倩云, 岳涛, 等. 湿式静电协同脱除SO_2、PM试验研究[J]. 工程热物理学报, 2015, 06: 1365–1370.

[14] 万益. 湿式静电除尘水膜均布及细颗粒物强化脱除研究[D]. 杭州: 浙江大学硕士学位论文, 2014.

[15] 陈吉宁. "十二五" 生态环境保护成就报告 [R]. 中华人民共和国环境保护部, 2015.10

[16] 电力规划设计总院. 火电工程限额设计参考造价指标（2012年水平）[M]. 北京: 中国电力出版社, 2013.

[17] GB 13223－2011, 火电厂大气污染物排放标准[S]. 北京: 中国标准出版社, 2011.

[18] 中华人民共和国国务院. 国务院关于印发大气污染防治行动计划的通知[EB/OL]. [2013-09-12]. http://www.gov.cn/zwgk/2013-09/12/content_2486773.htm.

[19] 中华人民共和国国家发展和改革委员会. 煤电节能减排升级与改造行动计划（2014－2020年）[EB/OL]. [2014-09-19]. http://www.sdpc.gov.cn/gzdt/201409/t20140919_626240.html.

原文刊于中国工程科学，2015, 17(9)：49-55

Emerging energy and environmental applications of vertically-oriented graphenes†

Zheng Bo,‡[a] Shun Mao,‡[bf] Zhao Jun Han,‡[c] Kefa Cen,[a] Junhong Chen*[bf] and Kostya (Ken) Ostrikov*[cde]

Graphene nanosheets arranged perpendicularly to the substrate surface, *i.e.*, vertically-oriented graphenes (VGs), have many unique morphological and structural features that can lead to exciting properties. Plasma-enhanced chemical vapor deposition enables the growth of VGs on various substrates using gas, liquid, or solid precursors. Compared with conventional randomly-oriented graphenes, VGs' vertical orientation on the substrate, non-agglomerated morphology, controlled inter-sheet connectivity, as well as sharp and exposed edges make them very promising for a variety of applications. The focus of this *tutorial review* is on plasma-enabled simple yet efficient synthesis of VGs and their properties that lead to emerging energy and environmental applications, ranging from energy storage, energy conversion, sensing, to green corona discharges for pollution control.

Key learning points
(1) Unique morphology, structure, and properties of vertically-oriented graphenes (VGs).
(2) Plasma-enabled VG growth on a wide range of substrates from various precursors.
(3) Mechanisms and key features of the plasma-assisted VG growth.
(4) Energy applications of VGs: from electrochemical capacitors to rechargeable batteries and fuel/solar cells.
(5) Environmental applications of VGs: from gas- and bio-sensors to pollution control.

1. Introduction

Graphene, a lattice of sp^2 carbon atoms densely packed into a hexagonal structure, is a novel two-dimensional (2D) carbon material potentially suitable for a wide range of applications, *e.g.*, energy storage/conversion devices and biological/chemical sensors.[1,2] Although many promising applications of graphene-related materials have been demonstrated in the past decade, major challenges and needs still remain for the efficient use of their large surface areas and extraordinary electrical, chemical, optical, and mechanical properties.

Since oriented one-dimensional (1D) nanomaterials can outperform their non-oriented counterparts in specific applications,[3] the 'transition' of graphene orientation on a substrate from horizontal (randomly oriented or parallel to the substrate surface) to vertical (perpendicular to the substrate surface), *i.e.*, to form vertically-oriented graphenes (VGs), is promising. Compared with stacks of horizontal graphene from conventional chemical processes, VG networks normally produced by plasma-based approaches show many unique features,[4] such as vertical orientation on the substrate, non-agglomerated three-dimensional (3D) inter-networked morphology, controlled inter-sheet connectivity, as well as exposed ultra-thin and ultra-long edges. These unique features lead to advanced functional properties, *e.g.*, readily accessible surface areas, very long and thin reactive edges, easy examination by a scanning electron microscopy (SEM), and unusual surface functionalities that can potentially enable new applications. Conventional applications of VGs have mainly focused on electron field emitters, where the vertical orientation, sharp exposed edges, and high electrical conductivity of VGs have led to the exceptional field emission performance.[5,6] With the increasing demands from energy and environmental

[a] *State Key Laboratory of Clean Energy Utilization, Department of Energy Engineering, Zhejiang University, Hangzhou, Zhejiang 310027, China*
[b] *Department of Mechanical Engineering, University of Wisconsin-Milwaukee, Milwaukee WI, 53211, USA. E-mail: jhchen@uwm.edu*
[c] *CSIRO Manufacturing Flagship, P.O. Box 218, Bradfield Road, Lindfield, New South Wales 2070, Australia. E-mail: kostya.ostrikov@csiro.au*
[d] *Institute for Future Environments and School of Chemistry, Physics and Mechanical Engineering, Queensland University of Technology, Brisbane, Queensland 4000, Australia*
[e] *School of Physics, The University of Sydney, Sydney, New South Wales 2006, Australia*
[f] *NanoAffix Science LLC, Shorewood WI, 53211, USA*

† Electronic supplementary information (ESI) available. See DOI: 10.1039/c4cs00352g
‡ Contributed equally.

fields that are critical for a sustainable future, applications of VGs in these fields have recently been actively pursued.

This *tutorial* introduces VG's unique properties and state-of-the-art energy and environmental applications. The key aim here is to present insights into VG-based architectures and their potential for functional devices of major interest to a wider scientific community and emerging nanotechnology-based industries.

2. Vertically-oriented graphenes: growth, unique features and properties

2.1 VGs and their unique properties

VGs are made of a stack of graphene nanosheets arranged perpendicularly to the substrate surface. Despite the common hexagonal carbon networks, VGs differ from the conventional horizontal, randomly oriented graphenes in many aspects. Their distinctive morphology and structure over micro- and nano-meter scales are sketched in Fig. 1, in comparison with horizontal graphene. These features determine a number of unique mechanical, chemical, electronic, electrochemical, and optoelectronic properties, which could benefit a wide range of applications.

The first notable feature of VGs is the vertical orientation on the substrate which improves mechanical stability. Although VG networks can have petal-, turnstile-, maze-, and cauliflower-like morphologies,[7–10] each VG nanosheet usually represents a free-standing, self-supported rigid structure. This structure enables the mechanical stability of 2D graphene nanosheets, which would otherwise collapse and stack with each other in random directions, in part due to the strong van der Waals

Zheng Bo

Shun Mao

Zheng Bo is an associate professor at the State Key Laboratory of Clean Energy Utilization, Department of Energy Engineering, Zhejiang University (China). He received his PhD degree from Zhejiang University in 2008. During 2009 to 2011, he was a postdoctoral research associate at the University of Wisconsin-Milwaukee (USA). His research interests include energy and environmental applications of nanomaterials, flow and transport at the nanoscale, interfacial phenomena, and non-thermal plasmas. In 2011, he was awarded the National Excellent Doctoral Dissertation of China (top 100). He is currently a member of the editorial board of Scientific Reports.

Shun Mao received his PhD degree in Mechanical Engineering from the University of Wisconsin-Milwaukee (USA) in 2010 for the study of hybrid nanomaterials for biosensing applications. He is currently a research associate at the University of Wisconsin-Milwaukee and a principal investigator at the Industry-University Cooperative Research Center (I/UCRC) on Water Equipment and Policy, supported by the U.S. National Science Foundation (NSF). His research focuses on hybrid nanomaterials for environmental and energy applications.

Zhao Jun Han

Kefa Cen

Zhao Jun Han has been a Research Scientist at CSIRO Manufacturing Flagship since 2013. He graduated from Nanyang Technological University with both BE and PhD degrees in Electrical and Electronic Engineering. During 2009–2013, he was the Office of Chief Executive (OCE) Postdoctoral Fellow at CSIRO Materials Science and Engineering. His research topics include the synthesis and plasma nanofabrication of various nanomaterials and their applications in energy storage devices, water treatment, and biomedical engineering. He is the recipient of CSIRO's Julius Career Award (2014), Australian Research Council's Discovery Early Career Researcher Award (2013), and the Institute of Engineering Singapore Award (2007).

Kefa Cen is currently a professor at the State Key Laboratory of Clean Energy Utilization, Department of Energy Engineering, Zhejiang University. He received his PhD degree from Moscow Industrial Technology University (USSR) in 1962. He is currently the Director of the Institute for Thermal Power Engineering of Zhejiang University. His research mainly focuses on renewable energy, emission control, and nanotechnology. He became an Academician of the Chinese Academy of Engineering in 1995.

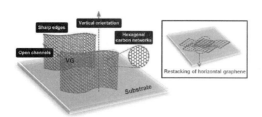

Fig. 1 A schematic representing VGs' structural and morphological features. Inset illustrates the restacking of horizontal graphene nanosheets.

interactions. For electronic, optoelectronic, and electrochemical applications, the alignment between highly conductive graphene planes and the direction of charge transport in the devices can lead to a higher device efficiency. The vertical orientation of VGs also facilitates SEM imaging due to their lateral dimensions that are much larger than their thicknesses.

Second, VGs feature a non-agglomerated morphology with a high surface-to-volume ratio and open channels between the sheets, making the entire VG surface area readily accessible by gases or liquids in sensing or electrochemical applications. The strong interest in graphene applications is due to its one to a few carbon-atom thickness and a high specific surface area. However, re-arrangement (e.g., stacking) of horizontal graphene nanosheets can easily lead to a significant decrease in graphene's available surface area. This problem can be minimized to a large extent by using VGs as an alternative. Depending on the plasma process and growth parameters, the inter-sheet spacing between the adjacent VG nanosheets varies from a few tens to several hundred nanometers and even larger.[6,11,12] Taking the advantage of this non-agglomerated structure, the specific surface area of the VG networks could reach a high value of $\sim$1100 m^2 g^{-1}.[13]

Third, VGs exhibit long exposed ultra-thin, reactive graphene edges, attractive for applications that rely on the edge activity. An individual nanosheet of VGs typically has a tapered shape, with a few graphene layers formed at the base and atomically thin carbon layers formed at the top.[10] The thin graphene layers, commonly with an interlayer (002) spacing of 0.34 to 0.39 nm,[14] can be stacked in the Bernal AB configuration. However, rotating and disordered stacking orders are more often found in few-layer graphenes.[10] It was recently revealed that most of the VG edges are made of folded seamless graphene sheets and only a relatively small fraction of the edges remain open during the plasma-based growth.[15] These active edges can boost the chemical and electrochemical activity of VGs for sensing applications.

These unique morphological and structural features make VGs very attractive for many emerging energy and environmental applications in addition to the common field emission devices.[6] For example, the large accessible surface area and high in-plane electrical conductivity can benefit VG's use in electrochemical capacitors (ECs, or so-called 'supercapacitors' and 'ultracapacitors'), batteries, fuel cells, and solar cells;[9,16–18] the high density of open edges with controlled structural defects can enhance the chemical and electrochemical activity in biosensors and gas sensors;[13,19] the high aspect ratio and electrical conductivity can facilitate generation of atmospheric corona discharges with lower power consumption and reduced emission of hazardous ozone.[20]

2.2 Plasma-enabled growth

Apart from the synthesis in arc discharges[21] and cutting from rolled graphene film stacks,[22] high-quality VGs are produced by a plasma-enhanced chemical vapor deposition (PECVD).[14] By tuning the plasma growth conditions, the VG can be synthesized in a low-temperature, highly-efficient, and catalyst-free manner, with controllable structures and properties. In this

Junhong Chen

Junhong Chen is a Professor of Mechanical Engineering, a Professor of Materials Science and Engineering at the University of Wisconsin-Milwaukee, and a Fellow of American Society for Mechanical Engineers (ASME). He is also the Director of the Industry-University Cooperative Research Center (I/UCRC) on Water Equipment and Policy, supported by the U.S. National Science Foundation (NSF) and water-based industrial partners. His research interests lie in nanocrystal synthesis and assembly; nanocarbons (i.e., graphene and CNTs) and hybrid nanomaterials; nanostructure-based gas sensors, water sensors, and biosensors; and nanocarbon-based hybrid nanomaterials for sustainable energy and environment (http://www.uwm.edu/nsee/).

Kostya (Ken) Ostrikov

Kostya (Ken) Ostrikov is a Science Leader, ARC Future Fellow, Chief Research Scientist with CSIRO and a Professor with Queensland University of Technology, Australia. His achievements include Pawsey (2008) medal of Australian Academy of Sciences, Walter Boas (2010) medal of Australian Institute of Physics, Building Future Award (2012), NSW Science and Engineering Award (2014), 8 prestigious fellowships in 6 countries, 3 monographs, and 430 journal papers. His research on nanoscale control of energy and matter contributes to solution of the grand challenge of directing energy and matter at the nanoscale, to develop renewable energy and energy-efficient technologies for a sustainable future.

section, recent progress in the PECVD growth of VGs as well as the growth mechanism are discussed.

2.2.1 VG growth on various substrates. PECVD growth of VGs can be conducted on virtually any substrates ranging from macro sized planar, foam like, and cylindrical shapes to micrometer- and even nanometer-sized structures. This ability, together with controlled VG structures, allows the easy fabrication of different VG-based devices for diverse applications.

PECVD growth of VGs on planar substrates has been extensively demonstrated.[14,23] One of the major advantages of the plasma-enabled growth is that it requires no catalyst, making the growth amenable to diverse materials such as dielectric SiO_2 and Al_2O_3, semi-conductive Si, and conductive carbon as well as various metals (e.g., Cu, Ni, W, Al, Ti, Pt, and stainless steel).[14] As an example, Fig. 2a and b show SEM images of VGs grown on a planar n-type Si(100) substrate.[23] A CH_4–H_2 gas mixture was used as the precursor and the synthesis was conducted in a 13.56 MHz radio-frequency (rf) PECVD reactor. Metallic impurity-free VG networks were obtained with very sharp graphitic edges (~1 nm) and a uniform height distribution (standard deviation of <10%).

Fig. 2c shows an SEM image of VGs grown on a porous nickel foam.[24] A commercially available nickel foam was first compressed and chemically etched to remove surface oxides. The substrate was further etched with H_2 plasmas and VGs were produced by using CH_4 (diluted in H_2) as the carbon feedstock gas through a microwave PECVD. The SEM image clearly shows that VGs with a few micrometers in height were grown perpendicularly around the Ni scaffold on both inner and outer surfaces. The oxygen impurity in VG nanosheets is also low due to the use of a high-purity carbon source gas and a low pressure plasma process.

VGs can also be grown on cylindrical substrates such as metallic wires. This was achieved using an atmospheric-pressure glow discharge in a CH_4–H_2O–Ar mixture.[20] The VGs synthesized by this method showed good uniformity in both circumferential and axial directions (Fig. 2d), in part due to the use of a rotating substrate stage. The transmission electron microscopy (TEM) image in Fig. 2e shows a solid contact at the interface of the VGs and the metal wire substrate.

VGs can also be produced on micrometer- and even nanometer-sized substrates. Fig. 2f and g show SEM images of VGs grown on carbon cloth,[25] by rf magnetron sputtering of a carbon target in the Ar–H_2 gas mixture at a temperature of 350 °C. The thickness and lateral dimension of VG nanosheets were about 5–10 nm and 300 nm, respectively. Similar to the nickel foam substrate, the porous and fibrous structure of the carbon cloth enabled a high density of VG networks with large surface areas. VG flakes can also form on the lateral surfaces of carbon nanotubes (CNTs), as shown in Fig. 2h.[18,26] Fig. 2i shows that VGs are seamlessly integrated into the outer walls of a multi-walled CNT by forming sp^2 covalent bonds.[18]

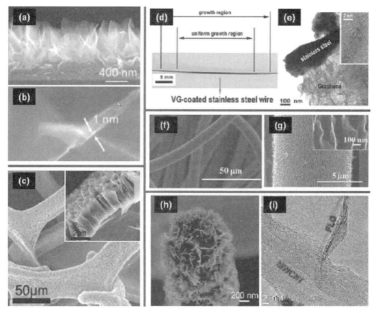

Fig. 2 VGs grown on various substrates. (a) cross-sectional SEM of VG networks and (b) magnified SEM image of an individual VG grown on planar Si.[23] (c) Top-view and cross-sectional (inset) SEM images of VGs on 3D porous Ni substrates.[24] (d) A photograph and (e) a TEM micrograph of VGs on a stainless steel wire.[20] (f) Low- and (g) high-magnification SEM images of VGs on carbon cloth (inset shows the VG edges).[25] (h) SEM and (i) TEM images of VG seamlessly integrated with a CNT.[18]

The catalyst-free and material/structure-independent growth of VG could lead to the direct integration of VGs into many functional devices.[18,20,24,27] For example, VGs grown on conductive planar metal substrates showed promising performance as anode materials in lithium-ion batteries;[27] growth of VGs on metallic cylindrical electrodes helped generating atmospheric corona discharges;[20] VG growth on nickel foams enabled 3D architectures that combine the benefits of vertical orientation and a high loading density of active materials for high-performance ECs;[24] and the 'fused' VG–CNTs hybrid structure exhibited better optoelectronic and gas sensing properties than the randomly mixed graphene and CNTs.[18] Nonetheless, it is noted that although the growth process is substrate-independent, the final structure and crystallinity of the VGs can vary among different substrates.[11] This may lead to the ability to tune and improve the performance of VG-based devices.

2.2.2 VG growth from different precursors. Gaseous precursors are the most common precursors used for the VG growth in PECVD processes, mainly including hydrocarbons (*e.g.*, CH_4, C_2H_4, C_2H_2), fluorocarbons (*e.g.*, CF_4, CHF_3, C_2F_6), and carbon monoxide/dioxide.[11,14] In addition, these carbon-containing gases are often diluted in Ar and/or H_2 or H_2O to improve the plasma stability and better control the VG structure and crystallinity. The fact that VGs can grow without the catalyst implies that precursor dissociation by the plasma plays an important role in the VG nucleation.[15] Unfortunately, due to the complexity in the plasma chemistry, it is very difficult to identify which species contribute most to the VG growth. A number of free radicals, ions, and other reactive species form as gas precursors undergo inelastic collisions with electrons and other species in the plasma. For instance, in C_2H_2 rf plasmas, H_2, $C_4H_2^+$, $C_4H_3^+$, $C_2H_2^+$, a C_2 dimer, a C_4H_2 neutral, a C_4H_3 radical, a C_2H radical and $C_{2n}H_2$ polyacetylenes are generated and contribute to the VG formation.[28]

The growth rate of VGs is closely related to the gas dissociation energy and the formation of reactive carbon dimers C_2 in the plasma.[14] For example, it is easier to produce C_2 dimers by dissociating a C_2H_2 precursor than CH_4 molecules, mostly because of the strength of the C≡C bond in C_2H_2 molecules. Consequently, in rf PECVD processes VGs grow much faster from C_2H_2 than from CH_4.[28] When CF_4 and C_2F_6 fluorocarbon precursors are used, VGs appear to be straighter and thicker than in a CH_4-based process.[11] In addition, a trace amount of oxidizing gases, such as oxygen or water, is often added to improve the crystalline structure and properties of VGs.[12]

Recent advances in using solid or liquid natural precursors to produce graphenes stimulate environmentally-friendly, low cost and large-scale production of this material.[29] VGs can also be synthesized from solid and liquid natural precursors, such as honey, sugar, butter, milk, cheese, and wax, and using a rapid reforming in low-temperature Ar/H_2 plasmas without any metal catalyst or external substrate heating.[30] VGs derived from these solid and liquid precursors show vertical orientation, open structure, and long reactive edges that are very similar to those produced using gaseous hydrocarbon precursors.[30]

One particular advantage of using solid and liquid precursors is that biomass even from natural waste could be potentially transformed into useful VG structures for device fabrication. Existing thermal- and chemical-based methods for transforming biomass are not only precursor-specific but also expensive and energy-, time-, and resource-consuming. In contrast, the plasma-based reforming represents an eco-friendly, energy-efficient, and inexpensive approach. Moreover, structural parameters such as height and open edge density, surface functional groups, adhesion, purities, and crystallinity of VGs can be controlled in PECVD by choosing different solid or liquid precursors, which may open up a new avenue for the large-scale production of VG networks for practical applications.

2.2.3 Growth mechanisms. Despite substantial recent efforts, the VG growth mechanisms remain elusive. Indeed, in addition to the substrate and precursor effects, other factors such as the plasma source and power, etching rate, surface temperature, and plasma pre-treatment may also affect the final structure.[14] The vapor–liquid–solid (VLS) or vapor–solid–solid (VSS) mechanisms widely used to explain the growth of 1D nanotubes or nanowires cannot be directly applied for VG growth as the process requires no catalyst. The nucleation mechanism for 2D thin film deposition is also of a limited relevance because it describes continuous layers rather than networks of vertically oriented wall-like structures such as VGs. Recent advances in time-resolved growth and microanalysis techniques allow in-depth understanding and several growth mechanisms have been put forward, as discussed below.

The VG growth process is likely to involve three steps: (i) first, a buffer layer is formed on the substrate surface with irregular cracks and dangling bonds, which serve as nucleation sites for the VG growth; (ii) then, graphene nanosheets grow vertically under the influence of stress and/or localised electric field, and carbon atoms are continuously incorporated into open edges; (iii) the growth of VG finally stops upon the closure of open edges determined by the competition of material deposition and etching effects in the plasmas.[31]

The buffer layer formed in the nucleation step is usually made of either amorphous carbon or carbide.[10,19] Amorphous carbon is formed due to the large mismatch between the lattice parameters of the substrate material and the graphite, while a carbide layer is formed when the substrate can react with (*e.g.*, dissolve) carbon atoms.[31] A planar or carbon onion-like graphitic layer can be present in between the amorphous carbon buffer layer and the vertical graphene nanosheets, as shown in Fig. 3a.[15] A suitable amount of H atoms or OH radicals can etch the amorphous carbon and help the growth of graphene nanosheets.[12,14] Once the buffer layer is formed and the graphene nanosheets start to grow, VGs no longer show any substrate-dependent features, accounting for the similar morphology on all substrates.[10,31]

The next essential question is why VG can grow vertically instead of forming thicker graphene films as seen in other carbon-based nanostructures. This is likely due to three reasons, *namely*, the electric field, the internal stress, and the anisotropic growth effects, as explained below.

Electric field effects. The electric field in the plasma sheath directs the growth of various oriented nanostructures (*e.g.*, CNTs)

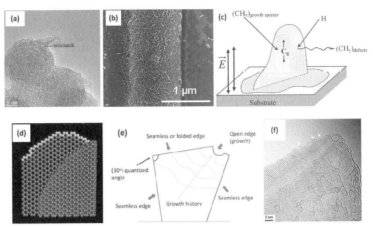

Fig. 3 VG growth mechanism: (a) TEM image of a carbon onion with mismatched graphitic layers at the surface, which may initialize the VG growth.[15] (b) SEM image of VGs grown on an Au stripe due to the electric field effect.[32] (c) A schematic of VG growth controlled by the electric field and carbon surface diffusion.[33] (d) Atomistic model of a curved vertical graphene with active growing edges (highlighted in color).[15] (e) Schematic representation of VG with folded/seamless and open edges.[10] (f) TEM image of a VG nanosheet with the tapered shape. Folded edges are shown by the arrows.[15]

in the vicinity of the substrate surface.[4] Hence, the VG growth direction and spatial distribution are affected by the electric field in the plasma sheath. In the case of grounded conductive substrates, the electric field is normal to the substrate surface and is stronger near the edges and sharp points. This localized electric field above the substrate can be used to control the density and orientation of the VG networks.[32] As shown in Fig. 3b, VGs grow at a high density above the Au stripe while neither VGs nor amorphous carbon were found on the neighbouring SiO_2 surface.[32] It can be explained by noting that the electric field above the Au strip especially near the edges is much stronger than that above the SiO_2 substrate. Customizing the surface electric field distribution thus opens up a way to pattern VGs for practical devices. On the other hand, when the substrate is non-conductive or 'floating' (disconnected from an external electric circuit) in the plasma, the relatively low electric field leads to much more irregular and random VG networks.[19]

Stress effects. Internal stresses arising from the temperature gradients, ion bombardment and lattice mismatch between the substrate material and the graphitic material may cause defects or buckling in the buffer layer, which serve as nucleation sites for the VG growth.[31] The initial planar growth of 2D graphitic layers eventually switches to upward growth of 'impinging' graphene sheets, which releases the stress accumulated during the initial growth phase. The dissociated carbon species in the plasma then continuously provide radicals, ions and neutrals to open sites of vertically growing hexagonal lattices of VGs.[19,28]

Anisotropic growth effects. The directional growth of VGs could also be due to the anisotropic growth effect. It was proposed that growth rates in the directions that are parallel and perpendicular to the graphene layer were different.[12,34] Moreover, the VGs oriented normally to the substrate usually grow faster than their randomly oriented counterparts, which is partly due to the surface diffusion of carbon atoms (Fig. 3c).[34] Carbon-containing species landing on the surface of a growing nanosheet rapidly move along the sheet surface, reach the upper edge, and covalently bond to the edge atoms before desorbing from the surface. In contrast, carbon-containing species diffusing to the substrate surface can be re-evaporated because of the weak adsorption of the species to the substrate. More carbon atoms can also be preferentially directed to growing edges of VGs due to their sharp features that produce stronger localised electric fields.[4,33] As a result, the growth rate in the vertical direction is higher as compared with the lateral direction.

Recently, a kinetic model supplemented with experimental results showed that the VG growth can be considered as a step-flow process where the nucleation takes place at the bottom.[10,15] According to this model, the VG nucleation is triggered by the mismatch of graphitic carbon layers at either the buffer layer or the carbon onions that form on the surface. The growth of individual nanosheets is then determined by the number of layers nucleated from the bottom and the diffusion rate of carbon atoms to each layer (Fig. 3d). Moreover, VG growth only occurs at open edges but not at folded or seamless edges, as sketched in Fig. 3e. As the neighboring layers can form a closure and cease the growth, tapered VG nanosheets may form (Fig. 3f).

While the VG growth mechanism from gaseous precursors has received much attention, there is presently no clear explanation for the growth from liquid or solid precursors. There are some apparent similarities in the growth kinetics when VGs are grown from liquid, solid or gaseous precursors.[30] The plasma first acts on the solid or liquid natural precursors by dehydrating them as a result of the plasma-related heating. The plasma then converts or decomposes the dehydrated natural precursors into smaller, more common carbon-containing species, regardless of the initial precursor, through interactions with the plasma-generated ions and radicals. These species then act as the basic building units for the VG growth.[30] Several questions remain, e.g., why VG from different precursors exhibit different adhesion to the substrate, and what exact surface reconstructions under the plasma exposure can cause the preferential growth of VG in the vertical direction. As the understanding of growth mechanism is fundamental to the controlled growth of VGs, and consequently their device performance, more studies are warranted in this direction to harness the full potential of such a material.

3. Energy applications of vertically-oriented graphenes

3.1 VGs for ECs

ECs are advanced electrochemical devices for energy storage that requires frequent charge–discharge cycles at a high power and over a short period of time. These devices also have a much higher capacitance than regular capacitors, in part due to the use of advanced nanomaterials.[34] Here we discuss recent progress in applications of VGs and VG-based hybrids/composites in ECs.

3.1.1 VG-based active materials for EDLCs. Typical ECs operate based on the electric double-layer (EDL) mechanism, where active materials are charged for the rapid separation and surface adsorption of ions with the opposite charge. Unlike batteries that store energy through Faradaic redox reactions, EDL capacitors (EDLCs) work in a direct, electrostatic way, thereby leading to a higher power density, shorter charge–discharge cycles, and a longer lifespan. Although EDLCs are available commercially, there is still a need for the further improvement of some critical characteristics such as the mass/volume-specific capacitance, rate capability, as well as power and energy densities.

Besides the ultrathin charge separation distance (the thickness of EDL), the 'super' capacitance of EDLCs is mainly attributed to the use of active materials with a large surface area. Consequently, the characteristics of active materials play a crucial role in the performance of EDLCs. An ideal EDLC active material should fulfill the following requirements: a high electrolyte accessible surface area for the effective adsorption of ions to obtain a high capacitance and a high energy density; a suitable structure for the easy diffusion of ions to realize high rate capability; as well as the minimum resistances within the material and at the contact interface of the material/current-collector for the fast electron transport to obtain the high power density.

The unique features of VGs enable them ideally satisfy the above criteria. First, VG networks present a non-agglomerated structure with exposed edge planes, which facilitates the surface utilization for charge storage. The graphene-based EDLC electrodes are commonly fabricated from reduced graphene oxide via chemical routes, followed by assembly of these active materials on current collectors using binders. Due to van der Waals interactions and the use of binders, commonly observed restacking of horizontal graphene nanosheets leads to a considerable reduction of the available surface areas for charge adsorption and electrochemical reactions. In contrast, VG's non-agglomerated morphology leads to a higher electrochemically-accessible surface area, and thus to a higher capacitance. Meanwhile, the dense edge planes of VGs can also enhance the charge storage capability, since the edge planes have a much larger area-specific capacitance than the basal plane surface.[13]

Second, when VGs are used as active materials, the large ionic resistance associated with the distributed charge storage in porous materials can be minimized, making it possible to use EDLCs in a high-frequency mode. Electrolyte access into the pores plays a major role in EDLC's rate capability and frequency response. Considerable ionic resistance is formed when the pores are small and tortuous, thus leading to a poor capacitive behavior, especially at relatively high frequencies.[13] This problem not only exists for activated carbons (ACs, the most common porous active materials for commercial EDLCs), but also for horizontal graphene stacks where pores mainly originate from the 2D inter-layer spacings. As for VGs (Fig. 4a), the vertical orientation and open inter-sheet channels can facilitate the ion migration between the layers and minimize the undesirable porosity effects especially at high frequencies.

Third, the EDLC series resistance can be reduced and the power and rate capabilities can be significantly enhanced by using VGs. The vertical orientation with the intrinsic good in-plane electronic conductivity of graphenes facilitates the charge transport within active materials. Meanwhile, the direct growth of VGs without a commonly used binder can reduce the contact resistance between active materials and the current collector. As such, series resistance as low as 0.05 Ω has been achieved in the VG-based EDLCs.[13]

A significant feature of VG-based EDLCs is the ultrafast dynamic response. In particular, the 120 Hz alternating-current (ac) line-filtering has become possible.[13] The smooth transition from 120 Hz ac to direct-current (dc) is required for line-powered electronics and relies on traditional electrolytic capacitors. Although EDLCs can provide a much higher specific capacitance, most devices made from porous materials (e.g., ACs and horizontal graphene stacks) fail to show the capacitive behavior at relatively high frequencies, which is due to the porosity effect discussed above.[13] As shown in Fig. 4b, AC-based EDLC features an impedance phase angle of $\sim 0°$ at 120 Hz, and behaves like a resistor. In contrast, the VG-based EDLC showed an impedance phase angle of $-82°$ at the same frequencies, which is close to the ideal capacitive behavior ($-90°$) and is suitable for the 120 Hz ac line-filtering application. Meanwhile, compared with the Al electrolytic capacitor owning an impedance phase angle of $-83°$ at 120 Hz, the VG-based EDLC presented a much higher volumetric energy density, which allows size reductions of the filtering system. The 0.6 μm thick VG layer stored ~ 1.5 and ~ 5.5 F V cm^{-3} with aqueous and organic electrolytes, respectively, significantly higher than that of an Al electrolytic capacitor (~ 0.14 F V cm^{-3}).[13]

Further, kilohertz ultrafast EDLCs were also reported with VGs grown on the nickel foam current collectors.[24] The use of a foam-type current collector instead of a foil-type counterpart

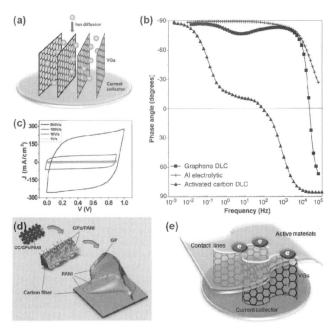

Fig. 4 Recent applications of VGs in ECs. (a) Schematic of ion diffusion within VGs. (b) Impedance phase angle data of a VG-based EDLC (labeled as 'Graphene DLC' in the figure), an AC-based EDLC (labeled as 'Activated carbon DLC' in the figure), and an Al electrolytic capacitor.[13] (c) Cyclic voltammetry curves of a VG-based EDLC electrode (VG grown on a nickel foam) at scan rates of 1, 10, 100, and 500 V s^{-1}.[24] (d) A hierarchical electrode composed of CC, VG (labeled as 'GP' in the figure), and PANI, for pseudo-capacitors.[35] (e) VG bridges connecting active materials and a current collector increase contact area and enhance charge transport.[9]

can lead to a higher mass loading of active materials, as VGs can fully cover the 3D metallic scaffold. For thin-film ECs, areal and volumetric capacitances are important metrics to evaluate their performance.[36] The proposed VG-based EDLCs exhibited a capacitance of ~0.32 mF cm^{-2} at 1 kHz, higher than any previously reported values of EDLCs at the same frequency.[24] It is noted that reports on most EDLC studies are conducted using the highest scan rate of 1 V s^{-1}.[24] As shown in Fig. 4c, with an increasing cyclic voltammetry (CV) scan rate from 1 V s^{-1} to 500 V s^{-1}, the shape of the curves remained quasi-rectangular (one of the indicators of the EDLC mode operation). These results demonstrate the high rate performance of VG-based EDLCs.

The capacitive behaviours of VGs can be tailored by tuning their morphology and structure in the PECVD growth, which could be realized through adjusting growth precursors, plasma sources, and growth parameters. For example, the graphitization degree and the density of edge planes strongly affect their electrochemical performance.[37] Specifically, thinner edge planes lead to a higher specific capacitance,[37] arising from the much larger area-specific capacitance of edge planes than basal surface planes.[13] Meanwhile, a higher sp^2 content also improves the charge storage capability, since the sp^3-bonded carbon only increases the charge transfer resistance and contributes little to the charge storage.[37] The optimized specific capacitance of VG-based EDLCs could reach a high value of 230 F g^{-1} (or ~23 mF cm^{-2}) at a CV scan rate of 10 mV s^{-1}.[37] The specific capacitance can be improved by using the VG-based hybrid structures, e.g., CNT-on-VG.[38] Such a combination of 1D and 2D nanostructures can further increase the surface area and enhance the electron transport within active materials, leading to a high specific capacitance (278 F g^{-1} or ~36 mF cm^{-2} at 10 mV s^{-1}) and good cycling stability (capacitance retention of >99% after 8000 charge–discharge cycles).[38]

3.1.2 VG-based active materials for pseudo-capacitors. In contrast to EDLCs based on physical adsorption of ions, pseudo-capacitors (also known as redox ECs) rely on pseudo-capacitance derived from reversible Faradaic-type charge transfer in electrodes.

Due to the presence of Faradaic redox reactions, pseudo-capacitors usually have a higher capacitance but a lower power density and poorer cycling stability compared with EDLC devices.[34] By combining merits of VGs and transition metal oxides or electrically conducting polymers (herein referred to as pseudo-species, enabling repeated Faradaic-type reactions), high performance pseudo-capacitors can be realized. VGs with both the high specific surface area and the high electrical conductivity can synergistically enhance the electrochemical performance of pseudo-species. Indeed, VGs increase the specific loading of the pseudo-species for a higher energy density, enhance the charge transport between the pseudo-species and the substrate for higher power and rate capabilities, and also improve the adhesion of the pseudo-species for the enhanced cycling stability.

As an example, electrochemical properties of pseudo-capacitors based on hybrid MnO_2–VG nano-architectures can compete with those of EDLCs.[37] VGs grown on a nickel foil were used as conductive templates for the deposition of MnO_2 nanoflowers. The exposed surface and high conductivity of VGs can enhance electrochemical properties of MnO_2. At a CV scan rate of 10 mV s^{-1}, the MnO_2/VG electrode presented a high specific capacitance of 1060 F g^{-1} (calculated based on the mass of MnO_2). Moreover, the MnO_2/VG electrode exhibited a remarkable capacitance retention (>97%) after 1000 cycles, which can be attributed to the strong adhesion between the MnO_2 and VGs.[37] Based on density functional theory calculations, the vertical orientation and sharp edge planes of VGs facilitate the ion diffusion with low energy barriers, while the covalent bonding between MnO_2 and graphene leads to the effective charge transfer.[39] Similar applications of VGs decorated with MnO_2 of diverse morphologies[39] and other transition metal oxides (such as NiO)[25] have also been reported. These studies demonstrate the critical roles of VGs in the reduction of internal resistance, the enhancement of specific capacitance, and the improvement of cyclic stability.

Hierarchical electrodes (Fig. 4d) composed of carbon cloth, VGs, and electrically conducting polyaniline (PANI) can further improve the EC performance.[35] The specific surface area of the carbon cloth, which served as a flexible and open scaffold, increased by a factor of ~3 with the decoration of VGs. At a scan rate of 2 mV s^{-1}, the mass-specific capacitance of this hierarchical electrode (referred to as CC/VGs/PANI) was also ~3 times higher compared with that of the CC/PANI electrode. The CC/VGs/PANI electrode presented a mass specific capacitance of 2000 F g^{-1} and an area-normalized specific capacitance of 2.6 F cm^{-2}. The presence of VGs also enhanced the electron transport, leading to energy and power densities higher than the previously reported values for PANI-based counterparts. A high-performance all-solid-state flexible EC based on the CC/VGs/PANI electrode and the PVA–H_2SO_4 polymer gel electrolyte was also demonstrated.[35] The EC presented an energy density ~10 times higher than that of a commercial 3.5 V/25 mF EC and comparable to the upper range of that of a 4 V lithium thin-film battery, yet the power density of which was ~2 orders of magnitude higher than that of the lithium thin-film battery.

3.1.3 VGs for bridging active materials and current collectors.

VGs are also used as bridges connecting active materials and current collectors in ECs for the fast transport of electrons. Due to the surface roughness, active materials and the current collector meet at a finite number of contact points, which induces a considerable contact resistance at the interface. This contact resistance can be reduced by VG bridges between the current collector and the active materials, as shown in Fig. 4e, leading to ECs with a high rate performance and a high power density.[9] Even with a conventional graphene film used as an active material in this proof-of-concept device, the VG-bridged ECs outperformed nearly all of the previously reported counterparts. This was evidenced by a capacitance retention of ~90% when the CV scan rate increased from 20 to 1000 mV s^{-1} or the galvanostatic charge–discharge current density increased from 1 to 100 A g^{-1}.

The exposed edges of VGs can provide dense contact points with active materials, thereby reducing the contact resistance. Meanwhile, due to the high in-plane electrical conductivity of graphene nanosheets, the vertical orientation of VGs enables the high-quality electrical contact at the interface between the current collector and the active material. Otherwise, inserting horizontal graphene films instead of VGs may cause a negative effect on the rate capability due to the increase of the internal resistance between planar graphene sheets.[9] This type of application of VGs could also be suitable for other electrochemical energy storage and conversion devices to advance their performance.

3.2 VGs for other energy applications

The unique features of VGs also make them suitable for other energy storage and conversion applications such as rechargeable batteries, fuel cells, and solar cells.

Rechargeable batteries. As one of the most common rechargeable batteries, lithium-ion batteries (LIBs) store charges through the reversible insertion–extraction of lithium ions between redox-active host materials (i.e., anode and cathode).[40] During charge–discharge processes, lithium ions are inserted into/extracted from the anode materials. Consequently, high reversible lithium storage capability and good cycling stability are desirable for the anode materials. VGs directly grown on bare[27] or graphene-coated[41] metal foil current collectors are promising anode materials for LIBs. VGs with the exposed graphene surface and edge planes can provide numerous sites for the capture of Li^+ ions. Meanwhile, the open inter-sheet channels, vertical alignment, and good electrical connection to the substrate of VGs can significantly reduce the transport resistance of Li^+ ions, the intrinsic resistance within the anode materials, and the contact resistance between the anode materials and the current collector, respectively. This is why LIBs employing VGs as anode materials present a very high reversible lithium storage capacity and good cycling stability.[27,41] Moreover, the combination of VG and lithium alloying materials (such as GeO_x) further improves the lithium capture.[17] As shown in Fig. 5a, VGs could work as fast electron transport channels and ensure smooth lithium diffusion pathways in a VG@GeO_x sandwich nanostructured anode. The VG@GeO_x anode presented a stable capacity of 1008 mA h g^{-1} at 0.5 C (retention of 96% capacity after 100 cycles), a capacity of 545 mA h g^{-1} at a high rate of 15 C, and a capacity retention of 92% when the rate recovered to 0.5 C (Fig. 5b). These characteristics are considered very competitive compared with other lithium alloying material-based LIBs.[17]

VGs are also used as the electrodes in vanadium redox flow batteries (VRFBs),[42] which employ active vanadium species in different oxidation states for electrochemical energy storage. The electrodes in VRFBs are used as the support for vanadium reactions, and a high surface area with active sites and a high electrical conductivity are desirable features. The dense exposed edge planes of VGs with oxygen-based functional groups act as

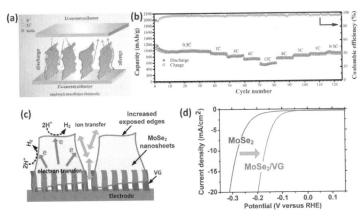

Fig. 5 (a) Schematic of the Li ion diffusion mechanism in the VG@GeO$_x$ sandwich nanoflakes-based electrode.[17] (b) Performance of the VG@GeO$_x$ electrode (70% GeO$_x$) at the rate of 0.5 C, 1 C, 3 C, 6 C, and 15 C.[17] 1 C rate means that at the discharge current the battery can discharge completely in one hour. For example, for a battery with a capacity of 10 A h, 1 C equates to a discharge current of 10 A. (c) Schematic of hydrogen evolution reaction and electron transport between the perpendicularly-oriented MoSe$_2$ nanosheets, VGs, and the electrode.[44] (d) Improved hydrogen evolution catalytic activity of the MoSe$_2$ nanosheets with a VG support.[44] The electron transfer at the electrode interface is greatly promoted through the highly-conductive VG, which smoothly bridges MoSe$_2$ nanosheets and the current collector due to the *in situ* growth of both VG and MoSe$_2$ nanosheets.

active sites for vanadium reactions. Moreover, the interconnected VG networks perpendicular to the substrate can facilitate the charge transfer. As a result, VG-based VRFB electrodes showed high performance in terms of low overpotential, high peak current density, fast electron transfer kinetics of V$^{4+/5+}$, and favourable long-term stability.[42]

Fuel cells. Fuel cells can convert chemical energy from fuels (*e.g.*, methanol and hydrogen) into electricity through a chemical reaction with oxygen or other oxidizing agents. In a typical fuel cell system, chemical reactions occur at the interfaces of three different segments, *i.e.*, the anode, the electrolyte, and the cathode. At the anode, a catalyst is normally employed to assist the oxidation of fuels and thus the catalytic behaviour strongly influences the fuel cell performance.

The unique features of VGs make them an ideal catalyst support for such anode reactions. The non-agglomerated morphology and open channels of VGs can enhance the deposition and dispersion of catalysts.[43] Moreover, along the in-plane direction of graphene nanosheets, fast electron transport between reaction sites and the current collector can be realized. Specifically for the methanol oxidation reaction in direct-methanol fuel cells, Pt nanoparticles supported by VGs exhibited the improved electrocatalytic performance in activity and stability than the counterpart employing vertically-aligned carbon nanofibers as the support.[16]

Moreover, transition metal dichalcogenides supported by VGs have also been demonstrated as effective catalysts for fuel generation. Electrolysis of water is one of the major sources to produce hydrogen. VG supported MoSe$_2$ nanosheets showed a greatly improved catalytic activity for hydrogen evolution reaction compared with bare MoSe$_2$ nanosheets (Fig. 5c).[44] A remarkable positive shift of the onset potential was found in the polarization curves of the catalysts, confirming that the hydrogen evolution reactions were catalyzed at a lower overpotential with the MoSe$_2$/VG catalysts (Fig. 5d).

Solar cells. Besides working as the catalyst support, VGs with a controlled number of oxygen functional groups can work as catalysts themselves for dye-sensitized solar cells (DSSCs). DSSCs use light-absorbing dye molecules to generate electricity from sunlight. In a typical DSSC operation, I$_3^-$ is reduced at the counter electrode for dye regeneration. Conventional Pt catalysts possess a high catalytic activity for I$_3^-$ reduction but suffer from high cost. Because of the large surface area and a low cost, VGs with oxygen functional groups (catalytic sites) were demonstrated as promising substitutes for Pt in the counter electrode of DSSCs.[8,45] Importantly, a VG-based DSSC counter electrode exhibited a charge transfer resistance of about 1% of the Pt electrode and has improved the power conversion efficiency of the cell.[45]

4. Environmental applications of vertically-oriented graphenes

4.1 VGs for biosensors and gas sensors

Graphenes have been used in various health and environmental applications including bio- and gas sensors.[2,46] Sensitive and selective detection of proteins, DNA and bacteria as well as gases, *e.g.*, NO$_2$, CO, and H$_2$, plays a critical role in improving environment (*e.g.*, water and air) and public health. Recently, VGs and VG/nanoparticle-based composites have been explored in these devices due to their unique structure and properties. In this section, representative VG-based sensors are discussed.

4.1.1 VGs for biosensors.
Depending on the specific working principle, graphene-based biosensors either use their electrical properties (*e.g.*, high carrier mobility), electrochemical properties

(e.g., high catalytic activity and electron transfer rates), or unique structure (e.g., atom-layer thickness and a high surface-to-volume ratio) for biomolecule detection. Different from conventional graphenes, VGs draw increasing attention in biosensor applications due to their unique vertical orientation and open structure. VGs are high-performing sensing materials as their surface can be fully accessible by analytes, and the high length of exposed edges and the high in-plane carrier mobility lead to superior sensor performance. A field-effect transistor (FET) biosensor can be fabricated by direct growth of VGs on the sensor electrodes, as shown in Fig. 6a and b.[47] The sensor contains patterned metal electrodes, VGs (sensing material), and gold nanoparticle (Au NP)–antibody conjugates (as a probe for analyte protein binding). When an analyte protein binds to the antibody, it causes a change in the electrical conductance of the VG nanosheet that can be measured by the external circuit/measuring system (Fig. 6c). The VGs showed p-type semiconducting characteristics in an ambient environment, which are quite similar to the characteristics of graphene or reduced graphene oxide (RGO). The sensor has a high sensitivity (down to 2 ng ml^{-1} or 13 pM for IgG antigens) and a fast response (on the order of seconds) to target proteins. By comparing the VG sensor with conventional flat graphene-based electronic sensors, the vertical orientation and open structure of VG increase the accessible area of the device to analytes, thereby increasing the sensitivity of the sensor. Moreover, direct PECVD growth of VGs on sensor electrodes could achieve higher stability and reproducibility than the drop-casting method that is commonly used in graphene biosensor fabrication, which is attributed to the stronger binding between the VG and the sensor electrode as compared with the drop-casting deposition.

In addition to electronic sensors based on the conductivity change in the sensing element, VG-based electrochemical sensors were also reported for biomolecule detection. This type of sensor uses the electrocatalytic activity of the sensing electrode in a redox system for analyte detection. Typically, the sensor records reaction fingerprints (e.g., oxidation or reduction peaks) of the analyte in the CV measurements and the intensity of current peaks can be related to the analyte concentration. Fig. 6d shows the CV curves of a VG-based electrochemical sensor for the simultaneous detection of dopamine (DA), ascorbic acid (AA), and uric acid (UA).[48] In particular, the detection of DA with various concentrations (1 to 100 μM) was demonstrated in the presence of common interfering agents of AA and UA, as shown in Fig. 6e, indicating high sensitivity and selectivity of the sensor. The exceptional sensor performance is attributed to the enhanced electron transfer from the VG electrode. In general, high electronic density of states (DOS) in the electrode leads to the increased electron transfer in a redox system.[48] Compared with the basal planes of graphene with a low DOS at the Fermi level, defects (such as kinks, steps, vacancies, etc.) on the edges of VGs can produce localized edge states between the conduction and the valence bands, resulting in a high DOS near the Fermi level. The investigation into the stability of the VG electrode showed that the morphology of VGs remained unchanged after long-term cycling. This research confirms the mechanical robustness of VGs and their good electrochemical stability.

VG-based electrochemical sensors were also used for DNA detection. For example, a highly-sensitive electrochemical biosensor with VGs was demonstrated for the detection of four bases of DNA (G, A, T, and C) by monitoring oxidation signals of individual nucleotide bases.[49] The VG electrode was able to detect a wide concentration (0.1 fM to 10 mM) of double-stranded DNA (dsDNA). The wide dynamic response window of the VG electrode at such high and low concentrations was attributed to the high surface porosity (open space in the VG network) and edge defects of VGs, which inhibited the electrode

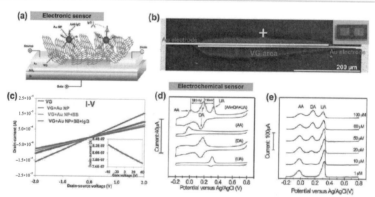

Fig. 6 (a) Schematic of the VG electronic sensor. The binding of the analyte protein (IgG) to the antibody on the VG causes a change in the electrical conductance of the sensor. (b) SEM image of a sensor fabricated by direct growth of VG between gold electrodes. Inset is a digital image of the sensor electrode. (c) Sensor conductivity changes with probe antibody labeling and analyte protein detection.[47] Inset is the FET transport characteristics of the VG sensor (d) CV profiles of the VG-based electrochemical sensor in the solution of 50 mM, pH 7.0 PBS with individual 1 mM AA, 0.1 mM DA, and 0.1 mM UA, and their mixtures. Each analyte has a characteristic pair of redox peaks in the CV and the mixture shows distinct peaks for each analyte. (e) Differential pulse voltammetric profiles of the VG electrode with 1 mM AA, 0.1 mM UA and different concentrations of DA from 1 to 100 μM. The DA peak current increasing with the DA concentration.[48]

fouling and accelerated the electron transfer between the electrode and dsDNA, respectively.

4.1.2 VGs for gas sensors. Air pollutants, *e.g.*, sulfur oxides, nitrogen oxides, carbon monoxide, ammonia, volatile organic compounds and particulates, are significant risk factors for a number of health conditions. The real-time monitoring of the air condition is critical for the reduction of harmful effects from air pollutants to humans; this monitoring generally relies on a gas sensor to detect specific species. Graphene/RGO-based materials have been widely studied for electronic gas sensor applications due to their large specific surface areas and their high sensitivity to electronic perturbations upon gas molecule adsorption. Micro-sized sensors made from graphenes are able to detect individual gas molecules, which change the local carrier concentration in graphene sheets. The gas-induced changes in conductivity have different magnitudes for different gases, and the sign of the change (increase or decrease in conductivity) indicates whether the gas is an electron acceptor (*e.g.*, NO_2) or an electron donor (*e.g.*, NH_3 and H_2). Similar to VG-based biosensors, the vertical and open structures of VGs offer large accessible surface areas for gas molecule adsorption and inhibit the agglomeration of graphene sheets during the sensor device fabrication.

The electric field distribution above the substrate guides the VG growth by PECVD and can be used for area-specific synthesis of VG-based FET gas sensors.[32] Due to the enhanced electrical field, VG sheets could be selectively grown on a gold sensor electrode with different patterns. The VG-based sensor was tested with three consecutive steps that include exposure of the device to an air flow to record a base value of the sensor resistance, then to the analyte gas to register a sensing signal, and subsequently to an air flow for sensor recovery. The VG based sensor can be operated at room temperature for NO_2 (100 ppm) and NH_3 (1%) detection (Fig. 7a). The sensitivities, defined as ratios of R_{air}/R_{NO_2} and R_{NH_3}/R_{air} (R_{air}, R_{NO_2}, and R_{NH_3} are the sensor resistances in air, in NO_2, and in NH_3, respectively) of the VG sensor to NO_2 (1.57) and NH_3 (1.13), were comparable with multilayer graphene-based devices. The VGs behaved like a p-type semiconductor in an ambient environment, which was confirmed by the measurement of FET transport characteristics shown in Fig. 7b. The sensing mechanism is based on the adsorbed NH_3 donating electrons and neutralizing holes, which decreases the VG conductance. On the other hand, NO_2 accepts electrons from VGs, which leads to the increased VG conductance. The VG sensor is promising for gas sensing as the device is suitable for large-scale fabrication and has a better stability than sensors fabricated by other methods such as drop-casting of RGO sheets.

Theoretical studies have shown that the graphene-gas molecule adsorption is strongly dependent on the graphene structure and the molecular adsorption configuration.[50] Gas molecules have much stronger adsorption on the doped or defective graphenes than that on the pristine graphene. Therefore, high adsorption

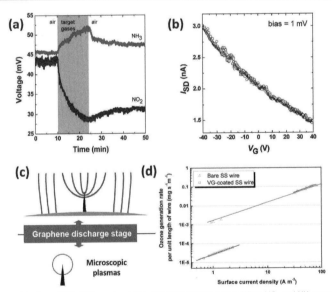

Fig. 7 (a) Room-temperature detection of NO_2 (100 ppm) and NH_3 (1%) using the VG sensor. When NO_2 and NH_3 gases are injected in the testing chamber, the sensor conductivity shows significant changes.[32] When NO_2 and NH_3 gases are shut off and an air flow is introduced, the sensor conductivity begins to recover. (b) The FET transport characteristics of a VG sensor. The sensor conductivity gradually decreases when the gate voltage changes from −40 to +40 V, showing that the VGs in the sensor behave like a p-type semiconductor.[32] (c) Schematic of electric field lines and the plasma region for the VG-based corona discharge.[20] (d) Coating of SS wires with VGs decreases the rates of ozone generation in the microscopic corona-type discharges by more than two orders of magnitude at a surface current density of <2 A m^{-2}.[20]

energy is expected for VGs with gas molecules due to a large number of defects, edges and a curved morphology of the sheets, which lead to higher sensitivities of the sensor compared with horizontal graphene-based devices. The demonstrated biosensor and gas sensor applications of VGs rely on the unique properties arising from the vertical orientation of graphene sheets. In addition, the sensing performance could be further improved by decreasing the thickness of VG sheets or engineering the VG structure with physical/chemical modifications.

4.2 VGs for other environmental applications

Corona discharge is a localized breakdown phenomenon in gases, which is typically created by an asymmetric electrode pair (*e.g.*, pin-to-plate and wire-to-plate). Corona discharges employing a microsized metallic wire as the discharge electrode are widely used for indoor electrostatic devices (*e.g.*, photocopiers and printers). However, ozone is emitted as a hazardous byproduct that poses serious health hazards to the human respiratory system.

A more health-benign corona discharge employing a VG-coated stainless steel (SS) wire as the discharge electrode was demonstrated.[20] Due to the good electrical conductivity and a high aspect ratio, the electric field near the VG edges can easily reach a critical value for electric breakdown, as shown in Fig. 7c. As a consequence, a VG-based corona discharge can be initiated and operated at a much lower voltage as compared with common metal electrodes. More importantly, for a given surface current density (<2 A m^{-2}), the ozone generation rate per unit length of the wire in VG-based corona discharges is only 1% of the amount of ozone produced using the micro-sized SS wire (Fig. 7d). Ozone generation in atmospheric-pressure corona discharges is affected by the size of the plasma region. Ionization is very effective in the plasma region, while electrons outside the plasma region are not energetic enough to generate ozone. Consequently, VG networks with ultrathin edges can generate microscopic corona discharges, further leading to much lower ozone emission rates.

5. Conclusions and outlook

Plasma-based approaches enable the direct synthesis of VGs on various substrates, employing gaseous, liquid, or solid precursors. In contrast to the conventional horizontal graphene stacks obtained by various chemical or physical routes, VGs show many unique features such as vertical orientation, non-agglomerated morphology, a high surface-to-volume ratio, and exposed sharp edges. Combined with the inherent good electrical, chemical, and mechanical properties of graphenes, VGs show great potential and several advantages in emerging applications ranging from energy storage (ECs and rechargeable batteries), energy conversion (fuel cells and solar cells), sensing (biosensors and gas sensors), to green corona discharges for pollution control (a schematic overview of VGs' unique properties and their energy and environmental applications is given in the ESI†).

Although many proof-of-concept studies have demonstrated VG's superior performance in a wide range of energy and environmental devices, there is no doubt that the potential of VGs for such applications has not been fully exploited. For example, the specific capacitance of VG-based EC electrodes still requires further improvement. The FET biosensors and gas sensors have limitations arising from the intrinsic electrical properties of VGs, *i.e.*, a very narrow bandgap and a low on–off current ratio. In addition, for practical purposes, critical factors such as reliability, selectivity, and robustness of sensors still require comprehensive assessments according to industrial standards.

A number of further studies are warranted to better utilize the VGs' unique features. The growth processes require a more precise control on height, lateral size, density, and crystalline structure of VGs through the adjustment of plasma parameters and operation conditions. Meanwhile, a more profound understanding of the VG growth mechanisms through advanced *in situ* characterization still remains a major challenge. For real-world applications, a better understanding of the functional performance of VGs is essential for the optimum material design/fabrication, performance optimization, as well as the development of scale-up processes of technological relevance. Examples of such mechanisms include liquid flow/mass transport in nanoscale space and charge storage at the interface of electrolyte/VGs for ECs, and electron transport during the interactions of VGs with bio- or gas molecules. Moreover, surface engineering and doping of VGs could also enhance the VG device performance.

The unique structures and intrinsic properties of VGs greatly expand the use of conventional graphene sheets, and more applications in diverse fields are anticipated in the near future. Finally, VG's outstanding performance will encourage and inspire additional studies on the exploration and use of other vertically-oriented 2D nanostructures.

Acknowledgements

Financial support from US National Science Foundation (EECS-1001039 and IIP-1128158), National Natural Science Foundation of China (No. 51306159), Research Growth Initiative Program of UWM, Australian Research Council (ARC) and CSIRO Science Leadership Program is acknowledged.

References

1. S. Guo and S. Dong, *Chem. Soc. Rev.*, 2011, **40**, 2644–2672.
2. Y. Liu, X. Dong and P. Chen, *Chem. Soc. Rev.*, 2012, **41**, 2283–2307.
3. J. Liu, G. Cao, Z. Yang, D. Wang, D. Dubois, X. Zhou, G. L. Graff, L. R. Pederson and J.-G. Zhang, *ChemSusChem*, 2008, **1**, 676–697.
4. K. Ostrikov, E. C. Neyts and M. Meyyappan, *Adv. Phys.*, 2013, **62**, 113–224.
5. A. N. Obraztsov, I. Y. Pavlovsky, A. P. Volkov, A. S. Petrov, V. I. Petrov, E. V. Rakova and V. V. Roddatis, *Diamond Relat. Mater.*, 1999, **8**, 814–819.

6 J. J. Wang, M. Y. Zhu, R. A. Outlaw, X. Zhao, D. M. Manos, B. C. Holloway and V. P. Mammana, *Appl. Phys. Lett.*, 2004, **85**, 1265–1267.
7 Y. Wu, P. Qiao, T. Chong and Z. Shen, *Adv. Mater.*, 2002, **14**, 64–67.
8 C. Yang, H. Bi, D. Wan, F. Huang, X. Xie and M. Jiang, *J. Mater. Chem. A*, 2013, **1**, 770–775.
9 Z. Bo, W. Zhu, W. Ma, Z. Wen, X. Shuai, J. Chen, J. Yan, Z. Wang, K. Cen and X. Feng, *Adv. Mater.*, 2013, **25**, 5799–5806.
10 K. Davami, M. Shaygan, N. Kheirabi, J. Zhao, D. A. Kovalenko, M. H. Rümmeli, J. Opitz, G. Cuniberti, J.-S. Lee and M. Meyyappan, *Carbon*, 2014, **72**, 372–380.
11 M. Hiramatsu, K. Shiji, H. Amano and M. Hori, *Appl. Phys. Lett.*, 2004, **84**, 4708–4710.
12 Z. Bo, K. Yu, G. Lu, P. Wang, S. Mao and J. Chen, *Carbon*, 2011, **49**, 1849–1858.
13 J. R. Miller, R. A. Outlaw and B. C. Holloway, *Science*, 2010, **329**, 1637–1639.
14 Z. Bo, Y. Yang, J. Chen, K. Yu, J. Yan and K. Cen, *Nanoscale*, 2013, **5**, 5180–5204.
15 J. Zhao, M. Shaygan, J. Eckert, M. Meyyappan and M. H. Rümmeli, *Nano Lett.*, 2014, **14**, 3064–3071.
16 C. Zhang, J. Hu, X. Wang, X. Zhang, H. Toyoda, M. Nagatsu and Y. Meng, *Carbon*, 2012, **50**, 3731–3738.
17 S. Jin, N. Li, H. Cui and C. Wang, *Nano Energy*, 2013, **2**, 1128–1136.
18 K. Yu, G. Lu, Z. Bo, S. Mao and J. Chen, *J. Phys. Chem. Lett.*, 2011, **2**, 1556–1562.
19 M. Cai, R. A. Outlaw, S. M. Butler and J. R. Miller, *Carbon*, 2012, **50**, 5481–5488.
20 Z. Bo, K. Yu, G. Lu, S. Cui, S. Mao and J. Chen, *Energy Environ. Sci.*, 2011, **4**, 2525–2528.
21 Y. Ando, X. Zhao and M. Ohkohchi, *Carbon*, 1997, **35**, 153–158.
22 Y. Yoon, K. Lee, S. Kwon, S. Seo, H. Yoo, S. Kim, Y. Shin, Y. Park, D. Kim, J. Y. Choi and H. Lee, *ACS Nano*, 2014, **8**, 4580–4590.
23 S. Wang, J. Wang, P. Miraldo, M. Zhu, R. Outlaw, K. Hou, X. Zhao, B. C. Holloway, D. Manos, T. Tyler, O. Shenderova, M. Ray, J. Dalton and G. McGuire, *Appl. Phys. Lett.*, 2006, **89**, 183103.
24 G. Ren, X. Pan, S. Bayne and Z. Fan, *Carbon*, 2014, **71**, 94–101.
25 H.-C. Chang, H.-Y. Chang, W.-J. Su, K.-Y. Lee and W.-C. Shih, *Appl. Surf. Sci.*, 2012, **258**, 8599–8602.
26 C. B. Parker, A. S. Raut, B. Brown, B. R. Stoner and J. T. Glass, *J. Mater. Res.*, 2012, **27**, 1046–1053.
27 X. Xiao, P. Liu, J. S. Wang, M. W. Verbrugge and M. P. Balogh, *Electrochem. Commun.*, 2011, **13**, 209–212.
28 M. Cai, R. A. Outlaw, R. A. Quinlan, D. Premathilake, S. M. Butler and J. R. Miller, *ACS Nano*, 2014, **8**, 5873–5882.
29 Z. Sun, Z. Yan, J. Yao, E. Beitler, Y. Zhu and J. M. Tour, *Nature*, 2010, **468**, 549–552.
30 D. H. Seo, A. E. Rider, Z. J. Han, S. Kumar and K. Ostrikov, *Adv. Mater.*, 2013, **25**, 5638–5642.
31 A. Malesevic, R. Vitchev, K. Schouteden, A. Volodin, L. Zhang, G. Van Tendeloo, A. Vanhulsel and C. Van Haesendonck, *Nanotechnology*, 2008, **19**, 305604.
32 K. Yu, P. Wang, G. Lu, K.-H. Chen, Z. Bo and J. Chen, *J. Phys. Chem. Lett.*, 2011, **2**, 537–542.
33 M. Zhu, J. Wang, B. C. Holloway, R. A. Outlaw, X. Zhao, K. Hou, V. Shutthanandan and D. M. Manos, *Carbon*, 2007, **45**, 2229–2234.
34 L. L. Zhang and X. S. Zhao, *Chem. Soc. Rev.*, 2009, **38**, 2520–2531.
35 G. Xiong, C. Meng, R. G. Reifenberger, P. P. Irazoqui and T. S. Fisher, *Adv. Energy Mater.*, 2014, **4**, 1300515.
36 Y. Gogotsi and P. Simon, *Science*, 2011, **334**, 917–918.
37 D. H. Seo, Z. J. Han, S. Kumar and K. Ostrikov, *Adv. Energy Mater.*, 2013, **3**, 1316–1323.
38 D. H. Seo, S. Yick, Z. J. Han, J. H. Fang and K. Ostrikov, *ChemSusChem*, 2014, **7**, 2317–2324.
39 G. Xiong, K. P. S. S. Hembram, R. G. Reifenberger and T. S. Fisher, *J. Power Sources*, 2013, **227**, 254–259.
40 A. D. Roberts, X. Li and H. Zhang, *Chem. Soc. Rev.*, 2014, **43**, 4341–4356.
41 H. Kim, Z. Wen, K. Yu, O. Mao and J. Chen, *J. Mater. Chem.*, 2012, **22**, 15514–15518.
42 Z. Gonzalez, S. Vizireanu, G. Dinescu, C. Blanco and R. Santamaria, *Nano Energy*, 2012, **1**, 833–839.
43 Z. Bo, D. Hu, J. Kong, J. Yan and K. Cen, *J. Power Sources*, 2015, **273**, 530–537.
44 S. Mao, Z. Wen, S. Ci, X. Guo, K. Ostrikov and J. Chen, *Small*, 2015, **11**, 414–419.
45 K. Yu, Z. Wen, H. Pu, G. Lu, Z. Bo, H. Kim, Y. Qian, E. Andrew, S. Mao and J. Chen, *J. Mater. Chem. A*, 2013, **1**, 188–193.
46 S. Mao, G. Lu and J. Chen, *J. Mater. Chem. A*, 2014, **2**, 5573–5579.
47 S. Mao, K. Yu, J. Chang, D. A. Steeber, L. E. Ocola and J. Chen, *Sci. Rep.*, 2013, **3**, 1696.
48 N. G. Shang, P. Papakonstantinou, M. McMullan, M. Chu, A. Stamboulis, A. Potenza, S. S. Dhesi and H. Marchetto, *Adv. Funct. Mater.*, 2008, **18**, 3506–3514.
49 O. Akhavan, E. Ghaderi and R. Rahighi, *ACS Nano*, 2012, **6**, 2904–2916.
50 Y.-H. Zhang, Y.-B. Chen, K.-G. Zhou, C.-H. Liu, J. Zeng, H.-L. Zhang and Y. Peng, *Nanotechnology*, 2009, **20**, 185504.

垂直取向石墨烯在能源与环境领域应用的最新进展

摘要：垂直于衬底表面排列的石墨烯纳米片，即垂直取向石墨烯（VGs），具有许多可以导致激发性质独特的形态和结构特征。等离子强化化学气相沉积可以使用气体、液体或固体前体在各种基底上生长 VGs。与传统的随机取向石墨烯相比，VGs 在衬底上的垂直取向、无团聚形态学、控制片间连通性，以及锋利和暴露的边缘使它们非常有前景的各种应用。本文聚焦等离子体激发的简单性高效合成 VGs 及其带来的新兴能源和环境应用特性，从能量存储、能量转换、传感器，到绿色电晕放电用于污染控制。

原文刊于 Chemical Society Reviews, 2015, 44: 2108-2121

Formation, Measurement, and Control of Dioxins from the Incineration of Municipal Solid Wastes: Recent Advances and Perspectives

Yaqi Peng, Shengyong Lu,* Xiaodong Li, Jianhua Yan, and Kefan Cen

ABSTRACT: Polychlorinated dibenzo-*p*-dioxins (PCDDs) and polychlorinated dibenzofurans (PCDFs) consist of 210 kinds of pollutants, 17 of which are highly toxic to humans. PCDD/F is mainly produced during combustion and incineration processes. This review focuses on the formation, measurement, and control of dioxins in municipal solid waste incineration. Typical PCDD/F emission concentrations and air pollutant control devices are introduced in the work. The formation mechanisms of PCDD/F are divided into homogeneous synthesis and heterogeneous synthesis, and some previous reviews and the latest work are introduced. Considering the complexity, high cost, and time lag of traditional offline detection methods based on high-resolution gas chromatography coupled with high-resolution mass spectrometry, fast and indirect measurement techniques have been recently developed, among which resonance-enhanced multiphoton ionization-time-of-flight mass spectrometry is a good option. The study of online/at-line in situ measurements and the correlation between PCDD/F and the indicator are reviewed in detail. Due to the strict policies regarding pollutants' emissions, state-of-the-art remediation technologies are used to reduce dioxin formation and emission from incineration. A comprehensive discussion about the control of dioxin emission is also reviewed in this work. Finally, challenges, future directions, and perspectives are proposed.

1. INTRODUCTION

1.1. Dioxins in Chemical Synthesis. PCDD/F is commonly used to designate polychlorinated dibenzo-*p*-dioxins and polychlorinated dibenzofurans, which are referred to as dioxins. The structure of PCDD/F is shown in Figure 1, two benzene rings connected by two ether bridges. There are 75 PCDD congeners and 135 PCDF congeners in total, differing in the chlorine number and location. Seventeen of them with chlorine substitutions in 2,3,7,8 positions are especially toxic.

Polychlorinated biphenyls (PCBs) are derived from biphenyl, and 12 of the 209 congeners are structurally "dioxins-like". Under some conditions, PCBs may form PCDF through partial oxidation. The chemical structures of PCDD/F and PCBs are given in Figure 1.

Dioxins are solid with a high melting point and are difficult to dissolve in water. They are stable in strong acids and alkalis and have strong chemical stability.[1] Dioxins can exist for a long time in the environment. With the degree of chlorination increasing, dioxins have reduced solubility and volatility. Microbial degradation, hydrolysis, and photolysis in the natural environment have little effect on the molecular structure of dioxins.

PCDD/Fs with chlorine atoms substituted at the 2,3,7,8-positions are toxic, of which 2,3,7,8-TCDD is the most toxic. The strength of the toxicity determines its strength in binding molecules in the human body. Dioxins present in the environment exist in the form of mixtures, and the evaluation of the potential health effects of exposure to these mixtures is not a simple sum of the contents. To evaluate the potential health effects of dioxin mixtures, the concept of toxic equivalence factor (TEF) is proposed.[2−4] 2,3,7,8-TCDD is given a TEF of 1, and the other 16 toxic congeners have values relative to 2,3,7,8-TCDD. In 1997, the World Health Organization (WHO) proposed the toxicity equivalent factor, WHO-TEF. The toxicity equivalent factor of 2,3,7,8-TCDD still remains 1, and the TEFs of the other 16 species of 2,3,7,8-chloro-substituted dioxins were increased or downgraded. The calculation of toxic equivalent quantity (TEQ) of a congener is to multiply the concentration by its TEF, and the TEQ of the mixture in the flue gas or fly ash is the sum of the individual TEQs.

1.2. Dioxins in Incineration Processes. The study on dioxins started since their identification in waste incineration fly ashes approximately in 1970s. The PCDD/F compounds were first found in Dutch and Swiss studies by Olie et al.[5] and Rappe et al.[6] In the 1980s, with the development of analysis methods, kinds of environmental samples were detected. Dioxins are released into the environment in various ways and in a wide range of quantities according to the source. The major identified sources released to the environment can be grouped into four categories by Kulkarni et al.: incineration, combustion, industrial, and reservoir.[7] Incineration, including municipal solid waste incineration (Figure 2), hazardous waste inciner-

Received: July 21, 2020
Revised: September 21, 2020
Published: September 22, 2020

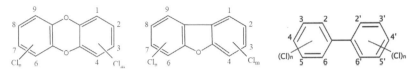

Figure 1. Chemical structures of PCDDs, PCDFs, and PCBs.

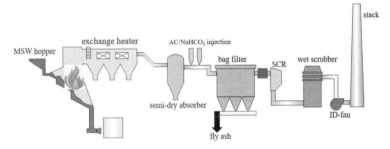

Figure 2. Schematic diagram of a typical municipal solid waste incinerator.

Table 1. Concentrations of PCDD/Fs in Different Incinerators

fuel	sampling point	APCDs	concn	district	year	ref
MSW	catalytic filter inlet	semidry absorber + catalytic filter	N.A.[a]	Taiwan, China	2014	13
RDF	furnace outlet, normal conditions	spray drier absorbers (SDA) + BF	27.1–108.2	U.S.	2007	14
	furnace outlet, start-up and shut down		64.5–124.4			
MSW	cyclone outlet, stack gas	cyclone dry sorbent injection (with activated carbon injection) BF	3.93, 0.17	Taiwan, China	2005	15
MSW	electrostatic precipitator, stack gas	EP + wet scrubber + SCR	7.89, 0.043	Taiwan, China	2005	15
MSW and coal (8:2)	stack gas	semidry scrubber and ACI + BF	0.0108	Mainland, China	2006	16
MSW and coal (8:2)	stack gas	semidry scrubber and ACI + BF	0.1961	Mainland, China	2006	16
MSW and Coal (8:2)	stack gas	semidry scrubber and ACI + BF	0.0054	Mainland, China	2006	16

[a]N.A. = not available.

ation, and sewage sludge incineration, is the largest emission source of dioxins released into the environment.[7] UNEP assessed the releases of dioxins according to ten source groups.[8] Open burning, as the largest source group, accounted for 45% of the total release to the environment. The second and third largest source groups were waste incineration (17%) and metal production and heat and power (both with 14%).

Numerous organic compounds such as chlorobenzenes (CBzs), chlorophenols (CPhs), and polycyclic aromatic hydrocarbons (PAHs) and PCDD/F are all already present in fresh solid waste,[9] and these compounds enter the combustion chamber with the waste. The operating, chemical, and catalytic factors have an important influence on dioxin formation.[10] The factors have different impacts, with multiple mutual interactions, and the importance varies with the operating conditions. Feeding too much at once or inadequate premixing result in poor combustion in the furnace, which leads to the production of PICs (products of incomplete combustion). In the postcombustion area, with the help of oxygen and catalyst in the fly ash, many dioxins can be formed from the PICs. Thus, some operating methods could be beneficial to the combustion of the waste and the minimization of dioxin formation. Buekens and Cen have given some suggestions, such as well-mixed, homogeneous waste, low rates of primary air, and homogeneous combustion.[10]

With the principle "3T+E", which means Temperature higher than 850 °C, residence Time longer than 2 s, Turbulence, and Excessive air in the furnace, the pollutants are mostly destroyed during the incineration of solid waste.[11] However, PCDD/Fs can be formed in the postcombustion zone and in the electrostatic precipitator with the flue gas cooling down. In the process, dioxins are formed along three pathways: high-temperature homogeneous synthesis, de novo synthesis, and precursor synthesis.[12] As the latter two routes happed on the surface of the fly ash, de novo and precursor syntheses are collectively referred to as heterogeneous synthesis.

Table 2. Summary of the PCDD/F Emission Regulations Worldwide

area	emission limit for MSW incinerators	baseline oxygen concn (%)	emission standards	measurement method
U.S.	13 ng/m^3	7	40 CFR Part 60	EPA method 23
EU	0.1 ng TEQ/m^3	11	Directive 2010	EN 1948
China	0.1 ng TEQ/m^3	11	Gb18485-2014	HJ 77.2-2008
Japan	0.1 ng TEQ/m^3 (>4 t/h); 1 ng TEQ/m^3 (2–4 t/h); 5 ng TEQ/m^3 (<2 t/h)	12	law concerning special measures against dioxins	JIS K0311-2008

The concentration of PCDD/F emission depends on the fuel composition, type of furnace, combustion conditions, and air pollution control devices (APCDs). The flue gas cleaning system is essential to control the emission of dioxins. In recent years, public and political requirements have put much more pressure on the control of pollutants' emission. The flue gas cleaning system can remove the following pollutants:

- Particulates and dust, including associated solid dioxins and heavy metals;
- Acid gases, such as SO_2, HCl, and HF;
- Nitrogen oxides, such as NO, NO_2, and N_2O;
- Semivolatile organic compounds, such CBzs, PCBs, PAHs, and dioxins.

Due to their high boiling point and hydrophobicity, dioxins can attach to the surface of the heat exchange tube and the fly ash. Approximately 85% of dioxins attached to the surface of the fly ash can be captured by the baghouse filter. While the concentration of the gas-phase dioxins in the flue gas still exceeds the regular limit (0.1 ng I-TEQ/m^3), some other operations should be taken. Activated carbon (AC) can adsorb the gas-phase dioxins in the flue gas effectively and then be arrested by the baghouse filter (BF), which is a commonly used dioxin control method.

Table 1 summarizes some representative PCDD/F concentrations for MSW incineration. Under stable operating conditions, the PCDD/F concentrations range from 27.1 to 124.4 ng I-TEQ/(Nm^3).

1.3. State-of-the-Art Formation, Measurement, and Control of Dioxins during MSW Incineration.

Ever since the discovery of dioxins in fly ash, lots of researchers have devoted efforts to identifying the dioxin formation pathways. With the strict emission regulars, measurement and remediation technologies are required to reduce the emission of dioxins to the environment. In recent years, much work has been done on the formation, measurement, and control of dioxins. Zhang et al. reviewed the dioxins' emission from biomass combustion and open burning.[17,18] As a major source of dioxins' emission, open burning has attracted much attention. The influence of fuel components, the combustion conditions, the phases of the fire (flaming vs smoldering), and other factors are reviewed and discussed in this work. Zhou et al. paid attention to the Cl-containing species and chlorine release behavior from MSW.[19] This work reviewed the modes of occurrence of chlorine of both organic and inorganic chlorine-containing species, which is helpful to reveal the role of chlorine in PCDD/F formation. Rathna et al. reviewed sources of dioxins' pollution and some innovative remediation technologies, such as bioremediation, plasma pyrolysis, and so on.[20] Wu et al. summarized the emission of PCDD/F and other unintentional persistent organic pollutants (UPOPs) in the metal industry.[21] Lei et al. reviewed the levels and profiles of PCDD/F in different environmental media, such as air, water, sediments, and soils.[22]

Any thermal process proceeding in the presence of C, Cl, H, and metal catalyst may produce dioxins. Dioxins may form at two temperature windows: the homogeneous synthesis taking place between 500 and 800 °C and the heterogeneous pathways proceeding mainly between 200 and 450 °C.[23,24] Given that the PCDD/F formation mechanisms are very complicated and important for the control of dioxins, many researchers have focused on the formation mechanisms; however, a universally accepted result, which can fully reveal the formation of dioxins, has yet to be obtained.[25] In the previous review,[12] the formation of chlorophenols in the MSWI and the formation of dioxins from CPh were reviewed in detail. In this work, data regarding CPh concentrations formed in the incineration process and the influence of APCDs on the emission of CPh were introduced. CPh is synthesized through several pathways, including oxidative conversion and hydrolysis of chlorobenzene, de novo synthesis from carbon matrix, or cyclization combined with the oxidation of reactively small molecules, decomposition, or oxidation combined with chlorination of selected PAHs. Another class of chlorinated compounds existing in the flue gas is chlorobenzene. This work reviewed the concentrations of CPh, CBz, and PCDD/F emissions in different incinerators. It showed that, under stable combustion conditions, the total concentration of CPh ranges from 800 to 46,600 ng/(N m^3), and CBz is of the same order of magnitude as CPh, which both are 2 or 3 orders of magnitude higher than that of PCDD/F. The formation rate of PCDD/F from the CPh route is some 2 orders of magnitude higher than other routes. On the basis of several studies, CPh preferentially formed PCDD, rather than PCDF. In this review, the homogeneous generation pathway and heterogeneous pathway, including de novo synthesis and formation from precursors, were fully discussed. The authors proposed a unified PCDD/F formation pathway.

The emission limit and measurement method for PCDD/F emission for MSWI has been established in many countries.[26−35] As shown in Table 2, most of the regulations limit the emission concentration as 0.1 ng TEQ/m^3 and require measurement of PCDD/F emission once or two times per year.

In these guidelines or standards, PCDD/F concentrations in the flue gas and fly ash are determined by high-resolution gas chromatography coupled with high-resolution mass spectrometry (HRGC/HRMS). Before the detection, the samples from the incinerators should be pretreated in the laboratory, which can take several days or weeks and is expensive. The costly and infrequent measurements can only obtain the PCDD/F emission values in the inspection periods. People living nearby the MSWI want to know the PCDD/F emissions in real time, similar to the other pollutants, including NO_x and HCl. Moreover, continuous PCDD/F emission data help to optimize plant operations. Cao et al. reviewed the advances regarding fast indirect measurement of PCDD/F TEQ emission.[36] Detecting the concentration of indicators to predict the PCDD/F emission is a much more promising method. CO, total hydrocarbons, PCBs, PAHs, CBz, and CPh are the potential indicators. In particular, CBz, as the precursor of PCDD/F formation, has a structure similar to those of PCDD/F and is easy to detect. Many researchers have found a strong correlation between CBz and PCDD/F, with a Pearson correlation coefficient of more than 0.95 in flue gas between PCDD/F and CBz.[37] In addition, the I-TEQ concentration of PCDD/F is strongly correlated with tetra- (TeCBz), penta- (PeCBz), and hexachlorobenzene (HxCBz), and the correlation coefficients are approximately 0.85–0.93.[38−40]

REMPI-TOFMS (resonance-enhanced multiphoton ionization-time-of-flight mass spectrometry) was employed for online measurement of the PCDD/F indicators in MSWI. In recent years, much effort has been paid to improve the stability and sensitivity of the instruments.[41,42] Cao et al.[41] applied LASTI (laser adsorption spectrometry tunable ionization)-TOFMS to detect 1,2,4-TrCBz and predicted the PCDD/F concentration in the flue gas. The average of the relative difference between PCDD/F TEQ predicted and measured by HRGC/HRMS was 18.9%.

With the strict environmental policies for dioxin emission worldwide, technologies for the inhibition of dioxin formation as well as the remediation in flue gas and solid residues were rapidly developed. Several measures were taken to reduce the formation of dioxins, such as reducing the supply of primary air, raising combustion and postcombustion temperatures, and improving mixing by judicious injection of secondary air.[43] Due to the strong adsorption characteristics of dioxins, sorbents, such as activated carbon, with high surface area and adapted pore structure were used to adsorb the gas-phase dioxins in the flue gas. The used activated carbon and the solid dioxins in the fly ash were captured by the baghouse filter. In recent years, conventional or modified SCR catalysts, such as V_2O_5–WO_3–TiO_2 honeycomb blocs or catalytic filters, were used to destroy the low-concentration dioxins.[44,45]

With the development of technology and society, it can be expected that the environmental protection policies will be tightened with time. The foreseeable aggravation of environmental regulations does not create principal problems; the challenge is more to optimize existing technologies.

1.4. Structure of This Review. Expanding on the previous review studies and some latest advancements, this work provides systematic, comprehensive overviews of the formation, measurement, and control of dioxins.

Homogeneous synthesis, precursor synthesis and de novo synthesis contribute to the formation of PCDD/F. This review deals with the relationship between the concentration of PCDD/F and combustion conditions in Section 2, as well as the formation of dioxins from CPh, CBz, and carbon matrix. PCDD/F sampling and analysis methods are presented in Section 3. Some fast online monitoring techniques, such as the work by the U.S. Environmental Protection Agency (EPA) and our group, are well-described in detail, including the study on the PCDD/F indicators and advanced instrumentations. Dioxins formed from incineration and combustion processes enter the environment through flue gas, fly ash, and even contaminated soil in the plant. Feedstock optimization, adjustment of incineration conditions, and addition of additives are summarized in detail in Section 4. Air pollution control devices have great effects on the abatement of dioxins, such as activated carbon injection with baghouse filter, catalytic oxidation, and wet scrubbers. The nonthermal and thermal treatments of fly ash are also described in this section. Despite the rapid development of incineration and combustion, measurement and emission control of dioxins still face some challenges. Detailed future research perspectives in PCDD/F formation, fast measurement, remediation, and reduction of dioxins are outlined in Section 5.

2. FORMATION MECHANISMS OF PCDD/F IN INCINERATION

Since the identification of dioxins in the fly ash from the Amsterdam incinerator, many researchers have studied the formation of dioxins in the combustion process. Some researchers proposed that dioxins are synthesized from related compounds, such as CPh and CBz, which formed the basis of the "precursor route". In 1985, Vogg and Stieglitz[46] proposed the "de novo" theory. Two elements stand central in de novo synthesis, oxygen and metal catalysis. Another important route is homogeneous synthesis, which proceeded in gas phase at high temperature without the catalysis of metals. Thus, three formation routes to the formation of dioxins have been described:

- Homogeneous synthesis is processed at high temperature (>500 °C) from a variety of reactive gases,[47] and the reactions do not need the catalytic process of fly ash. These include various pyrolysis and partial oxidation reactions, cyclization of reactive intermediates to aromatic compounds, self-condensation of precursors, and both chlorination and dichlorination reactions.[48]
- Precursor synthesis means PCDD/F formed from structurally similar compounds, such as the condensation of two CPh or CPh and CBz, oxidation of PCB, and chlorination of DD/DF.[48]
- De novo synthesis means the synthesis of PCDD/F from low-temperature catalytic chlorination and oxidative breakdown of the carbonaceous matrix.[48]

PCDD/F can be formed in the thermal process with the elements C, H, Cl, and O. The results showed that homogeneous synthesis plays a rather subordinate role in the PCDD/F generation process, but amounts of CPh and CBz were formed, which will supply sufficient precursors for PCDD/F synthesis in the lower temperature zone.[49,50] There is still an ongoing debate about whether precursor or de novo synthesis is the major route for PCDD/F formation. The precursors reacted fast to form dioxins, but de novo synthesis reproduces the thermal fingerprint found in MSWI.[12]

2.1. Homogeneous Pathway. The homogeneous pathway occurs at temperatures of 500–800 °C, and the elements of O and Cl are necessary. Previous research showed that the dimerization of chlorinated phenoxy radicals was the major pathway to form PCDD/F. CBz was oxidized to CPh, and CPh condensed to PCDD/F, which was one of the routes of formation. This subsection will be focused on the formation mechanisms, such as formation and oxidative coupling of chlorophenyl radicals, PCDD formation, and PCDF formation. PCDD/F formation mechanisms from CPh have already been excellently reviewed by Altarawneh et al.,[48] Stanmore,[23] and Peng et al.[12]

The formation and degradation of PCDD/F exist simultaneously in the furnace, and the degradation rate was higher when the temperature of the furnace was greater than 900 °C.[51] Altwicker[52] found that the homogeneous formation route contributes ca. 30% of the whole PCDD/F emission. With the development of furnace techniques and the "3T+E" principle in the operation guides, the contribution becomes lower.

Research found that the dimerization of chlorinated phenoxy radicals plays an important role in PCDD/F formation.[53,54] Qu et al.[55] found that the ortho—ortho coupling of 2,4,5-TCP radicals is a spontaneous reaction to form PCDD/F and that 2-MCP radicals should cross a potential energy of 8.27 kcal/mol, which means that the number of chlorines of chlorophenol can heavily affect the ortho—ortho coupling of CPh radicals. Fadli et al.[56] discovered that, at a high temperature (575–825 °C), CBz formed CPh and then converted to PCDD/F.

The PCDD/F formation through homogeneous synthesis can be divided into several parts.

2.1.1. Formation of Chlorophenoxy Radicals. Altarawneh et al.[48] divided the generation of chlorophenol radicals into three types, as shown in Figure 3:[12] initiation, oxidation of MCP, and

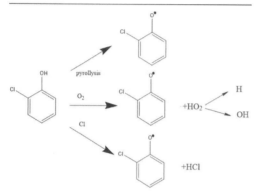

Figure 3. Three formation modes of phenoxy radicals. Reprinted with permission from ref 12. Copyright 2009 Elsevier, Inc.

H abstraction. Initiation means the abstraction of a hydrogen atom from the CPh molecule, which can occur in pyrolytic conditions.[57,58] In the oxidation process the phenolic hydroxyl CPh reacts with oxygen and forms chlorophenoxy and HO$_2$ radicals.[59] Since the concentration of Cl in the flue gas is higher than that of OH and H, H abstraction can easily occur.

2.1.2. Oxidative Coupling of Phenoxy Radicals. Phenoxy radical character is localized on the phenolic oxygen atom, as well as on the para- and ortho-carbon atoms.[60] Results from the calculation showed that phenolic oxygen, para-carbon, and ortho-carbon accounted for 41.3%, 39.1%, and 19.6%, respectively. The self-reaction of the phenoxy radicals through the three radical sites can generate six different products. On the basis of the frontier molecular orbital theory,[61] coupling is preferred at sites that possess the highest spin density, and the coupling prefers the para—para dimer, while the experiment results showed that the ortho—ortho coupling product dominates due to the lower barrier for tautomerization of ortho—ortho to the final 2,2′-dihydroxybiphenyl. It was postulated that kinetic factors associated with the structure of the transition states determine the final coupling product.[62]

2.1.3. Mechanistic Steps for the PCDD/F Formation. Experiments have shown that 2,2′-dihydroxybiphenyl, the ortho—ortho coupling product, is the key intermediate to PCDF formation, while the formation of PCDD differs from the synthesis of PCDF because they involve kinetically favored radical/radical, molecule/molecule, and molecule/radical reacting coupling types. PCDD formation needs only one ortho-chlorine atom, while PCDF formation needs two chlorophenol

molecules, each with an available ortho-hydrogen,[48] which lead to the high PCDD/PCDF ratio when CPh acts as precursor in heterogeneous synthesis.

Due to the low contribution of the homogeneous pathway to PCDD/F emission, the study on dioxin formation shifts to heterogeneous synthesis.

2.2. Heterogeneous Pathway. *2.2.1. Surface Catalytic Precursor Pathway.* The reaction steps of the precursor pathway are as follows: adsorption, adsorption of a second CPh molecule (E—R model), condensation of the adsorbed molecules to form a PCDD/F molecule, and finally desorption of the PCDD/F molecule.[12]

Due to adequate amounts of exposed metals, the fly ash supplies necessary active sites for CPh adsorption and condensation. The calculated results show that there are approximately $(2.8-5.85) \times 10^{18}$ adsorption sites.[63,64] The amount of CPh adsorbed on fly ash decreased with increasing temperature, and the CPh concentration in the gas phase got higher.

CPh shows a high conversion rate to PCDD/F through heterogeneous synthesis. Milligan and Altwicker[65] found a yield of PCDD from 2,3,4,6-TeCP (10% O$_2$, 350 ng/mL, 350 °C) of 4%, and the higher the concentration, the lower the conversion efficiency. In the formation process, the catalyst plays an important role; notably, the transition metals gave a clear promoting effect.[66] In particular, divalent oxides of copper showed a marked enhancement of the yields of PCDD/Fs in comparison with other oxides.

The formation of PCDD/Fs from CPh resembles Ullmann-type reactions, with coupling of aryl groups catalyzed by metals. Dellinger's and Altarawneh's groups have performed a large amount of work. Lomnicki and Dellinger[67] found that approximately 0.5% of the 2-chlorophenol was converted into PCDD/F in the absence of O$_2$ in the atmosphere. The oxygen was assumed to have originated from the copper oxide according to the Mars—van-Krevelen theory. After the adsorption of the CPh molecule on the catalyst surface oxide and hydroxyl sites, chlorophenolate was formed through elimination of H$_2$O or HCl. Then, the chlorophenolate is a donor of an electron to the metal cation site to form a surface-associated chlorophenoxyl radical.[68]

The Eley—Eideal mechanism means the formation reaction taking place between a gaseous species and a species adsorbed on the catalyst. The Langmuir—Hinshelwood mechanism means the reaction occurs between two adsorbed species. Lomnicki and Dellinger,[67] Altarawneh et al.,[48] and Peng et al.[12] have clearly stated the formation process.

2.2.2. De Novo Pathway. De novo synthesis refers to the formation of PCDD/F in a temperature window between 200 and 400 °C from the carbon, biomass, soot, or other organic structures, including small molecules, such as ethylene, acetylene, and PAHs. This process includes the transfer of inorganic chloride to the macromolecular structure of residual carbon under catalysis and the partial oxidation of the chlorinated carbon matrix. The de novo route has been studied by several groups, and the related work has been repeatedly reviewed. Stieglitz et al.[69,70] reviewed the de novo synthesis reactions in the laboratory. Addink and Olie,[71] Buekens and Huang,[72] McKay,[11] and Stanmore[23] reviewed the dioxin formation in the MSWI. Altarawneh et al.[48] reviewed the de novo process, especially regarding the dynamic change of copper during the de novo process and the kinetic modeling. Peng et al.[12] proposed a unified PCDD/F formation pathway involving both the de novo route and the precursor route. Zhang and Buekens[47] reviewed the de novo synthesis and the influencing factors.

The de novo synthesis of dioxins involves the low-temperature catalytic gasification of carbon and catalytic chlorination of carbon. The transition metal plays an important role in the process and can be reduced. The catalyst can be reoxidized by the oxygen adsorbed on the surface of the fly ash. The chloroaromatic it is still the key mechanistic step, in which, some precursors, including CPh, CBz, and PCBs, are formed. Dickson et al.[63] added ^{13}C-PCP to the model fly ash and found that no scrambled PCDD was detected in the de novo process, meaning that the CPh molecule would not react with the carbon matrix to form PCDD. However, Hell et al.[73] found that approximately 40% of the total PCDD was formed through condensation of de novo created chlorophenols. Peng et al.[12] ascribe the difference to the capacities of the various kinds of fly ash.

The de novo kinetics study found that the maximum rate is situated at 300−350 °C. Some characteristic studies proved the dramatic change in the chemical forms when the fly ash was heated to 300 °C.[74] At room temperature, $CuCl_2 \cdot 3Cu(OH)_2$, CuCl, and $Cu(OH)_2$ account for 75%, 14%, and 11%, while CuCl transferred to 45% when heated. The noticeable change is consistent with the amount of formation of PCDD/F in this temperature zone.[74] Changes of the copper chemical form were attributed to the oxychlorination of carbon in the fly ash through oxidative degradation. When using $CuCl_2 \cdot 2H_2O$ in the model fly ash, the contents changed to 60% CuCl, 20% Cu_2O, and 20% Cu. The reduction of $CuCl_2$ was consistent with the formation of carbon−chlorine bonds, while, for CuO, the best temperature to catalyze the de novo reaction is approximately 400 °C.[74]

3. SAMPLING AND ANALYSIS

3.1. Direct Measurement of PCDD/F. *3.1.1. Determination of PCDD/F by Isotope Dilution HRGC/HRMS.* Isotope dilution HRGC/HRMS has been used as the internationally recognized method for the determination of PCDD/F emission. Different standard methods such as EPA method 23,[35] EN 1948,[33] JIS K0311,[34] and HJ77.2-2008[27] are applied for offline determination of PCDD/F emissions from stationary sources such as MSWI in the U.S., the European Union, Japan, and China.

The general procedure of PCDD/Fs analysis with HRGC/HRMS consists of sample collection, sample extraction, sample cleanup and fractionation, analysis, and data processing.[75] The absorbent material of the sample device consists of XAD-II resin in the absorption chamber and a fiber filter in the heating filter, which is used to absorb particulate-phase sample and gas-phase sample, respectively. Before sampling, quantitative isotopically (^{13}C) labeled PCDD/Fs should be added to the XAD-II resin, which can guarantee the accuracy of measurement results. Samples from the flue or stack gas must be collected isokinetically and then stored in a dark environment and at low temperature until analysis. The two-step Soxhlet extraction is applied with dichloromethane and toluene to obtain semivolatile organic compounds, including CBz, CPh, PAH, and PCDD/F. In this process, internal standard stock solution was added in the toluene. Before subsequent cleanup processing, ^{13}C-1,2,3,7,8,9-HxCDD in the alternate recovery stock solution was added for the assessment of purification efficiency. Finally, samples are concentrated and ^{13}C labeled recovery standard stock solution was added before the analysis by HRGC/HRMS.

HRGC/HRMS has been used as the official method for trace analysis of dioxins and can provide accurate measurements for homologues of PCDD/F. However, this traditional method is time-consuming due to the complicated procedures, which include sampling, preprocessing, and analysis. The time lag, typically several days to weeks, prevents the plant from connecting emissions to the condition of the incinerator, which cannot meet the requirement for industrial plant control.

In Europe, dioxins stack emission determinations are based on a sampling period of 6−8 h according to Directive 2010/75/EU. In China, the standard (HJ77.2-2008) stipulates that each sample time cannot be less than 2 h, while sampling once per year or per season can cover less than 2‰ of the total operation time, which was usually performed at the best operation conditions.[76] The general wish is to have online monitoring equipment, which is not feasible with the current technology. Continuous sampling or long-term sampling covering the whole plant operation time, including the several conditions such as start-up and shut-down procedures, is a practical choice.

Since 2000 long-term sampling systems have been developed which are commercially available. Reinmain et al.[77] introduced the AMESA system in detail, and the functionality and stability have been demonstrated over 20 years. Belgium is the first country to require all hazardous and municipal waste incinerators to install the continuous sampling system. In the first 2 years, some facilities struggled to continuously meet the emission standard. And, in this period, the operators learned from experience to operate the furnace and the air pollution control devices better. Average emission concentration is about 0.02 ng TEQ/(N m^3), and excessive emissions rarely occur.

Abad et al. evaluated the long-term monitoring system for the analysis of PCDD/F concentrations from stationary sources in a series of studies.[76,78−80] In their research a long-term sampling system named DioxinMonitoringSystem (Monitoring Systems, Bad Voeslau, Austria) was used according to method EN-1948. The device allows continuous isokinetic sampling of the flue gas from several hours up to months. In the sampling process relative humidity, filter temperature, oxygen concentration, and flue gas volumes are recorded. The adsorbent media, including the particulate filter, polyurethane form, and XAD-2 resin were sent to the laboratory for the following pretreatment after the long-term sampling. In a 2 year time period, the absolute levels varied from 0.011 to 0.077 ng TEQ/(N m^3) (mean value, 0.030 ng TEQ/(N m^3); RSD = 70%; n = 16), showing a good system performance over a wide range of dioxins' concentrations.[78] Compared to the short-term samples which varied from 0.007 to 0.021 ng TEQ/(N m^3), both the mean and maximum values are obviously higher for the continuous monitoring.

Another advantage of continuous sampling is that the data obtained can be used to calculate the emission factors. After more than a year of sampling in a cement factory in Spain, Conesa et al. calculated the PCDD/F emission factor to be 8.5 ng I-TEQ/(ton of clinker) and 3.2 ng of WHO-TEQ/(ton of clinker) for PCBs.[80] And at low emission levels, only long-term sampling can provide complete information on dioxins' emission, making it possible to study the behavior of low emissions. Slight differences in the emission profile could be assessed from the continuous sampling results.[79]

3.1.2. Determination of PCDD/F by Bioanalytical Detection Methods. The principle of bioanalytical detection methods (BDMs) is the ability to identify the unique structural properties of the dioxin-like compounds by hinge biomolecules and based on special reactions to dioxin-like compounds by cells or

Table 3. Comparison of Online/At-Line Measurement Equipment for PCDD/F

equipment	LOD^a	measured compd	R^2 with PCDD/F (I-TEQ)	measurement cycle	application
VUV-SPI-IT-TOFMS	ppq	2,3,4,7,8-PeCDF	0.9	2–6 h	universality
VUV light ionized IT- TOFMS	ppb	TrCBz		20 s	site specificity
RIMMPA-TOFMS	pptv	14 PCDD/F congeners			universality
TD-GC-REMPI-TOFMS	ppb	organ indicators (CBz)	0.73	5 min	site specificity
REMPI-TOFMS	ppb	PAHs, chlorobenzene, and some aromatic compounds	0.82		site specificity
jet-REMPI-TOFMS	ppb	chlorobenzene	0.74	5 min	site specificity
TD-GC-LASTI-TOFMS	ppb	chlorobenzene	0.89	15 min	site specificity

^{a}LOD = limit of detection.

organisms.[81] Compared to the traditional methods using HRGC/HRMS, BDMs are rapid and low cost,[82] which mainly include the enzyme immunoassay (EIA) and bioassays based on the molecular function of aryl hydrocarbon receptors (AhRs).

EIA is based on clonal antidioxin antibodies, which can particularly bind PCDD/F. A competitor enzyme is added to compete for limited antibody-specific binding sites with PCDD/F.[81] EIA becomes a promising alternative to the traditional method, as it is relatively less time-consuming and is easy to operate, though it is lacking in sensitivity. Cost reduction can be an advantage of EIA for its support to inexpensive sample cleanup procedures.[83]

AhR-based bioassays can be classified into two categories by the AhR sources: endogenous AhR-based and exogenous AhR-based.[84] Ethoxyresorufin-O-demethylase (EROD) assay and chemically activated luciferase expression (CALUX) are two of the most widely used methods for detecting dioxin-like compounds. EROD can indirectly reflect the ability of binding with Ah receptors for the test compounds by measuring the activity of EROD enzyme. In recent years, the EROD assay has been used for identifying humic soil samples and water samples, which reflect dioxin-like toxicity successfully.[85,86] CALUX is based on a modified recombinant cell line that has been transfected with luciferase reporter gene plasmids, and the expression of the reporter genes responds to the dioxin-like compounds.[81] Due to its wider detection limit and higher sensitivity than EROD, it has been validated by the U.S. EPA to become an official method for the rapid screening of dioxin-like chemical activity in soils and sediments.[87] The minimal detection limit has been updated with brand new, third-generation (G3) CALUX cell lines developed by He.[88] Brennan et al.[89] used sediment samples and chemical libraries including 176 compounds to make a comparison between different G3 CALUX cell lines (mouse, rat, human, and guinea pig) and found that the rat hepatoma G3 CALUX cell line is the best option for filtering dioxin-like compounds. The utilization of exogenous AhR molecules is uncommon in the detection of dioxin. However, it will be a promising method for the development of techniques in cloning and manipulation of AhR.[84]

3.2. Determination of PCDD/F by Indicators. The traditional method for measurement of dioxin using HRGC/HRMS is very expensive and requires high skills to operate. In addition, it is time-consuming to execute the complicated procedures, including sampling, pretreatment, and analysis. Therefore, measurement of PCDD/F through indicators, whose concentrations are 2 or 3 orders of magnitude higher than that of PCDD/F, has been developed.[25] The indicators of PCDD/F can be classified as (1) chlorobenzenes (CBzs), chlorophenols (CPhs), and polycyclic aromatic hydrocarbons (PAHs) and (2) other indicators.

3.2.1. CBz, CPh, and PAHs as Indicators. CBzs have always been considered as the most common indicators of PCDD/F. Lavric et al.[90] summarized the experimental studies of correlations between dioxin and surrogates and found that CBzs had been widely used as dioxin surrogates. Kato and Urano[91] stated that TeCBz, PeCBz, and HxCBz have good correlations with TEQ value, and PeCBz is better than TeCBz and HxCBz due to its lack of isomers, low toxicity, and high recovery in the pretreatment procedure.[92] Wang et al.[93] discussed different indicators of PCDD/F TEQ and PCDD/F concentrations in MSWI in China and found that 1,2,3-TrCBz, 1,2,3,4-TeCBz, and PeCBz show good correlations with PCDD/F TEQ concentration. In addition, correlations between PCDD/F with low chlorinated chlorobenzenes have been studied, with coefficients of 0.80–0.83.[94,95] In brief, highly chlorinated chlorobenzenes show better correlations with PCDD/F than those of low chlorinated chlorobenzenes.

CPhs are one of the most important precursors of PCDD/F. Blumenstock et al.[96] used data from HWI and the literature to demonstrate that only a few chlorophenols can be indicators of PCDD/F TEQ only in flue gas, and there is no correlation between chlorophenols and PCDD/F TEQ in stack gas, probably due to the influence of the flue gas cleaning system. Kaune et al. found a fair correlation ($R^2 = 0.81$) between I-TEQ and CPh in the flue gas of HWI and $R^2 = 0.83$ downstream of the electrostatic precipitator.[40,97] Chlorophenols were sensitive to the condition of the incinerator and had relatively poor correlations with PCDD/F TEQ and concentration.[93] It is obvious that chlorophenols are undesirable choices for PCDD/F indicators compared to chlorobenzenes.

PAHs form PCDF isomers in one of the de novo formation paths.[98] A similar trend of the concentration of PCDF and PAHs, especially benzo(a)anthracene, was found by Li et al.[99] through an incineration experiment with a mixture MSW and coal so that benzo(a)anthracene might be a typical indicator of PCDF. Furthermore, Yan et al.[100] reported on the relevance between PAHs and PCDD/F on fluidized bed incinerators and mechanical stocker incinerators in China and established a multiple regression analysis with naphthalene, fluorene, and phenanthrene to indicate PCDD/F TEQ, which showed a high coefficient of determination ($R^2 = 0.85$). However, later research observed that the correlations between PAHs and PCDD/F were poorer than those between CBzs and PCDD/F, especially in the start-up and shut-down periods of MSWI.[93]

3.2.2. Other Indicators. Some common compositions of flue gas, such as CO, O_2, HCl, and NO_x, have also been used as indicators of PCDD/F. However, the correlations between these common pollutants and PCDD/Fs are not as definite as those

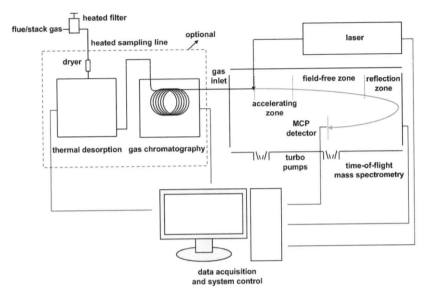

Figure 4. Schematic of a typical REMPI−TOFMS instrument. Reprinted with permission from rev 101. Copyright 2019 Springer Nature.

between CBzs and PCDD/F. Obviously, they have nothing to do with PCDD/Fs based on structure similarity or reaction mechanisms.[101]

Oh et al.[102] demonstrated that CO concentration was positively correlated with PCDD/Fs in 4 different incinerators of Korea, while O_2 and HCl were unrelated to PCDD/F. CO concentration in stack gas of a fluidized bed municipal solid waste incinerator had a weak positive correlation with total PCDD/F concentration.[16,103] Low CO concentration (<10 ppm) was a prerequisite for the low emission level of PCDD/Fs.[103]

However, HCl, O_2, and NO_x cannot be used as indicators of PCDD/Fs because there were no noticeable correlations between HCl, O_2, NO_x, and PCDD/Fs.[16,103]

3.3. Online/At-Line Measurement of PCDD/F. 3.3.1. Determination of PCDD/F by PI-TOFMS Techniques. Indicators of PCDD/F are more likely to be online/at-line measured than PCDD/F due to concentration. However, challenges still exist in operating online/at-line in situ measurements in flue gas, which have complex composition. To overcome these problems, specific kinds of photon ionization time-of-flight mass spectrometers (PI-TOFMS) have been developed for the online/at-line measurement of indicators of PCDD/F (Table 3).

Tsuruga et al.[104] conducted a sensitive and robust vacuum ultraviolet (VUV) single-photon ionization (SPI) ion trap (IT) time-of-flight mass spectrometer (VUV-SPI-IT-TOFMS) system, which can monitor the PeCDF for several months continuously. The device can measure the PeCDF concentration every 2−6 h. Morii et al.[105] applied vacuum ultraviolet light ionized ion trap time-of-flight mass spectrometry (VUV light ionized IT-TOFMS, hereafter abbreviated as VUV analyzer) to solve the problem. Online measurement of PeCDF and real-time measurement of TrCBz were successfully completed.

In 1999, Heger et al.[106] redesigned and constructed the REMPI-TOFMS instrument (Figure 4) to monitor the PAH concentrations in the hot flue gas. The UV laser was fixed at 266 or 248 nm to detect the PAH precisely, and the detection limit for naphthalene is as low as 10 ppt concentration range. The group also used the instrument to detect chlorobenzene in an industrial hazardous waste incinerator, and dioxins were detected by laboratory GC-MS to build the correlation model.[94] The correlation coefficients were 0.82 and 0.76 for the flue gas and stack gas, respectively. In this study, the detection limit of chlorobenzene was achieved as 10 ppt. Differently from previous research, they tuned the laser system to online detect more aromatic compounds, such as phenol and benzene. On the basis of a fused-silica capillary with an integral nozzle, Hafner et al. developed a new supersonic jet inlet system for the REMPI-TOFMS.[107] With the new inlet system, excellent jet beam qualities and high spatial overlap of sample and laser beam are achieved with good adiabatic cooling properties of analyte molecules. Compared to a conventional system, the new inlet system has the advantage of higher sensitivity and selectively and reduced degree of fragmentation. With the new instrument, they found that non-chlorinated and chlorinated aromatic compounds behave differently to the combustion conditions.[107]

The Gullett group from the Environmental Protection Agency has also worked for many years on online measurement of PCDD/F indicators.[108−110] Continuous measurements of chlorobenzene congeners with GC coupled to REMPI-TOFMS after the boiler chamber were compared to 5 min sampling for PCDD/F by standard methods. With the GC ahead of the REMPI-TOFMS and tunable laser wavelength, distinction of chlorobenzene isomers was achieved.[110] In the process, 1,4-DCBz, 1,2,4-TrCBz, 1,2,3-TrCBz, and 1,2,4,5-TeCBz were detected, and the correlation coefficients were calculated. The results illustrated that 1,2,4-TrCBz has the best

Table 4. Correlations between I-TEQ Values and Indicators That Can Be Detected by Photo-ionization Mass Spectrometry ($R^2 >$ 0.7)

authors	indicator	R^{2a}	type of incinerator	sampling locations	no. of samples
Zimmerman et al.[94]	MCBz	0.76	HWI	stack gas downstream of the wet electrostatic dust precipitator	7
Blumenstock et al.[96]	MCBz	0.82	HWI	flue gas in the boiler	16
	2,4-DCP	0.8	HWI	flue gas upstream of ESP	12
	2,4,6-TrCP	0.82	HWI	flue gas upstream of ESP	12
	PeCP	0.79	HWI	flue gas upstream of ESP	12
	MCBz	0.85	HWI	flue gas upstream of ESP	24
	1,2-DCBz	0.78	HWI	flue gas upstream of ESP	24
	1,2,3-TrCBz	0.76	HWI	flue gas upstream of ESP	24
Gullett et al.[111]	1,2,4-TrCBz	0.709	HWI	flue gas upstream of the air pollution control system (APCS)	36
	1,3,5-TrCBz	0.703	HWI	flue gas upstream of APCS	36
Yan et al.[100]	NAP; FLU; PHA	0.8452	MSWI	stack gas downstream of bag filter	9
Wang et al.[93]	1,3-DCBz	0.70	MSWI	stack gas downstream of bag filter	13
	1,2-DCBz	0.77; 0.75	MSWI	stack gas downstream of bag filter	7; 13
	DCBzs	0.73; 0.75	MSWI	Stack gas downstream of bag filter	7; 13
	1,3,5-TrCBz	0.71; 0.77	MSWI	Stack gas downstream of bag filter	7; 13
	1,2,4-TrCBz	0.76; 0.83	MSWI	Stack gas downstream of bag filter	7; 13
	1,2,3-TrCBz	0.86; 0.90	MSWI	Stack gas downstream of bag filter	7; 13
	TrCBzs	0.83; 0.88	MSWI	Stack gas downstream of bag filter	7; 13
	2-MCP	0.85	MSWI	Stack gas downstream of bag filter	6
Cao et al.[41]	1,2,4-TrCBz	0.751; 0.908; 0.892	MSWI	Stack gas downstream of bag filter	8; 8; 16

$^a R^2$ = coefficient of determination.

ability to predict PCDD/F TEQ emission with the correlation coefficient value of 0.852. Furthermore, a linear regression model was formed to predict the PCDD/TEQ, with an average relative difference from HRGC/HRMS values of 72%.[110]

Cao et al.[41] from Zhejiang University applied LASTI (laser absorption spectrometry tunable ionization)-TOFMS for at-line measurement of 1,2,4-TrCBz to predict the PCDD/F emission in the stack gas of an incinerator. Eight samples were taken in each of the two test days, and the operation conditions changed frequently in day 1, while it was quite stable in day 2. From the results, the concentrations of 1,2,4-TrCBz were 3 orders of magnitude higher than those of I-TEQ. The correlation coefficients were obtained between 1,2,4-TrCBz and I-TEQ with r = 0.867, 0.953, and 0.944 in unstable, stable, and integrated conditions, respectively.[41] The authors applied the linear correlation models between 1,2,4-TrCBz and I-TEQ to predict the I-TEQ concentrations, and the results showed that the predicted I-TEQ was generally in line with measured values (Table 4).

3.3.2. Determination of PCDD/Fs by Soft Sensors. Soft sensors are predictive models of process industry that are able to achieve online prediction of variables that cannot be easily measured directly due to technological or economic reasons.[112] Linear regression models are widely used due to their simplicity and directness. Chang and co-workers[113] used data collected from several testing programs held in North America to present three multiple linear regression models for different types of incinerators using monitoring variables and operating variables (CO, O_2, and combustion chamber temperature, etc.) to predict PCDD/Fs' concentrations. Most of the results showed the t-ratios at the 5% level of significance and satisfactory R^2 and F values. Then, they put forward nonlinear models with the help of genetic algorithms using the same database and made progress compared to previous research.[114] Bunsan et al.[115] selected five input variables through principal component analysis (PCA)

and performed a three-layer back-propagation neural network analysis to forecast the dioxin emission of a municipal solid waste incinerator in Taiwan successfully. Lack of modeling samples will lead to the inaccuracy of soft sensors. To solve this problem, a virtual sample generation (VSG) method was used by Tang et al.,[116] which combined linear interpolation and prior knowledge to generate virtual samples based on real samples. Soft sensors are quite promising methods to realize online measurement of dioxin; however, they should be integrated with the formation mechanism of dioxin.

3.4. Conclusion. A comprehensive investigation of different sampling and analysis methods of dioxins from incineration of wastes is provided in this section (Table 5). Isotope dilution HRGC/HRMS is the widely accepted "gold standard" for trace analysis of dioxins, which can provide an accurate measurement for dioxin homologues. However, it requires a complicated process and expensive apparatuses. BDMs are considered to become substitutes or collaborators to the traditional HRGC/HRMS-based methods due to their relative ease of operation, low cost, and short time interval. However, the lack of international quality criteria and cross-validation studies between similar or different bioassays still remains a serious problem. Measurement of PCDD/Fs through indicators can reduce the detection difficulty, and some homologues of chlorobenzenes might be the best indicators due to their high correlations with PCDD/Fs. TD-GC-REMPI-TOFMS has been confirmed to be a good option to online/at-line detect indicators of PCDD/Fs in flue gas of waste incinerators with low limits of detection. Soft sensors combined with machine learning do not require either complex processes or expensive equipment, but their accuracy and generalization ability are unsatisfactory.

For future work, it is promising to integrate dioxin indicators detected by TD-GC-REMPI-TOFMS and soft sensors to obtain high-accuracy and low-generalization-error models of dioxin. In addition, the lack of samples urgently needs to be resolved.

Table 5. Comparison of Different Measurement Methods of PCDD/F

detection methods of dioxins	complexity	time consumption	LOD[a]	advantages	disadvantages
isotope dilution HRGC/HRMS	complicated procedure includes sampling, preprocessing, and analysis	weeks	pptv	provide an accurate measurement for dioxin homologues	time-consuming, labor-consuming, and expensive
bioanalytical detection methods	relatively easy to operate	days	pptv	short procedure time, low cost, rapid screening of large-scale environmental samples, beneficial to the assessment of toxicity and health risks	lack of cross-validation study, degree of reliability doubted, lack of international quality criteria
indicators (REMPI-TOFMS)	easy to operate based on the established detection systems	minutes	ppb	online/at-line measurement, provide feedback and optimize combustion condition	reliability and universality need further study
soft sensors	easy to operate based on the established model	seconds	/	do not need chemical process, online measurement, low cost	relatively low accuracy due to the lack of sample, poor generalization ability of models, low correlation with PCDD/Fs

[a]LOD = limits of detection.

4. DIOXIN PREVENTION, REMEDIATION, AND REDUCTION

4.1. Prevention Methods. *4.1.1. Adjustment of Waste Composition and Properties.* PCDD/Fs cannot be formed without chlorine.[117] The presence of poly(vinyl chloride) (PVC) in municipal wastes increased the chlorine content in the flue gas, promoting the formation of PCDD/F. A positive correlation is observed between PVC and PCDD/F.[118,119] Similar emission patterns of PCDD/Fs were found for PVC combustion in the absence of HCl and for PVC-free combustion in the presence of HCl.[117] Hence, there is possibly no correlation between PVC and PCDD/F on this issue. Specifically, it was determined that PVC was unlikely to be a major source of PCDD/F in house fires.[120] Additionally, PVC was not the only chlorine source in MSW, and the separation of PVC from MSW cannot significantly reduce PCDD/F formation because of the presence of inorganic chlorine sources in MSW, such as metal chlorides.

In addition, PCDD/Fs are mainly generated from precursor pathways or de novo synthesis,[121,122] both of which are catalyzed by some metal catalysis, via the cyclization of the carbon chain, the coupling of precursors and the chlorination of carbon.[66,123] MSW could also be processed into refuse-derived fuels (RDFs) in a waste treatment plant. MSW could be separated, recovering large amounts of the plastics and metals, and the rest of the MSW may be transferred into RDFs with lower contents of metals compared with MSW. It is possible that fewer PCDD/Fs would be synthesized from RDF incineration due to the low metal content.

4.1.2. Optimization of the Combustion Process. Combustion conditions have considerable influences on PCDD/Fs formation in MSW incinerators. Among numerous combustion parameters, the most important factors have been accepted as "3T": temperature, residence time, and turbulence. Specifically, a gas residence time in the combustion zone greater than 1 s at 1000 °C and a residence time of 2 s at 850 °C are recommended for ensuring the complete destruction of MSW. For better combustion, advanced combustors were also designed, such as fluidized-bed, vortex flow, and swirl combustors.[72]

In addition to the 3T principle, oxygen concentration affected the emission of PCDD/Fs. Incomplete combustion can occur due to oxygen deficiency, while excess oxygen can promote the synthesis of PCDD/Fs. Utilization of a 3–6% (v/v) excess oxygen level would be recommended for ensuring complete combustion.[124]

4.2. Treatment of the Flue Gas. *4.2.1. Chemical Inhibition.* Chemical inhibition has emerged as one of the promising technologies for in-process prevention of PCDD/F formation. Generally, the chemical inhibitors can be divided into the following groups: sulfur-based compounds, nitrogen-based compounds, S–N-containing compounds, and alkaline compounds.

As early as 1986, Griffin et al.[125] first proposed that higher sulfur content in coal was the cause of low PCDD/F emission by Cl_2 consumption during coal combustion. Subsequently, a significant number of studies on sulfur-based inhibitors have been conducted, especially for SO_2. The significant inhibition efficiency of SO_2 has been observed not only in laboratory experiments[126−128] but also in full-scale municipal solid waste incinerators.[129−131] In addition to SO_2, sulfur trioxide (SO_3), sulfur element, coal with sulfur, sulfur-containing compounds such as sodium sulfide (Na_2S), and carbon sulfide (CS_2) were also used as inhibitors and could successfully inhibit PCDD/F formation.[132−136] However, the presence of sulfur, the molar ratio of sulfur to chlorine, and the reaction temperature could all influence the inhibition effects.

Moreover, many researchers have tried to reveal the inhibitory mechanism with which sulfur-based inhibitors prevent PCDD/F formation, and the following mechanisms were proposed. Cl_2, which is mostly derived from the Deacon reaction, is regarded as the main chlorine source for substitution reaction of the aromatic structure, further influencing the formation of PCDD/Fs, while SO_2 can inhibit PCDD/F formation by converting Cl_2 into HCl[132,133] through the reaction 1:

$$Cl_2 + SO_2 + H_2O \rightarrow 2HCl + SO_3 \quad (1)$$

Since metal catalysts can catalyze the formation of PCDD/Fs, various studies have supported that sulfur-based inhibitors could react with metal catalysts to form sulfate with lower catalytic ability.[126,137,138] In addition, Tuppurainen et al.[122] verified that sulfur-based inhibitors can cause sulfate dioxin precursors to form sulfonates.

Urea and ammonia have also been widely recognized as effective nitrogen-based inhibitors.[139−141] For instance, the combustion of refuse-derived fuel (RDF) containing 10% (w/w) urea resulted in PCDD/F emission lower than 8 ng I-TEQ/(g of RDF), much lower than that of 17 ng I-TEQ/g without urea.[142] Hajizadeh et al.[143] found that the injection of gaseous ammonia into the flue gas could decrease the concentrations of PCDD/F by 34−75% from the solid phase and by 21−40% from the gas phase. Moreover, compounds containing both sulfur and nitrogen tend to function synergistically rather than competitively. The inhibition effects of thiourea and ammonium thiosulfate on both the experimental scale and MSWI are tested systematically. The inhibition efficiency was more than 99% in the model fly ash in the addition of 1.0% thiourea at 300 °C.[144] Chen et al.[145] also found that thiourea as an inhibitor in two full-scale MSW incinerators showed a significant inhibitory effect in the formation of PCDD/Fs.

Numerous studies have been conducted to investigate the suppression effects of alkaline compounds on PCDD/F formation.[146−150] Liu et al.[148] found that CaO could inhibit more than 90% of PCDD/F formation from PCP. CaO could consume acidic compounds such as HCl and Cl_2 resulting from a low concentration of chlorine and weakening the chlorination of PCDD/Fs. In addition, the ability to absorb precursors of PCDD/Fs in the combustion zone has been confirmed.[135]

4.2.2. Adsorption. Activated carbon adsorption is the most widely applied technology in the abatement of dioxin from flue gas. The dioxin is absorbed by activated carbon featuring a large specific surface area and a well-developed pore structure to limit emissions. In China, activated carbon injection combined with baghouse is the main applied form in MSWI, divided into single- and double-bag systems. The basic principle of the single-bag system is that various types of particulate matter and dioxins in flue gas contact with the injected activated carbon and become adsorbed. After the flue gas passes through the bag filter, activated carbon becomes enriched to form a filter cake on the side of the bag. Furthermore, the filter cake can continuously adsorb dioxin and other pollutants.[151] The double-bag system is a combination of a two-stage bag filter in series; activated carbon is sprayed from the inlet of the second bag and then circulated to the first bag. Compared with the single-bag system, the air resistance of the double-bag system is greater, but the removal efficiency of dioxin is also significantly increased. Lin et al.[152] determined that the double-bag system only consumed 40% of the activated carbon in the single-bag system, but increased the dioxin removal efficiency from 97.6% to 99.3%, with 0.03 ng I-TEQ/(N m^3) of dioxin emission concentration. Generally, activated carbon adsorption technology is easily applied in industry and exhibits an ideal dioxin removal effect. However, the gas-phase dioxins are only transferred to the solid-phase adsorbent by the technology, which requires further treatment of captured adsorbent and increases the application cost.

4.2.3. Catalytic Oxidation. Instead of transferring dioxin to liquid or solid phase, catalytic oxidation technology is marked as a highly promising technology of dioxin decomposition, which decomposes dioxin into harmless molecules and avoids additional treatment of fly ash. In the present research, catalysts for dioxin degradation can be divided into noble metal catalysts and transition metal oxide catalysts based on the active component.

Noble metals, including Au, Pd, Pt, Ru, and Rh, show excellent catalytic activity in the degradation of a chlorinated aromatic compound. Noble metal catalysts differ from transition oxide catalysts in catalytic oxidation of volatile organic compounds. Taking Pd as an example, in the process of catalytic degradation of the target molecule, the metal element is first oxidized by oxygen in the atmosphere to form $(Pd^{2+}O^{2-})$. After the target organic molecule is oxidized by $(Pd^{2+}O^{2-})$, Pd^{2+} is reduced to Pd. Therefore, the oxidation ability of noble metals is more important than the reducing ability.[153] Okumura et al.[154] used a multicomponent noble metal catalyst to degrade PCDD/Fs. The results show that the Ir catalyst presents rather poor catalytic activity, but the catalyst activity is greatly improved by combining Ir with other noble metal catalysts. Au/Fe_2O_3–Ir/La_2O_3, Pt/SnO_2–Ir/La_2O_3, and Ir/La_2O_3–Pt/SnO_2–Au/Fe_2O_3 show degradation efficiencies at 423 K of 78.8%, 69.9%, and 98.0%, respectively. It is evident that synergistic and multifunctional effects exist among the components of the multicomponent catalyst. Taralunga et al.[155] developed Pt/HFAU, Pt/Al_2O_3, and Pt/SiO_2 catalysts for the degradation of chlorobenzene, ranked on the basis of activity from high to low as Pt/HFAU > Pt/Al_2O_3 > Pt/SiO_2. The complete combustion temperature of chlorobenzene on Pt/HFAU is 350 °C, and highly chlorinated byproducts are formed. In general, some drawbacks impose restrictions on the industrial application, including high cost, poor chlorine resistance,[156] sintering at high temperature,[155,157] and highly chlorinated byproducts.[157,158]

Transition metal oxide catalysts usually contain one or more transition metal oxides as active components, which are evenly distributed on TiO_2, Al_2O_3, activated carbon, and other carriers. Compared with other TiO_2-based transition metal oxide catalysts (such as CrO_x, CuO_x, CeO_x, MnO_x, CoO_x, WO_x, MoO_x, and FeO_x), V_2O_5 as an active component features high activity, high selectivity, and strong chlorine resistance.[159−161] The V=O group in the vanadium oxide can achieve nucleophilic adsorption of dioxins through the C−Cl bond, and then an oxidation reaction occurs to form a series of intermediate products. The intermediate products are further oxidized in the process of adsorption and desorption and eventually form CO_2, H_2O, and HCl. SCR systems originally used for NO_x reduction with NH_3 are found to show effective removal and destruction ability of dioxin.[162−164] Yu et al.[165] show that appropriately elevated vanadium content over a TiO_2 support presents better catalytic capability of PCDD/Fs with 5% optimal vanadium loading. However, increased V content could promote the oxidation of SO_2, weakening sulfur resistance of the catalyst. Therefore, the V content on commercial SCR catalysts is generally controlled below 1%. Moreover, the SCR system is generally installed after the baghouse in MSWI, where flue gas needs to be reheated to meet the activity temperature of the catalyst above 220 °C. A risk of dioxin formation through de novo reaction exists. Thus, varieties of catalyst modifications are investigated to improve the catalytic activity of the catalyst under low-temperature conditions, including active component doping,[159,166] carrier optimization,[167] and O_3 coupling.[168−170] Yu et al.[159] select cerium as the best second-active component based on a V_2O_5/TiO_2 catalyst that attains the highest PCDD/F destruction efficiency of 92.5% over VO_x−CeO_x/TiO_2 catalyst. Debecker et al.[171] observe that sulfated TiO_2 used as the support of VO_x facilitates the catalytic activity in PCDD/F degradation due to the larger specific surface area and increased surface acidity.

The catalysts for dioxin destruction can be designed as a honeycomb, plate, or corrugated plate in the form of a fixed bed when applied in the MSWI. Generally, the catalytic bed system is fixed behind the baghouse in MSWI, before where the flue gas needs to be reheated. Therefore, the risk of dioxin formation through de novo reaction is a great possibility. Moreover, the higher cost of equipment investment and extra heat energy is required. Additionally, the catalytic bag filter system combines the function of dust removal and catalytic degradation, achieving simultaneous removal of particle-phase and gas-phase PCDD/Fs. The catalytic filter only replaces the raw filter installed at the existing baghouse, requiring no new equipment or operation procedures. Moreover, the removal efficiency with catalytic filter is higher than the combination of activated carbon injection and baghouse filter in pilot tests.[172] Chang et al.[173] determined that the removal efficiencies achieved with the CF in the pilot test at 180 °C are 99.50%. Thus, the catalytic filter system can decrease the PCDD/F content in fly ash and decrease the burden of additional treatment of fly ash, which is considered as hazardous waste.

4.2.4. Wet Scrubbers. Wet scrubbers are used to reduce acidic gases and particulate matter simultaneously. The control of PCDD/Fs by wet scrubbers has attracted attention in MSWI for many years. Wevers et al.[174] found that wet scrubbers not only could remove acidic gases such as HCl, HF, and SO_2 but also could mediate the 70% reduction of PCDD/Fs (21.4 ± 21.8 ng of TEQ/(N m^3) at the inlet and 6.3 ± 5.7 ng of TEQ/(N m^3) at the outlet). Lehner et al.[175] verified that PCDD/Fs as well as fine dust, HCl, and HF could be highly separated in an advanced compact wet scrubber. The performance of the venutri scrubber and bag filter on reducing PCDD/F emissions was characterized, and the results showed that the removal efficiency of venutri scrubbers on the total PCDD/F and the total TEQ were 46.0% and 44.5%, respectively.[176] Significant removal efficiencies for PCDD/Fs were detected in the seven wet scrubbers in Korean waste incineration.[177]

Despite these positive results, higher PCDD/F emissions have been found following the use of wet scrubbers. Kim et al.[162] found that the TEQ of PCDD/Fs in the flue gas was 60 times higher after wet scrubbers. This phenomenon was also observed by several other researchers[178−181] and was caused by the so-called "memory effect", where PCDD/Fs that have been highly absorbed on the plastic packing materials in wet scrubbers could be desorbed under high temperature.

4.3. Treatment of the Fly Ash. The harmless disposal technology of fly ash can be roughly divided into thermal disposal and nonthermal disposal technologies (Table 6).

Table 6. Advantages and Disadvantages of Common MWSI Fly Ash Disposal Methods

classification	disposal technology	mechanism	advantages	disadvantages
thermal disposal technologies	sintering and melting	melt solid fly ash into a glassy, ceramic-like substance, or hard sintered body at high temperature	effective disposal of heavy metals and dioxins	the cost is very high, secondary fly ash is formed, and the flue gas needs to be treated
	low-temperature thermocatalysis	in inert atmosphere, dioxins decomposition at specific temperature (300−400 °C)	disposal of dioxins	high requirements for inert atmosphere environment, high energy consumption, no heavy metal stabilization
	hydrothermal	zeolite-like substance synthesized in the hydrothermal process with stabilization of heavy metals, ion adsorption, ion exchange, precipitation, and physical packaging	disposal of heavy metals and dioxins at the same time, with great potential for resource utilization	high equipment requirements, high energy consumption, and high cost
nonthermal disposal technologies	cement curing	formation of C−S−H gel, enveloping heavy metals	low cost, partially solidified heavy metals	large compatibilization, poor Cr curing effect, untreated dioxin
	stabilization of chemical agents	chemical reaction turning heavy metals into poor solubility, low-mobility, and low-toxicity substances	less capacity increase conducive to large-scale processing	cost of chemical agents too high and dioxin not treated
	biological/chemical extraction	microorganisms leaching heavy metals from solid phase to liquid phase, electrochemical recovery of heavy metals chemical reagents leaching heavy metals from solid phase to liquid phase, electrochemical recovery of heavy metals	complete harmlessness and resource utilization of heavy metals	long cycle, high cost of microbial cultivation, difficult to achieve recovery, untreated dioxin cost of large amount of acid and complexing agent, low concentration of heavy metals in fly ash
	mechanochemical methods	inducing chemical reactions through mechanical force	great reaction conditions and industrial application possibilities	long reaction time

Thermal disposal technologies mainly include sintering, molten vitrification, low-temperature thermal disposal, hydrothermal disposal, and supercritical water oxidation. Nonthermal disposal methods include cement curing, agent stabilization, biological/chemical leaching, and mechanochemical disposal.

4.3.1. Thermal Disposal Technologies. 4.3.1.1. Traditional Heat Treatment Methods. The traditional heat treatment method refers to the use of high temperatures to treat fly ash, making it a stable substance in the environment.[182,183] The advantage of the method is that the volume of fly ash is greatly reduced. In addition, because the porosity of the product is very small, the leaching of heavy metals is greatly reduced. At the same time, dioxins in fly ash are efficiently degraded due to the high temperature.[184] Thermal disposal of fly ash is considered to be one of the best methods for degrading dioxin in fly ash. There are reports that after thermal disposal of fly ash, more than 95% of the dioxin is degraded.[72] In general, the traditional heat treatment methods can be divided into the following three categories: (a) sintering, with the temperature used for treatment being usually 900−1000 °C [at this temperature, the crystal-phase boundary in the fly ash partially melts, removing most of the pores from the fly ash to form a dense and hard sintered body]; (b) vitrification, with fly ash and additives (glass precursors) being melted at a high temperature of 1100−1500 °C and contaminants such as heavy metals being enclosed in a crystal network (aluminosilicate Minerals); and (c) melting, with the temperature used for melting being the same as that for vitrification, except that no additives are added to the melting, so the glass body formed is relatively uneven. Equipment commonly used for heat treatment includes electric heating furnaces, microwave heating furnaces, incinerators, electric heating rotary kilns, and plasma melting furnaces.[72]

The products obtained by traditional heat treatment can be used for foundations, roadbeds, and other building materials, but the high-temperature treatment conditions result in high energy consumption and high cost of the technology and are only widely used in developed countries such as Japan. In addition, during the heat treatment process, volatile heavy metals such as Pb and Cd will enter the flue gas or secondary fly ash, and some of the decomposed dioxins will be generated again in the tail flue gas, so during the heat treatment process, the exhaust and secondary fly ash need to be further disposed.

4.3.1.2. Low-Temperature Heat Treatment. Low-temperature heat treatment refers to the harmless treatment technology of fly ash that uses a much lower temperature than traditional heat treatment to degrade dioxin in fly ash. Vogg and Stieglitz[46] found that more than 95% degradation of dioxins can be achieved at 600 °C for 2 h in an oxidizing atmosphere. In addition, a 90% degradation rate of dioxin can be obtained at 300 °C within 2 h in an inert atmosphere.

Studies have shown that some of the dioxins decomposed at high temperature will be generated again in the low-temperature region (300−500 °C): (1) organic carbon, oxygen, and chlorine atoms undergo heterogeneous reactions under the catalysis of the metal catalysts Cu and Fe to generate dioxins (commonly known as de novo synthesis[185,186]), and (2) other dioxins such as CPh, CBz, and PCBs are converted to produce dioxins (precursor generation).[11] Therefore, low-temperature heat treatment not only greatly reduces energy consumption but also effectively eliminates the problem of dioxin regeneration. It is a very good fly ash dioxin degradation technology, and its degradation effect has been confirmed by many research results,[187] but the technology needs to ensure an inert atmosphere, which increases the difficulty of actual operation.

4.3.1.3. Disposal by Hydrothermal Treatment. Hydrothermal treatment is a method of disposing fly ash in a high-temperature and high-pressure aqueous solution; simultaneous degradation of dioxins and efficient stabilization of heavy metals are efficiently achieved by the method. Under the conditions of high temperature and high pressure, the aqueous solution is in a subcritical or even supercritical state. The aqueous solution in this state is similar to an organic solvent and can be used as a good reaction medium. Therefore, the dioxins in the fly ash can be dissolved into the solution and degraded efficiently. In addition, under hydrothermal conditions, the silicon aluminum and other substances in the fly ash can form zeolite-like minerals with the aid of external additives, by which heavy metals in the ash are stabilized. Yamaguchi et al.[188] degraded dioxin in fly ash for the first time by hydrothermal method. They put fly ash in a NaOH solution containing methanol and reacted the mixture at 300 °C for 20 min. The dioxin concentration in fly ash decreased from 1100 to 0.45 ng/g with the method. Considering the toxicity of methanol, Hu et al.[189] used iron sulfate to assist in the degradation of dioxins and found that the temperature is the most important parameter during the hydrothermal degradation of dioxins, and the degradation efficiency of dioxins is increased by the addition of iron sulfate and ferrous sulfate. Ma et al.[190] and others found that the introduction of O_2 during the reaction can greatly accelerate the degradation rate of dioxin, decrease the reaction temperature, and shorten the time. They also found that the stabilization efficiency of each heavy metal exceeded 95% when fly ash from the fluidized bed was reacted in a 0.5 mol/L NaOH solution at 150 °C for 12 h.

Compared with sintering and vitrification, the hydrothermal method is more energy-saving, which further reduces the disposal cost. After disposal, fly ash can be used in the cement industry.[191] However, the promotion and application of the hydrothermal method also face problems. Under high temperature and high pressure, the reactor is easily corroded by alkali and chloride ions; additionally, the waste liquid after disposal is very alkaline, and the chloride ion concentration is also very high, making the disposal of waste liquid tricky. For fly ash with low contents of silicon and aluminum (such as grate fly ash), the effect of stabilizing heavy metals after disposal is not good.

4.3.1.4. Supercritical Water Oxidation. Supercritical water oxidation is similar to hydrothermal treatment, except that the former requires the aqueous solution to reach a supercritical state (>374 °C, >22.1 MPa). Due to its special dissolution and physical properties, supercritical water is a unique reaction medium for organic pollutants. When organic matter and oxygen are dissolved in supercritical water, the two react quickly in a homogeneous medium. There is no restriction between the phases, and the reaction rate is also very fast due to the ultrahigh temperature. Supercritical water oxidation has been shown to have very good degradation efficiency for dioxins in fly ash. Sako et al.[192] placed fly ash in an aqueous solution and added H_2O_2 at 400 °C and 30 MPa for 30 min to achieve a 99.7% dioxin degradation rate. Later, they developed another fly ash dioxin removal process based on supercritical oxidation. They first used activated carbon under supercritical conditions to absorb, extract, and concentrate the dioxins in the fly ash and then completely degraded the dioxins in the activated carbon through supercritical oxidation.

4.3.2. Nonthermal Treatment Technology. 4.3.2.1. Cement Curing. Cement solidification is the most common method of

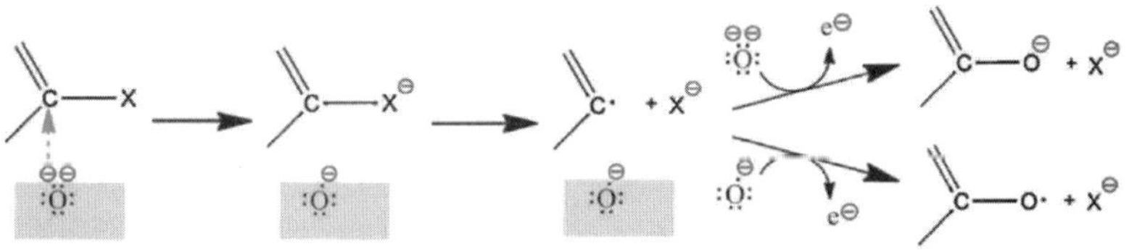

Figure 5. CaO dehalogenation mechanism in a mechanochemical degradation reaction (X refers to a halogen atom). Reprinted with permission from ref 194. Copyright 2016 Elsevier, Inc.

fly ash disposal, and it has been widely used in fly ash disposal processes around the world. First, the fly ash is mixed with cement or other hardening materials according to a certain ratio during the simple cement curing process. Then, water is added, and the mixture is stirred and removed from the mold for curing. During the hydration reaction of cement, heavy metals will eventually stay in the cement hydration product (such as hydrated silicate) in the form of hydroxide or complex through adsorption, ion exchange, surface complexation, chemical reaction, or precipitation. At the same time, the cement solid formed by the hydration reaction of fly ash and cement wraps the heavy metals in the fly ash to prevent it from leaching, and the addition of cement also provides an alkaline environment for heavy metals to inhibit their leaching. Disposal of various heavy metals in fly ash (except Cr) can achieve effective solidification. However, cement solidification has its own defects, such as the obvious increase in compatibilization and the poor long-term stability of heavy metals such as Cd, Cr^{6+}, and Zn. After the chloride salt in fly ash is dissolved, the structure of the solidified body is loose, and the risk of heavy metal leaching increases. Dioxins have no degradation effect; the disposal of fly ash can only be achieved by landfill disposal and does not have the conditions for recycling.

4.3.2.2. Stabilization of Chemical Agents. The chemical agent stabilization method involves changing the heavy metal ions from soluble and leaching forms to insoluble and nonleaching forms, such as a heavy metal polymer complex or an inorganic mineral salt, by adding a chemical reaction with the heavy metal compound in the fly ash, thereby achieving heavy metal stabilization. The chemical agent stabilization process is simple, and it is easy to achieve large-scale disposal of fly ash and has the advantage of less or no compatibilization compared to cement curing. Improving the structure and characteristics of the agent can also improve the long-term stability of the product. Fly ash after disposal also has the conditions for resource utilization. Organic agents mainly form chelate compounds with heavy metal ions through coordination groups to stabilize heavy metals. Compared with inorganic agents, they have the advantages of small dosage and strong resistance to acid leaching, but they have the problem of high price. Agents are usually selective for the stabilization of different heavy metals; thus, several agents are often used in combination to achieve full stabilization of heavy metals in fly ash, or chelating agents are used in conjunction with cement curing to achieve double solidification and stabilization of heavy metals.

4.3.2.3. Biological/Chemical Extraction. Biological/chemical leaching removes heavy metals from fly ash to achieve fly detoxification of gray heavy metals. In the long run, heavy metal leaching is more harmless to fly ash and heavy metals. It can eliminate the environmental pollution of heavy metals in fly ash after disposal. In addition, the heavy metals obtained by leaching can be reused by electrochemical recovery, and fly ash has the conditions for resource utilization. The methods of heavy metal leaching are mainly divided into biological leaching and chemical leaching. Biological leaching is a leaching method that uses the metabolism of specific microorganisms to convert insoluble heavy metals into soluble heavy metals and then transfer from the solid phase to the liquid phase. For example, *Thiobacillus ferrooxidans* adsorbs heavy metal sulfides through its own exogenous polymers and relies on its own internal catalytic enzymes to oxidize heavy metals into soluble sulfates at the same time. The sulfide is oxidized, which promotes the dissolution of heavy metals. The leaching effect of biological leaching is affected by many factors, including bacterial species, temperature, oxygen and CO_2 concentration, pH, mineral components, and inhibitory factors. Its leaching cycle is long, and the cost of microorganism cultivation is high. This method is still in the research stage.

Chemical leaching is a method for extracting heavy metals in fly ash through chemical agents. The leaching agents used include inorganic acids (nitric acid, hydrochloric acid, and sulfuric acid), organic acids (acetic acid, formic acid, and oxalic acid), alkalis (NaOH and KOH), and complexing agents {ethylenediaminetetraacetic acid (EDTA) or ethylenediaminetetraacetic acid disodium salt (EDTA-2Na), and nitrilotriacetic acid (NTA series)}. Among them, nitric acid and hydrochloric acid have the best extraction effects and can extract most heavy metals. Sulfuric acid can extract other heavy metals except Pb, and acetic acid has a very good extraction effect on Pb.[193] NaOH and KOH are mainly used to extract Zn and Pb and other amphoteric metals. Due to the high content of alkaline substances in fly ash, chemical leaching usually consumes a large number of drugs such as acids and complexing agents. The extraction, separation, and purification of heavy metals in the solution after leaching, including the disposal of the final waste liquid, will greatly increase the disposal cost.

4.3.2.4. Disposal by Mechanochemical Methods. Mechanochemistry (MC) modifies solid reactants through multiple modes of mechanical force, such as collision, compression, shear, and friction. The reactivity is increased by inducing changes in their physicochemical properties, thereby activating or accelerating the chemical reaction between solids. The mechanochemical degradation of halogenated persistent organic pollutants is usually accompanied by dehalogenation and other reactions. The other reactions include the formation of inorganic halides, the decomposition of benzene rings or carbon chains, the production of small molecular hydrocarbons, the polymerization of carbon molecules, and the oxidation of carbon molecules (air or oxidizing atmosphere). The dehalogenation reaction is usually dominant among these reactions (Figure 5). Halogenated pollutants are mainly degraded by dehalogenation reactions, and the main function of additives is dehalogenation. In the ball milling process of halogenated pollutants and additives, the pollutant molecules are first adsorbed on the surface of the additives under the action of ball milling. Then, the additives are gradually activated under the action of mechanical energy, and the activated additives generate free electrons or free radicals to be transferred to contaminants, resulting in dehalogenation.

Mechanochemical methods can not only degrade POPs but also stabilize heavy metals. Montinaro et al.[195] used dry ball milling to repair sand, bentonite, and kaolin contaminated with Cd, Pb, and Zn. It was found that the three heavy metals in the soil could be effectively solidified. After the disposal, the leaching concentration of heavy metals in the soil even meets the EPA Standard for potable water. The mechanism is that heavy metals in soil are gradually irreversibly adsorbed to the newly formed surface of the soil particles and the crystal network structure during the ball milling process, thereby being solidified. Nomura et al.[196] found that the additive CaO can not only achieve the degradation of dioxin in fly ash but also stabilize Pb. The 93% leaching concentration of Pb in fly ash is decreased after the treatment. Li and others have also achieved the solidification of Pb in fly ash without the use of additives through wet mechanochemical treatment, with a curing rate of 96%. The solidification mechanism is explained as heavy metal Pb being continuously broken and aggregated in the fly ash particles. In the process, it is wrapped inside the particles, making it difficult to leach.

5. CHALLENGES, FUTURE DIRECTIONS, AND PERSPECTIVES

Developing reliable mechanisms for the PCDD/F formation is challenging work. Numerous reactions happen in a short time, and it is difficult to detect and quantify the available species and intermediates. Quantum chemical calculation appears to be a potential method to address many of these deficiencies. And with the development of the detection technology, some intermediates can be measured in the near future to verify some new theory.

The online measurement of PCDD/F by indicators consists of two parts: the indicator measurement system and the mapping model between indicators and dioxin. The problems of the method include the high equipment complexity and difficult maintenance at industrial sites. At the same time, high equipment costs also make the method difficult to promote. Moreover, the accuracy and stability of the mapping model are challenging to guarantee, considering the difficulty involving multiple operating conditions obtaining a large amount of effective observation data and the construction of a reliable mapping model. The volatility of raw material properties in different MSWI processes and the differences in exhaust gas treatment equipment make it difficult to maintain a stable indicator/correlation. At present, most mapping models are only established by a single indicator. How to construct a fusion model from the perspective of multiple indicators to adapt to different industrial realities is a problem worthy of in-depth study. Additionally, the feasibility and maintainability of the method need to be further strengthened. The key parameters that affect dioxin emissions are determined to establish a feedback adjustment mechanism for the incinerator operation of the dioxin online rapid detection system.

Though the application of inhibitors achieved a significant inhibitory effect on the formation of PCDD/Fs, the improper use of traditional inhibitors can lead to several problems. For instance, utilization of sulfur- or nitrogen-based inhibitors would increase the concentration of SO_2 or NO_x in the flue gas, leading to an increased burden on air pollution control devices.[197] Moreover, sulfur-based inhibitors will produce sulfate, which adheres to the surface of the heat exchanger, causing corrosion during the operation.[198] It is extremely important to clarify the parameters (e.g., temperature and amount) of the application of an inhibitor. Additionally, the development of new, economical, and environmentally friendly inhibitors needs attention. Moreover, the memory effect is still a problem, increasing the PCDD/Fs in wet scrubbers. Hence, more study about the factors influencing the memory effect is needed for further optimization of the design of wet scrubbers.

Treatment of the flue gas after the combustion zone is the final guarantee for dioxin emission compliance. To control the emission of pollutants such as fly ash, acid gases (NO_x, SO_x, and so on), heavy metals, and dioxin generated during the waste incineration process, the waste incineration system will be equipped with APCDs in the tail combustion area. Generally, APCDs include (1) deacidification devices, including wet, semidry, and dry deacidification devices; (2) denitrification devices, including selective noncatalytic reduction (SNCR) and selective catalytic reduction (SCR) with a synergetic effect on dioxin degradation; (3) activated carbon adsorption devices, which are used to adsorb mercury vapor and organic pollutants such as dioxin in flue gas; and (4) dust removal devices, including bag dust removal and electrostatic dust removal devices. However, the electrostatic dust removal device has an additional effect on the regeneration of dioxin; currently, baghouse is mostly used in waste incineration systems. Most modification methods of activated carbon have been well-investigated to obtain activated carbon with a large specific surface area and pore volume, which are both key characteristics of the adsorption ability. Since the size of dioxin molecules is approximately 1−2 nm, it is generally believed that mesopores are the key to the adsorption of dioxin by activated carbon.[199] The directional design of mesopore structure and pore size distribution of activated carbon through modification is one of the research hotspots. In addition, because mercury vapor in the flue gas is easily captured by AC, synergistic removal of dioxins and heavy metals is an important direction for the application of activated carbon. The future direction of catalytic oxidation includes simultaneous removal of NO_x and PCDD/Fs, which are both present in flue gas, promoting low-temperature catalytic activity of the catalyst and improving the resistance to water vapor and chlorine.

Although the thermal treatment method can effectively dispose of dioxins and heavy metals in MSWI fly ash, a series of problems exists in practical industrial applications. For example, all thermal disposal methods have the disadvantage of high energy consumption. When sintering and melting methods are used to treat fly ash, the flue gas generated in the process needs to be disposed of, and the flue gas usually carries secondary fly ash, with higher contents of heavy metals and dioxins. In China, there is a national standard (GB 30485-2013) for the coordinated disposal of fly ash by cement kilns. However, due to the need to ensure the quality of cement, the amount of fly ash added is very low, and the disposal capacity is limited. A sintering plant is an important link in the whole production chain of comprehensive iron and steel enterprises. Low-cost iron-bearing sinter is provided for blast furnaces. Lots of solid waste is disposed of, containing iron, carbon, and fly ash produced in the processes of iron making, steel making, and steel rolling. Meanwhile, a back-end flue gas treatment device controls the emission of sulfur oxide, nitrogen oxide, and dioxins to the standard. Therefore, the coordinated disposal of fly ash from MSWI with existing equipment in industrial production seems to be a feasible route.

Nonthermal disposal methods have the advantages of low energy consumption but cannot degrade dioxins. The cement solidification method results in capacity increase, which is not conducive to transportation and landfill. Moreover, the subsequent products do not have the conditions for resource utilization. As a new treatment method of fly ash, the mechanized method can dispose of heavy metals and dioxins at the same time with a small number of additives. Additionally, there are low requirements for equipment and atmosphere, among other factors. The subsequent products can be recycled, which may become an important treatment method for fly ash in the future. However, additives are the most important parameters in the mechanochemical method. Currently, the practicability, degradation effect and cost of additives are different. Screening out efficient, safe, and low-cost additives will be the focus of future research on the treatment of fly ash by the mechanochemical method.

■ AUTHOR INFORMATION

Corresponding Author

Shengyong Lu — *Institute for Thermal Power Engineering, Zhejiang University, Hangzhou 310027, China;* orcid.org/0000-0003-2684-3498; Email: lushy@zju.edu.cn

Authors

Yaqi Peng — *Institute for Thermal Power Engineering, Zhejiang University, Hangzhou 310027, China;* orcid.org/0000-0002-5331-5968

Xiaodong Li — *Institute for Thermal Power Engineering, Zhejiang University, Hangzhou 310027, China*

Jianhua Yan — *Institute for Thermal Power Engineering, Zhejiang University, Hangzhou 310027, China*

Kefan Cen — *Institute for Thermal Power Engineering, Zhejiang University, Hangzhou 310027, China*

Complete contact information is available at:
https://pubs.acs.org/10.1021/acs.energyfuels.0c02446

Notes

The authors declare no competing financial interest.

■ ACKNOWLEDGMENTS

This work was supported by the National Key Research and Development Project (Grant No. 2017YFE0107600) and the National Natural Science Foundation of China (Grant Nos. 51976192 and 51676172).

■ ABBREVIATIONS

AC = activated carbon
AhR = aryl hydrocarbon receptors
APCD = air pollution control devices
BDMs = bioanalytical detection methods
BF = baghouse filter
CALUX = chemically activated luciferase expression
CBz = chlorobenzenes
CPh = chlorophenols
DCBz = dichlorobenzene
EDTA = ethylenediaminetetraacetic acid
EDTA-2Na = ethylenediaminetetraacetic acid disodium salt
EIA = enzyme immunoassay
EP = electrostatic precipitator
EROD = ethoxyresorufin-*O*-demethylase
FF = fabric filter
HWI = hazardous waste incinerator
HxCBz = hexachlorobenzene
IT = ion trap
MC = mechanochemistry
MCB = monochlorobenzene
MSWI = municipal solid waste incinerator
NTA = nitrilotriacetic acid
PAH = polycyclic aromatic hydrocarbons
PCA = principal component analysis
PCB = polychlorinated biphenyls
PCDD = polychlorinated dibenzo-*p*-dioxins
PCDF = polychlorinated dibenzofurans
PeCBz = pentachlorobenzene
PI = photon ionization
PIC = products of inadequate combustion
PVC = poly(vinyl chloride)
RDF = refuse-derived fuel
REMPI = resonance-enhanced multiphoton
SCR = selective catalytic reduction
SDS = semidry scrubber
TCDD = tetrachloro dibenzo-*p*-dioxins
TD = thermal desorption
TeCBz = tetrachlorobenzene
TEF = toxic equivalence factor
TEQ = toxic equivalent quantity
TOFMS = time-of-flight mass spectrometer
TrCBz = trichlorobenzene
VUV = vacuum ultraviolet

■ REFERENCES

(1) Lohmann, R.; Jones, K. C. Dioxins and furans in air and deposition: a review of levels, behaviour and processes. *Sci. Total Environ.* **1998**, *219* (1), 53−81.

(2) van Zorge, J. A.; van Wijnen, J. H.; Theelen, R. M.; Olie, K.; van den Berg, M. Assessment of the toxicity of mixtures of halogenated dibenzo-p-dioxins and dibenzofurans by use of toxicity equivalency factors (TEF). *Chemosphere* **1989**, *19* (12), 1881−1895.

(3) Bellin, J. S.; Barnes, D. G. *Interim procedures for estimating risks associated with exposures of mixtures of chlorinated dibenzo-p-dioxins and-dibenzofurans (CDDS and CDFS)*; U.S. Environmental Protection Agency (EPA): Washington, DC, USA. 1987.

(4) Barnes, D. *Interim procedures for estimating risk associated with exposures to mixtures of chlorinated dibenzo-p-dioxines and dibenzofurans*; Risk Assessment Forum, U.S. Environmental Protection Agency (EPA): Washington, DC, USA, 1989.

(5) Olie, K.; Vermeulen, P.; Hutzinger, O. Chlorodibenzo-p-dioxins and chlorodibenzofurans are trace components of fly ash and flue gas of some municipal incinerators in the Netherlands. *Chemosphere* **1977**, *6* (8), 455−459.

(6) Rappe, C.; Marklund, S.; Bergqvist, P. A.; Hansson, M. Polychlorinated dioxins (PCDDs), dibenzofurans (PCDFs) and other polynuclear aromatics (PCPNAs) formed during PCB fires. *Chem. Scr.* **1982**, *20* (1−2), 56−61.

(7) Kulkarni, P. S.; Crespo, J. G.; Afonso, C. A. Dioxins sources and current remediation technologies—a review. *Environ. Int.* **2008**, *34* (1), 139−153.

(8) Fiedler, H. Release inventories of polychlorinated dibenzo-p-dioxins and polychlorinated dibenzofurans. *Dioxin and Related Compounds*; Springer: Cham, Switzerland, 2015; pp 1−27, DOI: 10.1007/698_2015_432.

(9) Grosso, M. Post-combustion PCDD/F formation and destruction mechanisms: experiences in a full scale waste incineration plant. Ph.D. Thesis, Politecnico di Milano, Milan, Italy, 2000.

(10) Buekens, A.; Cen, K. Waste incineration, PVC, and dioxins. *J. Mater. Cycles Waste Manage.* **2011**, *13* (3), 190−197.

(11) McKay, G. Dioxin characterisation, formation and minimisation during municipal solid waste (MSW) incineration. *Chem. Eng. J.* **2002**, *86* (3), 343−368.

(12) Peng, Y.; Chen, J.; Lu, S.; Huang, J.; Zhang, M.; Buekens, A.; Li, X.; Yan, J. Chlorophenols in municipal solid waste incineration: a review. *Chem. Eng. J.* **2016**, *292*, 398−414.

(13) Hung, P. C.; Chang, S. H.; Chang, M. B. Removal of chlorinated aromatic organic compounds from MWI with catalytic filtration. *Aerosol Air Qual. Res.* **2014**, *14* (4), 1215−1222.

(14) Oh, J.-E.; Gullett, B.; Ryan, S.; Touati, A. Mechanistic relationships among PCDDs/Fs, PCNs, PAHs, ClPhs, and ClBzs in municipal waste incineration. *Environ. Sci. Technol.* **2007**, *41* (13), 4705−4710.

(15) Chi, K. H.; Chang, M. B.; Chang-Chien, G. P.; Lin, C. Characteristics of PCDD/F congener distributions in gas/particulate phases and emissions from two municipal solid waste incinerators in Taiwan. *Sci. Total Environ.* **2005**, *347* (1−3), 148−162.

(16) Yan, J.; Chen, T.; Li, X.; Zhang, J.; Lu, S.; Ni, M.; Cen, K. Evaluation of PCDD/Fs emission from fluidized bed incinerators co-firing MSW with coal in China. *J. Hazard. Mater.* **2006**, *135* (1−3), 47−51.

(17) Zhang, M.; Buekens, A.; Li, X. Dioxins from biomass combustion: an overview. *Waste Biomass Valorization* **2017**, *8* (1), 1−20.

(18) Zhang, M.; Buekens, A.; Li, X. Open burning as a source of dioxins. *Crit. Rev. Environ. Sci. Technol.* **2017**, *47* (8), 543−620.

(19) Zhou, S.; Liu, C.; Zhang, L. Critical Review on the Chemical Reaction Pathways Underpinning the Primary Decomposition Behavior of Chlorine-Bearing Compounds under Simulated Municipal Solid Waste Incineration Conditions. *Energy Fuels* **2020**, *34* (1), 1−15.

(20) Rathna, R.; Varjani, S.; Nakkeeran, E. Recent developments and prospects of dioxins and furans remediation. *J. Environ. Manage.* **2018**, *223*, 797−806.

(21) Wu, G.; Weber, R.; Ren, Y.; Peng, Z.; Watson, A.; Xie, J. State of art control of dioxins/unintentional POPs in the secondary copper industry: A review to assist policy making with the implementation of the Stockholm Convention. *Emerging Contaminants* **2020**, *6*, 235−249.

(22) Lei, R.; Liu, W.; Wu, X.; Ni, T.; Jia, T. A review of levels and profiles of polychlorinated dibenzo-p-dioxins and dibenzofurans in different environmental media from China. *Chemosphere* **2020**, *239*, 124685.

(23) Stanmore, B. The formation of dioxins in combustion systems. *Combust. Flame* **2004**, *136* (3), 398−427.

(24) Tuomisto, J.; Vartiainen, T.; Tuomisto, J. T. *Synopsis on dioxins and PCBs*; National Institute for Health and Welfare (THL): Helsinki, Finland, 2011.

(25) Zhou, H.; Meng, A.; Long, Y.; Li, Q.; Zhang, Y. A review of dioxin-related substances during municipal solid waste incineration. *Waste Manage.* **2015**, *36*, 106−118.

(26) Nganai, S.; Dellinger, B.; Lomnicki, S. PCDD/PCDF ratio in the precursor formation model over CuO surface. *Environ. Sci. Technol.* **2014**, *48* (23), 13864−13870.

(27) Ambient air and waste gas determination of polychlorinated dibenzo-p-dioxins (PCDDs) and polychlorinated dibenzofurans (PCDFs) isotope dilution. *Method Standard*, HRGC-HRMS HJ77.2-2008; Ministry of Ecology and Environment: Beijing, People's Republic of China, 2008.

(28) Standard for pollution control on the municipal solid waste incineration. *Solid Wastes Pollution Control Standard*, GB 18485-2014; Ministry of Ecology and Environment: Beijing, People's Republic of China, 2014.

(29) *Emission guidelines for existing small municipal waste combustion units*, Final Rule; 40 CFR Part 60; U.S. Environmental Protection Agency (EPA): Washington, DC, USA, 2000.

(30) *New source performance standards for new small municipal waste combustion units*, Final Rule, 40 CFR Part 60; U.S. Environmental Protection Agency (EPA): Washington, DC, USA, 2000.

(31) U.S. EPA. *Standards of performance for new stationary sources and emission guidelines for existing sources: Large municipal waste combustors*, Final Rule, 40 CFR Part 60; U.S. Environmental Protection Agency (EPA): Washington, DC, USA, 2006.

(32) *Law concerning special measures against dioxins*, Law No. 105 of 1999; Environment Agency of Japan: Tokyo, Japan, 1999.

(33) Stationary source emissions: Determination of the mass concentration of PCDDs/PCDFs and dioxin-like PCBs. *European standard*, EN-1948 Parts 1−3; European Committee for Standardization: Brussels, Belgium, 2006.

(34) Method for determination of tetra-through octachlorodibenzo-p-dioxins, tetra-through octachlorodibenzofurans and dioxin-like polychlorinatedbiphenyls in stationary source emissions (Amendment 1). *Japanese Industrial Standard*, JIS K0311-2008; Japanese Standards Association: Tokyo, Japan, 2005.

(35) U.S. EPA. *Method 23—determination of polychlorinated dibenzopdioxins and polychlorinated dibenzofurans from stationary sources.* U.S. Environmental Protection Agency (EPA): Washington, DC, USA, 2017.

(36) Cao, X.; Lu, S.; Stevens, W. R.; Zhong, H.; Wu, K.; Li, X.; Yan, J. Fast indirect measurement of PCDD/F TEQ emission from municipal solid waste incineration: A review. *Waste Dispos. Sustainable Energy* **2019**, *1*, 39−51.

(37) Öberg, T.; Bergström, J. G. Emission and chlorination pattern of PCDD/PCDF predicted from indicator parameters. *Chemosphere* **1987**, *16* (6), 1221−1230.

(38) Kaune, A.; Lenoir, D.; Nikolai, U.; Kettrup, A. Estimating concentrations of polychlorinated dibenzo-p-dioxins and dibenzofurans in the stack gas of a hazardous waste incinerator from concentrations of chlorinated benzenes and biphenyls. *Chemosphere* **1994**, *29* (9−11), 2083−2096.

(39) Kaune, A.; Schramm, K.; Kettrup, A.; Jaeger, K. Indicator parameters for PCDD/F in the flue gas of the hazardous waste incinerator at Leverkusen, Germany. *Organohalogen Compd.* **1996**, *27*, 163−166.

(40) Kaune, A.; Lenoir, D.; Schramm, K.-W.; Zimmermann, R.; Kettrup, A.; Jaeger, K.; Rueckel, H.; Frank, F. Indicator parameters for polychlorinated dibenzodioxins and dibenzofurans: comparison of different incinerators. *Organohalogen Compd.* **1998**, *36*, 37−40.

(41) Cao, X.; Stevens, W. R.; Tang, S.; Lu, S.; Li, X.; Lin, X.; Tang, M.; Yan, J. Atline measurement of 1, 2, 4-trichlorobenzene for polychlorinated dibenzo-p-dioxin and dibenzofuran International Toxic Equivalent Quantity prediction in the stack gas. *Environ. Pollut.* **2019**, *244*, 202−208.

(42) Heger, H.; Zimmermann, R.; Blumenstock, M.; Kettrup, A. On-line real-time measurements at incineration plants: PAHs and a PCDD/F surrogate compound at stationary combustion conditions and during transient emission puffs. *Chemosphere* **2001**, *42* (5−7), 691−696.

(43) Buekens, A. *Incineration technologies*; Springer Science & Business Media: New York, 2013; DOI: 10.1007/978-1-4614-5752-7.

(44) Lin, T.; Hu, Z.; Zhang, G.; Li, X.; Xu, W.; Tang, J.; Li, J. Levels and mass burden of DDTs in sediments from fishing harbors: the importance of DDT-containing antifouling paint to the coastal environment of China. *Environ. Sci. Technol.* **2009**, *43* (21), 8033−8038.

(45) Chiu, S.; Ho, K.; Chan, S.; So, O.; Lai, K. Characterization of contamination in and toxicities of a shipyard area in Hong Kong. *Environ. Pollut.* **2006**, *142* (3), 512−520.

(46) Vogg, H.; Stieglitz, L. Thermal behavior of PCDD/PCDF in fly ash from municipal incinerators. *Chemosphere* **1986**, *15* (9−12), 1373−1378.

(47) Zhang, M.; Buekens, A. De novo synthesis of dioxins: A review. *Int. J. Environ. Pollut.* **2016**, *60* (1−4), 63−110.

(48) Altarawneh, M.; Dlugogorski, B. Z.; Kennedy, E. M.; Mackie, J. C. Mechanisms for formation, chlorination, dechlorination and destruction of polychlorinated dibenzo-p-dioxins and dibenzofurans (PCDD/Fs). *Prog. Energy Combust. Sci.* **2009**, *35* (3), 245−274.

(49) Altwicker, E.; Schonberg, J.; Konduri, R. K.; Milligan, M. Polychlorinated dioxin/furan formation in incinerators. *Hazard. Waste Hazard. Mater.* **1990**, 7 (1), 73–87.

(50) Olie, K.; Addink, R.; Schoonenboom, M. Metals as catalysts during the formation and decomposition of chlorinated dioxins and furans in incineration processes. *J. Air Waste Manage. Assoc.* **1998**, 48 (2), 101–105.

(51) Kim, K.-S.; Hong, K.-H.; Ko, Y.-H.; Kim, M.-G. Emission Characteristics of PCDD/Fs, PCBs, Chlorobenzenes, Chlorophenols, and PAHs from Polyvinylchloride Combustion at Various Temperatures. *J. Air Waste Manage. Assoc.* **2004**, 54 (5), 555–562.

(52) Altwicker, E. R. Some laboratory experimental designs for obtaining dynamic property data on dioxins. *Sci. Total Environ.* **1991**, 104 (1–2), 47–72.

(53) Khachatryan, L.; Asatryan, R.; Dellinger, B. An elementary reaction kinetic model of the gas-phase formation of polychlorinated dibenzofurans from chlorinated phenols. *J. Phys. Chem. A* **2004**, 108 (44), 9567–9572.

(54) Sidhu, S.; Edwards, P. Role of phenoxy radicals in PCDD/F formation. *Int. J. Chem. Kinet.* **2002**, 34 (9), 531–541.

(55) Qu, X.; Wang, H.; Zhang, X.; Shi, X.; Xu, F.; Wang, W. Mechanistic and kinetic studies on the homogeneous gas-phase formation of PCDD/Fs from 2,4,5-trichlorophenol. *Environ. Sci. Technol.* **2009**, 43, 4068–4075.

(56) Fadli, A.; Briois, C.; Baillet, C.; Sawerysyn, J.-P. Experimental study on the thermal oxidation of chlorobenzene at 575–825 C. *Chemosphere* **1999**, 38 (12), 2835–2848.

(57) Xu, Z.; Lin, M.-C. Ab initio kinetics for the unimolecular reaction C6H5OH· CO+ C5H6. *J. Phys. Chem. A* **2006**, 110 (4), 1672–1677.

(58) Zhu, L.; Bozzelli, J. W. Kinetics and Thermochemistry for the Gas-Phase Keto- Enol Tautomerism of Phenol↔ 2, 4-Cyclohexadienone. *J. Phys. Chem. A* **2003**, 107 (19), 3696–3703.

(59) Evans, C. S.; Dellinger, B. Mechanisms of dioxin formation from the high-temperature oxidation of 2-chlorophenol. *Environ. Sci. Technol.* **2005**, 39 (1), 122–127.

(60) Linstrom, P. J.; Mallard, W. G. The NIST Chemistry WebBook: A chemical data resource on the internet. *J. Chem. Eng. Data* **2001**, 46 (5), 1059–1063.

(61) Libit, L.; Hoffmann, R. Detailed orbital theory of substituent effects. Charge transfer, polarization, and the methyl group. *J. Am. Chem. Soc.* **1974**, 96 (5), 1370–1383.

(62) Armstrong, D. R.; Cameron, C.; Nonhebel, D. C.; Perkins, P. G. Oxidative coupling of phenols. Part 8. A theoretical study of the coupling of phenoxyl radicals. *J. Chem. Soc., Perkin Trans. 2* **1983**, 2 (5), 575–579.

(63) Dickson, L.; Lenoir, D.; Hutzinger, O. Quantitative comparison of de novo and precursor formation of polychlorinated dibenzo-p-dioxins under simulated municipal solid waste incinerator post-combustion conditions. *Environ. Sci. Technol.* **1992**, 26 (9), 1822–1828.

(64) Milligan, M. S.; Altwicker, E. R. Chlorophenol reactions on fly ash. 2. Equilibrium surface coverage and global kinetics. *Environ. Sci. Technol.* **1996**, 30 (1), 230–236.

(65) Milligan, M. S.; Altwicker, E. R. Chlorophenol reactions on fly ash. 1. Adsorption/desorption equilibria and conversion to polychlorinated dibenzo-p-dioxins. *Environ. Sci. Technol.* **1996**, 30 (1), 225–229.

(66) Qian, Y.; Zheng, M.; Liu, W.; Ma, X.; Zhang, B. Influence of metal oxides on PCDD/Fs formation from pentachlorophenol. *Chemosphere* **2005**, 60 (7), 951–958.

(67) Lomnicki, S.; Dellinger, B. Formation of PCDD/F from the pyrolysis of 2-chlorophenol on the surface of dispersed copper oxide particles. *Proc. Combust. Inst.* **2002**, 29 (2), 2463–2468.

(68) Bandara, J.; Mielczarski, J.; Lopez, A.; Kiwi, J. 2. Sensitized degradation of chlorophenols on iron oxides induced by visible light: comparison with titanium oxide. *Appl. Catal., B* **2001**, 34 (4), 321–333.

(69) Stieglitz, L. Selected topics on the de novo synthesis of PCDD/PCDF on fly ash. *Environ. Eng. Sci.* **1998**, 15 (1), 5–18.

(70) Stieglitz, L.; Jay, K.; Hell, K.; Wilhelm, J.; Polzer, J.; Buekens, A. Investigation of the formation of polychlorodibenzodioxins/-furans and of other organochlorine compounds in thermal industrial processes. *Wiss. Ber. - Forschungszent. Karlsruhe* **2003**, 6867, 48.

(71) Addink, R.; Olie, K. Mechanisms of formation and destruction of polychlorinated dibenzo-p-dioxins and dibenzofurans in heterogeneous systems. *Environ. Sci. Technol.* **1995**, 29, 1425–1435.

(72) Buekens, A.; Huang, H. Comparative evaluation of techniques for controlling the formation and emission of chlorinated dioxins/furans in municipal waste incineration. *J. Hazard. Mater.* **1998**, 62 (1), 1–33.

(73) Hell, K.; Stieglitz, L.; Dinjus, E. Mechanistic aspects of the de-novo synthesis of PCDD/PCDF on model mixtures and MSWI fly ashes using amorphous 12C-and 13C-labeled carbon. *Environ. Sci. Technol.* **2001**, 35 (19), 3892–3898.

(74) Takaoka, M.; Shiono, A.; Nishimura, K.; Yamamoto, T.; Uruga, T.; Takeda, N.; Tanaka, T.; Oshita, K.; Matsumoto, T.; Harada, H. Dynamic change of copper in fly ash during de novo synthesis of dioxins. *Environ. Sci. Technol.* **2005**, 39 (15), 5878–5884.

(75) U.S. EPA, Method 23-Determination of Polychlorinated Dibenzo-p-dioxins and Polychlorinated Dibenzofurans from Stationary Sources, 2017, https://www.epa.gov/sites/production/files/2017-08/documents/method_23.pdf.

(76) Rivera-Austrui, J.; Martínez, K.; Adrados, M.; Abalos, M.; Abad, E. Analytical approach and occurrence for the determination of mass concentration of PCDD/PCDF and dl-PCB in flue gas emissions using long-term sampling devices. *Sci. Total Environ.* **2012**, 435, 7–13.

(77) Reinmann, J.; Weber, R.; Haag, R. Long-term monitoring of PCDD/PCDF and other unintentionally produced POPs—concepts and case studies from Europe. *Science China Chemistry* **2010**, 53 (5), 1017–1024.

(78) Rivera-Austrui, J.; Borrajo, M.; Martinez, K.; Adrados, M.; Abalos, M.; Van Bavel, B.; Rivera, J.; Abad, E. Assessment of polychlorinated dibenzo-p-dioxin and dibenzofuran emissions from a hazardous waste incineration plant using long-term sampling equipment. *Chemosphere* **2011**, 82 (9), 1343–1349.

(79) Rivera-Austrui, J.; Martínez, K.; Marco-Almagro, L.; Ábalos, M.; Abad, E. Long-term sampling of dioxin-like substances from a clinker kiln stack using alternative fuels. *Sci. Total Environ.* **2014**, 485, 528–533.

(80) Conesa, J. A.; Ortuño, N.; Abad, E.; Rivera-Austrui, J. Emissions of PCDD/Fs, PBDD/Fs, dioxin like-PCBs and PAHs from a cement plant using a long-term monitoring system. *Sci. Total Environ.* **2016**, 571, 435–443.

(81) Behnisch, P. A.; Hosoe, K.; Sakai, S.-i. Bioanalytical screening methods for dioxins and dioxin-like compounds—a review of bioassay/biomarker technology. *Environ. Int.* **2001**, 27 (5), 413–439.

(82) Tavakoly Sany, S. B.; Narimani, L.; Soltanian, F. K.; Hashim, R.; Rezayi, M.; Karlen, D. J.; Mahmud, H. An overview of detection techniques for monitoring dioxin-like compounds: Latest technique trends and their applications. *RSC Adv.* **2016**, 6 (60), 55415–55429.

(83) Harrison, R.; Eduljee, G. Immunochemical analysis for dioxins—progress and prospects. *Sci. Total Environ.* **1999**, 239 (1–3), 1–18.

(84) Otarola, G.; Castillo, H.; Marcellini, S. Aryl hydrocarbon receptor-based bioassays for dioxin detection: Thinking outside the box. *J. Appl. Toxicol.* **2018**, 38 (4), 437–449.

(85) Mertes, F.; Mumbo, J.; Pandelova, M.; Bernhöft, S.; Corsten, C.; Henkelmann, B.; Bussian, B. M.; Schramm, K-W. Comparative study of dioxin contamination from forest soil samples (BZE II) by mass spectrometry and EROD bioassay. *Environ. Sci. Pollut. Res.* **2018**, 25 (5), 3977–3984.

(86) Shi, P.; Zhou, S.; Xiao, H.; Qiu, J.; Li, A.; Zhou, Q.; Pan, Y.; Hollert, H. Toxicological and chemical insights into representative source and drinking water in eastern China. *Environ. Pollut.* **2018**, 233, 35–44.

(87) U.S. EPA. SW-846 Test Method 4435: Screening For Dioxin-Like Chemical Activity In Soils And Sediments Using The Calux Bioassay And Toxic Equivalents (TEQs) Determinations, https://www.epa.gov/sites/production/files/2015-12/documents/4435.pdf.

(88) He, G.; Tsutsumi, T.; Zhao, B.; Baston, D. S.; Zhao, J.; Heath-Pagliuso, S.; Denison, M. S. Third-generation Ah receptor-responsive luciferase reporter plasmids: Amplification of dioxin-responsive

elements dramatically increases CALUX bioassay sensitivity and responsiveness. *Toxicol. Sci.* 2011, *123* (2), 511–522.

(89) Brennan, J. C.; He, G.; Tsutsumi, T.; Zhao, J.; Wirth, E.; Fulton, M. H.; Denison, M. S. Development of species-specific Ah receptor-responsive third generation CALUX cell lines with enhanced responsiveness and improved detection limits. *Environ. Sci. Technol.* 2015, *49* (19), 11903–11912.

(90) Lavric, E. D.; Konnov, A. A.; De Ruyck, J. Surrogate compounds for dioxins in incineration. A review. *Waste Manage.* 2005, *25* (7), 755–765.

(91) Kato, M.; Urano, K. Convenient substitute indices to toxic equivalent quantity for controlling and monitoring dioxins in stack gas from waste incineration facilities. *Waste Manage.* 2001, *21* (1), 55–62.

(92) Kato, M.; Urano, K. A measuring method of chlorobenzenes as a convenient substitute index of dioxins in stack gas from waste incineration facilities. *Waste Manage.* 2001, *21* (1), 63–68.

(93) Wang, T.; Chen, T.; Lin, B.; Lin, X.; Zhan, M.; Li, X. Emission characteristics and relationships among PCDD/Fs, chlorobenzenes, chlorophenols and PAHs in the stack gas from two municipal solid waste incinerators in China. *RSC Adv.* 2017, *7* (70), 44309–44318.

(94) Zimmermann, R.; Heger, H. J.; Blumenstock, M.; Dorfner, R.; Schramm, K. W.; Boesl, U.; Kettrup, A. On-line measurement of chlorobenzene in waste incineration flue gas as a surrogate for the emission of polychlorinated dibenzo-p-dioxins/furans (I-TEQ) using mobile resonance laser ionization time-of-flight mass spectrometry. *Rapid Commun. Mass Spectrom.* 1999, *13* (5), 307–314.

(95) Blumenstock, M.; Zimmermann, R.; Schramm, K.-W.; Kettrup, A. Influence of combustion conditions on the PCDD/F-, PCB-, PCBz- and PAH-concentrations in the post-combustion chamber of a waste incineration pilot plant. *Chemosphere* 2000, *40* (9), 987–993.

(96) Blumenstock, M.; Zimmermann, R.; Schramm, K.-W.; Kettrup, A. Identification of surrogate compounds for the emission of PCDD/F (I-TEQ value) and evaluation of their on-line real-time detectability in flue gases of waste incineration plants by REMPI-TOFMS mass spectrometry. *Chemosphere* 2001, *42* (5–7), 507–518.

(97) Kaune, A.; Lenoir, D.; Schramm, K.-W.; Zimmermann, R.; Kettrup, A.; Jaeger, K.; Rückel, H.; Frank, F. Chlorobenzenes and chlorophenols as indicator parameters for chlorinated dibenzodioxins and dibenzofurans in incineration processes: influences of various facilities and sampling points. *Environ. Eng. Sci.* 1998, *15* (1), 85–95.

(98) Iino, F.; Imagawa, T.; Takeuchi, M.; Sadakata, M. De novo synthesis mechanism of polychlorinated dibenzofurans from polycyclic aromatic hydrocarbons and the characteristic isomers of polychlorinated naphthalenes. *Environ. Sci. Technol.* 1999, *33* (7), 1038–1043.

(99) Li, X.; Yin, X.; Lu, S.; Gu, Y.; Yan, J.; Ni, M.; Cen, K. The correlation between PAHs and dioxins formation during coal and municipal solid waste co-incineration process. *J. Eng. Thermophys.* 2006, No. 04, 691–694.

(100) Yan, M.; Li, X.; Zhang, X.; Liu, K.; Yan, J.; Cen, K. Correlation between PAHs and PCDD/Fs in municipal solid waste incinerators. *J. Zhejiang Univ. (Eng. Sci.)* 2010, *44* (06), 1118–1121 + 1132.

(101) Cao, X.; Lu, S.; Stevens, W. R.; Zhong, H.; Wu, K.; Li, X.; Yan, J. Fast indirect measurement of PCDD/F TEQ emission from municipal solid waste incineration: a review. *Waste Disposal & Sustainable Energy* 2019, *1* (1), 39–51.

(102) Oh, J.-E.; Lee, K.-T.; Lee, J.-W.; Chang, Y.-S. The evaluation of PCDD/Fs from various Korean incinerators. *Chemosphere* 1999, *38* (9), 2097–2108.

(103) Weber, R.; Sakurai, T.; Ueno, S.; Nishino, J. Correlation of PCDD/PCDF and CO values in a MSW incinerator–indication of memory effects in the high temperature/cooling section. *Chemosphere* 2002, *49* (2), 127–134.

(104) Tsuruga, S.; Suzuki, T.; Takatsudo, Y.; Seki, K.; Yamauchi, S.; Kuribayashi, M.; Morii, S. On-line monitoring system of P5CDF homologues in waste incineration plants using VUV-SPI-IT-TOFMS. *Environ. Sci. Technol.* 2007, *41* (10), 3684–3688.

(105) Morii, S.; Suzuki, T.; Seki, K.; Okada, M.; Okuno, S.; Tanaka, M. Dioxins and their precursors On-line Monitoring Method for Incineration Plants. *Proc. Symp. Environ. Eng.* 2004, *2004.14*, 231–234.

(106) Heger, H. J.; Zimmermann, R.; Dorfner, R.; Beckmann, M.; Griebel, H.; Kettrup, A.; Boesl, U. On-line emission analysis of polycyclic aromatic hydrocarbons down to pptv concentration levels in the flue gas of an incineration pilot plant with a mobile resonance-enhanced multiphoton ionization time-of-flight mass spectrometer. *Anal. Chem.* 1999, *71* (1), 46–57.

(107) Hafner, K.; Zimmermann, R.; Rohwer, E. R.; Dorfner, R.; Kettrup, A. A capillary-based supersonic jet inlet system for resonance-enhanced laser ionization mass spectrometry: principle and first on-line process analytical applications. *Anal. Chem.* 2001, *73* (17), 4171–4180.

(108) Oudejans, L.; Touati, A.; Gullett, B. K. Real-time, on-line characterization of diesel generator air toxic emissions by resonance-enhanced multiphoton ionization time-of-flight mass spectrometry. *Anal. Chem.* 2004, *76* (9), 2517–2524.

(109) Gullett, B.; Touati, A.; Oudejans, L.; Tabor, D. Real time monitoring of PCDD/PCDF for transient characterization and process control. *Proceedings, Hazardous Waste Combustors Conference and Exhibition*, Charleston, SC, USA, Mar. 13–14, 2007; Air & Waste Management Association (AWMA): Pittsburgh, PA, USA, 2007; pp 13–14.

(110) Gullett, B. K.; Oudejans, L.; Tabor, D.; Touati, A.; Ryan, S. Near-real-time combustion monitoring for PCDD/PCDF indicators by GC-REMPI-TOFMS. *Environ. Sci. Technol.* 2012, *46* (2), 923–928.

(111) Gullett, B.; Oudejans, L.; Touati, A.; Ryan, S.; Tabor, D. Verification results of jet resonance-enhanced multiphoton ionization as a real-time PCDD/F emission monitor. *J. Mater. Cycles Waste Manage.* 2008, *10* (1), 32.

(112) Kadlec, P.; Gabrys, B.; Strandt, S. Data-driven soft sensors in the process industry. *Comput. Chem. Eng.* 2009, *33* (4), 795–814.

(113) Chang, N.-B.; Huang, S.-H. Statistical modelling for the prediction and control of PCDDs and PCDFs emissions from municipal solid waste incinerators. *Waste Manage. Res.* 1995, *13* (4), 379–400.

(114) Chang, N. B.; Chen, W. Prediction of PCDDs/PCDFs emissions from municipal incinerators by genetic programming and neural network modeling. *Waste Manage. Res.* 2000, *18* (4), 341–351.

(115) Bunsan, S.; Chen, W.-Y.; Chen, H.-W.; Chuang, Y. H.; Grisdanurak, N. Modeling the dioxin emission of a municipal solid waste incinerator using neural networks. *Chemosphere* 2013, *92* (3), 258–264.

(116) Tang, J.; Qiao, J.; Gu, K.; Yan, A. Dioxin soft measuring method in municipal solid waste incineration based on virtual sample generation. *2017 Chinese Automation Congress (CAC)*; IEEE: Piscataway, NJ, USA, 2017; pp 7323–7328, DOI: 10.1109/CAC.2017.8244101.

(117) Hatanaka, T.; Kitajima, A.; Takeuchi, M. Role of Chlorine in Combustion Field in Formation of Polychlorinated Dibenzo-p-dioxins and Dibenzofurans during Waste Incineration. *Environ. Sci. Technol.* 2005, *39* (24), 9452–9456.

(118) Wirts, M.; Lorenz, W.; Bahadir, M. Does co-combustion of PVC and other plastics lead to enhanced formation of PCDD/F? *Chemosphere* 1998, *37* (8), 1489–1500.

(119) Bi, W.-z.; Zhao, R.-d.; Chen, T.-j.; Wu, J.-l.; Wu, J.-h. Study on the formation of PCDD/Fs in PVC chemical looping combustion. *Journal of Fuel Chemistry and Technology* 2015, *43* (7), 884–889.

(120) Carroll, W. F. The relative contribution of wood and poly(vinyl chloride) to emissions of PCDD and PCDF from house fires. *Chemosphere* 2001, *45* (8), 1173–1180.

(121) Zhou, H.; Meng, A. H.; Long, Y. Q.; Li, Q. H.; Zhang, Y. G. A review of dioxin-related substances during municipal solid waste incineration. *Waste Manage.* 2015, *36*, 106–118.

(122) Tuppurainen, K.; Halonen, I.; Ruokojarvi, P.; Tarhanen, J.; Ruuskanen, J. Formation of PCDDs and PCDFs in municipal waste incineration and its inhibition mechanisms: A review. *Chemosphere* 1998, *36* (7), 1493–1511.

(123) Fujimori, T.; Takaoka, M.; Takeda, N. Influence of Cu, Fe, Pb, and Zn chlorides and oxides on formation of chlorinated aromatic compounds in MSWI fly ash. *Environ. Sci. Technol.* 2009, *43* (21), 8053–8059.

(124) Liu, H.; Kong, S.; Liu, Y.; Zeng, H. Pollution control technologies of dioxins in municipal solid waste incinerator. *Procedia Environ. Sci.* 2012, 16, 661−668.

(125) Griffin, R. D. A new theory of dioxin formation Response: in municipal solid waste combustion. *Chemosphere* 1986, 15, 1987−1990.

(126) Pekarek, V.; Puncochar, M.; Bures, M.; Grabic, R.; Fiserova, E. Effects of sulfur dioxide, hydrogen peroxide and sulfuric acid on the de novo synthesis of PCDD/F and PCB under model laboratory conditions. *Chemosphere* 2007, 66 (10), 1947−1954.

(127) Ke, S.; Jianhua, Y.; Xiaodong, L.; Shengyong, L.; Yinglei, W.; Muxing, F. Inhibition of de novo synthesis of PCDD/Fs by SO2 in a model system. *Chemosphere* 2010, 78 (10), 1230−1235.

(128) Fujimori, T.; Nishimoto, Y.; Shiota, K.; Takaoka, M. Contrasting Effects of Sulfur Dioxide on Cupric Oxide and Chloride during Thermochemical Formation of Chlorinated Aromatics. *Environ. Sci. Technol.* 2014, 48 (23), 13644−13651.

(129) Gullett, B. K.; Bruce, K. R.; Beach, L. O. Effect of sulfur dioxide on the formation mechanism of polychlorinated dibenzodioxin and dibenzofuran in municipal waste combustors. *Environ. Sci. Technol.* 1992, 26 (10), 1938−1943.

(130) Lindbauer, R. L.; Wurst, F.; Prey, T. Combustion dioxin supression in municipal solid waste incineration with sulphur additives. *Chemosphere* 1992, 25 (7−10), 1409−1414.

(131) Ruokojärvi, P. H.; Halonen, I. A.; Tuppurainen, K. A.; Tarhanen, J.; Ruuskanen, J. Effect of gaseous inhibitors on PCDD/F formation. *Environ. Sci. Technol.* 1998, 32 (20), 3099−3103.

(132) Ogawa, H.; Orita, N.; Horaguchi, M.; Suzuki, T.; Okada, M.; Yasuda, S. Dioxin reduction by sulfur component addition. *Chemosphere* 1996, 32 (1), 151−157.

(133) Raghunathan, K.; Gullett, B. K. Role of sulfur in reducing PCDD and PCDF formation. *Environ. Sci. Technol.* 1996, 30 (6), 1827−1834.

(134) Stieglitz, B. R. L.; Zwick, G.; Will, R.; Roth, W.; Hedwig, K. Influence of elemental sulfur on the de-novo-synthesis of organochlorine compounds from residual carbon on fly ash. *Chemosphere* 1998, 37 (9−12), 2261−2278.

(135) Ma, H.; Du, N.; Lin, X.; Liu, C.; Zhang, J.; Miao, Z. Inhibition of element sulfur and calcium oxide on the formation of PCDD/Fs during co-combustion experiment of municipal solid waste. *Sci. Total Environ.* 2018, 633, 1263−1271.

(136) Wu, H.-L.; Lu, S.-Y.; Li, X.-D.; Jiang, X.-G.; Yan, J.-H.; Zhou, M.-S.; Wang, H. Inhibition of PCDD/F by adding sulphur compounds to the feed of a hazardous waste incinerator. *Chemosphere* 2012, 86 (4), 361−367.

(137) Ryan, S. P.; Li, X.-D.; Gullett, B. K.; Lee, C. W.; Clayton, M.; Touati, A. Experimental study on the effect of SO2 on PCDD/F emissions: Determination of the importance of gas-phase versus solid-phase reactions in PCDD/F formation. *Environ. Sci. Technol.* 2006, 40 (22), 7040−7047.

(138) Addink, R.; Altwicker, E. R. Role of copper compounds in the de novo synthesis of polychlorinated dibenzo-p-dioxins/dibenzofurans. *Environ. Eng. Sci.* 1998, 15 (1), 19−27.

(139) Takacs, L.; Moilanen, G. L. Simultaneous control of PCDD/PCDF, HCl and NOX emissions from municipal solid waste incinerators with ammonia injection. *J. Air Waste Manage. Assoc.* 1991, 41 (5), 716−722.

(140) Tuppurainen, K. A.; Ruokojärvi, P. H.; Asikainen, A. H.; Aatamila, M.; Ruuskanen, J. Chlorophenols as precursors of PCDD/Fs in incineration processes: correlations, PLS modeling, and reaction mechanisms. *Environ. Sci. Technol.* 2000, 34 (23), 4958−4962.

(141) Kuzuhara, S.; Sato, H.; Tsubouchi, N.; Ohtsuka, Y.; Kasai, E. Effect of nitrogen-containing compounds on polychlorinated dibenzo-p-dioxin/dibenzofuran formation through de novo synthesis. *Environ. Sci. Technol.* 2005, 39 (3), 795−799.

(142) Samaras, P.; Blumenstock, M.; Lenoir, D.; Schramm, K. W.; Kettrup, A. PCDD/F inhibition by prior addition of urea to the solid fuel in laboratory experiments and results statistical evaluation. *Chemosphere* 2001, 42 (5−7), 737−743.

(143) Hajizadeh, Y.; Onwudili, J. A.; Williams, P. T. Effects of gaseous NH3 and SO2 on the concentration profiles of PCDD/F in flyash under post-combustion zone conditions. *Waste Manage.* 2012, 32 (7), 1378−1386.

(144) Fujimori, T.; Nakamura, M.; Takaoka, M.; Shiota, K.; Kitajima, Y. Synergetic inhibition of thermochemical formation of chlorinated aromatics by sulfur and nitrogen derived from thiourea: Multielement characterizations. *J. Hazard. Mater.* 2016, 311, 43−50.

(145) Chen, Z. L.; Lin, X. Q.; Lu, S. Y.; Li, X. D.; Yan, J. H. Suppressing formation pathway of PCDD/Fs by S-N-containing compound in full-scale municipal solid waste incinerators. *Chem. Eng. J.* 2019, 359, 1391−1399.

(146) Gullett, B. K.; Lemieux, P. M.; Dunn, J. E. Role Of Combustion and Sorbent Parameters in Prevention of Polychlorinated Dibenzo-p-Dioxin and Polychlorinated Dibenzofuran Formation during Waste Combustion. *Environ. Sci. Technol.* 1994, 28 (1), 107−118.

(147) Qian, Y.; Zheng, M. H.; Liu, W. B.; Ma, X. D.; Zhang, B. Influence of metal oxides on PCDD/Fs formation from pentachlorophenol. *Chemosphere* 2005, 60 (7), 951−958.

(148) Liu, W. B.; Zheng, M. H.; Zhang, B.; Qian, Y.; Ma, X. D.; Liu, W. X. Inhibition of PCDD/Fs formation from dioxin precursors by calcium oxide. *Chemosphere* 2005, 60 (6), 785−790.

(149) Lu, S. Y.; Chen, T.; Yan, J. H.; Li, X. D.; Ni, Y. L. M. J.; Cen, K. F. Effects of calcium-based sorbents on PCDD/F formation from pentachlorophenol combustion process. *J. Hazard. Mater.* 2007, 147 (1−2), 663−671.

(150) Li, Q. Q.; Li, L. W.; Su, G. J.; Huang, X. C.; Zhao, Y. H.; Li, B. K.; Miao, X.; Zheng, M. H. Synergetic inhibition of PCDD/F formation from pentachlorophenol by mixtures of urea and calcium oxide. *J. Hazard. Mater.* 2016, 317, 394−402.

(151) Yan, J.-h.; Peng, Z.; Lu, S.-y.; Li, X.-d.; Cen, K.-f. Removal of PCDDs/Fs from municipal solid waste incineration by entrained-flow adsorption technology. *J. Zhejiang Univ., Sci., A* 2006, 7 (11), 1896−1903.

(152) Lin, W.-Y.; Wang, L.-C.; Wang, Y.-F.; Li, H.-W.; Chang-Chien, G.-P. Removal characteristics of PCDD/Fs by the dual bag filter system of a fly ash treatment plant. *J. Hazard. Mater.* 2008, 153 (3), 1015−1022.

(153) Tidahy, H. L.; Siffert, S.; Wyrwalski, F.; Lamonier, J. F.; Aboukaïs, A. Catalytic activity of copper and palladium based catalysts for toluene total oxidation. *Catal. Today* 2007, 119 (1), 317−320.

(154) Okumura, M.; Akita, T.; Haruta, M.; Wang, X.; Kajikawa, O.; Okada, O. Multi-component noble metal catalysts prepared by sequential deposition precipitation for low temperature decomposition of dioxin. *Appl. Catal., B* 2003, 41 (1−2), 43−52.

(155) Taralunga, M.; Mijoin, J.; Magnoux, P. Catalytic destruction of chlorinated POPs—Catalytic oxidation of chlorobenzene over PtHFAU catalysts. *Appl. Catal., B* 2005, 60 (3), 163−171.

(156) Hashimoto, Y.; Uemichi, Y.; Ayame, A. Low-temperature hydrodechlorination mechanism of chlorobenzenes over platinum-supported and palladium-supported alumina catalysts. *Appl. Catal., A* 2005, 287 (1), 89−97.

(157) Everaert, K.; Baeyens, J. Removal of PCDD/F from flue gases in fixed or moving bed adsorbers. *Waste Manage.* 2004, 24 (1), 37−42.

(158) De Jong, V.; Cieplik, M. K.; Louw, R. Formation of Dioxins in the Catalytic Combustion of Chlorobenzene and a Micropollutant-like Mixture on Pt/γ-Al2O3. *Environ. Sci. Technol.* 2004, 38 (19), 5217−5223.

(159) Yu, M.-f.; Li, W.-w.; Li, X.-d.; Lin, X.-q.; Chen, T.; Yan, J.-h. Development of new transition metal oxide catalysts for the destruction of PCDD/Fs. *Chemosphere* 2016, 156, 383−391.

(160) Cho, C.-H.; Ihm, S.-K. Development of New Vanadium-Based Oxide Catalysts for Decomposition of Chlorinated Aromatic Pollutants. *Environ. Sci. Technol.* 2002, 36 (7), 1600−1606.

(161) Krishnamoorthy, S.; Rivas, J. A.; Amiridis, M. D. Catalytic Oxidation of 1,2-Dichlorobenzene over Supported Transition Metal Oxides. *J. Catal.* 2000, 193 (2), 264−272.

(162) Kim, S.-C.; Jeon, S.-H.; Jung, I.-R.; Kim, K.; Myung-Hee, K.-H.; Kwon, M.-H.; Kim, J.-H.; Yi, S.-J.; You, J.-C.; Jung, D.-H. Removal

efficiencies of PCDDs/PCDFs by air pollution control devices in municipal solid waste incinerators. *Chemosphere* 2001, 43 (4), 773–776.

(163) Goemans, M.; Clarysse, P.; Joannès, J.; De Clercq, P.; Lenaerts, S.; Matthys, K.; Boels, K. Catalytic NOx reduction with simultaneous dioxin and furan oxidation. *Chemosphere* 2003, 50 (4), 489–497.

(164) Wang, L.-C.; Lee, W.-J.; Tsai, P.-J.; Lee, W.-S.; Chang-Chien, G.-P. Emissions of polychlorinated dibenzo-p-dioxins and dibenzofurans from stack flue gases of sinter plants. *Chemosphere* 2003, 50 (9), 1123–1129.

(165) Yu, M.-f.; Lin, X.-q.; Li, X.-d.; Yan, M.; Prabowo, B.; Li, W.-w.; Chen, T.; Yan, J.-h. Catalytic destruction of PCDD/Fs over vanadium oxide-based catalysts. *Environ. Sci. Pollut. Res.* 2016, 23 (16), 16249–16258.

(166) Bertinchamps, F.; Grégoire, C.; Gaigneaux, E. M. Systematic investigation of supported transition metal oxide based formulations for the catalytic oxidative elimination of (chloro)-aromatics. *Appl. Catal., B* 2006, 66 (1–2), 10–22.

(167) Du, C.; Wang, Q.; Peng, Y.; Lu, S.; Ji, L.; Ni, M. Catalytic oxidation of 1,2-DCBz over V2O5/TiO2-CNTs: effect of CNT diameter and surface functional groups. *Environ. Sci. Pollut. Res.* 2017, 24 (5), 4894–4901.

(168) Wang, H. C.; Chang, S. H.; Hung, P. C.; Hwang, J. F.; Chang, M. B. Synergistic effect of transition metal oxides and ozone on PCDD/F destruction. *J. Hazard. Mater.* 2009, 164 (2–3), 1452–9.

(169) Wang, Q.; Tang, M.; Peng, Y.; Du, C.; Lu, S. Ozone assisted oxidation of gaseous PCDD/Fs over CNTs-containing composite catalysts at low temperature. *Chemosphere* 2018, 199, 502–509.

(170) Zhan, M. X.; Yu, M. F.; Zhang, G.; Chen, T.; Li, X. D.; Buekens, A. Low temperature degradation of polychlorinated dibenzo-p-dioxins and dibenzofurans over a VOx-CeOx/TiO2 catalyst with addition of ozone. *Waste Manage.* 2018, 76, 555–565.

(171) Debecker, D. P.; Delaigle, R.; Hung, P. C.; Buekens, A.; Gaigneaux, E. M.; Chang, M. B. Evaluation of PCDD/F oxidation catalysts: confronting studies on model molecules with tests on PCDD/F-containing gas stream. *Chemosphere* 2011, 82 (9), 1337–42.

(172) Hsu, W. T.; Hung, P. C.; Chang, M. B. Catalytic destruction vs. adsorption in controlling dioxin emission. *Waste Manage.* 2015, 46, 257–64.

(173) Hung, P. C.; Chang, S. H.; Lin, S. H.; Buekens, A.; Chang, M. B. Pilot tests on the catalytic filtration of dioxins. *Environ. Sci. Technol.* 2014, 48 (7), 3995–4001.

(174) Wevers, M.; De Fré, R.; Rymen, T.; Geuzens, P. Reduction of dioxin emission from a municipal waste incinerator by wet gas scrubbing. *Chemosphere* 1992, 25 (7), 1435–1439.

(175) Lehner, M.; Mayinger, F.; Geipel, W. Separation of Dust, Halogen and PCDD/F in a Compact Wet Scrubber. *Process Saf. Environ. Prot.* 2001, 79 (2), 109–116.

(176) Lee, W.-S.; Chang-Chien, G.-P.; Chen, S.-J.; Wang, L.-C.; Lee, W.-J.; Wang, Y.-H. Removal of Polychlorinated Dibenzo-p-Dioxins and Dibenzofurans in Flue Gases by Venturi Scrubber and Bag Filter. *Aerosol Air Qual. Res.* 2004, 4, 27–37.

(177) Choi, K.-I.; Lee, D.-H. PCDD/DF concentrations at the inlets and outlets of wet scrubbers in Korean waste incinerators. *Chemosphere* 2007, 66 (2), 370–376.

(178) Sierhuis, W. M.; de Vries, C.; Born, J. G. P. PCDD/F emissions related to the operating conditions of the flue gas cleaning system of MWI-Amsterdam. *Chemosphere* 1996, 32 (1), 159–168.

(179) Giugliano, M.; Cernuschi, S.; Grosso, M.; Miglio, R.; Aloigi, E. PCDD/F mass balance in the flue gas cleaning units of a MSW incineration plant. *Chemosphere* 2002, 46 (9), 1321–1328.

(180) Takaoka, M.; Liao, P.; Takeda, N.; Fujiwara, T.; Oshita, K. The behavior of PCDD/Fs, PCBs, chlorobenzenes and chlorophenols in wet scrubbing system of municipal solid waste incinerator. *Chemosphere* 2003, 53 (2), 153–161.

(181) Löthgren, C.-J.; van Bavel, B. Dioxin emissions after installation of a polishing wet scrubber in a hazardous waste incineration facility. *Chemosphere* 2005, 61, 405–412.

(182) Cheung, W. H.; Lee, V. K.; McKay, G. Minimizing dioxin emissions from integrated MSW thermal treatment. *Environ. Sci. Technol.* 2007, 41 (6), 2001–2007.

(183) Lundin, L.; Marklund, S. Thermal degradation of PCDD/F, PCB and HCB in municipal solid waste ash. *Chemosphere* 2007, 67 (3), 474–481.

(184) Sakai, S.-i.; Hiraoka, M. Municipal solid waste incinerator residue recycling by thermal processes. *Waste Manage.* 2000, 20 (2–3), 249–258.

(185) Addink, R.; Govers, H. A.; Olie, K. Kinetics of formation of polychlorinated dibenzo-p-dioxins/dibenzofurans from carbon on fly ash. *Chemosphere* 1995, 31 (6), 3549–3552.

(186) Wikström, E.; Ryan, S.; Touati, A.; Gullett, B. K. Key parameters for de novo formation of polychlorinated dibenzo-p-dioxins and dibenzofurans. *Environ. Sci. Technol.* 2003, 37 (9), 1962–1970.

(187) Addink, R.; Altwicker, E. Role of copper compounds in the de novo synthesis of polychlorinated dibenzo-p-dioxins/dibenzofurans. *Environ. Eng. Sci.* 1998, 15 (1), 19–27.

(188) Yamaguchi, H.; Shibuya, E.; Kanamaru, Y.; Uyama, K.; Nishioka, M.; Yamasaki, N. Hydrothermal decomposition of PCDDs/PCDFs in MSWI fly ash. *Chemosphere* 1996, 32 (1), 203–208.

(189) Hu, Y.; Zhang, P.; Chen, D.; Zhou, B.; Li, J.; Li, X.-w. Hydrothermal treatment of municipal solid waste incineration fly ash for dioxin decomposition. *J. Hazard. Mater.* 2012, 207, 79–85.

(190) Jin, Y.-q.; Ma, X.-j.; Jiang, X.-g.; Liu, H.-m.; Li, X.-d.; Yan, J.-h. Hydrothermal degradation of polychlorinated dibenzo-p-dioxins and polychlorinated dibenzofurans in fly ash from municipal solid waste incineration under non-oxidative and oxidative conditions. *Energy Fuels* 2013, 27 (1), 414–420.

(191) Ma, W.; Brown, P. W. Hydrothermal reactions of fly ash with Ca (OH)2 and CaSO4·2H2O. *Cem. Concr. Res.* 1997, 27 (8), 1237–1248.

(192) Sako, T.; Sugeta, T.; Otake, K.; Sato, M.; Tsugumi, M.; Hiaki, T.; Hongo, M. Decomposition of dioxins in fly ash with supercritical water oxidation. *J. Chem. Eng. Jpn.* 1997, 30 (4), 744–747.

(193) Karlfeldt Fedje, K.; Ekberg, C.; Skarnemark, G.; Steenari, B.-M. Removal of hazardous metals from MSW fly ash-An evaluation of ash leaching methods. *J. Hazard. Mater.* 2010, 173 (1–3), 310–317.

(194) Cagnetta, G.; Robertson, J.; Huang, J.; Zhang, K.; Yu, G. Mechanochemical destruction of halogenated organic pollutants: A critical review. *J. Hazard. Mater.* 2016, 313, 85–102.

(195) Montinaro, S.; Concas, A.; Pisu, M.; Cao, G. Remediation of heavy metals contaminated soils by ball milling. *Chemosphere* 2007, 67, 631–639.

(196) Nomura, Y.; Nakai, S.; Hosomi, M. Elucidation of degradation mechanism of dioxins during mechanochemical treatment. *Environ. Sci. Technol.* 2005, 39, 3799–3804.

(197) Duo, W.; Leclerc, D. Thermodynamic analysis and kinetic modelling of dioxin formation and emissions from power boilers firing salt-laden hog fuel. *Chemosphere* 2007, 67 (9), S164–S176.

(198) Boonsongsup, L.; Iisa, K.; Frederick Jr, W. J. Kinetics of the Sulfation of NaCl at Combustion Conditions. *Ind. Eng. Chem. Res.* 1997, 36 (10), 4212–4216.

(199) Nagano, S.; Tamon, H.; Adzumi, T.; Nakagawa, K.; Suzuki, T. Activated carbon from municipal waste. *Carbon* 2000, 38 (6), 915–920.

垃圾焚烧中二噁英的形成，测量和控制：最新进展与展望

摘要：多氯二苯并对二噁英(pcdd)和多氯二苯并呋喃(pcdf)共有 210 种污染物，其中 17 种对人体具有高毒性。PCDD/F 主要在燃烧和焚烧过程中产生。本文综述了城市生活垃圾焚烧中二噁英的形成、测量和控制。介绍了在运行中典型的 PCDD/F 排放浓度和大气污染物控制装置。PCDD/F 的形成机理分为均相合成和多相合成两大类，并介绍了前人的研究进展和最新研究成果。考虑到基于高分辨气相色谱法与质谱法相结合的传统的离线检测方法的复杂性、高成本和时滞性，一种快速、间接的测量技术近来得到发展，激光电离飞行时间质谱法是一种较好的选择。这项研究详细回顾了在线/实时现场测量的方法以及 PCDD/F 与指示物之间的相关性。由于严格的污染物排放政策，采用最先进的修复技术来减少二噁英的形成和焚烧产生的排放物。本文还对二噁英排放的控制进行了综述。最后，提出了二噁英排放的控制的挑战、未来方向和展望。

原文刊于 Energy & Fuels, 2020, 34: 13247-13267

Influences of Coal Type and Particle Size on Soot Measurement by Laser-Induced Incandescence and Soot Formation Characteristics in Laminar Pulverized Coal Flames

Jiahan Yu, Linghong Chen,* Jianfu Zhang, Jian Wu, Xuecheng Wu, Qingmin Zeng, and Kefa Cen

ABSTRACT: This paper discussed the limitation of the laser-induced incandescence (LII) method for the measurement of soot in different laminar pulverized coal flames and explored the soot formation characteristic. Multiple methods, including digital photography, thermophoretic sampling, elastic laser scattering (ELS), and LII methods, were adopted to laboratory-scale laminar pulverized coal flames with different coal types (anthracite, lignite, and bituminous coal) and coal particle sizes (61−90, 90−100, and 100−154 μm). The flame structure, the morphology of dominant particles in different combustion regions, and the spatial distribution of soot and micron-sized carbonaceous particles were obtained and analyzed. The optical measurement and sampling results showed that the soot yield was so low that the LII signal from soot did not dominate in the detection area for anthracite, lignite, and the larger size fraction bituminous coal flames. In particular, a comparison of LII and scattering signals indicated that the vast majority of the LII signal was excited from char and unburned coal for anthracite flame, and char combustion was predominant during the combustion process. Moreover, the coal particles were more dispersed for the larger particle size fraction bituminous coal flames, which favored soot oxidation but did not favor local fuel-rich combustion and the formation of soot. Therefore, in the present study, the soot yield was high enough only in 61−90 μm bituminous coal flame, and the interference LII signal from char or unburned coal particles could be neglected by carefully controlling the laser fluence. It was also found that the LII signal intensity increased as the particle size of the pulverized coal jet decreased in the noncombustion case, which could be explained by the smaller total emission surface area with larger particle size. In addition, the temperature of heated coal particles was also lower as the coal particle size increased, which resulted in the weaker LII signal from coal particles. On the basis of this, a new phenomenon of primary fragmentation of coal particles in pulverized coal flames was observed for the first time by LII measurement, which could provide a potential method for the study of primary fragmentation for coal particles.

1. INTRODUCTION

Numerous scientific studies have pointed out that the industrial combustion of fossil fuels is one of the important sources of atmospheric pollution.[1,2] Soot, as one type of particulate matter, is one of the major products of the incomplete combustion of fossil fuels, and it affects human health and the atmosphere in several ways.[3] On the other hand, compared with natural gas and oil, coal accounts for the greatest proportion of the overall energy needs in China, and it is unlikely to change the dominance of coal in the near future. In most coal-fired power plants, coal is pulverized into micrometer particles and dispersed into an oxidizer to form a jet flame. Hence, understanding the soot formation mechanism and emission characteristics in a pulverized coal flame has great significance to improve the efficiency of the coal combustion system and reduce the emission of pollutants.

Laser-induced incandescence (LII) is a noninvasive laser diagnostics technique with high sensitivity and spatiotemporal resolution, which has been widely used to measure the soot volume fraction and its spatial distribution in flames.[4] However, as far as we know, most of the reports using LII to measure the soot volume fraction were devoted to the gaseous and liquid hydrocarbon fuel flames.[5−8] For the pulverized coal flame, many experimental studies were focused on ignition behavior.[9,10] For instance, Liu et al.[10] investigated ignition delay of different pulverized coal streams using a scientific charge-coupled device (CCD) camera, and they concluded that the ignition delay time was sensitive to the coal particle size. In comparison to gas and liquid flames, besides soot, there are many other types of carbonaceous particles, such as unburned coal and devolatilized char, in the pulverized coal flame. Moreover, these particles can also be heated by a high-energy pulsed laser to emit the interference LII signal during soot-LII measurement. Some researchers have used the LII technique to measure the soot spatial distribution of pulverized coal combustion.[11−13] Balusamy et al.[11] employed the LII technique to obtain the spatial distribution of soot and discussed the characteristics of soot formation in turbulent pulverized coal flames with different coal loading rates. Ma et

Received: July 24, 2020
Revised: September 29, 2020
Published: November 2, 2020

al.[13] experimentally investigated the effects of sodium and iron on the formation of soot in pulverized coal combustion by measuring the soot-LII signal. However, these researchers simply selected a laser energy to acquire the maximum LII signal intensity but failed to consider the interference signal from other particles during soot-LII measurement. Hayashi et al.[14] detected the LII signal intensities under various incident laser fluences in a pulverized bituminous coal flame and pointed out that the interference from pulverized coal particles could be neglected through carefully controlling the laser fluence for the first time. They concluded that appropriate laser fluence was required for soot-LII measurement in pulverized coal flames, which was sufficiently high to bring all soot particles to the sublimation temperature but also sufficiently low to avoid the LII signal from pulverized coal particles. Recently, Wu et al.[15] have applied the identical selection strategy of laser energy and studied the evolution of soot during pulverized coal combustion in different oxygen concentrations.

The studies mentioned above only considered the impact of the coal mass flow rate and oxidizer/diluent concentrations on the soot-LII measurement but neglected the influences of the coal type and coal particle size. A few studies have proved that both coal type and coal particle size have effects on the characteristics of the combustion process and soot formation.[16−18] Using a high-speed camera to image the ignition process, Kim et al.[16] explored the ignition behavior of coal particles with different coal sizes and indicated that the particle with a size less than 45 μm underwent heterogeneous ignition for medium-volatile coal, whereas the particle in the size ranges of 150−200 and 75−90 μm underwent homogeneous ignition. McLean et al.[18] performed direct observation of devolatilizing pulverized coal particles, and results showed that a condensed phase, which was a soot-like material, was formed surrounding the bituminous coal, but similar observation did not appear when burning lignite. Whether LII could be applied in soot measurement for various pulverized coal flames is not yet known. Meanwhile, some researchers simply attributed the interference of the LII signal to the pulverized coal particle[12,14] but did not consider that the characteristics and morphology of particles in the pulverized coal flame varied with the combustion reaction region.

For all the abovementioned reasons, it is necessary to evaluate the feasibility and efficacy of the soot-LII method for pulverized coal flames with different coal types and particle size fractions, which requires to figure out the specific types and characteristics of the particle in different combustion areas and to consider the effects of the LII signal from these particles on soot-LII measurement. In this study, a combination of digital photography, thermophoretic sampling, elastic laser scattering (ELS), and LII methods was applied in different laminar pulverized coal flames to characterize the flame structure, the dominant particles in various combustion regions, and the two-dimensional distribution of soot and micron-sized carbonaceous particles. Based on the measurement results, the limitation of the LII method for the measurement of soot in laminar pulverized coal flames with different coal types and coal particle sizes was explored, and the soot formation characteristics in the corresponding flames were also discussed.

2. EXPERIMENTAL APPARATUS AND METHODS

2.1. Flow Reactor and Coal Samples. Experiments were conducted in a reactor system consisting of a non-premixed Hencken burner with a quartz tower, a gas supply system, and a feeder system, as shown in Figure 1. A detailed description of the reactor has been

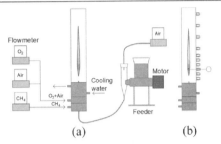

Figure 1. (a) Schematic diagram of the pulverized coal combustion system and (b) the sampling system.

provided in a previous study.[15] Identical honeycomb ceramic plates were placed inside the Hencken burner, and 574 steel pipes were inserted through the plates. The fuel gas (CH_4) was introduced through the pipes' spaced arrangement when the oxidizer (air and O_2) was supplied between the pipes. Therefore, a high-temperature environment that simulated a typical furnace condition could be supplied from a group of methane diffusion flames stabilized above the capillary pipe. The flow rates of CH_4, O_2, and air remained stable at 4, 5, and 53 standard liters per minute (SLPM) at 298 K and 101.3 kPa, respectively, leading to a concentration of 13% for the residual oxygen in the high-temperature environment. Measured with a 75 μm B-type thermocouple and corrected for radiative loss according to the study of McEnally et al.,[19] the temperature along the centerline reached over 1273 K at 50 mm above the burner and remained stable at about 1375 K below 200 mm. Pulverized coal particles were supplied by a screw pulverized coal feeder with a relatively stable feeding rate controlled by a motor speed. The particle feeding rate was set to 6.67 × 10^{-6} kg/s and calibrated before and after each experiment. Pulverized coal particles entrained by a low air gas flow were transported through a steel pipe which was inserted at the center of the burner. The coal stream was rapidly heated and ignited by the high-temperature environment to produce a laminar pulverized coal flame. The flow rate of the airstream was held constant at 0.6 SLPM with a flow velocity u_0 of 2.6 m/s. Reynolds' number ($Re = u_0 d_0/\nu$, where u_0, d_0, and ν were the mean velocity at the jet exit, the diameter of the jet exit, and the kinetic viscosity of the airflow at 298 K, respectively) at the tube exit was 373, as calculated. A quartz glass tower was fixed above the burner to stabilize the coal jet flame. In this experiment, all gas flows were controlled using mass flow meters (Alicat Scientific, Inc.) which had been calibrated by soap bubble flow meters.

Three typical coals were selected in this study: anthracite, bituminous coal, and lignite. The proximate analysis and calorific value of the investigated coals are listed in Table 1. Among the three coals, lignite has the highest volatile hydrocarbon content, bituminous coal is in the middle, and anthracite has the lowest content. However, this situation is opposite using the carbon content as a comparative metric. When comparing the ash content, it can be seen that anthracite > lignite > bituminous coal. The calorific value from high to low is bituminous coal, anthracite, and lignite. Many numerical and experimental studies[10,16,17] have reported that the coal particle size

Table 1. Proximate Analysis and Calorific Value of Three Coal Types

	M_{ad} (%)	A_{ad} (%)	V_{ad} (%)	FC_{ad} (%)	$Q_{b,ad}$ (kJ/kg)
anthracite	1.41	21.59	12.91	64.05	26,318
bituminous coal	5.38	3.99	32.01	58.62	29,585
lignite	7.34	12.53	36.82	43.31	21,981

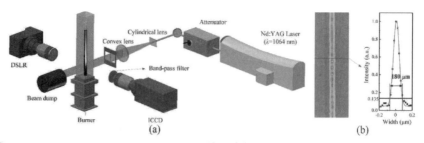

Figure 2. (a) Schematic illustration of the optical measurement system[15] and (b) cross section of the laser sheet and its width.

has a strong influence on the combustion characteristics of pulverized coal. Therefore, bituminous coal was sieved to three particle size fractions (61–90, 90–100, and 100–154 μm) in this study. Moisture was confirmed to have a significant impact on the ignition behavior of coal particles in previous research studies.[10,20] To eliminate the effect of moisture, the pulverized coal particles were fully dried in an oven at 378 K before use. Each experiment was carried out under the constant coal feeding rate, and the overall excess air ratio was above 26 for the complete burning of the coal jet, while the excess air ratio of the coal stream was less than 0.3.

2.2. Optical Measurement System. An experimental setup similar to our previous studies[7,15] was used to perform the LII and ELS measurements, as shown in Figure 2a. The fundamental output of 1064 nm from an Nd:YAG laser beam (Quantel, Brilliant B), with a frequency of 10 Hz and a duration of ~10 ns [full width at half-maximum (fwhm)], provided the laser beam for the LII measurements. On the other hand, the laser source was employed to operate at the second harmonic output of 532 nm for ELS measurements. The energy of the pulsed laser beam was adjusted using an attenuator (Newport, M-935-10). After passing through the attenuator, the laser beam was shaped into a laser sheet with a height of 50 mm using a cylindrical lens (focal length = −25 mm) and a convex lens (focal length = 500 mm) and then directed into the flame. The laser sheet was measured by a beam profile (Duma Optronics, BeamOn) to analyze its spatial intensity profile on the center axis of the burner. The laser sheet, with an intensity variation of less than 15% vertically, was a Gaussian sheet in the horizontal axis, and the thickness of it was around 180 μm ($1/e^2$ value), as shown in Figure 2b.

To collect the LII signal and ELS signal, a gated intensified CCD camera (ICCD, Princeton Instruments, PI-MAX4:1024i) connected with a UV lens (Nikon, UV Nikkor f = 105 mm) was located perpendicularly to the laser sheet. For LII measurements, a band-pass filter with a center wavelength of 450 nm (fwhm = 40 nm) was attached in front of the camera to avoid flame luminosity. The Mie scattering signal was measured at 532 nm using a band-pass filter centered at 530 nm (fwhm = 10 nm). The camera was synchronized with the laser pulse, and the image resolution was 0.0586 mm/pixel. Besides, one hundred shots were accumulated for each image to improve the signal-to-noise ratio and decrease the influences of laser energy fluctuation and the fluctuation in the coal feeding rate. The signals were detected at the peak of the pulsed laser, and the detection gate widths for LII and ELS measurements were set to 50 and 5 ns, respectively. The corresponding background was recorded in the case without the incident laser and subtracted from each LII or ELS image. Moreover, a digital single-lens reflex camera (DSLR, Nikon D700) was employed to record direct photographs of coal jet flames, and the exposure time was 1/20 s. The laser energy for ELS measurement was chosen as 6.5 mJ to ensure that there was no discernible LII signal being detected. Because of the length limitation of the laser sheet, signals from only part of the flame were detected during measurement. To obtain the complete flame signals, segmented images were taken every 40 mm in the vertical direction and then spliced into intact images.

2.3. Sampling System. A thermophoretic sampling of particles formed in pulverized coal flames was performed to examine particle characteristics in different combustion reaction regions. As shown in Figure 1, a quartz glass tower with several circular sampling ports along the heights was employed for sampling of particles at different heights. A 500-mesh stainless-steel grid attached to a probe was employed to collect particles. To reduce the disturbance to the flames, only one port allowed the probe to be inserted into the center of the flame and the rest of the ports were closed with rubber stoppers for each testing. The resident times of insertion were approximately 0.5 s. The morphological characteristics of samples were studied by scanning electron microscopy (SEM, FEI, SIRON). The SEM images were recorded at magnifications of 5000× and 20,000×.

3. RESULTS AND DISCUSSION

3.1. Digital Camera Images. Figure 3 presents the photographs of ignition and combustion of the pulverized coal

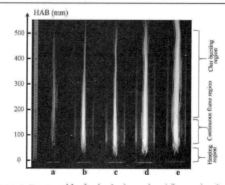

Figure 3. Pictures of five kinds of pulverized coal flames taken by the digital camera and three regions divided by the combustion condition. (a) 61–90 μm anthracite flame. (b) 61–90 μm lignite flame. (c) 61–90 μm bituminous coal flame. (d) 90–100 μm bituminous coal flame. (e) 100–154 μm bituminous coal flame. HAB stands for the height above the burner.

with different types and size fractions taken by the digital camera. According to previous studies,[9,21,22] three combustion regions are defined during the combustion process of pulverized coal jet flames based on the ignition characteristics. The jet flame of bituminous coal particles with 61–90 μm was taken as an example. The first one is a heating region, where pulverized coal particles are rapidly heated by the high-temperature environment and coal pyrolysis occurs. The image of the heating region presents no flame and only some isolated bright particles. The second region is a continuous flame region. Different from the first region, the image of the second region shows the bright white radiation. In this area, most of the pulverized coal particles are ignited, and the main body of the cloud flame looks like a continuous gaseous flame. A similar phenomenon was also observed by Xu et al.,[9] which could be caused by the combustion of volatiles. The accumulation of the volatile matter constantly increases as the number of ignited coal particles increases, resulting in a local fuel-rich environment. As the volatile matter is consumed, oxygen diffuses inward from the surroundings, leading to char combustion, and the flame gradually develops into the third region. The char particles burn slowly, the temperature of which is lower than the volatile combustion flame; therefore, the flame of char combustion presents apparently reduced brightness and accounts for the main portion during the entire combustion process, as shown in Figure 3.

From the flame photographs of different types of coal, it is clear that both lignite and bituminous coals exhibited much brighter radiation than anthracite. The bright white radiation region even is not formed in the flame of anthracite. Figure 3 also illustrates that the flame becomes larger in height and width as the size fraction of bituminous coal increases.

3.2. Analysis of Particles by SEM Images. To clarify the characteristics of dominant particles formed in various combustion regions of pulverized coal flames with different coal types, thermophoretic sampling was performed at different heights for each coal type flame, and the morphology of samples was visualized via SEM. Three different sampling heights were selected based on the three combustion reaction regions in the pulverized coal flame (HAB = 0, 80, and 200 mm; HAB stands for the height above the burner).

First, the particles sampled at HAB = 0 mm are undoubtedly pulverized coal particles in the noncombustion case, as shown in Figure 4a. Because the sampling location is at the nozzle tip position, the pulverized coal particles here are sampled before heating by high-temperature fuel gas. It can be seen from Figure 4a that pulverized coal particles are irregular lumps with particle size equal to or greater than 10 μm. It should be noted that because the morphology of anthracite or lignite particles is

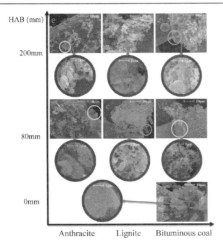

Figure 4. SEM images of particles at different heights (0, 80, and 200 mm) for three types of 61–90 μm pulverized coal flames (anthracite, lignite, and bituminous coal). Yellow circle: unburned coal particle; blue circle: ash; green circle: mineral particle.

similar to that of bituminous coal, only the SEM result of bituminous coal at 0 mm is given.

In comparison to Figure 4a, the size of the particles marked by a yellow circle in Figure 4b significantly decreases (1–10 μm), but the shape of the particles is basically unchanged. From this result, at a height of 80 mm, the dominant particles in anthracite flame are considered to be unburned carbon particles. Similar micron-sized particles are also observed for bituminous coal flame, indicated by a yellow circle in Figure 4d. However, Figure 4d shows that a large number of small particles aggregate around the surface of carbon particles to form a flocculent structure and even adhere to the stainless grid. It is clear that the particle size is smaller than 100 nm from the magnified SEM image. The morphological features of the small particles are quite similar to the SEM results of soot reported by Apicella et al.[23] and Xiao et al.[24] Therefore, it is reasonable to consider that soot is dominant among the particles in the continuous flame region of bituminous coal flame. Different from the other two groups, it is evident, in Figure 4c, that there is an aggregation with a size much larger than 10 μm. It can be seen from the magnified SEM image that many irregular particles with a size from 0.1 to 1 μm cluster on the surface of the aggregation. Moreover, as shown by a blue circle in Figure 4c, some ellipsoid-shaped particles of several micrometers in size are partially or totally embedded in the aggregation. The size and morphological features of the aggregation are consistent with the SEM images of ash and unburned char particles obtained by Zeng et al.[25] Thus, at a height of 80 mm, most of the lignite particles are burned into the aggregation of small ash particles and large unburned char particles. Interestingly, the morphological characteristics of aggregation between ash particles and char particles are similar to the soot aggregates based on the magnified SEM images. These aggregations can be distinguished from two aspects. On the one hand, the sizes of soot particles are far smaller and more uniform than those of ash particles. On the other hand, soot presents a more regular shape compared with ash particles.

As shown in Figure 4e–g, SEM results for three types of coal flames show almost no soot. The particles in Figure 4f,g exhibit a more hollow shape with a size much larger than 10 μm. This suggests that char particles further burn out at a height of 200 mm for lignite and bituminous coal flames. According to the research of Xiao et al.,[26] the spherical particles, marked by a green circle in Figure 4g, with a size of 1–10 μm can be inferred as mineral particles. As shown in Figure 4e, in addition to ash particles with a size of 0.1–1 μm, some unburned carbon particles marked by a yellow circle are also observed for anthracite flame.

3.3. Particle Distribution. In the current study, the ELS signal intensity can be determined with the following formula[27]

$$I_{sca} = \frac{1}{4} N_p k_{sca} \pi d_p^{2} \tag{1}$$

where k_{sca}, N_p, and d_p are the scattering efficiency, the number density, and the effective diameter of the particles, respectively. The scattering efficiency of micron-sized carbonaceous particles, such as pulverized coal and char, is weakly dependent on the particle effective diameter,[28] indicating that the ELS signal intensity signal is only related to the number density and the effective diameter of particles. Therefore, the scattering signal intensity represents the spatial distribution of particles in the flame.[29] Figure 5 presents the spatial measurements of ELS

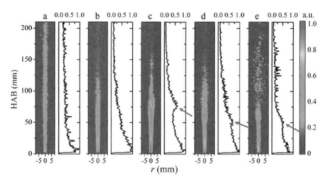

Figure 5. Two-dimensional ELS measurements and the normalized scattering intensity profiles of the central flame along the height for (a) 61−90 μm anthracite flame, (b) 61−90 μm lignite flame, (c) 61−90 μm bituminous coal flame, (d) 90−100 μm bituminous coal flame, and (e) 100−154 μm bituminous coal flame. For each map and curve, normalization was carried out with reference to its peak.

for the five tested pulverized coal flames and the normalized scattering intensity profiles along the centerline. Because the pulverized coal flame is not exactly symmetrical, the scattering signal intensity of the central flame at each height was an average value in the radial range of −0.5 to 0.5 mm from the centerline.

From Figure 5, it is clear that the scattering signal intensity is significantly greater at the flame center than at the flame edges. This is potentially because the smaller particles with higher mobility tend to move to the outside of the flame by the influence of heat expansion of the gaseous phase, while the particles with larger diameters tend to be concentrated in the central part of the flame.[30] Also, the coal particles in the edge position are easy to devolatilize, leading to the decrease in scattering signal intensity as the radial distance increases.

Overall, the scattering signal intensity gradually decreases with increasing height for each pulverized coal flame. This result can be interpreted by the fact that the devolatilization and combustion of coal particles result in a decrease in number and particle size, so the signal intensity decreases according to eq 1. However, as indicated by the red arrow in Figure 5c−e, there is an obvious second small peak in the normalized scattering intensity profiles for bituminous coal flames, at least in the case of 61−90 μm. Hayashi et al.[14] reported that the scattering signal appeared in the same region as high LII signal intensity for a pulverized coal flame, and they thought the existence of sufficient volatile matter for soot formation accounted for this phenomenon. Subsequently, Xu et al.[27] further confirmed that the scattering signal from soot would increase the overall scattering signal intensity and affect the Mie scattering measurement of coal particles based on the comparison of Mie scattering images under combusting and cold conditions. Considering these findings mentioned above and SEM analysis in the previous section, it is reasonable to believe that a large amount of soot formation is responsible for the second small peak of the scattering signal at a height of about 75 mm for 61−90 μm bituminous coal flame. Remarkably, the second tiny peak of the scattering signal intensity for 90−100 and 100−154 μm bituminous coal flames appears at a height of about 50 mm, where an obvious bright flame is hardly seen in the corresponding flame pictures. This means that soot formation does not seem to be the primary cause for the second tiny peaks of the scattering signal in bituminous coal flames with larger particle sizes, which will be detailed in Section 3.4.2.

3.4. LII of Pulverized Coal Flames. 3.4.1. Laser Fluence Curves. The laser pulse fluence value (F) should be optimized to avoid the interference LII signal from other carbonaceous particles before the LII measurement of soot in pulverized coal flames.[14] Based on the abovementioned experimental results and analysis, three regions with different heights in a pulverized coal flame (HAB = 0−5, 60−80, and 200−220 mm) were chosen as detection areas, which were denoted as H1, H2, and H3, respectively. The average LII signal intensities derived from different regions were recorded under various F. Among the three detection areas, H2 is the brightest area in the flame and the area where soot is most likely to form.

The normalized laser fluence curves derived from H1 in bituminous pulverized coal flames with three various particle size fractions are shown in Figure 6. Obviously, the LII signal from H1 is excited from pulverized coal particles in the heating region, which can be regarded as the signal obtained in a noncombustion case. As shown in Figure 6, the LII signal intensity increases with F, and there is a clear negative correlation between the LII signal intensity and the size fraction of bituminous coal at the same F. These results can be

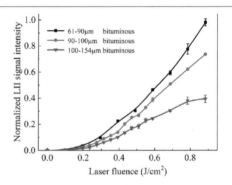

Figure 6. LII signal intensity as a function of laser pulse fluence. The LII signal was derived from H1 = 0−5 mm in bituminous pulverized coal flames with three different coal size fractions, and the signal intensity was normalized by the maximum value for simplicity of comparison.

explained in two aspects. First, the peak LII signal intensity $S_{max}(\lambda)$ recorded via the ICCD can be calculated by the following equation according to Planck's law

$$S_{max}(\lambda) = C_s \frac{12\pi hc^2 \varepsilon_\lambda \varphi_c}{d_c \rho_c \lambda^6 [\exp(hc/k T_{max}\lambda) - 1]} \quad (2)$$

where C_s is a constant calibration parameter of the detection system, d_c is the effective diameter of coal particles, ε_λ is the spectral emissivity of coal particles, T_{max} is the maximum temperature of particles by the laser, λ is the emission wavelength, φ_c is the mass concentration of coal particles, ρ_c is the density of coal, and h, c, and k represent Planck's constant, speed of light, and Boltzmann's constant, respectively. Assuming that the coal particle is spherical, the total emission surface area can be expressed as $6\varphi_c/d_c\rho_c$. It is obvious from eq 2 that the larger the particle size, the smaller the total emission surface area and the lower the LII signal intensity under otherwise identical conditions. Second, the results indicate that F in the present experiments is insufficient to increase the temperature of pulverized coal particles to the sublimation temperature. Therefore, the equation describing the energy balance for the interaction of a particle with a laser can be simplified as

$$\frac{1}{6}\pi d_c^3 \rho_c c_c \frac{dT_c}{dt} \approx \frac{\pi d_c^2 k_{abs}}{4} q(t) \quad (3)$$

where c_c is the specific heat capacity of the coal, T_c is the temperature of the coal particles, k_{abs} is the absorption efficiency of the coal particles, and $q(t)$ is the laser power density. Solving eq 3

$$T_{max} \approx T_0 + \frac{3k_{abs}}{2d_c\rho_c c_c} \int_0^{t_0} q(t)dt \quad (4)$$

k_{abs} can be treated as a constant because of the large size of the coal particles.[28] Thus, it follows from eq 4 that when F remains unchanged, the maximum temperature of coal particles would decrease with increasing particle size, resulting in a decrease in LII signal intensity.

Figure 7 presents the normalized laser fluence curves detected in the three different height regions of 61–90 μm

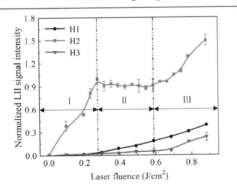

Figure 7. Laser fluence curves from three regions (H1 = 0–5 mm, H2 = 60–80 mm, and H3 = 200–220 mm) of 61–90 μm bituminous pulverized coal flame. The LII signal was normalized by the maximum value for simplicity of comparison. Each curve was divided into three regions by dashed lines (I–III).

bituminous pulverized coal flame. Based on the analysis of SEM images and ELS results, it is reasonable to speculate that the LII signal derived from H1, H2, and H3 is mainly excited from coal particles, soot, and char, respectively, for 61–90 μm bituminous flame. The LII signal intensity from H2 demonstrates three regions as a function of F. The first region ($F = 0$–0.3 J/cm^2) is the heating region, where the LII signal intensity from soot first increases rapidly with F because of the heating of soot particles by the laser sheet. The second region ($F = 0.3$–0.6 J/cm^2) is the plateau region, where the temperature of soot reaches the sublimation temperature and the LII signal intensity decreases slightly with increasing F because of the volatilization of soot particles. In the third region ($F = 0.6$–0.9 J/cm^2), the LII signal intensity increases again with F. It is also found that the LII signal intensity from H3 is close to zero in the first and second regions and only begins to increase with F in the third region. Therefore, the second increase in the laser curve derived from H2 is mainly caused by the fact that the LII signal from char is superposed on the LII signal from soot.

The optimal F should be selected to guarantee that the interference signal excited from micron-sized carbonaceous particles is as low as possible, and the LII signal from soot is sufficiently high. Based on the position of the plateau region of the laser curve, the optimal laser fluence of the 1064 nm incident laser is 0.3–0.6 J/cm^2 for soot-LII measurement in 61–90 μm bituminous coal flame. It should be noted that the optimal value of laser fluence is different in various research studies[12,14] because this value is related to many factors, such as the wavelength of the incident laser and the absorption function of soot.[31]

Figure 8 illustrates the laser fluence curves from H2 for the five groups of coal flames. Surprisingly, LII signal intensity

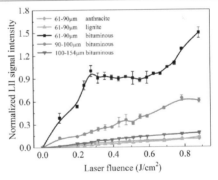

Figure 8. Laser fluence curves of the five kinds of tested coal jet flames. The LII signal was derived from a region located between H2 = 60–80 mm and normalized by the maximum value for simplicity of comparison.

monotonically increases with laser fluence for the other four kinds of coal flames, and the laser curves do not reach a plateau region over the fluence range studied. Therefore, the optimal laser fluence could not be chosen to reduce the interference signal excited from micron-sized carbonaceous particles for the other four kinds of coal flames. It is also found that the LII signal intensity detected in anthracite and lignite flames is much lower than that in bituminous coal flame under the same laser fluence. Considering the high excitation efficiency of soot particles, it is reasonable to infer that the soot yield in H2 of anthracite and lignite flames is very low. Moreover, the LII signals from H2 for anthracite and lignite flames are mainly excited from micron-sized carbonaceous particles, such as char and unburned coal particles, which is clearly supported by the results of SEM images in Section 3.2. It is further observed, in Figure 8, that the signal intensity detected in 90−100 and 100−150 μm bituminous coal flames is also lower than that in 61−90 μm bituminous coal flame, but obviously higher than that in anthracite and lignite flames. These results imply that although soot is formed in H2 for bituminous coal flame with larger particle size, the amount of soot is not dominant compared with other carbonaceous particles. Thus, if the laser fluence curve measured in the most luminous area of a pulverized coal flame does not reach a plateau region, it means that the soot yield is not enough to cause the LII signal from soot to be dominant, and there is no optimal laser fluence that can be used to ignore the interference signal excited from micron-sized particles.

3.4.2. LII Measurements. According to the analysis in the previous section, the optimal laser fluence for 61−90 μm bituminous coal flame is 0.3−0.6 J/cm^2, but there is no optimal laser fluence for the other four kinds of coal flames. In different pulverized coal jet flames, the laser fluence for soot-LII measurement was set at 0.4 J/cm^2.

Figure 9 shows the two-dimensional LII measurements for the five groups of pulverized coal flames. Evidently, the LII

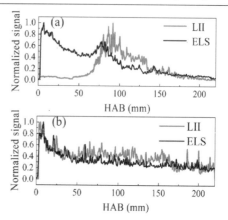

Figure 10. Comparison of the normalized LII and ELS signal intensity along the centerline for (a) 61−90 μm bituminous coal flame and (b) 61−90 μm anthracite pulverized coal flame. The signal intensity was normalized by the maximum value for simplicity of comparison.

higher. This result may be explained by the higher oxygen concentration at the flame edge, which leads to a faster oxidation rate of soot and a decrease in LII signal. Figure 10a shows the LII and scattering signal intensity profiles along the central flame for 61−90 μm bituminous coal flame. Comparing the two signal profiles, it is found that the second peak position of the scattering signal is roughly consistent with the soot-LII signal peak position, which further confirms that the second increase in scattering signal is due to the soot formation. The abovementioned analysis suggests that the region where soot or micron-sized carbonaceous particles exist can be well distinguished based on the combination of ELS and LII measurements.

Different from 61 to 90 μm bituminous coal flame, obvious LII signals are detected near the burner port, as shown in Figure 9a,b,d,e. This result shows that the interference from large particles cannot be avoided completely in the LII measurements of soot for the other four kinds of coal flames. In particular, the spatial distribution of the LII signal for anthracite jet flame is similar to that of the scattering signal. As shown in Figure 10b, the LII signal and scattering signal along the centerline exhibit almost the same trend, and both decrease with heights. This tendency implies that char combustion is predominant in anthracite flame, and the vast majority of the LII signal was excited from unburned coal and char particles, which coincides to a certain extent with the analysis of the corresponding flame photograph and SEM results. The results can be attributed to the lower volatile hydrocarbon content for anthracite. Thus, the LII method can even be employed to measure the spatial distribution of carbonaceous particles for 61−90 μm anthracite pulverized coal flame.

As shown in Figure 9b,d,e, LII signals present a similar distribution pattern for 61−90 μm lignite and 90−100 and 100−154 μm bituminous coal jet flames. The overall LII signal distribution can be divided into two distinct parts as follows. In the upstream region below 70 mm, the distribution of the LII signal is relatively uniform, and the signal is mainly concentrated near the flame centerline. Considering that the

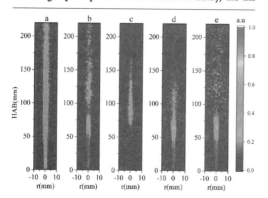

Figure 9. Two-dimensional LII measurements for (a) 61−90 μm anthracite flame, (b) 61−90 μm lignite flame, (c) 61−90 μm bituminous coal flame, (d) 90−100 μm bituminous coal flame, and (e) 100−154 μm bituminous coal flame. For each map, normalization was carried out with reference to its peak.

signal below 50 mm could be neglected compared to the signal in the continuous flame region, as shown in Figure 9c. The result indicates that the interference of the LII signal excited from micron-sized particles can be efficiently avoided in 61−90 μm bituminous pulverized coal jet flame under the current laser fluence. Thus, the spatial distribution of soot particles could be represented by the detected LII signal. The soot LII signal appears at around 70 mm and reaches a peak at 90 mm; then, the signal intensity decreases gradually with increasing height. The region where the soot LII signal appears coincides with the continuous flame region, which suggests that soot is formed in a local fuel-rich condition. Compared with the flame edge position, the LII signal intensity in the flame center is luminous emission of the flame is not strong in this region, it is reasonable to speculate that this part of the LII signal is excited from unburned coal particles. The pulverized coal particles entrained by a low air gas flow were ejected from the nozzle and then heated to devolatilization in the high-temperature gas environment. During the heating process, the coal particles were heated by an incident laser pulse to emit the LII signal. In the area above 70 mm, the LII signal exhibits a relatively dispersed and disordered distribution. Although it can be seen from the corresponding flame pictures that most coal particles were ignited, forming the bright white flame, according to the analysis of the laser fluence curve, few soot particles were formed in 61–90 μm lignite and 90–100 and 100–154 μm bituminous coal flames. Hence, part of the LII signal in the downstream region of the flame is emitted from micron-sized carbonaceous particles, while part is excited from soot. Moreover, it is hardly possible to distinguish the soot LII signal from other LII signals under current laser fluence.

The LII signal intensity along the centerline in the five groups of pulverized coal flames is presented in Figure 11. It is

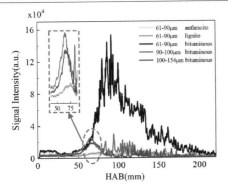

Figure 11. Comparison of the normalized LII signal intensity at the axis in the five tested pulverized coal flames.

apparent that the LII signal intensity detected in 61–90 μm bituminous coal flame is far larger than that in the other four coal flames. This finding also confirms that the soot yield of bituminous coal flame, everything else being equal, is higher than that of anthracite and lignite flames. It can be explained by the fact that bituminous coal volatiles contain large fractions of soot, producing heavy hydrocarbons, while lignite volatiles are largely composed of light hydrocarbons[18,20] and anthracite is characterized by low volatiles (see Table 1).

Similarly, the measurement results reveal that the soot formed in 61–90 μm bituminous coal flame is more than that in 90–100 and 100–154 μm bituminous coal flames, which seems to be contrary to the study of Feng et al.[32] They used a single coal particle combustion model to find that the soot yield for the larger coal particle is higher. The contradiction can be explained by the characteristics of a pulverized coal jet flame. For the pulverized coal jet flame, the larger the particle size fraction of coal, the smaller the number density of coal particles, the wider and longer the flame is under a certain particle feeding rate. It can be found in Figure 5c–e that the particles in bituminous coal flame with larger particle size are more dispersed than those in the flame with smaller particle size, which seems to favor soot oxidation. In addition, because of the slower internal heating rate and smaller specific surface for the larger coal particles, the evolution rate of volatile matter decreases, and it is difficult to form a local fuel-rich condition that is conducive to the formation of soot. As a consequence, the soot yield for the larger size fraction bituminous coal jet flame is lower compared with smaller size fraction bituminous coal. Similar conclusions were also drawn by Yu et al.[33]

Furthermore, for 90–100 and 100–154 μm bituminous coal jet flames, the LII signal intensity suddenly increased at HAB = 50 mm and reaches a peak at around 65 mm, the peak location of which coincides approximately with the position of the second peak of the scattering signal (indicated by the red arrow in Figures 5 and 11). A similar behavior of the LII signal also appears in lignite coal flame (not obvious in Figure 11 but clearly visible in Figure 9b). As mentioned in the analysis of two-dimensional LII measurements, the LII signal around 65 mm is mainly excited from unburned coal particles. As can be seen in eqs 1 and 2, the LII signal intensity from smaller-sized particles is much stronger than that from larger-sized particles, while the scattering signal intensity exhibits a quite opposite trend. Therefore, it can be inferred that the sudden increase in LII signal can be ascribed to the increase in coal particle number and the decrease in coal particle size. By comparing the flame photographs and intensity profiles of the LII signal, the position of the peak is located in the intermediate transition zone between the heating region and the continuous flame region. Moreover, the LII signal increases and decreases in a sudden manner rather than in a gentler manner similar to soot formation. Based on the present analysis, it is reasonable to speculate that the primary fragmentation of coal particles is responsible for the dramatic change in LII signal. The primary fragmentation of the coal particle is the breakage of a raw coal particle into smaller particles during devolatilization, which is considered to be caused by the stress made by thermal shock and the increase in pressure of the volatile matter in the pores of coal particles.[34,35] Compared to the entire combustion process of coal particles, primary fragmentation takes a short time based on the fragmentation model introduced by Senneca et al.[35] During the early stage of a combustion process of coal particles, the volatile matter caused by pyrolysis continued to accumulate with the increase in height. In the transitional zone between the heating region and the continuous flame region, the pressure stress generated by volatiles reached a critical threshold, and the primary fragmentation of most coal particles occurred rapidly and almost simultaneously, leading to the dramatic variation in the number of coal particles and particle size.

Primary fragmentation is commonly observed and studied in a fluidized bed combustion system. Many numerical and experimental studies have found that the extent of fragmentation is strongly dependent on the initial size of coal particle and coal ranks.[35–39] Various coal ranks can lead to differences in volatile-matter content, the hardness, and the strength of coal particles. Cui et al.[37] observed that the fragmentation of anthracite is far weaker compared to that of low-rank coal. The volatile-matter content of anthracite is so low that the inner pressure is lower during devolatilization, causing its fragmentation to be the least intense. This explains why the LII signal profile of anthracite does not increase suddenly at around HAB = 65 mm. Nevertheless, LII measurements cannot be used to determine whether primary fragmentation occurred for 61–90 μm bituminous coal jet flame because of the formation of a large amount of soot. As presented in the studies of Zhang[38] and Dacombe,[36] the larger the coal particle size is, the more intense the fragmentation is. This is because the coal particles with a larger size have a larger quantity of volatile matter, leading to a higher inner pressure, which is helpful for fragmentation. In contrast, the smaller particle size has the bigger specific surface and the evolution rate of volatile matter is higher, resulting in a lower inner pressure; therefore, fragmentation occurs less intensely. From the data mentioned above, compared with 61−90 μm lignite, 90−100 and 100−154 μm bituminous coal particles underwent much stronger primary fragmentation, resulting in the larger peak of LII signals. It is also potentially explained why there is no second peak of the scattering signal for 61−90 μm lignite. Based on the results mentioned above, it is reasonable to think that the primary fragmentation of coal particles can be observed using the LII method for a pulverized coal flame that only forms little or no soot.

4. CONCLUSIONS

Digital photography, thermophoretic sampling, ELS, and LII methods were successfully used to study the flame structure, the characteristics of dominant particles in various combustion regions, and the spatial distribution of soot and micron-sized carbonaceous particles for a lab-scale laminar pulverized coal flame. The influences of the coal type and particle size fractions on the measurement and formation of soot were also analyzed. From this experimental study, the following conclusions can be drawn:

(1) Careful control of the laser fluence was considered as a useful strategy for reducing the interference LII signal from micron-sized carbonaceous particles for the soot-LII measurement. Nonetheless, in this study, this strategy was effective only for 61−90 μm bituminous coal flame because for the other four flames, the soot yield was so low that the LII signal from soot did not dominate in the detection area. Therefore, not only the coal type but also the particle size fraction should be taken into consideration when the LII method was applied to the pulverized coal flame. For 61−90 μm bituminous coal flame, the optimal laser fluence of the 1064 nm incident laser was 0.3−0.6 J/cm^2.

(2) For 61−90 μm anthracite flame, the intensity of LII and scattering signals exhibited almost the same trend, which implied that the vast majority of the LII signal was excited from char particles and unburned coal, and char combustion was predominant during the combustion process. Therefore, as with the ELS method, the LII method could be employed to measure the spatial distribution of carbonaceous particles in anthracite coal jet flame.

(3) Two-dimensional ELS measurements showed that the coal particles in 90−100 and 100−154 μm bituminous coal flames were more dispersed than those in 61−90 μm bituminous coal flame, which seems to favor soot oxidation but not favor the formation of a local fuel-rich condition. Hence, the soot yield of the bituminous coal jet flame with larger particle size was lower.

(4) With all else being equal, LII signal intensity was a negative correlation with the particle size of pulverized coal jet in the noncombustion case. It could be interpreted as the larger the particle size, the smaller the total emission surface area and the lower the temperature of heated coal particles, which resulted in the weaker LII signal from coal particles. On the basis of this, a new phenomenon of primary fragmentation of coal particles in pulverized coal flames was observed for the first time by LII measurement. This finding provided a potential method for the study of primary fragmentation for coal particles.

■ AUTHOR INFORMATION

Corresponding Author

Linghong Chen − *State Key Laboratory of Clean Energy Utilization and Qingshanhu Energy Research Center, Zhejiang University, Hangzhou 310027, China*; Email: chenlh@zju.edu.cn

Authors

Jiahan Yu − *State Key Laboratory of Clean Energy Utilization, Zhejiang University, Hangzhou 310027, China*; orcid.org/0000-0003-1808-2297

Jianfu Zhang − *State Key Laboratory of Clean Energy Utilization, Zhejiang University, Hangzhou 310027, China*

Jian Wu − *State Key Laboratory of Clean Energy Utilization, Zhejiang University, Hangzhou 310027, China*

Xuecheng Wu − *State Key Laboratory of Clean Energy Utilization and Qingshanhu Energy Research Center, Zhejiang University, Hangzhou 310027, China*; orcid.org/0000-0001-9897-8776

Qingmin Zeng − *State Key Laboratory of Clean Energy Utilization, Zhejiang University, Hangzhou 310027, China*

Kefa Cen − *State Key Laboratory of Clean Energy Utilization and Qingshanhu Energy Research Center, Zhejiang University, Hangzhou 310027, China*

Complete contact information is available at:
https://pubs.acs.org/10.1021/acs.energyfuels.0c02483

Notes

The authors declare no competing financial interest.

■ ACKNOWLEDGMENTS

This research was supported by the Natural Science Foundation of China (no. 51876190), National Key Research and Development Program of China (no. 2018YFB0605200) and the Program of Introducing Talents of Discipline to Universities (no. BP0820002).

■ REFERENCES

(1) Pope, C. A.; Ezzati, M.; Dockery, D. W. Fine-Particulate Air Pollution and Life Expectancy in the United States. *N. Engl. J. Med.* 2009, *360*, 376−386.

(2) Bond, T. C.; Streets, D. G.; Yarber, K. F.; Nelson, S. M.; Woo, J.-H.; Klimont, Z. A technology-based global inventory of black and organic carbon emissions from combustion. *J. Geophys. Res. Atmos.* 2004, *109*, D14203.

(3) Kennedy, I. M. The health effects of combustion-generated aerosols. *Proc. Combust. Inst.* 2007, *31*, 2757.

(4) Schulz, C.; Kock, B. F.; Hofmann, M.; Michelsen, H.; Will, S.; Bougie, B.; Suntz, R.; Smallwood, G. Laser-induced incandescence: recent trends and current questions. *Appl. Phys. B: Lasers Opt.* 2006, *83*, 333−354.

(5) Bladh, H.; Johnsson, J.; Bengtsson, P.-E. On the dependence of the laser-induced incandescence (LII) signal on soot volume fraction for variations in particle size. *Appl. Phys. B: Lasers Opt.* 2008, *90*, 109−125.

(6) Snelling, D. R.; Link, O.; Thomson, K. A.; Smallwood, G. J. Measurement of soot morphology by integrated LII and elastic light scattering. *Appl. Phys. B* **2011**, *104*, 385−397.

(7) Wu, J.; Chen, L.; Bengtsson, P.-E.; Zhou, J.; Zhang, J.; Wu, X.; Cen, K. Effects of carbon dioxide addition to fuel on soot evolution in ethylene and propane diffusion flames. *Combust. Flame* **2019**, *199*, 85−95.

(8) Liu, H.; Zhang, P.; Liu, X.; Chen, B.; Geng, C.; Li, B.; Wang, H.; Li, Z.; Yao, M. Laser diagnostics and chemical kinetic analysis of PAHs and soot in co-flow partially premixed flames using diesel surrogate and oxygenated additives of n-butanol and DMF. *Combust. Flame* **2018**, *188*, 129−141.

(9) Xu, K.; Wu, Y.; Wang, Z.; Yang, Y.; Zhang, H. Experimental study on ignition behavior of pulverized coal particle clouds in a turbulent jet. *Fuel* **2016**, *167*, 218−225.

(10) Liu, Y.; Geier, M.; Molina, A.; Shaddix, C. R. Pulverized coal stream ignition delay under conventional and oxy-fuel combustion conditions. *Int. J. Greenhouse Gas Control* **2011**, *5*, S36−S46.

(11) Balusamy, S.; Kamal, M. M.; Lowe, S. M.; Tian, B.; Gao, Y.; Hochgreb, S. Laser diagnostics of pulverized coal combustion in O_2/N_2 and O_2/CO_2 conditions: velocity and scalar field measurements. *Exp. Fluids* **2015**, *56*, 108.

(12) Xu, K.; Li, Z.; Zhang, H.; Wu, Y.; Luo, K.; Lü, J. Laser Induced Incandescence Measurement of Soot Distribution in a Coal Jet Flame. *J. Combust. Sci. Technol.* **2018**, *24*, 232−237.

(13) Ma, P.; Huang, Q.; Gao, Q.; Li, S. Effects of Na and Fe on the formation of coal-derived soot in a two-stage flat-flame burner. *Fuel* **2020**, *265*, 116914.

(14) Hayashi, J.; Hashimoto, N.; Nakatsuka, N.; Tsuji, H.; Watanabe, H.; Makino, H.; Akamatsu, F. Soot formation characteristics in a lab-scale turbulent pulverized coal flame with simultaneous planar measurements of laser induced incandescence of soot and Mie scattering of pulverized coal. *Proc. Combust. Inst.* **2013**, *34*, 2435−2443.

(15) Wu, J.; Chen, L.; Bengtsson, P.-E.; Zhou, J.; Zhang, J.; Wu, X.; Cen, K. Optical investigations on particles evolution and flame properties during pulverized coal combustion in O_2/N_2 and O_2/CO_2 conditions. *Fuel* **2019**, *251*, 394.

(16) Kim, R.-G.; Li, D.; Jeon, C.-H. Experimental investigation of ignition behavior for coal rank using a flat flame burner at a high heating rate. *Exp. Therm. Fluid Sci.* **2014**, *54*, 212−218.

(17) Ryan, W.; Annamalai, K. Group ignition of a cloud of coal particles. *J. Heat Transf.-Trans. ASME* **1991**, *113*, 677−687.

(18) McLean, W. J.; Hardesty, D. R.; Pohl, J. H. Direct observations of devolatilizing pulverized coal particles in a combustion environment. *Symp. (Int.) Combust.* **1981**, *18*, 1239−1248.

(19) McEnally, C. S.; Köylü, Ü. Ö.; Pfefferle, L. D.; Rosner, D. E. Soot volume fraction and temperature measurements in laminar nonpremixed flames using thermocouples. *Combust. Flame* **1997**, *109*, 701−720.

(20) Shaddix, C. R.; Molina, A. Particle imaging of ignition and devolatilization of pulverized coal during oxy-fuel combustion. *Proc. Combust. Inst.* **2009**, *32*, 2091−2098.

(21) Yamamoto, K.; Murota, T.; Okazaki, T.; Taniguchi, M. Large eddy simulation of a pulverized coal jet flame ignited by a preheated gas flow. *Proc. Combust. Inst.* **2011**, *33*, 1771−1778.

(22) Du, X.; Gopalakrishnan, C.; Annamalai, K. Ignition and combustion of coal particle streams. *Fuel* **1995**, *74*, 487−494.

(23) Apicella, B.; Senneca, O.; Russo, C.; Heuer, S.; Cortese, L.; Cerciello, F.; Scherer, V.; Schiemann, M.; Ciajolo, A. Separation and characterization of carbonaceous particulate (soot and char) produced from fast pyrolysis of coal in inert and CO_2 atmospheres. *Fuel* **2017**, *201*, 118−123.

(24) Xiao, Z.; Tang, Y.; Zhuo, J.; Yao, Q. Effect of the interaction between sodium and soot on fine particle formation in the early stage of coal combustion. *Fuel* **2017**, *206*, 546−554.

(25) Zeng, X.; Zheng, S.; Zhou, H.; Fang, Q.; Lou, C. Char burnout characteristics of five coals below and above ash flow temperature: TG, SEM, and EDS analysis. *Appl. Therm. Eng.* **2016**, *103*, 1156−1163.

(26) Xiao, Z.; Shang, T.; Zhuo, J.; Yao, Q. Study on the mechanisms of ultrafine particle formation during high-sodium coal combustion in a flat-flame burner. *Fuel* **2016**, *181*, 1257−1264.

(27) Xu, K.; Zhang, H.; Shen, W.; Zhang, Y.; Wu, Y.; Lyu, J. Soot Formation and Distribution in Coal Jet Flames over a Broad Range of Coal Concentration. *Energy Fuels* **2020**, *34*, 7545−7553.

(28) Gogoi, A.; Choudhury, A.; Ahmed, G. A. Mie scattering computation of spherical particles with very large size parameters using an improved program with variable speed and accuracy. *J. Mod. Opt.* **2010**, *57*, 2192−2202.

(29) Gronarz, T.; Schnell, M.; Siewert, C.; Schneiders, L.; Schröder, W.; Kneer, R. Comparison of scattering behaviour for spherical and non-spherical particles in pulverized coal combustion. *Int. J. Therm. Sci.* **2017**, *111*, 116−128.

(30) Hwang, S. m.; Kurose, R.; Akamatsu, F.; Tsuji, H.; Makino, H.; Katsuki, M. Application of Optical Diagnostics Techniques to a Laboratory-Scale Turbulent Pulverized Coal Flame. *Energy Fuels* **2005**, *19*, 382−392.

(31) Cléon, G.; Amodeo, T.; Faccinetto, A.; Desgroux, P. Laser induced incandescence determination of the ratio of the soot absorption functions at 532 nm and 1064 nm in the nucleation zone of a low pressure premixed sooting flame. *Appl. Phys. B* **2011**, *104*, 297−305.

(32) Feng, L.; Wu, Y.; Xu, K.; Zhang, H.; Zhang, Y.; Zhang, M. Coal-Derived Soot Behaviors in O_2/N_2 and O_2/CO_2 Atmospheres, Studied through a 1-D Transient Coal Combustion Model. *Energy Fuels* **2019**, *33*, 3620−3629.

(33) Yu, W. W.; Wu, Y.; Kailong, X.; Zhang, H.; Yang, H.; Lu, J. Experimental investigation on characteristics of coal particle stream ignition under turbulent condition. *J. Eng. Thermophys.* **2016**, *37*, 443−447.

(34) Chirone, R.; D'Amore, M.; Massimilla, L. Carbon attrition in the fluidized combustion of a petroleum coke. *Symp. (Int.) Combust.* **1985**, *20*, 1505−1511.

(35) Senneca, O.; Urciuolo, M.; Chirone, R. A semidetailed model of primary fragmentation of coal. *Fuel* **2013**, *104*, 253−261.

(36) Dacombe, P.; Pourkashanian, M.; Williams, A.; Yap, L. Combustion-induced fragmentation behavior of isolated coal particles. *Fuel* **1999**, *78*, 1847−1857.

(37) Cui, T.; Zhou, Z.; Dai, Z.; Li, C.; Yu, G.; Wang, F. Primary Fragmentation Characteristics of Coal Particles during Rapid Pyrolysis. *Energy Fuels* **2015**, *29*, 6231−6241.

(38) Zhang, H.; Cen, K.; Yan, J.; Ni, M. The fragmentation of coal particles during the coal combustion in a fluidized bed. *Fuel* **2002**, *81*, 1835−1840.

(39) Paprika, M. J.; Komatina, M. S.; Dakić, D. V.; Nemoda, S. Đ. Prediction of Coal Primary Fragmentation and Char Particle Size Distribution in Fluidized Bed. *Energy Fuels* **2013**, *27*, 5488−5494.

煤种和粒度对激光诱导白炽法测量烟灰的影响及层流煤粉火焰中烟灰的形成特性

摘要：本文讨论了激光诱导白炽法（LII）测量不同层流煤粉火焰中烟灰的局限性，并研究了烟灰的生成特性。采用数字摄影、热泳采样、弹性激光散射（ELS）和LII方法等多种方法，对实验室规模不同煤种（无烟煤、褐煤和烟煤）和不同煤粒径（61～90μm、90～100μm和100～154μm）的层流煤粉火焰进行了测量。获取并分析了火焰结构、不同燃烧区域中主要粒子的形貌以及烟灰和微米级碳质颗粒的空间分布。光学测量和采样结果表明，在无烟煤、褐煤和较大粒径分数的烟煤火焰中，烟尘产量很低，因此烟尘产生的LII信号并不占主导地位。特别的，LII信号和散射信号的比较表明，无烟煤火焰的绝大部分LII信号都是由焦炭和未燃烧的煤激发的，在燃烧过程中，焦炭燃烧占主导地位。此外，对于粒径较大的烟煤火焰，煤颗粒更分散，有利于烟灰氧化，但不利于局部富燃料燃烧和烟灰的形成。因此，在本研究中，只有在61～90μm烟煤火焰中烟灰产量才足够高，通过控制激光能量密度可以忽略焦炭或未燃煤颗粒对LII信号的干扰。本研究还发现，在不燃烧的情况下，LII信号强度随着煤粉射流粒径的减小而增大，这可能是因为粒径越大，总发射表面积越小。此外，随着煤粒粒径的增大，受热煤粒的温度也会降低，从而导致煤粒的LII信号变弱。在此基础上，通过LII测量首次观察到了煤粉火焰中煤粒一次破碎的新现象，为研究煤粒一次破碎提供了一种潜在方法。

原文刊于 Energy & Fuels, 2020, 34: 13740-13749

A solar micro gas turbine system combined with steam injection and ORC bottoming cycle

Gang Xiao[a,b,*], Jinli Chen[a,b], Mingjiang Ni[a,b], Kefa Cen[a,b]

[a]State Key Laboratory of Clean Energy Utilization, Zhejiang University, 38 Zheda Road, Hangzhou 310027, China

[b]Qingshanhu Energy Research Center, Zhejiang University, 1699 Dayuan Road, Hangzhou 311035, China

ARTICLE INFO

Keywords:
Distributed power system
Solar micro gas turbine
Thermodynamic model
Steam injection
Organic Rankine cycle

ABSTRACT

Micro gas turbine (MGT) is an important section in solar-driven power systems, as it can improve system efficiency and provide unique flexibility to meet quickly changing demands. Herein, a solar micro gas turbine (MGT) combined with steam injection and organic Rankine cycle (ORC) is proposed to improve efficiency and flexibility. A steam receiver that produces the steam injected into the MGT is placed around the air receiver aperture to increase the receiver intercept efficiency. A thermodynamic model including a heliostat field, a solar receiver, an MGT and an ORC is developed to investigate the performance of the proposed system, where each sub-model is validated with experimental/referenced data for the subsequence optimization of design and operating parameters. For a 100 kW MGT driven by a solar heliostat field, the simulation results show that the optimized aperture diameter of the air receiver is 1.1 m. On four typical days of a year, i.e. the spring equinox, the summer solstice, the autumn equinox, and the winter solstice, the average optical efficiency of the heliostat field is 0.672, 0.754, 0.672 and 0.597, respectively. Fuel consumption of the MGT during the four days is reduced by 14.8%, 24.6%, 22.4% and 3.7%, respectively. The solar energy share can reach up to 86.0% with a system power efficiency of 19.9% at 12:30 of the autumn equinox. By adding steam injection and ORC, the power output can be increased by 8.29 kW and 30.37 kW, respectively, bringing an increase of the total power output by 37.7% and improving the system efficiency and flexibility. The proposed system can benefit the distributed energy systems especially for remote areas or islands as a flexible option.

1. Introduction

Micro gas turbine (MGT) is an advanced technology with a simple structure and fast load response [1]. It is a good choice for the next-generation distributed power systems, where the fossil fuel is largely replaced by renewable energies such as the solar power [2]. Currently, there are two main issues for a solar MGT distributed power system, i.e. to further improve flexibility and efficiency, especially when combined with a distributed PV system with dynamic operating conditions [3].

Several solarized turbine projects were carried out to investigate the system performance. However, the power efficiency of solarized turbine system is still poor. The electrical efficiency of the AORA solarized microturbine system is 16.0% with a power output of 100 kW [4]. As to the SOLGATE solarized turbine, the electrical efficiency is approximately 18.2% with a power output of 230 kW when solar energy share is nearly 60.0% [5].

Solar air receiver is a key component to convert solar energy to produce high-temperature working fluids [6], and a cavity receiver is recommended to obtain a high thermal efficiency [7], where the receiver aperture should be matched with the heliostat field. One way to increase the receiver intercept efficiency is to place additional absorbing tubes around the air receiver aperture and produce low temperature working fluids (e.g. steam) for micro gas turbines to improve the power output. Ward et al. [8] investigated the impact of steam injection on the electrical efficiency of a T100 micro-turbine, and reported a rise of the electrical efficiency by 2.2% if 5.0% mass-share steam was injected. Jong et al. [9] illustrated that the best position of steam injection in MGT was the compressor outlet. Araki et al. [10] illustrated that the advanced humid air turbine could provide higher thermal efficiency and operational flexibility. Maya et al. [11] added an extra concentrating solar collector to heat the steam injected into the gas turbine but it would significantly increase the cost. Here, a steam receiver is suggested to put around an air receiver aperture of a solar MGT to increase the receiver intercept efficiency and produce steam utilized by the MGT.

In a solar MGT system, the cost of a heliostat field takes up to 29.6%

---

* Corresponding author at: State Key Laboratory of Clean Energy Utilization, Zhejiang University, 38 Zheda Road, Hangzhou 310027, China.
E-mail address: xiaogangtianmen@zju.edu.cn (G. Xiao).

Nomenclature

Latin symbols

A	area (m^2)
cosω	cosine efficiency
c_p	constant pressure specific heat (J/kgK^{-1})
c_{pave}	average of constant pressure specific heat (J/kgK^{-1})
D	diameter (m)
F	view factor
Gr	Grashof number
h	enthalpy (J/kg)
$\bar{h}$	heat transfer coefficient (W/m^{2}K^{-1})
L	length (m)
m	mass flow (kg/s)
Nu	Nusselt number
P	pressure (kPa)
Pow	power (kW)
Pr	prandtl number
Q	heat power (kW)
q	heat flux (W/m^2)
R	heat transfer resistance (m^{2}K/W)
S	heat exchange area (m^2)
T	temperature (K)

Greek symbols

β	reflectance of receiver
δ	heliostat cleanliness
γ	ratio of specific heats
ε	emissivity
η	efficiency
ρ	heliostat reflectivity
Φ	inclination
σ	error/constant

Abbreviations

CSP	concentrating solar power
DNI	direct normal irradiance (W/m^2)
LHV	low heating value of fuel (MJ/kg)
MGT	Micro Gas Turbine
ORC	Organic Rankine Cycle
PR	Pressure rate

Subscripts

a	air
abs	absorb
air	air receiver
ap	aperture
ast	astigmatic effect of reflected rays
bq	beam quality
cav	cavity
cc	combustor combustion
comp	compressor
cond	conduction
conv	convection
f	fuel
g	gas
gen	generator
h	heliostat
in	inlet
ins	thermal insulation layer
inv	invertor
m	mechanical transmission
out	outlet
rad	radiation
recup	recuperator
ref	reflect
sb	Stefan-Boltzmann constant
steam	steam receiver
sun	sun shape
sur	surrounding
t	tracking
tub	tube
turb	turbine
w	inner wall of cavity

of the total construction cost [12]. In a CSP plant with a Solar Multiple of 2.5, 14 h of TES, the heliostat field contributes approximately 41.2% of the total construction cost [13]. Therefore, it is important to improve the system power output and efficiency while reducing the system cost at the same time.

As for the MGT, its power efficiency is ~30.0% (such as T100 micro gas turbine) despite its high global efficiency (~80.0% of heat and power) [14], indicating a large energy waste if there is no heating demand. An organic Rankine cycle (ORC) is suitable for a bottoming cycle to increase the power output, the capacity of which ranges from a few dozen to hundreds of kW [15]. Clemente et al. [16] investigated the main features of the ORC to recover the waste heat from a 100 kW commercial gas turbine, where an 8.0% increase of the global electrical efficiency was achieved. Camporeale et al. [17] compared three different configurations of the topping cycle combined with a bottoming ORC. The results showed that the ORC system produced an additional electric output of 20.67 kW with the efficiency increasing from 30.1% to 36.6%.

To further improve the efficiency and flexibility, a solar MGT combined with steam injection and ORC bottoming cycle is proposed, and a thermodynamic model is put forward, which is validated with the experimental/referenced data. The model can be divided into five parts, namely the heliostat field, the air receiver, the steam receiver, the MGT and the ORC. Based on the proposed model, the design and performance parameters will be optimized, which is capable to provide useful guidance for a solar-driven distributed power system.

2. Thermodynamic model development

Fig. 1 illustrates a schematic of the proposed system including a heliostat field, an MGT, a solar receiver and an ORC system. An air receiver is employed to absorb concentrated solar energy and produce high-temperature air for MGT, and a combustor serves as a backup for unstable or sunset periods. A circle steam receiver is placed around the air receiver aperture to produce steam which is injected into the MGT to increase the power output. An ORC system is added to recover the waste heat from MGT. A water-cooled condenser is adopted to recycle the steam in the exhaust gas, which is pumped into the steam receiver. Each component is modeled by using Matlab software.

2.1. Heliostat field

The heliostat field is located in the Qingshanhu Energy Research Center of Zhejiang University, Hangzhou, China as shown in Fig. 2(a). Here, 40 mirrors out of 100 mirrors are selected for modeling, as shown inside the red lines of the trapezium in Fig. 2(b), where each mirror has a reflection area of 19.5 m^2.

The flux density distribution concentrated by mirrors on the image

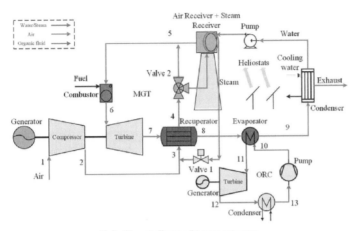

Fig 1. Schematic diagram of the proposed system.

(a) Solar tower and receivers

(b) Selected mirrors for modeling

Fig 2. Qingshanhu Energy Research Center of Zhejiang University.

plane can be expressed as [18]:

$$Flux_h(x,y) = \frac{Pow_h}{2\pi\sigma_h^2} e^{\frac{(x^2-y^2)}{2\sigma_h^2}} \qquad (1)$$

where Pow_h is the total solar energy reflected by the heliostat. It is determined by direct normal irradiance (DNI), heliostat surface area (A_h), heliostat reflectivity (ρ), heliostat cleanliness (δ) and the heliostat surface cosine efficiency (cosw). σ_h is the effective deviation of heliostat which is the result of a convolution of four errors [19], namely the sun shape error (σ_{sun}), the beam quality (σ_{bq}) associated with the mirror slope error, the astigmatic effect of reflected rays (σ_{ast}) and the tracking error (σ_t). These values are shown in Table 1 from the heliostat manufacturer.

Once the flux density distribution on the image plane has been calculated, the flux density distribution can be projected to the target surface which could be regarded as the boundary condition of the solar receiver [20].

2.2. Solar receiver system

The solar receiver system is composed of an air cavity tubular receiver and a steam external tubular receiver. The air receiver is located at the center of the solar focus spot, where the concentration ratio is high enough and the compressed air is heated up to over 800 °C before entering the combustion chamber. The steam receiver is located around the air receiver aperture (Fig. 3(a)), which is used to produce steam injected into the MGT. It is an effective way to increase the intercept efficiency of the receiver system.

The energy equilibrium of air receiver or steam receiver can be expressed as:

$$Q_{in} = Q_{abs} + Q_{ref} + Q_{rad} + Q_{conv} + Q_{cond} \qquad (2)$$

Table 1
Main assumptions for the performance calculation of heliostat field.

	Symbol	Value
Number of heliostats	–	40
Heliostat surface area, m²	A_h	19.5
Heliostat reflectivity	ρ	0.94
Heliostat cleanliness	δ	0.95
Sun shape error, mrad	σ_{sun}	2.24
Beam quality, mrad	σ_{bq}	3.8
Tracking error, mrad	σ_t	1.7

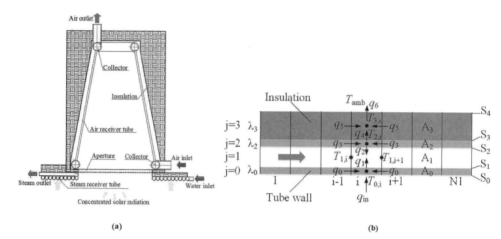

Fig 3. Solar receiver schematic (a) and receiver tube calculation model (b).

where Q_{in} is the total incident solar energy that reaches the air receiver aperture or steam receiver tube, Q_{abs} is the thermal energy absorbed by the working fluid, Q_{ref} is the reflective heat loss from air receiver cavity or steam receiver, Q_{cond} is the conductive heat loss through the receiver insulation layer, Q_{rad} and Q_{conv} are the radiative heat loss and convective heat loss, respectively.

The reflectance of the air receiver cavity to solar radiation β_{air} (Eq. (3)) can be calculated by:

$$\beta_{air} = 1 - \frac{\varepsilon}{1 - (1-\varepsilon)(1 - A_{ap}/A_w)} \quad (3)$$

where ε is the emissivity of cavity material or receiver tube, A_{ap} is the receiver aperture area, A_w is the total inner area of the cavity. As to the steam receiver, this value is calculated as:

$$\beta_{steam} = 1 - \varepsilon \quad (4)$$

The radiative heat loss of air receiver or steam receiver can be expressed as [21]:

$$Q_{rad} = \varepsilon \sigma_{sb} F_{tub-sur} A (T_{tub}^4 - T_{sur}^4) \quad (5)$$

where σ_{sb} is the Stefan-Boltzmann constant, $F_{tub-sur}$ is the view factor from the receiver tube wall to the surroundings and A is the radiative heat transfer area between the surrounds and the receiver tube wall.

The Nusselt number adopted to calculate the convective heat loss for air receiver or steam receiver is given as [22,23]:

$$Nu_{conv} = \begin{cases} 0.088 Gr^{1/3} \left(\frac{T_{tub}}{T_{sur}}\right)^{0.18} (cos\phi)^{2.47} \left(\frac{D_{ap}}{L_{cav}}\right)^s & \text{for air receiver} \\ 0.56 (GrPrcos\phi)^{1/4} & \text{for steam receiver} \end{cases} \quad (6)$$

$$s = 1.12 - 0.98 \left(\frac{D_{ap}}{L_{cav}}\right) \quad (7)$$

where Φ is the cavity inclination ranging from 0 to $\pi/2$, T_{tub} is the temperature of the receiver tube, T_{sur} is the surrounding temperature, D_{ap} is the diameter of the air receiver aperture, and L_{cav} is the length of the air receiver cavity. The convective heat loss of the air receiver or the steam receiver is calculated from the receiver tube wall to the surroundings.

The conductive heat loss through the insulation layer can be calculated using a thermal resistance network, ignoring the radiative heat loss. Thus, the total heat loss (Eq. (8)) through the insulation layer can be expressed as:

$$Q_{cond} = \frac{T_{tub} - T_{sur}}{R_{ins}} \quad (8)$$

where R_{ins} is the heat transfer resistance of the thermal insulation layer. The heat flux through the receiver tube is calculated according to the model below (Fig. 3(b)).

The calculation model of the air receiver tube or steam receiver tube is shown in Fig. 3(b). There are four calculation parts in the receiver tube model, namely the front surface of the receiver tube which absorbs the solar radiation (j = 0), the air flow (j = 1), the back surface of the receiver tube (j = 3) and the insulation (j = 4). Here, the heat conduction between front surface and back surface of the receiver tube and the heat conduction of the receiver tube along the air flow direction are calculated. The tube thickness is ignored in this model calculation. The general steady-state energy equations for the receiver tube model are shown as follows:

$$\sum q = 0 \text{ for } j = 0, 2, 3 \quad (9)$$

$$mc_{pf}\frac{dT_f}{dz} = 2\pi r_i \bar{h}(T_{tub} - T_{fluid}) \quad \text{for } j = 1 \quad (10)$$

In each calculation part, $q_{j=0,1,2,3,4,5,6}$ represent the heat fluxes transferred inside the receiver and can be calculated through the heat transfer equations. Besides, there is a two-phase region in the steam receiver calculation. In this region, the inner tube circumferential average working fluid side heat transfer coefficient ($\bar{h}$) can be written as [24]:

$$\bar{h} = 2.54(T_{tub} - T_{fluid})^3 e^{P_{fluid}/1551} \quad (11)$$

2.3. Micro gas turbine

A 100-kW micro gas turbine (T100) is employed for modeling. The governing equations are written in MATLAB and the air property data in REFPROP are used, where the rotational speed of MGT is set at 70,000 rpm and the turbine outlet temperature (TOT) is set at 645 °C.

For the compressor and turbine, the characteristic maps are adopted [25], respectively. The efficiency and the corrected mass flow rate are calculated with linear interpolation according to the maps based on the corrected speed and the pressure ratio. Then the compressor outlet temperature T_2 (Eq. (12)) can be calculated as per the equation:

$$T_2 = T_1\left[1 + \frac{(P_2/P_1)^{(\gamma-1)/\gamma} - 1}{\eta_{comp}}\right] \quad (12)$$

where γ is the ratio of air's specific heat at constant pressure and volume, T_1 is the inlet stagnation temperature, P_1 is the inlet stagnation pressure, P_2 is the outlet stagnation pressure, η_{comp} is the isentropic efficiency of the compressor.

The power consumed by the compressor (Eq. (13)) can be given according to the following equation:

$$Pow_{comp} = m_{comp}c_{pave}(T_2 - T_1)/\eta_m \quad (13)$$

where m_{comp} is the mass flow rate through the compressor, c_{pave} is the average specific heat of the air in the compressor, η_m is the mechanical transmission efficiency, i.e. 0.99 in this case.

In the same way, the turbine outlet temperature T_7 (Eq. (14)) and the power generated by the turbine P_{turb} (Eq. (15)) can be calculated respectively:

$$T_7 = T_6\left[1 - \eta_{turb}\left(1 - \left(\frac{P_6}{P_7}\right)^{-(\gamma-1)/\gamma}\right)\right] \quad (14)$$

$$Pow_{turb} = m_{turb}c_{pave}(T_6 - T_7) \quad (15)$$

where T_6 is the inlet stagnation temperature, P_6 is the inlet stagnation pressure, P_7 is the outlet stagnation pressure, η_{turb} is the isentropic efficiency of the turbine, m_{turb} is the mass flow rate through the turbine. The heat transfer process is considered between compressor and turbine, as well as their heat losses to surroundings.

For combustor simulation, the energy conservation equations (Eqs. (16) and (17)) are used to describe the compressor outlet conditions where m_a is the air mass flow rate and m_f is the fuel mass flow rate. It can be observed that the reference temperature of LHV is 25 °C. The combustion efficiency η_{cc} is 0.996.

$$(m_gh_g)_{out} = (m_ah_a)_{in} + m_fh_f + m_f\eta_{cc}LHV \quad (16)$$

$$m_g = m_a + m_f \quad (17)$$

The energy balance equation about the recuperator is given by Eq. (18), where Q_{recup} is the heat transferred from the hot steam to the cold steam in the recuperator. The inlet temperatures of the recuperator for hot and cold sides are previously calculated by using the compressor-turbine equations. Meanwhile, the Q_{recup} can be expressed from the convection heat transfer equation shown in Eq. (19) where h is the convective heat-transfer coefficient, A is the heat transfer area, T_f is the steam temperature and T_m is the heat exchange plate temperature in the recuperator.

$$Q_{recup} = mc_{p,3-4}(T_4 - T_3) = mc_{p,7-8}(T_7 - T_8) \quad (18)$$

$$Q_{recup} = \begin{cases} h_{hot}A_{hot}(T_{f,hot} - T_m) & \text{for hot steam side} \\ h_{cold}A_{cold}(T_m - T_{f,cold}) & \text{for cold steam side} \end{cases} \quad (19)$$

Eq. (20) shows the electric power output calculation of MGT. Pow_{elec} is a function of the power generated by the turbine (Pow_{turb}), the power consumed by the compressor (Pow_{comp}), the generator efficiency (η_{gen}) and the invertor efficiency (η_{inv}).

$$Pow_{elec} = (Pow_{turb} - Pow_{comp})\eta_{gen}\eta_{inv} \quad (20)$$

2.4. Organic Rankine cycle

An ORC system is applied as a bottoming cycle for recovering waste heat from the micro gas turbine. There are four components, i.e. a pump, an evaporator, a turbine and a condenser. The pump is used to increase the working fluid pressure, the evaporator is adopted to recover the waste heat from the exhaust gas of MGT to heat the organic working fluid, the turbine is employed to generate the power, and the condenser is used for cooling. Considering the exhaust gas temperature is around 270 °C [26], R141 b is selected as the organic working fluid, the critical temperature, normal boiling temperature and critical pressure of which are 204.4 °C, 32.1 °C and 42.1 bar, respectively.

The energy equation of the evaporator is similar to that of the MGT recuperator (Eq. (21)). The Eqs. (12) and (14) can be adopted to obtain the working fluid outlet temperatures of the turbine and pump. The Eqs. (13) and (15) are used to calculate the power.

$$Q_{ORC} = m_{MGT}c_{p,8-9}(T_8 - T_9) = m_{ORC}c_{p,10-11}(T_{11} - T_{10}) \quad (21)$$

To simplify the calculation of the organic Rankine cycle, the assumptions can be summed up as follows [27]:

- The minimum temperature difference within the evaporator is 30 °C;
- The condensing temperature of the working fluid is set at 30 °C;
- Isentropic efficiency of turbine and pump are 0.75 and 0.6, respectively;
- Heat and pressure losses are not considered.

In addition, a water-cooled condenser is adopted to recycle the steam in MGT exhaust gas for reducing water consumption. The minimum heat transfer temperature difference is 15 °C within the water-cooled condenser.

Fig. 4 illustrates the flow chart of the computational procedure to determine the system performance. The input data are time, ambient temperature, ambient pressure, ambient humidity, rotational speed (MGT) and DNI. Firstly, the heliostat field performance is calculated. It can determine the cosine efficiency, shading and blocking efficiency of the heliostat field as well as the solar energy incident to the steam receiver and air receiver. Through the steam receiver calculation, the steam mass flow rate heated by solar energy can be approached. In the MGT calculation, the compression rate of the compressor is assumed first. Then the mass flow through the compressor can be determined. Based on the mass flow rate through the compressor, the energy and pressure balance of MGT and air receiver are calculated. With the expansion rate of the turbine, the mass flow rate through the turbine is calculated. If the mass flow rates of compressor and turbine are converged, the calculation is finished. Otherwise, another compression rate of the compressor is assumed for a new iterative calculation. The ORC calculation is conducted after the MGT exhaust temperature is obtained, and then the whole system performance can be determined.

3. Model validation

The thermodynamic model is validated by experimental and referenced data. The validation of each sub-model is presented as follows.

The heliostat field performance is validated based on experimental data of the solar tower. The flux density distribution is measured by using the CCD-Lambert method [28]. For concentrated flux density distribution of 8 heliostats, the measurement has been compared with the simulation, the relative errors are within 2.13% for the maximum flux value and the total energy collected by the lambert plane (shown in Fig. 5(a)).

The performance of the steam receiver is validated based on reference data [29], and the validations of the solar air receiver model and MGT model are based on the authors' previous research [30]. For steam receiver model validation, the receiver efficiency is calculated, and the receiver outlet temperature is assumed to be the same as the experimental data. The relative error is within 2.35% for the receiver efficiency (see Table 2).

The simulation results of the steam-injection MGT are validated by adopting the reference data [8]. The simulated electrical efficiency of the MGT is higher than the experiment data, and it is the same tendency

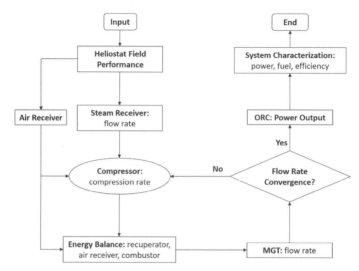

Fig 4. Flow chart of the computational procedure.

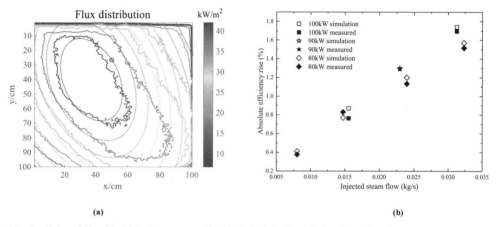

(a)   (b)

Fig 5. Results of heliostat field model validation (a), measurement (solid lines), simulation (dotted lines); Model validation for steam-injection MGT system (b).

Table 2
Results of the model validation for the steam receiver.

Outlet temperature, °C	Experiment efficiency, %	Simulation efficiency, %
177	94	93
283	92	92
502	85	87

as the simulation result reported in the literature (see Fig. 5(b)). The ORC cycle performance is validated by using the data reported in the literature as well [31]. The relative error of the ORC thermal efficiency is within 0.65%. Although the data for ORC model validation is simulated data from references, we believe the universal usage of ORC allows the assumption that these simulation results could be used for model validation.

To further estimate the accuracy of the whole system, we compared the system power output with and without considering the relative error ranges of the sub-models (as shown in Table 3). The simulated power output is 141.26 kW (as shown in Fig. 12), while the possible power output range is from 135.65 kW to 143.97 kW (as shown in Fig. 6). Thus, the relative error range of the whole system is concluded to be -3.97% to 4.12%.

Table 3
The relative error range of power output for each sub-model.

Power output	Heliostat field	Air receiver	Steam receiver	MGT	ORC
Relative error range	−0.22% to 2.13%	−1.85% to −2.18%	−1.06% to 2.35%	−0.20% to 0.34%	−0.23% to 0.65%

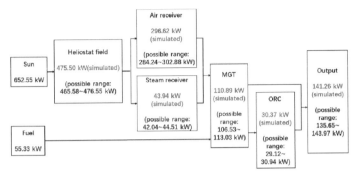

Fig 6. The energy flow of the whole system with simulated values and possible range.

4. Simulation results and discussions

This section shows simulation results and discussions on the proposed model, including the relevant parameters and the system performance.

4.1. Main parameters

There are four important parameters for system design and operation, including the air receiver aperture diameter, the injected steam temperature, the critical DNI value of receiver bypassed and the turbine inlet temperature of ORC. In this section, the simulation is carried out with 40 mirrors when DNI is 750 W/m².

4.1.1. Aperture diameter of air receiver

The air receiver aperture diameter is a key parameter to balance the solar energy entering the receiver and the thermal losses through the aperture. A cavity tubular air receiver is adopted in this system and the cavity is composed of receiver tubes. Here the cavity area is assumed to be constant and its geometric dimension is determined by the aperture diameter. The basic parameters of the air receiver tube are shown in Table 4.

Fig. 7 shows the simulation results of air receiver performance with different aperture diameters. Usually, the noon on the spring equinox is the design point for CSP plants. Three more time points are added in the simulation to provide a more representative performance of the air receiver [12]. The spring equinox and the autumn equinox are identified as the same condition because the sun altitude and azimuth are the same on these two days. The utilization efficiency is defined as the product of intercept efficiency and receiver thermal efficiency. The intercept efficiency of the receiver increases when the receiver aperture diameter increases. However, a larger receiver aperture causes a higher thermal loss, leading to a lower thermal efficiency. For example, at 12:00 am on the spring equinox, the receiver intercept efficiency increases from 70.4% to 95.8% when the receiver aperture diameter increases from 0.8 m to 1.4 m. But the receiver thermal efficiency decreases from 74.3% to 58.5%.

At 12:00 am on the spring equinox, the utilization efficiency is the highest when the aperture diameter is 1.1 m. If the concentrating performance of the heliostat field is poor (e.g. 9:00 am on the spring equinox, 12:00 am on the winter solstice), the aperture diameter should be increased from 1.1 m to 1.2 m to improve the intercept efficiency. Though the thermal efficiency of the air receiver decreases, the utilization efficiency increases. If the concentrating performance of the heliostat field is good enough (12:00 am on the summer solstice), a decrease of the aperture diameter (from 1.1 m to 1.0 m) can increase the thermal efficiency as well as the utilization efficiency. According to the simulation results, the aperture diameter of 1.1 m is selected for the air receiver in this system.

4.1.2. Injected steam temperature

The steam receiver is located around the air receiver aperture to increase the utilization efficiency of solar energy. The outer diameter of the receiver is 1.8 m and the inner diameter is 1.1 m. In this way, the total intercept efficiency of the air receiver and steam receiver can be up to 97.0%. The steam injection position of MGT is at the compressor outlet, so the injected steam pressure should be higher than 450 kPa. In the simulation, the steam injection pressure is 500 kPa. The simulation is carried out at 12:00 am on the spring equinox.

The surge margin is defined as Eqn. (21) according to the literature [32]. The steam injection of MGT will increase the discharge pressure of the compressor, and this will probably lead to a decrease in surge margin. The surge margin of the compressor should be larger than 1.2 for a safe operation.

$$SM = \frac{m_{operation}}{m_{surge}} \cdot \frac{PR_{surge}}{PR_{operation}} \tag{22}$$

where $m_{operation}$ and $PR_{operation}$ are the air mass flow rate and the pressure ratio of the compressor at the operation point; m_{surge} and PR_{surge} are the air mass flow rate and the pressure ratio of the intersection point of the surge line and the speed line.

The simulation results of the steam injected into the MGT are shown in Table 5. The steam receiver produces less steam with a higher steam outlet temperature. The lowest steam outlet temperature in Table 5 is a bit higher than the temperature of water evaporation when the pressure is 500 kPa.

After the steam injection, the electricity output of MGT increases from 102.70 kW to 110.30 kW with the generation efficiency increasing from 15.2% to 16.2% (the heliostat field efficiency is considered for the generation efficiency calculation). With the steam injected into the MGT, the mass flow rate through the compressor is reduced, causing a decrease in power consumed by the compressor. On the other hand, the turbine power generation and the electricity output are increased. It can be seen that when MGT operates with load regulation by adjusting fuel consumption, the steam injection can further increase the MGT

Table 4
Basic parameters of the air receiver tube.

Parameter	Value
Number	80
Length, m	3.2
Diameter, mm	17 + 2*3.35
Material	Inconel 625
Emissivity	0.875

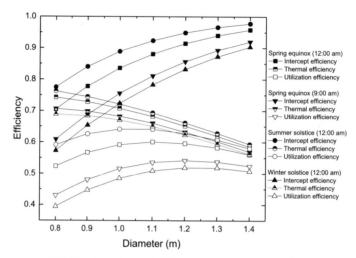

Fig 7. Air receiver performance for different aperture diameters at different times.

Table 5
Simulation results of the steam injected into the MGT.

T_{steam}, °C	m_{steam}, kg/s	Pow_{MGT}, kW	η	Surge margin	Pow_{com}, kW	Pow_{tur}, kW	m_{com}, kg/s	m_{tur}, kg/s
–	0	102.7	0.1524	1.378	160.5	283.5	0.8121	0.8139
155	0.0155	110.3	0.1618	1.355	159.1	290.2	0.8030	0.8204
175	0.0152	110.2	0.1617	1.355	159.2	290.1	0.8032	0.8203
195	0.0149	110.0	0.1615	1.356	159.2	289.9	0.8034	0.8202
215	0.0147	109.9	0.1613	1.356	159.2	289.8	0.8035	0.8201

flexibility. As shown in Table 4, the best steam injected temperature is 155 °C when the surge margin is always above 1.2.

4.1.3. Critical DNI value of receiver bypassed

When solar irradiation is inadequate, the air receiver should be bypassed to avoid a reduction of MGT generation efficiency. In this situation, the pressure loss between the compressor and turbine reduces, resulting in an increase of efficiency. When the inlet temperature of the air receiver is around 590 °C, the minimum incident solar energy to the air receiver should be greater than the thermal loss of the receiver. In addition, the conversion efficiency from fuel to electricity should be not less than the system efficiency when the air receiver is bypassed.

The critical DNI value of the air receiver bypassed is different with time (see Fig. 8(a)). The lowest value occurs on the summer solstice, and the highest occurs on the winter solstice because the solar altitude is low in winter with a poor heliostat field efficiency. The lowest values during

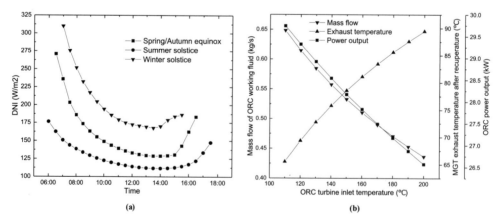

Fig 8. Critical DNI value of air receiver bypassed (a); Simulation results of ORC performance (b).

these four typical days all occur between 13:30 and 14:00. This is because the direction of the solar tower is 25 degrees south by east. On the winter solstice, the DNI cut-in value is between 167.7 W/m^2 and 309.9 W/m^2. On the summer solstice, this value is between 111.6 W/m^2 and 176.5 W/m^2. This can serve as guidelines for the operation of the air receiver.

4.1.4. Turbine inlet temperature of ORC

The key parameter of ORC is the turbine inlet temperature. The exhaust temperature of MGT is around 280 ℃ with a mass flow rate of 0.8 kg/s. The mass flow rate of organic working fluid increases when the ORC turbine inlet temperature decreases. Meanwhile, the MGT exhaust temperature after the evaporator reduces. The simulation results of ORC performance are shown in Fig. 8(b).

The ORC power output rises with the increase of the ORC working fluid, despite the lower inlet temperature of the ORC turbine. When the mass flow rate increases from 0.4362 kg/s to 0.6484 kg/s, the power output of ORC increases from 26.57 kW to 29.74 kW. In the meanwhile, the MGT exhaust temperature after the evaporator decreases from 89.4 ℃ to 65.5 ℃. The evaporation pressure of the working fluid is 800 kPa, while the corresponding evaporation temperature is around 107.6 ℃. Therefore, 110 ℃ is selected as the turbine inlet temperature of ORC.

4.2. System performance

The system performance includes the heliostat field performance and the system energy distribution. The heliostat field performance shows the different kinds of efficiency of the heliostat field at different times. The system energy distribution illustrates the thermal loss of the system, thus providing guidelines for the improvement of system performance. The simulation results are carried out during four typical days (i.e., the spring equinox, the summer solstice, the autumn equinox and the winter solstice).

4.2.1. Heliostat field performance

The parameters that describe the heliostat field performance are cosine efficiency, shading efficiency, blocking efficiency and optical efficiency. The optical efficiency is the ratio of the solar energy that reaches the solar tower to the total solar irradiation incident to the heliostat field. The simulation results are shown in Fig. 9. The simulation results of heliostat field on the spring equinox and the autumn equinox are the same.

The summer solstice has the highest cosine efficiency thus having the highest optical efficiency (0.815). While the highest optical efficiency on the winter solstice is only 0.629, indicating the large effect of cosine efficiency on the optical efficiency. After 14:00, the optical efficiency reduces considerably due to the reduction of shading efficiency. The blocking efficiency during these four days is close to 1.0 while the average optical efficiency during these typical days are 0.672, 0.754, 0.672 and 0.597, respectively.

4.2.2. Energy distribution

The energy distribution of the system illustrates the relationship among the energy collected by the heliostat field, the energy provided by fuel, the MGT power output, the ORC power output and the whole system power output. Fig. 10 shows the simulation results of the system energy distributions on four typical days. The actual DNI values and ambient temperatures of these four days are adopted for simulation.

In the case without solar energy share, the system fuel consumption is 358.80 kW with a total power output to be 138.38 kW, in which the MGT and ORC power output are 107.00 kW and 31.38 kW, respectively. When the DNI becomes higher than the critical value, the solar energy share increases and the fuel consumption decreases. The solar energy collected by the heliostat field has a similar tendency with the DNI fluctuation. During these four typical days, the fuel consumption is reduced by 14.8%, 24.6%, 22.4% and 3.7%, respectively. On the winter solstice, the poor DNI condition and low solar altitude lead to the lowest solar energy share of the system.

Fig. 11 shows the system efficiency of four typical days. The system efficiency decreases with the increase of solar energy share. This is because the utilization efficiency of solar energy in this system is significantly lower than the fuel utilization efficiency. With the solar energy share reaching its highest (86.0%) at 12:30 on the autumn equinox, the system efficiency decreases to only 19.9%. However, on the winter solstice, the system efficiency is always above 24.5% with a low solar energy share.

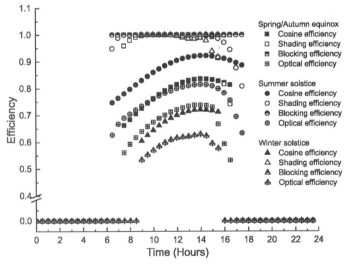

Fig 9. Heliostat field performance of the system on four typical days.

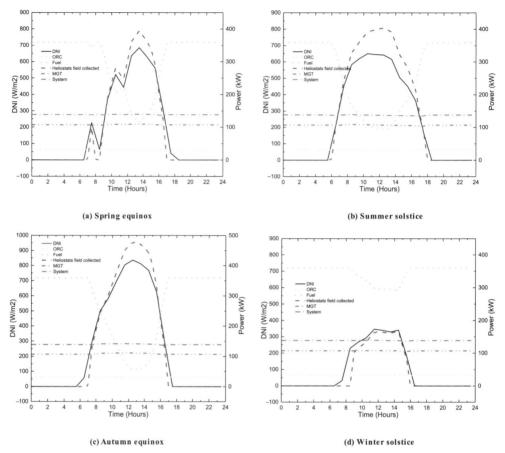

Fig 10. Energy distribution of the system on four typical days.

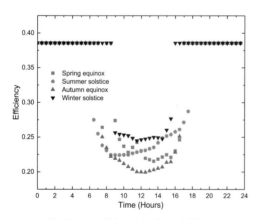

Fig 11. System efficiency during four typical days.

Fig. 12 shows the system energy flow at 12:30 on the autumn equinox with the maximum solar energy share. The total input energy is 707.88 kW with a total power output of 141.26 kW. The maximum thermal loss is the power system loss which accounts for 34.3% of total input energy. The thermal loss of ORC accounts for over 60% of the power system loss. It is because the ORC adopted here is a simple cycle. To reduce the power system loss, an ORC with the recuperation system can be adopted. In addition, the heliostat field loss and the receiver loss are the other main thermal losses. The air receiver loss accounts for over 95% of receiver loss because the thermal efficiency of the air receiver at this moment is 69.7%. The air receiver efficiency can be improved by reducing the receiver aperture through optimizing heliostat field performance to improve the concentration ratio of the heliostat field. In this case, the addition of steam injection and ORC system increases the system power output by 8.29 kW and 30.37 kW, respectively, thus bringing an increase of total system power output by 37.7%. The steam injection and ORC system increase the system construction cost by 29.5% [33] which is worth considering the total increase of system power output.

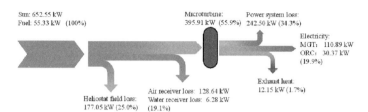

Fig 12. The energy flow of the system at 12:30 on autumn equinox (DNI = 836 W/m^2).

5. Conclusion

A novel system combining the solar-micro gas turbine with steam injection and the ORC bottoming cycle is proposed for distributed energy applications. A thermodynamic model is developed and validated by experimental/referenced data, and used to study the operating parameters for the investigation and design of the system performance. The relative errors of the sub-models are calculated, based on which the relative error range of the whole system is estimated to be −3.97% to 4.12%.

In regards of the concentration characteristics of the heliostat field as well as the balance between the intercept efficiency and the receiver thermal efficiency, the aperture diameter of the air receiver is optimized to be 1.1 m, which is also the design parameter for air receiver. To achieve the maximum power output, the injected steam temperature and the turbine inlet temperature of ORC are designed to be 155 °C and 110 °C, respectively. In operation, the critical DNI value of the air receiver bypassed changes with time, where the value is between 167.7 and 309.9 W/m^2 on the winter solstice, and decreased to between 111.6 and 176.5 W/m^2 on the summer solstice.

The system performances in representative days, including spring equinox, summer solstice, autumn equinox and winter solstice are investigated. The heliostat field performance is various with the changing time while the system efficiency is affected by both the time and the DNI. The highest optical efficiency of the heliostat field is up to 0.815 at 14:00 on the summer solstice, which has the highest average optical efficiency of the heliostat field. The system fuel consumption is considerably reduced with the utilization of solar energy, where 24.6% of the fuel is saved on the summer solstice. However, the system efficiency decreases with the increase of the solar energy share, where the system efficiency only reaches 19.9% with an 86.0% solar energy share at 12:30 on the autumn equinox. 8.29 kW and 30.37 kW power outputs are increased with the addition of steam injection and ORC bottoming cycle, respectively, raising 37.7% of the total power output. This method will increase the system flexibility in dealing with complex loading conditions.

Despite the benefits brought by steam injection and the ORC bottoming cycle, a more complex system is expected, leading to additional costs. The impact of the steam injection on the MGT operation life is still unknown, and more studies related to economic analysis are needed.

CRediT authorship contribution statement

Gang Xiao: Conceptualization, Methodology, Investigation, Resources, Data curation, Writing - review & editing. **Jinli Chen:** Methodology, Software, Validation, Formal analysis, Writing - original draft. **Mingjiang Ni:** Writing - review & editing, Visualization. **Kefa Cen:** Supervision.

Declaration of Competing Interest

The authors declare that they have no known competing financial interests or personal relationships that could have appeared to influence the work reported in this paper.

Acknowledgements

The authors would like to acknowledge the finical support from the National Natural Science Foundation of China (No. 51776186) and the Foundation for Innovative Research Groups of the National Natural Science Foundation of China (No. 51621005).

References

[1] Hohloch M, Zanger J, Widenhorn A, Aigner M. Experimental characterization of a micro gas turbine test rig. Proc Asme Turbo Expo 3: 2010; 671–81.
[2] Pramanik S, Ravikrishna RV. A review of concentrated solar power hybrid technologies. Appl Therm Eng 2017;127:602–37.
[3] Schwarzbözl P, Buck R, Sugarmen C, Ring A, Marcos Crespo MJ, Altwegg P, et al. Solar gas turbine systems: design, cost and perspectives. Sol Energy 2006;80(10):1231–40.
[4] Nelson J, Johnson NG, Doron P, Stechel EB. Thermodynamic modeling of solarized microturbine for combined heat and power applications. Appl Energy 2018;212:592–606.
[5] Heller P, Pfänder M, Denk T, Tellez F, Valverde A, Fernandez J, et al. Test and evaluation of a solar powered gas turbine system. Sol Energy 2006;80:1225–30.
[6] Pavlovic S, Loni R, Bellos E, Vasiljević D, Najafi G, Kasaeian A. Comparative study of spiral and conical cavity receivers for a solar dish collector. Energy Convers Manage 2018;178:111–22.
[7] Kanatani K, Yamamoto T, Tamaura Y, Kikura H. A model of a solar cavity receiver with coiled tubes. Sol Energy 2017;153:249–61.
[8] De Paepe W, Delattin F, Bram S, De Ruyck J. Steam injection experiments in a microturbine – a thermodynamic performance analysis. Appl Energy 2012;97:569–76.
[9] Jun Lee J, Sung Jeon M, Seop Kim T. The influence of water and steam injection on the performance of a recuperated cycle microturbine for combined heat and power application. Appl Energy 2010;87:1307–16.
[10] Araki H, Koganezawa T, Myouren C, Higuchi S. Experimental and analytical study on the operation characteristics of the AHAT system. ASME Turbo Expo, Vancouver; 2011.
[11] Livshits M, Kribus A. Solar hybrid steam injection gas turbine (STIG) cycle. Sol Energy 2012;86:190–9.
[12] Giostri A, Binotti M, Sterpos C, Lozza G. Small scale solar tower coupled with micro gas turbine. Renew Energy 2020;147:570–83.
[13] Mata-Torres C, Zurita A, Cardemil JM, Escobar RA. Cost allocation analysis of a CSP plus MED plant considering time-varying conditions. Aip Conf Proc 2126 2019.
[14] De Paepe W, Carrero MM, Bram S, Contino F. T100 Micro gas turbine converted to full humid air operation: test rig evaluation. Proceedings of the ASME turbo expo: turbine technical conference and exposition, 2014, vol. 3a; 2014.
[15] Mahmoudi A, Fazli M, Morad MR. A recent review of waste heat recovery by Organic Rankine Cycle. Appl Therm Eng 2018;143:660–75.
[16] Clemente S, Micheli D, Reini M, Taccani R. Bottoming organic Rankine cycle for a small scale gas turbine: a comparison of different solutions. Appl Energy 2013;106:355–64.
[17] Camporeale SM, Ciliberti PD, Fortunato B, Torresi M, Pantaleo AM. Externally fired micro gas turbine and orc bottoming cycle: optimal biomass/natural gas chp configuration for residential energy demand. Asme turbo expo: turbine technical conference and exposition, 2015, vol. 3. (2015).
[18] Salomé A, Chhel F, Flamant G, Ferrière A, Thiery F. Control of the flux distribution on a solar tower receiver using an optimized aiming point strategy: application to THEMIS solar tower. Sol Energy 2013;94:352–66.
[19] Besarati SM, Yogi Goswami D. A computationally efficient method for the design of the heliostat field for solar power tower plant. Renew Energy 2014;69:226–32.
[20] Sánchez-González A, Santana D. Solar flux distribution on central receivers: a projection method from analytic function. Renew Energy 2015;74:576–87.
[21] Xu L, Stein W, Kim J-S, Wang Z. Three-dimensional transient numerical model for the thermal performance of the solar receiver. Renew Energy 2018;120:550–66.

[22] Stine WB, McDonald CG. Cavity receiver convective heat loss. Proceedings of the International Solar Energy Society (ISES) SolarWorld Conference, Kobe, Japan; 1989.
[23] Fujii T, Imura H. Natural-convection heat transfer from a plate with arbitrary inclination. Int J Heat Mass Transf 1972;15:755–64.
[24] Jakob M. Heat transfer. New York: John Wiley & Sons Inc.; 1957.
[25] Caresana F, Pelagalli L, Comodi G, Renzi M. Microturbogas cogeneration systems for distributed generation: effects of ambient temperature on global performance and components' behavior. Appl Energy 2014;124:17–27.
[26] Hærvig J, Sørensen K, Condra TJ. Guidelines for optimal selection of working fluid for an organic Rankine cycle in relation to waste heat recovery. Energy 2016;96: 592–602.
[27] Invernizzi C, Iora P, Silva P. Bottoming micro-Rankine cycles for micro-gas turbines. Appl Therm Eng 2007;27:100–10.
[28] Xiao G, Guo K, Xu W, Ni M, Luo Z, Cen K. An improved method of Lambertian CCD-camera radiation flux measurement based on SMARTS (simple model of the atmospheric radiative transfer of sunshine) to reduce spectral errors. Energy 2014; 67:74–80.
[29] Zapata J, Lovegrove K, Pye J. Steam receiver models for solar dish concentrators: two models compared solar PACES 2010 conference; 2010.
[30] Chen J, Xiao G, Ferrari ML, Yang T, Ni M, Cen K. Dynamic simulation of a solar-hybrid microturbine system with experimental validation of main parts. Renew Energy 2020;154:187–200.
[31] Dai Y, Wang J, Gao L. Parametric optimization and comparative study of organic Rankine cycle (ORC) for low grade waste heat recovery. Energy Convers Manage 2009;50:576–82.
[32] Walsh PP, Fletcher P, editors. Gas turbine performance. 2nd Ed. Wiley; 2004.
[33] Javidmehr M, Joda F, Mohammadi A. Thermodynamic and economic analyses and optimization of a multi-generation system composed by a compressed air storage, solar dish collector, micro gas turbine, organic Rankine cycle, and desalination system. Energy Convers Manage 2018;168:467–81.

结合蒸汽注入和有机朗肯循环的太阳能微型燃气轮机系统研究

摘要：微型燃气轮机（MGT）是太阳能驱动发电系统的一个重要组成部分，因为它可以提高系统效率，并提供独特的灵活性来满足快速变化的需求。本文提出了一种结合蒸汽注入和有机朗肯循环（ORC）的太阳能微型燃气轮机，以提高效率和灵活性。将产生注入 MGT 蒸汽的蒸汽接收器放置在空气接收器口径周围，以增加接收器截获效率。开发了一个包括聚光镜场、太阳能接收器、MGT 和 ORC 的热力学模型，以研究所提出系统的性能，其中每个子模型都通过实验/参考数据进行了验证，以便后续优化设计和工作参数。对于由太阳能聚光镜场驱动的 100kW MGT，仿真结果显示，空气接收器的最佳口径直径为 1.1m。在一年中的四个典型日子，即春分、夏至、秋分和冬至，聚光镜场的平均光学效率分别为 0.672、0.754、0.672 和 0.597。MGT 在这四天的燃料消耗分别减少了 14.8%、24.6%、22.4%和 3.7%。在秋分 12:30，太阳能占比可达 86.0%，系统功率效率为 19.9%。通过添加蒸汽注入和 ORC，功率输出可分别增加 8.29kW 和 30.37kW，总功率输出增加 37.7%，提高了系统效率和灵活性。所提出的系统可以作为一种灵活的选择，在分布式能源系统中，特别是偏远地区或岛屿地区，带来裨益。

原文刊于 Energy Conversion and Management, 2021, 234: 114032

Astigmatic dual-beam interferometric particle imaging for metal droplet 3D position and size measurement

Yingchun Wu,[1,2] Zhu Zhuo,[1] Zhiming Lin,[1] Botong Wen,[1] Xuecheng Wu,[1,*] AND Kefa Cen[1]

[1]*State Key Laboratory of Clean Energy Utilization, Zhejiang University, Hangzhou 310027, China*
[2]*e-mail: wuyingchun@zju.edu.cn*
[*]*Corresponding author: wuxch@zju.edu.cn*

Received 5 February 2021; revised 7 March 2021; accepted 12 March 2021; posted 12 March 2021 (Doc. ID 421480); published 14 April 2021

We propose astigmatic dual-beam interferometric particle imaging (ADIPI) to simultaneously measure the three-dimensional (3D) position and size of spherical metal droplets. A theoretical model reveals that the orientation and spacing of the ADIPI fringes generated from the two reflections propagating through an astigmatic imaging system relate to the depth position and size, respectively. Proof-of-concept experiments on micron-sized gallium droplets are performed, and the tilted fringes in elliptical patterns are observed in the ADIPI interferogram, confirming theoretical predictions. Droplet 3D position and size are determined with ADIPI, and the relative discrepancies are within 5% and 2% compared to those with a dual-view digital inline holography system, demonstrating the feasibility and high accuracy of ADIPI. © 2021 Optical Society of America

https://doi.org/10.1364/OL.421480

Metal droplets are commonly encountered in a large number of applications, from metal robotics [1] at room temperature, molten droplets in three-dimensional (3D) metal printing [2,3], to burning aluminium [4,5] or other metal drops [6] in the harsh and high temperature flames of solid propellants. Although comprehensive investigations have been conducted on measurements of transparent droplets/bubbles with several advanced optical techniques developed, e.g., phase Doppler particle analyzer (PDPA) [7,8] and interferometric particle imaging (IPI) [9–13], unfortunately they do not fit the measurement of metal droplets well. IPI, originally known as interferometric laser droplet size measurement (ILIDS) [14–18], measures droplet size from the interference fringes generated by the reflection light on the droplet outer surface and the direct transmission light. IPI was originally proposed for two-dimensional (2D) planar measurement and then extended to volume measurement by using a cylindrical interferometric out-of-focus imaging configuration [19,20]. Digital holography has been applied to measure burning aluminum particles with both the Gabor inline configuration and the off-axis configuration [21–24], implying that two windows placed at opposite sides are needed for metal characterization in a chamber. To address this issue, Wu et al. [25–27] proposed an interference imaging method for opaque particle measurement, in which the particle interferogram is produced from the two reflections of two laser beams illuminations from two different angles, named dual-beam IPI (DIPI). In this letter, a technique called astigmatic DIPI (ADIPI) is proposed to simultaneously measure the 3D position and size of opaque metal droplets in a volume.

In ADIPI, two crossed laser beams illuminate metal particles in their crossing volume, which defines the measurement volume, as illustrated in Fig. 1. The light scattered by the particle is collected by an optical system with astigmatism comprising a spherical and cylindrical lens. Let us consider the Cartesian coordinate system with the origin at the crossing point of the optical axes of the two laser beams and the z axis along the optical axis of the astigmatic optical system. The reflection can be approximated as a spherical wave emitting from the incidence points (x_a, y_a) and (x_b, y_b) with an initial phase

$$U_0(x_0, y_0, 0) \cong u_1\delta(x_0 - x_a, y_0 - y_a)\exp(i\varphi_1) + u_2\delta(x_0 - x_b, y_0 - y_b)\exp(i\varphi_2), \quad (1)$$

where the function $\delta(x, y)$ is the Dirac function, u_1 and u_2, respectively, are the amplitudes of these two glare points [28] depending on the scattering angle θ_1 and θ_2, and φ_1 and φ_2 are the initial phases of the two incident laser beams at the incident points. Without loss of generality, take the planes Rp_1 and Rp_2 that are perpendicular to the incident beams and tangent to the droplet surface as the incident reference planes, where the two beams have fixed phases, and the planes Rp_{out} that are perpendicular to the outgoing rays and tangent to the droplet surface as the outgoing reference plane, as illustrated by the inset marked with dashed red rectangle in Fig. 1. The initial phase can be evaluated as

$$\varphi_1 = 2\pi L_{AC}/\lambda, \quad \varphi_2 = 2\pi L_{BD}/\lambda, \quad (2)$$

with the optical path $L_{AC} = R(1 - \cos\theta_1)$ and $L_{BD} = R(1 - \cos\theta_2)$. The distance between the two reflected lights is $a + b = D(\sin\theta_1 + \sin\theta_2)/2$, with D the droplet diameter.

The angle between the line between the two glare points and the cylindrical lens or the camera is α_0. Thus, $x_a = a \cos\alpha_0$, $y_a = a \sin\alpha_0$, $x_b = -b \cos\alpha_0$, and $y_b = -b \sin\alpha_0$.

Without considering the aperture of the optical system, the complex amplitude field at the camera sensor can be described by the generalized Huygens–Fresnel integral [29,30]:

$$U(x,y) = \frac{\exp(ikn_s l)}{i\lambda\sqrt{B_{t,x}B_{t,y}}} \iint U_0(x_0, y_0)$$
$$\times \exp\left[i\frac{\pi}{\lambda B_{t,x}}\left(A_{t,x}x_0^2 - 2x_0 x + D_{t,x}x^2\right)\right]$$
$$\times \exp\left[i\frac{\pi}{\lambda B_{t,y}}\left(A_{t,y}y_0^2 - 2y_0 y + D_{t,y}y^2\right)\right]dx_0 dy_0, \quad (3)$$

where the A, B, C, and D are elements of the transfer matrix from the glare point on the particle to the camera sensor. Substitute Eq. (1) into Eq. (3), and the light field can be resolved as

$$U(x,y) = \frac{\exp\left(i\frac{2\pi}{\lambda}n_s l\right)}{i\lambda\sqrt{B_{t,x}B_{t,y}}}\left[u_1\exp(i\varphi_{t1}) + u_2\exp(i\varphi_{t2})\right], \quad (4)$$

with the phases

$$\varphi_{t1} = \frac{\pi}{\lambda}\left[\frac{1}{B_{t,x}}\left(A_{t,x}x_a^2 - 2x_a x + D_{t,x}x^2\right)\right.$$
$$\left.+ \frac{1}{B_{t,y}}\left(A_{t,y}y_a^2 - 2y_a y + D_{t,y}y^2\right)\right] + \varphi_1,$$

$$\varphi_{t2} = \frac{\pi}{\lambda}\left[\frac{1}{B_{t,x}}\left(A_{t,x}x_b^2 - 2x_b x + D_{t,x}x^2\right)\right.$$
$$\left.+ \frac{1}{B_{t,y}}\left(A_{t,y}y_b^2 - 2y_b y + D_{t,y}y^2\right)\right] + \varphi_2. \quad (5)$$

Thus, the phase of the wavefront is

$$\varphi_t = \varphi_{t1} - \varphi_{t2} = \frac{2\pi}{\lambda}\left[\frac{(x_b - x_a)}{B_{t,x}}x + \frac{(y_b - y_a)}{B_{t,y}}y\right] + \varphi_c, \quad (6)$$

where φ_c is constant and therefore not considered below. Equation (6) implies that the orientation angle of the fringes, which is the equiphase line of the wavefront determined by a constant value of φ_t, is

$$\tan(\alpha) = -\frac{B_{t,y}(x_b - x_a)}{B_{t,x}(y_b - y_a)} = -\frac{B_{t,y}}{B_{t,x}}\cot(\alpha_0), \quad (7)$$

where α is the angle between the stripes and the x axis. α_0 is related to the glare points.

In practice, the optical system has a finite aperture, and the diffraction of each reflection by the spherical aperture during its propagation produces an elliptical pattern on the camera sensor. Note the ellipses' centroids as $(x_{s,a}, y_{s,a})$ and $(x_{s,b}, y_{s,b})$, respectively, and their relations with the glare points can be derived with the transfer matrix:

$$\begin{aligned} x_{s,b} - x_{s,a} &= A_{t,x}(x_b - x_a), \\ y_{s,b} - y_{s,a} &= A_{t,y}(y_b - y_a). \end{aligned} \quad (8)$$

The spatial spacing of the fringes can be evaluated from a variation of 2π along the normal vector of the equiphase line in the phase, with value of

$$\sigma = \frac{\lambda}{\sqrt{\left(\frac{x_b - x_a}{B_{t,x}}\right)^2 + \left(\frac{y_b - y_a}{B_{t,y}}\right)^2}}. \quad (9)$$

The defocus parameters $B_{t,y}$ and $B_{t,x}$ in Eq. (7) implicitly encode the particle depth position, and their ratio $B_{t,y}/B_{t,x}$ varies along the z axis since they have different values due to astigmatism, modulating the inclination angle of the fringe. In Eq. (8), the two magnification parameters, $A_{t,x}$ and $A_{t,y}$, also differ along the depth position, transforming the circular aperture into elliptical. This technique utilizes the astigmatism of interferometric fringes of two reflections to retrieve the particle 3D position and thus is called ADIPI. ADIPI reduces to the traditional DIPI when the astigmatism vanishes with the transfer matrix radially symmetric and interference fringes perpendicular to line segment of two glare points.

Figure 1 schematically illustrates the experimental setup of an ADIPI system for metal droplet measurement. A continuous vertically polarized laser beam with a wavelength of 532 nm was first spatially filtered, collimated into a parallel laser beam with a diameter of about 30 mm, and then split into two beams through a 50:50 beam splitter cube. Two laser beams were reflected by the mirrors. The laser beams refracted into a quartz cavity filled with deionized water and illuminated micron-sized droplets of gallium alloy with a low molten point from two orthogonal directions, with an experimentally calibrated value of 89.14°. The light scattered by the opaque gallium droplets from two incident laser beams was acquired by an astigmatic optical system laid at the symmetry axis of the two beams. The backward reflection around 135° was first collected by a spherical lens, with a focal length of 50 mm, and then transferred by a cylindrical lens with focal lengths $f_{cx} = 250$ mm and $f_{cy} = \infty$. Then, the light traveled to the camera with a pixel size of 3.45 μm and a resolution of 2048 × 2048 pixels and was captured as an ADIPI interferogram. In order to simplify the configuration, the cylindrical lens was rotated 45° with respect to the coordinates. Therefore, the fringe angle Θ in the camera was $\Theta = \alpha - \pi/4$. In order to determine the astigmatism of the optical system, a calibration plate was mounted and moved along the depth direction. As illustrated by the inset highlighted by the dashed blue rectangle in Fig. 1, the front focal planes in the x and y directions, where $B_{im,x}$ and $B_{im,y}$ were zeros, were located, and their transversal magnifications quantified with values of 0.637 and 0.708, respectively. The distance between the two focal planes was 8.56 mm. With these calibrated values, the ray transfer matrices at an arbitrary plane can be determined. Two cameras were placed perpendicular to the two illumination beams and recorded the digital inline holograms of the gallium droplet cloud with synchronization to the ADIPI system, forming a dual-view digital inline holography (DIH) system to verify the accuracy of ADIPI.

Figure 2 shows the relationship between the inclination of ADIPI fringe and the particle depth position, which refers to the stand-off distance z_1 from the particle to the first lens in this system. The inset figure shows that the region between the two focal planes in the $x - z$ and $y - z$ planes, which is located from 40 to 48 mm apart from the spherical lens, is more sensitive

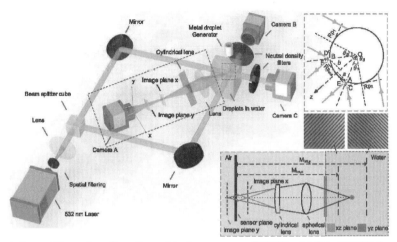

Fig. 1. Schematic diagram of the experimental setup for metal droplet measurement with ADIPI and dual-view DIH. The inset of the red rectangle shows the sketch of the optical path of the two incident and reflected lights. θ_1 and θ_2 are the reflection angles of the two incident light beams, θ_3 is the crossing angle between two laser beams. Point A and point B are marked as two glare points. The inset in the blue rectangle shows the ray transfer matrix of the astigmatic imaging system. $M_{ltn,x}$ and $M_{ltn,y}$ are the ray transfer matrices of the image system in the x and y direction, respectively.

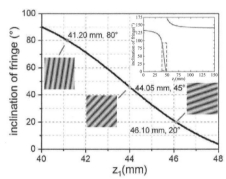

Fig. 2. Relationship between the inclination of fringe and the droplet depth position z_1 from 40 to 48 mm, and the inset shows the overall relationship between the inclination of fringe and z_1 from 0 to 150 mm.

than other regions, and thus this region is chosen as the measurement volume. The inclination angle in the measurement volume decreases from about 90° to nearly zero, with three simulated ADIPI interferograms exemplifying the dependence of the inclination angle on the depth position, and exhibits an approximated linear relationship versus the depth position, which facilitates the data processing. In this case, about ±1° resolution on the fringes orientation, $\Delta z = 0.1$ mm in particle depth position.

Figure 3 shows a representative experimental ADIPI interferogram of a cloud of gallium droplets. It can be observed that the aperture images are stretched to ellipses with different aspect ratios and orientations. A close examination shows that the ADIPI signals of each gallium droplet consist of two ellipses, which are generated by the diffractions of the two reflections by the aperture, and interference fringes are formed in the intersection of the two ellipses. The aspect ratio, as well as ellipse orientation, varies with the gallium droplet depth position. These ellipses are first detected, and their centroids are evaluated to determine droplet transversal position. The interference fringes are oblique, and orientations of fringes change with the particle depth position z_1, as exemplified by the two zoomed-in views of the interference pattern marked by the yellow and blue rectangles shown in Fig. 3(a). It is also observed that the interference fringes are not strictly lined segments, but present a sort of distortion as a cubic curve, especially in the outer region close to the boundary, and this is attributed to the existence of higher-order aberrations, which is mainly a spherical aberration and increases with the radial distance. Therefore, ADIPI signals in the center part of the ellipse are chosen for analysis. The frequency and inclination of the fringes are simultaneously quantified by performing a 2D Fourier transform on the interference fringes, and then the droplet size and depth position are determined with Eq. (9), and the relationship is formulated in Eq. (7) and plotted in Fig. 2.

The ADIPI in Fig. 3(a) is processed with the abovementioned methods. Seven elliptical patterns are detected, as marked by the solid circles filled with different colors in the projected ADIPI interferogram. The 2D Fourier spectrum of each interference fringe pattern is calculated, and the position of the dominant frequency is located to obtain the inclination and frequency of the ADIPI fringe. The orientation angles are calculated, with values of 27.4°, 42.0°, 19.5°, 33.2°, 23.3°, 18.0°, and 24.5°, respectively. According to the relationship in Fig. 2, droplet depth positions, z_1, can be obtained, with values of 45.43 mm, 44.28 mm, 46.15 mm, 44.96 mm, 45.79 mm, 46.29 mm, and 45.69 mm, correspondingly. The sizes of the seven droplets, respectively, are 214 μm, 208 μm, 210 μm, 208 μm, 208 μm, 230 μm, and 208 μm. The position of the droplet in the 2D direction is obtained by finding the center of

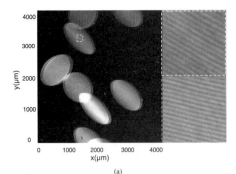

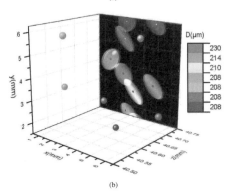

Fig. 3. (a) Typical cropped ADIPI interferogram of a cloud of gallium droplets recorded by camera A, with two zoomed-in views of the interference fringes marked by dashed rectangles. (b) 3D results of droplet diameter and 3D location of (a), with the sizes and colors of the scatterers proportional to droplet size, and the ADIPI interferogram is projected for comparison.

the elliptical pattern and determined by Eq. (8). The measured droplets are visualized in Fig. 3(b), with both the size and color of the scatterer proportional to the droplet size. Meanwhile, the droplets are measured with the dual-view DIH system for verification, where a comparison between the values with ADIPI and DIH shows that the average relative deviation in droplet size is 5%, and the relative deviation of depth position z_l is 2%. The comparison demonstrates the high accuracy of ADIPI in particle 3D position and size measurement.

In summary, the proposed ADIPI for metal droplet 3D position and size measurement has been theoretically elucidated and experimentally validated. Actually, applications of ADIPI can be extended to opaque spheres with sizes ranging from dozens to hundreds of microns with accuracy of several microns, and to curved containers, e.g., pipes [31–34], by considering its lens effect. The measurement volume is adjustable by the laser beam sizes and the astigmatic imaging system. Generally, the ADIPI signal exists at an arbitrary angle except the two laser beam axes, breaking the angular position constraint in IPI and Gabor inline holography. With this merit, the configuration of the laser illumination and image acquisition part can adapt to a variety of configurations, and thus facilitates its great potential in metal droplet characterization.

Funding. National Natural Science Foundation of China (52006193); Natural Science Foundation of Zhejiang Province (LQ19E060010); National Science and Technology Major Project (2017-V-0016-0069); National Key Research and Development Program of China (2020YFA0405700, 2020YFB0606200).

Disclosures. The authors declare no conflicts of interest.

Data Availability. Data underlying the results presented in this paper are not publicly available at this time but may be obtained from the authors upon reasonable request.

REFERENCES

1. X. Wang, R. Guo, and J. Liu, Adv. Mater. Technol. **4**, 1800549 (2019).
2. C. Ladd, J.-H. So, J. Muth, and M. D. Dickey, Adv. Mater. **25**, 5081 (2013).
3. L. E. Murr and W. L. Johnson, J. Mater. Res. Technol. **6**, 77 (2017).
4. D. R. Guildenbecher, M. A. Cooper, and P. E. Sojka, Appl. Opt. **55**, 2892 (2016).
5. Y. Chen, D. R. Guildenbecher, K. N. Hoffmeister, M. A. Cooper, H. L. Stauffacher, M. S. Oliver, and E. B. Washburn, Combust. Flame **182**, 225 (2017).
6. A. G. Korotkikh, O. G. Glotov, V. A. Arkhipov, V. E. Zarko, and A. B. Kiskin, Combust. Flame **178**, 195 (2017).
7. J. Ma, S. C. M. Yu, H. W. Ng, and Y. C. Lam, Plasma Chem. Plasma Process. **24**, 85 (2004).
8. C. Tropea, Annu. Rev. Fluid Mech. **43**, 399 (2011).
9. G. König, K. Anders, and A. Frohn, J. Aerosol Sci. **17**, 157 (1986).
10. A. Glover, S. Skippon, and R. Boyle, Appl. Opt. **34**, 8409 (1995).
11. T. Kawaguchi, Y. Akasaka, and M. Maeda, Meas. Sci. Technol. **13**, 308 (2002).
12. N. Semidetnov and C. Tropea, Meas. Sci. Technol. **15**, 112 (2003).
13. S. Dehaeck and J. Van Beeck, Exp. Fluids **42**, 767 (2007).
14. Y. Hardalupas, S. Sahu, A. M. Taylor, and K. Zarogoulidis, Exp. Fluids **49**, 417 (2010).
15. H. Shen, S. Coëtmellec, G. Gréhan, and M. Brunel, Appl. Opt. **51**, 5357 (2012).
16. A. Querel, P. Lemaitre, M. Brunel, E. Porcheron, and G. Gréhan, Meas. Sci. Technol. **21**, 015306 (2009).
17. M. Brunel and H. Shen, Particuology **11**, 148 (2013).
18. N. Damaschke, H. Nobach, T. Nonn, N. Semidetnov, and C. Tropea, in *11th International Symposium on Application of Laser Techniques to Fluid Mechanics* (2002).
19. H. Shen, S. Coetmellec, and M. Brunel, Opt. Lett. **37**, 3945 (2012).
20. H. Shen, S. Coetmellec, and M. Brunel, J. Quant. Spectrosc. Radiat. Transf. **131**, 153 (2013).
21. Y. Wu, Z. Lin, Z. Zhuo, S. Wu, C. Zhou, L. Yao, W. Ao, X. Wu, L. Chen, and K. Cen, "Particle burning behaviors of Al/AP propellant with high-speed digital off-axis holography," Proc. Combust. Inst. (to be published).
22. J. Gao, D. R. Guildenbecher, P. L. Reu, V. Kulkarni, P. E. Sojka, and J. Chen, Opt. Lett. **38**, 1893 (2013).
23. D. R. Guildenbecher, M. A. Cooper, W. Gill, H. L. Stauffacher, M. S. Oliver, and T. W. Grasser, Opt. Lett. **39**, 5126 (2014).
24. P. Memmolo, L. Miccio, M. Paturzo, G. Di Caprio, G. Coppola, P. A. Netti, and P. Ferraro, Adv. Opt. Photon. **7**, 713 (2015).
25. Y. Wu, Z. Lin, X. Wu, and K. Cen, Powder Technol. **356**, 31 (2019).
26. Z. Lin, Y. Wu, X. Wu, L. Chen, and K. Cen, Opt. Laser Eng. **129**, 106069 (2020).
27. Z. Lin, Y. Wu, Z. Zhuo, and X. Wu, Powder Technol. **382**, 505 (2020).
28. H. C. van de Hulst, *Light Scattering by Small Particles* (Dover, 1957).
29. C. Palma and V. Bagini, J. Opt. Soc. Am. A **14**, 1774 (1997).
30. H. T. Yura and S. G. Hanson, J. Opt. Soc. Am. A **4**, 1931 (1987).
31. M. P. Sentis, L. Bruel, S. Charton, F. R. Onofri, and F. Lamadie, Opt. Lasers Eng. **88**, 168 (2017).
32. S.-H. Lee, Y. Roichman, G.-R. Yi, S.-H. Kim, S.-M. Yang, A. Van Blaaderen, P. Van Oostrum, and D. G. Grier, Opt. Express **15**, 18275 (2007).
33. M. P. Sentis, F. R. Onofri, and F. Lamadie, Opt. Lett. **43**, 2945 (2018).
34. M. P. Sentis, F. R. Onofri, and F. Lamadie, Opt. Express **25**, 867 (2017).

用于金属液滴三维位置和尺寸测量的像散双光束干涉粒子成像技术

摘要：我们提出采用像散双光束干涉粒子成像(ADIPI)来同时测量球形金属液滴的三维(3D)位置和尺寸。理论模型表明，通过像散成像系统传播的两次反射所产生的 ADIPI 条纹的方向和间距分别与深度位置和尺寸有关。我们对微米级镓液滴进行了验证实验，在 ADIPI 干涉图像中观察到椭圆形图案的倾斜条纹，证实了理论预测。与双视在线数字全息系统相比，通过 ADIPI 确定液滴的 3D 位置和尺寸，相对差异分别在 5% 和 2% 以内，证明了 ADIPI 的可行性和高精度。

原文刊于 Optics Letters, 2021, 46(8): 1942-1945

Nonlinear Dynamic Characteristics of Turbulent Non-Premixed Acoustically Perturbed Swirling Flames

ZHOU Hao[*], TAO Chengfei, MENG Sheng, CEN Kefa

State Key Laboratory of Clean Energy Utilization, Zhejiang University, Hangzhou 310027, China

© Science Press, Institute of Engineering Thermophysics, CAS and Springer-Verlag GmbH Germany, part of Springer Nature 2022

Abstract: The main objective of this article was to experimentally investigate the dynamic response of diffusion flame under acoustic excitation in a laboratory-scale burner. Two parametric variations of the burner, the burner inlet length and variation of the airflow rate, were studied. Experimental results were analyzed through nonlinear time series analysis and several resonance characteristics were obtained. Results indicate that the flame-acoustic resonance only appears under certain frequencies together with the fuel tube vibration. Resonance characteristics of the combustion chamber and air inlet in the non-premixed burner indicate quasi-periodic or limit cycle oscillations, respectively. Flame-acoustic resonance would trigger the frequency and amplitude mode-transition in burners. Moreover, the intermittency of flame heat release was observed under variation of inlet length and airflow rate in the burner; the 445 mm case shows more frequency peaks and fluctuations than the 245 mm one. Four typical flame forms were examined during the flame-acoustic resonance conditions, evolves from wrinkled flames to diverged flames, then evolves to reattached flames and finally to blow-off flames. This study proposed the practical application of nonlinear time-series analysis method as a detection tool for flame-acoustic resonance in laboratory non-premixed burners, which could contribute to the detection and prevention of potential thermoacoustic instabilities or resonance structure failures of industrial boilers. Finally, this study demonstrates an alternative to conventional linear tool for the characterization of nonlinear acoustic resonance in industrial boilers.

Keywords: flame-acoustic resonance, non-linear time series analysis, non-premixed burners, intermittency, dynamic characteristics

1. Introduction

In combustion-based propulsion systems, thermoacoustic instability is an undesired phenomenon which can cause serious pressure fluctuations and serve damages to the combustor components [1, 2]. Thermoacoustic instability is an inherently nonlinear process and involves many complex factors in industrial burners which makes thermoacoustic modeling sophisticated [3]. Dynamic characteristics of burners with different geometry, emission, pilot burners, and fuel characteristics were widely studied [4–8]. The intermediate nonlinearity and sensitivity of burner contribute to the difficulty of understanding flame-acoustic resonance characteristics and damping of thermoacoustic instabilities [9]. However, the nonlinear time-series analysis tools are mainly used

Received: Oct 08, 2019   AE: YANG Bin   Corresponding author: ZHOU Hao   E-mail: zhouhao@zju.edu.cn

Nomenclature

AMI	average mutual information	L_2	air inlet length/mm
CH*	chemiluminence	P_1	section pressure/Pa
D	chamber width/mm	P_2	inlet acoustic pressure/Pa
d_E	embedding dimension	PMT	photomultiplier tube
FEM	Finite Element Method	**Greek symbols**	
FFT	Fast Fourier Transform	τ	optimum time delay/ms
L	chamber length/mm	Φ	equivalence ratio
L_1	air inlet length/mm		

to study the dynamics of premixed combustion [10–27]. Although time-frequency analysis has been applied to the fault detection of diesel engines [28], the dynamic characteristics of non-premixed combustion in industrial non-premixed combustors are not well studied. Mechanism of resonance behind flame-acoustic coupling interaction is seldom explored by nonlinear time series tools.

Combustion instability is a serious problem in industrial thermal power plants. One must identify the intermediate dynamic states of the system in order to fully comprehend the system and obtain complete characterization. Larry [10–12] investigated the periodic and aperiodic thermoacoustic oscillations by acoustic forcing. Sujith [13–20] studied combustion instabilities with the nonlinear time series analysis tools by exploring the domain of intermittency, bifurcation and blowout characteristics during thermoacoustic oscillations. Kabiraj [21] studied recurrence quantification to analyze the combustion noise and revealed that thermoacoustic instability exhibits several different dynamic states. Juniper [22] identified the bifurcations and chaos in self-excited thermoacoustic oscillations by phase portraits and Poincaré sections. With the concept of dynamical system theory, Gotoda [23–25] found that the combustion blowout can be detected and prevented through nonlinear time-series analysis. Unlike premixed flames, non-premixed flames can be hydro-dynamically unstable and oscillate at a low natural frequency [26]. Although nonlinear time-series analysis techniques were widely used to study thermoacoustic instabilities in premixed combustors [27–29], the resonance characteristics of non-premixed burners under acoustic excitations are not well studied. Through changing the flame position and airflow rate, Sen [30–31] obtained several interesting dynamic characteristics in self-excited inverse diffusion flames with recurrence analysis, such as limit cycles, intermittency, and homoclinic orbits. Guan [32] studied dynamic characteristics of high-frequency thermoacoustic oscillations in a solid rocket motor with nonlinear time-series analysis and observed transient mode switching between different unstable attractors.

Kim [33] examined nonlinear mode transition mechanisms of a self-excited Jet A-1 spray non-premixed flame with nonlinear time-series analysis tools. Zhou [34] explored dynamic characteristics of non-premixed swirl-stabilized flames to acoustic excitation through Poincaré Sections analysis and concluded that lock-in occurred more easily when the forcing frequency was greater than the natural frequency. The recent research [35] identified that burner geometry also affects non-premixed flame response under acoustic excitation, but the specific physical mechanism behind this nonlinear resonance phenomenon is still unknown. Further research with nonlinear analysis tools is needed.

The purpose of the present work is to explore the flame-acoustic resonance characteristics of a laboratory-scale non-premixed burner under acoustic disturbance with nonlinear dynamical system theory. With the nonlinear time series analysis tools [36–38] and COMSOL acoustic simulation techniques [39–42], the effects of burner inlet length and airflow rate on the non-premixed flame-acoustic resonance characteristics were investigated. Under flame-acoustic resonance conditions, the sound pressure fluctuation, heat release intermittency and flame morphology phenomena of the non-premixed burner were studied simultaneously. The resonance characteristics detection of a non-premixed burner through nonlinear time series analysis has not been studied before, therefore this study can lead to a better understanding of flame-acoustic resonance dynamic characteristics. As a result, providing a useful tool for detecting unsteady resonance characteristics among non-premixed burners can improve the operation safety of industrial boilers.

2. Experimental Setup and Analytical Methods

Experiments were performed on a laboratory-scale, swirl-stabilized non-premixed burner. Detailed experimental setup and indispensable instruments used in the present work are shown in Fig. 1 as a system diagram and the components of the experiment are connected in series by black lines. Combined with two symmetrically installed loudspeakers, the overall layout and measuring instruments of the burner are demonstrated in Fig. 2(a). The non-premixed flame was perturbed acoustically by the above loudspeakers, accompanied by amplified figures of the combustion chamber illustrated in Fig. 2(b) and swirlers in Fig. 2(c). It is presented in Fig. 1 and Fig. 2 that at the upstream of the burner, the geometry structure of the burner inlet channel comprises a cylindrical air inlet section and a fuel tube. An optically-accessible combustion chamber with three quartz square windows is arranged in the middle of the burner. The quartz windows were rubbed by ethyl alcohol before experiments for a better transmittance. Fig. 2(c) and (d) show the swirler of the burner. The axial and azimuthal velocities of the burner are assumed to be uniform and the vanes are very thin [3]. The diameter of

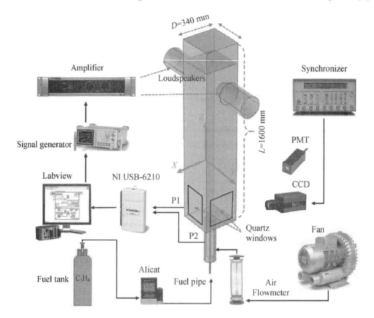

Fig. 1 System diagram of the acoustically perturbed non-premixed swirling burner (P1, P2: Position of sound pressure tap)

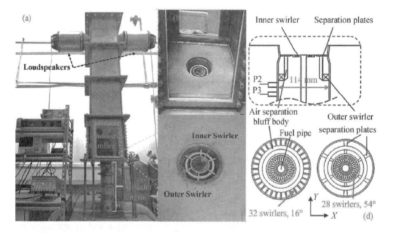

Fig. 2 (a) The layout of the non-premixed combustor with 445 mm and 225 mm air inlet length; (b) Magnified figure of the combustion chamber; (c) Magnified figure of swirlers; (d) The detailed configuration of the burner with separation plates, inner and outer swirlers (P2, P3: Position of sound pressure tap)

the burner inlet section is 114 mm. Inner swirler has 28 swirl vanes with thickness of 0.5 mm, aligned at 54° to the flow direction. Outer swirler has 32 swirl vanes with thickness of 1 mm and 16° angle.

The cross-sectional area of the combustor chamber is 340 mm×340 mm, labeled in Fig. 1. Settings of the air inlet length elected with two different values, L_1=245 mm and L_2=445 mm. The length of the combustion chamber and exhaust section is fixed at L=1600 mm; the width of the combustion chamber is D=340 mm, while two loudspeakers (150 W each, HiVi, D10G) locates at the 1400 mm downstream of the combustor and connects to a power amplifier (Yamaha, P5000S) through two audio signal wires (Zoguo, 3.2 mm×6.4 mm). The input voltage level of two loudspeakers keeps the same during excitation experiments. The burner is excited by the signal generator with the frequency of 30–400 Hz. The sweeping frequency gradient is set to 10 Hz during the period of 30–400 Hz, and 1 Hz for the flame-acoustic resonance region.

Sound pressure of the combustion chamber is gathered through a pressure transducer; the pressure transducers are installed at the wall of the chamber [34]. Flame heat release signal is measured with a photomultiplier tube; the sound pressure and CH* chemiluminescence signal is recorded simultaneously using a multi-channel signal recorder with a 20 kHz sampling rate [35]. The experimental data are transmitted and processed by using Labview2012 and Matlab2017a commercial software. In Fig. 1, along with the PMT (photomultiplier tube), the non-premixed flame images during the resonance process are recorded synchronously using a high-speed camera. The sweep frequency of the loudspeaker is controlled by the arbitrary signal generator; the amplitude of the sound excitation is kept constant during the experiments [34, 35].

Air for combustion is conveyed through two radially opposed apertures at the upstream of the burner. In the present work, the combustor is operated at atmospheric pressure and ambient temperature. Propane is delivered through the central stainless fuel pipe to the combustion chamber in this study. The inner diameter of the fuel tube is 13 mm. In Fig. 1, volume flow rate of the fuel is adjusted with two mass flowmeters and the thermal power of the combustor is fixed, with the propane flow rate fixed at 9.6 L/min. The propane is provided with fuel tanks. The air is supplied with an industrial fan and the flow rate is controlled by a volumetric flowmeter. During the combustion, the global equivalence ratio of 0.3 and 0.18 are selected. Combustion conditions of the flame-acoustic resonance process explored are summarized in Table 1.

Commercial FEM software, COMSOL Multiphysics [39], has been taken to investigate the acoustic mode of non-premixed burners with different inlet lengths. COMSOL Multiphysics has been combined with nonlinear polynomial flame response models to study the thermoacoustic oscillations in the simple laboratory-scale longitudinal combustor and an annular combustion chamber [40, 41]. In COMSOL acoustic modules, linearized potential flow equation, linearized-Euler equation or linearized equation can be solved. The Navier-Stokes equations are used to study the aeroacoustics [42]. Before acoustic FEM computation of the burner, working fluid medium of the non-premixed burners was regarded as ideal gases; the speed of sound in air was set as 343 m/s; density of air was set as 1.2 kg/m³; air temperature in the inlet section was defined as 25°C; air temperature in the combustion chamber was defined as 1270°C. Nonlinear time series analysis methods contain phase space reconstruction and poincaré section, optimal time delay τ and embedding dimension d_E, recurrence qualification analysis [29–38].

Table 1 Combustion conditions of acoustically perturbed non-premixed flame studied in the experiment

Air inlet length/ mm	Chamber length/ mm	C_3H_8 flow rate/ L·min⁻¹	Airflow rate/ L·min⁻¹	Global Φ —	Forcing frequencies/ Hz
245	1600	9.6	760/1270	0.30/0.18	30–400
445	1600	9.6	760/1270	0.30/0.18	30–400

3. Results and Discussion

3.1 Sound pressure resonance characteristics analysis

To find the given frequency that flame-acoustic resonance occurs in the burner, the nonlinear dynamic characteristics of non-premixed flames under acoustic excitation were studied in our previous research [35]. The peak pressure response of flame in the combustion chamber is recorded.

To further elucidate the flame-acoustic resonance characteristics, the acoustic resonance of the burner was explored by FEM (Finite Element Method) in COMSOL software [39]. Fig. 3 and Fig. 4 present the results of acoustic simulation of the burner with COMSOL software. For Case L_1=245 mm, air inlet and fuel tube of the non-premixed burner resonate simultaneously at 60 Hz; the maximum vibration emerges in the fuel tube; P is the non-diamensional value of the sound pressure by FEM. At 144 Hz, the combustion chamber and air inlet no longer resonate with the fuel tube, but the maximum sound pressure emerges in the fuel tube. At the response frequency of 190 Hz, the fuel tube of the burner no longer resonates, but the maximum vibration emerges in the air inlet. In Fig. 4, the resonance characteristics of L_2=445 mm length resemble that present in Case L_1=245 mm. The numerical results presented in Fig. 3 and Fig. 4 are conformed to the experimental results presented in Ref. [35]. Under conditions of flame-acoustic resonating (L_1: 144 Hz, L_2: 134 Hz), the vibrating fuel pipe causes the fluctuation of heat release rate.

We could draw from the COMOSOL simulation results in Fig. 3 and Fig. 4 that the resonance modes in both 445 mm and 245 mm cases are mixed resonance mode of the fuel pipe, combustion chamber and air inlet section. The dynamic characteristics of non-premixed flame under acoustic resonance will be studied in the next section.

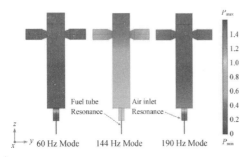

Fig. 3 FEM acoustic simulation of the burner with 245 mm inlet length compared with the results in Ref. [35]

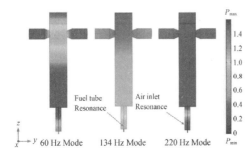

Fig. 4 FEM acoustic simulation of the burner with 445 mm inlet length compared with the results in Ref. [35]

In order to study the dynamic characteristics of sound pressure and flame in the case of resonance, the nonlinear time series analysis method is used [29–38]. Eight different conditions of experiments were designed through calculating the minimum delay time τ and embedding dimension d_E of time series in the combustion chamber and burner inlet parts. The AMI (average mutual information) approaches minimum when time delay τ=0.5. The embedding dimension used for nonlinear dynamics research is d_E=10. With the above optimum delay time τ and embedding dimension d_E, the time series

of pressure in the combustion chamber and burner inlet sections are analyzed sequentially with phase space reconstruction, poincaré sections, and recurrence qualification analysis. The sampling frequency was set as 2000 HZ; the time series of pressure was set as 0.2 s for analyzing. The phase space figure of Fig. 5(a) and Fig. 5(b) denotes that the sound pressure in combustion chamber is in chaotic states at Φ=0.18 and evolves to quasi-periodic states at Φ=0.30. This is compatible with the acoustic simulation results demonstrated in Fig. 4, which suggests that the resonance states in the combustion chamber are complex and in conformity. The poincaré section graph in Fig. 5 and Fig. 6 shows a scattered distribution, which means that the dynamic characteristics of sound pressure in the combustion chamber are in combustion noise states at Φ=0.18 and evolves to periodic limit cycle states at Φ=0.30. It can be seen that in Fig. 5(b), Fig. 6(a) and Fig. 6(b), the recurrence qualification of the periodic oscillating system appears along the main diagonal with equal intervals. This diagram indicates chaotic states of the sound pressure combustion chamber. Fig. 5(a-1) shows the phase space reconstruction of sound pressure. The phase space diagrams were composed of chaotic points. Fig. 5(a-2) shows the poincaré sections of phase space. The points are sparsely distributed near the diagonal line. As seen in Fig. 5(a-3), the recurrence qualification graph is composed of a large number of isolated points and some short diagonals, showing a chaotic structure with inherent certainty. This structure indicates quasi-periodic states of the combustion chamber.

Accordingly, by analyzing the dynamic characteristics of the sound pressure of the burner inlet section presented in Fig. 7 and Fig. 8, it can be seen that the burner inlet section is in a periodic limit cycle oscillation states under acoustic resonance, which is different from results shown in Fig. 5 or Fig. 6. The computational domain in Fig. 7 and Fig. 8 was set as 0.2 s with 400 sampling points. Compared with the acoustic simulation results in Fig. 3 and Fig. 4, under flame-acoustic resonating (144 Hz, 134 Hz), the combustion chamber and air inlet of the burner show different strengths of resonance, which matches the dynamic characteristics of the combustion chamber and air inlet presets in Figs. 5–8.

The phase space figures of Fig. 7 and Fig. 8 denote that the sound pressure in the combustion chamber is in the turbulent limit cycle states at Φ=0.18 and Φ=0.30. This is consistent with the acoustic simulation results demonstrated in Fig. 4, which indicates intense periodic resonance states in the air inlet section. The poincaré section graphs in Fig. 7 and Fig. 8 are in compacted distribution, which means that the dynamic characteristics of sound pressure in air inlet are in a stable limit-cycle states at Φ=0.18 and Φ=0.30. It can be seen that in Fig. 7 and Fig. 8, the recurrence qualification of the sound pressure in air inlet appears along the main diagonal with equal intervals. This structure indicates that under flame-acoustic resonance, the sound pressure in the air inlet shows periodic states.

3.2 Flame resonance characteristics analysis

Fig. 9 and Fig. 10 present the recurrence qualification analysis of PMT (photomultiplier tube) signal under flame-acoustic resonance, with different equivalence ratios and inlet lengths. There are sudden fluctuations of flame heat release under acoustic-flame resonance. Dynamic characteristics of higher equivalence ratio in both cases lead to higher flame heat release fluctuation. The fluctuation phenomena in Fig. 9 and Fig. 10 conforms to the COMSOL acoustic simulation results of the fuel inlet pipes oscillation modes presented in Fig. 3 and Fig. 4. The sound pressure fluctuation in the 245 mm case is more intense than the 445 mm case. However, this is poles apart when referring to the heat release

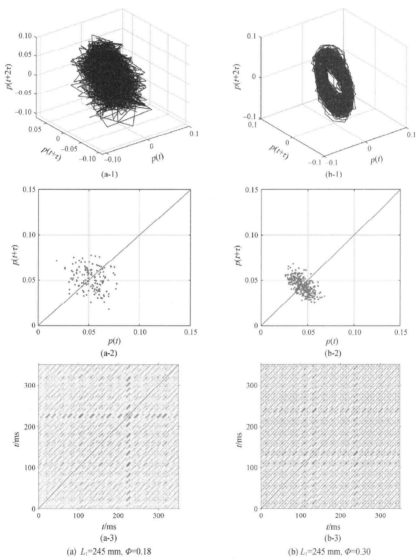

Fig. 5 Nonlinear time-series analysis for combustion section with inlet length of 245 mm and different equivalence ratio (Φ =0.18 or 0.30) ($p(t+2\tau)$, $p(t+\tau)$ and $p(t)$ denote the time-series of sound pressure.)

fluctuation in the 245 mm case compared with the 445 mm case.

Nonlinear dynamic characteristics of flame heat release rate are different from sound pressure in the combustion chamber. In Fig. 11, FFT analysis shows the 445 mm case exhibits higher heat release intensity than the 245 mm case. Furthermore, the heat release intensity gradually declines as the oscillation frequency increases. A relatively stable level with some sudden oscillations was maintained in the end. The sampling frequency was 2000 Hz; the time series of heat release rate was set as 2 s (with 4000 sampling points) for analyzing. Fig. 12 shows the time-series signal of flame heat release rates. More fluctuations emerge under the condition of 445 mm inlet lengths. These results conform to the experimental results presented in Fig. 9 and Fig. 10.

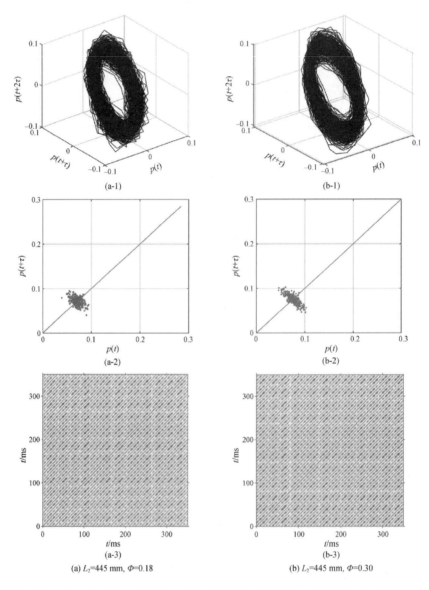

(a) L_2=445 mm, Φ=0.18   (b) L_2=445 mm, Φ=0.30

Fig. 6 Nonlinear time-series analysis for combustion section with inlet length of 445 mm and different equivalence ratio (Φ =0.18 or 0.30)

Flame intermittency was observed and accompanied by four kinds of forms during the mode transition. Images of flame were captured with the high-speed CCD camera under conditions of flame-acoustic resonance. The flame images under flame-acoustic resonance are captured using a digital high-speed camera (AOS S-PRI plus, 800×600 pixels image resolution, 14 μm pixel size) with a sampling rate of 1250 fps. For resonance frequency between 130 Hz (7.7 ms) and 150 Hz (6.7 ms), 10 pictures are used to get the average flame height in the article. As shown in Fig. 13, there are mainly four forms of unstable non-premixed flame, which are the results of the flame-acoustic resonance. The unexcited stable flame is set as Mode-A flame; the original height of the flame is near 300 mm. In the initial resonating stage, the sound field wrinkles the flame surface and fines the flame root

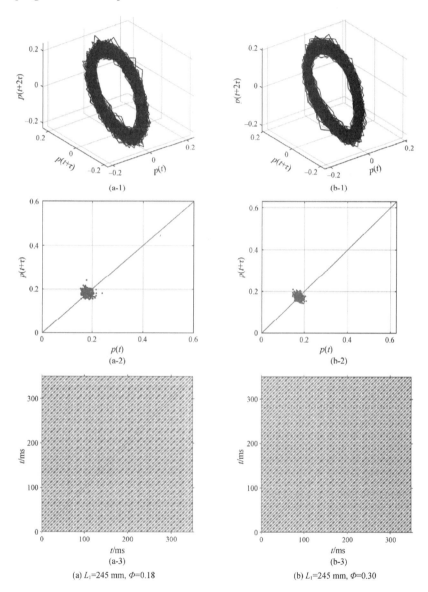

(a) L_1=245 mm, Φ=0.18   (b) L_1=245 mm, Φ=0.30

Fig. 7 Nonlinear time-series analysis for inlet air section with inlet length of 245 mm and different equivalence ratio (Φ=0.18 or 0.30)

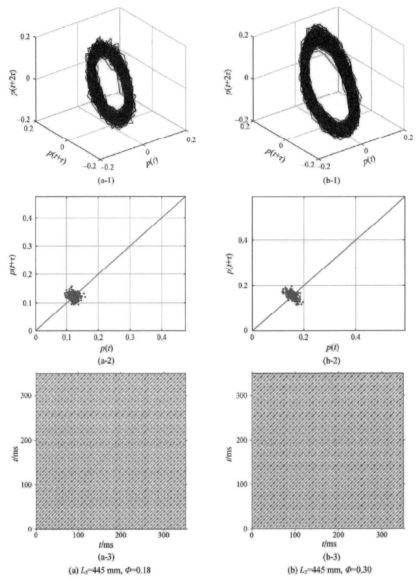

Fig. 8 Nonlinear time-series analysis for inlet air section with inlet length of 445 mm and different equivalence ratio (Φ =0.18 or 0.30)

(Mode B-Wrinkled flame). Then the flame is strongly coupled with the sound field, resulting in obvious faults and upward movement of the flame (Mode C-Diverged flame); the flame is stretched and becomes slimmer; the average height of the diverged based flame is close to 350 mm. Then the flame stretches severely, rises upward as a whole; the root of the flame disappears, and the flame breaks into several segments (Mode D-Transitional flame). The average height of the transitional reattached flame is close to 200 mm, and then flames become dimmer. Finally, the flame enters a strong turbulent regime and is on the verge of extinction; the flame becomes wider and shorter (Mode E-Lifted-off flame), and the average height of the lifted-off flame is near 100 mm. Rong Fung Huang [43-45] researched non-premixed flames subject to acoustic excitation at resonance conditions, and three different flame modes were observed.

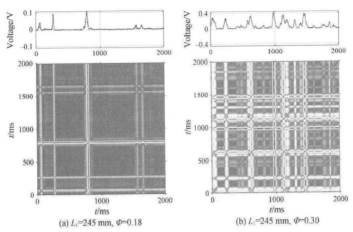

Fig. 9 Recurrence plots of flame heat release fluctuation with inlet length of 245 mm with an equivalence ratio of 0.18 and 0.30 respectively

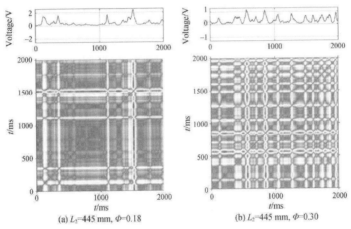

Fig. 10 Recurrence plots of flame heat release fluctuation with inlet length of 445 mm with an equivalence ratio of 0.18 and 0.30 respectively

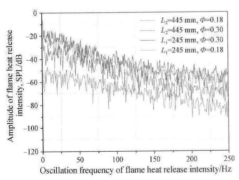

Fig. 11 FFT analysis for flame heat release intensity with inlet length of 245 mm and 445 mm and different equivalence ratio ($\Phi=0.18$ or 0.30)

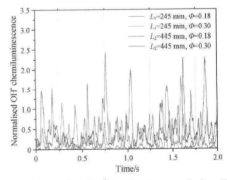

Fig. 12 Time-series of OH^* chemiluminescence for flame heat release rate with inlet length of 245 mm and 445 mm of different equivalence ratio ($\Phi=0.18$ or 0.30)

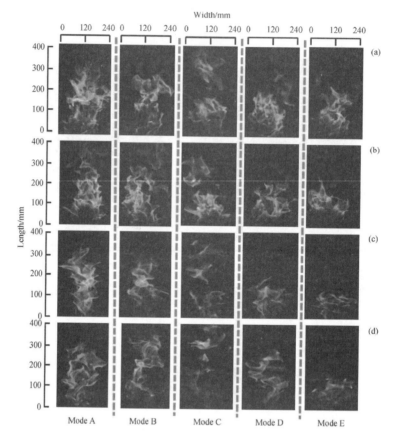

Fig. 13 Far field flame behaviors during resonance: with inlet length of 245 mm and 445 mm at different equivalence ratio. (a) L_1=225 mm, Φ=0.18, 144 Hz. (b) L_1=225 mm, Φ=0.30, 144 Hz. (c) L_1=445 mm, Φ=0.18, 134 Hz. (d) L_1=445 mm, Φ=0.30, 134 Hz.

4. Conclusions

The present work applies time series analysis to explore the nonlinear dynamics of non-premixed flame under flame-acoustic resonance. Two parametric variations of the burner, variation of the burner inlet length and combustion equivalence ratio, were performed. Several dynamic characteristics were revealed. This research provides a tool to study dynamic characteristics of non-premixed flame under resonance and may be of functional significance to the prevention of thermoacoustic instability and resonance structural failure of industrial boilers. Under the flame-acoustic resonance mode, the pressure fluctuation is in chaotic or quasi-periodic states of the combustor chamber, while the pressure fluctuation in the burner air inlet section is in turbulent limit cycle states. Sound pressure in the combustion chamber and air inlet shows frequency-mode-transition phenomena. The frequency and amplitude of the flame-acoustic vary when the inlet length changes. From time-series of the flame heat release rate recorded by PMT, non-premixed acoustically perturbed swirling flame exhibits intense pulsations under flame-acoustic resonance mode. Fuel equivalence ratio and burner inlet lengths have a great influence on flame nonlinear dynamics. During the transition stage, the non-premixed flame has five main forms under the resonance process. The proposed nonlinear time series analysis methods promote a better understanding of resonance dynamics observed in non-premixed combustion burners.

Acknowledgment

This work was supported by National Science Fund for Distinguished Young Scholars (51825605).

References

[1] Lieuwen T., Yang V., Combustion instabilities in gas turbine engines: Operational experience, fundamental mechanisms, and modeling. Progress in Astronautics and Aeronautics, 2005, AIAA, Inc. USA.

[2] Lieuwen T., Unsteady combustor physics. 2012, UK, Cambridge University Press, Cambridge.

[3] Huang Y., Yang V., Dynamics and stability of lean-premixed swirl-stabilized combustion. Progress in Energy and Combustion Science, 2009, 35: 293–364.

[4] Wang X., Lin Y., Zhang C., et al., Effect of swirl cup's secondary swirler on flow field and ignition performance. Journal of Thermal Science, 2015, 24: 488–495.

[5] Xing S., Fang A., Song Q., et al., Experimental investigation of dynamic and emission characteristics of a DLE gas turbine combustor. Journal of Thermal Science, 2013, 22: 180–185.

[6] Jacqueline O., Vishal A., Timothy L., Transverse combustion instabilities: Acoustic, fluid mechanic, and flame processes. Progress in Energy and Combustion Science, 2015, 49: 1–39.

[7] Lin J., Shi Z., Lai H., Numerical study of controlling jet flow and noise using pores on nozzle inner wall. Journal of Thermal Science, 2018, 27: 146–156.

[8] Zhao D., Ephraim G., Philip D., A review of cavity-based trapped vortex, ultra-compact, high-g, inter-turbine combustors. Progress in Energy and Combustion Science, 2018, 66: 42–82.

[9] Juniper M., Sujith R., Sensitivity and nonlinearity of thermoacoustic oscillations. Annual Review of Fluid Mechanics, 2018, 50: 661–689.

[10] Saravanan B., Larry K., Han. Z., et al., Nonlinear dynamics of a self-excited thermoacoustic system subjected to acoustic forcing. Proceedings of the Combustion Institute, 2015, 35: 3229–3236.

[11] Kashinath K., Larry K., Juniper M., Forced synchronization of periodic and aperiodic thermoacoustic oscillations: lock-in, bifurcations and open-loop control. Journal of Fluid Mechanics, 2018, 838: 690–714.

[12] Guan Y., He W., Murugesan M., et al., Control of self-excited thermoacoustic oscillations using transient forcing, hysteresis and mode switching. Combustion and Flame, 2019, 202: 262–275.

[13] Jegadeesan V., Sujith R., Experimental investigation of noise induced triggering in thermoacoustic systems. Proceedings of the Combustion Institute, 2013, 34: 3175–3183.

[14] Vishnu R., Sujith R., Aghalayam P., Role of flame dynamics on the bifurcation characteristics of a ducted V-flame. Combustion Science and Technology, 2015, 187: 894–905.

[15] Nair V., Sujith R., Intermittency as a transition state in combustor dynamics: An explanation for flame dynamics nearing lean blowout. Combustion Science and Technology, 2015, 187: 1821–1835.

[16] Lipika K., Sujith R., Nonlinear self-excited thermoacoustic oscillations: intermittency and flame blowout. Journal of Fluid Mechanics, 2012, 713: 376–397.

[17] Kabiraj L., Saurabh A., Wahi P., et al., Route to chaos for combustion instability in ducted laminar premixed flames. Chaos: An Interdisciplinary Journal of Nonlinear Science, 2012, 22: 023129.

[18] Nair V., Sujith R., A reduced-order model for the onset of combustion instability: physical mechanisms for intermittency and precursors. Proceedings of the Combustion Institute, 2015, 35: 3193–3200.

[19] Unni V., Sujith R., Flame dynamics during intermittency in a turbulent combustor. Proceedings of the Combustion Institute, 2017, 36: 3791–3798.

[20] Nair V., Sujith R., Identifying homoclinic orbits in the dynamics of intermittent signals through recurrence quantification. Chaos: An Interdisciplinary Journal of Nonlinear Science, 2013, 23(3): 033136.

[21] Kabiraj L., Saurabh A., Nawroth H., et al., Recurrence analysis of combustion noise. AIAA Journal, 2015, 53: 1199–1210.

[22] Karthik K., Waugh I., Juniper M., Nonlinear self-excited thermoacoustic oscillations of a ducted premixed flame: bifurcations and routes to chaos. Journal of Fluid Mechanics, 2014, 761: 399–430.

[23] Gotoda H., Nikimoto H., Miyano T., et al., Dynamic properties of combustion instability in a lean premixed gas-turbine combustor. Chaos, 2011, 21: 013124.

[24] Domen S., Gotoda H., Kuriyama T., et al., Detection and prevention of blowout in a lean premixed gas-turbine model combustor using the concept of dynamical system theory. Proceedings of the Combustion Institute, 2015, 35: 3245–3253.

[25] Kobayashi H., Gotoda H., Tachibana S., et al., Detection of frequency-mode-shift during thermoacoustic combustion oscillations in a staged aircraft engine model combustor. Journal of Applied Physics, 2017, 122: 224904.

[26] Li L., Juniper M., Lock-in and quasiperiodicity in hydrodynamically self-excited flames: Experiments and modelling. Proceedings of the Combustion Institute, 2013, 34: 947–954.

[27] Li Ping., Yang E., Song S., et al., Analysis of the dynamic characteristics of combustion instabilities in a pre-mixed lean-burn natural gas engine. Applied Energy, 2016, 183: 746–759.

[28] Taghizadeh A., Mahdavian A., Fault detection of injectors in diesel engines using vibration time-frequency analysis.

[29] Wu G., Lu Z., Guan Y., et al., Characterizing nonlinear interaction between a premixed swirling flame and acoustics: heat-driven acoustic mode switching and triggering. Energy, 2018, 158: 546–554.
[30] Mondal S., Mukhopadhyay A., Sen S., Dynamic characterization of a laboratory-scale pulse combustor. Combustion Science and Technology, 2014, 186: 139–152.
[31] Sen U., Gangopadhyay T., Bhattacharya C., et al., Dynamic characterization of a ducted inverse diffusion flame using recurrence analysis. Combustion Science and Technology, 2018, 190: 32–56.
[32] Guan Y., Liu P., Jin B., et al., Nonlinear time-series analysis of thermoacoustic oscillations in a solid rocket motor. Experimental Thermal and Fluid Science, 2018, 98: 217–226.
[33] Ahn B., Lee J., Jung S., et al., Nonlinear mode transition mechanisms of a self-excited Jet A-1 spray flame. Combustion and Flame, 2019, 203: 170–179.
[34] Zhou H., Huang Y., Meng S., Response of non-premixed swirl-stabilized flames to acoustic excitation and jet in cross-flow perturbations. Experimental Thermal and Fluid Science, 2017, 82: 124–135.
[35] Zhou H., Meng S., Tao C., et al., Study of burner geometry effects on non-premixed flame response under acoustic excitation. Journal of Low Frequency Noise, Vibration and Active Control, 2019, 38: 3–17.
[36] Kantz H., Schreiber T., Nonlinear time series analysis. Cambridge University Press, 2003, Cambridge, UK.
[37] Marwan N., Romano M., Thiel M., et al., Recurrence plots for the analysis of complex systems. Physics Reports, 2007, 438: 237–329.
[38] Zou Y., Donner R., Norbert M., et al., Complex network approaches to nonlinear time series analysis. Physics Reports, 2019, 787: 1–97.
[39] COMSOL Multiphysics. Acoustic Module Minicourse, 2020.
[40] Oh S., Shin Y., Kim Y., Stabilization effects of perforated plates on the combustion instability in a lean premixed combustor. Applied Thermal Engineering, 2016, 107: 508–515.
[41] Laera D., Campa G., Camporeale S., A finite element method for a weakly nonlinear dynamic analysis and bifurcation tracking of thermo-acoustic instability in longitudinal and annular combustors. Applied Energy, 2017, 187: 216–227.
[42] Kim S., Kim D., Cha D., Finite element analysis of self-excited instabilities in a lean premixed gas turbine combustor. International Journal of Heat and Mass Transfer, 2018, 120: 350–360.
[43] Zargar O., Huang R., Hsu C., Flames of swirling double-concentric jets subject to acoustic excitation at resonance. Journal of Thermal Science and Engineering Applications, 2019, 11(3): 031004.
[44] Zargar O., Huang R., Hsu C., Effect of acoustic excitation on flames of swirling dual-disk double-concentric jets. Experimental Thermal and Fluid Science, 2019, 100: 337–348.
[45] Loretero M., Huang R., Effects of acoustic excitation on a swirling diffusion flame. Journal of Engineering for Gas Turbines and Power, 2010, 132: 1113–1122.

湍流非预混声扰动漩涡火焰的非线性动态特性

摘要：本文的主要目的是在实验室规模的燃烧器中实验研究声激励下的动态响应扩散火焰。研究了燃烧器的两个参数变化，即燃烧器入口长度和风量的变化。通过非线性时间序列分析，对实验结果进行了分析，获得了若干共振特性结果表明，火焰声共振只在一定频率下与燃油管振动一起出现。非预混燃烧器的燃烧室和进气口的共振特性分别表示准周期或极限周期振荡。燃烧器中的声-火共振会引起频率和幅值模式的转变。此外，在燃烧器入口长度和气流速率变化下观察到火焰热释放的间歇性；445mm 的壳体显示出比 245mm 的更频繁的峰值和波动。在火焰声共振条件下，研究了四种典型的火焰形态，从褶皱火焰演变为分裂火焰，然后演变为重新附着火焰，最后演变为喷射火焰。本文提出了非线性时间序列分析方法作为实验室非预混燃烧器火焰声共振检测工具的实际应用，有助于工业锅炉潜在热声不稳定或共振结构失效的检测和预防。最后，本研究证明了一种替代传统线性工具的工业锅炉非线性声共振表征方法。

原文刊于 Journal of Thermal Science, 2022, 31(3): 882-894

第二篇　著　作

锅炉燃烧试验研究方法及测量技术

我国有上百万台电站锅炉和工业锅炉。为了节约燃料,适应煤种、负荷的变化,以及提高燃烧过程的经济性和稳定性,都需要对锅炉燃烧过程进行专门测量和试验研究,以便发现问题,寻求解决措施,提出合理的设计方案和较佳的运行方式。

全书共有十二章,系统地阐述了锅炉燃烧试验研究的各种方法、有关锅炉燃烧的测量技术以及测量数据的处理方法。

第一章探讨了锅炉热平衡试验的原理及测试方法。首先,简要概述了热平衡试验的目的与任务。锅炉设备的热平衡是指输入锅炉设备的燃料拥有热量与锅炉设备输出热量之间的平衡。热平衡试验是锅炉设备热工试验中最基本的一项实验,为锅炉设备运行调整提供了重要参考。通过热平衡试验可以确定锅炉的热效率、锅炉的各项热损失以及不同运行工况下的各项经济指标。热平衡试验的准备工作首先需要确定试验大纲、测量项目、测量方法及测定位置,其次针对不同的实验任务目标和要求,明确试验负荷、煤质以及试验前锅炉的运行状态,开展影响锅炉设备主要运行技术经济特性指标重要参数的测量,然后进行试验数据的整理、计算并完成试验报告。此外,分析和对比了反平衡法和正平衡法两种锅炉热平衡的定义、测量项目、适用范围及误差分析。最后,借鉴了国外锅炉技术的某些经验和特点,简要介绍了英国、美国等国的锅炉热效率的计算方法,以供参考。

第二章介绍了炉内模化技术及冷炉试验方法,包括炉内模化的目的及自模化区的确定方法、炉内冷态模化和热态模化原理以及气固两相模化理论基础。首先,在设计新型锅炉或新炉投入运行时,可通过模化试验或冷炉试验来了解掌握炉内的流动规律,从而验证和修改设计及运行方案。考虑到冷态等温模化技术是省时、省力、效率高、灵活性强的一种试验方法,通常用来确定锅炉燃烧系统的配风均匀程度,通过三次风布置位置、角度和所需风速等调节,探究燃烧系统及燃烧器的阻力特性,从而揭示燃烧器的流体动力运动规律。通过不同炉膛充满度的各种影响要素的研究,揭示炉内结渣的空气动力原因,从而寻求合理的运行方式,探索新的燃烧方式,优化炉膛结构。并且指出,冷态等温模化虽然不可能完全准确地描绘燃料在炉内燃烧的复杂物理化学过程,但能对炉内流动过程提供定性结果,具有一定的参考价值。本章的主要内容着重阐述如何使冷态等温模化试验尽可能接近真实情况。根据相似原理,进行炉内冷态等温模化试验时必须遵守以下三个原则:①模型与实物需几何相似;②保持气流运动状态进入自模化区;③边界条件相似。然后,针对炉内冷态、热态模化的工作原理以及相应的热态模化法、冷炉试验时近似热态模化法、气固两相模化等相关理论知识点进行了概述。重点分析了引起燃料颗粒在炉内气流运动的八种作用力,将其定义及物理概念进行了认真解读。在实际应用中需根据具体情况对两相模型设计予以简化,从而保证试验的顺利进行。当开展冷模及冷炉试验时,一般采用飘带法、示踪法、仪器测量法等方法对实验过程中的气流方向、速度等参量进行观测，从而全面掌握气流运动轨迹及其气流特性。

第三章介绍了旋流式燃烧器的试验研究方法，包括旋流式燃烧器试验特性参数及旋流强度确定、旋流强度对燃料着火过程影响试验方法、燃烧器空气动力特性对燃料燃烧影响试验方法以及旋流式燃烧器热态试验的总结方法。旋流式燃烧器是一种普遍应用的锅炉用燃烧器，燃烧器的二次风通常是利用蜗壳、轴向或切向叶片等部件产生旋转运动或部分旋转运动，一次风则可采用旋流或直流扩散锥式，以达到着火稳定、燃烧中心适中、火焰分布均匀、燃料燃烧干净、不出现炉内的积灰结渣等目的。为此旋流燃烧器的试验调试及参数优化稳定运行显得格外重要，以便获得良好的空气动力场。本章列举了燃烧器的重要特性参数，如旋流强度、实际旋流强度、综合旋转强度、燃烧器阻力系数、射流边界、射流扩展角、煤粉分配不均匀系数、烟气回流量、气流紊流强度、炉膛充满系数等计算公式及其定义，以及不同燃烧器型式的简图、旋流强度计算公式、特性曲线及主要结构数据，为燃烧器型式的选择提供了参考依据。由于旋流燃烧器的着火调试及稳定燃烧试验是最关键的步骤之一，首先根据燃料的特性及运行条件来确定其着火的温度，并根据煤中挥发物的含量、着火温度、一次风量及风温等参数来估算使其稳定着火所必需的最小烟气回流量。接着，通过调整旋流燃烧器的合适旋流强度以达到满足稳定着火所要求的烟气回流量。当给定最佳的旋流强度后，还必须对燃烧器的空气动力场进行完整调试。此外，书中充分详尽地介绍了如何开展各种参数对空气动力特性的影响试验，如燃烧器出口布风均匀试验、煤粉分布均匀试验、旋转射流火炬的扩展角、一二次风的混合点对着火的影响试验、火炬在炉膛内的充满程度、旋流燃烧器的旋转方向等。最后，介绍了应用物质平衡法开展旋流燃烧器热态试验的各种数据测量及数据的整理，计算综合分析能力利用效率及各项热损失。

第四章介绍了四角布置直流燃烧器的调试方法，包括四角布置直流燃烧器调试的一般原理、方法及燃用劣质煤的燃烧调试问题。从空气动力角度来讲，四角布置直流燃烧器实际上也是一个大型旋流式燃烧器，由于它对燃料适应性较广，在大容量炉子中较易布置，在国内外应用比较普遍。首先介绍了四角布置直流燃烧器炉膛的常用形式，四角布置直流燃烧器的切圆直径、炉内气流的速度分布等空气动力结构的设计步骤。本章引入了直流燃烧器的射流刚性的重要定义，是指射流喷入炉内后射流的轨迹偏离假想射流轴心线的程度，射流刚性越小，越容易偏离，并强调了产生偏离现象的三个具体原因以及解决方案。由于大型机组炉内热态试验具有较复杂、难度大的特点，有必要深入了解四角布置直流燃烧器炉内空气动力特性冷热态的差异性，才能恰如其分通过冷态试验来判断热态过程的规律，从而寻求解决问题的途径。接着，本章重点介绍了四角布置直流燃烧器调试过程中所需开展的各种试验，针对投产前或大修后开展一、二、三次风管道流量标定、四角直流燃烧器的安装角度符合设计要求、炉内冷空气动力场试验等。除此之外，包括四角配风均匀性试验、煤粉分布均匀性试验、煤粉细度对燃烧影响试验、炉内过量空气系数调整、缺角停投方式试验、热态试验调整等相关调试试验也进行了一一赘述。同时，针对燃烧劣质煤工况也给出了相对应的解决方案，如通过一、二次风的布置与风量分配、周界风和夹心风的调整、卫燃带的使用，来减缓结焦等现象的发生。

第五章介绍了炉内沾污、结渣、积灰的试验研究方法。首先简要阐述了炉内辐射受热面的沾污、结渣、腐蚀的定义、积灰结渣的类型以及研究炉内辐射受热面结渣、沾污的重要意义，揭示了炉内水冷壁沾污、结渣及腐蚀过程的详细机理，提出了结渣、沾污对炉内传热和运行影响的各种关键参数。并运用测量炉膛出口烟温的近似总结法、综合系数总结法、测量炉内温度场及出口烟温的方法、测量炉膛出口烟温及水冷壁热有效系数等多种工业试验方法来评判结渣、沾污对炉内传热和运行的影响规律。进一步通过专门性试验方法来研判沾污和结渣特性，如灰渣熔点及粘度试验、煤灰烧结强度来衡量煤灰沾污特性，根据煤灰成分来判断其结渣及沾污特性等进行了一一列举。最后，重点强调了炉内受热面经沾污、积灰腐蚀后，可采用吹灰、烟气再循环等一些有效的应对措施。

第六章介绍了气流速度测量技术，包括气力式和热电式速度测量方法、气流脉动速度测量原理及激光测速原理。气流速度是锅炉燃烧试验中最常遇到的测量参数之一。试验中需要测量气流速度的介质有冷空气、热空气、热烟气和高温火焰等。不同气体介质对速度测量装置有不同的要求，炉体内或旋流燃烧器出口则是三维旋转气流，不但需要测量气流速度的大小，还要测量气流的方向。因此，本章着重介绍在锅炉燃烧试验中经常用于速度大小及方向的各种气力式和热电式气流速度测量装置及其测量技术，并简单介绍了激光测速原理。首先，针对常用的几种气力式速度测量方法的原理及装置进行阐述，包括最常见的毕托管测速、放大测速装置、均速管、翼形测速装置、含尘气流、平面气流、空间气流、炉内高温气流等。然后，简要介绍了热电式速度测量的原理及方法，以及测量过程中气流温度、压力变化对测量精度的影响机制。最后，列举了一些较常用的激光多普勒等测速方法。

第七章介绍了气固两相浓度测量方法，包括等速取样原理、两相浓度表示方法及适用范围、煤粉及飞灰浓度测定方法、基于燃烧法的碳量快速测定法。在锅炉的调试和运行过程中，开展气固两相流的取样及测量尤为重要。相比于比色法、光电法、激光法等方法，虽然它们能快速测量两相流的浓度，但不够准确，而用通过炉内两相流动中对固体颗粒进行直接取样测定浓度的方法显得更加通用和直观。首先详细介绍了不等速取样、简单等速取样、静压零位等速取样的原理、测量方法、适用范围及其误差分析。考虑到实际气固两相流动工况常常是不稳定的，导致实际取样过程中，往往很难达到正确的等速取样的预期。当实际取样过程不等速情况下，选择合理的总结方法，可以使试验的误差减到最小的程度。为此，文中归纳了表示气固两相流固体颗粒浓度的三种方法。其次，系统阐述了如何利用等速取样的方法来达到煤粉及飞灰浓度测定的目的，及其整个取样工艺设计及流程。接着，对比了灰化法及燃烧法的优缺点，证明了燃烧法可在现场快速测定碳量，从而较快地获取锅炉的飞灰机械未完全燃烧热损失。

第八章介绍了高温测量技术，包括热电偶测量高温烟气的误差分析、抽气热电偶及效率、用于高温测量的组合热电偶。测量炉膛内的火焰温度或炉膛出口温度是锅炉燃烧调整测试试验，研究炉内燃烧过程的专门性试验时非常重要的关键测量参数。首先系统性地阐述了热电偶测量高温烟气时所存在的各种误差的定义及其计算方法，如导热误差、辐射误差、速度误差等，以及减少各类误差所采用的补偿措施。接着，针对近年来较常用的多种接触式高温测量仪器，如抽气热电偶、组合热电偶和气力高温计等的结构、原理、抽气装置、误差分析做了较为详细的介绍。为了能够准确地计算烟气的真实温度，作者引入了抽气热电偶的效率概念，即为抽气热电偶测定的温度与烟气真实温度的接近程度。文中指出通过简化物理模型、定时法、温度-速度曲线法、真实温度法或标定法等方法可求得抽气热电偶的效率，并计算出烟气的真实温度。针对高温测量用的组合热电偶，则是通过零直径外推法、双热电偶法、三热电偶法等方法来求得烟气的真实温度。

第九章介绍了烟气成分的分析方法，包括化学分析法、色谱分析法及烟气中有害气体成本的测量方法。在锅炉试验中，通过烟气成分的测定可用来评判燃料燃烧过程的完全程度、燃料燃烧的工况特性、炉内各区段燃料燃烧的特点、烟道漏风的大小及燃烧与污染物排放的关系。本章主要介绍锅炉燃烧试验中常用的烟气分析仪器来实现烟气成分的快速测量。首先介绍了如奥氏烟气分析仪的测试原理、构造和分析取样过程；接着详细阐述了气相色谱仪分析的原理、测色谱仪的使用和测量过程；概述了烟气中有害气体的组分，运用冷凝法可测量烟气中二氧化硫的浓度及其工作原理，用比色法可测定氧化氮浓度及其原理和化学分析方法。

第十章介绍了炉内热流的测量方法，包括导热式热流计、热流计热平衡及返现辐射测量方法、其他型式热流计及热流计的标定方法。炉内的高温烟气与受热面的热交换基本上以辐射和对流两种形式进行，由于炉内烟气流速很低，被污染的受热面温度又较高，对流传热所占的比例甚小，炉内主要以辐射换热为主，所以开展炉内辐射热流的测量对炉内换热的研究尤为重要。对炉内热流进行测量可以掌握锅炉炉膛内的热流分布情况，对锅炉运行特性做出评价，并可及时对局部过热区进行优化控制调整，保障锅炉的安全稳定运行。由于当前尚缺乏较公认的通用而完善的炉内热流测量装置，研究团队研制了各种简单实用、适用于特定条件的辐射热流计。首先简明扼要地介绍了辐射热流计的工作原理，并按其工作原理、结构和使用方法进行了分类归纳。根据导热元件的结构的不同，分成 5 种导热式热流计，原理性地介绍了相关构造的差异性及其应用范围。接着，基于导热式热流计的热量平衡原理，得到了导热系数、辐射热流量、散热损失、对流加热热流等关键参量。其次，对热量计式热流计、热容式热流计、测温式热流计、辐射式热流计的原理、特点及其标定的方法也进行了逐一阐述。

第十一章介绍了低温受热面腐蚀磨损及烟气露点的测量，包括影响低温受热面腐蚀磨损的因素、短期和长期腐蚀磨损速度的测量方法、酸沉积和露点测量的热质交换原理及炉烟露点的测量方法。首先阐述了造成低温受热面腐蚀磨损的主要原因。针对腐蚀磨损的测定分为长期和短期两种方法，当环境工况变动及寻求低温受热面的最佳运行方式时，则采用短期腐蚀磨损测量方法。本章也分析了多种典型测定管的原理、装置及其适用范围。为获取长期自动保持试片温度为定值时，则可采用管型长期腐蚀速度测量管、半工业性水冷多管式测管等方法。接着，系统性探究了酸沉积两相图，露点探头的传热传质原理，突出了探头表面酸沉积量的主要因素及其影响规律。为了使所测得的露点数值更能高效地衡量和解释受热面的腐蚀过程及规律，书中也提到了可以运用电阻法、电流法、冷却加热平均法、脉冲电流法、电力速度增长率为零的总结方法等有效的方法来实现这一目的。在此基础上，YL 型烟气露点测量仪也是试验过程中经常使用的测量手段之一。

第十二章介绍了燃烧试验数据微机处理，包括常用燃烧参数插值处理方法、试验数据准则方程多元回归分析方法、常用物性参数多项式求值方法、炉内燃烧气流速度场测量数据处理方法及锅炉燃烧过程中随机信号的数据处理方法。在燃烧试验中，许多测试仪器都是利用一次感受件将所需测量的参数转化成其他物理量，然后根据指示的物理量，借助实现标定的曲线或表格进行查取，得到所需的燃烧参数。本章首先概括了燃烧试验数据与电子计算机结合的重要性。其次介绍了一元三点插值、三次自然样条函数插值、二元三点等插值方法及计算机程序代码示例，应用计算机进行插值，不仅加快数据处理速度，而且也可以提高精确度。在此基础上，进一步介绍了试验数据准则方程多元回归分析、常用物性参数多项式求值、炉内燃烧气流速度场测量数据处理、锅炉燃烧过程中随机信号的数据处理等方法以及相应的计算机程序代码。借助计算机的数据计算优势，可有效减少计算工作量，减小计算误差，提高工作效率。

本文摘自《锅炉燃烧试验研究方法及测量技术》，水利电力出版社，ISBN: 7-120-00007-1，1987（1992年获全国优秀教材奖）。

工程气固多相流动的理论及计算

在工业生产过程中，经常碰到固体颗粒或液滴在气流中进行输送、反应或燃烧，以完成整个产品的制造或合成过程的问题，这种颗粒流动通常称为多相流动。气固多相流动的相关技术广泛应用于能源、化工、冶金、环保、建材、轻工等部门，并且往往成为设备的经济安全稳定运行，开发新的工艺流程，发展新型气固输送、反应、燃烧、分离等新装置的关键技术。

诸如在能源工程中固体和液体燃料在锅炉内的流动和燃烧过程；煤粉的制备、分离与管道输送；锅炉管内及核电站蒸汽发生器内的汽液两相流动及传热；燃气轮机燃烧室内的流动及燃烧；蒸汽轮机内湿蒸汽的流动及冷凝器的两相传热等等。在化学工程中，各种反应、萃取、分离装置也要牵涉到多相流动过程，如板内式塔、填料塔、固定床、气流床及流化床反应器等等；在冶金工业中诸如各种冶金炉，高炉煤粉喷吹等等；在采矿工业中煤炭的洗选、矿石的预加工、预处理、煤炭及矿粉的管道输送等等；在机械制造工业中的各种加热炉、热处理炉，也大多涉及到多相流动过程。在建材工业中水泥窑、玻璃窑、陶瓷隧道窑也离不开多相流动的处理工艺。在环保工程中各种除尘设备内的多相流动更是复杂，而且直接影响到除尘效率；烟气脱硫和脱硝装置内的多相反应；烟囱污染物排放与扩散的规律等等。在食品和运输工业中则把多相流动作为输送物料或制成产品的一种重要手段。由此可见，多相流动这种技术已经应用到国民经济的各个部门中。遗憾的是，过去与气固流动有关的计算、设计大多停留在经验分析为主的水平上。随着近代气固测试技术和数值计算方法的蓬勃发展，气固多相流动的研究达到了一个新的阶段，并成为当今国际上研究的前沿领域，也是目前国民经济所亟待解决的课题。可是有关气固多相流动的理论和计算方面的专著甚少，为促进我国气固多相流动的理论和技术的发展特写成此书。

多相流动中的"相"有两种含义，一种是泛指气、固、液相，另一种是指气-固或液-固流动过程中，颗粒相是由不同直径的固粒或液滴所组成，可以把每一组直径相同的颗粒称为一相，因而由不同直径组成的气-固或气-液流动也称为多相流动。本书主要讨论后一种情况的多相流，当然所涉及的原理和原则也可适用于由不同直径组成的气-液-固颗粒的多相流动，只不过此时各相间的物性参数、颗粒形状、受力情况和流动阻力均要用不同的系数或公式分别代入求解。

虽然多相流的应用领域很广，但是它的理论基础和物理模型都是相同的。为了得到一个比较完整的概念和计算方法，本书试图把各行业所出现的多相流动工艺加以归纳整理，把它们分解成典型的各种多相流动过程，而实际的多相流动工艺则是这些典型过程的相互组合。当然一本书不可能包罗万象，本书的重点是讨论各种典型的气-固多相流动。为了得出定量的计算结果，先从研究各种典型的气相流动入手，然后再考虑加入颗粒后的多相流动计算方法，再应用计算机辅助试验(CAT)的方法得到工程实际应用的回归方程计算公式。因此本书实际上是探讨了气相和多相流动的理论及其计算方法。

本书共十四章，前七章主要讨论各种单相射流的基本原理及其基本计算方法；后面七章主要讨论各种多相射流的理论及其计算方法。

第一章为引论，主要介绍了多相流动的工程应用范围和工程多相流动的空气动力学分类。

第二章概括性地介绍了气流的湍流模型及微分方程组。燃烧过程除了要进行一系列化学反应外，还与一些物理过程有密切关系，包括反应物和燃烧产物的扩散过程、高温的燃烧产物和温度较低的反应物之间的热交换过程、流体的流动过程、两相流动对燃烧过程的影响等物理过程。本章着重介绍与燃烧过程有关的单相流体流动的微分方程组，并讨论这些方程组的封闭性，提出了描述湍流运动的各种参数及目前常用的各种湍流模型，然后归纳成通用微分方程，提出了确定初始条件和边界条件的原则。本章的特色是对工程上广泛应用的 $K\text{-}\varepsilon$ 模型作了详细的推导。

第三章主要介绍了流体运动微分方程的数值求解方法。描述流体湍流运动的基本微分方程组构成了一个耦合的非线性系统，从数学上来说，要想用解析方法求解这一控制流场的守恒方程几乎是不可能的。随着计算机的发展和应用，数值计算成为求解非线性系统强有力的方法。目前流体力学中常采用的数值求解方法是有限差分法。本章围绕有限差分法，着重讨论了差分方程及其网格的划分、求解 Navier-Stokes 流动方程的 MAC 方法、求解湍流流动的涡量流函数方法和 SIMPLE 算法以及实际计算中的稳定性和收敛性问题。并从实用出发，根据物理概念，列出一些最终的公式和数据，而不计及这些公式的推导过程，以方便工程技术人员实际应用。

第四章讨论了直流射流的流动理论及数值计算。在燃烧技术中，直流射流是实现及组织燃烧过程的基本空气动力学方式之一，是电站锅炉中广泛应用的直流燃烧器的基本组成单元。对于燃烧技术工作者来说，详尽地掌握其湍流结构有着重大的现实意义：一方面，湍流结构直接影响到燃料的着火及火焰的稳定性问题；另一方面，它也决定着燃料的燃尽过程。

第五章阐述了气流绕非流线型物体流动时的理论及数值计算。稳燃器在燃烧技术中广泛应用，因其能点燃燃料射流，稳定燃烧火焰，其中装设在燃烧器出口的非流线型物体(又称为钝体)往往是稳燃器的主要部件之一。在燃烧设备的喷燃器中，常利用非流线型物体来形成回流，它一方面可将高温烟气卷吸到火炬的根部来提高气体和燃料混合物的温度，另一方面可通过强化湍流热、质交换来稳定燃烧过程。本章着重分析了工程上常碰到的绕流圆柱、管簇(横向及纵向冲刷)、各种形式的钝体以及带周界风的钝体等的流动工况及计算方法，使读者对绕钝体后的流场及回流区参数计算有一个清晰的概念。

第六章介绍了旋转射流的理论及数值计算。旋转射流目前在工程技术中得到广泛的应用，一方面因其有较大的喷射扩张角，使射程较短，能在较窄的炉室深度中完成燃烧反应过程；另一方面其较好的"抽气"能力可使大量高温烟气回流到火炬根部，以保证燃料顺利着火和火炬稳定燃烧。按照流体力学的分类方法，一般旋转射流可分为弱旋转射流和强旋转射流两类。本章分析了弱和强旋转射流的特性及其计算方法，还讨论了工程常用的旋转射流组、旋转和直流射流的组合、同轴双旋转射流的组合等的流动过程及其数值计算。

第七章叙述了燃烧射流的理论及计算方法。燃烧现象是物理和化学过程复杂的相互作用结果，包含流体流动、传热、传质和化学反应等分过程，其所涉及的物理过程，特别是能量、质量和动量交换，在燃烧系统中起着重要作用。本章主要介绍了气体湍流燃烧射流的基本理论和湍流燃烧数学模型，以及两类不同的燃烧射流(即扩散火焰和预混火焰)的数值计算方法，重点放在物理概念的讨论。鉴于目前国内、外已有一些气体燃烧射流方面的专著，因此本书在计算细节问题上不多加描述，只介绍一些在后文对多相燃烧射流讨论中需要应用的概念。

第八～十四章，主要研究了气固多相流动的理论及计算方法。

第八章着重讨论多相流的动量输运过程及颗粒受力分析。首先探索了各种气、固两相的流动特性表示方法，然后着重研究了多相流动过程中颗粒所受阻力、浮力、压力梯度力、虚假质量力、Basset 力、颗粒旋转所受的横向力、Saffman 升力、湍流脉动力、热泳力、静电力、颗粒相互碰撞的受力等等，并把它们综合在 Lagrangian 型运动方程中进行数值求解，得到在不同颗粒直径下所受各种力的数量大小，为工程应用提供了根据。本章的特点是对多相流工程中常碰到的旋转横向力、湍流脉动力、热泳力等作了详尽的分析，以使多相流动的计算更加接近实际。

第九章专门研究多相流动中最主要受力之一——阻力。在气固两相流动中，阻力对流动、传热、传质和燃烧起着十分重要的作用。在实际的多相流系统中，颗粒阻力的大小受到许多因素的影响，例如气流的不稳定流动和湍流运动、气体的稀薄效应和可压缩性、壁面的存在、颗粒和流体温度的不同、颗粒的燃烧及颗粒群的稠密度。另外由于燃料颗粒通常都不是圆形的，所以非球形的颗粒阻力系数和球形颗粒也不同。本章的重点放在决定各种情况下的阻力系数，通过精确求解或数值求解分别求出低雷诺数和高雷诺数时，颗粒稳定运动和不稳定运动情况，以及在湍流脉动气流下，有燃烧反应时、存在颗粒群的运动工况下，变形颗粒的运动阻力和在非牛顿流体下的颗粒运动阻力及阻力系数。通过比较理论计算和试验数据，推荐出工程应用的阻力系数计算公式。

第十章集中讨论了多相流动的数值计算模型及其微分方程组。为了能比较完善和有效地解决多相流动问题，首先必须建立多相流动理论模型，并相应地给出描述其运动规律的基本微分方程组，然后研究怎样来求解这些方程组。关于描述多相流体流动的理论模型综合起来有四种：无滑移连续介质模型、小滑移连续介质模型、滑移-扩散连续介质模型和分散的颗粒群轨迹模型。这里着重分析了这四种模型，以及颗粒半随机轨道模型、脉动频谱随机轨道模型等，讨论了各种模型的优缺点及其发展，并推荐其使用的各种条件。本章是后面讨论各种工程多相流动实际问题的基础。

第十一章讨论了在工程实际经常碰到的管内多相流动的理论和计算。在工程技术中，颗粒在管内的输运占有非常大的比重，许多动力被消耗用于燃料颗粒的运输。为了提高颗粒的输运安全性、效率和节约能源，必须对颗粒的管内流动问题进行深入的研究。这里重点放在管内多相流动的模型、理论和计算方法，统一从流体力学的观点加以探索。

首先讨论了多相管流的流动工况以及理论上颗粒最小输送速度的确定，然后分别研究水平管道、垂直管道、转弯管道等多相流动的近似计算和数值计算，并探索了颗粒在管内流动碰撞时产生的磨损问题。为进一步开展管内多相流动的研究提供了基础。

第十二章为多相射流的理论和计算。在工程技术中，经常碰到分散相颗粒或液滴喷至炉内空间进行反应。由于固相或液相颗粒的存在，对射流的发展规律带来了较大的影响，特别是当颗粒较粗或浓度较大时，其流动规律与单相射流相差很大，研究各种形式的多相射流的特性和计算方法对发展和完善两相流动相关的工程技术有着很大的意义。基于此，本章着重研究了各种单相射流中存在着不同直径的分散相颗粒时的流动过程和计算方法，着重讨论了低浓度和高浓度的多相自由射流、多相受限射流、环形多相射流、大速差多相流动、多相射流组等。并通过分析，提出了其近似或数值计算的方法，计算所得的结果能为工程实际所应用。

第十三章描述了绕钝体的多相流动问题。当煤粉气流喷入带钝体稳燃装置的燃烧室中，由于钝体的影响，使颗粒的运动情况不同于上章所述的多相直流射流。为了深入研究煤粉的回流燃烧，必须对钝体回流流场中的气固多相流动过程进行分析，并提出合理的计算方法。本章的重点放在工程常用的绕圆柱、绕钝体的多相流动，以及煤粉绕钝体后的数值计算方法。此外还利用多相绕流的理论阐述了等速取样的原理和误差计算方法，为取样探头的设计提供了依据。

第十四章研究了多相旋转射流的理论及其数值计算方法。在燃烧实践中，经常应用旋转气流，气固多相流从切向喷入燃烧室或旋风分离器内的流动都是多相旋转射流。这类流动的特点是，相间黏性作用较强，固体边界的黏性影响较少，颗粒在旋转运动中除了受到气流阻力和自身重力外，还将受到离心力的作用，且当气流高速旋转时，其离心力比颗粒重力大得多。本章着重讨论了颗粒在旋风分离器内，在旋流燃烧器的燃烧室内，在旋风炉内、在直流燃烧器四角布置切圆燃烧室中的多相流动过程，并在燃烧状态下所进行的数值计算方法及结果。尽管目前旋转燃烧多相射流的数值计算还包括了一些简化假定，所得的结果也是初步的，但有了这样的开端将为今后的发展提供良好的基础。

本文摘自《工程气固多相流动的理论及计算》，浙江大学出版社，ISBN:7-308-00406/TB.007，1990。本书获 1992 年高等学校出版社优秀专著优秀奖。

燃烧流体力学

我国现有几十万台电站锅炉、工业锅炉以及为数众多的工业窑炉和各种形式的燃烧装置，每年耗用大量的燃料。提高燃烧效率、设计出新的炉型、优化炉窑的运行特性、预防事故的发生、提高设备的可靠性和经济性的关键是对炉内的燃烧流体力学问题有深入的了解，能提出合理的计算方法与设计方案，并制定较佳的运行方式。炉内燃烧流体力学是一门交叉学科，它与物理学、化学、流体力学、燃烧学、传热传质学等学科交叉，涉及具有化学反应、不等温、变质量、变直径的复杂气固多相湍流流动过程，同时又和炉子结构有密切关系。尽管生产实践十分需要，国内外文献资料也不少，但燃烧流体力学的发展还不够成熟，并且缺乏系统的、由浅入深的、能结合实际工程问题进行分析计算的教材。本书作者从 60 年代初开始，为研究生、大学生编写了部分讲义，并开设燃烧流体力学有关课程和选修课。近年来，又开辟了燃烧流体力学的研究方向，发展了计算机辅助数值试验的方法，努力为锅炉及窑炉的设计、运行服务。在此基础上，根据国内外积累的有关资料，结合近年来作者的教学和科研实践，编写了本书。

本书共分为六章。第一章描述燃烧流体力学的基本理论和数值计算方法。首先介绍了炉内湍流燃烧流动的基本特点，即湍流流场是无数不同尺寸的涡旋相互掺混的流体运动场，每个流体微团的运动类似于分子运动，具有完全不规则的瞬息变化的运动特征；各种物理量都是随时间和空间变化的随机量；流体微团的随机运动在足够长的时间内服从某种数学统计规律。描述了湍流燃烧的基本方程组纳维-斯托克斯方程组的瞬时形式和时间平均形式，包含了连续性方程、动量方程、能量方程和化学组分方程。还介绍了常用的湍流模型，包括无方程模型、单方程模型和双方程模型等湍流燃烧数值计算的基本方法。

第二章描述燃料颗粒的受力分析和多相流动的理论基础。首先主要介绍并推导了颗粒在炉内所受的力，包括阻力、重力、浮力、压力梯度力、颗粒旋转时的马格纳斯(Magnus)升力、萨夫曼(Saffman)升力、虚假质量力、巴赛特(Basset)纳力、湍流脉动力、热泳力、颗粒间及颗粒与壁面相互碰撞产生的力以及燃料不均匀水分蒸发、挥发分析出和焦炭燃烧时所受的力，介绍了颗粒运动时的阻力系数的决定方法和计算公式。然后描述了燃料颗粒在气流中运动的特点：气体分子分布均匀，而燃料颗粒是分散相且直径大小不等；在燃烧装置中，通常颗粒浓度不太大，颗粒相不能作为连续介质；颗粒相有较大的惯性，气体和颗粒间有速度滑移存在；由于颗粒尺寸大小不一，且形状不同，每个颗粒均有不同的速度；在有速度梯度的流场中，颗粒经常处于加速或减速的不稳定状态；颗粒的湍流扩散系数和气体不同，通常认为小颗粒的扩散速率比大颗粒的大。接着介绍了多相流动的数学模型，包括无滑移连续介质模型、小滑移连续介质模型、滑移-扩散的多连续介质模型、颗粒群轨道模型。然后介绍了应用脉动频谱随机颗粒轨道模型计算气相场、颗粒相、颗粒源项、颗粒浓度和速度的方法及步骤。最后介绍了煤粉颗粒在湍流脉动气流中运动的方程式及影响煤粉颗粒运动轨迹的因素。

第三章针对炉内煤粉射流的流动及计算进行了讨论。射流是湍流中非常常见的一种构型，并广泛存在于各种锅炉中。本章对不同形式的射流的原理和计算方法进行了介绍，包括自由射流、气固多相自由射流、多相受限射流、喷入伴随流的气固多相射流、环形和同轴多相射流、矩形射流、平行射流组和交叉射流。自由射流的特点是气体射入一个相当大的空间，气流不受固体边界的限制可在此空间扩散，本章介绍了圆形射流的半径经验理论以及出口湍流度、出口速度对自由射流的影响，并给出了自由射流无因次速度、无因次温度差、无因次浓度、无因次边界宽度、射流扩展角等物理量的计算公式。当气相射流内含有煤粉或其他分散相的固体颗粒时，形成了气固多相自由射流，本章介绍了射流中存在颗粒时对气流及颗粒脉动的影响，气固多相射流的流动特性及气固多相射流的计算方法。当射流喷入尺寸有限的燃烧设备中时，形成了受限的射流，其特点是周围有一定壁面边界。与自由射流不同，受限射流喷出后要卷吸周围介质，而周围介质受壁面限制又不能无限供应被卷吸的流量，所以受限射流在喷出后会形成回流区。分别对轴对称受限射流、非轴对称受限射流和带有扩口的受限射流的计算方法进行了介绍。当射流往一个运动着的主气流（即伴随流）喷入时，即使射流喷射速度相同，但伴随流速度不同，也会对射流喷入后的流体运动特性带来很大影响，因此介绍了喷入伴随流的气固多相射流的特性和计算方法。因点火的需要，往往要在煤粉燃烧器中心放置油枪，煤粉和气流从燃烧器的环形通道内喷出，形成了环形射流。此外，燃烧器为了合理配风，除煤粉气流从一次风射流喷出还需在一次风嘴外的环形通道中喷出二次风，形成中心是煤粉气流、边缘是二次风气流的多相同轴射流，本章对环形和同轴的单相和多相射流的流通特性及计算进行了介绍。在切圆燃烧的电站锅炉中，直流燃烧器的横截面经常做成矩形，形成复杂的矩形射流，本章还介绍了矩形湍流射流的流动特性及计算方法。在实际燃烧装置中，往往使用的不是单个而是一列喷燃器，其最基本的空气动力结构就是一列相互平行的自由射流所组成的射流组。由于射流间的相互混合和影响，使射流组中每一个射流和单个的自由射流的流动规律有较大的差异。特别是当射流组中两个相邻射流在离喷嘴一定距离处汇合以后，相互的混合作用使速度场起了较大的变化，射流组的流动过程是很复杂的，书中对平行射流度的流动特性及计算方法进行了介绍。在燃烧工程中，经常碰到用二次风或三次风向炉内主气流喷射，以强化炉内燃料与氧气的混合，从而达到分段补氧、分段燃烧的目的，这种射流称为交叉射流或发展在横向动上的纵向射流。喷入的射流和主流的温度、速度、成分均明显不同，是一个复杂的三维流动，对交叉射流的流动特性及计算方法进行了介绍，并重点分析了相对速度、相对温度、喷嘴角、喷嘴形状、喷嘴节距对交叉射流流动特性的影响。

第四章主要介绍了旋流燃烧器的空气动力学特性及其计算。旋流燃烧器利用强烈的旋转气流产生强大的高温回流区，从而强化燃料的着火和燃烧，同时也计算了燃料和空气的混合。旋转气流有很多优点：旋转射流和气流绕钝体相比要产生相近宽度的回流区，钝体所需的尺寸比旋流器的尺寸要大得多；旋流燃烧器所产生的回流区的长度也比钝体更大；随着旋转强度的变化，回流区的尺寸及高温烟气回流量的调节方便。旋流燃烧器产生更长的回流区使煤粉火炬出口处形成强大的高温区，不但强化了着火，而且也加速了燃烧，缩短了火炬射程。本章首先介绍了旋流燃烧器的类型及其特点，包含蜗壳式、轴向叶片式、切向叶片式旋流燃烧器。然后描述了旋流燃烧器中重要的空气动力特性参数：燃烧器的旋流强度、动量矩保持系数、能量利用系数、实际旋流强度、综合旋流强度、组合旋转射流的平均参数、烟气回流量、燃烧器的阻力系数、气流出口不均匀系数、旋转射流的扩展角、炉膛充满系数和各燃烧器间空气分配不均匀系数。然后对旋转气流的速度，旋流强度及其对速度分布的影响，旋转射流的射程，旋转射流的回流区，旋转射流的卷吸量，旋转射流的动量和动量矩沿轴向变化规律，旋转射流的湍流参数分布规律等特性进行了阐述，并介绍了旋转射流的数值计算方法和模型。最后详细介绍了环状旋流燃烧器、直流和旋流组合的旋流燃烧器、共轴旋流组合燃烧器、有出口扩锥旋流燃烧器、多个平行旋转射流燃烧器的空气动力特性和计算方法。其中环形旋流燃烧器中心布置有喷油枪，一次风在油枪外的环形通道旋流喷入，形成环状旋转射流。直流和旋流组合的旋流燃烧器有两种形式：一种是一次风为直流射流而二次风为旋流射流；另一种一次风为旋转射流而二次风为直射流或部分直流、部分旋转。共轴旋流组合燃烧器在多股强旋转射流的作用下，形成强大的中心回流和外围回流两个回流区，不同轴中的旋流可以是同向的也可以是异向的。有出口扩锥旋流燃烧器的射流在喷入炉内前会经过带有出口扩锥的结构。多个平行旋转射流燃烧器由多个旋流燃烧器对称组合而成，在炉内形成多个组合的相互平行的旋转射流。

第五章主要介绍了煤粉火炬的稳燃方法及空气动力特性。首先阐述了煤粉火炬的稳燃理论。煤粉随一次风气流喷进炉膛后接受炉内对流热与辐射热而升温、着火。煤粉燃烧器喷出的煤粉与空气气流在炉内的流动过程是一个复杂的三维多相流动，同时还涉及到燃烧所引起的温度场和浓度场等复杂问题。影响煤粉气流着火的主要要素有煤的性质、一次风量和一次风速、一次风温和着火区域的烟气温度、炉内空气动力场和煤粉颗粒度，影响煤粉火炬火焰传播速度的主要因素有煤粉浓度、煤粉挥发分含量、煤粉灰分含量、煤粉颗粒度，书中利用静态热平衡理论、搅拌均匀反应热理论、考虑煤粉运动及燃烧特点的理论对煤粉稳燃机制进行讨论，影响煤粉火炬温度的主要因素有回流区大小、燃烧器出口煤粉气流速度、煤粉气流的初温、燃烧器区域的烟气温度、气流分配和燃料的性质。当气流流过非流线形物体运动时，在紧靠物体的下游区域将形成一个漩涡回流区，为燃料着火及火焰稳定创造了有利条件，形成了钝体燃烧器。在直流燃烧器一次风喷嘴内，安装可在一次风喷嘴内前后移动的船型钝体，以调整合适的稳燃工况，形成有火焰稳定船的直流燃烧器。煤粉预燃室依靠气流的旋转或经钝体绕流，在容积较小的前置燃烧室内形成强烈的回流，利用较小的点火能量点燃煤粉气流并使火焰稳定。本章详细地对钝体燃烧器、有火焰稳定船的直流燃烧器和煤粉预燃室的空气动力特性，稳燃机理和计算方法进行了阐述，接下来对大速差射流、不对称射流、反吹射流、高煤粉浓度高预热温度煤粉火炬、W型和Y型火焰稳燃技术及其动力特性、稳燃机理和计算进行了详细的讨论。其中大速差射流稳燃技术在圆柱形筒体内配直流燃烧器，与一次风喷嘴同一平面上布置数个喷嘴，通过高速空气或蒸汽以高速流动喷出，在预燃室内形成回流区。不对称射流稳燃技术中空气煤粉混合物通过下偏置或上偏置的一次风管，进入圆形或矩形截面的预燃室，在一次风口下方靠近预燃室底部的位置上有一股吹灰的平面射流，卷吸周围介质出现反向压力差,形成回流区。反吹射流稳燃技术中一次风为直流风,二次风切向引入,反吹射流的引入和卷吸使炉膛中心高温烟气随反吹射流一起倒流,形成中心回流区。高煤粉浓度高预热温度稳燃技术通过提高煤粉浓度和预热温度促进煤粉火炬的稳燃。W 型和 Y 型火焰稳燃技术通过形成复杂的回流运动和高温回流区来使火焰稳燃。

第六章介绍了四角切向燃烧锅炉的流体动力特性及其计算。四角切向燃烧广泛用于大中型燃煤或燃油锅炉,通常将直流式燃烧器布置在炉膛四角,出口气流几何轴线切于炉膛中心的假想圆,造成气流在炉内强烈旋转。当四股气流同时喷入炉内时,各股射流的相互影响形成一个强烈的炉内旋涡,实际的切圆直径变大。在四角切向燃烧器中四角射流着火后相交,相互点燃,使煤粉着火稳定。由于四股射流在炉膛内相交后强烈旋转,湍流的热量、质量和动量交换十分强烈,故能加速着火后燃料的燃尽程度。四角切向射流有强烈的湍流扩散和良好的炉内空气动力结构,炉膛充满系数高,炉内热负荷较均匀。切圆燃烧时每角均由多个一、二、三次风喷嘴所组成,负荷的变化时调节灵活对煤种的适应性强,控制和调节手段也较多。四角燃烧器炉膛结构较简单,便于大容量锅炉的布置,其采用摆动式直流燃烧器时,运行中改变上、下摆动角度即可改变炉膛出口烟温达到调节过热蒸汽温度的目的,并便于实现分段送风组织分段燃烧,从而抑制 NO 的排放等。书中对常见的四角切向燃烧器分类、数值计算方法以及四角切向射流的偏转和炉内实际切圆直径的计算方法进行介绍。然后对周界风直流燃烧器、夹心风直流燃烧器、摆动式直流燃烧器的空气动力特性及计算方法进行了阐述。其中周界风燃烧器在直流式燃烧器的一次风口四周包围一层速度较高的二次风辅助燃烧,防止煤粉从一次风气流中分离出来并冷却一次风喷口。夹心风燃烧器在一次风喷口中间竖直地插入一股二次高速风提高了气流的刚性,并为火焰及时补充氧气。摆动式燃烧器利用上下摆动的燃烧器来方便调节火焰中心位置、炉内热流分布、炉膛出口烟温和过热器再热器的蒸汽温度。本章还讨论了运行工况和结构参数对四角切向燃烧炉内的空气动力特性的影响。最后阐述了当煤粉在各燃烧器均匀分配的燃烧器和四角切向燃烧煤粉火炬燃烧器的空气动力特性。

本文摘自《燃烧流体力学》,水利电力出版社,ISBN: 7-120-01348-3/TK.229,1991。本书获第三届全国普通高等学校热能动力类专业部级优秀教材二等奖,1996 年。

锅炉和热交换器的积灰、结渣、磨损和腐蚀的防止原理与计算

我国是产煤大国，煤炭资源非常丰富，但是煤种品位高低不等，变化范围很宽。根据我国能源政策，火力发电厂应以煤作为主要燃料，且动力用煤应尽量使用低品位劣质煤，加之现有供煤和配煤系统有许多不完善之处，电站锅炉燃用煤质难以得到保证，锅炉燃煤多变且灰份有不断升高、热值有不断降低的趋势。燃煤总体品位下降使锅炉受热面产生磨损、积灰、结渣、腐蚀等一系列问题，受热面使用寿命降低，锅炉管子爆漏现象频繁。1992 年我国火电设备事故的统计结果表明，当年锅炉事故占全部发电事故的 56%，而锅炉四管爆破事故占到了全部锅炉事故的 64.2%，其中省煤器占 35.3%，过热器占 29.8%，水冷壁占 27.8%，再热器占 7.1%。产生事故的原因除管材和焊接质量问题外，主要是由于锅炉积灰、结渣、高低温腐蚀、磨损及振动所引起。由此可见，四管爆漏已成为阻碍提高机组利用率和安全性的重要因素，该问题在国际上也十分严重，诱发了不少发电事故。

国外对磨损、积灰、结渣、腐蚀和振动进行了大量的理论研究和实践探索，取得了丰富的研究结果和实践经验。国内科研单位也正在积极开展研究，取得了一些成果，但尚缺乏系统的归纳和总结。近年来，电力工业发展迅速，锅炉磨损、腐蚀、积灰、结渣等问题变得更为突出，明显了撰写相关专著的必要性和迫切性。为此，作者结合浙江大学热能工程研究所长期来在此领域研究所取得的成果，参考国内外研究资料，组织撰写了本专著。本专著系统论述了锅炉受热面磨损、积灰、结渣、腐蚀、振动的产生机理及防止措施。在计算分析锅炉受热面的磨损和结渣过程时，引入了作者所提出的计算机辅助优化数值试验方法(computer aided testing，CAT)，使各种参数变化对锅炉受热面的磨损和结渣的影响分析得以定量化。

全书共十章，系统地阐述了锅炉受热面及热交换器内烟侧飞灰的沾污、积灰、高低温腐蚀、磨损、空气动力振动过程的原理、计算方法及预防措施。

第一章主要探讨了煤及煤灰特性对受热面积灰、腐蚀及磨损的影响。首先，简要概述了积灰、腐蚀、磨损对锅炉运行的影响，当煤在炉膛中燃烧时，煤中的矿物质及金属有机物会转化为煤灰(灰渣)，煤灰会沉积在锅炉的各种受热面上，影响锅炉的正常运行。介绍了煤及其破碎过程中矿物成份的分布以及煤灰的矿物质成份及其微观结构。然后，分析了煤灰矿物成份对沾污、结渣的影响，概述了炉内矿物成分的结渣过程，详细分析了灰中铁含量、硫含量、碱金属含量、氯含量，以及硅酸盐含量对沾污和结渣的影响。接着，介绍了煤灰特性对受热面磨损的影响，分析了灰的熔化特性和黏度特性及其与成分的关系，研究了灰渣及灰垢的物理特性，用于计算结渣积灰过程，探索其对锅炉运行、传热过程的影响。

第二章主要研究了炉内燃烧时煤灰中各种矿物质的行为，灰粒沉积和黏结的受力分析以及受热面沾污、积灰、结渣过程的动态特性。首先，介绍了煤燃烧过程中硅酸盐、硫化物，以及碳酸盐类矿物质在高温中的行为，系统地分析研究煤在燃烧过程中各种矿物质的变化，准确地评价煤种在燃烧过程中可能引起的结渣和积灰的程度，以提高煤的燃烧效率和锅炉运行的安全性。然后，介绍了煤粉燃烧过程中微细飞灰的形成，飞灰中含有大量微细颗粒，除尘器难以分离并造成空气污染，同时这些微细颗粒往往是引起锅炉受热面沾污、结渣的根源之一。然后，分析了受热面积灰、结渣的特性及其影响，介绍了积灰、结渣的类型，飞灰对受热面沉积和结渣的过程以及结渣的机理，分析了影响飞灰沉积和结渣的各种因素。探究了灰粒向受热面沉积运动的受力以及灰粒在受热面上沉积和黏结过程及其黏结力。最后，详细介绍了受热面积灰和结渣厚度的计算方法以及锅炉受热面沾污、积灰、结渣的动态特性及其计算方法。

第三章主要讨论了积灰、结渣对炉内传热及对流受热面传热的影响。首先，简要介绍了炉膛传热过程的计算方法，研究了受热面积灰、结渣对炉膛传热的影响，阐述了炉膛壁面积灰、结渣的机理。积灰、结渣将造成传热热阻增加，水冷壁吸热量减少，导致锅炉出力下降，甚至影响锅炉的安全运行。然后，探究了火炬热流密度对膜式水冷壁温度场变化的影响，介绍了电站锅炉炉膛中膜式水冷壁温度分布的计算方法。随着锅炉容量的增大，出口烟囱面积增大、烟气温度不均匀性增加，仅仅计算炉膛出口烟温不能满足实际需求，需要发展三维模型的数值模拟计算方法以求解炉内烟温的空间分布，为此，介绍了区域法的传热计算原理，导出了基本的计算公式和计算步骤，并以实际锅炉的炉内传热计算作为实例，分析了积灰、结渣对炉膛传热的影响。基于统计学的蒙特卡罗方法是计算炉内烟温分布的另一种实用方法，介绍了蒙特卡罗方法用于炉膛辐射传热的计算公式和计算步骤，分析了积灰、结渣对炉膛传热的影响。烟道中的对流受热面主要以对流传热方式吸收热量，称为对流受热面，探究了积灰、结渣对锅炉对流受热面传热的影响。

第四章主要阐述了根据煤灰特性及运行特性对锅炉受热面结渣进行预测的方法。锅炉受热面结渣是经常发生而又难以彻底解决的问题，结渣影响了锅炉运行安全性、经济性及锅炉可用率。锅炉结渣受许多因素影响，其中煤本身结渣倾向、炉内温度水平及温度场，以及炉内空气动力工况等是主要影响因素。本章从煤的物理化学特性、运行中烟温变化趋势等方面对煤本身结渣趋势、实际运行中锅炉结渣状况进行预测，分析讨论了减少或降低结渣的措施以及吹灰器吹灰性能等问题，提出了根据煤灰物理特性和成份特性对受热面积灰、结渣、沾污进行预测的方法。然后，进一步考虑锅炉本身的特性及运行特性，在运行中判断结渣状况，综合地反映煤种潜在的结渣倾向及运行参数对结渣的影响。最后，详细介绍了采用气动方法、水射流方法以及振动方法对受热面进行吹灰的原理及装置，提出了采用合理的锅炉结构设计及合理的运行方式来减少锅炉受热面沾污、积灰、结渣的措施。

第五章主要探讨了锅炉受热面烟侧的高温腐蚀过程及其影响。首先阐述了煤燃烧过程中产生 SO_2、SO_3、H_2S、HCl 的化学反应机理及其对受热面腐蚀的影响，煤中的硫在燃烧过程中生成 SO_2，其中少量的 SO_2 转化为 SO_3，烟气中的 H_2S 主要由煤的缺氧燃烧导致，HCl 的形成则与煤中存在的 NaCl 密切相关。然后，简要概述了硫酸盐型和硫化物型的高温腐蚀，介绍了碱土金属和钒盐类对受热面的高温腐蚀以及水冷壁的高温腐蚀。最后，提出了防止高温腐蚀的措施，例如采用低氧燃烧技术、避免出现受热面壁温局部过高、提高各燃烧器间煤粉浓度分布均匀性，以及合理控制炉膛出口烟温等。

第六章主要分析了受热面的低温腐蚀过程及其计算方法。首先，阐述了 SO_3 的冷凝过程以及硫酸蒸气的热力学露点，详细介绍低温受热面区间烟气的 SO_3 浓度以及烟气露点温度的计算方法。然后，分析了硫酸蒸气向低温受热面沉积的热质交换过程，详细介绍了酸冷凝沉积量的计算方法。接着，阐明了低温受热面的积灰和腐蚀机理，介绍了烟气露点的测试方法，分析了各种运行参数对低温受热面积灰和腐蚀的影响。锅炉低温受热面的腐蚀较为严重，一切影响烟气中硫酸形成的因素、影响烟气中的硫酸蒸气冷凝在受热面上的因素、影响金属和硫酸溶液进行相互作用的因素均会影响低温受热面的腐蚀。总之，低温受热面的积灰、腐蚀主要原因是烟气中含有的 SO_3 以及受热面温度低于酸露点温度，烟气中的 SO_3 来自燃料中所含的硫，SO_3 的含量与燃料中含硫量条件有关。最后。提出了防止和减轻低温受热面的积灰、腐蚀的主要原则：燃料脱硫、改善燃烧方式以减少烟气中 SO_3 的含量、提高受热面壁温使之大于烟气的露点温度等。

第七章主要叙述了飞灰对受热面的磨损机理以及采用 CAT 计算磨损的方法。对流受热面严重的飞灰磨损将导致泄露、爆管等锅炉事故，导致电厂被迫停炉检修、停汽停电，给国民经济造成极大的直接和间接损失。本章基于磨损理论来指导对流受热面的磨损分析和计算，为此，首先介绍了对流受热面飞灰磨损的两种类型，即冲刷磨损及撞击磨损。然后，介绍了含灰气流对金属管壁的磨损理论，主要包括了冲蚀磨损的微切削理论、变形磨损理论，以及非刚体颗粒破碎的二次冲蚀理论，概述了金属管壁受飞灰冲蚀磨损的近似计算方法。在实际的飞灰气流冲击管束问题中，飞灰气流的流动情况十分复杂，为此，提出了基于 CAT 的飞灰冲击对流受热面磨损的数值计算原理及方法，建立了颗粒与管壁的碰撞-反弹模型和冲蚀模型，介绍了含灰气流对单个圆管冲蚀磨损的数值计算结果。最后，分析了各种空气动力参数对飞灰冲蚀磨损的影响，例如烟气速度、气流湍流度、灰粒直径等参数对冲蚀磨损的影响，同时分析了受热面壁温变化以及烟气成份对飞灰磨损的影响。

第八章主要讨论了锅炉各种受热面布置磨损过程的计算方法。首先，阐述了错列对流管束间的灰粒运动规律及磨损机理，概述了灰粒在错列管束间运动及磨损的近似计算方法，进一步介绍了基于 CAT 的烟气流过顺列和错列管束的磨损数值试验及研究结果，成功运用多相流理论并结合材料学磨损理论研究了锅炉受热面的磨损问题。然后，分析了烟气由上向下和由下向上流动对受热面磨损的影响，提出了对流管束安全长期运行允许烟速的确定方法，总结了流过管束间产生磨损的各种因素以及计算方法。管束和炉墙之间的间隙形成烟气走廊，烟气在烟气走廊中为加速运动，探究了烟气走廊引起对流管束的磨损机理及防止磨损的措施。接着，阐述了采用加翅片管束的防磨理论以及数值实验的方法和结果，从多个方面提出了横向冲刷对流受热面的防磨措施，例如，设计时选择合理的烟速、降低速度不均匀系数、安装炉内飞灰除尘器等。最后，分析了管式空气预热器烟侧的磨损机理，介绍了飞灰颗粒进入空气预热器的运动以及含灰气流对管壁的磨损计算方法，阐述了空气预热器的防磨机理及防磨措施。

第九章主要分析了锅炉受热面空气动力振动的原因、计算方法及防止措施。首先，介绍了炉膛及热交换器空气动力振动的分类，内部流动中引起振动的主要原因是管内流动的不稳定性和管道鞭击（水锤）现象，而外部流动中主要是空气动力学的不稳定性以及旋涡诱发振动。然后，依次详细阐述了气流旋涡诱发振动的原理、湍流颤振诱发振动的原理、流体弹性扰动诱发振动的原理以及声激振的原理。接着，详细介绍了换热器管束及烟道的固有频率及其阻尼的计算方法，提出了换热器振动分析的评价准则，探究了振动产生的疲劳应力，分析了影响管束材料疲劳性能的因素。进一步探讨了燃料燃烧诱发噪音、炉膛振动的原理及其预防措施，分析了燃烧诱发脉动的机理、四角布置切向燃烧时炉内压力的脉动机理、旋流燃烧器燃烧诱发脉动机理，以及各种炉内燃烧因素对脉动的影响，介绍了炉墙弹性计算以及燃烧诱发炉墙结构振动的计算方法。最后，全面探究了过热器、省煤器、空气预热器以及尾部烟风道的空气动力振动及其预防措施，简要分析了烟道和风箱的空气动力诱发振动现象。

第十章主要提出了四管爆漏及炉内燃烧事故的分析原理及预测。首先，分析了炉内严重结渣、大量掉渣砸坏水冷壁的可能性，分析了落渣时炉内工况的波动，对结渣砸落至水冷壁过程进行详细的受力分析。然后，分析炉膛外爆和内爆的主要原因，探究了炽热灰渣掉入渣池引起水蒸发以及炉内压力升高过程。水冷壁管爆破后，管内高压的饱和水将随破口迅速喷射出来，引起汽包水位发生变化，为此，提出了炉管破裂事故后汽包水位变化的计算方法。同时，炉管爆裂引起水汽混合物释放到炉膛中，与高温烟气迅速混合后引发膨胀积聚，促使炉内压力升高和温度降低，为此，分析了炉内压力升高的动态过程，介绍了炉内温度变化的计算模型。当炉内水冷壁上或冷灰斗内大量结渣时，饱和水喷至热渣上将导致水大量蒸发，为此，介绍了水的饱和蒸发速率的简化计算模型。最后，分析了煤粉和可燃物在炉内爆炸引起的超压，介绍了爆燃与缓燃的区别，提出了炉内爆管蒸汽物理爆炸能量的估算方法，分析了可燃混合气和煤粉爆炸的特性及其引发的超压。

本书汇集了作者多年来在该领域内的研究成果，同时还引用了近年来国内外最新的试验研究资料，力图使本书理论知识丰富并具备工程实用价值。本书在撰写过程中通过试验研究和计算机数值试验分析，尽可能做到理论结合实际，解决锅炉四管爆破的问题，但由于受热面沾污、积灰、结渣、磨损等问题极为复杂，很多问题目前尚未得到解决，所以本书所提出的某些分析或计算具有一定的探索性，未来仍需要进一步的全面分析和验证。本书可供动力、化工、冶金、建材、轻工、环保、石油等领域内从事炉、室及热交换器的设计和运行的科技工作者参考，也可作为高等院校热能工程、电厂热能动力、工程热物理等专业的研究生和本科生的教材或教学参考书。

本文摘自《锅炉和热交换器的积灰、结渣、磨损和腐蚀的防止原理与计算》，科学出版社，ISBN: 7-03-004129-1，1994。本书获教育部科学技术进步奖二等奖，2004。

煤浆燃烧、流动、传热和气化的理论与应用技术

我国能源以煤炭为主，煤炭燃烧利用后会排放出飘尘、二氧化硫、氮氧化物等，引发大气污染等问题。因此，寻求一种新型的煤基洁净燃料乃当务之急的重大课题。煤浆作为一种低污染、高效率、流动性强的代油新型清洁燃料，上世纪以来在国内外上得到迅速的发展。所谓煤浆就是由煤、水(或油、甲醇等)和少量添加剂按一定比例组成，通过物理加工处理，制成类似油一样的新型洁净流体燃料。

煤浆具备像燃料油那样易于装、贮、管道输送及雾化燃烧等特点，其技术应用将使我国煤炭的品质、运输、工业应用、环境效益发生根本性的改革。首先，煤炭的洗选与制浆相结合，使煤浆成为低灰、低硫、高品位的燃料。其次，在矿区制备好的煤浆可以不用传统的货车运输方式，而用封闭式、低损耗、洁净化的管道运输方式运向工业用户，使用户能得到稳定的、高质量的燃料供应，同时又能大幅度降低运输的投资与运行成本以及煤炭的运输损耗。再者，用户能得到易贮存、能泵送、无须在厂房内布置煤粉制备系统的洁净燃料，大大改善了用户端环境，降低了投资及使用成本。由此可见，煤浆技术的推广具有重大的技术经济效益和社会发展意义。更鉴于世界石油资源日趋减少，并主要用于化学工业和交通运输工业。因此，国际上如美国、日本、俄罗斯、意大利、瑞典等发达国家近20年来相继投入大量的人力、物力开展新型煤浆燃料的研究，并已达到大型工业化和商业化的应用阶段，而且每年均召开一次国际煤浆应用技术会议，至1996年已召开了21届。

浙江大学是我国最早研究煤浆燃烧技术的单位，在1978年开始在国家计划委员会和中国科学院组织领导下，开展了油煤浆技术的研究，并在鞍钢电厂100 t/h锅炉上加以实现，通过了国家鉴定。1981年3月，浙江大学向国家科学技术委员会、科学院、煤炭部建议开展水煤浆代油燃料的研究，之后由国家科学技术委员会和煤炭工业部分别列入国家"六五""七五""八五"科技攻关项目，并成立了国家水煤浆工程技术研究中心及华煤水煤浆技术联合中心，使我国的制浆技术、管道输运技术、煤浆燃烧技术与工业应用技术联成一体，从而使煤浆的研究和利用得到蓬勃的发展。这是一项新的技术，是物理化学、多相流体力学、燃烧学、传热学、机械学、材料科学等交叉学科相关的技术，目前虽有大量相关期刊论文和国际会议论文集，但国际上尚无一本全面、系统论述煤浆流动与燃烧方面的专著，因此作者根据浙江大学研究团队十余年的研究心得与理论实践成果，最终编写形成本专著。

我国开展水煤浆技术的研究以来，已进行了大量的基础研究和工业性试验，开发了工业锅炉和窑炉燃用水煤浆的关键技术，且已达到了工业应用的阶段。现正开发大中型电站锅炉、大型工业窑炉应用水煤浆的有关技术，并向大型化、系列化、经济合理化方向发展，尚需解决如下关键技术问题：一是各类低灰、中灰、高灰和低浓度水煤浆燃烧特性的研究及拟订评价方法；二是强化着火、燃烧的机理，低污染燃烧的机理，煤浆雾

化机理的研究；三是炉前煤浆输送管路、冲洗、过滤和控制的设计与研究；四是开发不同容量、不同种类的水煤浆喷嘴，特别是大型锅炉(3~5t/h 和 5~8t/h)水煤浆喷嘴及其耐磨材料；五是开发不同容量、不同类型的旋流燃烧器、直流燃烧器及其他各类新型的水煤浆燃烧器；六是水煤浆的少油或无油点火技术；七是强化着火和燃烧技术；八是炉内燃烧特性和传热特性的研究；九是受热面的腐蚀及磨损的研究；十是低 NO_x 和炉内脱硫技术的研究；十一是油炉改烧水煤浆少降负荷技术的研究；十二是锅炉的改造技术的研究；十三是锅炉改造设计计算方法的研究。

本书系统地阐述了煤浆燃料在管内流动、传热及在锅炉内燃烧、气化的原理，计算方法及工程应用技术，主要包括以下十一个部分的内容。

第一部分是煤浆的物理、化学特性及流变特性，为煤浆燃料的制备和应用提供基础；第二部分是煤浆在管内的流动特性及沉降特性，并探讨其在管内流动的理论方程与计算方法；第三部分是煤浆在炉前的贮存系统，为搅拌器与过滤器等装置的工业设计提供依据；第四部分是煤浆在管内流动时的传热原理，为加热系统提供理论求解与数值计算方法；第五部分是单滴煤浆的燃烧特性与反应机理，为煤浆的燃烧应用打下理论基础；第六部分是煤浆喷嘴和燃烧器的装置原理与结构设计，为工业应用提供关键装备保障；第七部分是煤浆火炬在炉内的燃烧特性及其数学模型，为大型锅炉煤浆燃烧应用提供理论基础；第八部分是改烧煤浆时的改炉技术及现有煤浆低污染燃烧新技术，为燃油锅炉改烧煤浆问题的解决与先进工艺的引进提供参考依据；第九部分是水煤浆气化原理及技术工艺，介绍水煤浆在我国煤化工及燃烧联合循环应用中新途径；第十部分是煤浆燃烧的试验方法和测量技术，以供基础研究和工程现场调试之用；第十一部分是煤浆在国内外工业领域的技术应用及发展前景，包括在电站锅炉、工业窑炉、燃气轮机、内燃机等应用前景。

不同的煤浆产品是根据煤与不同流体的混合来命名的：油煤浆是 50%煤粉和 50%油的混合物；煤油水浆是煤粉、油及 10%以上水的混合物；水煤浆是 60%~70%煤粉与40%~30%的水及少量添加剂的混合物(其中根据原煤的灰分高低又可分为超低灰、低灰、中灰和高灰煤浆)；煤-甲醇混合物是 60%煤粉和 40%甲醇或甲醇水的混合物。此外，还有石油焦浆(石油焦为低灰高热值的石油残渣)，石油焦浆又可分为石油焦与油混合的油焦浆和水与石油焦浆混合的水焦浆。

油煤浆的制备主要有干法制浆工艺和湿法制浆工艺两大类。所谓干法制浆即是用干法将煤磨制成煤粉，再与油强烈混合的工艺。国内主要采用干法制浆工艺，而湿法制浆是用煤和油按一定比例同时加入磨煤机，在磨煤机内采用湿磨的方法直接制备成油煤浆。国外湿磨和干磨都有采用，水煤浆可以像油一样贮存、输送和燃烧，能 100%代油。在我国由于要求压缩烧油、以煤代油的政策和煤与油的比价，使为代油燃料的水煤浆具有较强的竞争力。世界上第一个大型工业化水煤浆制浆厂的制备工艺流程采用瑞典 Carbogel提出的二段磨矿的方案，是低浓度湿磨、浮选、精煤过滤、脱水后调浆。它的优点是低浓度磨矿容易，粒度分布便于控制，但增加了能耗很大的过滤工艺。由于煤浆质量易于控制，所以得到了广泛的应用。另外，还有一种特殊的煤浆叫做煤水气混合物。即是在水中造成水泡沫，然后与煤粉混合，对煤粉起到悬浮介质的作用，从而可制备成可泵送、不沉淀的高浓度煤浆。这种煤浆的优点是浓度高，可达 75%～80%；稳定性高，可储存 4～5 天不沉淀和具有制造成本低等。

高浓度水煤浆的制备需要添加少量的添加剂以降低水煤浆的黏度，提高水煤浆的浓度、流动性及提高煤浆的稳定性。但煤浆的制备还与煤质特征，粒度分布及添加剂的种类有关。关于煤质特征对制浆性能的影响，成浆性能与下列因素有关：一是内在水分，内在水分高低有时会有几倍至几十倍之差，这些水分分布在煤粒的内表面，当煤浆的重量浓度相同时，势必要减少流动介质作用的水量，造成水煤浆的黏度高或难于获得高浓度的浆。二是比表面积和孔隙度，比表面积的差异也有几倍到一、二十倍的差别，比表面积能间接地反映煤的孔隙程度，发达的孔隙度是造成内在水分高的重要原因之一，此外高比表面积还会导致消耗更多的添加剂。三是氧碳比，含氧量高即提供的含氧官能团活性氧部分就越多，增加了煤表面的极性和与添加剂作用的复杂性，减少了与表面性质相适应的添加剂种类及增加了添加剂的消耗量。四是可磨性，当磨矿条件相同时，难于获得超细的颗粒，从而降低了煤粒的堆积效率，难于制出高浓度的煤浆，或者在保证必要的充填效率的条件下，将增大磨矿的能耗和磨制工艺的复杂性。

高浓度煤浆的煤重量浓度达 65%～70%，因而要求煤炭的粒度分布能达到较高的堆积效率。这就是通常所谓煤浆的粒度分布要有良好的级配。如果采用不同粒径来堆积，可提高堆积效率。通常煤粉的粒径分布规律符合 Rosin-Rammler 分布函数，而该函数是单峰分布的，影响了堆积效率的提高，而双峰分布的颗粒组成会有较高的堆积效率。有人提出了颗粒组成为双峰分布和多峰分布制浆的工艺。因而在工业制浆时应寻求合理的粒径分布，以得到较高的制浆浓度和较低的黏度。

煤粉颗粒表面是憎水性的，未完全湿润的煤粉会互相团聚、会使煤浆的黏度增加。因而未加添加剂的煤粉制成煤浆的浓度小于 60%。添加剂的功能在于提高煤粒表面的亲水性、整顿颗粒表面电荷密度，从而调整煤浆性能，高的表面电荷在水煤浆中建立了防止沉淀的三维结构，或者导入弱的絮凝倾向以减少平均粒径和最大的固体浓度。水煤浆的添加剂包括分散剂和稳定剂。分散剂作用是改善煤粉的润湿程度，提高煤粒表面的电位，使煤粒在水中能更好润湿和分散，降低黏度和改善流动性。稳定剂作用是防止煤浆在储存与运输过程中发生硬沉淀。添加剂的种类有非离子型、阴离子型和阳离子型三种。非离子型添加剂主要通过表面活性来降低煤浆液体的表面张力，使煤浆中煤粒表面润湿，控制表面的电荷来改变煤浆的性能。离子型的添加剂通过含极性基添加剂的静电吸附在煤浆颗粒上、降低颗粒表面的疏水性、控制颗粒表面的荷电性来改变煤浆的性能。当前国内外广为应用的分散剂是非离子型的萘磺酸盐，其优点是适用性强、价格便宜，降黏度效果较好。离子型添加剂由于通常价格太高而限制了它的应用。

煤气化发展至今已有百余年历史，按煤在气化炉的物理流态来分类，主要有固定（移动）床、流化床和气流床三种。工业实践证明，它们有各自比较适应的经济规模，移动床气化可以应用于非常小的容量（4～15MW 热量）规模，气流床气化较适用于大规模生产（>200MW 热量），流化床气化则介于中间。煤气化方法按技术水平和商业应用规模也可分为第一代、第二代、第三代气化技术。第一代煤气化技术一般是指目前已经商业化、在生产上具有较长期的成熟操作经验的气化装置，如固态排渣的鲁奇炉、K-T 炉和温克勒炉等。第二代煤气化方法是指把新的工程技术、煤化学的新理论及工程材料科学的新成就相结合产生的气化方法。部分目前正处于中试研究阶段和示范阶段，部分已进入商业应用。如采用固定床型的液态排渣鲁奇炉，流化床型的 HTW 炉、U-Gas 炉、KRW 法以及采用气流床型的德士古(Texaco)炉等。第三代煤气化方法是指比第二代煤气化技术更先进、综合效益更高的气化方法，大多数都正处于研究开发阶段，在技术上有所突破，但近期内还不会商业化。如催化气化、太阳能气化、核热气化、等离子气化和采用火箭技术的 Rockwell 气化法等。

相比于干法气化工艺，水煤浆气化技术研究较少，目前主要应用在气流床气化和流化床气化工艺中。特别是第二代煤气化法采用了高温高压运行、细粉给料，因此气化装置燃料转化率高，燃料适应性广，容量也更大；而水煤浆气化在满足上述工艺要求时显示了其独特的优越性：一是利用高压气化，输送给料方式简单；二是利用高温下水煤浆的水产生热分解促进气化反应进行；三是降低了制粉系统投资。目前已工业化的水煤浆气化技术是德士古气化技术，此外美国道化学公司开发的道气化技术，日本日立公司开发的油煤浆裂解气化法，也已进入示范阶段，浙江大学还提出了煤浆/煤气化燃气-蒸汽联产工艺。

在循环流化床煤浆/煤气化燃气-蒸汽联产工艺中，本书提出了煤或煤浆干馏和部分气化产生民用煤气，半焦送燃烧炉燃烬产汽发电方案，使煤中成分得到合理利用。该工艺中，气化室为常压鼓泡床，用水蒸气和再循环煤气作为气化剂，运行温度为 750~800℃，燃料给入气化室，首先受热裂解，析出高热值挥发分，半焦中部分碳和气化剂反应形成水煤气，气化吸热由燃烧室的高温循环物料来提供，气化后半焦随循环物料送入燃烧室燃烬，燃烧室为快速床，空气鼓风，运行温度为 900~950℃，燃用气化室来的半焦，产生热量，加热从气化炉来的低温循环物料变为高温物料再送至气化室提供气化吸热和产生水蒸气，从气化炉出来的高温煤气，经煤气冷却器冷却，净化器净化，除去灰、焦油、水后变为净化煤气。由燃烧室、煤气冷却器、烟气冷却器产生蒸汽除少量的供气化炉用汽外，大部分用于发电、供热，也可制冷。如此，实现了煤气热电冷多联产。

书中汇集了作者多年来在该领域内的研究成果，特别是总结了煤浆技术国家"六五""七五""八五"重大科技攻关项目及国家自然科学基金项目研究成果。同时介绍了在国内工业性生产中推广应用情况，及国外近年来最新试验研究资料，使本书在理论与工程应用相结合方面具有极高的实用价值。

本文摘自《煤浆燃烧、流动、传热和气化的理论与应用技术》，浙江大学出版社，ISBN: 7-308-01873-7/TK·006，1997。

循环流化床锅炉理论、设计与运行

作为重要的高效低污染燃烧技术，循环流化床锅炉技术可广泛应用于电站锅炉、工业锅炉和废弃物处理利用等领域。本书是根据十几年来浙江大学在循环流化床锅炉技术方面的研究成果组织编写而成的，本书对循环流化床的流动特性、传热特性、燃烧特性、分离回送特性、脱硫脱硝特性、分离器和回送装置等关键部件、设计计算方法、启动和运行等进行了系统的介绍。

1. 循环流化床锅炉原理和特点

循环流化床锅炉采用流态化的燃烧方式，即半悬浮燃烧方式。在一定的燃烧设备内，燃料在高速气流的作用下以沸腾悬浮状态进行燃烧。燃烧产物——烟气携带一部分固体颗粒离开炉膛进入物料分离器。物料分离器将固体颗粒分离出来返送回炉床内再燃烧，烟气排出进入烟道，形成循环流化床。循环流化床锅炉具有燃料适应性广、燃烧效率高、氮氧化物排放低、低成本石灰石炉内脱硫、负荷调节比大和负荷调节快、燃料预处理系统简单、易于实现灰渣综合利用等突出优点。

2. 循环流化床发展历史

循环流化床锅炉最早起源于20世纪50年代的德国，典型的是德国鲁奇公司(Lurgi)发展并运行的 Lurgi/VAW 氢氧化铝焙烧反应器，1976年，Yerushalmi 等首次提出了快速流态化的概念，从而引起了人们对循环流化床技术的日益重视。20世纪80年代技术的进步和环保意识的增强，循环流化床技术研究和开发受到高度重视，循环流化床锅炉在国际上开始得到广泛应用。

国外循环流化床锅炉发展到现在已有许多不同的流派和型式，其中较有代表性的是芬兰奥斯龙(Ahlstrom)公司的 Pyroflow 绝热分离器型循环流化床锅炉、美国福斯特惠勒(Foster Wheeler)公司的水冷分离器型循环流化床锅炉、德国鲁奇公司的外置式换热器型循环流化床锅炉、美国巴特利(Battelle)的多固体循环流化床锅炉及德国 Babcock 公司的 Circofluid 循环流化床锅炉等。

国内循环流化床燃烧技术的研究和开发是从80年代开始，经大专院校、科研单位、锅炉制造厂家和用户共同努力，开发了不同特色的循环流化床锅炉。为满足严格的环保法规和提高能源利用效率，循环流化床锅炉不断向高参数、大容量发展。浙江大学先后针对烟煤、石煤、煤泥、生物质、垃圾等不同燃料特性，开发了各具特色的循环流化床燃烧技术，包括洗煤泥异重床结团循环流化床燃烧技术、煤泥煤矸石混烧下排气循环流化床锅炉、生物质中温燃烧循环流化床锅炉等，先后在300多个电厂成功应用。

3. 循环流化床的流体动力特性

循环流化床气-固两相流体动力特性是 CFB 锅炉性能设计、炉内传热研究及锅炉运行调试的基础。循坏流化床的流体动力特性不仅取决于流化风速、固体颗粒循环流率、气固物性，而且受设备的结构尺寸，包括床径、床高、进出口结构以及运行参数（如温度、压力）的影响。

通常将床层从固定状态转变到流化状态（或称沸腾状态）时按布风板面积计算的空气流速称为临界流化速度即所谓的最小流化速度。对于工业应用的燃煤流化床锅炉，其正常运行的流化速度均要大于临界流化速度。浙江大学基于大量工业实验，提出针对宽筛分床料的临界流化风速计算公式，可广泛用于工业设计。

快速流化床被认为介于湍流流化床和气力输送状态之间的一个流型，在典型的快速流化床中可观察到不均匀的颗粒絮状物在非常稀相的上升气固流中随机地作上行或下行运动。快速流化床具有高的气固相对速度、颗粒絮状物的形成与解体、极好的混合条件等主要特征。另一个明显的特征是快速流化床中悬浮颗粒浓度不但沿轴向（高度）而且还沿径向位置有变化。

颗粒浓度很高的两相流系统常用到空隙率的概念，影响循环流化床内平均空隙率的因素很多，如运行风速、颗粒循环流率、颗粒物性、床层高度、循环流化床进出口结构等。根据理论和试验研究可以认为轴向空隙率分布大致分为 3 种基本类型包括单调指数函数分布、S 型分布和反 C 型分布。

4. 循环流化床的传热特性

循环流化床传热规律和传热系数的研究是一个十分重要的课题。在锅炉设计中，它决定着受热面的布置、受热面的数量及结构，如果传热系数选取不当，就难以达到稳定燃烧和最佳经济效益，甚至出现受热面损坏的现象。

在循环流化床锅炉下部的密相区中，固体颗粒浓度较高，颗粒对流传热起主导作用；在炉膛上部的稀相区，床层密度较小，传热主要由辐射控制。当锅炉满负荷运行时，炉膛中悬浮物料的热量吸收由对流和辐射两因素同时控制。为了配合锅炉在不同负荷运行，通过调节运行参数，使二次燃烧区或炉膛水冷区的床层密度减小，从而辐射在炉膛中起主导作用的区域增加，此时，对于炉膛中的传热对流分量仍起到一定的作用。将负荷减至 40%以下时，从而稀相区中的对流传热系数达到最小值，其传热主要由辐射控制。

浙江大学在流化床传热方面进行大量研究，提出了循环流化床换热系数计算方法，可指导工业循环流化床锅炉设计。

5. 循环流化床燃烧特性

煤粒在流化床内的燃烧是流化床锅炉内所发生的最基本而又最为重要的过程，它涉及流动、传热、化学反应及若干相关的物理化学现象。煤粒在送入循环流化床内迅速受到高温物料及烟气的加热。首先是水分蒸发，接着是煤中的挥发分析出并燃烧，以及焦炭的燃烧，其间还伴随着发生煤粒的破碎、磨损等现象，并且大量实验表明了挥发分的析出燃烧过程与焦炭燃烧过程有一定的重叠。

挥发分析出(或称热解)过程是煤粒受到高温加热后分解并产生大量气态物质的过程，热解产物由焦油和气体组成，在进行煤的热解特性实验研究的同时，不少研究者相继开发了一些热解过程数学模型。煤中挥发分在流化床条件下具有短时大量析出特点，使对床内燃烧的组织以及后续的焦炭燃烧过程产生较大的影响，单颗焦碳的燃烧速率受到外部对流传质，氧气穿过外包灰壳的灰层扩散传质及焦炭表面的非均相化学反应联合控制，依据流化床燃烧过程煤种反应性能不同，上述三个步骤对焦炭燃烧速度的影响程度亦各不相同。

流化床中细碳粒的扬析对燃烧效率及周围环境都很不利。而燃烧器中的细碳粒来源主要有三种：①原始给煤中的细碳粒；②煤在挥发分析出阶段破碎形成的细碳粒；③在燃烧的同时由于磨损造成的细颗粒。

6. 循环流化床脱硫脱硝特性

燃煤中的硫可分为有机硫、黄铁矿硫及少量硫酸盐硫，硫分在加热时析出，与氧气反应形成 SO_2。循环流化床由于中温燃烧的特点可以采用石灰石实现高效脱硫，影响循环流化床脱硫效率的主要因素包括 Ca/S(摩尔比)、床温、脱硫剂和燃料粒度、氧浓度、分段燃烧、床内风速、物料循环倍率、SO_2 在炉内停留时间、给料方式、运行负荷等。

煤在循环流化床中燃烧产生的 NO_x 可以分为热力型、燃料型和快速型，由于中温燃烧，循环流化床锅炉 NO_x 的排放水平一般为 50～150ppm，影响 NO_x 排放因素包括床温、过量空气系数、脱硫剂、燃料性质及床料、循环倍率等。循环流化床锅炉降低 NO_x 的措施包括：①低过量空气系数；②分级配风；③SCR；④SNCR；⑤烟气后燃等。

7. 循环流化床气固分离机构

循环流化床的分离机构是循环流化床中关键部件，其主要作用是将大量高温固体物料从气流中分离下来，送回燃烧室。以维持燃烧室的快速流态化状态，保证燃料和脱硫剂多次循环，反复燃烧和反应，这样才有可能达到理想的燃烧效率和脱硫效率。

循环流化床分离装置的种类很多，新的型式还在不断出现，但大致可分为两大类：①高温旋风分离器；②惯性分离器。

高温旋风分离器结构简单，分离效率高，广泛用于循环流化床锅炉中，典型结构有：① 耐火材料制成的高温旋风分离器。②水冷、汽冷高温旋内分离器。相对于旋风分离器，惯性分离器结构简单，易与整个锅炉设计相适应，种类也较多，但该类分离器分离效率普遍较低，影响燃烧效率和造成受热面磨损。

浙江大学开发的下排气旋风分离器具有阻力低、布置方便的优点，广泛用于各种工业循环流化床锅炉。

8. 固体物料回送装置

固体颗粒的回送装置的基本任务是将分离器分离的高温固体颗粒稳定地送回压力较高的燃烧室内，并且保证气体反窜进入分离器的量为最小。为满足上述的回送装置的基本要求，返料机构一般由立管和阀两部分组成，一般的流量控制装置可分为机械阀和非机械阀两大类，除了极少数炉型曾报道采用机械阀外，循环流化床锅炉中几乎全部采用非机械阀。

非机械阀可以依其功能分为三大类，第一类为可控型阀，主要型式包括L阀、J阀、H阀、V阀、换向密封阀等形式。这种阀不但可以将颗粒输送到主床，可以开启和关闭固体颗粒流量，而且可以控制和调节固体颗粒的流量。第二类称之为通流阀，主要型式包括密闭输送阀(sealpot)、流动密封阀(loopseal)、N阀、多点送风L阀等，这种阀主要是经过一压力的屏障，将固体颗粒从低压处送到高压处，而对固体颗粒的流量的调节作用很小，阀和立管依据自身的压力平衡关系自动地平衡固体颗粒的流量。浙江大学在大量实验基础上，提出了L阀设计方法。

9. 循环流化床关键部件

布风板作为重要的布风装置在流化床锅炉中的作用为支承静止的燃料层和均匀布风，对通过布风板的气流以一定的阻力，使在布风板上取得均匀的气流速度分布，维持沸腾床层的稳定，抑制沸腾床层的不稳定性。目前流化床锅炉采用的布风装置主要有两种型式，即风帽型和密孔板型。风帽型布风装置是由风室、花板、风帽和隔热层组成，通常把花板和风帽合称为布风板。密孔板型布风装置是由风室和密孔板构成。而在我国流化床锅炉中使用最广泛的是风帽式布风板。

给料装置作用是将经破碎后的煤和脱硫剂送入流化床，通常包括皮带输送、链板输送、埋刮板输送、气力输送以及盘式给料机和螺旋给料机(又称绞笼)等设备。

循环流化床锅炉点火启动就是将床料加热至运行所需的最低温度以上，以便实现投煤后能稳定运行。目前，工业上点火一般采用油气点火启动方式，包括床上点火和床下点火方式。

10. 循环流化床磨损

在循环流化床锅炉中，受热面的磨损与流经其表面的固体物料运动形式密切相关，因此要了解炉内受热面的磨损情况，不仅要分析流经受热面的固体物料的局部运动形式，也要分析循环流化床锅炉内物料总体循环形式。

炉内浓相区水冷壁与耐火材料交接处管壁的磨损是循环流化床锅炉有关材料方面问题之一，一种较新的方法是改变水冷壁管的几何形状，耐火材料结合简易弯管使卫燃带区域与上部水冷壁管保持平直，这样固体物料沿壁面平直下流。

对流受热面可采用以下防磨措施：①提高气固分离装置的分离效率，或在炉内装飞灰除尘器，这样可降低烟气中的飞灰浓度从而减轻对流受热面的磨损；②设计时应选择合理的烟速；③降低速度场和飞灰浓度场的不均匀性，如在烟道转弯处加装导向板等，以防止局部严重磨损；④受热面管束尽量采用顺列布置；⑤防止烟气走廊引起磨损；⑥尽可能采用上行烟气流动结构；⑦采用膜式省煤器或鳍片式受热面；⑧管束前加假管；⑨局部易磨处采用厚壁管；⑩采用管壁表面处理技术，如喷涂、渗氮等；⑪防止磨损和腐蚀同时发生。

11. 循环流化床锅炉的总体设计

循环流化床锅炉设计不仅仅是需要达到满足用户的设计要求，而且要求达到高效、低污染、锅炉制造费用较低、运行可靠、不需经常维修、运行费用较低等。

循环流化床锅炉设计首先要选择炉型，由于循环流化床锅炉的炉型种类很多，按不同的部件可以有不同的分类方法。循环流化床锅炉一般由炉膛、分离装置、回送装置、尾部受热面及外置式受热面等主要部件构成，其主要的区别在于分离器的位置、分离器的型式和外置式受热面等。

循环流化床锅炉的炉膛设计主要包括：①炉膛的结构设计，包括炉膛的截面尺寸、炉膛高度等；②炉膛内受热面的布置；③炉膛内各开孔的结构及位置；④循环流化床的布风装置等。在循环流化床锅炉设计中，水循环一般问题不太大，与常规的锅炉水循环相近。

对于采用高温分离型的循环流化床锅炉，其过热器的设计完全可以参阅常规的煤粉炉中过热器的设计方法，但对于中温分离或组合分离型的循环流化床锅炉，其过热器或再热器的设计有其特殊之处，其原因主要是由于过热器或再热器区域的固体颗粒浓度会很高，则带来的另一个问题是受热面的磨损，需要考虑受热面防磨的具体措施。

12. 循环流化床的启动与运行

循环床的点火启动一般通过床上和床下油气点火枪点火加热，使床层温度提高到并保持在投煤运行所需的最低水平上，从而实现投煤后的正常稳定运行。

影响循环流化床锅炉运行的因素众多，主要包括煤种特性、过量空气系数、风速、循环倍率、床温、负荷等。

煤种变化对循环锅炉运行的影响很大，煤种的变化主要指发热量和灰分、挥发分的变化，首先，当燃料发热量改变时，床内热平衡的改变将影响床温，这不仅会影响燃烧、传热和负荷，也会影响排放量。与此同时，煤中灰份还影响飞灰浓度，从而影响分离效率和受热面的磨损。此外，挥发份高的煤比挥发份低的煤更有利于达到很高的燃烧效率。对一定的运行风速，给料及床料粒度决定了颗粒在床内的行为。燃烧和脱硫效率都受粒度影响。

在一定范围内，提高过量空气系数可改善燃烧效率；采用分段燃烧的主要目的被认为是降低气体污染物排放。

流化风速是循环床运行的控制变量之一，它影响床内颗粒流动、燃烧和传热等。在考虑床层换热时，风速对传热系数的影响不是决定性的。

循环床燃烧技术对鼓泡床的优势之一是固体物料循环延长了细颗粒的停留时间，改善了燃烧效率，但提高循环倍率的同时增加了风机电耗，这意味着，锅炉系统存在一个能量最优倍率。

床温是运行中的控制变量之一。循环流化床的运行温度一般在 800~1000℃范围内，这样就能在保证很高的燃烧效率的同时，降低烟气污染物的排放。

本文摘自《循环流化床锅炉理论、设计与运行》，中国电力出版社，ISBN：978-7-801-25458-0，1998。

气固分离理论及技术

气固分离技术广泛应用于能源、环保、化工、建材、轻工、冶金等与国民经济密切相关的领域。由于工业流程的需要及科学技术的发展，先后出现了为数众多的气固分离技术，从常温到高温，从低气固浓度到高气固浓度，从单级到多级，从单用途到多用途。本书的目的是结合作者从事这一领域的研究结果，对现有的气固分离技术加以分类、总结，并遵循理论密切结合工程实践的原则，对典型气固分离技术的理论基础、计算设计原理及结构应用特点紧密地有机联系，全书共十七章，系统地阐述了气固分离的理论基础、分离机理、设计计算方法、各种分离技术及装置的原理、结构、运行影响参数，力图使之既有学术价值及相应的理论水平，又能指导工程实际应用。

第1章重点分析了影响气固分离的颗粒特性。分离器的设计、运行效果与所处理的固体颗粒及气体的性质有关，对气固分离效果影响最大的颗粒特性为颗粒尺寸。因此重点详述了颗粒当量粒径、颗粒群的平均粒径、颗粒粒径分布的表达方式及颗粒形状系数的确定等；同时分析了与气固分离过程有关的颗粒物理特性，主要有固体颗粒浓度、气体密度、颗粒密度和气固多相流密度、比表面积、气固多相流的黏度、气固多相流的比热和导热系数、颗粒的润湿性、固体颗粒的燃烧性及爆炸性、颗粒的荷电性和比电阻、颗粒沉降的安息角与滑动角及颗粒的凝聚性等。

第2章综述了气固分离方式及其机理。主要涉及重力沉降、惯性力分离、拦截分离、荷电分离、热泳力、扩散泳力等的分离机理，以及利用颗粒的凝聚特性、润湿特性等来提高或进行分离的机理。任何一种气固分离装置都是利用一种或数种气固分离的物理过程来达到分离气固混合物中固体颗粒的目的。如沉降室是利用固体颗粒的重力沉降机理达到气固分离的目的，电除尘器是利用固体颗粒的荷电分离机理来实现气固分离。常见的分离机理有重力沉降、惯性力分离、拦截分离、荷电分离等，此外还有热泳力、扩散泳力、磁场力和辐射力等作用与气固分离，人们还常常利用颗粒的凝聚特性、润湿特性等来提高或达到分离的目的。有许多参数会影响固体颗粒的分离特性，如上述不同分离机理所涉及的分离力、气固混合物的物理性质、流体动力特性、固体颗粒的粒径和分布以及固体颗粒密度等。目前，从理论上还不能导出计算气固分离效率的关系式，一般仅能以相似理论为基础建立起各种颗粒分离效率的无因次参数表达式。

第3章简要介绍了气固分离特性参数的测定和研究方法。气固分离装置的分离特性评价涉及的基本参数有气固两相流中颗粒尺寸、颗粒的速度及颗粒浓度。这些参数对各种分离装置的分离性能有很大的影响，而分离(或除尘)效率及分离器(或除尘器)的阻力则是评价分离性能的直观指标，另外诸如粉尘的比电阻测定则是静电除尘器设计的重要条件。

第4章阐述了气固两相流动边界层的流动规律及对气固分离的影响。气固两相流边界层对气固分离的影响主要基于边界层和颗粒—湍流相互作用两个中心概念。文中分析讨论了气固两相流中颗粒所受的各种力的作用，考虑半无限平板层流边界层和管内流动中颗粒相的浓度分布和沉降规律，继而考察湍流边界层内的流动规律以及颗粒-湍流相互作用，最后运用上述理论框架，研究旋风分离器内颗粒的运动轨迹及 Saffman 力、湍流边界层厚度、湍流强度等因素对分离效率的影响。

第 5 章分析了颗粒在管内流动及其沉降过程。颗粒在管内的流动和输送占有非常大的比重，如化工、机械、动力食品和医药等工程中均有广泛的应用。固体颗粒在管内随携带气流流动，会发生分离、沉降。阐述气固两相流在管内流动、分离和沉降的特性有助于了解气固两相流进入分离器前在管内的流动情形。

第 6 章介绍了重力分离理论及其分离器。在重力分离器中，气固两相流在流入一个较大空间后气速将降低，受重力的作用会发生颗粒沉降现象，利用颗粒的这一重力沉降作用，实现气固分离。本章重点阐述了重力沉降的基本理论、颗粒流动及沉降分离特性计算，介绍各种沉降室的结构形式和设计方法。

第 7 章主要介绍了转折式气固分离理论和技术。根据气流及颗粒的运动形式，惯性分离器可分为转折式和绕流撞击式两大类型。本章分析讨论了转折式惯性分离器的分离机理，从结构和分离特性等方面具体介绍 U 形惯性和百叶窗式惯性分离器。

第 8 章主要阐述绕流撞击式气固分离理论和除尘技术。绕流撞击式分离器是惯性分离器的另一种类型，此种分离器由于在分离空间内设置一些特殊形状阻挡件使气流多次急速转向，从而充分利用了气固两相的惯性差别，因此可获得足够高的分离效率，在许多场合得到了应用。本章还介绍典型撞击式惯性分离器如 U 形槽钢、鳍片管束及槽形半圆管等分离器的流动和分离特性，并对不同分离器的性能作了初步比较。

第 9 章介绍了气固旋风分离理论和除尘技术。旋风分离器是利用含尘气体旋转时所产生的离心力将粉尘从气流中分离出的一种干式气-固分离装置。旋风分离器用于工业生产以来，已有百余年历史。对于捕集、分离 5~10μm 以上的粉尘效率较高，被广泛地应用于能源、化工、石油、冶金、建筑、矿山、机械、轻纺等工业部门。本章介绍了旋风分离器的基本理论包括各种气固分离模型、旋风分离器内气相流动特性和颗粒运动规律，在此基础上对影响旋风分离器性能的各种因素和结构措施进行分析，并介绍各种典型旋风分离器和旋风分离器基本设计方法。

第 10 章重点介绍了气固高温气固分离及装置。本章概述了高温条件下气固旋风分离特点和影响高温气固分离的各种因素，介绍了常规高温上排气、下排气和方型上排气、下排气以及卧式高温旋风分离器等结构特征及其设计方法，同时也介绍了高温旋风气固内分离装置。

第 11 章分析讨论了袋式除尘理论及装置。重点阐述布袋式除尘器工作原理，性能和分类，随后对袋式除尘基本理论进行阐述，包括各种除尘机理和计算，在此基础上本章对袋式除尘器典型结构、滤袋材质要求、袋式除尘器设计和运行做一简要分析。

第 12 章介绍了颗粒层过滤式分离器。本章讨论了颗粒层除尘器的除尘特点，重点围绕颗粒层过来器的结构型式、颗粒层分离器的设计和颗粒层除尘器的运行特性进行阐述，同时还介绍了颗粒床高温燃气过滤器。

第 13 章主要介绍了高温陶瓷过滤器。本章首先讨论燃煤联合循环发电系统对高温燃气除尘的要求，接着介绍了陶瓷过滤器的工作原理，其中重点介绍棒式、管式、交叉流式陶瓷过滤元件，然后对影响陶瓷过滤器性能的因素进行分析讨论。最后结合实际应用，对三种高温陶瓷过滤器进行介绍。

第 14 章主要分析讨论了电除尘器。本章主要介绍电除尘器的主要特点，电除尘器的基本类型，着重叙述电除尘器的基本原理，分析各种因素如粉尘比电阻、温度、湿度、粉尘特性、烟气流速、烟气温度、烟气调质、使用的电压、电流等对电除尘器效率的影响，进而介绍了电除尘器的基本结构。

第 15 章主要分析了气固湿法捕集原理和装置。本章从原理出发对旋风水膜除尘器、管式水膜除尘器、填料塔湿式除尘器、文丘里除尘器、自激喷雾式除尘器、喷雾接触式除尘器、湍球塔湿法捕集装置和鼓泡接触型除尘器等常见装置的除尘机理和结构性能进行了详细的介绍，介绍了几种提高湿法捕集装置效率的新方法及其相关装置。

第 16 章主要分析讨论了复合式多级气固分离(除尘)技术。本章首先介绍了复合多级分离(除尘)装置的分类，介绍了复合多级分离(除尘)效率的计算方法，重点讨论了重力沉降室与旋风分离器组合、旋风分离器和袋式(含静电、湿法)除尘器组合、多级旋风分离器组合等复合除尘技术特征和优化建议。

第 17 章主要讨论了除尘脱硫结合的技术及装置。本章概述了除尘脱硫装置的基本特点及分类，讨论了除尘脱硫装置中气态污染物控制机理，重点介绍了湿法洗涤、活性炭干法、旋风式、高压脉冲电晕和布袋式等除尘脱硫装置等基本结构和系统优化建议，同时也讨论了脱硫对电除尘器性能的影响。

本文摘自《气固分离理论及技术》，浙江大学出版社，ISBN:7-308-02095-9，1999。

大型电站锅炉安全及优化运行技术

我国是产煤大国，煤炭资源非常丰富，但煤种特性变化范围很宽。根据我国的能源政策，火力发电厂大多以煤为主要燃料，而且根据可持续发展战略的要求，为获得高的能量转换效率和低的污染物排放率，大型燃煤发电机组日益成为我国火力发电厂的主力机组。但由于现有供煤和配煤系统存在许多不完善之处，电站锅炉燃用煤质难以得到保证，机组容量和参数的提高也对锅炉设计、调试、运行操作和维护提出了新的要求。大型电站锅炉的安全和优化运行技术成为提高机组利用率、安全性、经济性和环保性能的关键技术。影响大型电站锅炉经济、安全运行的因素错综复杂，涉及承压部件的应力分析和寿命问题，炉内燃烧过程的优化调整，炉内的积灰、结渣、腐蚀、磨损过程，炉内烟气温度、烟速偏差、汽温偏差，大型电站锅炉参与调峰所带来的新问题，制粉系统及其他辅助系统的安全经济运行技术，煤质的掺配技术及低污染运行技术等。国内外对大型电站锅炉经济、安全运行技术进行了大量的理论研究和实践探索，取得了丰富的研究成果和实践经验，近年来电力工业发展迅速，大容量机组纷纷投产，大型电站锅炉经济、安全运行技术变得尤为需要，包括了大型电站锅炉启动和停炉优化运行，四角切向燃烧锅炉和带旋流燃烧器的电站锅炉优化运行和调整技术，过热器、再热器的调温方式及防止汽温偏差的方法，锅炉受热面积灰、结渣及其防止措施，锅炉受热面磨损机理及防磨措施，锅炉受热面的高低温腐蚀及预防措施，制粉系统的优化运行，优化配煤及混煤燃烧技术，锅炉变负荷及调峰运行技术，锅炉安全经济运行技术、事故诊断技术和锅炉低污染运行技术。

锅炉启动和停炉过程是一个不稳定的变化过程。锅炉的工况变化很复杂，存在各种矛盾，如各部分的工作压力和温度随时在变化、启动时间的长短与启动费用的问题、启动时冷炉的燃烧稳定性、受热面内部的工质流动可靠性、热量回收等，即启动过程中的安全性和经济性两大问题。

启动过程中，由于各受热部件的加热不可能完全均匀，金属部件内存在温度差异，会产生热应力，特别是厚壁部件，要重视其温度场的不均匀性，以免产生过大的热应力而使部件损坏。启动过程中，各受热面内部的工质流动尚不正常，易引起局部超温，如水循环尚未正常时的水冷壁、未通汽或汽量很少时的再热器、断续进水的省煤器等，都存在管壁超温破坏的可能性。启动初期，炉膛温度较低，在点火后的一段时间内，燃料投入量少，燃烧不容易控制，容易出现燃烧不完全不稳定、炉膛热负荷不均匀等较突出的问题。启动过程中，锅炉利用热量一部分被工质吸收和加热各部件，另一部分则由排汽、放水带走。如何最大限度地利用热量，减少热损失，也是启动过程中的重要问题。

为优化燃烧经济性，需考虑以下几方面：①提升煤粉的燃尽程度以减少飞灰和炉渣的含碳量，从而提高锅炉效率。具体措施依煤种和炉内条件调整，如针对烟煤改善空气混合，对劣质煤则提升炉温和煤粉细度。②最佳的过量空气系数能平衡燃烧损失和排烟损失，通常通过试验找出，大型锅炉控制在 1.1～1.2 范围内。③煤粉细度的优化，通过筛选确定，需平衡燃尽程度和粉磨电耗，以找到经济性最佳点。④锅炉煤种和负荷适应性的提高是关键，因应能源政策和负荷需求变化，通过技术创新保持锅炉运行的经济性和安全性。

在提高锅炉安全性方面，重要的措施包括三个：①控制炉内结渣和积灰：炉内积累的结渣和积灰不仅影响锅炉效率，还可能导致设备损坏和安全事故。为了减少这些影响，需要采取包括优化燃烧技术、合理调整负荷和改善煤质等措施。②受热面磨损和腐蚀的控制：锅炉的受热面由于长期暴露于高温和腐蚀性环境下，容易发生磨损和腐蚀。应通过选用更耐磨和耐腐蚀的材料、优化设计和合理的运行维护策略来减轻这些问题。③四管爆漏的控制：为了减少和预防水冷壁、过热器、再热器和省煤器等关键部件的爆漏事故，需要深入理解爆漏的原因，并根据这些原因制定相应的预防措施和应对策略。定期的检查、预防性维护和对操作人员的培训是减少这类事故的关键。

旋流燃烧器利用强烈的旋转气流产生高温回流区，达到强化燃料着火和燃烧的目的。旋流燃烧器在锅炉中得到了较普遍的应用，相对于四角切圆燃烧方式，采用旋流燃烧器的锅炉主要有如下特点：①可以减轻四角切圆直流燃烧方式产生的炉膛出口扭转残余导致过热器区热偏差的现象；②燃烧器均匀布置于炉内，入炉热量比较均匀，可避免炉中部因温度过高而引起的结渣等现象；③各燃烧器单独组织燃烧，可通过调整旋流燃烧器的旋转强度，达到调节回流区大小的目的，相互间影响比较小；④对炉形状不如四角切圆直流燃烧方式要求严格，不必一定接近正方形，有利于尾部受热面的方便布置；⑤当锅炉容量增加时，单只燃烧器的功率不必相应增大，只需相应增加炉膛宽度和燃烧器个数。旋流燃烧器由旋流器产生燃烧器出口气流的旋转运动，根据旋流器的不同，可将旋流燃烧器分为蜗壳式、轴向叶片式和切向叶片式三大类。

过热器是将饱和蒸汽加热到额定过热温度的锅炉受热面部件，再热器则是将汽轮机高压缸（或中压缸）排汽重新加热到额定再热温度的锅炉受热面部件。设计锅炉的受热面时，规定了锅炉的燃料特性、给水温度、过剩空气系数和各种热损失等额定参数，但实际运行时由于有各种扰动，不能获得设计预定的工况，锅炉的蒸汽参数将发生变化。如果扰动是由锅炉设备本身的工作条件变化所引起，如受热面积灰、结渣、烟道漏风等因素，则称之为内扰；如果扰动是由锅炉外部的条件所引起，如用户对锅炉负荷需要的变化随时间而变化，则称之为外扰。锅炉汽温稳定是衡量锅炉运行质量的一个重要指标。汽温过高会引起锅炉和汽轮机金属材料的超温过热，加速管子金属的氧化，降低材料的使用寿命；而汽温过低会降低热力循环的效率，同时使汽轮机末级叶片处的蒸汽湿度增加，对叶片侵蚀作用增加，严重时甚至发生水冲击，影响汽轮机安全运行；再热汽温变化过大还会使汽轮机中压缸转子和汽缸之间的膨胀差变化，造成汽轮机剧烈振动。

炉内积灰、结渣会使炉内传热恶化，炉内辐射传热量减少，炉膛出口温度升高，对流受热面区域热负荷增加。对某台污染程度不同的 600MW 锅炉水冷壁进行的计算表明，沾污系数从 0.45 变为 0.25 时，炉膛出口烟温相差 65℃，且污染程度越严重，炉膛出口烟温越高。因此，当炉膛烟气流经大屏、再热器、过热器、省煤器、空气预热器时，各级的进口温度都将上升，越靠近炉膛，上升幅度越大。计算表明，大屏进口烟温上升

145℃，末级再热器进口烟温上升80℃，省煤器进口烟温上升12℃，空气预热器进口烟温上升5℃。可见积灰、结渣对高温受热面的影响较大。

含有硬颗粒的流体相对于固体运动，使固体表面产生的磨损称为冲蚀(或冲击损)。冲蚀有两种基本类型，一类叫冲刷磨损，另一类叫撞击磨损。这两类磨损的金属流失过程的微观形貌是不完全相同的。冲刷磨损时颗粒相对于固体表面冲击角较小，甚至接近于平行颗粒垂直于固体表面的分速度使它楔入被冲击的物体，而颗粒与固体表面相切的分速度使它沿物体表面滑动，两个分速度合成的效果即起一种刨削的作用。如果被冲击的物体经不起这种作用，即被切削掉一块。如此经过大量、反复的作用，固体表面就将产生磨损。而击磨损是指颗粒相对于固体表面冲击角度较大，或接近于垂直时，以一定的运动速度撞击后体表面使其产生微小的塑性变形或显微裂纹，在长期、大量的颗粒撞击下，逐渐使塑性变形层整片脱落而形成了磨损。一般在锅炉受热面的磨损中，煤灰颗粒与受热面的冲击角度在0°～90°，因此锅炉受热面的磨损是以上两类磨损基本类型的综合结果。

在实际锅炉运行过程中，煤燃烧产生污染物的同时对金属表面进行腐蚀。锅炉中受热面管常覆盖有附着层，经氧化作用分为金属基体、氧化层、浸润性内附着层和外附着层。浸润性内附着层主要由碱性硫酸盐组成的初始积灰层，外附着层主要由飞灰沉积造成。腐蚀是一个持续过程，取决于附着层的物理化学性质。根据致腐物质的迁移方式和腐蚀产物的脱离方式不同，腐蚀反应相态大致可分为纯气体腐蚀、熔盐腐蚀、在固相附着物参与作用下的气体腐蚀和在致腐气体作用下的熔盐腐蚀。

钢球磨煤机是在卧式旋转的钢制筒体内，利用钢球具有的能量通过撞击、挤压、研磨等作用将煤磨制成煤粉的机械设备。尽管钢球磨煤机有金属磨损量大、制粉电耗高、占用空间大和运行噪声大等缺点，但由于其具有煤种适应性广，运行安全可靠，维修方便，可以达到其他类型磨煤机难以达到的煤粉细度等优点，被国内许多燃用无烟煤、贫煤和劣质烟煤的电站所采用。对于煤种多变和煤质趋劣的电厂，选择钢球煤机是比较合适的。在我国燃煤电厂中，钢球磨煤机占到各类磨煤机总量的60%以上。钢球磨煤机有两大类：单进单出钢球磨煤机和双进双出钢球煤机，前者可简称为钢球磨煤机，后者简称为双式钢球磨煤机。

煤的着火和燃烧特性可以反映煤的燃烧化学反应能力，影响煤粉气流的着火难易程度并在很大程度上影响了煤粉在炉内燃烧的完全程度，即决定了飞灰可燃物含量的高低。一般而言，地质年代久远、碳化程度高的煤，化学反应能力较弱，着火和燃烧困难。反映煤的着火和燃烧特性的指标很多，可以分为煤的常规特性指标和非常规的实验室指标两大类，前者包括煤的挥发分、水分、灰分、发热量、灰熔融特性等，后者主要包括反应指数、可燃性指数、熄火温度等。

在配煤技术上，主要有两大类：一类是在电厂内部直接混合，另一类是建立配煤场。内部混合主要有仓混式、库混式、带混式或炉内直接混合等形式，只是将两种煤按发热量和挥发分要求进行一定比例的混合，虽然能满足一定的发热量、挥发分和灰分要求，但在燃烧效率、污染物排放和安全可靠性方面还存在一定差距。另一种方式是建立配煤场，这种方式目前大多还停留在对煤种进行机械混合，需要进行更深入的研究以提高配煤质量。动力配煤研究的目的就是要配出满足用户需求的最佳综合性能，包括混合系统和混合方法的研究，以及混煤着火燃烧性能、燃尽性能、结渣性能、污染物排放性能等方面的研究。

锅炉系统复杂，具有多个调节参数和调节作用，并且受多种扰动因素的影响。锅炉具有适应外界负荷变化的能力，其参数变化应在安全范围内。锅炉的运行特性包括静态特性和动态特性，各参数应处于设计的经济运行范围内。过热汽温随锅炉负荷变化，属于锅炉的静态特性。锅炉构造复杂，对各种扰动更加敏感，对运行调节系统的要求也更高。除了静态特性外，还应了解锅炉的动态特性，各工作参数随时间的变化规律。锅炉的蒸汽负荷不可能长期保持不变，应掌握负荷变动时锅炉各种参数的变化规律。单元机组的运行方式有定压运行和滑压运行两种，应充分掌握负荷变动时锅炉各种参数的变化规律。

可靠性管理是对表征设备使用可靠性的若干指标进行统计、分析和评价，以指导对设备的选择、运行、维修、改造直至报废的一种全过程管理。设备的可靠性是指一个部件或一个系统在规定的时间内和一定条件下完成预定功能的能力。对于发电设备来说，其可靠性是以统计时间为基准的表示机组所处状态的各种性能指标来表征的。

本文摘自《大型电站锅炉安全及优化运行技术》，中国电力出版社，ISBN：9787508310060，2003。

高等燃烧学

燃烧学是一门综合了物理、化学、力学、传热传质学和热力学等科学理论的综合交叉的一门科学。近年来，能源与环境的需求极大地促进和推动了燃烧科学的发展，《高等燃烧学》正是在这种形势下撰写和出版的。作为国家"十五"的重点图书和国家科学技术学术著作出版基金资助项目，本书系统介绍了燃烧理念的基本知识和最新研究成果，同时也系统地总结了浙江大学工程热物理学科近年来在煤与生物质的循环流化床燃烧、煤浆燃烧理论与技术、煤粉燃烧理论与技术、燃烧过程数值计算、煤的催化燃烧理论与应用和非线性理论研究燃烧过程等方面的研究成果和研究心得。上述内容汇集了作者大量承担的国家"六五"至"九五"攻关项目、国家自然科学基金、国家攀登计划等项目最新的具有国际先进水平的研究成果。

本书共计十四章，从燃烧化学动力学基础知识和反应机理出发，介绍了燃料着火和燃烧的基本理论和最新研究成果，尤其重点阐述了煤的热解、着火和燃烧等理论，充分体现了最新的研究和发展成果；本书还针对目前关注的洁净燃烧技术问题，对 SO_2、NO_x 和炭黑的形成机理进行了系统介绍；最后全书还介绍了催化燃烧和燃烧过程非线性现象的最新研究成果。

第 1 章首先介绍了燃烧科学的发展简史、燃烧科学的应用和燃烧科学的研究方法，然后主要介绍化学动力学的基础知识，包括各种化学反应速度及其各种因素(浓度、温度等)对反应速度的影响的确定；各种化学反应机理，即研究从反应物过渡到生成物所经历的途径。

第 2 章为燃料的着火理论，探讨了燃料混合物从初始状态转变为持续燃烧状态的复杂机理。主要介绍了经典的热力着火理论、点火理论和强迫着火理论，也引入了近年来对热力爆燃理论的最新发展的介绍。燃烧过程通常包含三个阶段——着火、燃烧和熄火，其中着火作为初始的过渡阶段，是混合物从非活性状态过渡到自维持放热反应的关键步骤。着火受多种因素影响，包括燃料性质(如化学组成、分子结构)、混合物浓度、压力、温度、湍流程度以及燃烧室几何形状等，这些因素共同决定了着火的难易程度和着火特性。

第 3 章为火焰传播与稳定理论，探讨了火焰传播的基础原理、影响因素及火焰稳定性的维持机制，为理解复杂燃烧现象及优化燃烧系统设计提供理论基础。火焰传播可分正常火焰传播、爆燃及爆震等不同传播模式，本章讨论了火焰传播速度与燃烧反应动力学、热传递及流动条件间的关系，解析了层流预混火焰传播的化学反应控制机理和热力学条件，以及湍流对火焰传播速度的增强效应，包括湍流混合对反应区结构的影响和湍流火焰的不稳定性问题。火影响火焰正常传播速度的主要因素包括过量空气系数、燃料特性、添加剂、混合可燃物初始温度、火焰温度、压力、惰性物质含量以及热扩散系数和比热等因素，实际测量可采用圆柱管法、定容球法、肥皂泡法(定压法)、粒子示踪法、平面火焰燃烧器法等火焰正常传播速度的测量方法。本章还讨论了火焰稳定特征、回火和吹熄的临界条件，提出了通过钝化回流区的 L/d 衡量是否会出现脱火的现象以及多种稳定火焰的方法。

第 4 章为湍流燃烧理论及模型，主要介绍湍流气流中火焰传播的表面燃烧模型、容积燃烧模型求解湍流火焰传播速度。湍流扩散火焰 $k\text{-}\varepsilon\text{-}g$ 模型探讨了利用几率分布函数方法来描述湍流场中化学组分的浓度分布，模拟湍流燃烧中的不确定性。湍流预混火焰模型描述了包括旋涡破碎模型、拉切滑模型、概率密度函数的输运方程模型分析在湍流流动条件下预混气体的燃烧过程。同时介绍了 Spalding 的 ESCIMO 湍流燃烧理论，包括"经历""统计""综合"三大部分，获得了活化能越低、点火越容易，体现为点火距离较短；拉伸速率越大，点火越困难等。

第 5 章为液体燃料的燃烧，介绍经典的油滴蒸发燃烧理论、液雾燃烧理论和近年发展起来的浆体燃烧的研究成果。阐述了液体燃料的物理和化学性质(如沸点、燃点、蒸发性、黏度等)对燃烧过程的影响。单颗液滴在静止气流中的燃烧是液体燃烧的基础，主要包含液滴在静止气流中的蒸发、扩散及燃烧。首先介绍了油滴蒸发的基础理论，包括蒸发的阶段划分，如初期的表面蒸发和后续的内部扩散控制阶段，以及蒸发速率受油滴大小、环境温度、饱和蒸汽压和周围气体的流动状态等因素的影响。其次详细描述了斯蒂芬流现象，燃料滴温度及蒸发浓度的确定，蒸发浓度效应对蒸发速率影响因素以及油滴蒸发过程中的传热和传质系数的确定方法。液滴在气流中的扩散主要包括折算薄膜理论，而液滴扩散燃烧的非稳态研究，表明液滴越细燃烧速度相对较快。本章进一步对雾化机理进行了分析，并对压力雾化机理和气力雾化机理进行了介绍，提出了衡量喷嘴的性能主要包括雾化角、液滴分布特性及雾化细度等指标。本章最后介绍了浆体燃料的燃烧过程，主要探讨了油/水煤浆代油燃烧技术，包括浙江大学对煤浆燃烧特性的主要研究成果。

第 6 章为煤的热解及挥发分的燃烧，主要研究在惰性气氛下和氧气气氛下的挥发分析出或称热解。煤的热解是煤燃烧的一个非常重要的过程。煤热解时产生挥发分及挥发分的燃烧对于整个煤的燃烧过程有着重要的影响，有时甚至是决定性的影响。本章主要介绍了煤的特性对热解的影响、煤的热解过程、煤热解产物的组成，以及各种因素(如温度、压力、加热速率等)对热解的影响。本章还介绍了各种热解模型及挥发分燃烧过程。研究表明，煤热解所产生的挥发分的着火是煤粉燃烧的主要着火机理，挥发分的燃烧对于煤粉火焰的稳定具有决定性的影响。加速煤燃烧时的煤热解过程能有效地提高总的燃烧效率。其次，煤的热解对于炭粒的进一步燃烧也有重要的影响。煤燃烧时所产生的污染物质主要有粉尘、氮氧化物、硫氧化物、一氧化物及各种有机碳氢化合物，而煤粉的热解过程又是这些污染物形成的最主要环节。

第 7 章为煤的着火理论，重点介绍了煤的着火理论和试验研究的发展。着火问题的核心是着火点的确定，本章首先介绍了控制颗粒试验法、自由运动法、群体爆炸法和连续流法四类着火温度的试验方法。大量试验研究表明，试验方法的不同如加热条件、环境温度控制、流动条件的变化都会对结果造成较大影响。进一步建立了煤的着火模式均相着火模型以及 Jundgen 通过实验提出的合着火方式。同时进行了煤粒的多相着火及其影响因素分析，包括辐射与对流散热、介质温度、煤种、煤粒直径、介质含氧量以及不同试验方法的影响。其次详细介绍了单颗煤粒着火的计算方法，包括煤粒非均相着火方法、大颗粒碳的着火分析、考虑挥发分燃烧的单颗煤粒的均相着火计算及单颗煤粒着火的随机模型计算等。本章还介绍了关于煤粉气流的着火特性的研究，主要包括浓度效应、颗粒群研究方法以及煤粉气流着火的数学模型方法。本章还探讨了煤粉气流着火的微分方程及求解过程，并分析了煤粉浓度、粒径、氧浓度、气流速度、掺杂等因素对煤粉气流着火的影响。浙江大学对煤粉着火建立了基于有限控制体假定的煤粉着火非稳态统一模型，既能反映颗粒的着火又反映颗粒群着火，既能研究其着火方式，又能确定着火的时间。

第8章为煤的燃烧理论(碳及煤焦的燃烧)，主要研究焦炭的非均相燃烧。半焦燃烧涉及物理化学过程主要讨论了煤焦反应的控制因素、燃烧速率、碳的形态和结构的变化过程、吸附过程、扩散过程及半焦燃烧机理。详细描述了碳的燃烧涉及的化学反应(包括碳与氧、二氧化碳、水蒸气和氢气等)的反应动力学及产物的二次反应。由于碳的反应机理既有表面反应又有空间反应，十分复杂，本章首先讨论了一些简单的仅考虑一次反应的数学模型，分别分析了温度较低或颗粒很小可略去时空间气相反应的情况下的碳球燃烧速度及碳球高温下的扩散燃烧。然后探讨了实际过程中碳燃烧的二次反应，指出由于一氧化碳在碳球表面附近燃烧，阻碍了氧气向碳球的扩散，使碳球燃烧模型起了很大变化，包括静止介质、流动介质中碳表面附近的燃烧、有一氧化碳空间反应时的碳球燃烧速率以及强迫对流条件下碳球燃烧速率。碳粒的燃烧不仅包括表面燃烧，同时对多孔性碳球内部的燃烧特性也进行了介绍。本章还系统分析了各种因素对焦炭燃烧的影响，包括煤中挥发物析出、灰分、氧浓度、总压力、气体温度和壁面温度及颗粒直径。

第9章为煤粉燃烧的数学模型，主要包括了单颗煤粒经历模型，煤燃烧过程中流动、气相反应过程模型，以及煤粉颗粒扩散及两相流模型。本章主要介绍了煤燃烧过程中气相湍流流动、颗粒加热、挥发分析出及着火、焦炭燃烧、气相湍流反应、辐射传热、颗粒的湍流扩散以及污染物形成方面的模拟研究工作。

第10章为燃烧过程中硫的反应动力学及燃烧的固硫机理，详细介绍了燃料燃烧过程中硫的析出以SO_2为主要产物，提出了SO_2析出的计算方法。当燃料处于高温热解条件时，有机硫会热解成H_2S的中间体；而对于无机硫其热解中间体则以S单质为主；燃烧区域内有氧富余时，部分SO_2还会转化为SO_3。针对不同的硫化物，本章对其生成的反应动力学进行了详细分析。石灰石是重要的脱硫剂，本章提出了石灰石煅烧反应动力学的研究，并对石灰石煅烧过程中孔隙结构变化、煅烧模型以及多种孔隙结构模型进行了详细介绍。进而详细描述了石灰石固硫机理，煅烧石灰石的硫盐化模型以及在工业应用中影响脱硫效率的主要因素。

第11章为燃烧过程中氮氧化物的生成及分解机理，详细阐述了燃烧过程中NO、NO_2的生成途径，主要包括热力型、快速型及燃料型，而N_2O则分为均相和多相生成。本章对于气体燃料、液体燃料及煤燃烧过程中各种影响因素对氮氧化物的生成及分解的机理的影响也进行了介绍，最后还介绍了在燃烧过程中降低NO、NO_2和N_2O的各种措施。降低NO_x排放的措施主要分为燃烧过程中脱硝技术包括空气分级、燃料分级、低氮燃烧、烟气再循环、浓淡偏差燃烧和非选择性催化还原法等措施，以及烟气脱硝包括选择性还原法脱硝和氧化法脱硝等。特别地对于 N_2O 的降解也提出了几种可能的方法，包括改变运行温度、低氧燃烧、再燃烧法和催化反应等措施。

第 12 章为燃烧过程中炭黑的形成机理，主要介绍了炭黑的生成机理与控制措施。燃料燃烧时会排放出炭黑，它是燃烧过程中析出的碳所产生。碳黑粒子在形成过程中会经历成核、表面增长和凝聚、集聚和氧化等一系列阶段，生成的碳黑粒子若不能在燃烧系统中完全氧化掉，则最终排入大气。在燃气轮机中，这些粒子的存在可能会严重影响透平叶片的寿命，然而在一些工业燃烧炉中，炭黑粒子的存在会加强辐射传热，因而明显地增加了换热效率。为了满足环境保护的要求，这些炭黑粒子希望在燃烧的后期与多余空气结合燃尽，这样既能提高换热效率，又不造成对环境的污染。

第 13 章为催化燃烧原理，主要介绍了催化燃烧的原理，包括催化燃烧的作用，催化脱硫脱氮，催化燃烧理论，气体、液体和煤的催化燃烧机理及催化作用等。将催化作用分为催化着火助燃、催化燃烧、催化低 NO_x 燃烧、催化脱硫和防止结渣的作用。接着介绍了燃烧催化剂的组成、性能指标及制备方法。介绍了气体燃料催化燃烧机理、燃烧动力学及应用。煤的催化燃烧原理研究大多集中在对碱金属、碱土金属、过渡金属元素催化剂的开发中。研究表明，影响煤催化着火燃烧的主要因素表现在煤种、煤中矿物质成分、压力及高灰分等。

第 14 章为非线性理论在燃烧领域中的应用，主要介绍在燃烧理论中应用较广的诸如混沌理论、分形理论、逾渗理论、小波分析和人工神经网络的应用情况。非线性科学是正在蓬勃发展中的前沿科学，燃烧理论中也出现了许多非线性的问题需要分析。

本文摘自《高等燃烧学》，浙江大学出版社，ISBN：7-308-02629-9/O·257，2002。

煤的热电气多联产技术及工程实例

煤在世界能源构成中占有重要的地位。我国是一个以煤炭为主要的能源结构的国家，而且在很长的一段时间内不会改变。实现煤炭的高效洁净利用是世界各国尤其是我国实现能源的可持续发展的关键。煤不仅仅是一种燃料资源，以煤为资源的生产工艺涉及各工业部门，如火力发电、动力、化工、冶金、建材及民用煤气生产等领域，同时煤炭是一种具有复杂组成、结构的物质。把煤炭作为单一用途来利用，存在着煤炭利用效率低、资源浪费、工艺复杂、生产成本高及污染物控制困难等问题。以煤炭为原料，通过把多种煤炭转化技术有机集成在一起，同时获得多种高附加值的化工产品、多种洁净的二次能源及其他产品，追求整个系统的资源利用、总体生产效益最大化和污染物排放最小化的多联产技术是煤炭利用技术的主要趋势之一。在一个系统中同时生产蒸汽、电力及可以作为后续工艺原料的煤气的热电气多联产技术是煤的多联产技术的主要实现方式。

煤的多联产技术是一个非常复杂的系统工程，它不是多种煤炭转化技术的任意简单的叠加，而是以煤炭资源合理利用为前提，建立在相关技术发展水平基础之上，以煤炭资源利用价值的提高、利用过程效率、经济效益及环境污染等为综合目标函数的多个子系统的优化集成，从而实现煤炭资源的分级利用、高利用效率、高经济效益及极低污染物排放。基于这一原则，多联产的主要技术方向可以分为如下几类：①以煤热解为基础的热电气多联产技术；②以煤部分气化为基础的热电气多联产技术；③以煤完全气化为基础的热电气多联产技术。

挥发份是煤组成中最活跃的组分，通常在较低的温度下就会析出，同时挥发份也是煤中比较容易进行利用的组分。以煤热解为基础的多联产技术针对这个特点，把煤先加入热解炉内经热裂解析出挥发份，所产生热解气可以作为化工合成原料和工业用气等。焦油则可以通过焦油加氢等生产燃料油及其他产品。热解煤气和焦油也可以通过进一步的工艺从中获得苯、奈、蒽、菲以及目前尚无法人工合成的多种稠环芳香烃类化合物及杂环化合物。热解所产生的煤半焦则可以直接被送到燃烧炉中作为燃料燃烧产生蒸汽，用于发电或供热。

以热载体热解为基础的热电气多联产技术是以热载体提供煤热解所需热量生产中热值热解煤气供民用或工业用的煤气化技术，该技术集煤热解、气化、燃烧分级转化于一体，同时产生热、电、煤气和焦油，方案结构简单，不需特殊的制氧设备，而且运行费用较低。热载体热解技术与循环流化床燃烧技术相结合的热电煤气焦油多联产系统得到广泛关注，即把热载体与煤在热解气化炉中混合进行热解，获得热解气，所产生半焦则直接送到循环流化床燃烧炉中燃烧产生蒸汽用于发电或供热。从循环流化床燃烧炉中排出的灰渣则可以综合利用，如作为建材原料、提取贵重金属等。根据气化反应装置、热载体性质的不同，该技术目前主要有以流化床煤热解为基础以及移动床煤热解为基础的热电气多联产技术，以焦热载体煤热解为基础的热电气多联产技术。

以流化床热解为基础的循环流化床热电气多联产技术主要工艺特点是利用循环流化床锅炉的循环热灰或半焦作为煤热解气化热源,煤先在流化床气化炉中热解析出挥发分,经净化除尘后获得煤气焦油,气化炉中半焦及放热后的循环灰一起送入循环流化床锅炉,半焦燃烧放出热量产生过热蒸汽用于发电、供热。浙江大学以及清华大学等单位开展了相关技术的研究开发工作。

浙江大学所开发的循环流化床热电气多联产技术主要工艺流程为:循环流化床锅炉运行温度在 900℃左右,大量的高温物料被携带出炉膛,经分离机构分离后部分作为热载体进入以再循环煤气为流化介质的流化床热解炉。煤经给料机进入热解炉和作为固体热载体的高温物料混合并加热(运行温度在 550~800℃)。煤在热解炉中经热解产生的粗煤气和细灰颗粒进入热解炉分离机构,经分离后的粗煤气进入煤气净化系统进行净化。冷却后煤气除作为热解炉流化介质的再循环煤气外,所产煤气则经脱硫等净化工艺后作为净煤气供民用或经变换、合成反应生产相关化工产品。收集下来的焦油可提取高附加值产品或改性变成高品位合成油。煤在热解炉热解产生的半焦、循环物料及煤气分离器所分离下的细灰(灰和半焦)一起被送入循环流化床锅炉燃烧利用,用于加热固体热载体,同时生产的水蒸汽用于发电、供热及制冷等。浙江大学在所建的 1MW 循环流化床热解燃烧分级转化中试试验装置上开展典型烟煤的试验研究,验证了以烟煤为原料的循环流化床热电气焦油联产技术的可行性,并获得了热解温度、石灰石以及煤种特性对热解煤气产率、煤气组分及焦油产率的影响特性。

在此工作基础上,浙江大学开发了 12MW 循环流化床热电气联产工业装置,由一台 75t/h 循环流化床锅炉和 1 台流化床热解炉组成。该 12MW 循环流化床热电气联产工业装置完成了 75t/h 循环流化床燃烧炉的建设和运行,运行结果表明,锅炉可以满负荷连续稳定运行,可以在 30~83t/h 稳定运行。煤种适应广,可燃用热值在 11.3~26kJ/kg 的煤种,可以满足热电气联产过程中半焦燃烧的需要。

以移动床热解为基础的循环流化床热电多联产技术的基本原理与以流化床为基础的热电多联产技术的原理基本相同,其主要差别在于热解炉。以移动床热解为基础的循环流化床热电多联产工艺的气化室采用移动床进行热解。国内进行这方面的研究有北京动力经济研究所及中国科学院等单位。

北京动力经济研究所热电气多联产工艺是在循环流化床锅炉一侧设置一个移动床干馏器,流化床的循环灰先被送入其中,同时,将锅炉给煤的一部分送到干馏器中。这样,循环热灰将作为热载体对煤进行干馏,析出其挥发分,而煤干馏形成的半焦和循环灰最后将被回送到锅炉进行循环燃烧。该技术完成了 150kg/h 的中试研究,验证了工艺的可行性,并获得运行特性。

以焦热载体热解为基础的多联产工艺的技术核心是以煤半焦作为固体热载体,并以流态化方式按气化过程所需热量来组织物料和热量的输送。国内外已有多家单位进行了该工艺的开发,主要有前苏联、鲁奇鲁尔公司、大连理工大学、清华大学等。大连理工大学所提出的褐煤固体热载体干馏技术的工艺流程为:原料褐煤从煤斗到混合器与热半焦相混合,由于混合强化和煤粒子较细且均匀分散,所以煤与半焦之间换热迅速,加热速率很快,从而发生快速热解。煤焦混合物由混合器去反应槽,在此完成干馏反应并析出挥发产物。半焦从反应槽去提升管下部与空气部分燃烧或由热的烟气加热并流化提升，热半焦回到集合槽再去混合器，完成循环利用。干馏挥发产物从反应槽导出后，经过滤除尘器，冷却冷凝出焦油和冷凝液。煤气经过干燥脱去水分，在30℃左右条件下冷风回收煤气中轻质油。大连理工大学在平庄建成了褐煤固体热载体干馏多联产中试试验装置，验证了以褐煤及长焰煤为原料的固体热载体热解联产煤气、焦油和半焦的可行性。

以煤的部分气化为基础的热电气多联技术依据煤中不同组分和不同转化阶段的反应性不同的特点，先在气化炉内将煤炭中容易气化的组分转化成合成气，没有被气化的半焦则进入燃烧炉燃烧利用产生蒸汽以发电、供热，而产生的合成气则可以作为多用途，如燃气-蒸汽联合循环发电、燃料气、其他化工产品的生产等。经过多年的发展，目前在国外主要表现为气化燃烧集成利用技术与联合循环技术相结合的先进燃煤发电技术和多联产技术。与其他先进技术相比（IGCC 等），这类技术具有系统简单、投资小、煤种适用性广的优点。以部分气化为基础的先进燃煤发电技术主要代表有美国 Foster Wheeler 公司开发的第二代增压循环流化床联合循环（2G-PFBC 或称 APFBC）。煤在增压部分气化炉中部分气化产生低热值煤气和半焦。煤气经过旋风分离器和陶瓷过滤器除去其携带的固体颗粒，并经过一个装有酸性白土的填充床除去碱金属后，进入燃气轮机前置燃烧室燃烧产生高温烟气驱动燃气轮机。部分气化炉的主要作用是将燃料转化成低热值煤气和半焦。部分气化炉产生的半焦、CaS，以及煤气中分离出来的固体颗粒经输送系统进入 CPFBC 进一步燃烧利用。在此基础上，Foster Wheeler 公司提出了部分气化模块（PGM），气化剂可以是空气、富氧空气，也可以是氧气，使用氧气作为气化剂时为控制 PGM 的温度可以添加适量水蒸气或者采用循环煤气，所产生的合成气既可以用于燃气轮机发电，也可以用于合成液体燃料和化学产品等。部分气化装置由增压循环流化床（PCFB）、旋风分离器和除尘器组成。煤、空气或氧气和蒸汽从 PCFB 反应器的底部加入，反应生成的合成气携带部分固体颗粒进入旋风分离器，旋风分离器将煤气中大部分的固体颗粒收集，并通过下降管重新返回反应器。经过旋风分离器的合成气经后续除尘净化后获得合成气。

浙江大学、东南大学和中国科学院煤炭化学研究所在国家重点基础发展规划项目支持下分别开展了常压气化燃烧、加压气化常压燃烧和加压气化常压燃烧集成利用技术的研究开发。浙江大学开展了将常压流化床部分气化和常压半焦循环流化床燃烧耦合起来的部分气化多联产技术的研发，煤在常压流化床气化炉中以空气或氧气/蒸汽为气化剂进行部分气化，产生的煤气经过高温净化后，供燃气轮机做功发电，排气经余热锅炉回收热量，也可以作为燃料气或化工原料；气化炉排出的半焦送燃烧炉燃烧，产生的蒸汽供发电供热。浙江大学开展了煤炭部分气化特性的实验研究及理论分析，并完成了中试试验验证。中国科学院山西煤化所则开展了将煤炭流化床加压部分气化和常压半焦循环流化床燃烧耦合起来的部分气化多联产技术的研发，并完成了加压流化床煤部分气化及半焦燃烧的实验研究，通过改变操作温度、压力及进料量等参数获得了在加压流化床上达到不同碳转化率的煤部分气化数据，得出压力、温度及进料量对煤部分气化的影响规律；验证了煤部分气化、燃烧集成系统的可行性和煤分级转化集成优化思想的正确性，为进一步验证奠定了良好的基础。东南大学开展了将煤炭流化床加压部分气化和加压半焦循环流化床燃烧耦合起来的部分气化多联产技术的研发,并完成了试验验证。

以煤完全气化为核心的多联产系统是将煤在气化单元内完全转化,将固相碳燃料转化为合成气,合成气则可以用于燃料、化工原料、联合循环发电及供热制冷,从而实现以煤为主要原料,联产多种高品质产品,如电力、清洁燃料、化工产品以及为工业服务的热力。一个以煤完全气化为核心的热电气多联产系统是很多技术模块的有机耦合,从煤等资源的给入到蒸汽、电力、燃料等产品的产出,涉及煤的气化、合成气净化、污染物排放控制、合成气转化等过程。煤炭等燃料气化、煤气净化、污染物控制、燃气轮机、燃料电池、合成气转化、气体分离等技术都是其关键技术,这些子单元的性能表现对整个多联产系统都有极其重要的影响。

以煤完全气化为基础的多联产系统是煤炭利用领域国内外研究的重点与热点之一。在代表美国 21 世纪能源生产领域重要研究方向的"展望 21(Vision 21)"计划中,以煤完全气化为核心的多联产系统是其重点之一,该系统具有多种先进技术的组合、燃料适应性广、近零排放等特点,可实现煤炭高效高价值利用。整个系统由三大部分组成:气化部分采用先进气化技术将原料转化为以 H_2 和 CO 为主的合成气;化学转化部分将部分合成气转化为超清洁的交通用燃料和高品质的化工产品;联合循环系统利用剩余部分合成气以及化学转化过程中未转化的合成气生产电能。其基本流程是把煤(或以煤为主同时混合生物质、石油焦等原料)在气化炉中进行氧气/水蒸汽气化,所产生的合成气以 H_2 和 CO 为主,同时含有少量的 H_2S、NH_3、CH_4 及 CO_2 等杂质,经高温除尘后进入高温净化装置,合成气在净化装置把所含的 H_2S、NH_3 以硫酸、硫磺及氨水等副产品的模式脱除。净化后的合成气一部分用于生产多种优质化工产品和交通用燃料;一部分合成气用于燃气-蒸汽联合循环发电系统发电、供蒸汽,另一部分则用于制氢,所得氢气可供燃料电池发电,也可以供其他以氢气为资源的用途。该多联产系统采用模块化的设计思想,整个系统由多个技术模块(或称为"能源岛")构成。具体技术模块的选择和各类产品的比例并不存在一个固定的模式,而是通过严密的市场需求分析来决定的,由此可使整个系统达到最大的效率和经济性。壳牌(Shell)公司提出的合成气园(Syngas Park)的系统则是一个以 Shell 气化技术为核心的多联产系统。

国内,中国科学院工程热物理所正在积极研究 IGCC 多联产系统,并与兖州矿务局合作建设多联产示范工程。中国科学院工程热物理研究所以 IGCC 动力系统为基础,从能的梯级利用概念出发,提出了新型的 IGCC 多联产总能系统,将合成气合成化学品和燃气轮机发电耦合,从系统高度研究动力系统与各生产过程的相互连接和耦合,实现能源资源的高效、洁净和经济利用,并进行示范工程的构建与建设。

清华大学、浙江大学等研究机构也相继构建了有自己特色的多联产系统。作为一个以现有较成熟工艺为基础而集成多联产系统工程,上海焦化总厂的热电气多联供工程为多联产的推广应用作了较好的示范。

本文摘自《煤的热电气多联产技术及工程实例》,化学工业出版社,ISBN: 978-7-502-55423-1,2004。

燃烧理论与污染控制

燃烧是物质剧烈氧化而发光、发热的现象,是人们利用能源的最主要方式。一方面,能源的需求还在不断地增加,另一方面,燃料中存在的有害物质在燃烧过程中会散发出来,包括烟尘、灰粒、炭黑粒子、氮氧化物、硫氧化物和一氧化碳等,这些排放物可能是影响全球环境的酸雨等的主要因素,而燃烧排放的二氧化碳会造成温室效应,甚至影响整个生态的平衡。因此,开展对燃烧污染物形成机理的研究、探索通过改变燃烧工艺、控制燃烧过程是减少或消除污染物排放的有效方法。近年来燃烧科学与技术的发展在很大程度上是这一需求的必然结果。本书系统介绍了燃烧理念的基本知识和最新研究研究成果,同时系统总结了浙江大学在煤与生物质的燃烧、煤浆燃烧、煤粉燃烧、燃烧过程数值计算、燃烧污染物控制等方面的研究成果,包括承担的国家973项目、国家自然科学基金、国家科技攻关和国家攀登计划等项目所取得的研究成果。

《燃烧理论与污染控制》从第一版出版到第二版经历了15年,作为一门科学,燃烧科学的基本理论和框架没有明显的变化,所以本书第二版在第一版的基础上保持科学理论的完整性及稳定性,同时结合了近年来燃烧科学与污染控制技术的最新研究成果与发展动态,增加了诸如极端条件下的燃烧、燃烧的数值模拟、燃烧污染物的控制以及燃烧过程中二氧化碳的分离等,进行了内容的更新与扩充,力图反映燃烧理论特别是燃烧技术上的最新发展成果。

理论深化与技术创新方面,本书在燃烧理论部分进一步深化了对燃烧化学动力学模型、复杂燃料的燃烧特性、以及先进燃烧模式(如低温燃烧、超临界燃烧等)的探讨,强调了燃烧效率与降低污染物排放的协同优化策略;污染控制技术进展方面,鉴于环保标准的日益严格,本书新增了近年来发展的高效污染控制技术,包括二氧化碳捕集与封存等减缓气候变化的技术。

本书内容主要分为以下10章。

第1章概述了燃烧科学的发展和应用情况,介绍了化学热力学和化学反应动力学基础知识。本章首先概述了燃烧科学的历史背景,包括燃烧研究的起源、发展简史和里程碑事件,如拉瓦锡的氧化理论、谢苗诺夫的链式反应理论和火焰传播理论。同时分析了燃烧产生的环境污染问题,包括酸雨、温室效应、雾霾等,以及这些问题如何促使燃烧科学向更环保的方向发展。化学反应的热效应数据对自然科学的研究、工业生产、燃料的利用及确定设备条件都是很重要的,因此,本章详细介绍了燃料燃烧反应的热效应,包括燃烧焓和生成焓等,以量化燃烧过程中的热效应,有助于设计高效燃烧系统,减少未燃尽燃料和降低不完全燃烧产生的污染物。本章还详细介绍了化学反应的速率及影响因素,通过精细调控反应速率,可以实现高效燃烧的同时最大限度地减少对环境的负面影响。

第2章为燃料的着火理论,详细阐述了燃料从初始加热到发生持续燃烧这一复杂物理化学过程的理论基础。本章首先明确了两种使可燃混合物着火的方法：热力着火和强迫着火。针对封闭或近乎封闭系统，详细介绍了谢苗诺夫的可燃气体混合物的热力着火理论，包括热力不稳定性的原理、临界着火条件和爆燃临界条件的定量关系；链爆燃理论包括链式反应机制和链爆燃条件等。由于煤的结构、种类极其复杂，其燃烧方式也繁多，因此，本章专门对其各种着火状况的理论和试验研究的发展情况进行了研究，包括煤的着火及其判据和煤的着火模式，并详细阐述了谢苗诺夫热力着火理论用于碳粒着火的分析的三种工况：动力燃烧工况、扩散燃烧工况及过渡燃烧工况，同时提供了辐射与对流散热、介质温度等因素对煤粒着火的影响。

第 3 章为火焰传播与稳定理论，主要研究火焰如何在燃料和氧化剂混合物中传播，以及如何维持火焰的稳定燃烧状态，以确保燃烧过程的安全、高效和环保。本章首先明确了两种稳定的火焰传播形式为正常火焰传播和爆燃，详细阐述了可燃气体的火焰正常传播理论，并分析讨论了过剩空气系数、燃料化学结构和添加剂等因素对火焰正常传播速度的影响。具体介绍了层流火焰正常传播速度的测量方法可采用圆柱管法、定容球法、肥皂泡法（定压法）、粒子示踪法、平面火焰燃烧器法等。详细介绍了层流火焰传播的机理，证明了层流火焰传播速度是可燃混合物物理化学性质的反映。另外，针对燃烧技术中要保证已着火了的燃料不再熄灭问题，介绍了火焰稳定的基本原理和方法，如钝体后回流区火焰稳定原理。

第 4 章为湍流燃烧理论及模型，首先以本生灯火焰为例说明了湍流燃烧及特点，分析了如何通过湍流燃烧模型处理平均化学反应速率，并详细介绍了研究湍流火焰的两类方法及其模型。一类为经典的湍流火焰传播理论，包括皱折层流火焰的表面燃烧理论与微扩散的容积燃烧理论。另一类是以计算湍流燃烧速度为目标的湍流扩散燃烧和预混燃烧的物理模型，包括概率分布函数的输运方程模型和 ESCIMO 湍流燃烧理论等。

第 5 章为液体燃料的燃烧，首先介绍了石油作为主要液体燃料的组成元素及化合物、燃料油品的物理和化学性能，如沸点、燃点等。其次，描述了浆体燃料的主要技术特性，如煤浆的黏度、流变性和稳定性等。单颗液滴在静止气流中的燃烧是液体燃烧的基础，主要包含液滴在静止气流中的蒸发、扩散及燃烧。然后介绍了油滴蒸发的基础理论，包括初期的表面蒸发和后续的内部扩散控制阶段，以及蒸发速率受油滴大小、环境温度、饱和蒸汽压和周围气体的流动状态等因素的影响。本章还详细描述了斯蒂芬流现象、燃料滴温度及蒸发浓度的确定。针对液体燃料的燃烧过程，分析了雾化喷嘴和雾化机理，提出了评定燃料雾化质量的主要指标包括雾化角，雾化液滴细度、雾化均匀度、喷雾射程和流量密度分布等。进一步探讨了相对静止环境和强迫气流中液滴的蒸发和扩散燃烧问题，同时针对工业喷雾燃烧的技术基础方面展开研究，简要介绍了常见的喷雾燃烧系统、提高雾化质量的基本措施和合理配风方式。

第 6 章为煤的热解及挥发分的燃烧。主要介绍了煤中有机化合物的特性、煤的特性对热解的影响、煤的热解过程、煤热解产物——挥发分的组成，以及各种因素（如温度、加热速度、压力、颗粒粒度、煤种、气氛等）的影响，同时详细分析了各种热解反应动力学模型以及挥发分的燃烧过程。介绍了浙江大学提出的综合考虑传热、扩散传质、热解动力学及二次裂解反应的热解模型，在建立综合热解模型时首先考虑传热因素，即整个粒子内部按不等温热解模式处理，对于大颗粒煤，采用"含尘气体"理论，模型计算结果与实验结果相符。

第 7 章为煤的燃烧理论(碳及焦炭的燃烧)主要研究焦炭的非均相燃烧。燃烧涉及的物理化学过程主要讨论了煤焦反应的控制因素、燃烧速率、碳的形态和结构的变化过程、吸附过程、扩散过程及碳燃烧机理。详细介绍了碳的燃烧化学反应的活化能包括碳与氧、二氧化碳、水蒸气、氢气的反应及产物在容积中的二次反应。由于碳的反应机理既有表面反应，又有空间反应，因此本章首先分析了一些简单分析的仅考虑一次反应求解模型，分析了温度较低或颗粒很小可略去时空间气相反应情况下的碳球燃烧速度以及碳球高温下的扩散燃烧。其次探讨了实际过程中碳燃烧的二次反应，即由于 CO 在碳球表面附近燃烧，阻碍了氧气向碳球的扩散，使得碳球燃烧过程起了很大变化，包括静止介质、流动介质中碳表面附近的燃烧、有 CO 空间反应时的碳球燃烧速率以及强迫对流条件下碳球燃烧速率。同时碳粒的燃烧不仅有表面燃烧，也可能有内部燃烧，本章也对多孔性碳球内部的燃烧特性进行了研究，分析了各种因素对焦炭燃烧的影响，包括煤中挥发分析出、灰分等因素。模化理论工作的开展着重介绍了包括湍流模型、两相模型和辐射传热模型，以描述煤燃烧的基本过程。

第 8 章为燃烧过程中氮氧化物的生成及分解机理，详细阐述了几种 NO_x 的生成机理，包括热力 NO_x、快速 NO_x 和燃料 NO_x 的生成，以及气体燃料燃烧、液体燃料燃烧和煤燃烧时 NO_x 的生成机理。在此基础上，介绍了降低 NO_x 排放的措施主要分为燃烧过程中脱硝技术和燃烧后脱硝技术，前者主要包括空气分级、燃料分级、低氮燃烧、烟气再循环、浓淡偏差燃烧和非选择性催化还原法等措施，后者主要包括选择性还原法脱硝和氧化法脱硝等。由于 N_2O 对大气环境的破坏作用，本章还简要介绍了燃烧过程中 N_2O 的危害、生成及分解机理和减少燃烧过程特别是流化床燃烧中 N_2O 的降解方法。

第 9 章为燃烧过程中硫氧化物、颗粒物及其他污染物的生成及脱除机理，简要介绍了燃料中硫的存在形态、硫在燃烧过程中的转化特性和 SO_3 的生成机理，分析了典型的气固两相反应石灰石燃烧固硫反应的基本过程、煅烧石灰石的固硫反应动力学、多孔介质内部气体扩散的数学模型和燃烧固硫总体模型，简述了燃烧脱硫技术包括炉内直接喷钙脱硫、循环流化床燃烧脱硫；烟气脱硫技术分为湿法脱硫包括石灰石石膏湿法脱硫、氨法脱硫、海水脱硫和镁法脱硫等；干法/半干法烟气脱硫技术包括烟气循环流化床脱硫 (CFB-FGD)、旋转喷雾干燥法 (SDA) 和增湿灰循环脱硫 (NID) 方法。炭黑粒子作为燃烧不彻底的产物对环境造成危害，燃烧过程中希望在燃烧后期与多余空气结合燃尽来提高换热效率避免炭黑的产生。本章节分析了炭黑形成的类型及性质，分别阐述了气体燃料、油和煤燃烧时，炭黑的生成机理、影响炭黑生成的因素以及降低炭黑排放的措施。煤燃烧过程中颗粒物的形成过程是一个复杂的物理化学过程，受诸多因素的共同影响。本章进一步分析了煤燃烧过程中颗粒物的形成机理、特性、影响因素、控制措施和脱除方法。浙江大学对燃煤电厂飞灰特性进行了测试研究，表明煤种、燃烧方式的不同会导致其飞灰的颗粒化学组分不同。燃烧过程中排放到环境中的有害痕量重金属及其化合物不能被降解，会对生物体产生严重的危害，尤其以重金属汞为代表。本章最后简要介绍了煤中汞含量的特征、燃煤烟气中汞的形态转化和控制机理。

第 10 章为燃烧技术的新进展，主要与清洁及低碳技术相关。首先介绍了富氧燃烧技术及其应用，提出了目前富氧燃烧技术商业推广面临的空气分离系统能耗高和设备长期运行后出现腐蚀等问题，下一步应更加关注系统长期运行的安全性、稳定性问题，以及各子系统优化组合，尤其是空气分离系统的合理配置对电厂经济性的影响等方面。化学链燃烧技术具有无需空气分离制氧设备、二氧化碳富集深度高以及可以与其他系统结合形成能源梯级利用等优点，但该技术存在系统复杂、载氧体寿命等问题。其次介绍了典型的催化燃烧技术的及其应用情况，采用催化燃烧技术可以促进劣质煤替代优质煤，减少高耗能行业的能耗，同时具有显著的环保效益。最后介绍了超焓燃烧技术及其应用情况，超焓燃烧特别适应于高温炉窑、低浓度燃气的能量回收利用或低浓度污染物的燃烧处置。

本文摘自《燃烧理论与污染控制》，机械工业出版社，ISBN：978-7-111-14022-1，2004；《燃烧理论与污染控制》第 2 版，机械工业出版社，ISBN：978-7-111-62944-3，2019。

洗煤泥及污泥焚烧技术与工程实例

煤炭是世界上最丰富的化石燃料资源。我国是煤炭生产和消费大国，在今后相当长的时间内，煤炭仍将是我国最主要的一次性能源。煤炭洗选是去除或减少原煤中所含的煤矸石、硫份等杂质，并按不同煤种、灰分、热值和粒度分成若干品种等级，以满足不同用户的需要。选煤是大幅度降低污染物排放、提高煤炭利用效率和节煤节运的最经济和有效的技术。选煤技术已在我国得到了广泛的应用。洗煤泥是煤炭洗选过程中生产的废弃物，如不加以妥善处理和利用，不仅造成环境污染，还导致能源的损失和浪费。因此洗煤泥的高效清洁利用是非常重要的。采用燃烧发电方法处理利用洗煤泥，能实现洗煤泥的规模化利用，既解决环保问题，又回收了能量，同时洗煤泥燃烧后的灰渣还可综合利用，完全解决了洗煤泥的处理利用问题。

污泥为水和废水处理过程所产生的固体沉淀物质。未经恰当处理处置的污泥进入环境后，直接给水体和大气带来二次污染，不但降低了污水处理系统的有限处理能力，而且对生态环境和人类活动构成了严重威胁。污泥焚烧具有减量效果明显、占地面积小等特点，在发达国家已得到广泛的应用，目前我国也在积极开展污泥焚烧技术的研究和应用工作。

本书主要介绍目前国内外洗煤泥及污泥焚烧技术的研究开发以及工程应用，共四章。

第一章论述洗煤泥及污泥的产生和特点，国内外洗煤泥及污泥处理处置的现状。

选煤技术在我国洁净煤技术体系中是最成熟、最可靠、最经济和最有效的技术，是洁净煤技术的关键和重点。洗煤泥是煤炭洗选过程生产的废弃物，洗煤泥的安全处置和高效清洁利用非常重要。选煤厂排出的煤泥的主要特点是：①粒度细，微粒含量多，尤其是小于200目的微粒占70%～90%。②持水性强，水分含量高。经圆盘真空过滤机脱水的煤泥含水一般在30%以上；折带式过滤机脱水的煤泥含水在26%～29%；压滤机脱水的煤泥含水在20%～24%。③灰分含量高，发热量较低。按灰分及热值的高低可以把煤泥分成三类：低灰煤泥，灰分为20%～32%，热值为12.5～20MJ/kg；中灰煤泥，灰分为30%～55%，热值为8.4～12.5MJ/kg；高灰煤泥，灰分＞55%，热值为3.5～6.3MJ/kg。④黏性较大。由于煤泥中一般含有较多的黏土类矿物，加之水分含量较高，粒度组成细，所以大多数煤泥黏性大，有的还具有一定的流动性。由于这些特性，导致了煤泥的堆放、贮存和运输都比较困难。尤其在堆存时，其形态极不稳定，遇水即流失,风干即飞扬。结果是不但浪费了宝贵的煤炭资源，而且造成了严重的环境污染，有时甚至制约了选煤厂的正常生产。大规模处理利用洗煤泥时，采用管道输送和其他形式的运输系统相比较，有着明显的优越性与可取之处。洗煤泥是一种高浓度、高黏度的浆状物料，其水份一般在25%～40%(即洗煤泥固体物含量60%～75%)，表观黏度在10泊以上，甚至大到几万泊，因此流动性很差。另外洗煤泥是由粒径小于1mm的细颗粒组成，其中小于0.2mm的组分占重量的80%上下，成份主要是黏土、砂石、煤粉，这样的洗煤泥表现出持水性好，同时洗煤泥灰份含量高，一般在 40%～70%。流变特性试验研究表明，洗煤泥属非牛顿流体，其流变特性符合下列本构方程：$\tau_w - \tau_0 = KS_w^n$，其中，τ_w 为剪切应力；τ_0 为屈服剪切应力；K 为幂定律系数；n 为幂定律指数；S_w 为剪切速率。传统的洗煤泥利用方法有生产民用型煤或烧砖、烧水泥、烧石灰；掺入中煤、混煤或原煤外销；作井下充填料或废弃在矸石山、直接排入环境等。上述利用煤泥的方法，由于受到多种因素的限制，其出路通常不稳定，利用量小，无法解决大量煤泥的出路。要想从根本上解决这一问题，还必须寻求高效益、大批量利用煤泥的新途径。近年来，探索和开发了一些合理利用煤泥的技术和工艺。主要有直接燃烧法处理和利用洗煤泥、生产锅炉用型煤和民用型煤、生产化工原料气或民用煤气等。

污水处理过程会产生沉淀物、颗粒物和漂浮物等，所产生的物质统称为污泥。污泥是一种由有机残片、细菌体、无机颗粒和胶体等组成的非均质体。它很难通过沉降进行彻底的固液分离。生活污水处理产生的混合污泥和工业废水产生的生物处理污泥是典型的有机污泥，其特性是有机物含量高(60%～80%)，颗粒细(0.02～0.2mm)，密度小(1002～1006kg/m³)，呈胶体结构，是一种亲水性污泥，容易管道输送，但脱水性能差。污泥性质主要包括：含水率与含固率、流变特性、脱水性能与污泥比阻、挥发性固体和灰分、湿污泥的密度与干污泥的密度、污泥的热值以及污泥的毒性与危害性等。污泥处置的最终目标是实现其减容化、稳定化和无害化，污泥处置的主要方法有以下几类：①污泥农用；②污泥高温堆肥；③污泥卫生填埋处理；④污泥焚烧等。和其他方法相比，焚烧法具有突出的优点：①大大地减少了污泥的体积和重量，因而最终需要处理的物质很少；②污泥处理速度快，不需要长期储存；③污泥可就地焚烧，不需要长距离运输；④可以回收能量用于发电和供热。

第二章讨论了洗煤泥流化床燃烧原理。

洗煤泥由于水分和灰分较高、发热量低，且物理性状特殊，因而采用传统技术燃烧难以获得满意的效果。为了达到稳定高效地燃用洗煤泥，必须发展与之相适应的新技术。流化床燃烧是 1960 年代开始发展起来的新型燃烧技术，最近几十年来得到了快速的发展。其优点如下：①燃料适应性广。它几乎可以燃烧一切种类的燃料并达到很高的燃烧效率。②能够在燃烧过程中有效地控制 NO_x 和 SO_2 的产生和排放。③燃烧热强度大，可以减小炉膛体积。④负荷调节性能好，负荷调节幅度大，其负荷调节范围可以从 100% 到 20%，在低负荷下也能保持燃烧稳定。⑤由于燃烧温度低，灰渣不会软化和黏结，燃烧的腐蚀作用也比常规锅炉小。此外，低温燃烧所产生的灰渣，具有较好的活性，可以用作制造水泥的掺合料或其他建筑材料的原料，有利于灰渣的综合利用。

但是常规的流化床燃烧技术并不能适应处理利用洗煤泥这种特殊燃料。浙江大学在分析总结国内外相关研究基础上，经过多年开发研究，形成了获国家技术发明二等奖(1997 年)的"洗煤泥异重流化床燃烧技术"。该洗煤泥高效低污染流化床燃烧技术的主要特点有：①采用燃料大粒度高位给料，利用洗煤泥的高温凝聚结团特性，使其在流化床内形成粒度较大的凝聚团，以减少燃料的扬析损失，提高燃烧效率。②采用异重流化床技术，防止大粒度凝聚团在流化床内的沉积，以保证流化床的稳定运行。③采用不排渣运行方式，在料层稳定的前提下减少大密度床料的消耗，并避免燃料的排渣损失，以进一步提高燃烧效率。④采用分级配风，降低 NO_x 排放量，即使床层维持在还原性气氛下，再辅之以二次风，抑制床内氮氧化物的生成，并加强悬浮空间对已生成的氮氧化物的还原反应，以达到燃烧脱硝的目的。⑤采用高效脱硫技术，将脱硫剂破碎后在燃料入炉前均匀混入洗煤泥，入炉后形成洗煤泥——脱硫剂凝聚团。既可采用较细的脱硫剂来提高脱硫剂利用率，又因为凝聚团在炉内有充分的停留时间，使脱硫剂完全反应，从而实现高效脱硫。

本章详细介绍了浙江大学获国家技术发明二等奖的"洗煤泥异重流化床燃烧技术"，包括异重流化床的流体动力特性、洗煤泥燃烧时的结团特性、洗煤泥的流化床燃烧过程、洗煤泥流化床燃烧中污染物的生成和控制、洗煤泥和煤矸石流化床混烧等。

第三章则结合前一章内容，重点介绍了洗煤泥流化床燃烧技术应用实例，洗煤泥煤矸石流化床混烧技术应用实例和洗煤泥输送给料技术的发展与应用。

洗煤泥和煤矸石同为煤炭洗选加工过程中排放的废弃物，但两者在物理形态和性质上差别很大。煤矸石通常呈较大的颗粒状，容易脱水，便于运输，在堆积状态下形态也比较稳定。洗煤泥由于其颗粒很细，水份高，黏度大，不易运输。为适应不同需求，在开发洗煤泥流化床燃烧技术基础上，浙江大学进一步开发了洗煤泥煤矸石流化床混烧技术。除保留原有特点，对于煤泥、煤矸石流化床混烧，主要通过对煤矸石给料粒度和煤泥给料粒度的合理选择，使之形成一个合适的异重流化床系统，既能使煤泥凝聚团能稳定充分地燃烧，也可使煤矸石尽可能地完全燃尽。对不同煤矿提供的煤泥和煤矸石进行了工业试验，表明该技术可适用灰份（干基）高达 70% 的各种煤泥和煤矸石，鼓泡流化床燃烧时燃烧效率达 92% 以上，高于相同条件下单烧煤矸石的燃烧效率，燃烧脱硫效率超过 80%（钙/硫摩尔比为 2），氮氧化物排放低于 200ppm（折算到烟气含氧量为 7%），实现了煤泥煤矸石的高效低污染混烧。

本章还介绍了技术应用实例。①兴隆庄煤泥热电厂。国家"七五"期间将《35T/H 煤泥流化床锅炉及煤泥燃料系统的研制》列为重点科技攻关项目，兴隆庄煤泥热电厂为该项目的配套工程，同时它也是全国第一家利用煤泥的热电厂，1990 年投运。其工艺过程包括：从选煤厂压滤车间来的煤泥卸至电厂煤泥棚，进入电厂煤泥燃料制备系统，在系统中经集料、转载、输送等技术环节，运至 21m 层铸石刮板机，经铸石刮板机分配进入炉顶立式煤泥给料机，煤泥在立式煤泥给料机中完成搅拌预碎、存贮挤压成型、给料等工艺过程，直接进入 35t/h 煤泥流化床锅炉炉膛燃烧。锅炉产生的中参数蒸汽通过主汽管路送至汽轮机，带动汽轮发电机组生产电力，汽轮机组可调整或非调整抽出做完功但仍有一定温度压力的蒸汽，经抽汽管路引至热网，供矿区、工厂区、家属区取暖。1991 年 12 月，国家"七五"重点科技攻关项目"35 吨/时洗煤泥流化床锅炉"通过国家级鉴定，鉴定结论认为洗煤泥流化床燃烧技术采用了"异重流化床，大粒度给料烧结成团和不排渣运行"等多项新技术具有开创性，具有国际领先水平。燃用发热量仅 2800 大卡/公斤的洗煤泥燃烧效率高达 96%，锅炉效率超过 83%。②东滩煤泥热电厂。东滩煤泥热电厂燃用东滩煤矿洗煤厂产生的洗煤泥。电厂设计规模 1×12MW + 2×15MW，配三台 75t/h 煤泥循环流化床锅炉，燃料根据全烧洗煤泥设计，东滩电厂与 1997 年 12 月 1# 炉投运至 1998 年 10 月形成三炉三机规模。③永荣矿务局电厂 35t/h 煤泥煤矸石混烧循环流化床锅炉。永荣矿务局电厂 35t/h 煤泥煤矸石混烧循环流化床锅炉于 1995 年 12 月 26 日并网发电。该炉不仅能全烧煤矸石，而且能混烧重量占 30%～92%的洗煤泥，煤矸石热值在 5.72～13.36MJ/kg，洗煤泥热值在 4.28～10.99MJ/kg 时都能保持稳定燃烧。热工测试表明，设计工况时，该锅炉热效率达 85.35%，燃烧效率达 95.12%。

第四章分析了污泥焚烧基本原理，介绍和对比了各种污泥焚烧方法及焚烧炉，并以目前国内外流行的流化床焚烧方式结合工程应用实例讨论了流化床焚烧过程、污染物排放及控制、流化床焚烧炉设计等。

污泥焚烧或燃烧是污泥中有机物的氧化过程，在产生稳定化飞灰的同时排放出一定量烟气。污泥中的 C、H、S 成分或可能包含的 NH_3 等可以进行燃烧化学反应，放出热量。污泥的焚烧已有 70 余年的发展历史，但直到 20 世纪 60 年代污泥焚烧处理才真正被广泛采用。污泥焚烧方法最常用的技术主要有多膛式、流化床和回转窑等，最新发展的焚烧技术还有熔融处理技术。分析比较表明，流化床污泥焚烧炉可广泛地用来处理各种固体废物及污泥。对于难于在多膛炉、炉排焚烧炉上焚烧的污泥采用流化床焚烧技术是非常适宜的。

本章还介绍了污泥焚烧在美国、日本、英国、印度、奥地利等国应用实例。浙江大学经多年的研究开发，形成了流化床污泥燃烧技术，成功应用于韩国清州 65t/d 高水份城市废水污泥流化床焚烧炉。该炉处理 65t/d 水分为 85.1%的清州污水污泥，同时处理废塑料和废橡胶各 2.5t/d，焚烧炉以废油为辅助燃料。焚烧锅炉由浙江大学设计，国内生产制造，于 1996 年 6 月制造完成并运往韩国。1997 年投入商业运行，得到了业主的好评。

本文摘自《洗煤泥及污泥处理焚烧技术与工程实例》，化学工业出版社，ISBN：978-7-502-57598-4，2004。

臭氧氧化烟气中多种污染物同时脱除技术

本书综合探讨了煤炭等能源利用过程中污染物控制技术的发展，包括脱硫、脱硝、汞去除以及挥发性有机化合物的处理，特别关注了臭氧在烟气多污染物同时去除中的应用。本书深入分析了臭氧与多种污染物之间的化学动力学和氧化机制，揭示了其在低于 300℃条件下的反应特性。试验结果表明，臭氧对氮氧化物、二氧化硫和汞等污染物具有高效的去除能力，且其去除效果受到多种因素的影响。此外，本书还探讨了臭氧与湿式洗涤器结合的多污染物去除方案，通过优化工艺参数，实现了对多种污染物的高效去除。最后，本书对臭氧多污染物控制技术的发展前景进行了展望，强调了技术创新在推动燃煤行业绿色转型中的关键作用。随着环保意识的提高和环保政策的加强，多污染物控制技术将在各能源行业中得到更广泛的应用和推广。本书的研究成果对于促进燃煤行业的可持续发展、保护生态环境具有重要意义，为相关领域的研究和应用提供了有益的参考和启示。

尽管氢燃料、风能、太阳能等清洁能源技术发展迅速，但在可预见的未来，大部分能源消费仍将依赖于使用化石燃料、废弃物和生物质的燃烧技术。众所周知，煤、石油、天然气、生物质和城市固体废物燃烧过程中排放的烟气污染物主要包括二氧化硫、氮氧化物、粉尘、汞、挥发性有机化合物和二噁英等。据预测，中国对煤炭的需求份额将在 2035 年下降至 45%以下，但彼时中国的煤炭消费量仍将占据全球的近一半。在这种情况下，煤炭燃烧产生的污染物排放突出。

到目前为止，各种减排方案包括石灰石-石膏法脱硫、低氮燃烧、选择性催化(非)催化还原脱硝、活性炭吸附、电除尘、布袋除尘、湿式静电除尘等已广泛应用于大型燃煤电站。近年来，烟气排放标准的进一步提高，控制范围的进一步扩大，以及超低排放的提出等对烟气污染治理提出了更高的要求。各种减排技术的叠加及增加负荷成为普遍的选择，但遗憾的是，单一种类污染物控制技术不仅投资高，而且会降低整个系统的可靠性。因此，在当前以煤为主要能源的情况下，积极开展燃煤多污染物协同高效脱除研究，探索先进、可靠、经济的多污染物协同脱除技术，是我国可持续发展的关键。此外，除大型燃煤电站外，工业锅炉和窑炉烟气排放贡献同样很大。据统计，小型的工业锅炉和窑炉消耗了 45%～50%的煤炭。但是，其具有的容量小、分布广、烟气成分复杂、温度低、湿度高、污染物浓度高等特点，迄今缺乏高效的污染物控制技术，更难以适应当下超低排放的要求。

臭氧前置氧化技术是目前最有前景的多污染物协同脱除技术之一，具有高效、节能、低成本等优点。臭氧喷入烟气中后，会产生·O_2、·O、·OH 等强氧化自由基，这些自由基能将几乎不溶于水的 NO 转化为具有较高溶解度的 NO_2 和更高溶解度的 N_2O_5，同时可将 Hg^0 转化为水溶性的氧化态(如 HgO 和 $HgCl_2$)。此外，VOCs 和二噁英等有机污染物同样可在氧化过程中完成降解。最后结合现有的烟气脱硫系统和特殊设计的碱吸收塔，同

时去除 NO_x、SO_2、Hg、VOCs 和二噁英等污染物。因为烟气污染物中，NO_x 是除 SO_2 外浓度最高的气态污染物，亦是目前重点关注的污染物，所以，臭氧多种污染物脱除技术的关键是 NO_x 的脱除。该技术突破了传统的还原法脱硝技术：SNCR 技术的脱硝效率较低，为 30%～50%，应用的温度窗口为 850～1100℃，而 SCR 技术的主要温度窗口为 300～400℃，脱硝效率可达 70%～90%。由于 SNCR 和 SCR 脱硝技术均需要特定的温度窗口，且钢铁烧结机、生物质焚烧炉、炭黑尾气炉、玻璃窑炉等工业排放过程中往往烟温偏低，富含金属、碱金属的飞灰易导致 SCR 催化剂中毒失效和堵塞等问题，其应用面临局限。相比 SCR 技术，臭氧氧化氮氧化物脱除技术是后发的先进脱硝技术，具有烟气成分和温度适应性强、改造简单、脱硝效率高和一塔多脱的优点。

浙江大学研究团队自 2004 年开始研究臭氧氧化多种污染物去除技术，本书系统介绍了过去几十年来在臭氧去除污染物领域大量的基础研究、重大的科学突破和多污染物协同脱除的潜力。此外，本书为读者提供与燃煤电厂相关的烟气污染控制技术领域的最新信息，由臭氧与烟气成分之间的详细反应机理组成的基础研究成果不仅来自于化学动力学建模，还来自实验室规模化实验研究。同时利用臭氧去除多种污染物技术在几种典型烟气治理领域的应用示范案例，在所获得的详细反应机理的帮助下，进行了经济性分析。简而言之，这本书在与臭氧去除多种污染物相关的这一关键领域的研究和开发中处于领先地位，给读者在烟气污染控制领域提供更丰富全面的信息，为烟气治理行业的研究者和工程师提供技术参考和选择依据。

本书内容主要分为 5 章。

第 1 章概述了燃煤污染控制技术的最新发展。本章首先对燃煤污染控制技术的当前发展状况进行了全面概述。燃煤过程中产生的二氧化硫、氮氧化物、汞、挥发性有机化合物(VOCs)以及二噁英等多种污染物对环境和人体健康造成了严重危害。因此，本章详细讨论了针对这些污染物的各种控制技术，包括它们的原理、应用效果优缺点。特别是，对于烟气中多种污染物的协同脱除技术，本章进行了深入分析和讨论。通过向上游烟气中喷射 CaO、$NaHCO_3$ 和 NH_3 的混合物，同时去除 NO_x、SO_2 和颗粒物，这种多污染物去除技术相对复杂且经济性较差，因此在实际炉膛中应用较少，需要探索先进、可靠、经济的多污染物协同脱除技术。

第 2 章介绍了臭氧的产生特性及去除多污染物的原理。首先，对臭氧的物理和化学性质进行了概述，包括其生成方法及臭氧合成技术的最新进展。特别介绍了等离子体放电臭氧合成技术的特点。臭氧放电技术转化气源为离子和中性物质所需的能量较少，降低了成本，而高能电子的产生有助于增强氧化反应的效率。介电层的存在可以消除带电粒子的扩散，使流注在介电表面均匀分布，同时可以消除阴极电子的发射，抑制弧光放电现象的发生，从而促进放电。尽管基于脉冲放电的臭氧发生器系统显示出良好的前景，但在实际应用中还需解决一些问题，如高压电源的寿命和成本，以及提高放电功率的适用方法。使用有机聚合物材料作为介电层可以提高系统的耐用性和经济性，如通过优化介电材料的厚度、直径和放电频率等参数，可以延长臭氧生成管的使用寿命，减少故障发生。

第 3 章详细探讨了臭氧氧化法脱除烟气中二氧化硫、氮氧化物等污染物的原理和应用效果。通过理论分析和试验验证，本章揭示了臭氧氧化法在烟气多污染物脱除中的独特优势和潜在应用价值。详细阐述了 O_3 与 $NO_x/SO_2/Hg$ 之间的化学动力学和氧化机理，该机理来源于化学动力学建模和实验室规模化试验研究。化学动力学模型考虑了烟气中的 40 种典型物种，共涉及 121 种基元反应。在烟气净化过程中，了解臭氧与污染物之间的反应速率对于优化去除效率至关重要。基于对化学动力学和氧化机理的理解，该章节还提出了优化臭氧去除多污染物技术的策略。这些策略包括调整反应条件（如温度、湿度、浓度等）、选择合适的催化剂以及优化反应器设计等，为臭氧去除多污染物技术的发展提供了重要的理论依据和实践指导。

第 4 章介绍了臭氧和湿法洗涤同时去除多种污染物的机理，主要探讨了利用臭氧与湿法洗涤塔相结合的技术，实现烟气中多种污染物的同时去除。该技术结合了臭氧的强氧化性和湿法洗涤塔的物理吸收能力，实现对烟气中氮氧化物、二氧化硫以及汞等多种污染物的有效去除。臭氧首先与烟气中的污染物发生化学反应，将其转化为更易溶于水或更易被吸附的形态。然后，通过湿法洗涤塔，利用水或吸收剂将转化后的污染物从烟气中分离出来。湿法洗涤塔中 NO_2/N_2O_5 的吸收对于实现最佳的 NO_x 脱除效率至关重要。实验研究表明，臭氧能高效去除烟气中的 NO_x、SO_2、Hg 等多种污染物，在与湿法联合脱除技术结合时，臭氧的引入显著提高了污染物的脱除效率。此外，本章还探讨了 pH 值、S(IV) 及初始 NO_2/SO_2 浓度等因素对 NO_2 去除的影响。

第 5 章关注臭氧多污染物控制技术的具体应用案例、实际运行效果分析、技术经济性评价等方面的内容。通过案例分析、数据分析和比较评估等方法，对臭氧多污染物控制技术的实际应用效果和潜在应用价值进行综合评价。重点介绍了臭氧多污染物控制技术与湿式洗涤塔系统的集成研究。通过实验研究和技术创新，成功开发出一种新型氧化-吸收系统，该系统能够有效脱除烟气中的多种污染物。本章还对该系统进行了经济分析，探讨了其商业化应用的可行性和潜力。该技术适用于各类工业烟气处理，如电厂、钢铁厂、化工厂等，可以有效降低烟气中的污染物排放，满足环保要求。结合臭氧氧化与碱液吸附技术的多污染物去除技术在理论上是可行的，并在实际应用中取得了一定的效果，虽然初期投资成本略高，但从长远来看，该技术可以为企业带来显著的经济效益和环境效益。未来研究可进一步探讨降低投资成本和运行成本的方法，提高技术的经济性和实用性。同时，还可以研究该技术与其他烟气处理技术的结合使用，以实现更高效、更环保的烟气处理效果。

2014 年起，作者团队研发的活性分子臭氧氧化燃烧烟气多种污染物一体化脱除技术开始工业化应用。从解决炭黑干燥炉尾气污染物治理问题到实现电站锅炉低负荷下的超低排放，臭氧多污染物控制技术日渐成熟，应用也趋向于多元化，已在炭黑尾气炉、热电锅炉、电站锅炉、生物质锅炉、钢铁烧结机等多行业上的 30 多个工程案例中取得成功应用，最大规模为 2×200MW 电站锅炉烟气超低排放改造项目。

本书全面梳理了燃煤过程中多污染物控制技术的发展现状，包括脱硫、脱硝、汞去除以及挥发性有机化合物的控制技术，特别阐明了臭氧在烟气多污染物去除中的关键作用。该书详细介绍了臭氧的特性及其在多污染物去除中的应用，通过化学动力学建模和实验数据揭示了臭氧与氮氧化物、二氧化硫及汞等污染物之间的反应机理。研究结果显示,臭氧在适当的条件下能有效去除烟气中的多种污染物,为燃煤污染控制提供了新的有效手段。此外,本书还关注了臭氧多污染物去除技术的实际应用和经济性,通过案例分析探讨了技术的可行性和优势。

本书的研究成果为煤炭利用过程中的污染控制提供了新的技术路径,为实现环保与经济效益的双赢提供了有力支持。随着技术的不断优化和推广应用,相信臭氧氧化多污染物去除技术将在各行业发挥更大的作用,推动煤炭等能源清洁高效利用的进程。本书内容涵盖了臭氧技术的理论基础、实验方法、实际应用以及未来发展趋势等方面,为读者提供了一份全面而深入的参考资料。本书不仅深入探索了多污染物协同减排的先进技术,更为我国的能源可持续发展贡献了卓越的智慧和力量。

未来,我们期待更多关于臭氧及其他先进技术在污染控制领域的研究,以不断推动技术创新和应用。同时,也需要关注技术的经济性和可行性,确保其在实际应用中能够发挥最大的效益。通过综合运用先进的烟气污染控制技术,我们有望为创造一个更加清洁、美丽的环境做出积极贡献,推动能源与环境可持续发展目标的实现。

本文摘译自《Simultaneous Multi-Pollutants Removal in Flue Gas by Ozone》,Springer出版社&浙江大学出版社;ISBN:978-7-308-13258-9,2014。

先进清洁煤燃烧与气化技术

本书是"中国煤炭清洁高效可持续开发利用战略研究"丛书之一。本书在系统调研国内外先进清洁煤燃烧与气化技术基础上提出了其发展战略及政策建议，按照存在问题及解决方案、典型案例分析、发展趋势与路线、具体措施建议的思路，对煤粉燃烧技术、循环流化床燃烧技术、工业锅炉燃烧技术、煤与生物质混合燃烧与气化技术、煤的气化技术、以发电为主的煤热解气化分级转化及灰渣综合利用技术、富氧燃烧及 CO_2 回收减排技术、化学链燃烧与气化和水煤浆燃烧等其他低污染燃烧与气化技术、煤炭地下气化技术 9 个技术方向开展战略咨询研究，并提出适用于我国国情的可大规模工业应用的关键技术的近、中期发展思路。

通过研究认为，煤炭利用应遵循"科学发展、战略需求、自主创新、重点突破"的重要原则，体现以下先进理念。

(1) 煤不仅是能源，而且是重要的资源，因此，煤的利用技术应该是分级转化综合利用、多级联产、烟气及煤炭灰渣近零排放，而且是有中国特色的新技术。

(2) 结合我国的国情和特色，发电以用煤为主，近几十年来不会改变，因此清洁煤分级转化技术，不单要能用于新设计发电机组，而且要对现有的 7 亿 kW 以上的现存煤发电机组也有可能因地制宜利用，较大幅度提高这些现有机组的节能减排效率，提高其产值和劳动生产率。

(3) 煤的燃烧、煤的气化和煤的分级转化为目前煤利用过程中 3 种主要的转化方式。煤燃烧发电时燃烧效率高，但污染排放高、发电效率和造价低，煤气化发电时煤的转化效率较低，但发电效率高，环保效率高，造价成本高。以煤的部分裂解气化制高级油品、半焦发电、灰渣综合利用为主要特点的煤分级转化技术，与现有煤燃烧与煤气化技术相比，在能耗、环保及经济性方面具有优越性，可以跨越式提高煤炭利用效率、环境效益和经济性，有望改变现有煤炭利用方式，促进传统产业的升级改造。

(4) 未来发展我国先进清洁煤燃烧与气化技术，除积极完善高效、低污染、适合我国国情的各种先进煤燃烧与气化技术的开发与应用外，更应建立煤分级转化技术创新体系，通过出台产业政策促进其推广应用，打造适合我国国情的煤炭利用新模式，从而推动形成煤分级转化战略性新兴产业链，解决我国煤炭的高效、洁净利用问题。

通过调研表明，煤的燃烧技术向大型化、清洁、高效、清洁燃料替代发展，煤的气化技术向大型化、高效率和环境友好发展，而将煤的燃烧与气化相结合形成新型煤炭转化方式，即煤的分级转化，有利于进一步提高煤炭综合利用与减排效率。传统燃煤方式忽视了煤的资源属性，将煤炭完全作为燃料燃烧，导致煤炭综合利用水平和效益不高。煤分级转化是基于"煤炭既是能源又是资源"的理念提出的煤炭转化利用的全新方向，可提高煤炭发电的综合效益，改变煤炭单一用于发电的产业结构；可形成基于煤炭资源化利用发电的新产业链，并缓解我国油气等资源的紧缺状况；对于改变和优化国家煤电产业结构、循环经济和节能减排具有重要意义。

采用技术经济比较和全生命周期分析表明，以煤的部分裂解气化制高级油品、半焦发电、灰渣综合利用为主要特点的煤分级转化综合利用技术，在能耗、环保以及经济性方面具有优越性。本书也分析了先进煤炭燃烧与气化技术的发展战略与目标，指出我国应积极发展先进煤粉燃烧技术、循环流化床燃烧技术、先进的工业锅炉燃烧技术、煤与生物质混合燃烧与气化技术、煤的先进气化技术、富氧燃烧及 CO_2 回收减排技术、先进煤炭地下气化技术、化学链燃烧与气化和水煤浆燃烧等其他低污染燃烧与气化技术。

根据"科学发展、战略需求、自主创新、重点突破"原则，研究认为，我国发电以用煤为主(每年约 18 亿 t)，今后新建机组宜采用清洁燃烧(超超临界)、完全气化(整体煤气联合循环发电系统)和煤分级转化综合利用等技术，现有电厂可采用超超临界结合煤分级转化技术进行低成本提效改造；兼顾 CO_2 减排问题积极发展富氧燃烧等技术，但应首先考虑低成本减排。今后我国需要发展高效、低污染、适合我国国情的未来先进清洁煤燃烧与气化技术，重点发展煤分级转化综合利用技术，以循环经济的全新模式新建或改造燃煤电厂，推动我国煤炭转化利用相关产业的产业结构升级转型。

我国今后发展先进煤燃烧与气化技术的战略目标如下。

(1)到 2020 年，实现 300~600MW 基于循环流化床技术的分级转化综合利用商业化应用，在较小的投资下，提高机组发电效率和煤炭利用效率。努力实现 8%左右的电力动力生产用煤采用煤炭热解气化半焦燃烧分级转化方案进行综合利用，预计每年可制取相当于约 210 亿 m^3 天然气或相当于约 1700 万 t 原油的油气替代产品。发展富氧燃烧和先进大型煤气化技术等以 CO_2 减排为特点的煤燃烧与气化技术，分别实现日处理 3000t 煤气化炉示范及 300MW 富氧燃烧示范。

(2)到 2030 年，发展基于煤粉燃烧技术的煤炭热解半焦燃烧分级转化、多联产及污染物和灰渣资源化利用相关关键技术，实现超超临界结合煤粉分级转化工程应用。该阶段预计可实现 25%左右的电力动力生产用煤采用煤炭热解气化半焦燃烧分级转化技术进行综合利用，预计每年可制取相当于约 675 亿 m^3 天然气或相当于 5400 万 t 原油的油气替代产品；以 CO_2 减排为特点的低成本煤燃烧与气化技术得到规模化应用。

通过本书研究，先进煤燃烧与气化技术的发展的保障措施及建议如下。

1)推动煤炭分级转化综合利用技术示范与应用

以煤的部分裂解气化制高级油品、半焦燃烧发电、灰渣综合利用为主要特点的煤分级转化综合利用技术将煤经温和的部分裂解气化，提取煤中轻质组分(挥发分和部分气化产物等)用于生产油、气和其他化工产品，硫等污染物资源化回收，而难于气化完全的半焦经高效燃烧发电，灰渣中有价元素根据赋存形式，先提取钒、铝等有价元素，然后进行综合利用。这种全新的煤炭发电方式有以下优点和创新：①工艺参数要求低，设备投资低。煤在常压低温无氧条件下热解气化，无需高压和制氧条件，对反应器及相关设备的材质要求低，设备投资低；②能耗增加不多，运行成本低。热解不需要纯氧作为气化剂，分级转化系统能量损耗低，与常规气化技术相比，过程热效率大幅度提高；③污染物易回收高值化利用。煤中所含硫大部分在热解过程中以 H_2S 形式析出，易于煤气净化系统脱硫，脱硫的副产品一般是硫磺，其利用价值高；④技术适用性好，可对现有燃煤电厂进行升级换代,又适用于新建电厂;⑤保证高效发电同时回收煤炭中油、气、铝、硫等资源,实现各种高附加值产品联产;⑥将各种单一产品为目标的电力、化工、冶金、建材和环保等工艺综合优化组合,在大幅提升煤炭利用综合效益的同时,大幅降低污染物和温室气体排放。

这种新的煤炭发电方式投资低,投入产出比高,可对现有燃煤电厂进行升级换代,又适用于新建电厂,可改变煤炭单一用于发电的产业结构,可形成基于煤炭资源化利用发电的新产业链,体现了"煤炭既是能源又是资源"的理念,对于改变和优化国家煤电产业结构和节能减排、循环经济具有重要战略意义。

2) 促进自主知识产权煤热解气化技术发展,提高煤热解气化比例

为了促进自主知识产权煤热解气化技术的健康、快速发展,建议采取以下政策措施:①国家应制定相关的产业政策加以扶持,凡是新上煤化工及相关项目,涉及煤热解气化技术的选型,如果国内已有成熟技术,应该优先选择国内技术。②继续加大对煤热解气化基础研究的投入,促进原创技术的形成,引导大型能源企业参与技术的研究与开发,进一步加强产学研用密切合作。③继续加大对煤热解气化技术工业示范的投入,开发有市场前景、符合煤炭高效清洁利用科学要求的新的煤炭热解气化技术(如分级气化技术、热解气化技术、催化气化技术、加氢气化技术、小型分散式气化技术等),在中试和产业化示范上给予政策和资金支持。④制定优惠政策,引导国内目前采用UGI等落后气化技术的部分合成氨企业和燃气企业进行技术改造,采用新一代的煤气化技术。⑤对煤炭资源的利用进行合理规划,对煤炭资源进行合理分类,摸清国内适宜气化的煤炭资源储量和区域分布,制定政策加以引导,凡是适应煤热解气化技术的煤,应该优先用于热解气化,以最大限度地提高煤炭利用效率。⑥发挥行业协会作用。行业协(商)会要充分发挥桥梁和纽带作用,在政府指导下,组织国内热解煤气化技术研发、示范、产业化应用的调研分析,及时反映技术发展情况、企业对技术的需求现状,促进科学和有序发展。⑦制定相关的财税政策,对进行新型热解煤气化技术示范的企业给予税收优惠或减免,降低企业风险。⑧加强煤炭热解气化领域基础研究和技术开发的人才队伍建设,对有影响的团队给予持续的支持。

3) 出台政策鼓励先进煤燃烧与气化技术示范与应用,减少燃烧煤耗

①出台法规性指导文件,鼓励研发、示范和推广先进燃烧和发电技术。根据各地区的环境容量,采取分步走战略,科学、有序、积极、稳健地控制煤炭燃烧和发电过程中污染物的排放,推动可实现资源化的污染物控制和脱硫脱硝一体化和汞排放的协同控制。建议出台煤生物质混烧激励机制和相应监管政策,推动生物质等可再生燃料在大型燃煤电站混烧,增加其替代比例,减少燃烧用煤炭的消耗。建议加大洗选煤在工业锅炉中的使用比例,提高其用煤质量。对中小工业锅炉燃料可加大天然气、生物质等低污染、可再生燃料替代比例。②加大煤炭分级转化综合利用技术的研发、示范和推广。根据国情科学地制定出煤炭分级转化综合利用的发展规划、确定重点技术方向和制定路线图。优先发展有中国特色、适应中国国情的煤炭利用技术,并在产业政策上予以支持。③建立多领域合作机制,重点突破多联产的关键工程技术和示范工程。组织多行业协同攻关,积极推进电力-化学品、电力-油/气、热解-气化-燃烧分级转化等煤基多联产工业示范。对

煤基多联产工业示范项目建设提供资金支持，对其生产的电力和化工产品、油品进行补贴。重点突破煤气化及煤炭/生物质共气化、电力和不同产品联产的集成设计与运行、与多联产系统匹配的二氧化碳捕集技术等关键单元技术及系统集成技术，建立一定数量和规模的煤基多联产示范工程。

4) 设立重大科技专项进行关键技术攻关，促进煤炭清洁高效开发利用

①设立重大科研专项，对煤炭清洁燃烧与气化技术的关键技术进行攻关。建议科技部和有关部门在"十二五""十三五"期间设立重大科技专项，对煤炭清洁燃烧与气化的重大技术方向和关键科技问题开展科技攻关。同时，企业要建立稳定、合理、长期的科技投入机制，并接受相关部门的监督和管理。②建立产学研结合的协同创新体系，推进关键技术的工程示范。充分利用高校和研究院所的研发基础和人才优势，以企业为推广主体，建立科技协同创新体系，加强科技成果的推广应用。在政策和资金给予支持，如煤分级转化、富氧燃烧、大型煤气化、地下煤气化等先进燃烧与气化技术等技术和装备的研发。

5) 建立产学研用联合培养机制，加强煤炭利用产业创新人才培养

①以提高自主创新能力为核心，加大高层次、紧缺人才的培养和引进力度。重点培养造就一批创新型领军人才，努力打造人才竞争优势；实施海外高层次人才引进计划和留学人员回国创业支持计划，完善留学人员回国服务体系，吸引留学人员以多种形式为国服务。在我国现有人才国际合作交流机制基础上，出台鼓励相关科研技术人员进行国际合作的相关倾斜政策，以促进我国煤炭清洁高效开发利用领域人才的快速成长。②建立科学合理的人才评价、激励和继续教育长效机制。积极拓展人才评价渠道，分类建立人才评价标准，形成以能力和成果为导向的人才评价机制；围绕煤炭清洁高效开发利用的重点方向，对做出突出贡献的人才给予特殊优惠政策和奖励。应加大专业技术人员继续教育投入力度，科学整合教育培训资源，有助于专业技术人员的知识和技能不断得到更新、补充、拓展和提高，从而不断完善知识结构，提高创新能力和专业技术水平。同时，充分发挥博士后制度的作用，聚集优秀人才，促进企业和科研机构的技术成果应用转化。③建立产学研用联合培养机制，加强我国煤炭高效清洁开发利用创新人才培养。充分发挥高校的多学科交叉和多种创新要素的集聚效果，通过有组织的合作创新活动和产学研用的有效分工协作，建立产学研用合作创新人才的培养体系，实现知识的创造、应用、分享、积累和增值，为我国煤炭开发与利用科学技术发展提供充分的人才支撑和智力保证是十分必要的。如煤炭热解气化半焦燃烧分级转化综合利用技术领域，涉及面广，高等院校、科研院所和产业化企业通过相关科研项目必须分工合作，协同作战，尽快完成煤炭分级转化综合利用技术的基础研究、技术创新及开发，为煤炭分级转化综合利用技术的推广应用提供保证。

本文摘自《先进清洁煤燃烧与气化技术》，科学出版社，ISBN 9787030403360，2014。

可燃固体废弃物能源化利用技术

我国拥有大量的可燃固体废弃物资源(城市生活垃圾、农林产业废弃物、工业有机废弃物和市政污水污泥等)，能源化利用潜力巨大。可燃固体废弃物能源化利用过程复杂，理论知识和工程技术上涉及可燃固体废弃物理化特性、多种能源化利用技术、以二噁英和重金属为代表的特殊污染物减排以及焚烧飞灰安全处置等，工程应用上涉及技术体系、设备制造和安装、运行水平、技术适应性和经济成本等。

固体废弃物处理是指通过物理、化学、生物等不同方法，使固体废弃物转化为适于资源化利用以及最终处置的一种过程。固体废弃物的物理处理包括破碎、分选、沉淀、过滤、离心等处理方式，化学处理包括焚烧、浸出等处理方法，生物处理包括好氧和厌氧分解等处理方式，固体废弃物最终处置或安全处置，是解决固体废弃物的归宿问题，如填埋等。

如何科学处置以城市生活垃圾为主的可燃固体废弃物已经成为社会经济可持续发展的一项重要内容。垃圾合理分类是对垃圾进行有效处置的一种科学管理方法，提高垃圾资源利用水平是实现垃圾减量化的重要途径。根据组分属性，通常将垃圾分为：纸类、织物类、木竹类、厨余类、塑橡类、金属类、玻璃类及其他物质。根据功能属性分类，可以将垃圾分为可燃物、不可燃物等。

从本质上来讲，垃圾是一种非均质多样物质的混合物，并不具有固有内部结构及外部特征，因而不存在特定的物理特性。物理性质会随结构中每种组分的物理特性及组分的具体含量变化而变化。一般而言，垃圾的物理特性一般包括容重、含水率、空隙率及粒度尺寸。只有将化学特性和物理特性有机地结合起来，全方面分析垃圾的基本特性，才能更好利用生活垃圾。在垃圾化学特性的主要特征参数包括元素分析(元素组成)、和工业分析表征(挥发分、灰分、可燃分等)、垃圾的灰熔点、热值等指标。

固体废弃物产量的预测方法和模型有很多，如聚类分析预测法、流行速率预测法、专家评估法、投入产出分析法、因果分析法、经验推断预测法(Delphi 法)、马尔可夫链分析预测法、回归分析法、神经网络预测模型、指数趋势模型、灰色系统预测模型和时间序列分析预测等。一般工程应用上常用的固体废弃物产量预测模型主要是回归分析法、时间序列分析预测、神经网络预测模型和灰色系统预测模型四种。

固体废弃物处理技术主要有填埋、堆肥、热处理三大类，其中热处理方法由于具有减量化程度高、资源利用率高、占地面积小等突出优点，近年来在世界各国得到了快速发展。一般来说，固体废弃物热处理技术包括热解、气化、焚烧等。

固体废弃物的热解是一个复杂的化学反应过程，包括大分子键的断裂、异构化和小分子的聚合等反应。热解的机理十分复杂，到目前为止，已经提出了许多不同的热解动力学模型，但尚没有公认的适用于任何热解物料的通用模型。目前用来描述热解过程的动力学模型当中最为完善的是由若干平行反应组成动力学模型。影响热解过程的主要因素有反应温度、加热速率、反应时间等。另外，固体废弃物的性质、含水率、反应器的类型等，都会对热解的结果产生影响。城市生活垃圾属于复杂组分燃料，其组分具有高度复杂性、多样性和不均匀性。垃圾中不同组分的物理特性和化学特性迥然不同，因而在热处理过程中各种组分之间必然会产生交互影响，其反应特性十分复杂，而不是其组分反应特性的简单线性叠加。

固体废弃物的气化过程实际上是固体废弃物在高温下的热化学反应过程。由于在气化炉内高温条件下发生多相反应，反应过程极为复杂，可能进行的化学反应很多。在气化过程中，不同的进气方式、气化段温度和过量空气系数对气化特性如气化气组成、气化气的热值、热转化率和碳转化率都会产生不同程度的影响。气化技术按照气化剂的不同，可分为空气气化、纯氧气化、水蒸气气化等。根据反应器的不同，气化技术可分为固定床、流化床、气流床及回转窑等。固体废弃物气化技术中还有一类典型工艺，称为气化熔融技术。

焚烧是通过高温下剧烈氧化反应处理废物的一种物理化学反应过程，常伴随光与热的现象。用于衡量焚烧效果的技术指标主要有可燃物减量比、热灼减量、燃烧效率、焚毁去除率和烟气排放浓度限制指标等。影响固体废弃物焚烧的主要因素有焚烧温度、湍流混合程度、烟气停留时间及过剩空气系数等。

垃圾焚烧过程中燃料型 NO_x 是由挥发分氮及半焦中的氮转化形成。垃圾中有机物含量增加，垃圾中含氮量相应增加，导致垃圾焚烧时 NO_x 排放浓度增加。垃圾粒径对 NO_x 排放影响也较大：垃圾粒径越细，垃圾被加热得越快，挥发分析出越多，垃圾燃尽率越高。因此，燃烧小粒径垃圾时 NO_x 的排放浓度要大于燃烧大粒径垃圾时的 NO_x 的排放浓度。少量水分对 NO 的转化率不会造成太大影响，在水分较少时甚至对 NO 生成起促进作用。

垃圾焚烧过程中由于垃圾成分的复杂性、多变性，会产生除常规污染物（SO_x、NO_x、CO、HCl）之外的毒性更强的无机污染物，如重金属。垃圾组分中所含有的重金属如 Hg、Pb、Cd、Cu、Zn、Ni 等会发生不同程度的挥发和迁移现象，从固相向气相中迁移，使烟气及飞灰中的重金属含量增加。而对于不同种类的重金属，其迁移程度不同。在垃圾焚烧前将垃圾分类，分拣出其中主要的重金属污染源，如回收电池、纸张等。然而垃圾分类只能减少入炉垃圾中重金属的含量。即便是去除了明显易生成重金属污染的垃圾源，焚烧后仍将有目标重金属存在，垃圾在焚烧处理后，其中所含的重金属最终将分布在焚烧炉底灰、飞灰、烟气及炉壁灰中。

自从 1977 年 Kees Olie 等在荷兰废弃物焚烧炉的飞灰中检出二噁英以来，二噁英相关研究及其控制就成了"热点"问题。各国研究人员对燃烧过程特别是垃圾焚烧过程中二噁英的来源、生成机理和主要的影响因素进行深入的研究。垃圾焚烧过程二噁英的生成途径主要有：①原生垃圾中固有的二噁英（固体废物和垃圾衍生燃料中含有痕量二噁英，在焚烧过程中由于未被完全地破坏或分解，从而残留在烟气、飞灰、残渣中）；②在炉膛高温段（500~800℃）发生的高温气相反应；③燃后低温段（200~500℃）发生的异相催化反应，包括从头合成（de novo 反应）和前驱物（precursor）合成。在特定的燃烧环境中，其生成量有所不同，一般情况下为：高温气相反应≤从头合成<前驱物生成。

现代垃圾焚烧炉设计时为控制二噁英的排放，均采用"3T"原则，即燃烧温度保持在 800℃以上(Temperature(温度))；在高温区送入二次空气，充分搅拌混合增强湍流度(Turbulence)；延长气体在高温区的停留时间(time(时间)＞2s)。上述措施实现后垃圾焚烧过程中二噁英可得到充分降解。

关于二噁英生成的关键影响因素认识主要有气相中的氧气对于二噁英"de novo"反应来说起着至关重要的作用，从头合成二噁英过程中，O_2是最重要的氧源。O_2由催化剂Cu_2Cl_2传递到飞灰中的途径是和Cu_2Cl_2合成络合物，合成的络合物氧化炭，同时被还原为Cu_2Cl_2，完成一个催化氧化循环。氯源是燃烧过程中二噁英生成的重要影响因素之一，氯源不仅是 de novo 反应的重要参数，也是二噁英形成必不可少的元素，氯在二噁英生成过程中的作用体现为氯化过程，对氯化过程的认识至今也未能有确切的结论。目前可以接受的氯化过程的机理认识有两种不同模型。一个即是 Deacon 氯化反应，另一个为直接 ligand 传递氯化反应。焚烧过程中的氯一般来自金属氯化物的分解、含氯化合物的不完全燃烧或气相中已存在的氯。飞灰中残碳含量对二噁英总量排放、有毒二噁英的排放及毒性当量排放有较大影响。从头合成反应中生成的 PCDD/Fs 的量与低温(C450)下可气化的碳有关，而不是总的残炭量，且与碳的来源无关，当反应温度为 300℃时，含碳量范围为 0%～4%，二噁英的生成与含碳量呈线性关系。金属催化剂是二噁英 de novo 合成反应必不可少的条件之一，催化剂在二噁英生成过程中的作用主要体现在以下 3 个方面：①从碳生成的从头合成反应；②环状物质的环化冷凝反应；③氯化和脱氯反应。催化剂的一个重要作用体现在影响氯化过程，即影响 Deacon 反应。由于 Cu 的存在，导致上述反应产生强氯化剂 Cl_2。从头合成二噁英过程中存在一个较宽的 C/Cu 摩尔比例范围"30～140"，即飞灰中 C/Cu 摩尔比在此范围内时的二噁英毒性当量排放是比较高的，典型垃圾焚烧炉飞灰中的 C/Cu(催化剂)比例刚好落在该范围内。

温度已被证明是影响二噁英生成或消解的最重要的影响因素之一，实践证明 800℃以下的燃烧温度可以促进二噁英的生成，当大于 850℃的燃烧温度完全可以分解二噁英，运行范围保持在 800～1050℃内即可。当焚烧烟气温度从 1100℃下降到 300℃时，PCDDs 的浓度增加 10 倍，PCDFs 的浓度增加 80 倍。已有研究表明 325℃时 PCDD/Fs 的生成量最高。

固体废弃物焚烧后产生的飞灰，多为圆柱状、环状的细小颗粒，根据原料以及焚烧炉型的不同，颜色可为白灰色、浅灰色、暗灰色，甚至可能为黑色。飞灰的物理特性主要包括：形状及颜色、堆积密度、含水率、热灼减率、熔融特性、pH、颗粒直径及粒径分布、比表面积和孔隙率等。焚烧飞灰化学成分相当复杂，其成分可以按酸碱特性的不同大致分为三类。

三类成分的大致分布对飞灰的熔融特性、技术选择、处理效果等有重要影响。酸性氧化物通常使飞灰熔点升高，增加熔融处理的难度，而碱性氧化物则使灰熔点降低。氯化物阻碍水泥水化过程，增加了固化体中重金属的可溶性。

焚烧法处理城市生活垃圾已有 100 多年的历史，欧洲最早采用焚烧技术处理生活垃圾。早在 1870 年，世界上第一台垃圾焚烧炉就已在英国投入运行，1895 年，德国汉堡建成了世界上首座固体废弃物发电厂。20 世纪 50 年代，西方国家伴随着经济的迅速发展，焚烧工艺不断完善，焚烧技术由固定炉排发展为回转窑、再到机械炉排和流化床。

垃圾焚烧技术按照运行方式分为连续运行焚烧炉和间歇运行焚烧炉。按照焚烧炉型分为固定炉排焚烧炉、机械炉排焚烧炉、流化床焚烧炉、旋转窑焚烧炉、等离子体焚烧炉等。目前我国内垃圾焚烧处理技术主要有两大流派：机械炉排炉和流化床。

20世纪流化床焚烧技术逐渐在日本、北美、欧洲等一些发达国家得到应用。国内在开发循环流化床垃圾焚烧技术方面投入力量较大，国内的科研单位也分别开发出有特色的流化床垃圾焚烧炉，典型的流化床垃圾焚烧炉有三种：①浙江大学的生活垃圾异重度循环流化床焚烧技术。特点为针对中国低热值高水分多组分的生活垃圾开发，异重度流化床使炉内实现稳定燃烧，特殊布风结合风帽布置方式，提高截面垃圾处理量；②北京中科通用能源环保公司的循环流化床垃圾焚烧技术。采用外置过热器技术，避免了氯化氢对过热器的高温腐蚀问题；③清华大学的炉排——循环床复合生活垃圾焚烧炉。

为了使在工业生产过程中二噁英排放达到要求，降低二噁英对生态环境的污染和对周边居民健康暴露风险，各国科学家都在努力开发不同的技术对二噁英进行控制排放。控制或减少二噁英排放的措施主要有以下三类：①控制垃圾的组分或促进其生成的组分；②抑制燃烧过程及燃后区域的二噁英生成；③在二噁英生成后，脱除或减少尾气和飞灰中二噁英的排放。同时抑制剂被认为是控制二噁英生成和排放的重要技术之一，具有低成本、高效益的优势，在国外被推荐作为一级控制措施。

为减少焚烧炉烟气中的粉尘排放，目前垃圾焚烧炉广泛采用布袋除尘器进行除尘，捕集烟气中的粉尘。布袋前采用喷射活性炭吸附剂的方法，吸附烟气中的二噁英，可以达到较好的二噁英减排效果。催化降解是一种工业烟气的末端处理技术，该方法能通过催化氧化还原反应去除烟气中的二噁英，并得到了良好的效果。

热降解飞灰中的二噁英一般有两种形式：在高温有氧气氛下进行降解；在低温无氧气氛进行降解。高温有氧气氛降解相比，低温热降解具有能耗低，脱除率高，且不会在冷却过程中再生成二噁英。该原理是通过对布袋飞灰的收集经过加热区域，在氮气气氛的保护、300~400℃的加热温度下将飞灰中的二噁英降解去除，加热的飞灰经冷却区域后排除，二噁英的去除效率可达98%。影响热解的因素包括：气体的流速、飞灰重量、反应温度及反应时间等。

熔融固化法就是把飞灰被加热到熔融温度（1200~1600℃），熔融后飞灰中的有机物热分解、燃烧、气化，而重金属因密度大而沉在熔炉的底部分离；硅酸盐类残渣浮熔融物上面，淬火后形成玻璃态物质，可作为建材使用，易挥发金属则在烟尘中被分离。垃圾焚烧灰渣熔融炉按照热源种类大致可分为燃料式熔融系统和电气式熔融系统两大类。

二噁英属于痕量级物质，浓度极低，现有的分析技术及仪器不易实现其直接测量。如果可以在线测量一种二噁英的替代物的浓度，且此替代物的浓度与二噁英的浓度之间存在着相应的关联关系，那么就可以间接地测量出二噁英的毒性当量浓度。二噁英的替代物一般也称之为指示物。指示物浓度相对较高、结构也相对简单，较容易实现在线测量。通过测量指示物不仅可以获得二噁英的浓度数据，还可以简化测量程序、节省测量成本。二噁英在线监测技术的研究集中在三方面：指示物的选择、关联模型建立、关联物在线测量技术。氯苯作为二噁英的指示物，对于优化燃烧过程，控制净化装置中烟气的分离效率，控制排放量可以起到很大作用。1985年，Oberg等首次在危险废物焚烧炉烟气中发现了六氯苯（HCBz）和二噁英存在良好的相关性。氯苯是目前研究得最多的二噁英替代物。通过对垃圾焚烧过程中二噁英替代物氯苯的快速检测可以达到间接监测二噁英的目的。

从尾部烟气中脱除二噁英是垃圾焚烧过程中必不可少的一个环节。燃烧设备特别是垃圾焚烧炉必须配置一定的尾气排放控制装置。一般尾气控制装置有：布袋除尘器、干法洗涤器、半干法洗涤器、湿法洗涤器、夹带流吸附装置、固定床吸附装置、移动床吸附装置以及选择性和非选择性催化器等。

目前比较常用的焚烧飞灰的处理方法，几乎都是根据如下三个处理原理中的一个或者多个发展起来的：物理或化学方法的分离技术、固化/稳定化技术及热处理技术。具体处理方法主要有：水、酸或其他溶剂洗提法；固化法，包括水泥固化法和热塑性材料固化法等；化学药剂稳定法；热处理法，包括烧结、熔融等；新型发展中技术，包括机械化学法、等离子体法等；以及这些处理程序的组合工艺。

危险废物管理是我国固体废物环境管理的重点。由于回转窑焚烧技术具有对物料适应性强，可以处理任何形态的固体、液体废弃物，焚烧处理时对入炉燃料的形状要求不高，不需要复杂的预处理过程等优点，在固体废弃物焚烧处理中得到了广泛的应用。回转窑式焚烧炉的发展已从过去单独设备上升为系统概念，典型的危险废物回转窑焚烧处理工艺包含废物预处理系统、焚烧系统、烟气处理系统等部分。废物预处理系统包括废物的预处理和进料工序；焚烧系统由回转窑和二燃室、出渣及控制系统组成；烟气处理系统由余热回收、急冷和除尘设备、酸性气体吸收组成。

固废的热解处理已吸引了越来越多的关注。因为热解相对于其他处理技术，不仅处置污染小，还可以回收得到热解油、气及焦炭等燃料。尤其是对高热值废物来说。为此选取了废旧轮胎、含油污泥和线路板等典型的高热值废物为例讨论说明其热解(技术)特点，并对废旧轮胎热解高质化利用实例进行分析，以说明热解技术对于高热值废物高质化利用的适用性。

本文摘自《可燃固体废弃物能源化利用技术》，化学工业出版社，ISBN：978-7-122-26197-7，2016。

推动能源生产和消费革命的支撑与保障

党的"十八大"报告中明确提出要"推动能源生产和消费革命",建设美丽中国。中央财经领导小组第六次会议上习近平总书记详细阐述了能源革命的内涵,提出了"四个革命"和"一个合作"的战略构想,充分表明党中央、国务院在对能源革命的态度和决心。为了保障能源生产和消费革命,我们应大力鼓励能源科技自主创新,建立新型的能源供需协调模式和市场、资源的保障模式,加快观念体制变革与创新;而首位重要的则是革命性的能源生产和消费理念以及顶层设计。依托中国工程院2013年5月启动的重大咨询项目"推动能源生产和消费革命战略研究",并根据第六课题组"能源生产和消费革命的支撑与保障"的研究,对如何支撑与保障推动能源生产与消费革命进行了系统的研究和分析。

研究旨在围绕促进和保障能源生产消费革命的体制机制问题,通过研究煤炭、石油、天然气、核能、可再生能源等典型能源生产行业的技术水平、政策法规、标准等现状和发展趋势,以及能源网(电、水、气、热、冷、废弃物)和火电、钢铁、建材、石化、化工等典型能源消费要素和行业的运行体制、能效、排放特征等,提出政策引导的内容与方式,市场运作的机制与保证标准及评价体系的完善与建立等,并在此基础上形成推动能源生产与消费革命综合技术创新与体制创新思路、基础建设与重大工程新方向以及政策法规及标准体系新建议。

主要研究和分析判断如下。

从体现战略性、方向性和可操作性要求出发,以2014～2020年、2021～2035年、2036～2050年三个不同发展时期为时间节点,按照能源发展需求→重要能源科技问题→重要能源技术方向→技术发展路线图→创新能源技术总体部署→保障体系建设的逻辑构思编制2050年我国能源生产与消费革命支撑保障战略及科技发展路线图。通过对能源需求、技术选择和各种技术路线的研究认为中国能源科技创新近、中、远期发展的阶段目标如下。

近期战略目标(2014～2020年),突破新型煤炭高效清洁利用技术,初步形成煤基能源与化工的工业体系;突破轨道交通技术、纯电动汽车,初步实现地面交通电动化的商业应用;在充分开发水力能源和远距离超高压交/直流输电网技术的同时,突破太阳能热发电和光伏发电技术、风力发电技术,初步形成可再生能源作为主要能源的技术体系和能源制造业体系。逐步提高核能、可再生能源和新型能源占总能的比重。

中期战略目标(2021～2035年),突破生物质液体燃料技术并形成规模商业化应用,突破大容量、低损失电力输送技术和分散、不稳定的可再生能源发电并网以及分布式电网技术,电力装备安全技术和电网安全新技术比重将达到90%,初步形成以太阳能光伏技术、风能技术等为主的分布式、独立微网的新型电力系统;突破新一代核电技术和核废料处理技术(ADS),为形成中国特色核电工业提供科技支撑。实现核能、可再生能源和新型能源的大规模使用。

远期战略目标(2036~2050年)，突破天然气水合物开发与利用技术、氢能利用技术、燃料电池汽车技术、深层地热工程化技术、海洋能发电等技术，基本形成化石能源、核能、新能源与可再生能等并重的低碳型多元能源结构。

路线图的实现需要在时序上、基础理论和技术应用的衔接上、技术竞争力和制造业的协同发展上全面积极推进，还必须调动广大能源科技人员的积极性和创造力，促进全社会能源科技资源的高效配置和综合集成，形成以政府主导、市场配置资源、产学研结合的能源技术创新体系，大幅度提高我国能源技术、能源产品、能源装备的自主创新能力。作为路线图实现的保障措施，必须加强政策、规则、标准的研究制定，加强人才队伍、科技平台及大科学装置的建设，加大必要的科技投入、执行，加快能源科技创新体系的建立。针对煤炭的洁净和高附加值利用技术、电网安全稳定技术、可再生能源规模化发电技术、氢能利用技术、新型核电与核废料处理技术等方面也提出不同时间节点的科技发展目标。

我国当前已成为世界上大气污染最为严重的国家之一，煤的燃烧生成了大量的二氧化硫(SO_2)、氮氧化物(NO_x)、重金属(Hg)和有机物等多种污染物，是大气污染物的重要来源。发展高效脱除与协同控制技术，实现燃煤烟气污染物的超低排放，达到燃气轮机的排放标准，是改善环境空气质量的有效途径。在此基础上，发展活性分子烟气深度净化技术和碳捕集与封存(carbon capture and storage，CCS)技术，实现燃煤烟气污染物的近零排放。同时开发活性半焦等资源化回收利用技术，满足国内硫磺、硫酸、硝酸钙等化工原料的需求。同步推进钢铁、有色、建材等行业污染物控制和资源化回收技术水平的提升，强化污染物控制装备运行的监管，保障污染物的长期稳定超低排放，为污染物的综合脱除与资源化回收提供支持与保障。

除此之外，同步推进产业结构和能源消费结构调整，加快传统产业技术创新，发展低能耗高附加值产业，严格控制高能耗低附加值产品出口。加大先进技术、工艺和装备的研发，加快运用高新技术和先进适用技术改造提升传统产业，促进信息化和工业化深度融合，支持节能产品装备和节能服务产业做大做强。支持优势骨干企业实施横向产业联合和纵向产业重组，通过资源整合、研发设计、精深加工、物流营销和工程服务等，进一步壮大企业规模，延伸完善产业链，提高产业集中度，增强综合竞争力。强化节能评估审查制度，提高行业节能准入门槛，控制高能耗产业增长速度，限制高资源消耗产品出口；提高节能环保市场准入门槛，严把土地、信贷两个闸门，严格控制新建高耗能、高污染项目。加大淘汰落后产能力度，地方各级政府要对限期淘汰的落后装备严格监管，禁止落后产能异地转移。要将"上大"与"压小"相结合，淘汰落后与新上项目相结合。

落实国家区域发展总体战略和主体功能区战略，根据资源能源条件、市场需求、环境容量、产业基础和物流配套能力，统筹沿海沿边与内陆、上下游产业及区域经济发展，优化产业布局，满足各地区经济社会发展需求。综合考察跨省区企业的产业链排放，并且建立区域间排放转移的补偿机制。在较不发达的中西部地区，应提高环境标准并严格执行。

以保护生态环境和人体健康为目标，加快环境保护标准制修订步伐，进一步完善国家环境保护标准体系。鼓励地方参与国家环境保护标准制修订，制定地方环境保护标准发展规划，制定实施较国家标准更为全面和严格的地方标准。坚持因地制宜，鼓励有条件的地区制订更严格的排放标准。严格执行节能、土地、环保等法律法规，综合运用差别电价、财政奖励、考核问责等法律手段、经济手段和必要的行政手段。

我国正处在快速发展时期，能源需求持续增长，能源和环境对可持续发展的约束越来越严重，因而发展清洁能源技术、加速本地化清洁能源的开发是必然选择。本书总结了我国太阳能发展的方向与途径，提出了目标建议与政策，并进行了案例分析。关于风能，介绍了国外不同形式的互补发电形式，指出了我国发展风能与其他能源互补发电技术的挑战和主要任务、发展方向及途径，提出了互补发电技术的发展目标和建议。生物质能被认为今后最有效的可再生能源，对生物质直燃/混燃典型工程技术的经济性和生物质电厂改造案例进行了分析，对生物质热裂解制取气/液高品位液体燃料进行了全生命周期评价，在此基础上提出了我国生物质能发展的目标和相关建议。提出了先进储能技术面临的挑战和发展建议，并进行了面向电力交通的储能动力系统建设的案例分析。最后，介绍了国内外的氢能发展现状，提出了我国发展氢能的战略定位、目标和综合建议。

在法律法规层面，研究并提出了常规能源法律政策现状、问题与完善建议，综合研究了国内外煤炭、油气、电力等法律政策支撑与保障，提出支撑与保障中国常规能源发展的综合建议；研究并提出了可再生能源法律政策现状、问题与完善建议，分析了国外可再生能源法律政策发展经验，并提出了支撑与保障中国可再生能源发展的综合建议；研究了国外节约能源法律政策发展经验并提出了中国节能减排法律政策现状、问题与完善建议；借鉴了国外能源安全法律政策的发展与启示，研究了能源安全法律政策现状、问题并提出相关建议。

在体制机制层面，强调了能源体制机制是支撑能源革命的重要保障，分析了现行能源体制机制存在的问题和成因，详细研究了能源体制机制的具体革命内容，并据此提出明确市场化方向，加强政府对环境的外部性管制、健全国家能源安全储备、预警与应急系统等方面的具体建议。结合国内外能源机制体制革命的借鉴研究和思考，提出了相应的改革思路和对策。

推动能源生产与消费各技术领域的综合建议如下。

加强煤分级转化与多级利用，加强有组织的合作创新活动和产学研用的有效分工协作，加强煤分级转化与多级利用科技创新与体制创新；加快煤分级转化与多级利用技术研发与工程应用；出台引导政策，调动企业创新技术研发与应用积极性；完善煤分级转化与多级利用标准体系。

推动污染物综合脱除与资源化回收，促进污染物控制从单一污染物治理向多种污染源物协同减排转变，实行分阶段污染物协同高效控制政策；鼓励污染控制技术向低消耗、低排放、高效益治理模式转变；大力推进燃煤污染物资源化利用；加强污染物共同减排政策与体制建设，促进多行业污染物共同减排。

推动海上天然气水合物开发，对天然气水合物调查和基础研究给予政策支持，建立天然气水合物技术标准体系，以规范、指导天然气水合物研究、试采、开采；积极探索天然气水合物研究的合作模式，加强天然气水合物研究与人才储备；对海域天然气水合物探测与资源评价技术、海上天然气水合物试采工程、天然气水合物环境效应研究等重点领域进行立项。

加大海洋能综合利用，通过建设初期给予信贷、上网电价的特殊支持、筹建国家级海洋能开发利用重点实验室等加快建立相关政策法规和技术标准体系；尽快出台我国海洋能资源开发的顶层设计，推进海洋能资源开发利用的发展；通过重大工程技术立项，开展全海域的海洋能资源普查与评价、加强高效能量转换和低成本的潮流能和波浪能装置开发、加强海洋能发电装置材料和防海生物附着技术研究、继续开展海洋能多能互补和并网技术研究。

推动油气供给与消费革命，改革现行的部分财税制度，合并相关资源类税费，实行从价计征，建立差别税率，启动及扩大对非常规天然气的财税政策支持，制定优惠税费政策，激励页岩气开发；建立适应社会主义市场经济特征，符合现代监管要求的法律体系，建立集中、统一、独立的监管委员会，并赋予其完善的职能，对自然和非自然垄断领域采取不同的管制方式，监管政策要鼓励对天然气基础设施的投资；推动油气供给与消费方面的科技创新。

推动核能革命性发展与利用，抓紧研究制订原子能法和核安全法，加强核安全法律法规建设，从加强核安全法律法规建设，全面加强核能发展管理，创新科研体制机制，加强人才队伍培养，积极提高核能公众接受度，努力开辟新的投资渠道、优惠政策。推动油气供给与消费方面的科技创新；

推动智慧能源网发展，是智慧能源网所带来的新型产业的落地规划；推行信息化能源管理平台和智慧水务网的建设；用分布式能源系统协调中央能源系统；大力发展新型的热电联供和冷热联供系统；大力推行中水回用技术和雨水收集回用技术；加强对绿色交通的投入力度；加强对于废弃物的监管力度。

推动重点耗能产业结构调整及工艺革新，支持高附加值、高技术含量产品研发力度，支持耗能行业清洁能源替代的相关研究；针对钢铁、煤电、水泥和有色金属等行业，优化区域布局；组织实施工业锅炉窑炉节能改造等九大重点节能工程；控制高耗能产业规模，限制高能耗产品出口，淘汰落后产能，加快兼并重组，实施严格污染物排放标准。

新能源与可再生能源的支撑与保障，加强太阳能、风能等行业监管，克服产能的过剩；推动风光等可再生能源间互补、常规能源与可再生能源互补的科技和体制机制创新；加大利用废弃物类原料，发展生物质制气和微藻能源等转化技术；明确储能产业发展整体框架和发展思路，大力发展先进高效储能技术，构建和完善储能设备评价标准，并建立和完善合理的优惠政策和补贴机制；制定有利于氢经济发展的优惠产业政策，尽快制定氢经济的行业和市场标准，成立国家发展氢能系统技术的组织领导委员会，建立国家级氢能研究平台。

本文摘自《推动能源生产和消费革命的支撑与保障》，科学出版社，ISBN：9787030523976，2017。

基于智能计算的燃烧优化

本书的论述重点是燃烧优化。首先简要讨论了燃烧参数对氮氧化物排放和碳燃尽率的影响。鉴于锅炉系统的复杂性,理论模型的建立到目前为止是极其困难的。为了解决这一问题,本书采用了包括支持向量回归(support vector regression,SVR)方法、人工神经网络(artificial neural networks,ANN)和计算流体力学(computational fluid dynamics,CFD)在内的人工智能方法,以构建替代模型。最后,本书回顾了在线燃烧优化系统的进展和设备,并对燃烧优化的未来展望。

火力发电是主要的电力来源,其发电量占据了全国总发电量的四分之三,其中超过90%依赖于燃煤。煤炭作为中国的主要能源,其种类繁多,特性各异。同时,随着可再生能源的利用日益增加,锅炉负荷的波动也变得更加频繁。如何在保证高燃烧效率的同时,实现低污染物排放,成为了一个亟待解决的问题。随着燃气轮机和燃气锅炉在中国的广泛应用,燃气燃烧同样受到广泛关注,并且面临类似的挑战。

煤燃烧过程中氮氧化物(NO_x)的排放是环境污染的重要来源。随着环境问题的日益严峻,氮氧化物排放问题也受到了更多关注。此外,粉煤灰中未燃碳的水平是影响粉煤锅炉效率的关键因素,尤其是在配备低氮氧化物燃烧器的锅炉中。粉煤灰含碳量高可能导致炉膛出口过热器和再热器管过热,甚至引起爆管。中国众多老旧设计的燃煤电站锅炉,这些锅炉热效率低,氮氧化物排放量高。虽然通过安装选择性催化还原(selective catalytic reduction, SCR)和选择性非催化还原(selective non-catalytic reduction, SNCR)烟气处理设备可以降低NO_x排放,但成本较高。近年来,通过精细调整锅炉运行参数,燃烧优化已被证实是降低燃煤锅炉氮氧化物排放和粉煤灰中未燃碳的有效方法。当前研究揭示了NO_x排放和碳燃烧并非孤立现象。在低NO_x燃烧器中实施的空气分配往往会导致飞灰中未燃碳浓度升高,因此,氮氧化物排放和碳燃尽率应被同时考虑。由于锅炉系统的复杂性,建立理论模型极为困难,而通过采用人工智能方法,如支持向量回归(SVR)、人工神经网络(ANN)和计算流体力学(CFD),可以构建替代模型。这些方法以其全局最优解、结构简单和泛化能力强等优点,已经受到了业界的广泛关注。

我国燃煤机组在燃烧设备和燃料特性的适应性方面面临诸多挑战。普遍存在的问题包括点火难度大、燃烧稳定性差、燃烧效率低、结垢和结渣现象、过热导致的爆管以及频繁的维护需求等。煤炭是一种有机矿物聚合物,其成煤时间与地质条件的差异导致煤的物理化学特性存在显著不同,从而表现出不同的燃烧性能。中国煤炭资源分布广泛但并不均衡,从褐煤到无烟煤,燃料煤的煤质差异显著。为了有效利用有限的煤炭资源,研究煤炭的燃烧特性显得尤为重要。在煤燃烧过程中,挥发性组分的分离和完全燃烧阶段仅占整个燃烧过程的10%,持续时间大约为0.2~0.5s。而焦炭的燃尽时间占据了整个过程的98%以上,其保持的热量释放占总热量的60%~95%,持续时间大约为1~2.8s。点火过程依赖于煤的干燥无灰基挥发物含量,而燃烧过程主要取决于焦炭的燃烧速率。

经验表明，煤的燃尽时间越长，其燃尽效率越低。粉状煤通过吸热升温，热源主要来自炉内 1300~1600℃的高温气体。这些气体通过对流、辐射和热传导的方式加热新燃料，提升其温度。煤炭中的水分首先被分离，随后煤炭干燥，水分蒸发使煤炭温度升高。在 120~450℃的温度范围内，煤中的挥发物发生分离，剩余的固体物质形成焦炭。

可燃挥发气体的点火温度低，如果氧气充足，挥发气体加热到 450~550℃时会被引燃，同时燃烧气体加热焦炭。焦炭从炉内挥发性气体和高温烟气燃烧产生的局部高温区吸收能量，温度升高，一旦达到点火点即发生着火，释放出一定量的热量。

在焦炭大部分燃尽后，内部灰分对燃烧过程产生影响。焦炭颗粒的内部灰分在燃料中均匀分布，在燃烧过程中从外向内逐渐迁移，外层的内部灰分会包裹内层焦炭，形成灰壳，甚至渣壳。这种外壳阻碍了氧气向焦炭表面的扩散，延长了燃尽过程。因此，灰分对燃烧过程的影响主要来自于内部灰分，而大部分外部灰分并不能独立阻碍可燃层的点燃。

煤的燃烧优化是一个多变量调控过程，涉及燃料特性和空气动力学条件的精细调整。通过精确的实验测量，包括炉内煤种选择、空气分布模式和燃烧角度，可以有效地控制和优化燃烧过程。这些实验结果为理论模型的构建提供了坚实的基础。以 NO_x 排放优化为例，研究者通过简化的化学动力学模型模拟了不同二次空气温度(350℃和 800℃)对 NO 生成的影响。模拟结果与实验数据的对比分析，揭示了高温空气能够减少 NO 的排放，为 NO_x 的减排提供了新的策略。然而，现有的优化模型尚显粗糙，且对风速和氧浓度测量的依赖性较强，这限制了其在复杂工况下的适用性。为了提高燃烧状态的诊断精度，专家系统被引入，利用模糊逻辑和预设模型来评估燃烧效率和稳定性。这些系统的核心包括知识库和推理机，它们集成了领域专家的经验和知识。尽管如此，专家系统在处理高度复杂的问题时，由于知识表示的局限性，其性能可能会受到显著限制。近年来，结合人工智能算法(如神经网络、支持向量机)与优化算法(如遗传算法 GA、粒子群优化 PSO、蚁群优化 ACO)的方法，已被证实在煤炭燃烧优化领域具有显著效果。这些方法通过处理大量的非线性数据和执行复杂的搜索任务，提高了燃烧效率和减少了污染物排放。

优化方法通常分为三个方向：①基于经验和实验的简化模型，用于控制燃料供给和空气分配；②根据模算法、遗传算法等人工算法计算出的模型进行优化；③基于 CFD 和化学机理的精确优化。第一种简单方法考虑的参数很少，在早期优化中占有优势，并与第二种新方法融合在一起。第二种优化方法可以处理大量的非线性输入，其计算和搜索能力保证了在线优化的启动。它已在许多方面得到了广泛的应用。第三种方法揭示了炉内精确的燃烧过程，但由于在简化机理和实际反应方面存在缺陷，其发展潜力有限，结果缺乏可靠性且耗时，目前尚处在研究阶段。

人工神经网络(ANN)模型是在现代神经学研究的基础上提出的，是一种简化和模拟生物神经网络模型。目前，人工神经网络已成功地应用于模式识别、语音识别、专家系统等。本书介绍了两种典型的神经网络算法：反向传播神经网络(BPNN)和广义回归神经网络(GRNN)。通过对 BPNN 和 GRNN 两种方法的原理和条件的比较分析，GRNN 在训练过程中采用了一些变量，GRNN 在训练速度和预测精度方面优于 BPNN。本书研究提出了将这两种方法结合的新策略，以提高燃烧过程的预测和控制精度。

基于结构风险最小化原理的支持向量机（SVM）作为一种重要的计算方法，它具有良好的泛化能力，在机器学习和数据挖掘领域得到了广泛的应用。SVM 的两种变体包括了支持向量分类（SVC）和支持向量回归（SVR），在燃煤发电领域的仿真和优化中发挥了重要作用。研究成果用于煤的鉴别、氮氧化物排放预测、粉煤灰中未燃碳预测等研究。发电厂锅炉煤燃烧过程中会向周围环境排放大量的氮氧化物，造成严重的危害。因此，氮氧化物的控制是燃煤锅炉运行中需要解决的重要问题之一。然而，NO_x 排放特性与运行条件之间的关系是非常复杂的。通过运行条件来估算 NO_x 排放量，寻找降低 NO_x 排放的最佳条件是十分困难的。因此，SVR 可以用于氮氧化物排放的建模和寻找最佳条件的全局搜索工具。

结合 SVM 和 ANN 的方法在实验中展现了显著的便利性和效率。这种综合方法利用了 SVM 在小样本、非线性问题上的优势以及 ANN 在处理复杂模式和函数逼近方面的能力。遗传算法（GA）是一种广泛应用的随机搜索技术，其灵感源自生物界的自然进化过程。GA 的核心思想是模拟达尔文提出的"适者生存"原则，通过选择、交叉（杂交）和变异这三种遗传操作来模拟自然进化中的关键机制，这些操作使 GA 能够有效地处理各种复杂的优化问题。蚁群优化算法（ACO）是一种新颖的元启发式算法，用于解决 NP-hard 问题，其设计灵感来源于蚂蚁觅食行为的观察。最初，ACO 主要用于解决旅行商问题（TSP）等离散优化问题，但随着算法的发展，它也被扩展到连续域问题的优化中。目前，已有多种 ACO 变体被提出用于处理连续域内的优化挑战。粒子群优化算法（PSO）是一种受鸟类群体觅食行为启发的随机全局优化方法。PSO 通过粒子间的位置和速度信息共享来引导搜索过程，通过速度更新来驱动位置更新，从而在解空间中搜索可能的最优解。该算法从一个随机初始化的粒子群开始，每个粒子在搜索空间中具有独特的位置和速度特性。这些算法的结合为解决燃煤锅炉燃烧优化问题提供了强大的工具。

氮氧化物排放的模拟和优化是改进燃烧过程的两个关键环节。首先，采用 SVR 方法对锅炉 NO_x 排放量与运行参数之间的函数关系进行建模。SVR 模型能够准确预测 NO_x 排放量，并且与基于遗传算法的方法相比，通过网格搜索优化的 SVR 模型在选择参数（C，γ）时，更适合在线建立最优 NO_x 排放预测模型。为了实现低 NO_x 燃烧，研究者提出了四种优化算法。实验结果表明，将 SVR 与优化算法（除了粒子群优化 PSO）结合的混合方法能有效地将燃煤锅炉的 NO_x 排放量降至低于国家法规要求的水平。这一发现意味着，通过调整锅炉运行参数，而不是依赖昂贵的选择性催化还原（SCR）和非催化还原（SNCR）烟气处理设备，可以在一定程度上实现 NO_x 排放的环境合规。此外，研究表明，基于 SVR 等人工智能方法和进化算法的燃烧优化技术为减少燃煤锅炉的 NO_x 排放提供了一种经济且有前景的途径。通过比较四种优化算法的结果质量和收敛速度，发现 PSO 算法的解质量最低，而蚁群优化算法的性能优于传统的遗传算法。

锅炉的燃烧是一个复杂的过程，与燃烧过程和设计因素有关。由于锅炉设计因素在运行中是固定的，因此，实现最优燃烧主要依赖于对燃烧操作参数的调整。影响燃烧状态的可调参数包括一次空气、二次空气、氧气供应、煤质、煤的细度等，这些参数需要根据锅炉负荷的变化进行协调调整。尽管运行人员通常依靠经验进行操作，但要实现锅炉的最佳燃烧状态几乎是不可能的。因此，采用燃烧调节试验来指导实际运行是提高锅炉运行质量的有效方法。为了满足实际燃烧过程中对优化的需求，在线燃烧优化系统提出了对实际燃烧优化的需求和局部优化的需求，如数据检测要求、快速准确要求、不同优化目标要求、在线自学习要求、参数优化极限要求、容错要求、报警要求、离线数据处理和优化的兼容性等。其次介绍了用于在线燃烧优化系统的仪表或传感器。然后，提出了在线 SVM 算法，主要涉及增量关系的推导、AOSVR 记账程序、R 矩阵的有效更新增量算法和减量算法的初始化。另外，在线燃烧优化系统主要有三大功能，它们分别是在线监测和报警功能、在线优化和自学习功能、离线建模和优化功能。最后讨论了在线燃烧优化系统的应用。

本书对基于计算智能的燃烧优化技术的未来发展潜力进行了深入探讨。该技术在电力行业中扮演着至关重要的角色，尤其是在电厂的氮氧化物减排和提高粉煤灰中碳的燃烧效率方面。尽管计算智能为燃烧优化提供了强大的工具，但仍存在一些挑战和不确定性，包括模型的煤炭特性、大型数据集的处理和特征选择，仍需要进一步的研究和开发。基于计算智能的燃烧优化技术是一项非常有价值和实用的技术，可用于产业变革的决策。基于计算智能的燃烧优化技术具有极大的应用价值，它不仅能够为工业决策提供支持，还有潜力引发产业变革。尽管存在挑战，但通过不断的技术创新和跨学科合作，这些挑战可以被克服。未来的研究将继续推动这一领域的发展，为实现更清洁、更高效的燃烧技术提供强有力的支持。

(1) 模型煤的特性。虽然煤质特性监测系统已成功地在电厂中投入使用，但燃烧优化模型中未包含实时煤质信息。煤质(如 Aar、Var、Qar、Net 和 Mar)被用于模型设置及优化参数集搜索。在计算过程中，煤质被视为一个常数，但它有时会发生两种或三种类型的变化。在电厂运行中，由于计算模型是以某一特定煤种为基础，煤的特性会发生不可预测的变化，导致优化参数的失效。在线煤特性监测信号延迟，DCS 中锅炉运行参数与煤特性监测无法同步匹配，增加了基于煤特性的燃烧建模难度。同时，煤粉细度也是煤的另一个重要特性。如何在优化模型中加入实时的煤粉特性和在线煤粉细度是一个亟待解决的问题。

(2) 处理大数据集。优化模型中使用的数据集在训练数据和测试数据方面都受到限制，大多数情况下其数据规模保持在 1000 以下。在各种机器学习问题中使用内核引起了很大的兴趣，其中 SVM 是燃烧优化中最突出的例子，处理大型操作数据中十分耗时，因此在发电厂应用中不可行。即使是训练数据也是从 DCS 中收集的，其中大部分数据都是人工筛选的，以避免在缩放数据计算时遇到障碍。工业数据在测量期间波动且不确定，人工筛选会损失大量的燃烧信息，所选数据是主观的或所谓的平均值，这会影响最终的结果。当前已研究了许多技术来减少时间和空间复杂度，流行的和激进的方法是低秩近似，例如贪婪近似、采样分解；缩放核方法，例如分块、复杂分解；其他方法是拉格朗日 SVM、减少 SVM、SMO、SVMlight、simpleSVM 等。将这些新的模型技术应用于处理大型燃烧数据，将保证最优结论的真实性。

(3) 特征选择。影响燃烧的因素很多，如煤质、锅炉负荷、气流组织、锅炉型式、燃烧器型式、炉膛温度、过剩空气系数、煤粉细度、风煤分布均匀性等。通过简化的数学公式对这些因素进行逻辑分析，这些因素都隐含着对残余元素的人为干预。燃烧是一个非常复杂的过程，受到大量因素的影响，DCS 数据库体现了这一点，优化模型会考虑一定数量的因素，这仅限于已经知道的影响燃烧的因素。这种选择过程简化了燃烧模型，将导致优化结果未知。在模型中引入足够多的因素的新算法正在研究中。

本文摘译自《Combustion Optimization Based On Computational Intelligence》，Springer 出版社&浙江大学出版社，ISBN：978-981-10-7873-6，2018。

太 阳 能

本书重点介绍太阳能光热发电、光热利用和光伏发电的基本原理、工程案例、研究进展以及相关的新概念和新方法；分析了太阳能资源和利用的总体概况；介绍了太阳光追踪的定位、太阳辐射在大气中的传递与测量、聚光方法与吸热介质、热力循环与分析软件等。针对太阳能光热发电技术及系统，包括槽式、线性菲涅尔式、塔式和碟式等系统，分析了各类聚光器、吸热器和发动机的设计与运行特点，还专门讨论了适合太阳能热发电的储热技术。介绍了中低温热水器与集热器、太阳能制冷与空调的基本原理，以及建筑一体化的设计原则等。以单晶硅光伏电池为例，主要介绍了光伏发电的基本原理与系统，分析了光热和光伏发电互补及稳定并网的技术前景。探讨了太阳能热光伏、光子增强热发射、热电材料、热/光化学转化及新型光热利用技术。

本书内容主要分为 6 章。

第 1 章为概论。该章首先介绍了我国太阳能资源的分布概况，表明我国 2/3 以上地区的年日照时数大于 2000h，太阳能资源分布具有高原大于平原、西部干燥区大于东部湿润区等特点。而全球的太阳能资源分布差异较大，一般而言，中国西部和北部、澳大利亚、阿拉伯地区、非洲的南部和北部、南欧、美国西南部、智利、巴西等地区的太阳能资源较好，本章小节末附上了一系列数据库网址供读者查找最新的全球太阳能资源分布情况。然后总体性地介绍了目前太阳能利用的情况。太阳能利用主要分为发电、热利用、化学转换三大类。太阳能发电主要包括光伏和光热两种发电方式。光伏发电是指利用半导体材料的光生伏特效应，将光能直接转换为电能的技术。光热发电是指通过聚光集热装置将太阳辐射收集起来，先转换为热能，提高工质的温度，再通过热功转换装置，将热能转换为电能的系统。该章还给出了国内外光伏、光热发电的发展历史以及全球光热发电的累计装机容量和市场分布情况。热利用主要是中低温供热，也包括太阳能干燥、海水淡化、制冷、空调、除湿及在建筑中的使用等。该章在热利用方面给出了太阳能热利用发展过程中发生的重要事件，并通过图表厘清了全球太阳能热水器装机容量数据，以及太阳能热利用对我国节能减排的贡献。太阳能化学转换主要包括储能、制氢、燃料及化学品等，该章主要阐述了太阳能制氢技术，包括太阳能发电电解水制氢、太阳能热分解水制氢、太阳能光化学分解水制氢、太阳能光电化学电池分解水制氢以及太阳光络合催化分解水制氢。

第 2 章为太阳能利用技术中的光学原理和热工基础。现代太阳能利用通常涵盖收集、吸收、转换、存储、传输和利用等过程中的两个或多个过程。该章主要介绍光学和热工学两大部分的基本理论和方法。第一部分介绍了常用的坐标体系。首先介绍了天球的概念。假定不同的点为天球中心，可确立不同的天球坐标系：①地平坐标系，天球中心与观察者位置重合。②赤道坐标系，天球中心与地球中心重合。③黄道坐标系，天球中心与太阳中心重合。④银河坐标系，天球中心与银河系中心重合。在太阳能工程中，最方便且常用的是地平坐标系,其中包含日地距离、赤纬角、时角、视太阳时间、钟表时间等重要参数。在地平坐标系中,地球上的观察者可通过太阳高度角和太阳方位角来描述任意时刻、任意位置上太阳相对于其的位置。第二部分介绍了太阳辐射及其测量和太阳辐射的聚光原理。太阳辐射方面主要介绍了平面上太阳辐射强度的计算,包括直射辐射强度和散射辐射强度。辐射测量原理是将入射到测量仪器特制受光面上的太阳辐射能全部吸收,使之转换为其他某种形式的能量并进行检测。主要测量仪器分为直接辐射表、总辐射表、散射辐射表、直射辐射表、反射辐射表等。从光学原理上看,聚光方式可以分为两大类:折射式聚光和反射式聚光。根据太阳辐射聚焦在接收器上形成点还是面可分为成像聚光和非成像聚光。前者是指通过聚光器将太阳辐射聚焦在接收器上形成焦点,以获得高能流密度的太阳辐射。后者的光线则会聚到一块接收器的面上以获得高强度太阳能。最后一部分介绍了新型吸热介质以及太阳能利用中的热力循环,主要分析了纳米流体、功能流体、液态金属的传热特性,并介绍了水/蒸汽朗肯循环、有机朗肯循环、斯特林循环、布雷顿循环、卡琳娜循环等的工作原理。在章节的最后罗列了太阳能利用中的常用分析软件。

第3章为太阳能光热发电系统与工程。该章首先介绍了我国太阳能热发电技术的发展现状,然后详细讲解了不同光热发电系统,主要包括槽式光热发电系统、线性菲涅尔式光热发电系统、塔式光热发电系统、碟式光热发电系统。对于槽式光热发电系统,该章主要围绕系统的聚光器的设计和光学性能、吸热器的参数和选择性涂层展开叙述,并给出了槽式光热发电系统的热力学分析以及技术发展展望。对于线性菲涅尔式光热发电系统则集中介绍了聚光器设计、吸热器设计及工质换热分析。塔式光热发电系统重点讲述了聚光器、吸热器、储热装置及二次反射技术,并在最后补充了新型塔式吸热器以及该项技术的未来发展趋势。碟式光热发电系统采用旋转抛物面聚光器,主要介绍了其斯特林发电系统、液态金属吸热器、燃气轮机发电系统、朗肯循环发电系统。其次对太阳能热发电系统中的储热技术进行了更全面的讲解。按照储存的能量形式,主要分为物理储热和化学储热。物理储热包括显热和潜热储热:显热储热是指通过储能介质温度的变化来实现储能的方式,潜热储热是指通过储热材料发生相变时吸收或释放热量来进行能量的储存和释放的储热方式。化学储热是指利用化学键的断裂重组实现能量的存储和释放的储热方式。该章主要讲解了各储热方式的材料选择、系统组成及储热量分析计算。最后该章介绍了其他光热发电技术,主要包括空间站太阳能发电技术、太阳能热气流发电、太阳能池热发电技术。空间太阳能发电是指在地球外层空间建立太阳能发电基地,通过微波将电能传输回地球,再将微波转换成电能供人类使用。太阳能热气流发电,又称太阳能烟囱发电,是利用太阳能加热吸热器内的空气,热空气在烟囱内产生上升的推力驱动风机做功发电的技术。太阳能池热发电则是以太阳能池底的高温盐水为热源,通过热交换器加热工质,驱动热机发电。

第4章为太阳能中低温利用基本原理及工程。该章主要是对太阳能热利用进行进一步的深化,第一部分是太阳能集热器,分别介绍了平板型集热器的组成结构以及不同类型真空管集热器的技术原理,之后按不同的组合方式,列举了太阳能热水器循环系统的分类情况以及太阳能集热器热性能测试方法,最后详细给出了设计太阳能热水器系统的设计步骤。第二部分则是讲了太阳能制冷，按循环类型分类，可分为吸附式制冷循环、吸收式制冷循环及喷射式制冷循环等。在该节里对吸收式制冷循环的基本原理、制冷工质及与太阳能相结合的制冷机组工作原理进行了详细阐述。对于吸附式制冷循环，给出了其工作原理及系统分类。除此之外，还有太阳能喷射式制冷系统及太阳能除湿空调系统，并讲解了各技术的基本原理。第三部分则是讲了太阳能在海水淡化方面的应用及太阳能建筑一体化的概念。对于利用太阳能产生热能以驱动海水相变过程的海水淡化系统，通常称为太阳能蒸馏系统。按是否存在任何利用电能驱动的动力原件和利用太阳能集热器等部件进行加热，太阳能蒸馏系统可分为被动式太阳能蒸馏系统和主动式太阳能蒸馏系统两大类。该节对这两类系统进行了详细介绍。在太阳能建筑一体化方面，首先给出了其基本概念，然后将其细分为太阳能光热利用与建筑一体化和太阳能光伏利用与建筑一体化，并分别列举了已经应用的真实案例及技术原理。

第 5 章为太阳能光伏发电原理与系统。该章第一部分介绍了太阳能光伏发电的原理。分析了半导体材料的能带结构、材料特性及材料制备，并讲解了能带理论和半导体的基本特性及物理机制。在发电原理方面，介绍了 p-n 结的形成机制、工作原理及光生伏特效应。光线照射在太阳能电池上，能量大于半导体材料禁带宽度的光子将进入 p-n 结区，这会将电子从价带激发，并在 p-n 结附近将产生"电子-空穴"对，电子和空穴在内建电场的作用下定向迁移形成电流，最后介绍了硅基太阳能电池的基本结构及工作原理。按光伏技术的发展将太阳能电池进行分类，即第一代晶硅太阳能电池、第二代薄膜太阳能电池和第三代新型高效电池。第二部分则是介绍了太阳能光伏发电系统，按照其是否并网可以将系统分为离网式光伏发电系统和并网式光伏发电系统。离网式光伏发电系统一般由光伏电池、充放电控制器、蓄电池、逆变器及负载等构成。并网式光伏发电系统一般由光伏电池阵列、并网控制器、蓄电池及逆变器等构成。并网系统还可分为分布式和集中式光伏发电系统。从发电出力特性、最大功率点跟踪控制等方面介绍了光伏电站的运行特性。最后一部分介绍了太阳能光伏/光热综合利用技术，分析了光伏电池技术与太阳能集热技术结合的方法，利用冷却介质带走电池组件中多余的热量并加以利用，从而提高系统效率。根据冷却介质的不同可分为光伏/热水综合利用、光伏/热空气综合利用、光伏/热泵综合利用以及光伏/光热发电综合利用技术。

第 6 章为新型太阳能发电与利用技术。该章主要介绍了一些新技术及发展趋势，包括太阳能热光伏发电技术、太阳能热电子发电技术、太阳能温差发电技术、太阳能化学转化技术等。太阳能热光伏是将太阳能辐照到吸收体，吸收体受热后以红外辐射的形式，将能量传输至近邻的光伏电池，再由光伏电池转换为电能的发电技术。与传统的光伏发电技术相比，太阳能热光伏转换技术适用于较高聚光比和较高集热温度的场合，并可通过改变吸收体的集热温度来调制发射光谱，匹配光伏电池的禁带宽度，从而提高发电效率。太阳能热电子发电技术是利用聚光太阳能辐射给金属加热，当金属的温度升高到一定值时，大量电子获得大于表面逸出功的动能，从金属内部逸出到真空。此时高温金属成为电子发射源，在近邻区域放置金属板接收真空中的电子，即可在两极板极之间形成电流。其还可细分为太阳能热致热电子发电和太阳能光子增强热电子发电。太阳能温差发电基于塞贝克效应，太阳辐射经聚光器聚焦并被温差发电材料的热端吸收转化为热能，导致其热端和冷端产生温度差。在热激发作用下，p-n 节中处于热端的空穴和电子浓度大于冷端，在浓度梯度驱动下，空穴和电子向冷端扩散，从而在 p-n 节两端形成电势差，电子流经外部负载发电。太阳能温差发电技术装置结构简单，体积小，制造工艺简单，运行成本低，寿命长，工作时无噪音，对环境污染较小。太阳能化学转化技术可分为热化学转化技术和光化学转化技术。太阳能热化学转化技术主要是指利用热化学反应将聚焦的太阳能转化为易于储存和运输的碳氢燃料的化学能的过程。光化学转化包括光电催化制氢、光敏化降解有机污染物、光合生物反应等。其他太阳能利用技术还包括太阳能直接产生水蒸气技术、太阳能空气取水技术等。

本书涵盖太阳能跟踪与测量、聚集与收集、传递与储存、转换与转化、发电与利用、系统及工程等多方面内容，结构体系完整，内容翔实，逻辑性强。面向能源动力类专业学生，重点介绍了太阳能光热发电和光热利用的基本原理，并通过案例分析加强工程理解。本书专门介绍了适合太阳能热发电的各种储热技术，突出了其在太阳能热发电技术中的重要地位，拓展介绍了光热和光伏利用的新技术和新方法，以期为未来太阳能技术研发和规模化应用启迪新思路。

本书还为读者提供了课后学习的数字资源，涵盖目前全球光热发电工程的清单以及典型案例、全球太阳能资源分布数据库、太阳能利用中的常用分析软件等。此外，该教材还配套了虚拟仿真教学平台，网址为 https://www.ilab-x.com/details/page?id=11369&isView=true，以期强化学生对相关知识的理解，提升应用与创新能力。

本文摘自《太阳能》，中国电力出版社，ISBN: 978-7-5198-2752-6，2019。

烟气多种污染物高效脱除技术原理及应用

煤炭曾在世界工业发展历程中扮演着极其重要的角色。然而，煤炭燃烧过程排放的颗粒物（PM）、二氧化硫（SO_2）、氮氧化物（NO_x）、汞（Hg）、三氧化硫（SO_3）等多种污染物，是导致区域灰霾等大气污染问题的重要原因。随着人们对大气环境质量要求的不断提高，欧美等发达国家的大气污染防治政策法规日趋严格，对燃煤电厂等大气污染源提出了更严的排放限值要求。为应对这些排放限值，欧美等发达国家一方面通过"煤改气"减少燃煤机组的数量来削减污染物排放，另一方面也持续投入研发更先进的燃煤烟气污染治理技术。但如何实现复杂煤质、多变工况条件下的多种污染物的高效协同脱除，达到更严格的排放限值要求，仍是亟待解决的难题。

我国是全球最大的煤炭消费国，煤炭消费量约占全球一半以上，且短时间内以煤为主的能源消费结构难以根本改变。据预测，2030年煤炭占我国能源消费总量的占比仍超40%。煤电消费了我国一半以上的煤炭，也曾是我国工业部门最主要的大气污染物排放源。因此，大幅减少燃煤机组烟气污染物排放量，是改善我国大气环境质量的关键。但是美英等国"煤改气"治霾的成功经验依赖于丰富的天然气资源，这在煤电机组装机规模大而天然气短缺的中国难以复制。因此，亟须研发及推广应用更先进的燃煤烟气污染物高效脱除技术，实现燃煤烟气污染物超低排放，满足燃机机组排放标准限值的要求。针对燃煤烟气多种污染物超低排放的迫切需求，本书分为五章，框架结构如图1所示，

图1　本书的框架结构

从多种污染物催化脱除强化、静电脱除强化、吸收脱除强化、多种污染物脱除系统及智能控制等方面,详细介绍了从单一污染物治理技术到多种污染物高效协同脱除技术研发、装备研制和工程应用最新进展。

第 1 章详细探讨了化石能源利用过程中产生的各种大气污染物及其治理进展。总结了发电和其他工业部门污染物的主要类型和来源,并讨论了这些污染物对环境和人体健康造成的不利影响;梳理了国际社会为解决空气污染问题出台的相关政策和实施的行动;从多种污染物催化脱除、静电脱除、吸收脱除等方面回顾了污染物脱除技术的最新进展,并提出了未来技术展望。

第 2 章针对多种污染物催化脱除过程,首先介绍了选择性催化还原(SCR)脱硝技术的基本原理,并详细讨论了高效 SCR 脱硝技术中的关键问题。重点阐述了通过稀土、过渡金属、类金属等元素的掺杂改性,提升催化剂的抗中毒、单质汞氧化能力及低温区间的反应活性,开发适合我国复杂多变煤质特性的高效抗碱金属/碱土金属/重金属等中毒、宽温度窗口的系列催化剂配方,解决燃用复杂煤质、污泥等锅炉的高效催化脱硝及锅炉低负荷脱硝难题;研发脱硝催化剂再生改性工艺技术及装备,在实现 NO_x 高效脱除的同时可协同控制重金属汞等其他污染物,并解决了废旧催化剂再生与处置问题;开发反向双旋涡强化的还原剂分配和烟气混合系统,解决了不同炉膛结构及燃煤锅炉大截面、短行程、突变/多变截面烟道等复杂条件下还原剂氨与烟气混合难题;针对变负荷条件下 NO_x 高效脱除难题,提出了宽温度窗口催化剂耦合智能化精准调控实现 NO_x 全负荷高效脱硝的新思路。在此基础上,开发适应工况多变的全负荷高效脱硝技术,可实现脱硝效率达 90%以上,在锅炉 35%~100%BMCR 负荷时 NO_x 排放浓度均低于 50mg/m^3,氨逃逸小于 3ppm,且大幅减小 SCR 系统出口 NO_x 浓度波动,解决燃煤机组低负荷运行下 NO_x 高效脱除及氨逃逸的难题。

第 3 章针对污染物静电强化脱除过程,以颗粒物为例,详细讨论了燃煤烟气颗粒物物理化学特性和颗粒物静电脱除原理,总结了影响颗粒物静电脱除的主要影响因素及颗粒的凝聚、长大、荷电、迁移等过程对颗粒物静电脱除强化效果,并介绍了典型的细颗粒物静电强化脱除技术。梳理总结了包括湿式静电除尘、低低温电除尘、高效凝并、烟气调质、高效供电电源等多种污染物静电脱除强化关键技术。基于颗粒荷电理论,通过高频电源、脉冲电源及三相电源,有效提升细颗粒物的荷电量,解决细颗粒荷电难从而导致的捕集效率低的难题;基于静电场中细颗粒物"凝结-团聚-荷电-迁移"的多元强化机制,研制高稳定放电电极、二次扬尘抑制收尘板、增均流预荷电器、凝并团聚强化装置等部件,解决宽温/宽湿烟气放电易击穿、颗粒荷电效率及脱除效率低难题;结合湿法烟气脱硫塔后采用新型湿式静电除尘技术,研发脱硫塔前除尘、脱硫塔内除尘及脱硫塔后除尘的细颗粒物高效脱除系统,可实现颗粒物排放浓度小于 1mg/m^3,且具有很好的 SO_3 协同脱除效果。

第 4 章针对 SO_2 等污染物吸收脱除过程,详细介绍了 SO_2 等气态污染物吸收脱除的基本原理,讨论了影响 SO_2 等气态污染物吸收脱除的关键因素。针对 SO_2 等污染物高效脱除过程中的溶解、吸收等关键步骤,提出浆液分区、添加剂与构件等多元强化 SO_2/Hg 吸收及颗粒物捕集的单塔高效脱硫协同除尘脱汞的新思路。通过研究加装不同构型、开孔率等构件后塔内持液特性、压力特性及气液传质特性的变化规律,得到强化传质塔内构件的关键设计参数,有效提高气液两相在塔内分布的均匀性与有效接触面积;总结梳理了液相多脱副产物 SO_3^{2-}/SO_4^{2-}、NO_2^-/NO_3^-、Hg/Hg^{2+} 等的氧化、再释放路径,为湿法脱硫实现多种污染物协同脱除提供了支撑;针对脱硫副产物的品质特性,通过开展温度、pH、溶解氧等对亚硫酸根氧化的影响规律研究,获得石膏结晶的运行参数以及浆液组分在结晶过程的作用规律,保证高脱硫效率的同时提升了石膏品质、减少了系统结垢;在此基础上,形成适应不同煤种、不同负荷等系列 SO_2 高效脱除及 PM、Hg、SO_3 等污染物的协同脱除技术,脱硫效率可达99%以上,SO_2 排放浓度可低于 20 mg/m³,且具有较好的颗粒物、Hg、SO_3 的协同脱除效果。

第 5 章详细介绍了多种污染物高效协同脱除的超低排放系统及智能调控方法。通过 NO_x、SO_2、PM、Hg 等多种污染物高效脱除与协同控制关键技术及设备集成方法,建立多种污染物高效协同脱除超低排放系统,形成了适合不同炉型、容量、煤质的烟气污染物高效稳定低成本超低排放系列化解决方案,如图 2 所示。研发烟气多污染物脱除系统的智能调控关键技术,建立烟气多污染物脱除系统的智能调控平台,通过污染物排放精准预测、脱除过程协同优化、运行成本综合评估、碳排放精准核算,显著提升了"双碳"目标下燃煤多污染物脱除系统运行的稳定性、可调性与经济性。

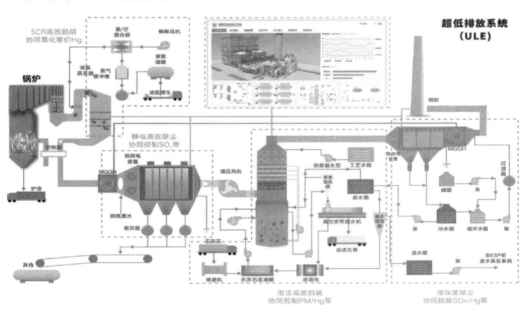

图 2 典型燃煤电厂超低排放系统流程图

本书中所介绍的多种污染物催化脱除强化、静电脱除强化、吸收脱除强化、多种污染物脱除系统及智能调控等超低排放关键技术成果率先在1000MW燃煤机组上成功示范应用,如图 3 所示,实现主要烟气污染物排放指标优于中国天然气燃气轮机组排放标准限值,污染物排放浓度与国家火电厂污染物排放标准相比,PM 排放降低 60%以上,SO_2 排放降低 80%以上,NO_x 排放降低 50%以上。该超低排放技术示范应用电厂被国家能源局授予"国家煤电节能减排示范电站",为国务院政府工作报告中的"燃煤电厂超低排放"

提供了典型案例。介绍了超低排放技术成果的推广情况,成果在浙能、国家能源集团、华能、大唐等大型发电集团十几个省市的 300 多台 50～1000MW 燃煤机组及中小型热电机组上得到规模化工程应用,显著提升了燃煤污染治理技术及装备水平,开启了我国燃煤电厂进入超低排放的新阶段。

图 3　1000MW 燃煤机组超低排放示范工程

同时,本书对我国超低技术成果应用后的污染物减排效果进行综合评估分析。2013～2019 年我国电力行业污染物排放情况显示,自超低排放政策出台以来,NO_x、SO_2 和 PM 的总减排率分别高达 92.6%、93.1%和 95.8%。2019 年,火电行业 SO_2、PM、NO_x 排放总量相比于 2013 年分别下降 19%、29%、9%;同时我国电力行业 SO_2、PM、NO_x 排放总量占比从 2013 年的 25%下降到 2019 年的 10%,超低排放技术推广减排效果显著。

最后,对未来燃煤电厂污染物高效脱除技术发展进行了展望。近年来,随着燃煤机组超低排放技术的大规模应用,燃煤烟气硫氧化物、氮氧化物和颗粒物排放量大幅降低,但为了应对环境空气质量持续改善的重大需求,还需进一步控制三氧化硫、氨和重金属等非常规污染物;同时,国家"双碳"目标对燃煤机组低碳清洁发展提出了新的要求。我国燃煤机组炉型/容量多样、燃用燃料复杂、工况波动大,如何在现有超低排放基础上,实现常规/非常规污染物高效协同脱除及 CO_2 协同减排已成为支撑煤电行业绿色低碳高质量发展的关键。未来,针对"双碳"目标下燃煤电站清洁低碳化的技术需求,重点突破燃煤烟气多种污染物与二氧化碳协同减排、烟气污染物与二氧化碳资源化利用、重金属等非常规污染物高效协同控制、工业废水低成本零排放、固体废弃物资源化利用、新一代低成本低能耗二氧化碳捕集利用与封存(CCUS)等关键技术,通过不断提高碳污协同治理效率,降低污染治理和碳捕集成本,实现煤炭利用过程更清洁低碳,为推动煤电行业绿色低碳转型,助力构建清洁低碳、安全高效的新型能源体系提供关键科技支撑。

本文摘自《Multi-Pollutant Control for Flue Gases—Principles and Applications》,Springer 出版社&浙江大学出版社,ISBN:978-981-16-1518-4,2021。

循环流化床锅炉数值优化设计与运行

本书是一本利用数值方法解决大型多颗粒循环流化床锅炉设计、运行中实际问题的专著。在循环流化床气固流动、传热、磨损、燃烧、燃烧产物、受热面沾污结渣模型与模拟，大型循环流化床锅炉三维整体数值模型与二维当量快算数值模型基础上，以不同循环流化床炉型，包括300MW、330MW、600MW、660MW和1000MW等循环流化床锅炉为示例，针对设计、运行中的问题开展数值模拟、变参数数值试验，如大型循环流化床大炉膛双布风板支腿密相区颗粒混合、床料平衡与控制、高炉膛气固浓度分布、二次风穿透、悬吊受热面表面颗粒流动、多分离器循环回路物料平衡、炉膛水冷壁、中隔墙和悬吊受热面热流分布、超临界水动力特性、回料系统受阻对炉膛运行影响等设计和运行中至关重要和必须解决的问题，给出问题的数值回答与讨论，为循环流化床锅炉设计、运行提供参考和思路。本书具有以下特点：①针对目前热点超/超临界参数大容量循环流化床锅炉研发中的新问题；②数值计算用于大型循环流化床锅炉设计与应用最新成果；③浙江大学近十余年大型循环流化床数值模型研发与数值试验研究结果与积累；④基于计算机辅助实验探讨大型循环流化床锅炉研发中的热点、难点问题。

浙江大学循环流化床研究团队自20世纪80年代初以来一直开展循环流化床理论和技术研究。1991年组织中国循环流化床工业界培训会议，1994年翻译了世界上第一本关于循环流化床的专业书 Circulating Fluidized Bed Boilers Design and Operations（《循环流化床锅炉的设计与运行》），1998年撰写专著《循环流化床锅炉理论设计与运行》，2003年编写循环流化床设计导则，2005年组织承办第8届国际循环流化床会议。近十余年来结合国家项目和市场需求，积极开展大型超/超临界循环流化床锅炉的理论和技术工作，特别在长期试验和理论研发基础上，针对大型循环流化床锅炉现场试验困难问题，发展适用于大型炉膛的循环流化床锅炉综合数学模型，采用数值模拟计算机辅助试验方法解决大型循环流化床锅炉研发、优化运行过程中的新问题。本书为浙江大学近十余年大型循环流化床数值模型研发与数值试验研究结果与积累。

本书内容主要分为8章。

第1章介绍了循环流化床锅炉发展历程和发展中遇到的问题。循环流化床锅炉是一种高效、低污染的燃烧设备，其核心技术是流化床燃烧原理。通过控制气流速度，使固体燃料颗粒悬浮在气流中，形成流化状态，从而提高燃料与空气的混合程度，促进完全燃烧。循环流化床锅炉具有燃料适应性广、燃烧效率高、污染排放低等优点，广泛应用于电力、化工、冶金等行业。循环流化床锅炉技术起源于20世纪70年代，经过几十年的发展，已成为成熟的燃烧技术，应用范围从处理工业废物和劣质燃料扩展到广泛的能源领域。并且，随着循环流化床锅炉大型化的发展，高参数循环流化床炉膛中的流动、燃烧、传热，与其在化工领域快速流态化反应器、小型流化床锅炉中存在差异，这些差异将影响大型循环流化床锅炉的设计和运行。随着循环流化床锅炉的大型化，设计运行中呈现了各种问题。在试验困难、试验方法局限性条件下，采用数值模拟方法针对问题开展模拟计算、变参数分析，不失为一种有效的方法。

第 2 章简要讨论了循环流化床锅炉气固流动、传热、磨损模型与数值模拟方法。首先，本章介绍了半经验燃烧模型和数值计算燃烧模型各自的优势和不足。数值模拟是研究循环流化床锅炉内复杂物理过程的重要手段。通过建立数学模型和计算机仿真，可以深入了解锅炉内的流动、传热、燃烧和化学反应等过程。常用的数值模拟方法包括欧拉-拉格朗日方法、欧拉-欧拉方法和离散单元法等。建立这些模型需要考虑气-固两相流动、化学反应、传热传质等多个因素，通常包括动量守恒方程、能量守恒方程、组分守恒方程和颗粒动力学方程等。通过这些方程，可以准确描述锅炉内的流体流动、温度分布、组分变化和颗粒运动等过程。最后本章对前文在循环流化床气固流动中的欧拉双流体模型、传热模型与磨损模型方面的工作进行了整理和总结，为建立大型循环流化床燃烧模型提供研究思路和方法。

第 3 章详细介绍了循环流化床燃烧、燃烧产物和受热面沾污结渣的数学模型与数值模拟。固体燃料在循环流化床锅炉中燃烧，气固流动燃烧反应主要发生在炉膛内，包括气固流动、煤颗粒干燥、挥发分析出和燃烧、颗粒破碎、焦炭燃烧及气固传热等一系列物理化学变化，是一种十分复杂的燃烧过程。通常煤颗粒进入循环流化床锅炉炉膛后，随着炉内气固流动经历下述一系列物理化学过程：①煤颗粒受高温床料加热，当温度达到 100℃左右时，开始析出水分。②当温度达到 300~400℃后，煤发生热分解，开始析出挥发分，主要为气态的碳氢化合物，同时生成焦和半焦。③伴随干燥和热解过程，煤颗粒由于热应力和热解作用，发生一次破碎。④当加热到约 500℃后，挥发分首先发生着火，随后焦炭开始着火。⑤挥发分燃烧，焦炭燃烧。挥发分的燃烧过程为气-气同相化学反应，燃烧速度快，从析出到基本燃尽所用的时间约占煤全部燃烧时间的 10%。⑥焦炭继续燃烧，同时发生二次破碎和磨损等物理变化，直到燃尽。焦炭的燃烧为气-固异相化学反应，燃烧速度慢，燃尽时间长。在燃烧过程中，NO_x 和 SO_2 是两种典型气相污染物。虽然它们基本上不直接影响锅炉地燃烧性能和运行，但会对人体产生较大危害。因此，本章详细介绍了 NO_x 和 SO_2 的生成反应机理。最后，本章对碱金属迁移和沾污结渣机理进行了总结，主要原因可归结为两方面：①碱金属蒸气遇到温度相对较低的受热面发生凝结，形成一层具有黏性的初始层，导致受热面对飞灰颗粒的黏附性增强；②灰颗粒中富含 Na、K 的矿物成分易与其他矿物成分之间形成低温共熔体，导致飞灰熔融温度降低，易黏附在受热面上。

第 4 章针对大型循环流化床锅炉建立了三维整体数值模型与二维当量快算数值模型。针对大型循环流化床锅炉的大尺度超/超超临界循环流化床锅炉整体数值模型，是以欧拉两相流模型和多尺度最小能量(EMMS)曳力模型的气固两相流动流场为基础，综合考虑煤燃烧过程中水分蒸发、挥发分析出、焦炭燃烧、气相反应以及燃烧产物生成等组分反应模型建立的。壁面传热计算基于颗粒团更新传热模型和离散坐标辐射模型。当锅炉工质为超/超超临界参数时，模型还耦合了水冷壁管内工质热工水动力和炉膛壁面传热计算。锅炉三维整体数值模型的计算目标，能够在企业级服务器上模拟获得超/超超临界循环流化床锅炉内的气固流动、温度、传热热流和燃烧产物三维分布结果，提供 2D 与

3D 动态数据显示，方便变参数计算，分析和解决锅炉设计与运行中存在的问题，为大型循环流化床锅炉精细、精准优化、设计、运行和调整提供预测知识和直观动态显示。同时，考虑到三维模拟耗时较长、计算时间成本较大，因此本章最后在三维计算模拟的基础上，提出了二维当量快算法，可有效缩短计算时间，提高计算效率。并将碱金属钠迁移的变参数作为算例，验证了二维模型的准确性。

第 5 章基于大型循环流化床锅炉三维整体数值模型，针对典型大型超/超超临界循环流化床锅炉(包括炉膛两侧墙布置 6 分离器循环流化床锅炉、炉膛单侧墙布置 4 分离器循环流化床锅炉和环形炉膛循环流化床锅炉)炉膛开展了数值计算研究，报告了循环流化床锅炉炉膛气固流场、主循环回路气固流场、二次风穿透、多分离器气固均匀性、中隔墙对气固流动的影响，悬吊屏气固流场和防磨梁对炉膛气固流场的影响。

第 6 章讨论了大型循环流化床锅炉受热面传热数值计算的相关结果，研究对象为 330MW 亚临界、350MW 超临界 600MW 超临界、660MW 超超临界和 1000MW 超临界循环流化床锅炉炉膛中的受热面，涉及内容包括水冷壁、中隔墙、悬吊屏及水冷壁防磨梁设置对传热的影响。

第 7 章探讨了大型循环流化床锅炉在燃烧过程中污染物的生成与控制的数值计算结果。锅炉炉膛内部的复杂环境使实际测量颇具挑战，通常假设温度场和各燃烧产物在炉膛中均匀分布。然而，数值模拟揭示了炉膛内温度和产物分布的实际不均匀性，这对于理解和控制污染物生成至关重要。模拟结果不仅提供了关于如何优化锅炉设置的具体建议，还探讨了不同操作变量对炉膛内温度场和产物场的影响，这对于设计更高效和环保的循环流化床锅炉具有指导意义。

第 8 章针对近年来大型循环流化床锅炉研发、运行中出现的一些问题基于数值模型开展计算研究，给出了数值回答，包括双布风板炉膛翻床机理，床存量、粒径对炉膛压降的影响，多分离器循环回路回料失衡对炉气固流场的影响，布风板布风对回料率的影响，受热面磨损与防磨梁布置，高碱煤燃烧受热面沾污问题。数值模拟结果的体现包括针对问题的数值和变化趋势。考虑问题分析时条件不同，具体结果的数值存在误差，但结果的变化趋势特别是局部区域分布和变化特性对锅炉的设计和运行具有指导意义。

总的来说，本书是一部系统、全面、深入的专业著作。书中的内容涵盖了循环流化床锅炉的基本原理、数值模拟方法、优化设计、运行管理、污染控制以及未来的发展方向，为相关领域的研究和实践提供了宝贵的参考资料。本书可供从事循环流化床锅炉和热能动力研究、设计、运行和管理人员参考，也可作为高等院校有关专业研究生、大学生的参考资料。通过学习本书，读者可以深入了解循环流化床锅炉的各个方面，掌握先进的技术和方法，从而在工程实践中提高锅炉的设计和运行水平。本书不仅具有重要的学术价值，也具有很高的实际应用价值。希望本书能够帮助更多的读者深入了解和掌握循环流化床锅炉技术，为推动能源技术进步和环境保护事业贡献力量。在能源紧张和环境保护的双重压力下，循环流化床锅炉技术的推广和应用无疑将为实现可持续发展的目标做出重要贡献。通过本书的学习，读者不仅可以获得理论知识，还可以在实际操作中得到指导，促进循环流化床锅炉技术的进一步发展和应用。本书在理论与实践之间架起了桥梁，既有助于推动循环流化床锅炉技术的进一步发展，也为实现清洁、高效的燃烧技术提供了宝贵的经验和指导。希望通过本书的广泛传播，更多的工程师、研究人员和学生能够掌握循环流化床锅炉的核心技术和最新进展，为实现清洁、高效的能源利用和环境保护目标做出积极贡献。

本文摘自《先进热能工程丛书——循环流化床锅炉数值优化设计与运行》，化学工业出版社，ISBN：9787122410634，2023。

第三篇 报 告

研究型大学要培养多层次的现代工程师

岑可法(中国工程院院士、浙江大学教授)

我国研究型大学已经成为高级人才和创新思想的摇篮、科技发展的中坚,并且走到社会中心,为推动经济和社会进步做出越来越多的贡献。高等学校正在成为基础研究的主力军,高新技术研发的重要方面军和成果转化与产业化的强大生力军。社会经济发展和科学技术产业化需要研究型大学培养工程师。

从当前社会发展需求和学科定位目标来看,我国研究型大学还需要培养具有较强创新意识、具有国际交流能力的现代工程师。主要有以下原因。

我国的工程本科教育还是以培养面向生产第一线的高级技术人才为目标。我国本科教育有着独特的历史背景,经过几次调整,本科专业设置还是以二级学科划分为主,不像国外以一级学科设置本科专业。国内最好的研究型大学一半以上具有强大的工科背景,工科招生人数和培养目标都是以面对生产第一线为主。

我国的科研体制也比较特殊,科研力量主要集中在中国科学院和高校系统,中国科学院和大学系统的科研人员占了大多数,企业的科研力量十分薄弱。企业自己具有知识产权的产品还不多。因此,企业以生产为主,自主开发和研究能力不足。虽然我国培养的工程师人数占全世界第二位,但高层次的不多,结构也十分不合理。

当今高科技迅猛发展,像能源这样对国家可持续发展具有支撑意义的领域,需要实行跨越式发展的模式。各种高技术彼此渗透、高度融合,为新技术的产生和新的主导产业的形成提供了原始动力,而且,解决高技术产业化的瓶颈问题,也需要多种技术的配套。

对外开放的需要。随着我国不断开放,以及 WTO 对中国开放市场要求提高,我国企业将直接面对国际市场,具有国际交流能力的现代工程技术人才十分紧缺。

因此,我国研究型大学的工程专业需要培养现代工程师。

作为研究型大学,其工程专业应该面向生产第一线培养学生,培养的目标应该是企业的研究人员。换言之,我们要为企业培养更多的工程师,但已经不是原来意义上的工程师,而是具有创新能力和市场开拓能力的研究人员。研究型大学工程专业培养的人才中,既要有本科生层次的生产技术型人员,也要有硕士生、博士生层次的研究型人员,在培养结构上以培养博士学位的现代工程师为主。

这里,以热能专业的本科教育为例,来讨论一下现代工程师应该具有什么内涵。根据对现代工程师的理解和学校的具体情况,我们提出了多元化的培养模式,将专业培养规格定位于"研究型"或"技术型"。研究型人才培养模式按照热能与动力工程专业的自身学科特点进行专业教育,培养的学生具有扎实的热科学专门知识,可以进一步深造,成为热能与动力学科的专门研究人才和师资,也可在工程领域中从事与热能和动力相关的工程设计、技术开发及技术管理。技术型人才培养模式依托于某个工程领域，学生除掌握热能与动力工程专门知识以外，还要熟悉该工程领域的有关专业知识。毕业生应能够在该工程领域中从事与热能和动力工程相关的工程设计、技术开发及技术管理工作。从素质结构来看，要求思想素质、专业素质、文化素质、身心素质全面发展。从能力结构来看，需要有较强的获取知识、应用知识的能力和创新能力。

面对新世纪，要努力提高教学效果及教学质量，走出一条培养高层次跨世纪现代工程师的新途径。

在研究生教育方面，我认为应做好以下工作。

(1) 博士研究生学位论文的指导工作应由老、中、青教授相结合的导师组承担，尤其是由在全国乃至国际上著名的教授担任导师，会引起国家评审委员会的重视。导师组中每位导师的职责有分工，有的侧重于把握整个研究方向，有的侧重于指导基础理论研究，有的侧重于指导试验和工业应用研究，使之相辅相成、相得益彰。

(2) 要积极创造条件，鼓励研究生参与国家重大科研项目。如我院程军博士在三年攻博期间，在导师组的精心安排下，曾参加了国家"八五"重点科技攻关项目、国家"973"项目、国家"863"项目、国家自然科学基金项目、国家高技术产业化示范工程项目，以及省级基金和攻关项目等12项纵向研究课题，在重大项目的严格磨砺中，其理论和实践水平获得很大提高。

(3) 导师应该重用有发展潜力的博士生，创造条件让他们勇于挑担子，鼓励他们在国际SCI核心期刊尤其是在相关领域的国际权威杂志上发表论文。如能源领域国际最高级别的期刊 *Progress in Energy and Combustion Science* 的主编、Carnegie Mellon 大学的著名科学家 N.A.Chigier 了解到我们研究所在高温脱硫方面取得了突出的研究成果，就特别约稿，请我们撰写一篇关于高温脱硫的研究评述。为此，导师组特别安排程军博士总结我们研究所十几年来的研究成果，并汇聚了各位导师及相关师生研究实践中的心得体会，撰写了一篇长达25页的集大成的评论性文章，发表在该国际顶尖杂志上 (SCL/EI 双收录，2002年SCI影响因子3.06)，获得国际同行权威专家的广泛好评，这也是最近10年以来中国作者在该杂志上发表的首篇论文，为中国能源领域的科研人员赢得了更高的国际声誉。

(4) 对于工科博士生，应特别强调理论与实践相结合，在深化理论研究的同时应该争取在工程实践中取得重要的应用成果。如针对目前炉内燃煤过程中脱硫率低的难题，导师组提出了两段脱硫的研究方向，并指导程军博士进行了深入系统的基础理论研究和大、中、小各类规模的试验研究，在提高高温脱硫率的实践上取得重要突破，获得国家科技进步二等奖、省科技进步一等奖和教育部科技进步二等奖，并获国家发明专利授权。强调研究成果在工程实践中的应用性是今年改进优秀博士论文评选的一个鲜明特色。

培养人才要从基础抓起，培养高级人才要从大学本科抓起。为此，我院本科生教育正在专业教学计划、课程建设、创新实践、教学内容与方法等方面进行改革。

(1) 参考欧美著名大学机械类课程设置，结合我国国情，以拓宽专业基础、拓宽专业知识面为宗旨，制订教学计划，与欧美大学接轨。接轨"大机械"模式，增设专业基础课，包含机械、电学、力学、热学、材料、自动控制、系统工程等课程；公共基础课增设工程化学课程。根据学科发展与国民经济需求，以模块化方式设置。同时，设置专业课模块。能源与环境系统工程专业先设置能源生产及自动化、能源环境及自动化、人工环境及自动化、低温环境与低温生物四个模块。为适应不同层次学生的要求，专业课程中设置限选课、任选课与本硕结合课程。本硕结合课程高年级本科生与研究生同时上课，所修学分可带入研究生学习阶段。

(2)把教师最新的科研成果直接应用于教学。增设有关能源环境和系统工程的新课程，如热力环境控制、环境流体力学、基于循环经济的能源环境系统、能源与环境系统工程概论、微尺度传热学、微尺度流体力学等课程。对原有基于产品的专业课程进行改造，如能源转换、流体机械、热力系统工程、热工信号处理技术、能源生产过程自动控制、能源与环境实验。对于上述新增和改造的课程组织教材编写工作，形成一批内容新颖、知识面宽的专业课教材。

(3)以创新为目标，积极开设各种创新实践，培养学生的创新思维和创新能力。对欧美大学进行的专题调查发现，我国高等学校实践课占总学分的比例只有欧美大学的一半。为克服实践环节薄弱的现状，专业开设课外创新实验，拟开设太阳能汽车、自动控制系统等竞赛性实验，培养学生创造性思维、实践动手能力以及对专业的兴趣。提倡"以学生为中心"的探究式教学方法，引导学生主动学习，减少课内讲授学时，增加课内的讨论内容和课外实践学时。

(4)加强与国外知名高校的交流，努力提高学生的国际交流能力。

原文发表于《高等工程教育研究》2005(1): 16-18。

培养高质量博士的探索与实践*

岑可法

摘要：介绍了浙江大学工程热物理学科培养高质量博士生的经验：组成导师团队，发挥集体指导的优势；博士生研究方向要与国家重大需求相结合；理论联系实际，引导博士生把基础研究变成生产力；为博士生提供良好的物质条件；鼓励博士生参加国际会议和交流。

关键词：工程热物理学科；博士生培养；团队指导

作者简介：岑可法，浙江大学机械与能源工程学院教授，中国工程院院士，杭州 310027

浙江大学工程热物理学科是全国首批博士点、首批国家重点学科，20多年来为国家培养了175位博士，其中45人晋升教授，任职于清华大学、中国科学院、上海交通大学、浙江大学、大连理工大学、天津大学等全国著名高校和科研机构。这些人里有3位是教育部长江学者奖励计划特聘教授，3位是国家杰出青年科学基金获得者，4位国家"百千万人才工程"第一、二层次人选，5位全国优秀博士学位论文奖获得者，6位教育部跨世纪、新世纪人才，2位"973计划"首席专家，9位浙江省"151"人才梯队第一层次人才。在2003年全国一级学科评估中本学科在人才队伍和人才培养两项指标上排名均居全国第一。本学科所在研究所两次被评为省劳动模范集体和省级先进党支部。在教学成果方面，"培养高水平工学博士的新机制"获1997年国家级优秀教学成果二等奖；"瞄准能源学科前沿，构建一流导师群体，培养一流创新人才"获2005年度国家级教学成果二等奖。

工程热物理学科的这些成绩，是经过学科点全体成员艰苦不懈的奋斗才取得的。从1962年到1983年我一直在做学科负责人陈运铣教授的助手。1983年，陈运铣教授不幸突然逝世，工程热物理学科的发展面临很多困难。当时校领导让我负责这个学科的工作，要求把整个学科发展起来。我们只能靠人和、靠团结、靠团队精神、靠集体的力量去克服困难，经过不懈的努力，20多年来终于慢慢走出了一条具有浙江大学工程热物理特色的学科发展道路。

在总结培养高质量博士生的经验之前，我想指出我们培养博士生所遵循的两个原则。一个是科学技术是没有国界的，要相互学习。第二是我们的制度和培养方法要符合中国实际情况，要符合建设有中国特色的社会主义的方针。前一个强调科学技术是不分国界的共同的真理和学问，后一个强调科学技术要为国家的经济建设服务。下面具体谈一下工程热物理学科培养博士生的经验。

一、组成导师团队，发挥集体指导的优势

我们认为，传统单元式博士生培养模式（导师-博士生的一对一模式）有好的一面，也存在一些不足之处：①单元式培养方法使博士生的研究范围受到限制；②不利于导师和博士生充分发挥各自的长处，导师和学生各自的专长在这种培养模式下很难结合得很好；③不利于各分支学科的导师之间以及学生之间的学术交流和研究积累，很难形成一个凝聚力很强的研究群体；④不利于发挥集体力量完成国家重大科研项目及高水平的学位论文和科研成果，不利于不同学科导师交叉合作来指导博士生。

所以我们采取了团队指导的方式。本学科博士生指导教师团队的组成有如下特点：①由著名教授做学术带头人，导师组整体学术水平较高；②导师组由老、中、青教授相结合组成；③导师分工合作，相辅相成，有的侧重于把握整个研究方向，有的侧重于指导基础理论研究，有的侧重于新的试验方法及新的试验路线的拟定，有的侧重于指导试验和工业应用研究，有的侧重于大规模数值计算，这样自然能使学生得到全面培养；④强调学科交叉，形成交叉学科研究团队，这样有利于承担国家重大科研项目，锻炼博士生的创新思维。

二、博士生研究方向要与国家重大需求相结合

选择正确的研究方向，对于保证和提高博士生培养质量非常重要。因此，我们非常重视博士生研究方向的选择。我们认为，工科博士生的研究要有创新性，要有为国家工程服务的能力。所以我们尽可能地把博士生的课题与国家建设的需求结合起来，尽可能把博士生推向国家及省部级重大科研项目的第一线，比如国家自然科学基金重点项目、国家杰出青年科学基金、国家 973 计划项目、国家 863 计划项目、国家科技攻关项目、国家高技术产业化示范项目、国家经贸委技术创新项目、省部级重大重点项目、国际合作项目、重大横向项目等，将大课题分解成几个子课题交给博士生来做，并作为博士生的学位论文选题。我们先后有近 200 名博士生参加了 100 余项重大科研项目，在科研实践中锻炼成才。全国博士学位论文质量评审重点考查两点：选题是不是国家重大需求？是不是科学前沿课题？这两点是前提。本学科有意识地安排优秀的博士生结合国家重大需求或重大理论创新选择研究方向，先后培养了池作和、邱利民、程军、周昊、罗坤等 5 位全国优秀博士论文奖获得者。程军博士参与的研究项目不但获得国家科技进步二等奖，而且发表了我国学者在国际顶尖能源期刊上的首篇综述性文章。

博士生尤其是工科博士生的研究，只有考虑了国家建设的需求，与重大科研项目相结合，才能取得重大的科研成果并在实践中加以推广。以水煤浆代油技术课题为例。水煤浆是一种高效清洁的代油燃料，我们从 1981 年提出这个新课题，因为看到了我国能源紧缺及环境污染的现状。从 2005 年到 2020 年，我国进口石油占需求总量的百分比将从 43%增长到 75%；如果按照现在的石油价格计算，我们到 2025 年需要在进口石油上面花费将近 5000 亿美元。经济问题之外，还存在（能源）危机问题、安全问题。假如这 75% 的石油进不来，我们的工业就要全部停顿。所以我们 1981 年就提出，能不能利用我们现有的资源，把煤变成油。我们和美国合作，建成了目前高等院校最大的实验台架，也是

全国最大的煤粉/水煤浆热态实验锅炉(3.52MW)，投资500多万元来进行这方面的研究。项目申请之前有专家质疑，煤可以燃烧，但是煤和水混合怎么可能燃烧呢？所以我们自己先进行了两年研究，确定水煤浆可以燃烧了，才向国家申请立项。国家将这个项目列入"六五""七五""八五"攻关项目来大范围推广。我们研制成功的全世界最大的670t/h全烧水煤浆专用锅炉(200MW)已于2005年在广东南海发电厂建成并成功投入运行，现在正在建设第二台。从基础理论、专利、应用实践、关键设备到最后建成大型发电厂，这个研究过程非常漫长，经过了25年努力，直到今年才通过省级鉴定。由浙江大学负责新建或改造并且实际应用的水煤浆锅炉达100余台(最大670t/h)，其中电站锅炉20余台，工业锅炉20余台，工业窑炉20余台。到目前为止，按锅炉容量折算我们每年为国家节约石油150万吨。现在世界上有很多国家在和我们合作，如意大利、俄罗斯、菲律宾等。

第二个例子是洗煤泥燃烧发电，也是基于国家重大需求。我国铁路运输紧张，火车运力不够，而60%的火车用来从内蒙古、山西等地运输煤到上海、北京、广州等沿海地区。粗煤纯度低，有石头、泥沙等混合物，所以运送之前需要先将煤洗干净，避免污染沿海环境，但是洗完煤之后剩下的煤泥等同样污染了当地环境，流到河流和田地里生物都要死亡。这是个国际难题，当时没有一个国家能够解决，所以国家科委给我们这个任务，希望我们解决这个国家重大需求。经过两年多的艰苦试验，国家科委、煤炭部的领导看到我们的实验结果，马上把它立为国家攻关项目，在兖州煤矿试点，建成了全世界第一座全烧煤泥的发电厂——兖州矿务局电厂。目前全国已有几十座煤泥发电厂，基本上都采用我们的技术。美国能源部在对世界各国流化床燃烧技术发展的研究报告列出的值得重视的技术中，对中国只列出了本项技术。"煤水混合物异重床结团燃烧技术"获得了1997年国家技术发明二等奖。这项技术从无到有的过程中也培养了很多优秀博士，其中有一位博士凭毕业论文得到了"有突出贡献的博士学位获得者"称号。

三、理论密切结合实际，引导博士生把基础研究变成生产力

国家现在越来越重视科研成果在实际生产中的应用，连重大基础研究973计划的导向也有所转变。我是973计划项目评审专家，开始讨论的时候有人认为重大基础研究就是数、理、化、天文、地理、生物，另外一些专家认为重大基础研究是数、理、化、天文、地理、生物为国家建设服务。最后国家同意了后一种意见，将基础研究分为六大领域：农业、能源、生命科学、环境、信息、材料。后来补充了两个，一个是重大学科交叉，一个是科学前沿，一共八个领域。说明国家慢慢将科学研究的重点从基础理论研究转变为重大实际问题服务。973计划项目验收的时候专家们也会提出有多大的可能性为国家重大实际问题服务，而不仅仅是看发表了几篇论文。我们从很早开始就注意这个问题，要求研究生"能文能武"，文是指基础理论，武是指工业应用，这是工科学生应该做到的。

以废弃物能源化利用技术为例，我们导师组将重大项目"垃圾焚烧发电"分解成6个关键性问题，指导博士生分工协作、共同攻关。这个项目得到了生活垃圾发电技术第一个国家奖，并且培养出一大批优秀博士。由此我们还成立了超洁净二噁英实验室。因为垃圾焚烧释放的二恶英毒性很强，比氰化钾毒1000倍，焚烧过程必须把它消除掉才能

保证安全。浙江省环境监测中心站联合比利时 SGS 二噁英实验室对焚烧炉烟气排放监测表明，烟气中有害物质排放优于国家排放标准，尤其是二恶英排放大大低于国家标准和欧盟标准。这样从基础研究到应用一条龙就完成了。浙江大学垃圾焚烧技术在国内得到大力推广，已运行的垃圾焚烧电厂 11 座，正在建设的有 9 座。目前采用浙江大学流化床技术的焚烧厂的市场占有率居全国第一，达 26.3%，导师组与博士生共同攻关，在垃圾清洁焚烧领域取得了重大的科研成果。

类似的例子还有很多，比如，我国近 100% 的煤泥发电运用的是浙大技术；水煤浆代油技术经过 25 年的努力，建成世界上最大的 20 万千瓦全烧水煤浆发电机组，2007 年获浙江省科技进步一等奖；洁净煤及低负荷稳燃技术已在 100 座电站机组推广，获国家科技进步二等奖；劣质煤高效燃烧技术分别获国家发明三等奖、四等奖；优化配煤催化燃烧脱硫技术在每年 200 万吨配煤厂推广，获国家科技进步二等奖；废弃物发电技术已建成和在建 21 座电厂，在全国市场占有率第一，获国家科技进步二等奖；根据西部地区缺水的实际情况研究的半干法烟气脱硫技术已在 66 个电厂推广，获浙江省科技进步一等奖；污泥焚烧处理成套技术已成套出口到韩国清州日处理量为 65 吨的污水处理厂，正在向全国推广；经过近 10 年的基础研究建成的全国首台最大生物质循环流化床发电机组（3 万千瓦）已在江苏省宿迁市建成运行；烟气脱硝技术已成功在多台 30 万和 12.5 万千瓦发电机组中应用；烟气脱硫技术成功应用于 30 万、60 万千瓦发电机组。这些研究在投入实际应用、转化为生产力，产生重大经济效益的同时，也培养出了多位全国优秀博士学位论文奖得主。

四、要为博士生提供良好的物质条件

有了好的导师队伍、好的课题之后，还要有良好的物质条件作保障。我们先后承担了 5 项"973 计划"项目，其中包括一级课题 1 项、二级课题 7 项、三级课题 26 项，共计课题经费 3000 多万元，加上"863 计划"、国家支撑计划、国家重点基金、杰出青年基金等 2600 万元资助，平均每位指导教师有 200 多万元的基础研究经费，2000 年至 2006 年总科研经费达 2.11 亿元，2007 年到目前为止已到款科研经费 4800 万元。同时，我们建设了一大批特色鲜明的大中小型实验台架，拥有 40 台/套大型试验台架和先进仪器，培养博士生使用现代化仪器的能力，为博士生科研创新提供了具有国际先进水平的实验平台。3.5MW 大型煤粉燃烧实验台是我们与美国 B&W 公司合作建设的，是目前国内高校最大的试验台。多位博士生利用该实验台研究煤粉稳燃及低污染燃烧技术，从基础研究到工业性实验，取得了重大的工业应用成果，获得了国家科技进步二等奖，池作和博士荣获全国优秀博士学位论文奖。80 户生物质中热值气化中试装置是在英国壳牌（Shell）基金支持下建成的。10 吨/天多功能垃圾焚烧中试试验台，是用"211 工程"经费建设的。这是固体废弃物回转窑热解/焚烧中试装置，已成功应用于浙江湖州的医疗垃圾处理中心。还有半干法烟气脱硫实验台、二恶英分析中心实验室等。充足的科研经费和高水平的实验平台，使每位博士生均能有充足的物质条件大展身手，开展探索性的重大基础研究及应用基础研究。

五、要鼓励博士生参加国际会议和国际交流

这一点非常重要,假如我们自己把自己封闭起来,就不可能成为一流。我们组织研究生参加了中日韩三国的定期交流。通过交流,学生们不但科研水平提高了,组织能力提高了,外语水平也提高了。我们还进行了中美、中法、中瑞交叉培养。除了"送出去",我们还"请进来"。我们已与国外 10 余所能源领域实力比较强的知名大学和研究机构签订了合作培养研究生协议。我们已培养国外进修人员 15 人,与国外大学联合培养国内博士生 15 名。我们与法国联合培养的博士生周昊回国后取得了突出的科研成绩,攻读博士学位期间发表 SCI 论文 7 篇、EI 论文 20 篇;获得国家自然科学二等奖 1 项、国家科技进步二等奖 1 项、浙江省科技进步一等奖 1 项、中国高校自然科学奖 1 项;与导师合作出版专著 1 部;获国家发明专利 2 项;荣获全国优秀博士学位论文奖。他自己认为,他在国外的经历对他帮助很大,能看到其他国家是怎么做的,同时增强了自己的信心。我们和大企业也有合作,这使得博士生有机会参加有大量经费来往的实质性国际科研合作,接受高水平锻炼。近五年来,我们与欧美、日本等国家的相关机构开展了 9 项国际合作科研项目,总经费达 900 万元,50 余名博士生在国际合作项目中受到高水平锻炼。

最后我想讲一点。我认为研究生培养应该创新,对于管理者来讲也应该进行管理创新。比如前面讲过的导师组方案一度不能通过,就不利于交叉学科合作,当然现在已经大大改变了。另外培养优秀学生应该考虑不同学科的不同要求,工科的学生更应该培养团队精神,这就要求给导师更大的自由度,要相信导师们的智慧和创新性,让导师们发挥自由能动性。过多的行政规定是不利于发展的。全校一盘棋,但不能要求一刀切、用同一个模式评价所有学科。每个学科都没有棱角没有特色是不行的。创新难,管理创新最难!

*本文根据中国工程院院士岑可法教授 2007 年 11 月 30 日在浙江大学首期"求是导师学校"的讲座录音整理而成。

原文发表于《学位与研究生教育》2008(5): 1-4。

附录 大事记

1952 年 9 月　考入华南理工大学。
1953 年 8 月　经院系调整至华中工学院。
1954 年 9 月　毕业于华中科技大学动力系,毕业时通过了国家派遣留学苏联研究生考试。
1955 至 1956 年　在浙江大学热工教研组实习工作。
1958 年 12 月　在莫斯科包曼高等工业大学读研究生,师从国际著名燃烧学专家 г.ф.КHoppe 教授。
1962 年 7 月　被授予副博士学位,选择回国,回到浙江大学热工教研室任教。
1963 年　发表 3 篇学术论文。
1976 年　担任燃烧理论与技术教研室主任。
1978 年 9 月　晋升为副教授。
1978 年　获全国科学大会奖。
1978 年　主持"油煤混合燃料(简称"COM")的制备与小型燃烧试验研究"技术攻关。
1979 年　获"浙江省先进科技工作者"称号。
1979 年　"低热值石煤沸腾炉的研制及应用"获浙江省科技进步奖三等奖。
1979 年　"石煤沸腾焙烧提取五氧化二钒(中试)"获浙江省科技进步二等奖。
1981 年　任油煤混合燃料工业性(中间)实验总体技术组组长。
1982 年　"油煤混合燃料制备输送燃烧小型试验研究"获中国科学院科技成果一等奖。
1983 年　洗选煤泥流化床燃烧技术攻关项目取得成功,这一成果为国内首创,达到国际先进水平。
1983 年 3 月　晋升教授。
1984 年　任热物理工程学系主任。
1984 年 5 月　加入中国共产党。
1986 年　获国家级有突出贡献的中青年科技工作者、浙江省优秀教师称号。
1986 年　获教育部博士生导师资格。
1987 年　获全国"五一"劳动奖章、全国先进科技工作者、浙江省劳动模范。
1988 年　主持国家"七五攻关项目"50t/h 燃油工业锅炉应用水煤浆代油燃烧技术联合攻关"。
1989 年　担任浙江大学热能工程研究所所长。
1990 年 12 月　荣获全国高等学校先进科技工作者称号。
1990 年 12 月　"低热值石煤的预热层燃技术及装置"获国家发明奖四等奖。

1991 年　国家七五攻关项目"50T/H 燃油工业锅炉应用水煤浆代油燃烧技术"获国家科委、计委、财政部联合颁发的"七五"重大科技成果奖。

1992 年 11 月　教材《锅炉燃烧试验研究方法及测量技术》获第二届普通高等学校优秀教材—全国优秀奖。

1993 年 11 月　获光华科技基金奖一等奖。

1993 年　《循环流化床流动、燃烧、脱硫的机理研究》一书获国家教委科技进步奖二等奖。

1993 年　"计算机辅助优化数值试验(CAT)理论及在工程气固多相流动研究中的应用"获国家教委科技进步奖一等奖。

1993 年　"用 CAT 解决韶关电厂 8#炉锅炉结渣"获广东省科技进步奖一等奖。

1993 年　"水煤浆代油燃烧技术"获国家教委科技进步奖三等奖。

1995 年 9 月　获全国优秀教师称号。

1995 年　当选为中国工程院院士，属能源与矿业工程学部。

1995 年　"用 CAT 解决大型电站锅炉结渣问题"获国家科技进步奖三等奖。

1995 年　"煤水混合物流化床燃烧技术"获国家教委科技进步二等奖(乙类)。

1996 年 12 月　获第二届"孺子牛金球奖"。

1997 年 10 月　"培养高水平工学博士的新机制"获浙江省教学成果一等奖。

1997 年 12 月　"煤水混合物异重床结团燃烧技术"获国家发明二等奖。

1998 年 6 月　"高浓度有机废液和污泥流化床焚烧技术"获浙江省科技进步奖二等奖。

1998 年 10 月　获何梁何利基金会"何梁何利科学技术奖"。

1998 年　"煤浆燃烧、流动、传热和气化的理论与应用技术"获浙江省科学技术进步奖一等奖。

1999 年　任浙江大学机械与能源工程学院院长。

2000 年　"湍流式烟气除尘脱硫一体化装置的开发及应用"获浙江省科技进步二等奖。

2000 年　"35t/h 煤矸石流化床锅炉混烧洗煤泥技术"获国家煤炭科技进步奖一等奖。

2000 年 11 月　"催化洁净燃烧及优化配煤技术研究"获教育部科技进步奖二等奖。

2001 年　任 2001～2005 年教育部高等学校能源动力学科教学指导委员会副主任委员，以及热能与动力工程专业教学指导分委员会主任委员。

2001 年 12 月　"煤的优化配制、催化洁净燃烧及产业化应用"获 2001 年国家科技进步奖二等奖，岑可法为第一获奖人。

2002 年 2 月　"城市生活垃圾清洁焚烧的研究"　获中国高校科学技术奖二等奖。

2002 年 4 月　"锅炉受热面管束磨损和防磨措施的数值试验研究"获教育部科技成果奖。

2002 年 7 月　"ZDM 型洁净煤添加剂研究"获教育部科技成果奖，"工程气固多相两相流动中若干关键性问题的研究"获教育部科技成果奖。

2003 年　获浙江省科学技术重大贡献奖。

2004 年　《高等燃烧学》获第十四届中国图书奖。

2004 年 12 月　"基于循环流化床技术的低品质燃料洁净综合利用"获教育部科技成果奖。

2005 年 1 月　"基于循环流化床技术的低品位燃料洁净综合利用"获教育部提名国家科技进步奖。

2005 年 1 月 15 日　把荣获的浙江省科学技术重大贡献奖的一半奖金 25 万元捐献出来，成立浙江大学热能工程研究所岑可法教育基金，用于奖励和资助优秀的教师和学生。

2005 年 4 月　获国务院颁发的"全国先进工作者"称号。

2005 年 9 月　"瞄准能源学科前沿，构建一流导师群体，培养一流创新人才"获国家级教学成果奖二等奖。

2005 年 11 月　"工程气固多相两相流动中若干关键基础问题的研究"获国家自然科学奖二等奖。

2005 年　任 2006-2010 年教育部高等学校能源动力学科教学指导委员会及热能与动力工程专业教学指导分委员会的主任委员。

2006 年 1 月　"城市生活垃圾异重流化床焚烧集成处理装置"和"循环悬浮式半干法烟气净化装置"分获中国专利优秀奖。

2006 年　能源清洁利用国家重点实验室通过国家科技部建设验收，任学术委员会主任。

2007 年 2 月　"生活垃圾循环流化床清洁焚烧发电集成技术"获国家科学技术进步奖二等奖。

2007 年　被任命为国家重点基础研究发展计划(973 计划)第四届专家顾问组成员。

2008 年 11 月　浙江大学工学部成立，任工学部学术委员会主任。

2008 年 12 月　"电厂锅炉多种污染物协同脱除半干法烟气净化技术"获国家技术发明奖二等奖。

2009 年　"水煤浆代油洁净燃烧技术及产业化应用"获国家科学技术进步奖二等奖。

2009 年　"生物质热化学转化制取高品位燃料的基础研究"获浙江省科学技术奖一等奖。

2009 年　"结合国家重大需求，创建能源与环境复合型人才培养新体系"获国家级教学成果奖二等奖。

2009 年 11 月　获 "International Conference on Power Engineering"特别贡献奖。

2010 年 1 月 15 日　出资设立"浙江大学岑可法教育基金"，该基金接受社会各界的捐赠，总额达到 1000 万元。

2010 年 7 月 26 日至 29 日　获 The 6th International Conference on Combustion, Incineration/Pyrolysis and Emission Control, i-CIPEC "杰出贡献奖"（Outstanding Performance Award）。

2011 年 6 月　当选浙江省"十大之江先锋"，获浙江省优秀共产党员。

2012 年 12 月　"湿法高效脱硫剂硝汞控制一体化关键技术与应用"获国家科学技术进步奖二等奖。

2013 年 5 月 "水煤浆代油洁净燃烧技术与产业化应用"项目获浙大十大学术进展。

2014 年 9 月 "强化节能减排意识，提升创新实践能力，创建与推进全国大学生节能减排竞赛"获国家级教学成果奖二等奖。

2014 年 12 月 "面向能源学科前沿与国家重大需求，团队式国际化培养创新人才的实践"获中国学位与研究生教育成果一等奖。

2015 年 "燃烧过程的场参数实时检测、在线诊断和优化控制技术"获浙江省科学技术进步奖一等奖。

2016 年 12 月 "浙江大学能源清洁利用创新团队"获国家科技进步（创新团队）奖。

2017 年 2 月 "生物质热化学定向转化分级制取高品位液体燃料"获中华人民共和国教育部自然科学奖一等奖。

2017 年 9 月 获 2017 年"浙江省杰出人才创新奖"。

2017 年 12 月 "燃煤机组超低排放关键技术研发与应用"获国家技术发明一等奖。

2017 年 "微生物转化生物质制油气燃料的能质传递强化机理"获浙江省自然科学奖一等奖。

2017 年 "锅炉烟气臭氧氧化脱硝方法"获中国专利优秀奖。

2017 年 "大型油气锅炉燃烧振动控制技术"获高等学校科技进步奖一等奖。

2018 年 获得全国大学生节能减排社会实践与科技竞赛委员会荣誉主任。

2022 年 可调谐激光光谱结合飞行时间质谱在线监测二噁英的方法获中国专利金奖。

2023 年 获中国工程热物理学会终身成就奖。